KAC-MOODY AND VIRASORO ALGEBRAS

Advanced Series in Mathematical Physics
Vol. 3

KAC-MOODY AND VIRASORO ALGEBRAS

A Reprint Volume for Physicists

Edited by
Peter Goddard
Department of Applied Mathematics and Theoretical Physics
University of Cambridge
David Olive
The Blackett Laboratory
Imperial College of Science and Technology

World Scientific
Singapore • New Jersey • Hong Kong

ADVANCED SERIES IN MATHEMATICAL PHYSICS

Published*

Vol. 29: *Bombay Lectures on Highest Weight Representations of Infinite Dimensional Lie Algebras (Second Edition)*
by Victor G Kac, Ashok K Raina and Natasha Rozhkovskaya

Vol. 28: *An Introduction to the Mathematical Structure of Quantum Mechanics (Second Edition)*
A Short Course for Mathematicians
by F. Strocchi

Vol. 27: *An Introduction to the Mathematical Structure of Quantum Mechanics*
A Short Course for Mathematicians
by F Strocchi

Vol. 26: *Soliton Equations and Hamiltonian Systems (Second Edition)*
by L A Dickey

Vol. 25: Basic Methods in Soliton Theory
by I Cherednik

Vol. 24: *The Mathematical Beauty of Physics*
edited by J M Drouffe and J B Zuber

Vol. 23: *Quantum Theory and Global Anomalies*
by R A Baadhio

Vol. 22: *W-Symmetry*
by P Bouwknegt and K Schoutens

Vol. 21: *Multidimensional Hypergeometric Functions and Representation Theory of Lie Algebras and Quantum Groups*
by A Varchenko

Vol. 20: *The State of Matter*
edited by M Aizenman and H Araki

Vol. 19: *Under the Spell of the Gauge Principle*
by G 't Hooft

Vol. 18: *Exactly Solvable Models and Strongly Correlated Electrons*
by V Korepin and F H L Ebler

*For the complete list of volumes in this series, please visit
www.worldscientific.com/series/asmp

Published by

World Scientific Publishing Co. Pte. Ltd.

5 Toh Tuck Link, Singapore 596224

USA office: 27 Warren Street, Suite 401-402, Hackensack, NJ 07601

UK office: 57 Shelton Street, Covent Garden, London WC2H 9HE

British Library Cataloguing-in-Publication Data
A catalogue record for this book is available from the British Library.

Advanced Series in Mathematical Physics — Vol. 3
KAC-MOODY AND VIRASORO ALGEBRAS
A Reprint Volume for Physicists

ISBN-13 978-9971-5-0419-9
ISBN-10 9971-5-0419-7
ISBN-13 978-9971-5-0420-5 (pbk)
ISBN-10 9971-5-0420-0 (pbk)

PREFACE

Mathematicians and physicists share the common aim of finding unified explanations of the phenomena with which they deal. In the late 1970's each community found a theory holding the promise of such a unification for a wide class of phenomena. Physicists found the "dual resonance" model which soon developed into "string theory", which could encompass as special cases the gauge theories of particle interactions as well as Einstein's theory of gravity and introduced a new principle, supersymmetry, leading later to supergravity. Mathematicians discovered the existence of a class of infinite-dimensional Lie algebras, called affine Kac-Moody algebras, with properties which generalised in a relatively straightforward way those of the familiar finite-dimensional simple algebras. The root systems of these algebras, although infinite, were finite-dimensional and possessed geometric properties leading to a systematic understanding of certain combinatorial identities involving infinite products and sums, known in some cases from the nineteenth century, such as Jacobi's triple product identity.

About ten years later it emerged that certain of the irreducible representations of these algebras could be constructed rather explicitly using the concept of a "vertex operator", one of the key ingredients in string theory. The forging of this vital link initiated a unification of ideas from mathematics and physics which is still in full spate and enriching each subject. It is these developments which provide the occasion for the present collection of reprints.

Before describing other applications in physics, we shall briefly explain why, in retrospect, it is natural for infinite-dimensional Lie algebras to be relevant. Physics deals with the regularities or symmetries of nature and the symmetries of any given physical system form a group. Finite or discrete groups are relevant to objects such as crystals but, because space and time are continuous, we should expect continuous groups, i.e. Lie groups, to enter physics. These are infinite groups but the original examples were finite-dimensional (i.e. their elements could be thought of as depending on a finite number of continuous parameters), such as the group of rotations in three dimensions, $SO(3)$, used in the study of atomic spectra and elsewhere. Quantum mechanics furnished the vector spaces on which these groups act naturally and, since that theory has traditionally been formulated in terms of hermitian observables, physicists have been led naturally to consider the Lie algebra of generators of the Lie group. For $SO(3)$ this consists of the angular momenta.

The theory of elementary particles combines quantum mechanics and special relativity as well as other ideas. In the early 1960's the theory of "current algebras", motivated by data on radio-active (i.e. "weak") decays held sway, later to be supplanted by two apparently different theories, string theory and the unified theory of gauge interactions in the late 1960's. In the latter theory, forces were associated with the generators of the "gauge group", a compact Lie group chosen on the basis of experimental data, and the plethora of "elementary" particles was fitted into families, i.e. irreducible representations, of the gauge group. The weak and electromagnetic forces were successfully described by the theory proposed by Weinberg and Salam, which took the group $U(2)$, consisting of 2×2 matrices, as gauge group. The strong forces, responsible indirectly for the forces that hold nuclei together, were described by an $SU(3)$ gauge group.

The desire to fit these groups together, as subgroups of a larger simple group, in as economical a way as possible, led in the 1970's to an interest in larger groups, such as $SU(5)$, $SO(10)$, and even the exceptional group E_6. This programme was called "grand unification". These and other developments persuaded many theoretical physicists that a general understanding of the Lie algebras was necessary, not only as a working tool but, more significantly, also as a guide and language for the formulation of concepts.

The affine Kac-Moody algebras constitute, in a well-defined sense, the simplest kind of infinite-dimensional algebra, with properties naturally generalising those of finite-dimensional Lie algebras. Automatically associated with such an algebra is a Virasoro algebra, another infinite-dimensional algebra which was first identified and studied in string theory. During the last twenty years there has been a growing number of situations in which one or both of these two types of infinite-dimensional Lie algebras (i.e. affine Kac-Moody or Virasoro) have played a rôle at least comparably important to those assumed by finite-dimensional Lie algebras.

These situations are ones in which the physical systems involved are in some essential respect two-dimensional. For example, it is the string world-sheet, the surface swept out by the string in its motion in space-time, which is two-dimensional. The elementary particles are quantum excitations of the string. The success of string theory in unifying all known basic symmetry principles, including gauge invariance, explains why it is currently regarded as the most promising candidate for the unified theory of all the interactions, including gravity.

A surprising and rewarding arena for the theory concerns the behaviour of statistical physics models consisting of, say, spin variables at the sites of a two-dimensional lattice interacting with their nearest neighbours. At a certain critical temperature, the system can make a "second order" phase transition whereby it changes its state (as, for example, ice melts into water). The scale given by the lattice spacing effectively diverges and therefore becomes irrelevant, leaving a scale invariant theory controlled by the Virasoro algebra, whose representation theory determines the "critical exponents" of the transition. These exponents specify certain power law behaviours in spatial separations which are measurable in the laboratory, where two-dimensional systems can indeed be constructed and are of considerable current technological interest. This result, that the possible critical behaviour of such models is thus largely independent of the details of the model, gives explicit examples of the concept of "universality" introduced in solid state physics in the 1960's.

Yet another manifestation of these algebras is in the theory of solitons and integrable systems. Often integrable non-linear equations of physical or engineering significance can be regarded as "zero curvature" conditions for connections taking values in an affine Kac-Moody algebra. The sine-Gordon equation is a prime example and the vertex operator construction was first proposed for its soliton by Skyrme at the beginning of the 1960's. He had in mind applications to a unified theory of elementary particles but the concept of soliton also has applications in nerve fibres, optical fibres and many other areas of science and engineering.

A framework which can encompass so many studies at the frontiers of research in physics and engineering is obviously worthy of study and must be expected to be of importance in mathematics, as had indeed been realised independently as we mentioned earlier.

The fact that so many areas are interrelated ought to stimulate new developments in each of the constituent subjects and indeed this has already happened. For example, recent developments in incorporating gauge symmetries into string theory are in part due to mathematical stimulus (see chapter 2). There seems every reason to believe that this fruitful interrelation will continue and that, in particular, future developments in string theory will exploit and stimulate progress in the theory of infinite-dimensional Lie algebras.

Prompted by the increasing interest in this area amongst physicists, Dr. K.K. Phua suggested to us that we might produce a volume of reprints, including our recent review article published in International Journal of Modern Physics A. In selecting papers for this volume, we have tried to choose those which form good sources for the current developments in physics and those likely to be relevant in the future. Such criteria are, of course, subjective and result in the exclusion of many papers of crucial importance historically. Additionally, we have had it in mind that we wished to produce a book of no more than 500 to 600 pages.

ACKNOWLEDGEMENTS

The publisher and editors thank the authors and the following publishers for permission to reproduce published papers in this volume:

Academic Press, Inc. (*J. Funct. Anal.*)
American Mathematical Society (*Bull. Amer. Math. Soc.*)
American Physical Society (*Phys. Rev. Lett.*)
North-Holland Physics Publishing (*Phys. Lett. B* and *Nucl. Phys. B*)
Plenum Publishing Corp. (*Funct. Anal. Appl.*)
Springer-Verlag, Heidelberg (*Invent. Math.* and *Commun. Math. Phys.*)

CONTENTS

KAC-MOODY AND VIRASORO ALGEBRAS

CHAPTER 1

INTRODUCTION

Reprinted Papers

1. R.V. Moody, "Lie Algebras Associated with Generalised Cartan Matrices", Bull. Amer. Math. Soc. **73** (1967) 217–221.

2. P. Goddard and D. Olive, "Kac-Moody and Virasoro Algebras in Relation to Quantum Physics", Int. J. Mod. Phys. **A1** (1986) 303–414.

The study of what came to be called Kac-Moody algebras began in 1967, with the work of V.G. Kac [1], R.V. Moody [*Reprinted Paper* #1] and I.L. Kantor [2], about the same time that the seeds of string theory (for an account see ref.[3]), with which its development has been intertwined, were sown in the study of the "duality" of different descriptions of the strong interactions of elementary particles. Accounts of the development of the theory of Kac-Moody algebras can be found in the book of Kac [4], and in the Introduction, by J. Lepowsky, to the Proceedings of the Conference on Vertex Operators held in Berkeley in November 1983 [5]. In this first chapter, the paper of Moody is reprinted to illustrate the mathematical context of the origins of the Kac-Moody algebras, and a review paper by the editors of this volume [*Reprinted Paper* #2] is included in order to introduce Kac-Moody and Virasoro algebra in a way which it is hoped is accessible to physicists, and which relates aspects of their study to contexts in theoretical physics. For other accounts of Kac-Moody algebras and their physical applications see the reviews of L. Dolan [6] and B. Julia [7].

Interest in Kac-Moody algebras was much stimulated by the realization of their connection with the Macdonald identities [8] through the Weyl-Kac character formula [9]; an account of this is given in the review of I.G. Macdonald [10]. An exposition of the theory of infinite-dimensional Lie algebras is given in the book of Kac [4]. A global approach, discussing the corresponding groups, is given in the book of A. Pressley and G. Segal [11].

The approach adopted in the review reproduced here [*Reprinted Paper* #2] is closer in spirit to the book of Kac [4] than to that of Pressley and Segal [11] in that it deals with Lie algebras rather than the global properties of the corresponding groups. However it differs from [4] in that it concerns itself immediately with affine Kac-Moody algebras rather than more general possibilities and the theme of the article is an insight gained from the theory of current algebras, which is used as a unifying thread for the treatment of various constructions of representations which are of interest in physics. This insight is a precise version of the Sugawara form for the energy-momentum tensor, which in this context provides a construction of the

Virasoro algebra that holds whenever we have a representation of an affine Kac-Moody algebra built up from certain "ground states" in a sense to be explained.

References

[1] V.G. Kac, "Simple Graded Algebras of Finite Growth", Funct. Anal. Appl. **1** (1967) 328–329.

[2] I.L. Kantor, "Simple Graded Infinite-dimensional Lie Algebras", Sovt. Math. Dokl. **9** (1984) 409–412.

[3] M.B. Green, J.H. Schwarz and E. Witten, *Superstring Theory*, two volumes (Cambridge University Press, 1987).

[4] V.G. Kac, *Infinite Dimesional Lie Algebras* (Second Edition, Cambridge University Press, 1985).

[5] J. Lepowsky, S. Mandelstam and I.M. Singer (editors), *Vertex Operators in Mathematics and Physics*, MSRI Publication #3 (Springer, Heidelberg, 1984).

[6] L. Dolan, "Kac-Moody algebras and exact solvability in hadronic physics", Phys. Rep. **109** (1984) 1–94.

[7] B. Julia, "Supergeometry and Kac-Moody algebras", in *Vertex Operators in Mathematics and Physics*, MSRI Publication #3 (Springer, Heidelberg, 1984) 51–96.

[8] I.G. Macdonald, "Affine Root Systems and Dedekind's η-function", Invent. Math. **15** (1972) 91–143.

[9] V.G. Kac, "Infinite-dimensional Lie Algebras and Dedekind's η-function", Funct. Anal. Appl. **8** (1974) 68–70.

[10] I.G. Macdonald, "Affine Lie Algebras and Modular Forms", in *Seminaire Bourbaki Exp. 577, Lecture Notes in Mathematics* Vol. **901** (Springer Verlag, New York, 1981) 258–265.

[11] A. Pressley and G. Segal, *Loop Groups* (Cambridge University Press, 1986).

LIE ALGEBRAS ASSOCIATED WITH GENERALIZED CARTAN MATRICES

BY R. V. MOODY[1]

Communicated by Charles W. Curtis, October 3, 1966

1. **Introduction.** If $\mathfrak{L}$ is a finite-dimensional split semisimple Lie algebra of rank n over a field Φ of characteristic zero, then there is associated with $\mathfrak{L}$ a unique $n\times n$ integral matrix (A_{ij})—its Cartan matrix—which has the properties

M1. $\quad A_{ii} = 2, \quad i = 1, \cdots, n,$

M2. $\quad A_{ij} \leqq 0, \quad$ if $i \neq j,$

M3. $\quad A_{ij} = 0, \quad$ implies $A_{ji} = 0.$

These properties do not, however, characterize Cartan matrices.

If (A_{ij}) is a Cartan matrix, it is known (see, for example, [4, pp. VI-19–26]) that the corresponding Lie algebra, $\mathfrak{L}$, may be reconstructed as follows: Let $e_i, f_i, h_i, i=1, \cdots, n$, be any $3n$ symbols. Then $\mathfrak{L}$ is isomorphic to the Lie algebra $\tilde{\mathfrak{L}}((A_{ij}))$ over δ defined by the relations

$$\left.\begin{aligned}[h_ih_j] &= 0,\\ [e_if_j] &= \delta_{ij}h_i,\\ [e_ih_j] &= A_{ji}e_i, \quad [f_ih_j] = -A_{ji}f_i,\end{aligned}\right\}\text{for all } i \text{ and } j$$

$$\left.\begin{aligned}e_i(\operatorname{ad} e_j)^{-A_{ji}+1} &= 0,\\ f_i(\operatorname{ad} f_j)^{-A_{ji}+1} &= 0,\end{aligned}\right\}\text{if } i \neq j.$$

In this note, we describe some results about the Lie algebras $\tilde{\mathfrak{L}}((A_{ij}))$ when (A_{ij}) is an integral square matrix satisfying M1, M2, and M3 but is not necessarily a Cartan matrix. In particular, when the further condition of §3 is imposed on the matrix, we obtain a reasonable (but by no means complete) structure theory for $\tilde{\mathfrak{L}}((A_{ij}))$.

2. **Preliminaries.** In this note, Φ will always denote a field of characteristic zero. An integral square matrix satisfying M1, M2, and M3 will be called a *generalized Cartan matrix*, or *g.c.m.* for short. Z will denote the integers, and in any Lie algebra we will use the symbol $[l_1, l_2, \cdots, l_n]$ to denote the product $[\cdots[[l_1l_2]\cdots]l_n]$.

[1] These results were obtained in my dissertation at the University of Toronto under the supervision of Professor M. J. Wonenburger.

Reprinted from *Bulletin of the American Mathematical Society* (1967), "Lie Algebras Associated with Generalised Cartan Matrices," R. V. Moody, Volume 73, pages 217–221, by permission of The American Mathematical Society.

Let (A_{ij}) be a g.c.m. and let $\mathfrak{L}=\tilde{\mathfrak{L}}((A_{ij}))$ be the Lie algebra over Φ which is obtained by using the method of § 1. Many of the customary features of finite-dimensional split semisimple Lie algebras appear in $\mathfrak{L}$. For example, $\mathfrak{L}$ is graded by $Z\times\cdots\times Z$ (taken n times), and if we denote the subspace of elements of degree $(d_1,\cdots,d_n)$ by $\mathfrak{L}(d_1,\cdots,d_n)$, we find that $\mathfrak{H}=\mathfrak{L}(0,\cdots,0)$ is the subspace generated by $h_1,\cdots,h_n$, and $\mathfrak{L}(d_1,\cdots,d_n)=(0)$ unless $d_1,\cdots,d_n$ are all nonnegative or all nonpositive. Furthermore, if $d_1,\cdots,d_n$ are all nonnegative (respectively all nonpositive) and not all zero, then $\mathfrak{L}(d_1,\cdots,d_n)$ is spanned by the elements of the type $[e_{i_1},\cdots,e_{i_r}]$ (respectively $[f_{i_1},\cdots,f_{i_r}]$) where each e_j (respectively f_j) appears precisely $|d_j|$ times.

Let $\mathfrak{A}$ be an n-dimensional vector space over Φ with a basis $\alpha_1,\cdots,\alpha_n$, and define a mapping, $\sim$, of $\mathfrak{A}$ into $\mathfrak{H}^*$, the dual space of $\mathfrak{H}$, by putting $\tilde{\alpha}_i(h_j)=A_{ji}$ for $i,j=1,\cdots,n$. Then, if $a\in\mathfrak{L}(d_1,\cdots,d_n)$, $[a\ h]=(\sum_{i=1}^n d_i\alpha_i)^\sim(h)a$ for all $h\in\mathfrak{H}$. If $\mathfrak{L}(d_1,\cdots,d_n)\neq(0)$, we call $\beta=\sum d_i\alpha_i$ a *root* and write $\mathfrak{L}_\beta$ for $\mathfrak{L}(d_1,\cdots,d_n)$. We use the words positive and negative for the nonzero roots in the usual way.

A g.c.m., (A_{ij}), is said to be *decomposable* if, after a suitable permutation of the rows together with the corresponding permutation of the columns, it takes a diagonal block form. Clearly, any block obtained in this manner is a g.c.m. (A_{ij}) is called *indecomposable* if it is not decomposable. If (A_{ij}) decomposes into indecomposable blocks $B_1,\cdots,B_k$, then $\tilde{\mathfrak{L}}((A_{ij}))\cong\tilde{\mathfrak{L}}(B_1)\times\cdots\times\tilde{\mathfrak{L}}(B_k)$. Consequently, we restrict our attention to indecomposable g.c.m.'s.

At this point, we impose a strong condition on our g.c.m.'s. Further results on the algebras $\tilde{\mathfrak{L}}((A_{ij}))$ when no further restrictions are placed on the matrix are discussed in a forthcoming note in this journal by Daya-Nand Verma.

3. G.C.M.'s of Type (1) and (2), and their classification. Let (A_{ij}) be an indecomposable g.c.m. which satisfies the following condition: If $\xi_1,\cdots,\xi_n$ are any nonnegative rational numbers such that $\sum_{i=1}^n\xi_iA_{ji}\leq 0$, $j=1,\cdots,n$, then $\sum_{i=1}^n\xi_iA_{ji}=0$, $j=1,\cdots,n$. Such a matrix will be called a g.c.m. of *type* (2) if there exist nonnegative rationals $\xi_1,\cdots,\xi_n$, *not all* zero, such that $\sum_{i=1}^n\xi_iA_{ji}=0$, $j=1,\cdots,n$, and a g.c.m. of *type* (1) otherwise. From now on all g.c.m.'s will be of type (1) or (2).

A real, symmetric $n\times n$ matrix is called an *a-form*, [1, p. 175], if every entry off the diagonal is nonpositive. It is called *connected* if it is indecomposable in the sense above.

THEOREM 1. *If* (A_{ij}) *is a g.c.m. of type* (1) *or* (2), *then there exist unique positive rational numbers* $\omega_1, \cdots, \omega_n$ *such that* (i) $\omega_i A_{ij} = \omega_j A_{ji}$ *for* $i, j = 1, \cdots, n$, *and* (ii) $\{\omega_i A_{ii} | i = 1, \cdots, n\}$ *is a set of integers with no common factor. The matrix* $(a_{ij}) \equiv (\omega_i A_{ij})$ *is a connected a-form. It is positive definite or positive semidefinite according as* (A_{ij}) *is of type* (1) *or* (2), *and* $A_{ij} = 2a_{ij}/a_{ii}$. *Conversely, if* (a_{ij}) *is a positive definite (respectively positive semidefinite) connected a-form such that* $A_{ij} \equiv 2a_{ij}/a_{ii}$ *is an integer for each* i *and* j, *then* (A_{ij}) *is a g.c.m. of type* (1) *(respectively type* (2)).

Let $\mathfrak{A}_0$ be the rational space spanned by $\alpha_1, \cdots, \alpha_n$, and define a bilinear form σ on $\mathfrak{A}_0$ by $\sigma(\alpha_i, \alpha_j) = a_{ij}$. σ is positive definite or positive semidefinite according as (A_{ij}) is of type (1) or (2), and in the latter case, $\mathfrak{A}_0$ contains precisely one isotropic line, and it is the radical of σ.

On $\mathfrak{A}_0$ we define S_i to be the reflection determined by the hyperplane orthogonal to α_i ($i = 1, \cdots, n$). The group $\mathfrak{W}$ generated by the S_i is called the *Weyl group* of $\tilde{\mathfrak{L}}((A_{ij}))$. Let $\Gamma = \{\beta \in \mathfrak{A}_0 | \beta = \alpha_j T$ for some $j = 1, \cdots, n$, and some $T \in \mathfrak{W}\}$.

THEOREM 2. (i) $A_{ij}A_{ji} = 0, 1, 2, 3,$ *or* 4 *for all* i, j.

(ii) $(S_i S_j)^{p_{ij}} = 1$ *where*

$$p_{ij} = \pi/(\mathrm{Cos}^{-1}((A_{ij}A_{ji})^{1/2}/2)), \qquad \text{if } i \neq j$$
$$= 1, \qquad \text{if } i = j.$$

(iii) $a_{ij}/(a_{ii}a_{jj})^{1/2} = -\cos(\pi/p_{ij})$ *for all* i *and* j.

The positive definite and positive semidefinite matrices of the type $(-\cos(\pi/p_{ij}))$, where the p_{ij} are integers such that $p_{ii} = 1$, $p_{ij} > 1$ if $i \neq j$, and $p_{ij} = p_{ji}$ for all i and j, have already been classified [1, Chapter 11], and Theorem 2 (iii) provides us with a link by which we may classify the g.c.m.'s of type (1) and (2). These matrices may be diagrammatically described in the way customary for Cartan matrices —namely by drawing a dot for each number $1, \cdots, n$, joining the ith and jth dots by $A_{ij}A_{ji}$ lines, and writing the *weight* $\sigma(\alpha_i, \alpha_i) = a_{ii}$ over the ith dot. The g.c.m.'s of type (1) turn out to be precisely the indecomposable Cartan matrices, and the Lie algebras $\tilde{\mathfrak{L}}((A_{ij}))$ obtained from them are, of course, the corresponding finite-dimensional split simple Lie algebras over Φ. The diagrams for the type (2) matrices are, except for the weights, basically those given in [2, p. 142]. The only change required is the replacement of any line with a number m appearing over it by $4\cos^2(\pi/m)$ lines. Each type (2) matrix is designated by the letter attached to its diagram by

Coxeter. As in the case of Cartan matrices, different matrices may have the same diagram with only the weight distribution differing. We use a second subscript to distinguish matrices with the same diagram. With this notation, the complete list of type (2) g.c.m.'s is: P_n, $n>2$; $S_{n,1}$, $S_{n,2}$, $n>2$; $R_{n,1}$, $R_{n,2}$, $R_{n,3}$, $n>2$; Q_n, $n>4$; T_7, T_8, T_9; $U_{6,1}$, $U_{6,2}$; $V_{3,1}$, $V_{3,2}$; $W_{2,1}$, $W_{2,2}$.[2]

REMARK. If (A_{ij}) is a g.c.m. of type (1), then $\mathfrak{W}$ is the group defined by the relations of Theorem 2 (ii). We do not know whether this result holds for the g.c.m.'s of type (2).

4. The structure theory. If (A_{ij}) is of type (2), then $\tilde{\mathfrak{L}}((A_{ij}))$ has a centre, $\mathfrak{C}$. $\mathfrak{C}$ is a homogeneous ideal, and $\mathfrak{L}((A_{ij}))\equiv\tilde{\mathfrak{L}}((A_{ij}))/\mathfrak{C}$ still decomposes into a direct sum of root spaces. The image, $\mathfrak{H}$, of $\tilde{\mathfrak{H}}$ in $\mathfrak{L}((A_{ij}))$ is of dimension $n-1$. $\mathfrak{L}((A_{ij}))$ has no centre. We often designate $\mathfrak{L}((A_{ij}))$ by the symbol for the matrix (A_{ij}). The algebras $\mathfrak{L}((A_{ij}))$, when (A_{ij}) is a g.c.m. of type (2), are called *tiered* algebras

THEOREM 3. *If $\mathfrak{L}$ is tiered, then $\beta\in\mathfrak{a}_0$ is a nonisotropic root if and only if $\beta\in\Gamma$, and for such roots* dim $\mathfrak{L}_\beta=1$. *There exists a positive isotropic root ζ such that the set of isotropic roots is precisely $Z\zeta$. There exists a positive integer r such that if β is a root, then $\{\beta+Zr\zeta\}$ are all roots (clearly, then, $\mathfrak{L}$ is infinite dimensional). The minimum r for which this is true is called the tier number of $\mathfrak{L}$, and $\mathfrak{L}$ is said to be r-tiered. The tier number is always 1, 2, or 3. In fact, the algebras P_n, $S_{n,1}$, $R_{n,1}$, Q_n, T_7, T_8, T_9, $U_{6,1}$, $V_{3,1}$, and $W_{2,1}$ are 1-tiered, and the remaining ones are 2-tiered with the exception of $V_{3,2}$ which is 3-tiered.*

THEOREM 4. *If $\mathfrak{L}$ is r-tiered, and is treated as an $\mathfrak{L}$-module relative to its adjoint representation, then there exists an $\mathfrak{L}$-module automorphism of $\mathfrak{L}$ (denoted by ′) such that $\mathfrak{L}_\beta\rightarrow\mathfrak{L}_{\beta+r\zeta}$ for all roots β.*

The mapping of Theorem 4 is called the *shift* mapping, and plays a fundamental role in proving the remaining theorems. If $l\in\mathfrak{L}$, we define $l^{(i)}$, $i=0, 1, 2, \cdots$ inductively by $l^{(0)}=l$, $l^{(i)}=(l^{(i-1)})'$ for $i>0$.

THEOREM 5. *If $\mathfrak{I}$ is a nonzero ideal of the tiered Lie algebra $\mathfrak{L}$, then $\mathfrak{I}$ is generated by a single element of the form $\sum_{i=0}^{s}\lambda_i h_1^{(i)}$ with $\lambda_0\neq0$, and $\lambda_s=1$. The correspondence between nonzero ideals and elements of this type is bijective.*

Let $\Phi\langle x\rangle$ denote the ring of polynomials of the form $\sum_{i=-\infty}^{\infty}\mu_i x^i$ with almost all of the $\mu_i=0$.

[2] I am grateful to Professor G. B. Seligman for pointing out that the matrices of type W were omitted in my original classification.

THEOREM 6. *The lattice of ideals of a tiered Lie algebra over* Φ *is isomorphic to the lattice of ideals of* $\Phi\langle x\rangle$.

If $\mu\in\Phi$, $\mu\neq 0$, the ideal generated by $h_1-\mu h_1^{(1)}$ is maximal. We denote its quotient in $\mathfrak{L}$ by $\mathfrak{L}(\mu)$.

THEOREM 7. $\mathfrak{L}(\mu)$ *is finite-dimensional and central simple.* dim $\mathfrak{L}(\mu)$ *is the same for all nonzero* μ.

The next theorem tells us that the 1-tiered algebras are not really anything new.

THEOREM 8. *If* $\mathfrak{L}$ *is* 1*-tiered, then* $\mathfrak{L}\cong\mathfrak{L}(1)\otimes_{\Phi}\Phi\langle x\rangle$. *The relation between* $\mathfrak{L}(1)$ *and* (A_{ij}) *is given by the table:*

(A_{ij})	$\mathfrak{L}(1)$	(A_{ij})	$\mathfrak{L}(1)$
P_n	A_{n-1}	T_8	E_7
$S_{n,1}$	B_{n-1}	T_9	E_8
$R_{n,1}$	C_{n-1}	$U_{5,1}$	F_4
Q_n	D_{n-1}	$V_{3,1}$	G_2
T_7	E_6	$W_{2,1}$	A_1

THEOREM 9. *If* (A_{ij}) *is an* $n\times n$ g.c.m. *of type* (2), *and* dim $\mathfrak{L}(\mu)=m^2$ *and if* $(\mu_1\mu_2^{-1})^{m-n+1}$ *is a nonsquare in* Φ, *then* $\mathfrak{L}(\mu_1)\not\cong\mathfrak{L}(\mu_2)$.

This theorem is difficult to apply because we do not know the dimension of $\mathfrak{L}(\mu)$ in general. However, a low dimensional survey reveals that if Φ contains nonsquares there are at least two non-isomorphic algebras of each of the forms $S_{6,2}(\mu)$, $R_{4,2}(\mu)$, $R_{6,2}(\mu)$, and $R_{4,3}(\mu)$, and they are of the types (in the sense of [3, p. 299]) A_7, D_4, D_6, and A_8 respectively.

Added in proof: The problem raised in the "Remark" has been answered in the affirmative.

BIBLIOGRAPHY

1. H. S. M. Coxeter, *Regular polytopes*, 2nd ed., Macmillan, New York, 1963.

2. ———, *Generators and relations for discrete groups*, 2nd ed., Ergebnisse der Mathematik und ihrer Grenzgebiete, Band 14, Springer-Verlag, New York, 1965.

3. N. Jacobson, *Lie algebras*, Interscience Tracts 10, Interscience, New York, 1962.

4. J.-P. Serre, *Algèbres de Lie semi-simples complexes*, Benjamin, New York, 1966.

UNIVERSITY OF SASKATCHEWAN

International Journal of Modern Physics A, Vol. 1, No. 2 (1986) 303–414

KAC-MOODY AND VIRASORO ALGEBRAS IN RELATION TO QUANTUM PHYSICS

PETER GODDARD

Department of Applied Mathematics and Theoretical Physics, University of Cambridge, Silver Street, Cambridge CB3 9EW, U.K.

and

DAVID OLIVE

Blackett Laboratory, Imperial College, London SW7 2BZ, U.K.

Received 13 May 1986

Contents

1. Introduction

1.1. *Background*

Symmetry plays a central role in physics, in the analysis of physical systems and the formulation of physical laws. Group theory is in essence the mathematical study of symmetry and it follows that it provides mathematical tools appropriate for theoretical physics. Indeed the use of finite groups (e.g., in crystallography) and continuous infinite groups which are finite dimensional (e.g., the rotation group or internal symmetry groups) is well established. An important feature of nature is locality or causality; combining locality with symmetry leads to the possibility of infinite dimensional symmetries where the symmetry operations are functions defined over space-time, a prime example of this being gauge invariance in theories of interactions between fundamental particles.

In general the study of representation theory, etc. of infinite-dimensional groups such as the group of all possible gauge transformations has not progressed very far, except for those relevant to 1-dimensional and certain 2-dimensional systems (and systems in higher dimensions which in some essential respects are 1- or 2-dimensional). Here the theory of affine Kac-Moody algebras and their associated Virasoro algebras, provides an extremely powerful yet natural framework for unifying the concepts of symmetry and locality. These Lie algebras are the infinitesimal version of certain infinite-dimensional groups, which we shall define in Subsec. 1.2, but at present it is easier to study the algebras. We can regard this as a second stage generalization of angular momentum theory, beyond the theory of ordinary finite-dimensional Lie algebras.

The Virasoro algebra arises as the algebra of the conformal group in one or two dimensions where, unlike three or more dimensions, it is infinite dimensional. (In two dimensions the algebra consists of two commuting Virasoro algebras.) The Virasoro algebra $\hat{v}$, has a basis consisting of generators L_m, $m \in \mathbb{Z}$, the set of integers, together with a central element, c, i.e., an element $\mathbb{Z}$ such that $[L_n, c] = 0$, satisfying

$$[L_m, L_n] = (m - n)L_{m+n} + \frac{c}{12}m(m^2 - 1)\delta_{m,-n}, \quad m, n \in \mathbb{Z}. \tag{1.1.1}$$

Classically, $c = 0$ but, as we shall see, it plays a crucial role in any quantum mechanical application. We shall be interested in unitary representations, i.e., those satisfying the hermiticity conditions $L_n^\dagger = L_{-n}$ and if such a representation is irreducible, c will have to take a constant value in it and we can think of it as a real number. The Virasoro algebra was first studied extensively in the early 1970s in the context of string theories. It is relevant in any theory in 2-dimensional space-time which possesses conformal invariance.

Almost at the same time as dual resonance models, which evolved into string theories were formulated by particle physics, Victor Kac and Bob Moody independently initiated the study of a class of infinite-dimensional Lie algebras including the (untwisted) affine Kac-Moody algebras, which form the central topic of this article.[1] The untwisted affine Kac-Moody algebra, $\hat{g}$, associated with a compact finite-dimensional Lie algebra g, is defined by the commutation relations

$$[T_m^a, T_n^b] = if^{abc}T_{m+n}^c + km\delta^{ab}\delta_{m,-n}, \tag{1.1.2}$$

where $m, n \in \mathbb{Z}$; a, b, c run over the values 1 to dim g; k is a central term, commuting with all the T_m^a; and f^{abc} are a set of totally antisymmetric structure constants for g (so that the generators $\{T_0^a\}$ form a subalgebra isomorphic to the original finite-dimensional compact algebra, g). As will be explained in Subsec. 3.2, it proved possible to construct root-systems for affine Kac-Moody algebras (associated with a simple g) rather analogous to those for finite-dimensional simple algebras. Further, there is also a parallel theory for an important class of representations, including character formulas which generalize Weyl's famous expression for finite-dimensional representations of g.[a] Just as in Weyl's finite-dimensional case, the denominator of the Kac-Weyl formula can be written in two alternative ways, one as a product and the other as a sum, now both infinite. By equating these alternatives various combinatorial identities and relations between theta functions can be obtained. Thus the theory of Kac-Moody algebras has deep links with combinatorics and the theory of modular forms and theta functions. Beyond this there are profound connections with other topics in mathematics such as finite simple groups and areas of topology.

[a] See Ref. 128.

The Virasoro and Kac-Moody algebras are not unrelated structures. In a sense the latter is a more detailed structure than the former because with each Kac-Moody algebra (1.1.2) there is a Virasoro algebra associated in a natural way (1.1.1) so that together they form a semi-direct product,

$$[L_m, T_n^a] = -nT_{m+n}^a, \quad [L_m, k] = 0. \tag{1.1.3}$$

The untwisted affine Kac-Moody algebras (1.1.2) are really special cases of algebras studied earlier in theoretical physics; they are obtained from current algebras in the case in which space-time is 2-dimensional and space itself is compact, i.e., a circle (see Secs. 3 and 5). In this context the Virasoro algebra corresponds to the (traceless) energy-momentum tensor and the relation between the two algebras corresponds to the construction of the energy momentum tensor out of currents, as proposed by Sugawara and others (see Secs. 2 and 4).

We shall be considering a class of irreducible unitary representations of Kac-Moody and Virasoro algebras, whose definition is motivated by physical considerations (which are reviewed in Sec. 2), called highest weight representations. It is not surprising that these representations are also of particular mathematical interest. In such representations the values of the central elements c and k are not arbitrary and a question of considerable mathematical and physical interest, whose complete answer has only been given very recently, is the determination of these permitted values. This is discussed in Subsecs. 1.4 and 3.4.

From about 1978 onwards, much progress has been made in constructing highest weight representations of affine Kac-Moody algebras and this work incorporates various ideas and constructions from developments in theoretical physics of the 1960s and early 1970s. Among these is the vertex operator construction, which played a central role in dual resonance theory[2] and is discussed in Secs. 6 and 7. This provided the first construction of highest weight representations. Another ingredient was Gell-Mann's quark model construction of current algebra (applied in two dimensions). This leads to the fermionic construction discussed in Sec. 5. The equivalence of these constructions is a manifestation of the boson-fermion equivalence of the sort that Skyrme[3] established by constructing a fermionic quantum field operator for the Sine-Gordon soliton. This is the subject of Sec. 7.

In the last couple of years there has been a considerable feedback into apparently diverse areas of theoretical physics from the mathematical progress made after the mid-1970s. These areas include the theory of integrable field equations and solitons (particularly in the work of the Kyoto school; see Ref. 4), the theory of critical phenomena in 2-dimensional statistical systems (see Subsec. 2.5) and the construction of consistent and, potentially, physically realistic quantum string theories treating all particle interactions in a unified way (see Subsec. 7.5).

Subsection 1.2 is devoted to obtaining the algebras we are to study starting from the infinite-dimensional groups to which they correspond. So obtained, the algebras lack the central elements c and k and are called loop algebras. In Subsec. 1.3 we discuss central extensions of Lie algebras and their role in quantum theory, indicating why the forms they take in Eqs. (1.1.1) and (1.1.2) are essentially the most general in these cases. In the fourth and final subsections we define a class of unitary representations of the algebras which is particularly relevant in physics and outline results on the quantization of c and k in such representations.

The material presented here is based in part on previous reviews that we have written,[5,6,7,8] the main theme being the central importance of the Sugawara construction of the Virasoro algebra. For other reviews of Kac-Moody algebras and their physical applications consult Refs. 9, 10 and 11. For a synopsis of the mathematical development see Ref. 12 and for more details consult other articles in the volume of the Proceedings of the 1983 Berkeley meeting on *Vertex Operators in Mathematics and Physics* (see Ref. 12), the review by MacDonald[13] and the book by Kac.[14]

1.2. *Algebras and groups*

A more geometric picture of the algebras just introduced can be obtained by constructing the infinite-dimensional groups whose Lie algebras are $\hat{g}$ and $\hat{v}$, with their centres omitted, i.e., $\hat{g}_0$ and $\hat{v}_0$, respectively.

Suppose g is the Lie algebra of our ordinary finite-dimensional compact connected Lie group, G. To obtain the infinite-dimensional group $\mathscr{G}$ with the untwisted affine Kac-Moody algebra $\hat{g}_0$ associated with g as its Lie algebra, we take $\mathscr{G}$ to be the set of (suitably smooth) maps from the circle, S^1, to G. We shall represent S^1 as the unit circle in the complex plane

$$S^1 = \{z \in \mathbb{C} : |z| = 1\} \tag{1.2.1}$$

and denote a typical map by

$$z \to \gamma(z) \in G. \tag{1.2.2}$$

The group operation is defined on G in the obvious way, by pointwise multiplication, i.e., given two such maps $\gamma_1, \gamma_2 : S^1 \to G$, the product of γ_1 and γ_2 is $\gamma_1 \cdot \gamma_2 : S^1 \to G$ where

$$\gamma_1 \cdot \gamma_2(z) = \gamma_1(z)\gamma_2(z). \tag{1.2.3}$$

Clearly this makes $\mathscr{G}$ into an infinite-dimensional Lie group. It is called the loop group of G.

We shall now calculate the Lie algebra $\hat{g}_0$ of $\mathscr{G}$. We start with a basis T^a, $1 \le a \le \dim g$, for g, with

$$[T^a, T^b] = if^{ab}{}_c T^c, \tag{1.2.4}$$

where $f^{ab}{}_c$ are the structure constants of g (not necessarily assumed totally antisymmetric at this stage). A typical element of G (assumed connected) is then of the form

$$\gamma = \exp[-iT^a\theta_a], \tag{1.2.5}$$

the θ_a, $1 \le a \le \dim g$, being parameters for G. A typical element of $\mathscr{G}$ (or rather the connected component containing the identity, consisting of maps $\gamma : S^1 \to G$ which are topologically trivial, i.e., can be continuously deformed to the constant map $\gamma(z) = 1$) can then be described by $\dim g$ functions $\theta_a(z)$ defined on the unit circle,

$$\gamma(z) = \exp[-iT^a\theta_a(z)]. \tag{1.2.6}$$

For elements near the identity,

$$\gamma \approx 1 - iT^a\theta_a, \quad \gamma(z) \approx 1 - iT^a\theta_a(z). \tag{1.2.7}$$

Making a Laurent expansion of $\theta_a(z)$,

$$\theta_a(z) = \sum_{n=-\infty}^{\infty} \theta_a^{-n} z^n, \tag{1.2.8}$$

we see that if we introduce generators

$$T_n^a = T^a z^n \tag{1.2.9}$$

for G, the θ_a^n, $1 \le a \le \dim g$, $n \in \mathbb{Z}$, provide an infinite set of parameters for G, with

$$\gamma(z) \approx 1 - i\sum_{n,a} T_{-n}^a \theta_a^n \tag{1.2.10}$$

near the identity. From Eq. (1.2.9) we see that $\mathscr{G}$ has the Lie algebra

$$[T_m^a, T_n^b] = if^{ab}{}_c T_{m+n}^c. \tag{1.2.11}$$

This is the untwisted affine Kac-Moody algebra $\hat{g}_0$, the algebra of the group of maps $S^1 \to G$.

Note that the operators T_0^a, $1 \le a \le \dim g$, generate a subalgebra of $\hat{g}_0$ isomorphic to g. It corresponds to the subgroup of $\mathscr{G}$ consisting of constant maps $S^1 \to G$, which is clearly isomorphic to G itself.

If G is a compact group and $\{T^a\}$ a basis of Hermitian generators for g.

$$T^{a\dagger} = T^a, \tag{1.2.12}$$

then for $|z| = 1$, $z^* = z^{-1}$ so that

$$T_n^{a\dagger} = T_{-n}^a. \tag{1.2.13}$$

A representation of $\hat{g}$ satisfying this hermiticity condition will be called unitary because the representatives of $\gamma(z)$ will then be unitary (for real parameters θ_a^n and $|z| = 1$).

To construct the infinite dimensional group corresponding to the Virasoro algebra, consider the group $\mathscr{V}$ of smooth one-to-one maps $S^1 \to S^1$ with the group multiplication now defined by composition

$$\xi_1 \circ \xi_2(z) = \xi_1(\xi_2(z)). \tag{1.2.14}$$

Notice this is different from the group we would obtain if we regard S^1 as U(1), the Lie group of complex numbers of unit modulus under multiplication, and construct the corresponding loop group $\mathscr{G}$ of maps $S^1 \to \mathrm{U}(1)$, because the fact that U(1) is Abelian implies that $\mathscr{G}$ is also whilst the composition (1.2.14) is clearly not commutative.

To calculate the Lie algebra of $\mathscr{V}$, consider its faithful representation defined by its action on functions $f: S^1 \to V$, where V is some vector space,

$$D_\xi f(z) = f(\xi^{-1}(z)). \tag{1.2.15}$$

For an element $\xi \in \mathscr{V}$ close to the identity

$$\xi(z) = ze^{-i\varepsilon(z)}, \tag{1.2.16}$$

$$\xi^{-1}(z) \approx z + iz\varepsilon(z), \tag{1.2.17}$$

so that

$$D_\xi(f) \approx f(z) + i\varepsilon(z)z\frac{d}{dz}f(z). \tag{1.2.18}$$

Making a Laurent expansion of $\varepsilon(z)$,

$$\varepsilon(z) = \sum_{n=-\infty}^{\infty} \varepsilon_{-n} z^n, \tag{1.2.19}$$

we are led to introduce generators

$$L_n = -z^{n+1}\frac{d}{dz}, \quad n \in \mathbb{Z}. \tag{1.2.20}$$

These satisfy the Lie algebra $\mathscr{V}_0$

$$[L_m, L_n] = (m - n)L_{m+n}. \tag{1.2.21}$$

A representation of this algebra will be unitary if the hermiticity condition

$$L_n^\dagger = L_{-n} \tag{1.2.22}$$

holds.

An important finite-dimensional subgroup of $\mathscr{V}$ consists of Möbius transformations,

$$z \to \xi(z) = \frac{az + b}{b^* z + a^*}, \tag{1.2.23}$$

which clearly maps $S^1 \to S^1$ and, provided that $|a|^2 > |b|^2$, this map $\xi(z)$ covers S^1 once positively as z goes round S^1 positively. We can rescale a and b so that $|a|^2 - |b|^2 = 1$; then

$$\begin{pmatrix} a & b \\ b^* & a^* \end{pmatrix} \in \mathrm{SU}(1,1). \tag{1.2.24}$$

The generators of this SU(1, 1) subgroup are easily seen to be $\{L_{-1}, L_0, L_1\}$. Actually there is an infinity of SU(1, 1) subgroups which can be obtained by doing Möbius transformations on z^n:

$$z \to \xi(z) = \left(\frac{az^n + b}{b^* z^n + a^*}\right)^{1/n}. \tag{1.2.25}$$

The generators of this subgroup are $\left\{\frac{1}{n}L_{-n}, L_0, \frac{1}{n}L_n\right\}$.

There is a natural interrelation between the groups $\mathscr{V}$ and $\mathscr{G}$ and, consequently, the algebras $\mathfrak{v}_0$ and $\mathfrak{g}_0$. To see this, consider a faithful representation of G in a vector space V. A map $\xi \in \mathscr{V}$ is represented on functions $f: S^1 \to V$ as in (1.2.15) by

$$\xi f(z) = f(\xi^{-1}(z)) \tag{1.2.26}$$

whilst $\gamma \in g$ is represented by

$$\gamma f(z) = \gamma(z) f(z). \tag{1.2.27}$$

With this action $\mathscr{V}$ and $\mathscr{G}$ form the factors of a semidirect product; we define $(\xi, \gamma) = \xi \circ \gamma$ so that

$$(\xi, \gamma) f(z) = \gamma(\xi^{-1}(z)) f(\xi^{-1}(z)), \tag{1.2.28}$$

and

$$(\xi_1, \gamma_1)(\xi_2, \gamma_2) = (\xi_1 \circ \gamma_2, (\gamma_1 \circ \xi_2) \cdot \gamma_2). \tag{1.2.29}$$

Since the generators of $\mathscr{V}$ and $\mathscr{G}$ are represented by the operators

$$L_n = -z^{n+1}\frac{d}{dz}, \quad T_n^a = z^n T^a, \tag{1.2.30}$$

their semi-direct product has the algebra,

$$[T_m^a, T_n^b] = if^{ab}{}_c T_{m+n}^c, \tag{1.2.31a}$$

$$[L_m, T_n^a] = -nT_{m+n}^a, \tag{1.2.31b}$$

$$[L_m, L_n] = (m-n)L_{m+n}. \tag{1.2.31c}$$

The infinite dimensional groups $\mathscr{V}$ and $\mathscr{G}$ may be thought of as the groups of general coordinate and gauge transformations, respectively, defined on the 1-dimensional space S^1.

1.3. *Central extensions*

The algebra of Eqs. (1.2.31) differs from that of Subsec. 1.1 by the absence of the central elements c and k. As will be explained in Subsec. 1.4 and Sec. 2 these are crucial in the quantum mechanical applications of these infinite-dimensional algebras and to see something of their significance we consider the role of central extensions in quantization. If in a classical theory one has a group of transformations with Lie algebra

$$[t^a, t^b] = if^{ab}{}_c t^c, \tag{1.3.1}$$

the corresponding generators will, under suitable circumstances, satisfy the Poisson bracket relations,

$$[t^a, t^b]_{\text{P.B.}} = f^{ab}{}_c t^c. \tag{1.3.2}$$

Following Dirac's quantization procedure, the corresponding quantum commutation is

$$[t^a, t^b] = i\hbar f^{ab}{}_c t^c + O(\hbar^2), \tag{1.3.3}$$

where $\hbar$ is Planck's constant. The unspecified terms of order $\hbar^2$ have to satisfy the Jacobi identities. The simplest possibility to contemplate is that the $O(\hbar^2)$ are c-numbers, i.e., multiples of the identity. What we have then is a central extension of the original Lie algebra (1.3.1), rescaled by a factor $\hbar$.

In general, a central extension of (1.3.1) is an algebra with basis $\{t^a\}$ with central elements k^j, $1 \le j \le M$, of the form

$$[t^a, t^b] = if^{ab}{}_c t^c + id^{ab}{}_j k^j, \tag{1.3.4}$$

$$[t^a, k^j] = [k^i, k^j] = 0. \tag{1.3.5}$$

The additional structure constants $d^{ab}{}_j$ are constrained by Jacobi identities,

$$[[t^a,t^b],t^c] + [[t^b,t^c],t^a] + [[t^c,t^a],t^b] = 0, \tag{1.3.6}$$

which imply

$$f^{ab}{}_e d^{ec}{}_j + f^{bc}{}_e d^{ea}{}_j + f^{ca}{}_e d^{eb}{}_j = 0 \tag{1.3.7a}$$

and, additionally, we have

$$d^{ab}{}_j = -d^{ba}{}_j. \tag{1.3.7b}$$

Equations (1.3.7) constitute a set of linear equations and the dimension of its set of solutions gives the number of independent central extensions. However, some of these are, in a sense, trivial because they can be removed by a redefinition of the generators

$$t^a \to t^a - \xi^a_j k^j, \tag{1.3.8}$$

under which

$$d^{ab}{}_j \to d^{ab}{}_j + if^{ab}{}_e \xi^e_j. \tag{1.3.9}$$

In the quantum mechanical context such a redefinition is by a term of order $\hbar$, which is insignificant in the classical limit, when $\hbar$ tends to zero. Thus one really wants to know the space of solutions of Eqs. (1.3.7) modulo transformations of the form (1.3.9).

For compact semi-simple finite-dimensional Lie algebras (1.3.1), by taking a basis in which the structure constants f^{abc} are totally antisymmetric, and such that $f^{abc}f^{abd} = \delta^{cd}$, we can use (1.3.7) and the Jacobi identity for f^{abc} to show that

$$d^{ab}{}_j = f^{abe}f^{cde}d^{cd}{}_j. \tag{1.3.10}$$

Thus in this case $d^{ab}{}_j$ can be set to zero by a transformation of the form (1.3.9) and all central extensions are essentially trivial. This argument is easily amended to establish the same result for a noncompact semi-simple finite-dimensional Lie algebra. In the case of the untwisted affine Kac-Moody algebra $\hat{g}$, a general central extension would have the form

$$[T^a_m, T^b_n] = if^{abc}T^c_{m+n} + d^{ab}_{mnj}k^j. \tag{1.3.11}$$

Provided that g is semi-simple, we can establish an equation of the form of (1.3.10) for d^{ab}_{m0j} and so use a transformation of the form (1.3.9) to set it to zero. Then $\{T^a_0\}$ closes on the algebra of g and $\{T^a_n\}$ transforms under its adjoint representation. It follows that d^{ab}_{mnj} is an isotropic tensor under this action of G. In the case where g is compact and simple. d^{ab}_{mnj} must be of the form $\delta^{ab}d_{mnj}$ (in a basis in which $f^{ab}{}_c$ is totally antisymmetric) and Eqs. (1.3.7) then imply that there is only one linearly independent solution for d^{ab}_{mnj}, modulo the transformations (1.3.9), namely that proportional to $m\delta^{ab}\delta_{m,-n}$. This gives the algebra (1.1.2). (If g is compact but not simple we can have an independent central extension to each U(1) or simple factor.)

To determine the possible central extensions of (1.2.21),

$$[L_m, L_n] = (m-n)L_{m+n} + c_{m,n}, \tag{1.3.12}$$

note that Eqs. (1.3.7) take the form

$$(m-n)c_{l,m+n} + (n-l)c_{m,n+l} + (l-m)c_{n,l+m} = 0, \tag{1.3.13a}$$

$$c_{m,n} = -c_{n,m}. \tag{1.3.13b}$$

Using the transformation $L_n \to L_n + c_{n,0}n$, $n \neq 0$, $L_0 \to L_0 + \frac{1}{2}c_{1,-1}$, we have afterwards $c_{n,0} = c_{0,n} = c_{1,-1} = 0$. Then putting $l = 0$ we deduce that $c_{m,n} = 0$ unless $m = -n$ and putting $l = -m-1, n = 1$, we find that the general solution is of the form

$$c_{m,n} = \frac{c}{12}m(m^2-1)\delta_{m,-n}, \tag{1.3.14}$$

as in (1.1.1).

1.4. *Highest weight representations*

In the next section, when we discuss some of the applications of infinite-dimensional Lie algebras in physics, we shall see that it is a particular class of representations that is relevant. This class has many properties in common with the unitary representations of finite-dimensional compact Lie algebras. The criterion for their selection comes from the requirement that, in physical applications, the spectrum of eigenvalues of L_0 must be non-negative, or at least bounded below. In Sec. 2 we shall define 2-dimensional quantum field theories with two commuting Virasoro algebras $\{L_n\}$ and $\{\bar{L}_n\}$ with, up to positive constants,

$$L_0 \approx H - P \quad \text{and} \quad \bar{L}_0 \approx H + P, \tag{1.4.1}$$

where H is the total energy and P the total momentum, so that both their eigenvalues need to be non-negative (see Eq. (2.1.38)). In string theory, the operator L_0 determines the mass spectrum and again we wish it to be intrinsically non-negative. In the applications in 2-dimensional statistical physics we again have two commuting Virasoro algebras but this time $L_0 + \bar{L}_0$ corresponds to D, the generator of dilatations but, again, absence of singularities at the origin requires that the spectrum of L_0 and $\bar{L}_0$ be non-negative.

Now the commutation relations (1.1.1) and (1.1.3) imply that the action of T^a_n or L_n

on an eigenvector of L_0 with eigenvalue λ gives an eigenvector of L_0 with eigenvalue $\lambda - n$, if it is not annihilated. Since the spectrum of L_0 is bounded below it has some lowest eigenvalue h and any corresponding eigenvector $|\Psi_0\rangle$ must satisfy

$$T_n^a|\Psi_0\rangle = 0, \quad n > 0, \tag{1.4.2a}$$

$$L_n|\Psi_0\rangle = 0, \quad n > 0. \tag{1.4.2b}$$

A representation for which L_0 is bounded below will be called a *highest weight representation* (for perverse historical reasons). If we further impose the unitarity conditions

$$T_n^{a\dagger} = T_{-n}^a, \tag{1.4.3a}$$

$$L_n^\dagger = L_{-n}, \tag{1.4.3b}$$

these are the representations of physical interest in Sec. 2 [In the applications in statistical physics, the unitarity conditions (1.4.3) do not always obtain.] Such representations are not necessarily irreducible but they can be decomposed into irreducible ones.

Not all representations are highest weight ones. For example the defining representations of Subsec. 1.2 have $c = 0$ and $k = 0$ and it is easy to see that L_0 is unbounded above and below. (Indeed if we relax the highest weight condition the restrictions on the values of c and k, which we wish to discuss, disappear.)

It is possible to enumerate all the irreducible unitary highest weight representations of both the Virasoro algebra (1.1.1) and the affine Kac-Moody algebra (1.1.2). It is clear that, in such a representation, all the states can be built up from "vacuum states," satisfying (1.4.2), by repeated application of the operators T_{-n}^a, L_{-n}, $n > 0$.

In the case of the Kac-Moody algebra, the vacuum states must form a representation of the finite-dimensional algebra $g = \{T_0^a\}$ and it is easy to see that an irreducible vacuum representation will generate an invariant subspace. Thus if the representation is irreducible with respect to $\hat{g}$ so is this vacuum representation with respect to g. Then the whole representation of $\hat{g}$ is characterized by this vacuum representation together with the value of k in (1.1.1) because, using this algebra and Eqs. (1.4.2a) and (1.4.3a), all of the inner products between states of the representations can be calculated. In Subsec. 3.4 we shall determine which vacuum representations are possible for a given value of k, but we note here, that, as will be apparent in the examples of Sec. 2, the values of k are quantized,

$$2k/\psi^2 \in \mathbb{Z}, \tag{1.4.4}$$

where ψ is a long root of g (see Sec. 3 for further explanation).

If we consider the Virasoro algebra alone, there is only one vacuum state $|h\rangle$ in any given irreducible highest weight representation and it satisfies

$$L_0|h\rangle = h|h\rangle, \quad L_n|h\rangle = 0, \quad n > 0. \tag{1.4.5}$$

Again all of the inner products between the states which are obtained by repeated application of L_{-n}, $n > 0$, can be calculated using (1.1.1), (1.4.3) and (1.4.5). Thus, in particular, if $n > 0$,

$$\begin{aligned}\|L_{-n}|h\rangle\|^2 &= \langle h|L_nL_{-n}|h\rangle \\ &= \langle h|[L_n, L_{-n}]|h\rangle \\ &= \left\{2nh + \frac{c}{12}n(n^2-1)\right\}\||h\rangle\|^2.\end{aligned} \tag{1.4.6}$$

Thus

$$2nh + \frac{c}{12}n(n^2-1) \geq 0 \quad \text{for all } n > 0 \tag{1.4.7}$$

and so we see, by taking $n = 1$ and then n large,

$$h \geq 0 \quad \text{and} \quad c \geq 0 \tag{1.4.8}$$

necessarily for an irreducible unitary highest weight representation. Such representations can then be characterized by the pair of non-negative numbers (c, h). Not all such values of (c, h) are permitted. The restrictions on them were found by Friedan, Qiu and Shenker[15] who established the following theorem:

In order for there to be a unitary representation of the Virasoro algebra corresponding to given values of c and h, it is necessary that *either*

$$c \geq 1 \quad \text{and} \quad h \geq 0, \tag{1.4.9}$$

or

$$c = 1 - \frac{6}{(m+2)(m+3)} \tag{1.4.10}$$

and

$$h = \frac{[(m+3)p - (m+2)q]^2 - 1}{4(m+2)(m+3)}, \tag{1.4.11}$$

where $m = 0, 1, 2, \ldots$; $p = 1, 2, \ldots, m+1$; $q = 1, 2, \ldots, p$.

This gives $\frac{1}{2}(m + 1)(m + 2)$ values of h for each value of c. The possibilities for c resemble the allowed energies in a quantum mechanical potential problem, with a threshold at $c = 1$. The first four terms in the sequence are:

$$c = 0, \quad h = 0;$$

$$c = \frac{1}{2}, \quad h = 0, \frac{1}{16} \text{ or } \frac{1}{2};$$

$$c = \frac{7}{10}, \quad h = 0, \frac{3}{80}, \frac{1}{10}, \frac{7}{16}, \frac{3}{5}, \text{ or } \frac{3}{2};$$

$$c = \frac{4}{5}, \quad h = 0, \frac{1}{40}, \frac{1}{15}, \frac{1}{8}, \frac{2}{5}, \frac{21}{40}, \frac{2}{3}, \frac{7}{5}, \frac{13}{8} \text{ or } 3. \tag{1.4.12}$$

It has now been established that all the representations permitted by the theorem indeed exist and they have been explicitly constructed.[16,17,18]

To indicate how the restrictions on the values of (c, h) are established, we consider a potential basis for the corresponding unitary highest representation. It is certainly spanned by vectors of the form

$$L_{-1}^{n_1} L_{-2}^{n_2} \ldots L_{-r}^{n_r} |h\rangle. \tag{1.4.13}$$

Each of these is an eigenvector of L_0 with eigenvalue, say

$$h + \sum_j j n_j \equiv h + N, \tag{1.4.14}$$

For different values of N, these vectors are orthogonal. For a given value of N the matrix of scalar products between the various vectors (1.4.13) can be calculated as a function of (c, h) using the Virasoro algebra, (1.4.2b) and (1.4.3b), up to a factor of $\langle h|h\rangle$ which we take to be 1. To illustrate this, consider the first three values of N.

$N = 0$:

$$\langle h|h\rangle = 1; \tag{1.4.15}$$

$N = 1$:

$$\langle \Psi|\Psi\rangle = \langle h|L_1 L_{-1}|h\rangle = 2h, \tag{1.4.16}$$

where

$$|\Psi\rangle = L_{-1}|h\rangle;$$

$N = 2$:

$$\begin{pmatrix} \langle\Psi_2|\Psi_2\rangle & \langle\Psi_1|\Psi_2\rangle \\ \langle\Psi_2|\Psi_1\rangle & \langle\Psi_1|\Psi_1\rangle \end{pmatrix} = \begin{pmatrix} 4h + \frac{1}{2}c & 6h \\ 6h & 8h^2 + 4h \end{pmatrix}, \tag{1.4.17}$$

where

$$|\Psi_1\rangle = L_{-1}^2|h\rangle, \quad |\Psi_2\rangle = L_{-2}|h\rangle.$$

The determinant of the matrix (1.4.17) is

$$4h(\tfrac{1}{2}c + (c-5)h + 8h^2). \tag{1.4.18}$$

If the representation is unitary, each of these matrices must be positive semi-definite. Conversely, if the matrices of scalar products of the vectors (1.4.13) are positive for each N, they can be used to define an inner product on the space with basis (1.4.13) with respect to which $L_n^\dagger = L_{-n}$. (If any of the matrices is positive semi-definite, but not actually positive definite, i.e. has some zero eigenvalues, the corresponding space can be consistently converted into a positive definite space by taking the quotient by the subspace of null states.)

This problem has been analysed.[15] The matrices corresponding to $N \leq 2$ will be positive semi-definite if $h \geq 0$ and

$$\tfrac{1}{2}c + (c-5)h + 8h^2 \geq 0. \tag{1.4.19}$$

Further we know from (1.4.8) that we shall need $c \geq 0$. These conditions leave us with the region of the first quadrant of the (c, h) plane on and outside the curve given by equality in (1.4.19); this is illustrated in Fig. 1.

To progress beyond the first few levels one needs a general formula and Friedan, Qiu and Shenker exploited the formula given by Kac[19,20] for the determinant of the matrix $M_N(c, h)$ of the scalar products of the vectors for a given value of N. The number $\pi(N)$ of such vectors grows quickly with N; it is given by the partition function

$$\prod_{n=1}^{\infty} (1-q^n)^{-1} = \sum_{N=0}^{\infty} \pi(N) q^N. \tag{1.4.20}$$

The formula of Kac for this $\pi(N) \times \pi(N)$ determinant is

$$\det M_N(c, h) = \prod_{k=1}^{N} \eta_k(c, h)^{\pi(N-k)}. \tag{1.4.21}$$

Here

$$\eta_k(c, h) = \prod_{pq=k} [h - h_{p,q}(c)], \tag{1.4.22}$$

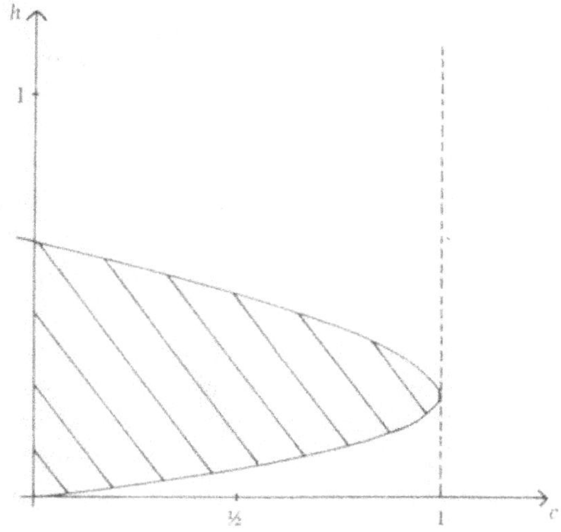

Fig. 1.

where p and q range over the positive integers and $h_{p,q}(c)$ is given by the expression for h in Eq. (1.4.11), with m related to c by Eq. (1.4.10). The formula (1.4.21) was proved by Feigin and Fuchs;[21] more recently other proofs have been given by Thorn[22], using techniques from string theory, and by Kac and Wakimoto[23] and by Kent[24] using techniques developed from those of Sec. 4.

The determinants of (1.4.21) are easily seen to be positive if $c > 1$ and $h \geq 0$. For large h, the matrices $M_N(c, h)$ are manifestly positive definite. It follows that they are positive semi-definite throughout $c \geq 1, h \geq 0$, as in (1.4.9). The only remaining region is $0 \leq c < 1$ and $h \geq 0$ and one might be tempted to try to reject all of this, except for $c = 0, h = 0$, which corresponds to the trivial representation $L_n = 0$. There are however simple and well-known representations for $c = \frac{1}{2}$ and $h = 0$, $\frac{1}{16}$ or $\frac{1}{2}$. (These correspond to taking $N = 1$ in Subsec. 2.1 and defining L_n by (2.1.29) with a single fermion field $H(z)$. Then $c = \frac{1}{2}$ from (2.1.31) and the NS case gives $h = 0$ and $\frac{1}{2}$ and the R case $h = \frac{1}{16}$.) The careful analysis of Friedan, Qiu and Shenker[15] showed that $c = 0$, $\frac{1}{2}$ were the first terms in the series, and the existence of representations for the other values of c in the sequence is established in Sec. 4.

We shall see, in Subsec. 4.5 and after, that the result, which comes as a by-product of the analysis of Friedan, Qiu and Shenker,[15] that a unitary representation of the Virasoro algebra in which $c = 0$ is trivial if L_0 is bounded below (i.e, if it is a highest weight representation) is important. Because their analysis is complicated, it is desirable to have a self-contained proof of it. We now give an argument due to Gomes.[25] Consider calculating the determinant of the matrix of (1.4.17) with $|\Psi_2\rangle = L_{-2N}|h\rangle$, $|\Psi_1\rangle = L_{-N}{}^2|h\rangle$ and $c = 0$. The result is

$$4N^3h^2(4h - 5N) \tag{1.4.23}$$

and for $h \neq 0$ this is clearly negative for large N. Hence if $c = 0$ we must have $h = 0$ and thus $L_n = 0$ for a unitary representation (since the norm of any state is clearly zero).

2. Physical Applications

2.1. *Current algebra*

There are a number of superficially disparate areas of theoretical physics in which infinite-dimensional algebras, of the sort introduced in Sec. 1, occur. Of these, the oldest is the current algebra approach to a local theory of symmetries of elementary particles, formulated in the 1960s. (See Ref. 26 for reprints of many of the original articles.) Next comes the string theories of particle interactions developed in the late 1960s and early 1970s.[2] More recent, and perhaps more surprising, is the theory of the behaviour of 2-dimensional statistical models, such as systems of spins on lattices, near the critical temperature for the system at which a second order phase transition occurs. Another class of theories recently discovered to possess such algebraic structures are certain 2-dimensional σ-models.

The feature common to all these theories is the connection with conformally invariant quantum field theories in two dimensions. The fact that we are restricted to talking about 2-dimensional systems, rather than 4-dimensional ones, is the price paid for having an algebraic structure which is tractable with our present knowledge. Of course, as this progresses, higher dimensional systems may become more accessible. Moreover it is not as severe a limitation as it might naively seem: string theories formulated in ten or more dimensions, but involving in an essential way 2-dimensional conformally invariant structures, are currently thought to offer the best prospects for the construction of a realistic unified theory of particle interactions; and experimentalists can readily make and study 2-dimensional substances described by the statistical models possessing the algebraic structures we are discussing.

The general idea that the Virasoro algebra (1.1.1) is relevant to conformally invariant quantum field theory in two dimensions has a long history.[27,28,29,30] More recently it has received persuasive advocacy from Friedan[31] and Belavin, Polyakov and Zamolodchikov.[32]

We shall say more about the relation of Kac-Moody algebras to σ-models in Subsecs. 2.3 and 2.4 and to statistical systems in Subsec. 2.5. In this section we shall discuss the connection between 2-dimensional current algebras and the algebras of Sec. 1. To see this connection we introduce the Kac-Moody field

$$T^a(z) = \sum_n T^a_{-n} z^n, \tag{2.1.1}$$

where the sum is over all $n \in \mathbb{Z}$. We then change from a field $T^a(z)$ defined on the unit circle S^1 to a field $J^a(\xi)$ defined on the real line but periodic with period L,

$$J^a(\xi) = \frac{\hbar}{L} T^a(e^{2\pi i \xi/L}). \tag{2.1.2}$$

In terms of $J^a(\xi)$, the Kac-Moody algebra (1.1.2) takes the form

$$[J^a(\xi), J^b(\eta)] = i\hbar f^{abc}J^c(\xi)\delta(\xi - \eta) + \frac{i\hbar^2}{2\pi}k\delta^{ab}\delta'(\xi - \eta) \tag{2.1.3}$$

(where the delta function δ and its derivative, δ', are understood as being periodic with period L). This equation is recognizable as a current algebra[26] and the c-member term occurs as the derivative of a delta function; this is known as a Schwinger term.[33] Notice that the period L does not occur explicitly in Eq. (2.1.3), leaving us free to consider the (infrared) limit $L \to \infty$. The appearance of the square of Planck's constant in the Schwinger term emphasizes that this term is a second order quantum effect, as discussed in Subsecs. 1.4 and 5.2, and, in calculational terms, corresponds to a 1-loop Feynman diagram. The integral suffix in (1.1.2) is seen to be $L/2\pi\hbar$ times the momentum conjugate to ξ, quantized in this way because of the periodicity.

The simplest physical theory in which we can find the current algebra (2.1.3) is that of free massless "quarks" (fermions) in one space and one time dimension. At first sight, this seems to be a rather trivial example because it is a free theory but we shall see that this conclusion is rather superficial. We start by considering a single real fermion field. To describe such a field, we use a real representation of γ-matrices in two dimensions:

$$\gamma^0 = \begin{pmatrix} 0 & 1 \\ 1 & 0 \end{pmatrix}, \quad \gamma^1 = \begin{pmatrix} 0 & -1 \\ 1 & 0 \end{pmatrix}, \quad \gamma^5 = \gamma^0\gamma^1 = \begin{pmatrix} 1 & 0 \\ 0 & -1 \end{pmatrix} \tag{2.1.4}$$

and a field ψ with two real components which, following Witten,[34] we write as

$$\psi = \begin{pmatrix} \psi_- \\ \psi_+ \end{pmatrix}, \tag{2.1.5}$$

for reasons that will become apparent. The system is described by the action

$$\mathcal{L} = \frac{1}{2}\int d^2x\bar{\psi}i\gamma^\mu\partial_\mu\psi, \tag{2.1.6}$$

where $\bar{\psi} = \psi^T\gamma^0$. This leads to the Dirac equation of motion

$$\gamma^\mu\partial_\mu\psi = 0, \tag{2.1.7}$$

where, as usual, $\partial_\mu = \partial/\partial x^\mu$, $x^0 = t$, $x^1 = x$. When written out in components, the equation of motion becomes

$$(\partial_0 + \partial_1)\psi^- = (\partial_0 - \partial_1)\psi^+ = 0. \tag{2.1.8}$$

These Weyl equations say that ψ^+ and ψ^- are functions only of $t + x$ and $t - x$, respectively,

$$\psi^+ \equiv \psi^+(t + x), \quad \psi^- \equiv \psi^-(t - x). \tag{2.1.9}$$

Each of ψ^+ and ψ^- is both Weyl (i.e., an eigenvector of γ^5) and a Majorana (i.e. real) spinor. Spinors with both these properties only exist in space-time dimensions of the form $8n + 2$, where n is an integer: see Ref. 35.

Independently, for each of the Weyl components, we have canonical anticommutation relations,

$$\{\psi(x), \psi(y)\} = \hbar\delta(x - y) \tag{2.1.10}$$

for $\psi \equiv \psi^+$ or ψ^- with $\{\psi^+(x), \psi^-(y)\} = 0$. Actually we wish to consider an elementary elaboration of this theory in which we have internal symmetry index i taking values from 1 to N, that is we take N non-interacting copies of the free real single fermion theory. Then the anticommutation relations (2.1.10) are replaced by

$$\{\psi_i(x), \psi_j(y)\} = \hbar\delta(x - y)\delta_{ij}. \tag{2.1.11}$$

Now to construct the current algebra (2.1.3) and so a representation of the untwisted affine Kac-Moody algebra $\hat{g}$ of (1.1.2), we take a real (not necessarily irreducible) representation of g under which $T^a = iM^a$, where M is an $N \times N$ real antisymmetric matix satisfying

$$[M^a, M^b] = f^{abc}M^c. \tag{2.1.12}$$

If g has such a (faithful) representation, the corresponding group G is a subgroup of the O(N) symmetry group of the theory. Associated with this symmetry group G are conserved currents

$$J^a_\mu = \frac{1}{2\sqrt{2}}\bar{\psi}M^a\gamma_\mu\psi \tag{2.1.13}$$

(where we have introduced a convenient normalization factor). We introduce light-cone coordinates for vectors $V = (V^0, V^1)$ by

$$V^\pm = (V^0 \pm V^1)/\sqrt{2}, \quad V_\pm = (V_0 \pm V_1)/\sqrt{2}, \tag{2.1.14}$$

so that $V^\pm = V_\mp$ and the scalar product

$$V^\mu W_\mu = V^+W^- + V^-W^+ = V_+W_- + V_-W_+. \tag{2.1.15}$$

Then the light-cone components of the currents are

$$J^a_\pm = \frac{i}{2}\psi^T_\pm M^a\psi_\pm \tag{2.1.16}$$

so that J_+^a is a function only of x^+ and J_-^a is a function only of x^-. Again we can consider $J_+^a(x^+)$ and $J_-^a(x^-)$ independently; they will commute with one another because ψ_+ and ψ_- anticommute.

Denoting either component by $J^a(x)$, the canonical anticommutation relations (2.1.10) imply the current algebra (2.1.3) with a specific value of k. To specify this value, note that we can choose the basis of generators $\{T^a\}$ of g so that

$$\mathrm{tr}(T^a T^b) = y\delta^{ab}, \tag{2.1.17}$$

where the constant y depends on the representation chosen. (If g is simple any basis in which f^{abc} is totally antisymmetric will do; for a general compact g the relative normalization of the generators of various simple or $u(1)$ factors needs to be adjusted, in a representation-dependent way, to achieve this orthonormality.)

The value of k is given by

$$k = \tfrac{1}{2}y_M, \tag{2.1.18}$$

where

$$\mathrm{tr}(M^a M^b) = -y_M\delta^{ab}. \tag{2.1.19}$$

The constant y_M is related to the Dynkin index of the representation and this is explained in more detail in Subsec. 5.2, where the current algebra (2.1.3) is carefully established, from the basic anticommutators for the Fermi fields, in the periodic case. Thus we have two current algebras one for each light cone component.

Note that (2.1.17) can be used to define a scalar product on g. If $X = X^a T^a$, $Y = Y^b T^b \subset g$, we define their scalar product by

$$\langle X, Y\rangle = \frac{1}{y}\mathrm{tr}(XY) = X^a Y^a. \tag{2.1.20}$$

This scalar product depends only on the scale (or scales if g is not simple) set by the choice of representation and not on the particular representation used. This scalar product is invariant under the action of G,

$$\langle \gamma X\gamma^{-1}, \gamma Y\gamma^{-1}\rangle = \langle X, Y\rangle, \tag{2.1.21}$$

which is equivalent to

$$\langle [Z, X], Y\rangle + \langle X, [Z, Y]\rangle = 0, \quad X, Y, Z \in g. \tag{2.1.22}$$

The required periodicity of J^a,

$$J^a(x) = J^a(x + L), \tag{2.1.23}$$

will follow if ψ is either periodic or antiperiodic, i.e.,

$$\textit{either} \quad \psi(x+L)=\psi(x) \qquad \textit{or} \quad \psi(x+L)=-\psi(x). \tag{2.1.24}$$

Just as Eq. (2.1.2) defines a dimensionless field $T^a(z)$ from $J^a(z)$, we can obtain a dimensionless Fermi field $H(z)$ from $\psi(x)$,

$$\psi_i(x)=\left\{\frac{\hbar}{L}\right\}^{1/2} H_i(z), \quad z=e^{2\pi i x/L}. \tag{2.1.25}$$

The periodicity assumptions mean that we have an expansion

$$H_i(z)=\sum_r b^i_{-r} z^r, \tag{2.1.26}$$

where in the case where ψ is periodic, the Ramond (R) case, $r \in \mathbb{Z}$ and in the case where ψ is antiperiodic, the Neveu-Schwarz (NS) case, $r \in \mathbb{Z}+\frac{1}{2}$.[36,37,59] The canonical anti-commutation relations (2.1.10) are then equivalent to

$$\{b^i_r, b^j_s\}=\delta^{ij}\delta_{r,-s} \tag{2.1.27}$$

where *either* $r, s \in \mathbb{Z}$ (R) *or* $r, s \in \mathbb{Z}+\frac{1}{2}$ (NS). These fermionic annihilation and creation operators satisfy $b^{i\dagger}_r = b^i_{-r}$ and define a Fock space by the application of the b^i_r, $r>0$, to a vacuum state $|0\rangle$ satisfying

$$b^i_r|0\rangle = 0, \quad r>0. \tag{2.1.28}$$

In the NS case we take the vacuum to be a single non-degenerate state; in the R case it must provide a representation for the Clifford algebra

$$\{b^i_0, b^j_0\}=\delta^{ij} \tag{2.1.29}$$

and so the vacuum states form a $2^{[N/2]}$-dimensional irreducible representation space for this algebra, where $[x]$ denotes the integral part of x. In terms of these operators.

$$T^a_n=\frac{1}{2}\sum_r b^i_r M_{ij} b^j_{n-r}, \tag{2.1.30}$$

where the sum is *either* over $r \in \mathbb{Z}$ (R case) *or* over $r \in \mathbb{Z}+\frac{1}{2}$ (NS case). This fermionic "quark model" representation of the untwisted affine Kac-Moody algebras is discussed in detail in Sec. 5.

To obtain a Virasoro algebra associated with this Kac-Moody algebra, we consider the symmetric energy momentum tensor of the theory.

$$\theta^{\mu\nu} = \tfrac{1}{4}\{\bar{\psi}\gamma^{\mu}\overleftrightarrow{\partial}^{\nu}\psi + \bar{\psi}\gamma^{\nu}\overleftrightarrow{\partial}^{\mu}\psi\}, \tag{2.1.31}$$

with an implicit sum over the internal index i. This tensor is traceless because of the equation of motion (2.1.7),

$$2\theta_{+-} = \theta^{\mu}_{\mu} = 0. \tag{2.1.32}$$

This is necessary for the conformal invariance of the theory. Thus the only nonzero components of $\theta_{\mu\nu}$ are θ_{++} and θ_{--}. These are proportional to $\psi_+\overleftrightarrow{\partial}_+\psi_+$ and $\psi_-\overleftrightarrow{\partial}_-\psi_-$, respectively. Again let us fix our attention on one light-cone component. Then θ_{++} is proportional to $\frac{1}{4}\left[z\frac{dH}{dz}, H\right]$ but for this to be well defined we need to introduce normal ordering with respect to the fermionic oscillators to avoid divergences,

$$\begin{aligned} {}^\circ_\circ b^i_r b^j_s {}^\circ_\circ &= -b^j_s b^i_r \qquad \text{if } r > 0, \\ &= \tfrac{1}{2}[b^i_r, b^j_s] \quad \text{if } r = 0, \\ &= b^i_r b^j_s \qquad \text{if } r < 0. \end{aligned} \tag{2.1.33}$$

We set

$$L(z) \equiv \sum_n L_{-n} z^n = \frac{1}{2}{}^\circ_\circ z\frac{dH}{dz} H {}^\circ_\circ + \varepsilon N, \tag{2.1.34}$$

where $\varepsilon = 0$ in the NS case and $\varepsilon = \frac{1}{16}$ in the R case. The term εN reflects a particular choice of the additive ambiguity introduced by normal ordering. With the definition (2.1.34) we obtain the Virasoro algebra

$$[L_m, L_n] = (m-n)L_{m+n} + \frac{N}{24}m(m^2-1)\delta_{m,-n}, \tag{2.1.35}$$

which is just (1.1.1) with $c = \frac{1}{2}N$. This commutator will be proved in Subsec. 5.4. We have further that

$$[L_m, T^a_n] = -nT^a_{m+n}, \tag{2.1.36}$$

providing a representation of the whole semi-direct product algebra, or rather two commuting copies, one for each light cone direction. We also have that

$$[L_m, b^i_r] = -\left(\frac{m}{2} + r\right)b^i_{m+r}. \tag{2.1.37}$$

The energy-momentum vector is

$$P^\mu = \int_0^L \theta^{\mu 0}\, dx\,, \tag{2.1.38}$$

so that, as we stated in (1.4.1) L_0 and $\bar{L}_0$ are equal to a positive constant times $H - P$ and $H + P$, respectively, where $\bar{L}_n$ is defined with respect to θ_{--} in the same way L_n is defined for θ_{++}. Hence we require the spectra of L_0 and $\bar{L}_0$ to be non-negative leading to the requirement for highest weight representations discussed in Subsec. 1.4. Similar considerations will apply in Subsec. 2.2.

The importance of the commutation relations of the components of the energy-momentum tensor was first realized by Dirac[38,39] and Schwinger,[40,41] who found an analogue of (2.1.35) in four space-time dimensions.

2.2. *Scalar fields in two dimensions*

We have seen that when space-time has two dimensions and the quark model currents are periodic in space, then their current algebra consists of two commuting copies of an affine Kac-Moody algebra $\hat{g}$. Surprisingly, it is considerably more difficult to construct the same algebraic structure from spinless boson fields unless g is Abelian. For example, if g is simple, the current bilinear in real Bose fields,

$$J^a_\mu = \partial_\mu \phi^i M^a_{ij} \phi^j\,, \tag{2.2.1}$$

possesses a commutator of its space components which vanishes instead of yielding an equal time delta function, multiplied by a structure constant and the time component, as in the quark model. Further, although J^a_μ is conserved, its dual $\varepsilon^{\mu\nu} J^a_\nu$ is not.

On the other hand if g is Abelian, the current linear in Bose fields:

$$J^a_\mu = \beta \partial_\mu \phi^a\,, \quad a = 1, \ldots, \text{rank } g\,, \tag{2.2.2}$$

yields the desired commutation relations. Furthermore it is conserved if the fields ϕ are free and massless, and its dual, $\varepsilon^{\mu\nu} J^a_\nu$ is conserved trivially. The currents (2.2.2) therefore yield two copies of the affine Kac-Moody algebra based on $U(1)^r$.

The energy-momentum tensor for these free massless scalar fields is traceless and yields a Virasoro algebra (1.1.1) with c-number equal to rank g, the number of Bose fields. Combining this result with (2.1.35) of the previous section, we have that the Virasoro algebra arising from the energy-momentum tensor of a theory of n_B free massless bosons and n_F free massless fermions has a c-number,

$$c = n_B + \tfrac{1}{2} n_F\,. \tag{2.2.3}$$

This result is very familiar in string theory.[2] This shows us that c has a physical significance, and provides us with a criterion as to whether or not a conformally invariant theory in two dimensions is composed of free fields.

If g is simple and simply-laced it is possible to extend (2.2.2) by applying it to the Cartan subalgebra h of g and constructing the currents corresponding to the step operators of g as exponentials in ϕ, via the vertex operator construction. This is a purely quantum mechanical construction, and will be explained more in Secs. 4 and 6.

Another possibility is to consider spinless fields confined to the manifold of a simple (or possibly semi-simple) Lie group G whose algebra is g. The natural G-invariant action for this theory, the principal σ-model associated with G, is

$$\mathscr{A}_0 = -\frac{1}{4\lambda^2}\int d^2x\psi^2\langle\gamma^{-1}\partial_\mu\gamma, \gamma^{-1}\partial^\mu\gamma\rangle, \tag{2.2.4}$$

where $\gamma(x,t)$ takes values in G. The bracket $\langle\,,\rangle$ contains an expression bilinear in the generators of g and is defined by Eq. (2.1.20). The factor ψ^2 denotes the square of the length of the highest root of g, and is included with $\langle\ ,\ \rangle$ to make it intrinsic to g, and independent of the basis chosen. Then the variation of $\mathscr{A}_0$ corresponding to some small variation $\delta\gamma$ of γ is proportional to

$$\begin{aligned} 2\lambda^2\delta\mathscr{A}_0 &= -\int d^2x\langle\gamma^{-1}\partial_\mu\gamma, \gamma^{-1}\partial^\mu\delta\gamma\rangle + \int d^2x\langle\gamma^{-1}\partial_\mu\gamma, \gamma^{-1}\delta\gamma\gamma^{-1}\partial^\mu\gamma\rangle \\ &= \int d^2x\langle\gamma^{-1}\delta\gamma, \partial_\mu(\gamma^{-1}\partial^\mu\gamma)\rangle, \end{aligned} \tag{2.2.5}$$

where we have discarded the surface terms which come from integrating by parts. Thus the action $\mathscr{A}_0$ leads to the equations of motion

$$\partial^\mu(\gamma^{-1}\partial_\mu\gamma) = 0. \tag{2.2.6}$$

Now, if we set

$$\gamma^{-1}\partial_\mu\gamma = J_\mu = \sum_{a=1}^{\dim g} J_\mu^a(x,t)T^a, \tag{2.2.7}$$

where T^a are, as usual, a set of generators of g, Eq. (2.2.6) amounts to the conservation equation

$$\partial^\mu J_\mu^a = 0. \tag{2.2.8}$$

But J_μ also satisfies a consistency condition coming from Eq. (2.2.7), which states that J_μ, regarded as a gauge potential, is a pure gauge. Hence the associated curvature vanishes:

$$\partial_\mu J_\nu - \partial_\nu J_\mu + [J_\mu, J_\nu] = 0. \tag{2.2.9}$$

We again find that the commutator of the space components of J_μ^a vanish instead of yielding the quark model result, which lead to the Kac-Moody algebra structure.

Further, although J^a_μ is conserved by Eq. (2.2.8) its dual, $\varepsilon^{\mu\nu}J^a_\mu$ is not, by virtue of Eq. (2.2.9).

In the free fermion case both the quark current J^a_μ and its dual $\varepsilon^{\mu\nu}J^a_\mu$ were conserved, so this meant that then

$$\partial_- J^a_+ = \partial_+ J^a_- = 0. \tag{2.2.10}$$

Finding how to obtain quantities J_+ and J_- with these properties (2.2.10) is the clue to obtaining quantities satisfying a commuting pair of Kac-Moody algebras $\hat{g}$, as we shall now see.

2.3. *Sigma-models with Wess-Zumino term*

It was Witten[34] who first found bosonic currents satisfying (2.2.10) and the Kac-Moody commutation relations (2.1.3) for any choice of simple g and level $x = 2k/\psi^2$. He first noticed that if

$$J_+ \equiv \gamma^{-1}\partial_+\gamma \tag{2.3.1}$$

were to satisfy the first of Eqs. (2.2.10), then, since

$$\begin{aligned}\partial_-(\gamma^{-1}\partial_+\gamma) &= \gamma^{-1}\partial_-\partial_+\gamma - \gamma^{-1}\partial_-\gamma\gamma^{-1}\partial_+\gamma \\ &= \gamma^{-1}\partial_+(\partial_-\gamma\gamma^{-1})\gamma,\end{aligned}$$

it is desirable to redefine J_- to be

$$J_- = \partial_-\gamma\gamma^{-1} = \gamma(\gamma^{-1}\partial_-\gamma)\gamma^{-1} \tag{2.3.2}$$

as it then automatically satisfies the second of Eqs. (2.2.10). It is this need to treat the two light-cone components of J_μ differently that had eluded the authors of previous treatments of σ-models and their relations to Kac-Moody algebras, or, equivalently, current algebras.

Equations (2.2.10) with the definitions (2.3.1) and (2.3.2) constitute perfectly good local equations of motion but they no longer follow from the action $\mathcal{A}_0$, (2.2.4). Consequently, it is necessary to modify this action by the addition of a new term, which Witten[34] identified as the so-called Wess-Zumino[42] term:

$$\mathcal{A} = \mathcal{A}_0 + K\Gamma(B), \tag{2.3.3}$$

where

$$\Gamma(B) = \frac{1}{24\pi}\int_B \varepsilon^{\lambda\mu\nu}\psi^2\langle\gamma^{-1}\partial_\lambda\gamma, \gamma^{-1}\partial_\mu\gamma\gamma^{-1}\partial_\nu\gamma\rangle\, d^3y \tag{2.3.4}$$

and K is a constant of the same dimensions as action.

What is meant by the integral $\Gamma(B)$ requires some explanation. First note the use again of the bracket (2.1.20) multiplied by ψ^2. Although the contents of the bracket appear to be trilinear in g, the antisymmetry assures, by the Lie algebra commutation relations, that it is actually bilinear as needed. It follows that $\Gamma(B)$ is independent of any choice of representation for g. Nevertheless we can think of the bracket as being like a trace as far as cyclic symmetry properties are concerned. Secondly note that $\Gamma(B)$ is not local in the ordinary sense; it is expressed as a 3-dimensional rather than a 2-dimensional integral.

To understand this suppose that, by analytic continuation, we may work in a Euclidean space-time which is a large 2-dimensional sphere, S^2. We take this S^2 to be the boundary of a spatial ball B in $\mathbb{R}^3$, whose coordinates are y, and extend γ in an arbitrary smooth fashion through B. Such a definition of $\Gamma(B)$ is not obviously unique. Indeed the difference between any two values of $\Gamma(B)$ obtained in this way can be represented as an integral $\Gamma(S^3)$ of the same form as (2.3.4), but with B replaced by some 3-dimensional sphere S^3. However, $\Gamma(S^3)/2\pi$ is an integer for any γ and (unless $G = SO(3)$) there exists a γ for which this integer is unity. This is because the integral in (2.3.4) is a sort of Jacobian and hence $\Gamma(S^3)$ is a winding number. $\Gamma(S^3)/2\pi$ is unity if we use the homeomorphism, expressed by Euler angles relating S^3 to SU(2), to map S^3 into an SU(2) subgroup of G associated with any long root.

This ambiguity in the definition of $\Gamma(B)$ introduces a discrete ambiguity into the definition of the action (2.3.3) of the form $2\pi nK$. This discrete ambiguity in $\mathscr{a}$ causes no problems for the derivation of the classical motion because each of the possibilities for $\Gamma(B)$ will be stationary for the same paths. Quantum mechanically it potentially has a significance because there $\exp(i\mathscr{a}/\hbar)$ has to be single-valued. This leads to the condition

$$K = x\hbar, \quad x \in \mathbb{Z}. \tag{2.3.5}$$

Note that this constraint on K is quantum mechanical rather than classical.

To see how $K\Gamma(B)$ modifies the equation of motion we calculate

$$\delta\Gamma(B) = \frac{\psi^2}{8\pi}\int_B \partial_\lambda \varepsilon^{\lambda\mu\nu}\langle\gamma^{-1}\delta\gamma, \gamma^{-1}\partial_\mu\gamma\gamma^{-1}\partial_\nu\gamma\rangle\, d^3y$$

$$= -\frac{\psi^2}{8\pi}\int_{S^2} \varepsilon^{\mu\nu}\langle\gamma^{-1}\delta\gamma, \gamma^{-1}\partial_\mu(\gamma^{-1}\partial_\nu\gamma)\rangle\, d^2x \tag{2.3.6}$$

as the boundary of B is the Euclidean space-time in which we are working. Combining this with (2.2.5) and (2.3.3), we obtain

$$\partial_\mu\{\gamma^{-1}\partial^\mu\gamma - \frac{\lambda^2 K}{4\pi}\varepsilon^{\mu\nu}\gamma^{-1}\partial_\nu\gamma\} = 0. \tag{2.3.7}$$

[Here $\varepsilon_{\mu\nu} = -\varepsilon_{\nu\mu}$, $\varepsilon^{01} = 1$.]

If we consider the special value of the coupling constant

$$\lambda^2 = 4\pi/K, \tag{2.3.8}$$

the equation of motion (2.3.7) reduces to the desired form

$$\partial_-(\gamma^{-1}\partial_+\gamma) = 0, \tag{2.3.9}$$

i.e., Eqs. (2.2.10) with the definitions (2.3.1) and (2.3.2).

If, on the other hand, we choose the special value $\lambda^2 = -4\pi/K$ we obtain instead

$$\partial_+(\gamma^{-1}\partial_-\gamma) = 0, \tag{2.3.10}$$

yielding Eqs. (2.2.10) with J^μ now defined by

$$J_+ = (\partial_+\gamma)\gamma^{-1}, \quad J_- = \gamma^{-1}\partial_-\gamma. \tag{2.3.11}$$

We shall now fix on the first possibility. (2.3.8). Putting together this with condition (2.3.5) we have the action

$$a = x\hbar[\mathscr{I}/16\pi + \Gamma(B)], \tag{2.3.12}$$

where

$$\mathscr{I} = -\int d^2x\psi^2\langle\gamma^{-1}\partial_\mu\gamma, \gamma^{-1}\partial^\mu\gamma\rangle.$$

This has only one free parameter, the integer x. Notice that in the theory (2.3.12), which we call the Wess-Zumino theory, it is not possible to take the classical limit, $\hbar \to 0$. The situation is similar to that with magnetic monopoles when the electron and magnetic charges q, g have to satisfy $qg/2\pi\hbar \in \mathbb{Z}$.

We now wish to quantize this theory (2.3.12) to see whether $J_\pm$ will indeed furnish commuting Kac-Moody algebras as in Subsec. 2.1. To do this we calculate the Poisson brackets in canonical fashion and then postulate the quantum version with commutator brackets. First let us absorb various dimensional constants by redefining

$$\sum_a T^a J^a_+ = -i\gamma^{-1}\partial_+\gamma\frac{x\hbar\psi^2}{4\pi}\sqrt{2}. \tag{2.3.13}$$

We obtain the Poisson bracket relations

$$[J^a_+(\xi), J^b_+(\eta)]_{\text{P.B.}} = f^{abc}J^C_+(\xi)\delta(\xi-\eta) + \frac{x\hbar\psi^2}{4\pi}\delta^{ab}\delta'(\xi-\eta). \tag{2.3.14}$$

There is a similar equation for J^a_- defined by

$$\sum_a T^a J^a_- = -i(\partial_-\gamma)\gamma^{-1}\frac{x\hbar\psi^2}{4\pi}\sqrt{2} \tag{2.3.15}$$

and

$$[J^a_+(\xi), J^b_-(\eta)]_{\text{P.B.}} = 0. \tag{2.3.16}$$

The transition to quantum mechanics is made by replacing the Poisson bracket by the quantum commutator bracket divided by $i\hbar$. The result is the current algebra (2.1.3) satisfied by J^a_+ and J^a_-, which mutually commute, with

$$k = \frac{x\psi^2}{2}, \quad \text{i.e., } x = \frac{2k}{\psi^2}. \tag{2.3.17}$$

To obtain the Kac-Moody algebra (1.1.2) we need to impose periodic boundary conditions again, as in Eq. (2.1.19), and define

$$J^a_n = \hbar T^a_n = \int_0^L J^a(\xi) z^n \, d\xi, \tag{2.3.18}$$

where

$$z = e^{2\pi i\xi/L}, \tag{2.3.19}$$

thereby inverting Eqs. (2.1.1) and (2.1.2), for each of $J^a_\pm$.

Notice that the Schwinger term in (2.3.14) arose classically. That it had the correct dependence on Planck's constant $\hbar$ was only possible because it occurred in the classical action (2.3.12), associated with the "quantization condition" (2.3.5) implied by the requirement of single valuedness. The only free parameter in the action (2.3.12), the integer x, occurs only in the Kac-Moody algebra c-number term (2.3.17) and is identified as what we call the level of the representation.[34]

The requirement that it be an integer arose for topological reasons in the above discussion, but, as we shall see in Subsec. 3.4, it will arise for algebraic reasons in the representation theory of Kac-Moody algebras.

2.4. *Conformal invariance of the Wess-Zumino model and non-Abelian bosonization*

Witten[34] showed that, in low orders of perturbation theory, the quantum Wess-Zumino model (2.3.12) is conformally invariant, unlike the more general theory (2.3.3) with generic values of λ and K not satisfying (2.3.8). This result can also be established exactly by constructing the energy-momentum tensor explicitly with the algebraic properties appropriate to conformal symmetry as we now show.

In the σ-model (2.3.3) the canonical classical energy-momentum tensor is traceless and has the current-current, or Sugawara, form,[43,44]

$$\theta_{\mu\nu} = \sum_a \left\{ J^a_\mu J^a_\nu - \frac{1}{2} g_{\mu\nu} J^a_\lambda J^{a\lambda} \right\} \Big/ k. \tag{2.4.1}$$

The nonzero components are

$$\theta_{++} = \sum_a J_+^a J_+^a / 2k \tag{2.4.2}$$

and θ_{--}, for which there is a similar expression. In the quantum theory, these expressions are potentially divergent and they need to be regularized by some sort of normal ordering procedure. We can do this in the special case when (2.3.8) holds, since then the currents are free fields satisfying the commuting Kac-Moody algebras. The definition of normal ordering is in terms of the T_m^a (2.3.18):

$$\begin{aligned} {}^{\times}_{\times} T_m^a T_n^a {}^{\times}_{\times} &= T_m^a T_n^a \,, m < 0 \,, \\ &= T_n^a T_m^a \,, m \geq 0 \,. \end{aligned} \tag{2.4.3}$$

The Fourier components of (2.4.2) are then, up to a factor of $(\hbar/L)^2$,

$$\mathscr{L}_m = \frac{1}{2k} \Sigma {}^{\times}_{\times} T_n^a T_{m-n}^a {}^{\times}_{\times} \,, \tag{2.4.4}$$

These operators do not themselves satisfy the Virasoro algebra (1.1.1). We need to introduce a multiplicative renormalization and replace $\mathscr{L}_m$ by $\mathscr{L}_m = \mathscr{L}_m / \beta$ where the renormalization constant

$$\beta = 1 + Q_\psi / 2k \,, \tag{2.4.5}$$

and Q_ψ is the quadratic Casimir operator for the adjoint representation of the group G, in the normalization implied by Eq. (1.1.2). The operators $\mathscr{L}_m$ then satisfy the Virasoro algebra with

$$c = \frac{2k \dim g}{2k + Q_\psi} \,. \tag{2.4.6}$$

Furthermore

$$[\mathscr{L}_m, T_n^a] = -n T_{m+n}^a \,. \tag{2.4.7}$$

These results, explained further in Sec. 4, mean that we have succeeded in constructing a conformally invariant quantum Wess-Zumino model, since our commutation relations guarantee the conformal invariance and the correct Heisenberg equation of motion.

Because the currents are free fields, the Wess-Zumino theory is in some sense a trivial theory, some say a free theory. However, the value of the Virasoro c-number (2.4.6) does not support this latter conclusion. We saw, in Eq. (2.2.3), that in a free theory c must be an integer or half-integer, yet as we see in more detail in Subsec. 4.3, expression

(2.4.6) is always rational but not necessarily integral or half-integral. There are interesting special cases when it is, however. For example, in the case of $G = SO(N)$, with $x = 1$, $Q_\psi = (N-1)\psi^2$, $\dim g = \frac{1}{2}N(N-1)$, so that, by (2.3.17) and (2.4.6), $c = \frac{1}{2}N$; and so this theory could be a free of N massless fermions with energy-momentum tensor (2.1.31). Indeed in that theory it is possible to construct from the N fermions a commuting pair of $SO(N)$ Kac-Moody algebras with level $2k/\psi^2 = y/\psi^2$, which is indeed 1 in this case. Thus the algebraic structures, two commuting copies of the semidirect product of a Virasoro algebra with an $SO(N)$ Kac-Moody algebra, coincide in the two theories: the theory of N massless-free fermions and the $SO(N)$ Wess-Zumino model (2.3.12) with coefficient $x = 1$.

Witten[34] pointed this out and argued that the two theories were therefore quantum equivalent, yielding what he called a non-Abelian bosonization. Before this conclusion can be substantiated two further points must be settled:

(i) whether the fermion theory, like the Wess-Zumino theory, possesses an energy-momentum tensor of the Sugawara form (2.4.1); and

(ii) the extent to which the isomorphism between the structures actually identifies them.

Since we have had to suppose that the fermions are free and massless question (i), stated more precisely in terms of Virasoro algebra generators, is whether L_n, specified by Eq. (2.1.30) (the free fermion construction), equals $\mathscr{L}_n = \mathscr{Q}_n/\beta$, specified by Eqs. (2.4.4), (2.4.5) and (2.1.26), the Sugawara construction. Evidently this equality is impossible if the c-numbers $N/2$ and (2.4.6) are unequal and in any case it looks unlikely since one expression is bilinear in fermions and the other quadrilinear. Nevertheless, in the case cited above with fermions in the defining representation of $SO(N)$, the c-numbers are equal and the Virasoro generators indeed coincide owing to a somewhat miraculous cancellation, which is purely quantum mechanical and has no classical analogue.

This result has been known for a long time in the physics literature and was used by Witten[34] to establish the isomorphism between the algebraic structures when $G = SO(N)$ and $x = 1$.

The question then arises as to whether the miraculous equality of the Sugawara and free fermion constructions can happen in other cases; Goddard, Nahm and Olive established a necessary and sufficient criterion for it, which is explained in Subsec. 5.5.[45]

Given this criterion, guaranteeing the isomorphism between the two algebraic structures, question (ii) asks to what extent the structures can be equated. This is like asking whether two angular momenta can be equated because they satisfy the same commutation relations. Evidently they cannot because they may well describe different spins. Thus we have to develop a representation theory of Kac-Moody algebras, analogous to that of angular momenta. This can be done if we assume L_0 is positive, as indeed it is here since it is proportional to $(H \pm P)$ which is positive on physical grounds. This is described in Subsec. 3.4, where it is established that, for a given value of $x = 2k/\psi^2$, there exist a finite number of inequivalent irreducible representations.

The Hilbert space of the Wess-Zumino theory has not yet been fully determined by our quantum procedure but that of the quark model is simply the fermion Fock space. Since the value of $x = 2k/\psi^2$ is fixed by the quark model as an integer (Eq. (2:1.17)), we know into which irreducible representations of $\hat{g}$ it may decompose but not the multiplicities. In general infinite multiplicities are possible but are excluded precisely when the criterion of Goddard, Nahm and Olive is satisfied.[45] Thus, in this case only, the quark model Fock space decomposes into a finite number of irreducible representations of $\hat{g}$.

The final question is whether the Wess-Zumino quantum theory decomposes into these same components, i.e., in physical terms, whether it has the same vacuum structure. One possibility is that, since there an ambiguity in our quantization of that theory, we can choose the desired vacuum structure at will. A second possibility is that the global structures of the relevant Kac-Moody groups will further constrain the vacuum structure. In any case, we can assert the quantum equivalence of the two quantum field theories up to a finite ambiguity when the criterion of Goddard, Nahm and Olive is met.[45]

2.5. *Statistical systems*

One of the most intriguing applications of infinite-dimensional algebras is to the critical behaviour of certain 2-dimensional statistical mechanical systems. For such systems, there exist critical points near which they are invariant under local scale transformations. In any many cases, the systems correspond to 2-dimensional Euclidean quantum field theories at these critical points, unitarity of the quantum field theory corresponding to a property called reflection positivity in the statistical system. The importance of conformal invariance in providing information about critical behaviour was realized by Polyakov.[46] Later Belavin, Polyakov and Zamolodchikov[32] developed this idea, showing how the conformal group could be used to provide a detailed analysis of conformally invariant 2-dimensional field theories.

The conformal group in 2-dimensional Euclidean space is infinite-dimensional and has an algebra consisting of two commuting copies of the Virasoro algebra. To see this, we have to introduce complex coordinates $z = x + iy$ and $\bar{z} = x - iy$, where x and y are Cartesian coordinates. (This is to be contrasted with the identification of z with light cone coordinates in Sec. 2.1, whereby $z = \exp\{\sqrt{2}\pi i(t + x)/L\}$.) Any analytic mapping of the complex plane into itself is conformal. A basis of generators for such mappings is provided by the infinitesimal transformations

$$z \to z + \varepsilon z^{n+1}, \quad \bar{z} \to \bar{z} + \overline{\varepsilon z}^{n+1}, \tag{2.5.1}$$

with corresponding generators L_n and $\bar{L}_n$, respectively. In this application, these generators occur as the coefficients in the Laurent expansion of the components of the stress-energy tensor, $T_{\mu\nu}(x, y)$, which must be conserved and traceless,

$$T_{\mu\nu}\, dx^\mu\, dx^\nu \equiv T(z)\, dz^2 + \bar{T}(\bar{z})\, d\bar{z}^2 ; \tag{2.5.2}$$

thus,

$$T(z) = \sum_{n \in \mathbf{Z}} z^{-n-2} L_n, \quad T(\bar{z}) = \sum_{n \in \mathbf{Z}} \bar{z}^{-n-2} \bar{L}_n. \tag{2.5.3}$$

In the quantum field theory, the L_n satisfy the Virasoro algebra (1.1.1) and the $\bar{L}_n$ satisfy the same algebra with the same value of the central term, c, which is positive and a characteristic of the theory.

The conformally invariant system is described by the correlation functions of certain primary fields $\phi(z, \bar{z})$ satisfying commutation relations

$$[L_n, \phi] = z^{n+1} \frac{\partial}{\partial z} \phi + h(n+1) z^n \phi, \tag{2.5.4a}$$

$$[\bar{L}_n, \phi] = \bar{z}^{n+1} \frac{\partial}{\partial \bar{z}} \phi + \bar{h}(n+1) \bar{z}^n \phi. \tag{2.5.4b}$$

The vacuum of the theory is invariant under the SU(1, 1) subgroup (1.2.24) of the conformal group (which corresponds to the theory being quantized radially, postulating fundamental commutation relations for the value of the fields on a circle $|z| = r$, constant, rather than on y constant, say). In fact, we have

$$L_n |0\rangle = 0, \quad n \geq -1. \tag{2.5.5}$$

Then, if ϕ is single-valued, it follows from (2.5.4a) that

$$L_0 \phi(0)|0\rangle = h\phi(0)|0\rangle \tag{2.5.6}$$

and that the 2-point function is given by

$$\langle 0|\phi(z_1)\phi(z_2)|0\rangle = r^{-2(h+\bar{h})} e^{-2i\theta(h-\bar{h})} \langle 0|\phi(0)\phi(1)|0\rangle, \tag{2.5.7}$$

where

$$z_2 - z_1 = re^{i\theta},$$

so that $h + \bar{h}$ is the scaling dimension of the field ϕ and $h - \bar{h}$ is its spin. The critical exponents in the theory are linear combinations of these scaling dimensions. Notice that, since correlation functions must decay with separation, (2.5.7) implies that $h + \bar{h}$ is positive.

Friedan, Qiu and Shenker[15] demonstrated how restrictive unitarity and conformal invariance are when $c < 1$. The states of the theory must fall into irreducible representations of the Virasoro algebra L_n (and of $\bar{L}_n$). It follows from their theorem, quoted in Subsec 1.4, that, in this case, c must take one of the values (1.4.10). Further, associated with each single-valued primary ϕ is a highest weight state,

$$|h\rangle = \phi(0)|0\rangle, \tag{2.5.8}$$

satisfying (1.4.5), and so h and $\bar{h}$ must be amongst the values listed in (1.4.11) for the value of m corresponding to c. By matching the permitted rational values of the scaling dimensions and spins with the known values in various 2-dimensional systems, Friedan, Qiu and Shenker[15] were able to deduce to corresponding values of c, or, equivalently, m in (1.4.10):

$c = \frac{1}{2}$, Ising model;

$c = \frac{7}{10}$, Tricritical Ising model;

$c = \frac{4}{5}$, 3-state Potts model;

$c = \frac{6}{7}$, Tricritical 3-state Potts model.

Subsequently, Huse[47] identified a series of models, solved a little earlier by Andrews, Baxter and Forrester[48], with the higher values of m (≥ 5).

The only nontrivial value of c in common between the discrete sequence (1.4.10) and the values (2.2.3) for a free theory is $\frac{1}{2}$, corresponding to a single, real fermion field. The above association of the Ising model at its critical temperature with $c = \frac{1}{2}$ suggests that it can be reformulated in terms of a single real fermion field, as indeed Onsager had discovered in 1944.[129]

For a recent review of the role of conformal invariance in critical phenomena see Ref. 49.

3. Roots and Weights

3.1. *Simple finite-dimensional Lie algebras*

In this section we describe aspects of the theory of roots, weights and Dynkin diagrams for affine Kac-Moody algebras. We begin in this section by recapitulating in summary the theory for the finite-dimensional case. Specifically, we consider the Lie algebra g,

$$[T^a, T^b] = if^{abc}T^c \tag{3.1.1}$$

of a compact simple Lie group G. (The results extend immediately to the case of a general compact Lie group; the Lie algebra of a such a group is the direct sum of simple Lie algebras and Abelian Lie algebras.) The basis $\{T^a\}$ is assumed to satisfy the orthonormality condition (2.1.17).

Throughout this article we shall adopt the convention of using lower case letters to refer to Lie algebras and upper case letters to refer to the corresponding groups. Thus su(n) denotes the Lie algebra of SU(n) and su(2) = so(3), etc.

A standard way of choosing a basis for g is first to seek a maximal set of commuting Hermitian generators, H^i, $i = 1, 2, \ldots, r$,

$$[H^i, H^j] = 0, \qquad 1 \leq i, \quad j \leq r. \tag{3.1.2}$$

The Abelian subalgebra of g generated by the H^i, $1 \le i \le r$, is called a *Cartan subalgebra.* It can be shown that any two such maximal Abelian subalgebras are conjugate under the action of G, so that the dimension r is the same for each Cartan subalgebra (CSA) and any other CSA has a basis H'^i $1 \le i \le r$ such that: $H'^i = \gamma H^i \gamma^{-1}$ for some $\gamma \in G$. The dimension r is called the *rank* of g (or G).

Given a choice of CSA, we extend it into a basis for the whole of g by taking complex combinations E^α of the generators of g, such that

$$[H^i, E^\alpha] = \alpha^i E^\alpha, \qquad 1 \le i \le r. \tag{3.1.3}$$

The real nonzero r-dimensional vector α is called a *root* and E^α the *step operator* corresponding to α. It can be shown that, for each root α, there is, up to scalar multiplication, just one step operator E^α satisfying (3.1.3) and further that the only multiples of a root α which are roots are $\pm\alpha$. It is easy to see that $-\alpha$ is a root, with step operator $E^{-\alpha} = E^{\alpha\dagger}$, by Hermitian conjugating (3.1.3). We denote the set of roots of g by Φ.

The basis of g thus obtained,

$$H^i, \quad 1 \le i \le r, \qquad E^\alpha, \quad \alpha \in \Phi, \tag{3.1.4}$$

is one in which the generators in the CSA are simultaneously diagonal. The r-dimensionl root vectors are the nonzero simultaneous eigenvalues of the $\{H^i\}$ and these simultaneous eigenvalues are nondegenerate. The zero vector occurs as an r-fold degenerate simultaneous eigenvector. The number of roots $|\Phi| = \dim g - \operatorname{rank} g$.

To complete the statement of the algebra g in this basis we need to consider $[E^\alpha, E^\beta]$ for each pair of roots α, β. It follows from the Jacobi identity that

$$[H^i, [E^\alpha, E^\beta]] = (\alpha^i + \beta^i)[E^\alpha, E^\beta] \tag{3.1.5}$$

so that, if $\alpha + \beta \notin \Phi$ and $\alpha + \beta \ne 0$, $[E^\alpha, E^\beta]$ must vanish. If $\alpha + \beta$ is a root, the commutator turns out to be nonzero and hence a multiple of $E^{\alpha+\beta}$. If $\alpha + \beta = 0$, it is easy to show, using (2.1.22), that $[E^\alpha, E^{-\alpha}]$ is orthogonal to $\xi \cdot H$ under the scalar product (2.1.20) if $\xi \cdot \alpha = 0$. Thus $[E^\alpha, E^{-\alpha}]$ is proportional to $\alpha \cdot H$. We can summarize these results by

$$[E^\alpha, E^\beta] = \varepsilon(\alpha, \beta) E^{\alpha+\beta}, \quad \text{if } \alpha + \beta \text{ is a root}, \tag{3.1.6a}$$

$$= 2\alpha \cdot H/\alpha^2, \quad \text{if } \alpha = -\beta, \tag{3.1.6b}$$

$$= 0, \quad \text{otherwise}, \tag{3.1.6c}$$

where the normalization of E^α is fixed by (3.1.6b). [The constants $\varepsilon(\alpha, \beta)$ in (3.1.6a), evidently antisymmetric in α and β, can be arranged to be ± 1, by choice of the phases of the E^α, if all of the roots α have the same squared length $\alpha^2 = \alpha^i\alpha^i$, i.e. g is simply-laced.] The basis (3.1.4), with the commutation relations (3.1.2), (3.1.3) and (3.1.6), is a modified version of what is called called a *Cartan-Weyl basis.*

For each root α, $E^\alpha, E^{-\alpha}$, $2\alpha\cdot H/\alpha^2$ form an su(2), subalgebra (in a complex basis), isomorphic to $I_+, I_-, 2I_3$, where

$$[I_+, I_-] = 2I_3, \quad [I_3, I_\pm] = \pm I_\pm \tag{3.1.7}$$

and satisfying the hermiticity conditions $I_+^\dagger = I_-$, $I_3^\dagger = I_3$. In consequence the eigenvalues of $2\alpha\cdot H/\alpha^2$ in any unitary representation are integral, as the eigenvalues of $2I_3$ are. Its eigenvalues in the adjoint representation are $2\alpha\cdot\beta/\alpha^2$, $\beta \subset \Phi$, together with zero r times. Thus

$$\frac{2\alpha\cdot\beta}{\alpha^2} \in \mathbb{Z} \quad \text{for all roots } \alpha, \beta. \tag{3.1.8}$$

Further, in the adjoint representation, the step operators of the form $E^{\beta+m\alpha}$, $m \in \mathbb{Z}$, must form an su(2) multiplet for E^α, $E^{-\alpha}$, $2\alpha\cdot H/\alpha^2$. Thus, in particular, since there must be a member of this multiplet with the opposite helicity to E^β,

$$2\alpha\cdot\beta/\alpha^2 + 2m = -2\alpha\cdot\beta/\alpha^2$$

for some value of m for which $\beta + m\alpha$ is a root. Then

$$\beta + m\alpha = \beta - 2(\alpha\cdot\beta)\alpha/\alpha^2 \equiv \sigma_\alpha(\beta) \tag{3.1.9}$$

is hence a root for each pair of roots α, β. The linear operator σ_α corresponds to reflection in the hyperplane normal to α. Thus, we have seen that each of these reflections permutes the roots, Φ. These reflections generate a finite group $W(g)$ called the *Weyl group* of g.

The number of root pairs $\pm\alpha$ in general exceeds the rank r of g and it is convenient to select a set of roots which forms a basis for the r-dimensional space of which they are members. It can be shown that a basis $\alpha_{(1)}, \alpha_{(2)}, \ldots, \alpha_{(r)}$ can be chosen in such a way that any root

$$\alpha = \sum_{i=1}^{r} n_i\alpha_{(i)}, \tag{3.1.10}$$

where each $n_i \in \mathbb{Z}$ and *either* $n_i \geq 0$, $1 \leq i \leq r$, *or* $n_i \leq 0$, $1 \leq i \leq r$. In the former case α is said to be positive ($\alpha > 0$), in the latter negative ($\alpha < 0$). Such a basis is called a basis of *simple roots.* Clearly, applying any transformation in the Weyl group to a basis of simple roots yields another such basis. In fact all possible bases of simple roots can be obtained in this way (and, in particular, the Weyl group applied to a basis of simple roots yields all the roots). Further, the Weyl group is generated by the reflections $\sigma_{\alpha_{(i)}}$ in the hyperplanes normal to the elements of a given basis of simple roots.

The scalar products

$$K_{ij} = 2\alpha_{(i)} \cdot \alpha_{(j)} / \alpha_{(j)}^2, \qquad 1 \le i, \ j \le r, \tag{3.1.11}$$

which have to be integers, form an $r \times r$ matrix called the *Cartan matrix* of g. (It is determined, up to simultaneous permutation of its rows and columns, by g.) Its diagonal elements are automatically 2 and its off-diagonal elements are all negative integers or zero. Clearly from knowledge of the Cartan matrix K we can reconstruct a basis of simple roots $\alpha_{(i)}$, $1 \le i \le r$, up to scale and overall orientation (whose ambiguities correspond to scaling or performing an orthogonal transformation on the CSA). From these we can generate the Weyl group and hence all the roots, and thus obtain all the structure constants in the Cartan-Weyl basis, though we have not given here the rules for determining $\varepsilon(\alpha, \beta)$. Hence we can reconstruct g from K.

The information in K can be coded into the *Dynkin diagram* which is constructed as follows: it consists of a point for each simple root $\alpha_{(i)}$ with points $\alpha_{(i)}$ and $\alpha_{(j)}$ being joined by $K_{ij}K_{ji}$ lines, with an arrow pointing from $\alpha_{(j)}$ to $\alpha_{(i)}$ if $\alpha_{(j)}^2 > \alpha_{(i)}^2$. The Dynkin diagrams for all the simple Lie algebras are given in Fig. 2. To reconstruct K from the Dynkin diagram note (i) that the possible values of $K_{ij}K_{ji}$ are 0, 1, 2, 3, 4 and (ii) if $i \neq j$, $K_{ij}K_{ji} \le 3$, and further if $\alpha_{(i)}^2 = \alpha_{(j)}^2$, $K_{ij} = K_{ji} = -1$ or 0 whilst if $\alpha_{(i)}^2 > \alpha_{(j)}^2$ and $K_{ij} \neq 0$ then $K_{ji} = -1$ and $K_{ij} = -2$ or -3. So if $\alpha_{(i)}$ and $\alpha_{(j)}$ are not orthogonal and $\alpha_{(i)}^2 \ge \alpha_{(j)}^2$, then $\alpha_{(i)}^2/\alpha_{(j)}^2 = 1, 2$ or 3. Since g is assumed simple, its roots cannot be divided into two mutually orthogonal sets and it can be shown that there can be at most two root lengths, which are called *long* and *short*. If the long and short lengths coincide, g is said to be simply-laced. Because g can be reconstructed from K and K can be retrieved from its Dynkin diagram, we conclude that the latter encodes all information about the structure of g.

Consider the finite-dimensional representations of g. We can take a basis $\{|\mu\rangle\}$ in which the CSA, H^i, $1 \le i \le n$, is diagonal,

$$H^i|\mu\rangle = \mu^i|\mu\rangle, \tag{3.1.12}$$

and the r-dimensional vector μ of eigenvalues is called a *weight* vector. For each root α, $2\alpha \cdot H/\alpha^2$ must take an integral value when acting on $|\mu\rangle$, so that

$$\frac{2\alpha \cdot \mu}{\alpha^2} \in \mathbb{Z} \quad \text{for all roots } \alpha. \tag{3.1.13}$$

The condition (3.1.13) defines a lattice $\Lambda_W(g)$ called the *weight lattice* of g. (It has as a sublattice the *root lattice* $\Lambda_R(g)$, the lattice generated by taking integral combinations of the roots.) Further, the states $\{|\mu\rangle\}$ must fall into su(2) multiplets under the action of E^α, $E^{-\alpha}$, $2\alpha \cdot H/\alpha^2$. If follows that the weights of a given representation must be mapped into one another by σ_α, for each root α, and so by the whole Weyl group $W(g)$.

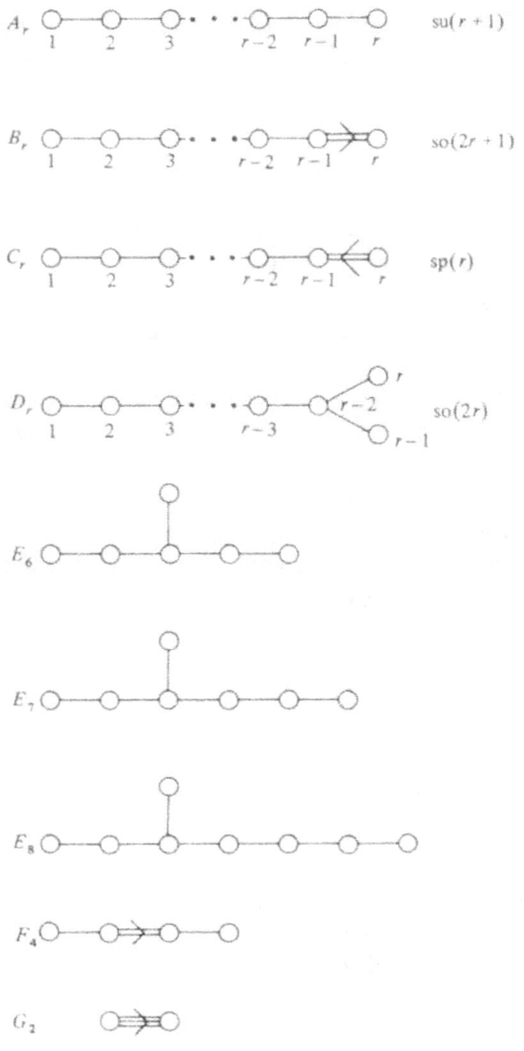

Fig. 2. Dynkin diagrams for finite-dimensional simple Lie algebras.

It is enough that condition (3.1.13) should hold for a given basis of simple roots (because all the roots are generated from simple roots by successive reflections in these roots). Consequently we can take a basis for the weight lattice $\Lambda_W(g)$ consisting of *fundamental weights* $\lambda_{(i)}$ defined by

$$2\lambda_{(i)} \cdot \alpha_{(j)}/\alpha_{(j)}^2 = \delta_{ij}, \qquad 1 \le i, j \le r. \tag{3.1.14}$$

Then any weight $\lambda \in \Lambda_W(g)$ is of the form

$$\lambda = \sum_{i=1}^{r} n_i \lambda_{(i)}, \qquad n_i \in \mathbb{Z}. \tag{3.1.15}$$

If all the integers $n_i \geq 0$, the weight λ is called *dominant*. Clearly a weight λ is dominant if and only if

$$\alpha_{(i)} \cdot \lambda \geq 0, \qquad 1 \leq i \leq r. \tag{3.1.16}$$

Every weight can be mapped into a unique dominant weight by the Weyl group. The particular dominant weight

$$\rho = \sum_{i=1}^{r} \lambda_{(i)} \tag{3.1.17}$$

has the property that $\rho \cdot \alpha > 0$ if α is positive and $\rho \cdot \alpha < 0$ if α is negative. It can be shown that ρ is equal to half the sum of positive roots.

In a given finite-dimensional representation of g, we can find a *highest weight* state $|\mu_0\rangle$, that is, one for which $\rho \cdot \mu_0$ is largest. It follows that

$$E^{\alpha}|\mu_0\rangle = 0, \qquad \alpha > 0, \tag{3.1.18}$$

because otherwise this state would have weight $\mu_0 + \alpha$ which would be higher than μ_0. Clearly, the subspace generated from $|\mu_0\rangle$ by the action of g consists of the linear span of states of form

$$E^{-\beta_1}E^{-\beta_2}\dots E^{-\beta_m}|\mu_0\rangle, \tag{3.1.19}$$

where each $\beta_i > 0$, each of which has a lower weight. This must be the whole space if the representation is irreducible. Thus an irreducible unitary representation of g has a unique highest weight state $|\mu_0\rangle$ and the other weights μ of the representation have the property that $\mu_0 - \mu$ is a sum of positive roots. Since, for each Weyl transformation σ, $\sigma(\mu)$ is a weight of the representation if μ is, we see that the weights μ of the representation have the property that $\mu_0 - \sigma(\mu)$ is always sum of positive roots. In fact it can be shown that all weights μ with this property will be weights of the representation with highest weight μ_0. The highest weight of a representation is necessarily a dominant weight and, conversely, for each dominant weight μ_0, there is a unique irreducible representation with this as its highest weight, consisting of states of the form (3.1.19), where $|\mu_0\rangle$ is defined by the conditions (3.1.18) together with

$$H^i|\mu_0\rangle = \mu_0^i|\mu_0\rangle. \tag{3.1.20}$$

The weights of the adjoint representation are the roots. Its highest weight is called the *highest root*, ψ. It thus has the property that, if α is any other root, $\psi - \alpha$ is a sum of positive roots.

In this subsection we have only been able to give a brief summary of the theory of roots, weights and Dynkin diagrams for a compact simple Lie algebra. A more extensive treatment, written with the objective of leading into Kac-Moody algebras, is given in the lecture notes of Olive.[50] For a full treatment the reader might consult the work of Humphreys.[51] Our treatment in the rest of this chapter of the generalization of these results to affine Kac-Moody algebras is based at least in part on Macdonald[13] and Kac.[14]

3.2. *The root system of $\hat{g}$.*

In the last subsection we discussed how to set up a modified Cartan-Weyl basis

$$[H^i, H^j] = 0, \tag{3.2.1}$$

$$[H^i, E^\alpha] = \alpha^i E^\alpha, \tag{3.2.2}$$

$$[E^\alpha, E^\beta] = \varepsilon(\alpha,\beta) E^{\alpha+\beta}, \quad \text{if } \alpha+\beta \text{ is a root}, \tag{3.2.3a}$$

$$= 2\alpha\cdot H/\alpha^2, \quad \text{if } \alpha = -\beta, \tag{3.2.3b}$$

$$= 0, \quad \text{otherwise}, \tag{3.2.3c}$$

where $1 \le i, j \le r$ and α, β are roots, for a compact simple algebra of rank r. We can now rewrite the untwisted affine Kac-Moody algebra $\hat{g}$ (1.1.2) in this basis

$$[H^i_m, H^j_n] = km\delta^{ij}\delta_{m,-n}, \tag{3.2.4}$$

$$[H^i_m, E^\alpha_n] = \alpha^i E_{m+n}, \tag{3.2.5}$$

$$[E^\alpha_m, E^\beta_n] = \varepsilon(\alpha,\beta) E^{\alpha+\beta}_{m+n}, \quad \text{if } \alpha+\beta \text{ is a root}, \tag{3.2.6a}$$

$$= \frac{2}{\alpha^2}\{\alpha\cdot H_{m+n} + km\delta_{m,-n}\}, \quad \text{if } \alpha = -\beta, \tag{3.2.6b}$$

$$= 0, \quad \text{otherwise}, \tag{3.2.6c}$$

$$[k, E^\alpha_n] = [k, H^i_n] = 0, \tag{3.2.6d}$$

where the normalization of the c-number term in Eq. (3.2.6b) follows from

$$\langle E^\alpha, E^{-\beta}\rangle = \frac{2}{\alpha^2}\delta_{\alpha,-\beta}, \tag{3.2.7}$$

which is a consequence of (2.1.22). We have the hermiticity conditions

$$H^{i\dagger}_n = H^i_{-n}, \quad E^{\alpha\dagger}_n = E^{-\alpha}_{-n}, \quad k^\dagger = k. \tag{3.2.8}$$

We now try to imitate the discussion of Subsec. 3.1, constructing an analogue of the CSA, roots, etc. We start with the CSA, H^i_0, $1 \le i \le r$, for the subalgebra generated by

H_0^i, $1 \le i \le r$, E_0^α, $\alpha \in \Phi$, which is isomorphic to g. To this we can obviously append the central element k. With this $(r+1)$-dimensional Abelian subalgebra, all of the E_n^α are step operators,

$$[H_0^i, E_n^\alpha] = \alpha^i E_n^\alpha, \tag{3.2.9a}$$

$$[k, E_n^\alpha] = 0, \tag{3.2.9b}$$

so that each of the roots $(\alpha, 0)$ is infinitely degenerate; moreover this Abelian subalgebra is not maximal because $[H_0^i, H_n^j] = 0$. To cure these problems we extend $\hat{g}$ by appending an extra element d designed to distinguish between E_n^α for different n. It has the defining properties

$$[d, T_n^a] = nT_n^a, \qquad [d, k] = 0, \tag{3.2.10a}$$

$$d^\dagger = d. \tag{3.2.10b}$$

Any $\hat{g}$ given by (1.1.2) can be extended in this way because (1.1.2) taken together with (3.2.10) is consistent with the Jacobi identities. In fact we can take $d = -L_0$, so that we are just using part of the semi-direct product structure (1.1.3).

Using as CSA, H_0^i, $1 \le i \le r$, k, d, we have as step operators:

$$E_n^\alpha \quad \text{corresponding to the root } a = (\alpha, 0, n); \tag{3.2.11}$$

$$H_n^i \quad \text{corresponding to the root } n\delta = (0, 0, n). \tag{3.2.12}$$

So we have a basis for the extended $\hat{g}$ consisting of this CSA together with corresponding step operators, the roots being $a = (\alpha, 0, n)$, $\alpha \in \Phi$, $n \in \mathbb{Z}$, each of which is non-degenerate, and $n\delta = (0, 0, n)$, $n \in \mathbb{Z}$, $n \ne 0$, each of which is r-fold degenerate. This degeneracy is the first strikingly new feature, compared with finite-dimensional Lie algebra theory.

The root system of $\hat{g}$ is thus infinite but spans an $(r+1)$-dimensional space. We can divide $\hat{g}$ into positive and negative roots by calling

$$(\alpha, 0, n) > 0 \quad \text{if } n > 0 \textit{ or } \text{if } n = 0 \text{ and } \alpha > 0. \tag{3.2.13}$$

This corresponds to taking as a basis of simple roots for $\hat{g}$

$$a_{(i)} = (\alpha_i, 0, 0), \quad 1 \le i \le r,$$

where $\alpha_{(i)}$, $1 \le i \le r$ is the chosen basis of simple roots for g, together with

$$a_{(0)} = (-\psi, 0, 1), \tag{3.2.14}$$

where ψ is the highest root of g; for then an arbitrary root of $\hat{g}$,

$$a = \sum_{i=0}^{r} n_i a_{(i)} \tag{3.2.15}$$

is positive if $n_i \geq 0, 0 \leq i \leq r$, and negative if $n_i \leq 0, 0 \leq i \leq r$, and these are the only two possibilities.

3.3. *Scalar product, Dynkin diagrams and the Weyl group*

To develop the theory further, and construct Cartan matrices, and hence Dynkin diagrams, we need to find an analogue for $\hat{g}$ of the invariant scalar product (2.1.20) on g. It must have the properties of being symmetric and satisfying the invariance property (2.1.22) for $X, Y, Z \in \hat{g}$. Because the nontrivial representations of $\hat{g}$ are infinite dimensional it is no longer completely straight-forward to define this scalar product by a trace. However it is fixed (up to a constant) by the properties we have specified. Because of its importance, we shall give the argument for this in detail.

By taking $Z = d$, $X = T_m^a$, $Y = T_n^b$ in (2.1.22), we see that

$$\langle T_m^a, T_n^b \rangle = 0 \quad \text{unless } m = -n, \tag{3.3.1}$$

and, taking $Z = T_0^a$ instead, we see that, when $m + n = 0$, $\langle T_m^a, T_n^b \rangle$ is an isotropic tensor with respect to g, and so proportional to δ^{ab}. By taking $Z = T_m^a$, $X = T_n^b$ and $Y = k$, we see that

$$\langle k, T_n^c \rangle = 0, \tag{3.3.2}$$

$$\langle k, k \rangle = 0. \tag{3.3.3}$$

Now with $Z = T_1^a$, $X = T_m^b$, $Y = T_{-m-1}^c$, we see that the coefficient of proportionality between $\langle T_m^a, T_{-m}^b \rangle$ and δ^{ab} is independent of m, so that, by a suitable choice of normalization,

$$\langle T_m^a, T_n^b \rangle = \delta^{ab}\delta_{m,-n}. \tag{3.3.4}$$

Finally, taking $X = T_m^a$, $Y = d$, $Z = T_n^b$ we have

$$m\delta^{ab}\delta_{m,-n} = if^{abc}\langle d, T_{m+n}^c \rangle + m\langle d, k \rangle \delta^{ab}\delta_{m,-n}$$

so that

$$\langle d, T_m^a \rangle = 0 \tag{3.3.5}$$

and

$$\langle k, d \rangle = 1. \tag{3.3.6}$$

The only unconstrained scalar product is, say

$$\langle d, d\rangle = x. \tag{3.3.7}$$

Actually it is clear that it is not possible to determine x because the algebra is unchanged if we replace d by $d' = d - wk$ for any number w: choosing $w = \frac{1}{2}x$ sets $\langle d', d'\rangle$ to zero and we adopt this choice. So hence forth, by convention,

$$\langle d, d\rangle = 0. \tag{3.3.8}$$

Let us consider the signature of this scalar product in a Hermitian basis. Such a basis is provided by

$$T_0^i, \quad (T_m^i + T_{-m}^i)/\sqrt{2}, \quad (T_m^i - T_{-m}^i)/\sqrt{2}i, \quad (k+d)/\sqrt{2}, \quad (k-d)/\sqrt{2}. \tag{3.3.9}$$

In fact this basis is orthonormal, the norm of all its vectors being 1 except $(k - d)/\sqrt{2}$ which has norm -1. This single negative value indicates that the scalar product is Lorentzian in the sense that it resembles the metric of space-time in the theory of relativity.

When the invariant scalar product on $\hat{g}$ is restricted to the CSA H_0^i, $1 \le i \le r$, k, d, it is still Lorentzian in character. We can use this to define the scalar product of two vectors of simultaneous eigenvalues of the CSA $m = (\mu, \mu_k, \mu_d)$, $m' = (\mu', \mu_k', \mu_d')$:

$$m \cdot m' = \mu \cdot \mu' + \mu_k \mu_d' + \mu_d \mu_k'. \tag{3.3.10}$$

With this definition, the scalar product of two roots

$$a = (\alpha, 0, n), \quad a' = (\alpha', 0, n)$$

is

$$a \cdot a' = \alpha \cdot \alpha', \tag{3.3.11}$$

so that the roots (3.2.11) have length squared $a^2 = \alpha^2 > 0$ while the roots (3.2.12) have length $(n\delta)^2 = 0$, and are orthogonal to all other roots. Roots of type (3.2.11) are usually called "real roots" in the literature whilst roots of type (3.2.12) are called "imaginary roots." We shall refer to the former as *space-like* and the latter as *light-like*, as this nomenclature seems more appropriate. As we have seen the space-like roots are non-degenerate whilst the light-like ones have an r-fold degeneracy, just as a photon, propagating in a Lorentzian space with the dimension of the Cartan subalgebra, would carry light-like momentum and possess r linearly independent transverse polarization states. [This resemblance is not accidential; in a covariant approach to the vertex operator construction of the algebra $\hat{g}$, the step operators for the light-like roots are represented by the vertices for massless vector particles (photons or gauge bosons) of string theory.[5,127]]

The Cartan matrix of $\hat{g}$ is defined as in (3.1.11) as the $(r+1) \times (r+1)$ matrix

$$K_{ij} = 2a_{(i)} \cdot a_{(j)}/a_{(j)}^2, \quad 0 \le i, j \le r. \tag{3.3.12}$$

It is thus formed from the Cartan matrix for g by the addition of a first row and column, K_{i0} and K_{0i}, which can be calculated using (3.1.11) with $\alpha_0 = -\psi$. To do this note that, from the defining property (3.1.14) of the fundamental weights,

$$\psi = -\sum_{i=0}^{r} K_{0i}\lambda_i . \tag{3.3.13}$$

Now ψ is necessarily a long root of g so that $\psi^2 \ge \alpha_{(i)}^2$, $1 \le i \le r$. Hence

$$K_{i0} = -1 \quad \text{if } K_{0i} \ne 0, \tag{3.3.14a}$$

$$= 0 \quad \text{if } K_{0i} = 0 \tag{3.3.14b}$$

(provided that ψ is not itself a simple root, which only happens for su(2)).

This enables us to calculate K for $\hat{g}$ and construct its Dynkin diagram. This is obtained from the Dynkin diagram for g by appending an extra point corresponding to $\alpha_{(0)}$, joined by $-K_{0i}$ lines to those points $a_{(i)}$, identified with the $\alpha_{(i)}$ for which $K_{0i} \ne 0$. If $-K_{0i} > 1$ an arrow points towards $a_{(0)}$. For example, for su$(r+1)$, $r \ge 2$

$$\psi = \lambda_1 + \lambda_r , \tag{3.3.15}$$

as is familiar for su(3) in the particle physics context, as it corresponds to the fact that an octet can be formed from the product of a triplet and anti-triplet (quark-anti-quark bound state). Hence the Dynkin diagram for su(n) is obtained from that for su(n) by adding an extra point joined by two single lines to the two extremities, as can be seen by comparing A_r in Fig. 2 with $\hat{A}_r$ in Fig. 3. As another example, for so(n), $n \ge 5$, $\psi = \lambda_2$, so that the extra point is connected by a single line to the second point of the Dynkin diagram for so(n).

The Dynkin diagrams for the untwisted affine simple algebras are given in Fig. 3. In each diagram, the point corresponding to $\alpha_{(0)}$ is marked by a zero. The distinction between the open and closed points is not important for the Dynkin diagram itself; its significance is explained in Subsec. 3.4. The diagram for $\hat{A}_1$ needs some explanation. For this algebra, there are only two simple roots $a_{(0)} = (-\alpha, 0, 1)$, $a_{(1)} = (\alpha, 0, 0)$, so that $K_{01} = K_{10} = -2$, a possibility not allowed in the Euclidean case; we have two roots

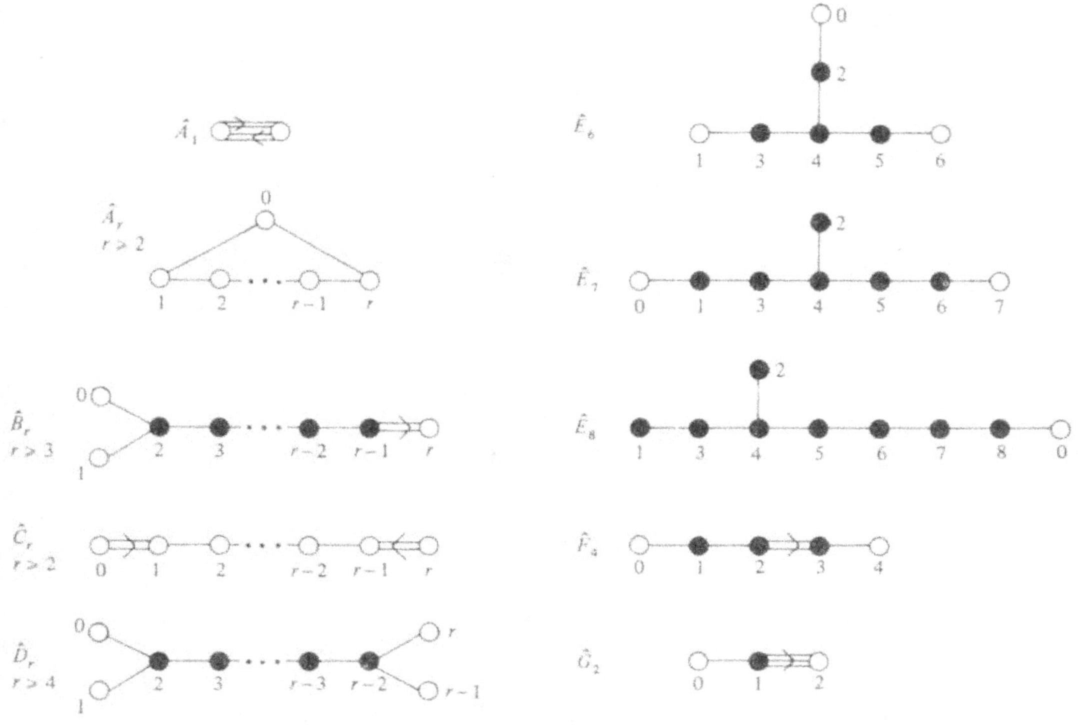

Fig. 3. Dynkin diagrams for untwisted affine simple Kac-Moody algebras.

of equal length joined by more than one line, four in fact, and the arrows point both ways to indicate the equality of $a^2_{(0)}$ and $a^2_{(1)}$.

[Notice that the extended Dynkin diagram describing $\hat{g}$, formed by adding the point 0, often has more symmetry than the original diagram describing g, and it never has less. The gain in symmetry can be related in a precise way to the centre of the simply connected group whose Lie algebras is g: this centre is isomorphic to a normal subgroup of the symmetry group of the extended Dynkin diagram, and the corresponding quotient group is isomorphic to the symmetry group of the original Dynkin diagram.[52] For example, SU$(r + 1)$ has centre $\mathbb{Z}_{r+1}$ and this is the cyclic symmetry of the diagram for $\hat{A}_r$, whose full symmetry is the $2(r + 1)$ dihedral group; the quotient is $\mathbb{Z}_2$, the symmetry of the A_r diagram. Likewise the group E_6 has centre $\mathbb{Z}_3$ which is a subgroup of the S_3 symmetry group of the Dynkin diagram for $\hat{E}_6$ and the quotient is again $\mathbb{Z}_2$, the symmetry of the E_6 diagram.]

Finally in this subsection we shall consider the structure of the Weyl group $\hat{W}$ of $\hat{g}$. As in Subsec. 3.1 this is defined as being generated by reflections

$$\sigma_a(x) = x - 2(x \cdot a/a^2)a \tag{3.3.16}$$

in the hyperplanes normal to roots a except that now we must use the Lorentzian scalar product (3.3.10) and we are restricted to using space-like roots $a = (\alpha, 0, n), \alpha \neq 0$, because the light-like roots $n\delta = (0, 0, n)$, $n \neq 0$, would give $a^2 = 0$. Note that any light-like root is left invariant by (3.3.16)

$$\sigma_a(n\delta) = n\delta, \tag{3.3.17}$$

because $\delta \cdot a = 0$ for any root a. Hence each light-like root is left invariant by all the elements of the Weyl group $\hat{W}$. On the other hand, if $a = (\alpha, 0, n)$, $a' = (\alpha', 0, n')$ are two space-like roots,

$$\sigma_a(a') = (\sigma_\alpha(\alpha'), 0, n - 2(\alpha \cdot \alpha'/\alpha^2)n') \tag{3.3.18}$$

so that $\hat{W}$ permutes the space-like roots.

We shall now show that $\hat{W}$ has the structure of a semi-direct product,

$$\hat{W} = W \ltimes \Lambda_R^v, \tag{3.3.19}$$

the semi-direct product of Weyl group of g and the coroot lattice of g, just as the Euclidean group is the semi-direct product of the rotation group and the translation group. The coroot lattice $\Lambda_R^\vee$ is the lattice generated by the *coroots* $\alpha^\vee = \alpha/\alpha^2$. (The coroots $\alpha^\vee$ form the root system of a Lie algebra $g^\vee$, dual to g, obtained by interchanging root lengths. Thus all simply laced algebras A_n, D_n, E_n are self-dual, as also are F_4 and G_2, whilst $B_n^\vee = C_n$, $C_n^\vee = B_n$. $\Lambda_R^\vee$ is the root lattice of $g^\vee$.)

To establish (3.3.19) we calculate the action of σ_a on $x = (\xi, \kappa, p)$ for $a = (\alpha, 0, n)$.

$$\begin{aligned} \sigma_a(x) &= (\xi - 2[\alpha \cdot \xi + n\kappa]\alpha/\alpha^2, \kappa, p - 2[\alpha \cdot \xi + n\kappa]n/\alpha^2) \\ &= (\sigma_\alpha(\xi'), \kappa, p + [\xi^2 - \xi'^2]/2\kappa) \end{aligned} \tag{3.3.20}$$

for $\xi' = \xi + 2n\kappa\alpha/\alpha^2$.

Hence

$$\sigma_a = \sigma_\alpha(t_\alpha)^n, \tag{3.3.21}$$

where

$$t_\alpha(x) = (\xi + 2\kappa\alpha/\alpha^2, \kappa, p + [\xi^2 - (\xi + 2\kappa\alpha/\alpha^2)^2]/2\kappa). \tag{3.3.22}$$

We can think of t_α as translation in the coroot $\alpha^\vee$; t_α, t_β commute for two roots α, β so that the t_α can be thought of as generating the translation group $\Lambda_R^\vee$. Further

$$\sigma_\beta t_\alpha \sigma_\beta = t_{\alpha'}, \quad \text{for} \quad \alpha' = \sigma_\beta(\alpha) \tag{3.3.23}$$

from which it follows that $\Lambda_R^\vee$ is a normal subgroup of $\hat{W}$ and hence that $\hat{W}$ has the structure (3.3.19). Thus $\hat{W}$ is isomorphic to the group of Euclidean transformations preserving the lattice $\Lambda_R^\vee$, whilst W is the subgroup of $\hat{W}$ which fixes any given point of the lattice $\Lambda_R^\vee$. Such groups were classified long ago by Coxeter using a Dynkin diagram notation; see Ref. 53.

3.4. *Highest weight representations of affine algebras*

As we saw in Sec. 2, in physical applications of affine Kac-Moody algebras $\hat{g}$, the operator $-d$, or equivalently L_0, is identified with a physical quantity such as the energy or the scale operator, whose spectrum must be bounded below for physical reasons. A representation in which the spectrum of $-d$ is bounded below is called a highest weight representation. In Subsec. 1.4 we remarked that the whole of a unitary highest weight representation can be built up from vacuum states $|\Psi\rangle$ satisfying

$$T_n^a|\Psi\rangle = 0, \quad n > 0. \tag{3.4.1}$$

If the representation is irreducible, these states form an irreducible representation of the finite dimensional algebra g and the irreducible representation of $\hat{g}$ is characterized by this vacuum representation of g, or, equivalently its highest weight μ_0, and the value of the central term k.

There is a simple set of necessary and sufficient conditions for there to be a unitary highest weight representation of $\hat{g}$ in which k takes a particular value and the vacuum representation of g has highest weight μ_0; it is

$$2k/\psi^2 \in \mathbb{Z}, \tag{3.4.2a}$$

and

$$k \geq \psi \cdot \mu_0 \geq 0, \tag{3.4.2b}$$

where ψ is the highest root of g. The non-negative integer $2k/\psi^2$ is called the level of the representation of $\hat{g}$.

To show that conditions (3.4.2) are necessary we only need to consider the su(2) subalgebra of $\hat{g}$ associated with the root $(-\alpha, 0, 1)$

$$E_1^{-\alpha}, \quad E_{-1}^{\alpha}, \quad \frac{2}{\alpha^2}(k - \alpha\cdot H_0), \tag{3.4.3}$$

which is isomorphic to I_+, I_-, $2I_3$, satisfying (3.1.7). Firstly the eigenvalues of $\frac{2}{\alpha^2}(k - \alpha\cdot H_0)$ must be integral. Applying this example to a state $|\mu\rangle$ satisfying

$$H_0|\mu\rangle = \mu|\mu\rangle, \tag{3.4.4}$$

we obtain

$$\frac{2k}{\alpha^2} - \frac{2\alpha\cdot\mu}{\alpha^2} \in \mathbb{Z} \tag{3.4.5}$$

and, since $2\alpha\cdot\mu/\alpha^2 \in \mathbb{Z}$, for any weight μ and root α of g, we have established (3.4.2a). If $|\mu\rangle$ is actually a vacuum state it obeys in addition

$$E_1^{-\alpha}|\mu\rangle = 0 \tag{3.4.6}$$

and thus

$$\begin{aligned} \|E_{-1}^{\alpha}|\mu\rangle\|^2 &= \langle\mu|E_1^{-\alpha}E_{-1}^{\alpha}|\mu\rangle \\ &= \langle\mu|[E_1^{-\alpha}, E_{-1}^{\alpha}]|\mu\rangle \\ &= \frac{2}{\alpha^2}(k - \alpha\cdot\mu)\|\,|\mu\rangle\|^2. \end{aligned} \tag{3.4.7}$$

Hence

$$k \geq \alpha\cdot\mu \tag{3.4.8}$$

for all roots α and vacuum weights μ.

To see that (3.4.2b) is the most stringent case of (3.4.8) note that the latter condition is invariant under Weyl transformations on μ and we can use this to take μ to be dominant. Then the root for which $\alpha\cdot\mu$ is largest is $\alpha = \psi$. Since the highest root has positive inner product with all the simple roots, $\psi\cdot\mu$ is largest for $\mu = \mu_0$ the highest weight of the vacuum representation.

Condition (3.4.2b) can be interpreted as specifying a minimum level at which the vacuum representation of g can be that with highest weight μ_0. To spell this out, expand μ_0 in terms of the fundamental weights for g,

$$\mu_0 = \sum_{i=1}^{r} n_i \lambda_{(i)}, \tag{3.4.9}$$

where the n_i are integers ≥ 0, and define integers $m_i > 0$ by

$$\psi/\psi^2 = \sum_{i=1}^{r} m_i \alpha_{(i)}/\alpha_{(i)}^2. \tag{3.4.10}$$

Then, by the definition (3.1.18) of $\lambda_{(i)}$, the level of the representation is

$$x = \frac{2k}{\psi^2} \geq \sum_{i=1}^{r} n_i m_i . \tag{3.4.11}$$

In other words the possible representations of $\hat{g}$ are those with levels x and vacuum representations with highest weight μ_0 where $x \geq \sum n_i m_i$. If $x = 0$ the only possible representation has $\mu_0 = 0$ and so it is trivial.

Actually we can reformulate this analysis of the possible unitary irreducible highest weight representation of $\hat{g}$ so that it exactly parallels that given in Subsec. 3.1 for g. There is a unique state in such a representation $|\mu_0\rangle$ for which

$$E_n^\alpha |\mu_0\rangle = 0 , \tag{3.4.12a}$$

if *either* $n > 0$ *or* $n = 0$ and $\alpha > 0$ and

$$H_n^i |\mu^0\rangle = 0 \quad \text{if} \quad n > 0 , \tag{3.4.12b}$$

namely the highest weight state of the vacuum representation of g, from which all other states follow by the application of step operators for negative roots. With respect to the CSA $\hat{H} = (H, k, d)$, this state has weight vector $\hat{\mu}_0 = (\mu_0, k, \nu_0)$ and the weight vector $\hat{\mu} = (\mu, k, \nu)$ of any state of the representation has the property that $\hat{\mu}_0 - \hat{\mu}$ is a sum of positive roots. The representation can be labelled by the highest weight $\hat{\mu}_0$ though the value of ν_0, the lowest eigenvalue of d, is really irrelevant, a matter of convention.

For each space-like root $a = (\alpha, 0, n)$ of $\hat{g}$, we have an su(2) subalgebra like (3.4.3) consisting of

$$E^a = E_n^\alpha , \quad E^{-a} = E_{-n}^{-\alpha}$$

and

$$2a \cdot \hat{H}/a^2 = 2(\alpha \cdot H + nk)/\alpha^2 . \tag{3.4.13}$$

The states of any unitary representation of $\hat{g}$ must fall into multiplets of this su(2) subalgebra and this implies that its set of weights $\hat{\mu} = (\mu, k, \nu)$ must be mapped into itself by the Weyl reflections (3.3.16), which can be realized in the representation by

$$\sigma_a = \exp\left(\frac{i\pi}{2}(E^a + E^{-a})\right) . \tag{3.4.14}$$

Hence $\hat{\mu}$ must satisfy the condition that $\hat{\mu}_0 - \sigma(\hat{\mu})$ is a sum of positive roots of $\hat{g}$ for any $\sigma \in \hat{W}$; this is also a sufficient condition. This condition applied to $\hat{\mu} = \hat{\mu}_0$ for $\sigma = \sigma_a$, a positive, gives

$$2a \cdot \hat{\mu}_0/a^2 \in \mathbb{Z} \quad \text{and non-negative}. \tag{3.4.15}$$

Taking $a = a_{(0)} = (-\psi, 0, 1)$ we immediately obtain (3.4.2).

To analyze (3.4.2) or, equivalently, (3.4.15) further, we introduce a set of fundamental weights for $\hat{g}$, $l_{(i)}$, $0 \le i \le r$, defined as in (3.1.14),

$$2l_{(i)} \cdot a_{(j)}/a_{(j)}^2 = \delta_{ij}, \quad 0 \le i, j \le r. \tag{3.4.16}$$

Then the general solution to condition (3.4.15) is

$$\hat{\mu}_0 = \sum_{i=0}^{r} n_i l_{(i)}, \tag{3.4.17}$$

where the $n_i \in \mathbb{Z}$ and non-negative (because a/a^2 can be expanded in terms of $a_{(i)}/a_{(i)}^2$), apart from the indeterminate component in the d direction. Using (3.1.18) and (3.4.10) we find

$$l_{(i)} = (\lambda_{(i)}, \tfrac{1}{2} m_i \psi^2, 0), \tag{3.4.18}$$

$$l_{(0)} = (0, \tfrac{1}{2} m_0 \psi^2, 0), \tag{3.4.19}$$

where $m_0 = 1$. It follows that the n_i in (3.4.9) and (3.4.17) are the same for $1 \le i \le r$. From (3.4.17) we deduce that the level is

$$x = \sum_{i=0}^{r} n_i m_i, \tag{3.4.20}$$

from which we can again deduce (3.4.11) as $n_0 m_0 \ge 0$.

The possible level 1 representations of $\hat{g}$ are those with highest weights $l_{(i)}$ for those i, $0 \le i \le r$, with $m_i = 1$. Thus the possible vacuum representations for level 1 representations are the scalar and those with highest weights $\lambda_{(i)}$ where $m_i = 1$. The corresponding points of the extended Dynkin diagrams are indicated by open rather than closed points in Fig. 3. We shall see in Sec. 6 that, at least for the simply-laced groups, the level 1 representations have a special significance as we can construct vertex operator representations for them.

Because the m_i, $0 \le i \le r$, satisfy

$$\sum_{i=0}^{r} m_i \alpha_{(i)}/\alpha_{(i)}^2 = 0 \tag{3.4.21}$$

for $\alpha_{(0)} = -\psi$, we have that $(m_0, m_1, \ldots, m_r)$ is a right null vector of the Cartan matrix:

$$\sum_{j=0}^{r} K_{ij} m_j = 0. \tag{3.4.22}$$

It follows from this that the positioning of the open dots in Fig. 3 must preserve the basic symmetry of the extended Dynkin diagram (since K has this symmetry). Thus any point related to the point 0 by such a symmetry will have $m_i = 1$. This serves to determine all the open dots for the simply-laced algebras.

In the case of the classical algebras A_r, B_r, C_r, D_r the value of m_i for a point represented by a closed dot is 2. For the exceptional groups the vector $(m_0, m_1, \ldots, m_r)$ is as follows:

$$\begin{array}{ll} \hat{E}_6 & (1,1,2,2,3,2,1), \\ \hat{E}_7 & (1,2,2,3,4,3,2,1), \\ \hat{E}_8 & (1,2,3,4,6,5,4,3,2), \\ \hat{F}_4 & (1,2,3,2,1), \\ \hat{G}_2 & (1,2,1), \end{array} \tag{3.4.23}$$

where the points are ordered as in Fig. 3: see, e.g., Ref. 51.

3.5. *Twisted affine Kac-Moody algebras*

So far in this article we have dealt with untwisted affine Kac-Moody algebras but for completeness we shall treat in this section the remaining affine Kac-Moody algebras, namely the twisted ones. To define one of these we need a compact finite-dimensional Lie algebra g,

$$[T^a, T^b] = i f^{abc} T^c, \tag{3.5.1}$$

and an automorphism τ of g, i.e.

$$[\tau(T^a), \tau(T^b)] = i f^{abc} \tau(T^c) \tag{3.5.2}$$

which has finite order N, say, specified by the smallest positive integer for which

$$\tau^N = 1. \tag{3.5.3}$$

Then by taking a complex linear combinations of the Hermitian generators we can divide g into eigenspaces $g_{(m)}$ of τ:

$$\tau(T) = e^{2\pi i m/N} T \quad \text{if } T \in g_{(m)}, \tag{3.5.4}$$

$0 \le m \le N-1$. Then, by (3.5.1),

$$[T, T'] \in g_{(m+n)} \quad \text{if } T \in g_{(m)}, \quad T' \in g_{(n)}, \tag{3.5.5}$$

that is we can assign a grade m to T if $T \in g_{(m)}$ and these grades add mod N on commutation.

Choosing the basis $\{T^a\}$ for g to consist of eigenvectors of τ (so that it is not Hermitian in general) we define the twisted affine algebra $\hat{g}^\tau$ to have a basis consisting of $\{T_r^a\}$, where

$$r \in \mathbb{Z} + m/N \quad \text{if } T_r^a \in g_{(m)}, \tag{3.5.6}$$

together with the central element k, and to have the commutation relations

$$[T_m^a, T_n^b] = if^{abc} T_{m+n}^c + km\delta^{ab}\delta_{m,-n} \tag{3.5.7}$$

formally as before in (1.1.2), but with the suffices m, n now being fractional in the way indicated. We can again extend the algebra further by introducing a derivation d satisfying

$$[d, T_m^a] = mT_m^a, \quad [d, k] = 0. \tag{3.5.8}$$

A simple example is given by taking $g = \text{su}(2)$.

$$[T^i, T^j] = i\varepsilon_{ijk} T^k, \tag{3.5.9}$$

and defining τ by $\tau(T^i) = -T^i$, $i \neq 3$, $\tau(T^3) = T^3$. Then $\tau^2 = 1$ and $\hat{g}^\tau$ has a basis consisting of T_m^i, $m \in \mathbb{Z} + \frac{1}{2}$, $i = 1, 2$; T_m^3, $m \in \mathbb{Z}$; k: and d. Actually this is a rather trivial example as we can see by changing to a basis of g consisting of the CSA T_3 and step operators $T^\pm = T^1 \pm iT^2$. Then

$$[T_m^3, T_n^\pm] = \pm T_{m+n}, \quad m \in \mathbb{Z}, n \in \mathbb{Z} + \tfrac{1}{2}, \tag{3.5.10a}$$

$$[T_m^+, T_n^-] = 2T_{m+n}^3 + km\delta_{m,-n}, \quad m, n \in \mathbb{Z} + \tfrac{1}{2}. \tag{3.5.10b}$$

So, if we define

$$S_n^\pm = T_{n\pm 1/2}^\pm, \quad S_n^3 = T_n^3 + \frac{k}{4}\delta_{n,0}, \tag{3.5.11}$$

$$d' = d - \tfrac{1}{2}T_3. \tag{3.5.12}$$

S_n^i satisfy the untwisted affine Kac-Moody algebra $\hat{\text{su}}(2)$ with central term k and derivation d'. Thus, in this case, the twisted algebra is isomorphic to the untwisted one and the twisting is, in a sense, trivial. This is not always the case. Where it is, this is a symptom of the fact that the automorphism τ is an inner automorphism, i.e., one of the form

$$\tau(T) = \gamma T \gamma^{-1} \tag{3.5.13}$$

for some $\gamma \in G$, a compact group with Lie algebra g. In our su(2) example we can take

$$\gamma = \exp(i\pi T_3). \tag{3.5.14}$$

We shall now show that a twist corresponding to an inner automorphism can always be removed by a shift of the form of Eq. (3.5.11). Clearly the automorphisms corresponding to conjugate elements γ_1, γ_2 of G produce isomorphic twisted algebras. By conjugation $\gamma \to \gamma_0 \gamma \gamma_0^{-1}$, we can map γ into the maximal torus generated by a given CSA, i.e., we may write

$$\gamma = \exp(i\chi \cdot H). \tag{3.5.15}$$

Then (3.5.3), i.e., $\gamma^N = 1$, is equivalent to

$$N\chi \cdot \alpha \in 2\pi\mathbb{Z} \quad \text{for all roots } \alpha \text{ of } g. \tag{3.5.16}$$

Writing g in a modified Cartan-Weyl basis (3.2.1)–(3.2.3), we see that

$$\tau(H^i) = H^i, \quad \tau(E^\alpha) = e^{i\chi\cdot\alpha} E^\alpha, \tag{3.5.17}$$

so that a basis for $\hat{g}^\tau$ consists of H^i_m, E^α_n, where $m \in \mathbb{Z}$ and $n \in \mathbb{Z} + \chi \cdot \alpha/2\pi$, together with k, and these operators satisfy an algebra which is formally the same as (3.2.4)–(3.2.6) but with the possible values of the suffices reinterpreted. To undo the "twist," we set

$$F^\alpha_n = E^\alpha_{n+\chi\cdot\alpha/2\pi}, \quad I^i_n = H^i_n + k\chi^i \delta_{n,0}/2\pi, \tag{3.5.18}$$

$$d' = d - \chi \cdot H/2\pi, \tag{3.5.19}$$

paralleling (3.5.11) and (3.5.12). F^α_n, I^i_n satisfy the untwisted algebra $\hat{g}$, the "twist" having been undone by a "shift."

Whilst this shows the mathematical isomorphism of the twisted algebra $\hat{g}^\tau$ and the untwisted algebra g for an inner automorphism τ, the physical interpretation may give more significance to one way of writing the algebra (e.g., the twisted) than the other. In particular the derivation d for $\hat{g}^\tau$ is different from that for $\hat{g}$, d'. The vacuum states in a given representation of $\hat{g}^\tau$, i.e., those with the lowest value of d, will form a representation of $g_{(0)}$, the subalgebra g commuting with $\chi \cdot H$, rather than g itself, whilst in the case of $\hat{g}$ the vacuum states must form a representation of g. Hence the twisting provides a way of breaking the symmetry from g to $g_{(0)}$. (This is true whether τ is an inner automorphism or not.)

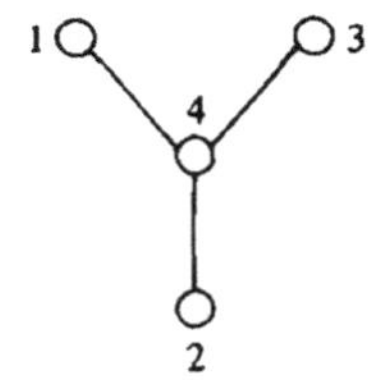

Fig. 4. Dynkin diagram for D_4.

Now let us consider outer automorphisms of g, that is ones which are not inner in the sense of (3.5.13). Such automorphisms exist if and only if the Dynkin diagram for g has nontrivial symmetries, that is when g is one of the algebras A_r $(r \geq 2)$, D_r $(r \geq 4)$ and E_6. Let us denote such a symmetry by τ. Clearly $\tau^2 = 1$ unless $g = D_4$ in which case we can find a τ with $\tau^3 = 1$ because the symmetry group of the Dynkin diagram is S_3, the permutation group on three elements (see Fig. 4). We can define a linear transformation on the roots by

$$\tau(\alpha_{(i)}) = \alpha_{\tau(i)}, \quad 1 \leq i \leq r \tag{3.5.20}$$

and this will be orthogonal because τ is a symmetry of the Dynkin diagram. This extends to the CSA in an obvious way, so that

$$\tau(\alpha)\cdot\tau(H) = \alpha\cdot H. \tag{3.5.21}$$

We then extend τ to an automorphism of g by

$$\tau(E^\alpha) = E^{\tau(\alpha)} \tag{3.5.22}$$

for each simple root α. It follows that, for a typical root,

$$\tau(E^\alpha) = \varepsilon_\alpha E^{\tau(\alpha)}, \tag{3.5.23}$$

where $\varepsilon_\alpha = \varepsilon_{-\alpha} = \pm 1$. It must be possible to assign these signs consistently because the algebra is determined by the Dynkin diagram, which is unchanged by τ. Any automorphism of g is equal to the product of an inner automorphism (3.5.13) and an outer automorphism, associated with a symmetry of the Dynkin diagram of g, of the sort just described. Thus, if we leave aside in the present discussion the potentially physically significant changes introduced by a shift of the form (3.5.18), it remains to consider $\hat{g}^\tau$ where τ is one of these outer automorphisms. The corresponding twisted algebras are denoted $A_2^{(2)}: A_r^{(2)}$, $r \geq 4$: $E_6^{(2)}$, each corresponding to an automorphism of order 2: and $D_4^{(3)}$ corresponding to one of order 3. [Note that, as $A_3 = D_3, A_3^{(2)} = D_3^{(2)}$, and we use the latter notation as it fits in better with that series.] The untwisted affine Kac-Moody algebra considered in previous sections correspond to the identity automorphism, which has order 1. Hence in the present notation we write $A_r^{(1)}$ for $\hat{A}_r$, etc.

Consider first the typical case where τ has order 2. The unbroken algebra $g_{(0)}$ consists of the eigenspace of g on which $\tau = 1$, whilst $g_{(1)}$ consists of the eigenspace on which $\tau = -1$; $g_{(1)}$ provides a representation of the algebra $g_{(0)}$. To see how the CSA divides between $g_{(0)}$ and $g_{(1)}$ we consider what happens to $\alpha_{(i)} \cdot H$. Since τ maps positive roots to positive roots. $\tau(\alpha) \neq -\alpha$ for any root α. It follows that

$$\alpha_{(i)} \cdot H \in g_{(0)} \quad \text{if} \quad \tau(i) = i \tag{3.5.24a}$$

and

$$\tfrac{1}{2}\{\alpha_{(i)} + \alpha_{\tau(i)}\} \cdot H \in g_{(0)}, \quad \tfrac{1}{2}\{\alpha_{(i)} - \alpha_{\tau(i)}\} \cdot H \in g_{(1)} \quad \text{if} \quad \tau(i) \neq i. \tag{3.5.24b}$$

The distinct operators amongst $\frac{1}{2}\{\alpha_{(i)} + \alpha_{\tau(i)}\} \cdot H$, $1 \leq i \leq r$, form a basis for the CSA of $g_{(0)}$, which is the subspace of the CSA of g formed by $\xi \cdot H$ with $\tau(\xi) = \xi$. For such an ξ

$$[\xi \cdot H, E^{\alpha} \pm E^{\tau(\alpha)}] = \tfrac{1}{2}\xi \cdot \{\alpha + \tau(\alpha)\}(E^{\alpha} \pm E^{\tau(\alpha)}). \tag{3.5.25}$$

Now

$$\tau(E^{\alpha} \pm E^{\tau(\alpha)}) = \pm\varepsilon_{\alpha}(E^{\alpha} \pm E^{\tau(\alpha)}). \tag{3.5.26}$$

From this it follows that, if $\alpha = \tau(\alpha)$,

$$E^{\alpha} \in g_{(0)} \quad \text{if} \quad \varepsilon_{\alpha} = 1$$

while

$$E^{\alpha} \in g_{(1)} \quad \text{if} \quad \varepsilon_{\alpha} = -1, \tag{3.5.27a}$$

and, if $\alpha \neq \tau(\alpha)$,

$$E^{\alpha} + \varepsilon_{\alpha} E^{\tau(\alpha)} \in g_{(0)}$$

while

$$E^{\alpha} - \varepsilon_{\alpha} E^{\tau(\alpha)} \in g_{(1)}. \tag{3.5.27b}$$

Thus the roots of $g_{(0)}$ are $\{\alpha : \tau(\alpha) = \alpha, \varepsilon_{\alpha} = 1\}$ together with $\{\frac{1}{2}\alpha + \frac{1}{2}\tau(\alpha) : \alpha \neq \tau(\alpha), \varepsilon_{\alpha} = 1\}$. It follows that a basis of simple roots consists of vectors of the form $\frac{1}{2}\alpha_{(i)} + \frac{1}{2}\alpha_{\tau(i)}$. If $r - s$ is the number of values of i for which $i \neq \tau(i)$ the rank of $g_{(0)}$ is s. We denote a basis of simple roots of $g_{(0)}$ by β_i, $1 \leq i \leq s$. We calculate its Dynkin diagram, and hence $g_{(0)}$, from our knowledge of the simple roots. These Dynkin diagrams can be found from Fig. 5 by deleting the points marked 0.

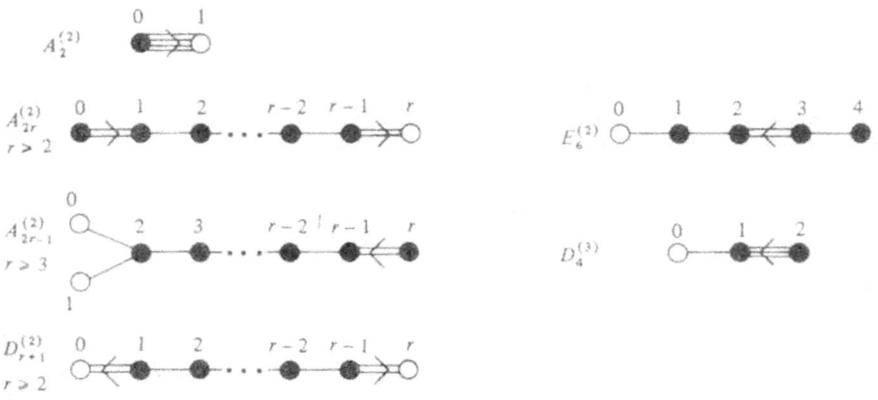

Fig. 5. Dynkin diagrams for twisted affine simple Kac-Moody algebras.

The roots of $\hat{g}^{\tau}$ are

$$(\alpha, 0, n), \quad \alpha = \tau(\alpha), \quad \varepsilon_\alpha = 1, \quad n \in \mathbb{Z}, \tag{3.5.28a}$$

$$(\alpha, 0, n), \quad \alpha = \tau(\alpha), \quad \varepsilon_\alpha = -1, \quad n \in \mathbb{Z} + \tfrac{1}{2}, \tag{3.5.28b}$$

$$(\tfrac{1}{2}\alpha + \tfrac{1}{2}\tau(\alpha), 0, n), \quad \alpha \neq \tau(\alpha), \quad 2n \in \mathbb{Z}, \tag{3.5.29}$$

$$(0, 0, n), \quad n \neq 0, \quad 2n \in \mathbb{Z}, \tag{3.5.30}$$

where α is a root of g. The space-like roots (3.5.28) and (3.5.29) are non-degenerate whilst the light-like roots (3.5.30) have degeneracy s for $n \in \mathbb{Z}$ and degeneracy $r - s$ for $n \in \mathbb{Z} + \frac{1}{2}$.

The vector we have to add to the s distinct vectors $\beta_{(i)}$ of the form $\frac{1}{2}\alpha_{(i)} + \frac{1}{2}\alpha_{\tau(i)}$ to go from a basis of simple roots for $g_{(0)}$ to one for $\hat{g}^{\tau}$ is

$$a_{(0)} = (-\phi, 0, \tfrac{1}{2}), \tag{3.5.31}$$

where ϕ is the highest vector of the form $\{\frac{1}{2}\alpha + \frac{1}{2}\tau(\alpha) : \alpha \neq \tau(\alpha)\}$ or $\{\alpha : \alpha = \tau(\alpha), \varepsilon_\alpha = -1\}$. Given this, we find the Dynkin diagrams as in Fig. 5, defining them as in Subsecs. 3.1 and 3.3.

We can perform a similar analysis for $D_4^{(3)}$ where $\tau^3 = 1$. The CSA of $g_{(0)}$ again is the subspace of the CSA of g consisting of $\xi \cdot H$ where $\tau(\xi) = \xi$. In this case $g_{(0)}$ has rank 2. It has roots $\{\alpha : \tau(\alpha) = \alpha\}$ and $\{\frac{1}{3}\alpha + \frac{1}{3}\tau(\alpha) + \frac{1}{3}\tau^2(\alpha) : \tau(\alpha) \neq \alpha\}$. Since $\tau(\alpha) \cdot \alpha = 0$ if $\tau(\alpha) \neq \alpha$, we see that the roots of $g_{(0)}$ have squared lengths in the ratio $1:3$. Hence $g_{(0)} = G_2$. A basis of simple roots for $g_{(0)}$ consists of α_4 and $\frac{1}{3}\alpha_1 + \frac{1}{3}\alpha_2 + \frac{1}{3}\alpha_3$, with the notation of Fig. 3. The simple root we must add to these to form a basis for $D_4^{(3)}$ is $(-\phi, 0, \frac{1}{3})$ where ϕ is the highest vector of the form $\frac{1}{3}\alpha + \frac{1}{3}\tau(\alpha) + \frac{1}{3}\tau^2(\alpha)$, where $\alpha \neq \tau(\alpha)$. Then $\phi = \frac{1}{3}\alpha_1 + \frac{1}{3}\alpha_2 + \frac{1}{3}\alpha_3$, and we obtain the diagram in Fig. 5.

In all these cases except $A_{2m}^{(2)}$ $(m \geq 2)$, $g_{(1)}$ (and, for $D_4^{(3)}$, $g_{(2)}$ as well) is the representation corresponding to the short roots of $g_{(0)}$. In the case of $A_{2m}^{(2)}$ the weights of $g_{(i)}$ are all the roots of $g_{(0)}$ together with twice the short roots.

We can adapt the discussion of Subsec. 3.4 to deal with representations of the twisted algebras $\hat{g}^\tau$. Now the vacuum states (i.e., those for which d takes its lowest value) form a representation of $g_{(0)}$. An irreducible representation of $\hat{g}^\tau$ is labelled by an irreducible representation of $g_{(0)}$, or equivalently its highest weight μ_0, and a value of k. We write $\hat{\mu}_0 = (\mu_0, k, 0)$. As before we have that

$$2a \cdot \mu_0 / a^2 \in \mathbb{Z} \text{ and non-negative} \tag{3.5.32}$$

for any root a of $\hat{g}^\tau$. To analyze this, we again introduce fundamental weights $l_{(i)}$ defined by (3.4.16) where

$$a_{(i)} = (\beta_i, 0, 0), \quad 1 \leq i \leq s, \tag{3.5.33a}$$

$$a_{(0)} = \left(-\phi, 0, \frac{1}{N}\right), \tag{3.5.33b}$$

where N is the order of τ. Then

$$l_{(i)} = \left(\lambda_i', \frac{Nm_i'}{2m_0'}\phi^2, 0\right), \quad 1 \leq i \leq s, \tag{3.5.34a}$$

$$l_{(0)} = \left(0, \frac{N}{2}\phi^2, 0\right), \tag{3.5.34b}$$

where λ_i', $1 \leq i \leq s$, are fundamental weights for $g_{(0)}$ and the m_i' are integers such that

$$m_0' \frac{\phi}{\phi^2} = \sum_{i=1}^{s} m_i' \frac{\beta_i}{\beta_i^2}. \tag{3.5.35}$$

We choose the m_i' to have no common factor. Now $m_0' = 1$ except for $A_{2s}^{(2)}$ for which $m_0' = 2$. The general solution to (3.5.32) is then as in (3.4.17). In particular,

$$\frac{2k}{\psi^2} = M \sum_{i=0}^{s} n_i m_i', \tag{3.5.36}$$

where

$$M = \frac{N\phi^2}{m_0'\psi^2}. \tag{3.5.37}$$

Except for $A_{2s}^{(2)}$, $\psi^2 = N\phi^2$ and $m_0' = 1$, so that $M = 1$. For $A_{2s}^{(2)}$, $r \geq 2$, $\phi^2 = 2\psi^2$, $m_0' = 2$, so that $M = 2$. For $A_2^{(2)}$, $\phi^2 = 4\psi^2$, $m_0' = 2$ so that $M = 4$. Thus $\frac{2k}{\psi^2}$ can take any integral value, except for $A_{2s}^{(2)}$ when it has to be even for $s \geq 2$ and a multiple of 4 for $s = 1$. This condition replaces (3.4.2a), and (3.4.2b) is replaced by

$$k \geq N\phi \cdot \mu_0 . \tag{3.5.38}$$

As in (3.4.21) and (3.4.22), it follows in the twisted case that the column vector (m_i') is a right null vector of the Cartan matrix defined as in (3.3.12). The points of the Dynkin diagram corresponding to level 1 representations, i.e. those with $m_0' = 1$, are indicated by open dots in Fig. 5. Again, they must occur symmetrically in the Dynkin diagrams. Note that for $A_{2s}^{(2)}$ the representation with a scalar vacuum is not level 1.

4. Sugawara's Construction of the Virasoro Algebra

4.1. *Sugawara's construction for g simple*

We have written down, in Eqs. (1.1.1) to (1.1.3), the semi-direct product of the Kac-Moody and Virasoro algebras as if the generators were totally independent, but in fact there is a construction of the Virasoro algebra in terms of bilinears in the Kac-Moody generators to which we referred in Subsec. 2.2, which it will be the purpose of this section to explain and develop. The idea arose naturally in the theory of current algebras,[54,55,26] where it was argued that the full dynamics of the theory should be formulated in terms of currents. This means that the energy-momentum tensor should be expressed in terms of currents and it was realized by Sugawara[43] and Sommerfield[44] that this was possible if this tensor was bilinear in currents and if the Schwinger[33] term (often ignored hitherto) was taken into account. This idea was originally applied in four space-time dimensions but it was soon realized that it worked most neatly in two dimensions. Since the currents correspond to the T_m^a, and the energy-momentum tensor to L_m, Sugawara's construction translated into the present language reads

$$\mathscr{L}(z) = \sum_{n\in Z} z^{-n}\mathscr{L}_n = \frac{1}{2k + Q_\psi} {}^{\times}_{\times} \sum_{a=1}^{\dim g} T^a(z)T^a(z) {}^{\times}_{\times} \tag{4.1.1}$$

or, equivalently

$$\mathscr{L}_n = \frac{1}{2k + Q_\psi} \sum_{m\in Z} {}^{\times}_{\times} \sum_{a=1}^{\dim g} T_{m+n}^a T_{-m}^a {}^{\times}_{\times} . \tag{4.1.2}$$

The first statement emphasizes that two quantum field operators $T^a(z)$ (see (2.1.1)) are multiplied together at the same point. In order to avoid a singularity and obtain a quantity with finite matrix elements in a highest weight representation it is necessary to introduce a normal ordering denoted by the double crosses defined by Eq. (2.4.3) whereby the T^a_m with positive suffixes are moved to the right of those with negative suffixes. The singularity is exhibited by the Wick contraction following from this definition and (1.1.2):

$$T^a(z)T^a(\zeta) = {}^\times_\times T^a(z)T^a(\zeta){}^\times_\times + \frac{kz\zeta}{(z-\zeta)^2}, \quad |z| > |\zeta|. \tag{4.1.3}$$

Notice that the prefactor is not $1/(2k)$ (as Sugawara thought), but is subtly "renormalized" to be $(2k + Q_\psi)^{-1}$ where Q_ψ is the quadratic Casimir in the adjoint representation of g (so that ψ denotes its highest weight). The necessity for this prefactor is seen by calculating the L_m, T^a_n commutator, paying attention to the normal ordering. Given the bilinear form of $\mathscr{L}$ in terms of T^a_m a simple determination of the coefficient was given by Knizhnik and Zamolodchikov.[56] Let the highest weight states (1.4.2a) belong to a representation of g with matrix representation t^a. Then

$$T^a_0|\Psi\rangle = |\Psi\rangle t^a. \tag{4.1.4}$$

If $1/\beta$ is the prefactor, we have by (1.4.2a)

$$\mathscr{L}_{-1}|\Psi\rangle = \sum_{a=1}^{\dim g} \beta^{-1} T^a_{-1}|\Psi\rangle t^a.$$

Acting on this equation with T^b_1 and using its commutation relations (1.1.1) and (1.1.3) with T^a_{-1} and $\mathscr{L}_{-1}$ yields $\beta = 2k + Q_\psi$.

Finally it has to be checked that $\mathscr{L}$ indeed satisfies (1.1.1) and that the c number is actually

$$c_g = \frac{2k \dim g}{2k + Q_\psi} = \frac{x \dim g}{x + \tilde{h}(g)}, \tag{4.1.5}$$

where x is the level $2k/\psi^2$ mentioned in Subsec 1.4 and 3.4 and $\tilde{h}(g) = Q_\psi/\psi^2$ is called the dual Coxeter number for g. We shall prove that it is an integer and list its values later in Subsec. 4.3.

The check is done by using (1.1.3) in the explicit form (4.1.2) for $\mathscr{L}_n$. The c-number arises in restoring the normal ordering after the commutation, using (1.1.2).

Notice that Sugawara's construction is automatically unitary in the sense (1.4.3) if the Kac-Moody generators are. Further,

$$\mathscr{L}_0 = \sum_{a=1}^{\dim g} \left(T^a_0 T^a_0 + 2\sum_{n=1}^{\infty} T^a_{-n} T^a_n \right) \Big/ (2k + Q_\psi) \tag{4.1.6}$$

and hence is positive. In fact, as we discussed in Subsec. 1.4, the states of an irreducible representation of $\hat{g}$, annihilated by T^i_n, $n > 0$, (and hence by L_n, $n > 0$,) form an irreducible representation of g whose quadratic Casimir operator is given by $Q_t = \sum(t^a)^2$ (see (4.1.4)). Thus within the irreducible representation of $\hat{g}$, $\mathscr{L}_0$ is bounded below by its eigenvalue on these states, namely

$$\mathscr{L}_0 > Q_t/(2k + Q_\psi) \geq 0. \tag{4.1.7}$$

The results described above form the culmination of a long series of papers in the physics literature (Sugawara;[43] Sommerfield;[44] Callan, Dashen and Sharp;[57] Coleman, Gross and Jackiw;[58] Bardacki and Halpern;[59] Dell'Antonio, Frishman and Zwanziger;[60] and Dashen and Frishman[61]) ending with the correct prefactor for level-1 representations of SU(n). More recently the correct general formula (4.1.1) has been given by Knizhnik and Zamolodchikov,[56] Goddard and Olive[16] and Todorov.[62] Applications in string theory have been discussed by Nemeschansky and Yankielowicz[63] and Jain, Shankar and Wadia.[64] The Sugawara formula has also appeared in the mathematical literature.[14,65,66,67]

4.2. *Sugawara's construction for g not simple*

If the Lie algebra g is semi-simple i.e. $g = g_1 \oplus g_2 \oplus \cdots$ then the relevant construction is simply

$$\mathscr{L}^g = \mathscr{L}^{g_1} + \mathscr{L}^{g_2} + \cdots, \tag{4.2.1}$$

where $\mathscr{L}^{g_i}$ denotes (4.1.1) evaluated for g_i. The c-number for $\mathscr{L}^g$ is also obtained additively:

$$c_g = c_{g_1} + c_{g_2} + \cdots \tag{4.2.2}$$

The result is valid even if g is not semi-simple. Thus if g is Abelian its structure constants vanish, so $Q_\psi = 0$ and we have simply

$$c_g = \dim g = \operatorname{rank} g. \tag{4.2.3}$$

A physical example of this case is the original construction of Virasoro in string theory.[68] The quantities p^a_m of string theory generate the Kac-Moody algebra:

$$[p^a_m, p^b_n] = mk\delta^{ab}\delta_{m+n,0} \tag{4.2.4}$$

(usually with $k = 1$ chosen) and Sugawara's construction (4.1.1) reduces to Virasoro's construction

$$L(z) = \frac{1}{2k} {}^{\times}_{\times} \sum_{a=1}^{\operatorname{rank} g} p^a(z)^2 {}^{\times}_{\times}. \tag{4.2.5}$$

Here the group G is just real space $\mathbb{R}^d$ with rank and dimension d. It was Weis who pointed out that, in this context, the algebra (1.1.1) has a c-number and that its value is given as $c = d$.[69] This result also applies if space is compactified to a torus $\mathbb{R}^d$/lattice since it is still an Abelian group.

4.3. *Properties of c_g for g simple*

We shall show that c_g is a rational number lying between dim g and rank g, and that it attains its lower bound, rank g, if and only if g is simply-laced (i.e. has roots all of the same length) and a level 1 representation of $\hat{g}$ is considered. To do this we shall find an expression for Q_ψ in terms of the roof system of g. Q_ψ was the quadratic Casimir in the adjoint representation of g and hence is defined in terms of structure constants of g:

$$Q_\psi \delta^{ab} = \sum_{c,d=1}^{\dim g} f^{acd} f^{bcd}. \tag{4.3.1}$$

It will be useful for later work to consider a more general real representation of g than the adjoint. Such a representation possesses real antisymmetric generators M^a satisfying

$$[M^a, M^b] = f^{abc} M^c, \quad M^{a^*} = M^a. \tag{4.3.2}$$

Since g is simple (and compact)

$$\mathrm{Tr}(M^a M^b) = -y_M \delta^{ab} = -x_M \psi^2 \delta^{ab} \tag{4.3.3}$$

for some real, positive y_M. Putting $a = b$ and summing from 1 to dim g yields

$$Q_M \dim M = y_M \dim g, \tag{4.3.4}$$

where Q_M is the quadratic Casimir $-\sum_{a=1}^{\dim g} M^a M^a$. Of course, in the adjoint representation Q_M equals Q_ψ. By summing instead over the rank g generators of the Cartan subalgebra we find

$$y_M = \sum \mu^2/(\mathrm{rank}\, g), \tag{4.3.5}$$

where the sum is over the weights μ of the representation M (with multiplicities included). The weights of the adjoint representation are the roots and when g is simple they have at most two distinct lengths. Let there be n_L long roots and n_S short roots respectively. Then

$$\dim g = n_L + n_S + \mathrm{rank}\, g \tag{4.3.6}$$

and the ratio of the square of the lengths of long and short roots is

$$(L/S)^2 = 1, 2, \text{ or } 3.$$

By (4.3.5)

$$\tilde{h}(g) \equiv Q_\psi/\psi^2 = (n_L + (L/S)^{-2}n_S)/\text{rank}\, g \tag{4.3.7}$$

so that the dual Coxeter number $\tilde{h}$ is certainly rational and hence, from (4.1.5), so is c_g.

Obviously c_g is less than dim g and we now show what is less obvious, that it is greater than or equal to rank g. By (4.1.5) (4.3.6) and (4.3.7),

$$c_g - \text{rank}\, g = \frac{n_L(x-1) + n_S(x - (L/S)^{-2})}{x + \tilde{h}} \geq 0 \tag{4.3.8}$$

as we consider level $x \geq 1$ and $(L/S)^{-2} \leq 1$ by definition.

What is more important is that c_g equals rank g if, and only if, $(L/S)^2 = 1$, i.e. g is simply-laced, and $x = 1$, i.e., the $\hat{g}$ representation has level 1. This conclusion remains true if g is semi-simple in the sense that each component g_i is to be simply-laced. These conditions are precisely those for the validity of the "vertex operator" representation of $\hat{g}$ and we shall show in Subsecs. 4.5 and 6.3 that this is no coincidence; there must be such a construction when c_g and rank g coincide.

Finally, introducing more Lie algebra theory (see Sec. 3.1 or Ref. 51), we shall show that the dual Coxeter number $\tilde{h}$ is an integer. If M generates an irreducible representation of g it has a unique highest weight λ, say, and then

$$Q_M = \lambda(\lambda + 2\rho) \tag{4.3.9}$$

where ρ is half the sum of positive roots of g. Hence

$$\tilde{h}(g) = Q_\psi/\psi^2 = 1 + 2\rho\cdot\psi/\psi^2 \tag{4.3.10}$$

as ψ is the highest weight of the adjoint representation. Now ψ/ψ^2 is a co-root of g and can be expanded as an integer linear combination of the simple co-roots of g:

$$\psi/\psi^2 = \sum_{i=1}^{\text{rank}\, g} m_i\alpha_i/(\alpha_i)^2 . \tag{4.3.11}$$

By Eqs. (3.1.14) and (3.1.17), $2\rho\cdot\alpha_i/(\alpha_i)^2 = 1$, and hence

$$\tilde{h}(g) = 1 + \sum_{i=1}^{\text{rank}\, g} m_i \tag{4.3.12}$$

and is clearly an integer. Its value can easily be calculated for simple groups: $A_n : n+1$, $C_n : n+1$, $E_6 : 12$, $E_7 : 18$, $E_8 : 30$, $F_4 : 9$, $G_2 : 4$ and $\text{SO}(n) : n-2$, $(n \geq 5)$.

4.4. *The situation where g has a subalgebra $h \subset g$*

In the first instance let us suppose that g and h are both simple. We can choose an orthonormal basis for g which includes as a subset an orthonormal basis for h, let us say $a = 1, 2, \ldots, \dim h$. We automatically obtain a Kac-Moody algebra $\hat{h}$ inheriting the same central term k as $\hat{g}$. But the $\hat{h}$ level max differ from that of $\hat{g}$ because the highest roots of g and h, ψ and ϕ, say, may have unequal lengths.

We may suppose $\hat{g}$ has level $2k/\psi^2$ equal to 1. Then the level of $\hat{h}$, $2k/\phi^2$, must equal 1, 2 or 3, etc. Hence ψ^2/ϕ^2 equals a positive integer. In general the $\hat{h}$ level must be greater than or equal to the $\hat{g}$ level.

Sugawara's construction can be applied to both g and h to obtain Virasoro generators $\mathscr{L}^g$ and $\mathscr{L}^h$ respectively. Of course they have different prefactors and different c-numbers in general. We have, by (1.1.3),

$$[\mathscr{L}^g_m, T^a_n] = -nT^a_{m+n}, \quad a = 1, \ldots, \dim g.$$

$$[\mathscr{L}^h_m, T^a_n] = -nT^a_{m+n}, \quad a = 1, \ldots, \dim h.$$

Hence, subtracting

$$[\mathscr{L}^g_m - \mathscr{L}^h_m, T^a_n] = 0, \quad a = 1, \ldots, \dim h, \tag{4.4.1}$$

and so, by (4.1.2),

$$[\mathscr{L}^g_m - \mathscr{L}^h_m, \mathscr{L}^h_n] = 0. \tag{4.4.2}$$

Thus $\mathscr{L}^g_m$ has been broken into two mutually commuting pieces, i.e.,

$$\mathscr{L}^g_m = \mathscr{L}^h_m + K_m, \tag{4.4.3}$$

where K_m commutes with the $\hat{h}$ Kac-Moody algebra by (4.4.1) and can be thought of as relating to the coset G/H. Further, since by (4.4.2),

$$[\mathscr{L}^g_m, \mathscr{L}^g_n] = [\mathscr{L}^h_m, \mathscr{L}^h_n] + [K_m, K_n],$$

we deduce that K_m, like $\mathscr{L}^g_m$ and $\mathscr{L}^h_m$, satisfies a Virasoro algebra and that the c-number is, by virtue of (4.1.5),

$$c_K = c_g - c_h = \frac{2k \dim g}{2k + Q_\psi} - \frac{2k \dim h}{2k + Q_\phi}. \tag{4.4.4}$$

Since the eigenvalues of $\mathscr{L}^g_0$ are bounded below (Eq. (4.1.7)), so are those of K_0. Therefore the highest weight representation of $\hat{g}$ must decompose into highest weight representations of the K_m Virasoro algebra so that in particular we must have c_K positive, i.e.

$$c_K \geq 0, \tag{4.4.5}$$

with c_K vanishing if, and only if, K_m vanishes, according to the theorem of Subsec. 1.4. Of course if c_K is less than unity it must take one of the values in the discrete sequence (1.4.10), and in Subsec. 4.6 we shall make choices of $h \subset g$ so as to obtain the complete sequence (1.4.10) with explicitly unitary representations. Thus by subtracting two Virasoro algebras whose c numbers exceed unity (Subsec. 4.3), we can obtain a Virasoro algebra with $0 \leq c \leq 1$.

These results were due to Goddard and Olive[16] and Goddard, Kent and Olive.[17] Particular examples of related algebraic structures occur in the earlier work of Bardacki and Halpern[59] and Mandelstam.[70,b]

Generalization to the cases when g and h are not simple is easily made using the results of Subsec. 4.2.

4.5. *A quantum equivalence theorem and the vertex operator construction*

A second use of the preceding argument concerns the deduction that the single numerical condition $c_g = c_h$ implies that

$$\mathscr{L}^g = \mathscr{L}^h. \tag{4.5.1}$$

This follows from the result of Subsec. 1.4 whereby K_n must vanish if $c_K = 0$ and K_0 is bounded below (as it must be in order not to contradict the positivity of $\mathscr{L}_0^g$, Eq. (4.1.6)). We call (4.5.1) a quantum equivalence theorem since it establishes the equality of two apparently different operators which, in a conformally invariant quantum field theory, correspond to components of the energy-momentum tensors of two apparently different theories. Two theories with the same energy-momentum tensor have the same Hamiltonian and so are indeed quantum mechanically equivalent. We stress that the result is exact to all orders in Planck's constant (if it is explicitly introduced) and hence nonperturbative. Indeed it will be seen in applications that nontrivial topological effects are included.

This is a remarkably powerful result which we shall return to in Sec. 5 when we study fermions. As an immediate application consider the Cartan subalgebra t of g which exponentiates to a maximal torus T subgroup of G. T is an Abelian group, isomorphic to $\mathbb{R}^{\operatorname{rank} g}/\Lambda_R(g)$ where $\Lambda_R(g)$ is the root lattice of g. We saw in Sec. 4.3 that $c_K = c_g - c_t = c_g - \operatorname{rank} g$ vanishes if and only if g is simply laced, and a level 1 representation of g is considered. Thus, when these two conditions are satisfied, the quantum equivalence theorem (4.5.1) states that, choosing $\psi^2 = 2$,

$$\frac{\operatorname{rank} g}{2\dim g}\sum_{a=1}^{\dim g} {}^{\times}_{\times}T^a(z)T^a(z)^{\times}_{\times} = \frac{1}{2}\sum_{i=1}^{\operatorname{rank} g} {}^{\times}_{\times}T^i(z)T^i(z)^{\times}_{\times}\,. \tag{4.5.2}$$

Now the $T^i(z)$, $i = 1, 2, \ldots, \operatorname{rank} g$, appearing on the right-hand side of (4.5.2) and corresponding to the Cartan subalgebra of g satisfy (4.2.4) (with $k = 1$), according to

[b] See also the work of M. Halpern, *Phys. Rev.* **D4** (1971) 2398 and M. Halpern and Thom, *Phys. Rev.* **D4** (1971) 3084.

(1.1.2). Thus the quantum equivalence theorem (4.5.2) states that when g is simply-laced and the level $x = 1$, the Sugawara construction (4.1.2) equals Virasoro's construction (4.5.2) for a string moving on the maximal torus T of G. This suggests that it must be possible to construct the Kac-Moody generators E^α_m corresponding to the step operators E^α of g from the T^i_m ($i = 1, \ldots, \text{rank}\, g$) and indeed this is precisely what the vertex operator construction achieves. One obtains the Fubini-Veneziano[71] vector $Q^i(z)$ by integrating

$$iz\frac{dQ^i}{dz} = T^i(z), \qquad i = 1, \ldots, \text{rank}\, g\,.$$

Then the E^α_m are the Laurent coefficients in the expansion of the vertex operator

$$\sum_{m \in Z} z^{-m} E^\alpha_m = z\!:\!e^{i\alpha\cdot Q(z)}\!:\, c_\alpha(T^i_0), \tag{4.5.3}$$

where the normal ordering is with respect to the bosonic oscillators which are the Laurent coefficients of $T^i(z)$, $i = 1, \ldots, \text{rank}\, g$, and $c_\alpha(T^i_0)$ is a Klein transformation needed to correct some signs in commutation relations. See Sec. 6 for more information about this construction due to Frenkel and Kac[72] and Segal.[65] In the case of su(N), this construction was anticipated by Halpern,[95] using the vertex operator construction for fermions, discussed in Sec. 7. The explicit verification of (4.5.2), due to Frenkel,[66] is given in Subsec. 6.3.

Two commuting copies of the above result (4.5.2) can be interpreted in terms of two dimensional conformally invariant quantum field theory. In Subsecs. 2.3 and 2.4 we saw that an interesting example was the "Wess-Zumino" theory, with a field confined to the manifold of a Lie group G, and described by an action (2.3.12) consisting of the usual kinetic term plus a "Wess-Zumino" term, each normalized so that the energy momentum tensor was indeed bilinear (*à la* Sugawara) in conserved currents satisfying the $\hat{g}$ Kac-Moody algebra. Witten[34] showed that the level of $\hat{g}$ was equal to the only free parameter left in the action, the overall coefficient which had to be an integer to ensure the single valuedness of $\exp\{i\, \text{Action}/\hbar\}$.

Equation (4.5.2) shows that if g is simply-laced and the level is unity then the Wess-Zumino model is quantum equivalent to a free scalar field theory on T, the maximal torus of G. At first sight this is highly surprising because the two equivalent theories possess different numbers of independent fields. Yet the total number of degrees of freedom is infinite in each case. The scalar field on the torus T "feels" the non-Abelian structure of G through the periodicity structure of the torus which is specified by the root lattice of g ($T = \mathbb{R}^d/\Lambda_R(g)$).

The Nambu action for the string can be regarded as a σ-model on the group $\mathbb{R}^d$ defined by flat space.[73] The ghost free nature of the theory depends on the conformal symmetry of this action but then holds only in 26 dimensions (10 for the fermionic string). Actually it is the c-number of the Virasoro algebra which is critical and in flat space this equals the dimension, (4.2.3). If we try to follow the Kaluza-Klein philosophy

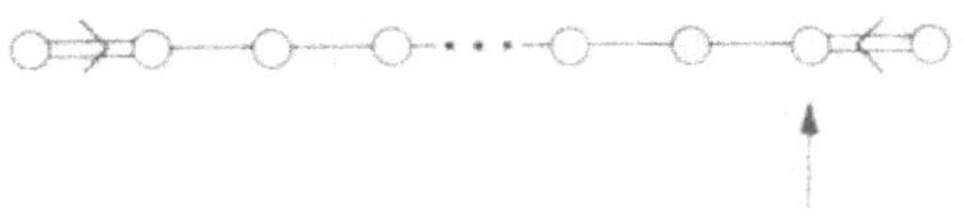

Fig. 6. Extended Dynkin diagram for sp(m + 1).

and let the string move on the manifold of $G \times \mathbb{R}^d$ we must add the Wess-Zumino term to the action in order to retain conformal symmetry and hence the no ghost theorem. Since we must choose g simply-laced and level 1 because of the string vertex operators, we have, by the above results, that the critical c number (26) equals d plus the rank of g rather than its dimension. Note that this differs from the recent attempts to use Kaluza-Klein theory in supergravity because the seven extra dimensions were apparently used in a different way.

The connection between the vertex operator construction of a Kac-Moody algebra g and the motion of a string on the maximal torus of G is discussed by Goddard and Olive[5] and the comments on the relation between string theories and σ-models are due to Witten, as reported by Freund, Oh and Wheeler.[74] Nemeschansky and Yankielowicz elucidated the point further.[63]

4.6. *The discrete sequence of Virasoro c-numbers less than unity*

We saw (Eq. (4.3.8)) that Sugawara's construction always yielded a Virasoro c-number c_g exceeding the rank of g and hence never less than unity. Nevertheless c less than unity could result from a judicious choice of $h \subset g$ (and level) in the construction $K = \mathcal{L}^g - \mathcal{L}^h$ of Subsec. 4.4. We now see that the complete sequence $c = 1 - 6/(m+2)(m+3)$ results from the choice $G/H = \mathrm{Sp}(m+1)/\mathrm{Sp}(m) \times \mathrm{Sp}(1)$—with level 1. That $\mathrm{Sp}(m) \times \mathrm{Sp}(1)$ is a subgroup of $\mathrm{Sp}(m+1)$ and inherits the same level can be seen from the extended Dynkin diagram shown in Fig. 6. Deletion of the arrowed point leaves the Dynkin diagram for the desired subgroup. Since the deleted point corresponds to a short root it is evident that the highest roots of each factor of h have the same lengths as that of g. Hence, as explained in Subsec. 4.4, the levels are all the same, and hence all of level 1 if we choose that in the first place. Then by (4.1.5)

$$c_{\mathrm{sp}(m)} = \frac{\dim \mathrm{sp}(m)}{1 + \tilde{h}(\mathrm{sp}(m))}. \qquad (4.6.1)$$

$\mathrm{sp}(m)$ has long roots $\pm 2e_i$, $i = 1, 2, \ldots, m$, short roots $\pm e_i \pm e_j$ $(i \neq j)$, and rank m. Hence $n_L = 2m$, $n_S = 2m(m-1)$ and $\dim \mathrm{sp}(m) = (2m+1)m$. Thus by (4.3.7) the dual Coxeter number $\tilde{h}$ equals

$$\tilde{h}(\mathrm{sp}(m)) = (n_L + n_S(L/S)^{-2})/m = (2m + m(m-1))/m = m + 1, \qquad (4.6.2)$$

as quoted at the end of Subsec. 4.3. Hence, inserting this in (4.6.1)

$$c_{\mathrm{sp}(m)} = \frac{m(2m+1)}{m+2} = 2m - 3 + 6/(m+2), \qquad (4.6.3)$$

and so, as desired[17]

$$c_K = c_{\mathrm{sp}(m+1)} - c_{\mathrm{sp}(m)} - c_{\mathrm{sp}(1)} = 1 - 6/(m+2)(m+3). \tag{4.6.4}$$

We shall see in the next section that level 1 sp(m) Kac-Moody algebra representations can be constructed in a unitary way by considering fermion fields in the defining representation of sp(m). It is possible to show that there exist within the fermionic Fock space all the highest weight states corresponding to (1.4.11).[24] Some of these states can be found explicitly quite easily.[75] The first proof that all the values of h in (1.4.11) occur in unitary representations was found using a reformulation of the above construction in terms of SU(2) groups following from results of the next Section.[18] The relationship between the construction given in this subsection and the lattice models (see Subsec. 2.5) with the same value of c is not yet clear but it is interesting that the defining representation of sp(m) used is quaternionic and thus related to Pauli spin matrices. It is also possible that the reformulation mentioned will aid the physical interpretation.

5. The Quark Model

5.1. *Kac-Moody generators bilinear in real Fermi fields*

In Secs. 3 and 4 we have been talking in general about unitary highest weight representations of (affine untwisted) Kac-Moody algebras without specifying any details of their construction. As was clear, the state space of these representations somewhat resembled a Fock space of a quantum field theory with the "highest weight state" corresponding to the vacuum. We saw in Subsec. 4.5 that the "vertex operator" representations (applicable at unit level when g is simply laced) acted in the Fock space of rank g real scalar Bose fields. In this section we construct representations in the Fock space of real Fermi fields, taking as our point of departure the "quark model" of the 1960's discussed in Subsec. 2.1 whereby currents are represented bilinearly in Fermi fields:

$$T^a(z) = \sum_{n \in \mathbf{Z}} z^{-n} T^a_n = \frac{i}{2} H^i(z) M^a_{ij} H^j(z). \tag{5.1.1}$$

The real, antisymmetric matrices M^a satisfy the commutation relations (4.3.2). The Fermi fields $H^i(z)$ are correspondingly real in the sense explained below. Thus we are using fermions to build a representation of $\hat{g}$ from a real representation of g. The extension to complex representations of g follows in Subsec. 5.3.

Because of the sum over integers in (5.1.1), the right-hand side must be single valued when the complex variable z encircles the origin. Thus the dim M Fermi fields $H^i(z)$ must be either periodic or antiperiodic under this operation, as in (2.1.24),

$$H^i(ze^{2i\pi}) = H^i(z) \quad : \text{periodic}, \quad (\mathrm{R}), \tag{5.1.2a}$$

$$= -H^i(z) \quad : \text{antiperiodic}, (\mathrm{NS}), \tag{5.1.2b}$$

corresponding respectively to the Ramond[36] and Neveu-Schwarz[37] Fermi fields of string theory.

As we shall see in Subsec. 5.6 it is essential to retain both possibilities since the representations (5.1.1) of $\hat{g}$ with different choices (5.1.2) are in general inequivalent. The Fermi fields (5.1.2a or b) can be expanded in integral or half integral powers of z, respectively, as in (2.1.26),

$$H^i(z) = \Sigma z^{-r} b_r^i, \tag{5.1.3}$$

where *either* $r \in \mathbb{Z}$ (R) for all r *or* $r \in \mathbb{Z} + 1/2$ (NS) for all r. These operators satisfy the reality condition

$$b_r^{a\dagger} = b_{-r}^a, \tag{5.1.4}$$

and the anticommutation relations specified by Eqs. (2.1.27).

Thus, when r is distinct from zero, b_r^i and b_{-r}^i are equivalent to a single fermionic harmonic oscillator. This applies to both Ramond and Neveu-Schwarz fields, but in the former case there is, in addition, a zero mode satisfying Eq. (2.1.29), which we can express in terms of Euclidean Dirac gamma matrices, γ^i, commuting with the b_r^i, $r \neq 0$, by

$$b_0^i = \frac{1}{\sqrt{2}}\gamma^i(-1)^{N_b}, \tag{5.1.5}$$

where N_b is the number operator for the b oscillators,

$$N_b = \sum_{r=1}^{\infty} \sum_{i=1}^{\dim M} b_{-r}^i b_r^i. \tag{5.1.6}$$

The extra phase factor (Klein transformation) is needed to ensure anticommutation with the nonzero modes. The space of states is a Fock space built up from vacuum states annihilated by the b_r^i with positive suffices $r > 0$ as in (2.1.28). The vacuum can be seen to be unique in the Neveu-Schwarz case but has to have a $2^{[\dim M]}$-fold degeneracy in the Ramond case owing to a Dirac spinor component for the zero modes. ($[p]$ is the integral part of p.) Relative to this vacuum a normal ordering operation denoted by open dots is defined by Eq. (2.1.33) whereby destruction operators are moved to the right of creation operators (with the inclusion of appropriate signs for Fermi statistics). Then normal ordered products of Neveu-Schwarz fields will have finite matrix elements in the Fock space and will be totally antisymmetric in these fields. We require the same to be true for Ramond fields and this means that we extend the definition of normal ordering to the zero modes so as to guarantee antisymmetry:

$$\circ b_0^i b_0^j \circ = (b_0^i b_0^j - b_0^j b_0^i)/2 = (\gamma^i\gamma^j - \gamma^j\gamma^i)/4 .$$

Using these definitions we can define a Wick[76] contraction function $\Delta(z,\zeta)$:

$$H^i(z)H^j(\zeta) = {}^\circ_\circ H^i(z)H^j(\zeta){}^\circ_\circ + \delta^{ij}\Delta(z,\zeta), \qquad |z| > |\zeta|, \tag{5.1.7}$$

and evaluate it as

$$\Delta(z,\zeta) = \sum_{r=1/2}^{\infty} (\zeta/z)^r = \frac{\sqrt{z\zeta}}{z-\zeta} \qquad \text{(NS)}, \tag{5.1.8a}$$

$$= \sum_{r=1}^{\infty} (\zeta/z)^r + \frac{1}{2} = \frac{1}{2}\frac{(z+\zeta)}{z-\zeta} \qquad \text{(R)}, \tag{5.1.8b}$$

in the Neveu-Schwarz and Ramond cases, respectively. Note that Eq. (5.1.7) makes explicit the pole singularity in the operator product of the two Fermi fields and that the residue is the same in each case. The convergence of the summations in (5.1.8) determines the inequality $|z| > |\zeta|$ in (5.1.7).

5.2. *The Kac-Moody algebra commutators*

Now we wish to check the Kac-Moody commutation relations (1.1.2) paying particular attention to how the value of k (the central term) depends upon the choice of quark representation generators, M^a. The product of two generators (5.1.1) is a product of four Fermi fields and we shall use Wick's theorem[76] to express this in normal ordered form with respect to the fermionic normal ordering with the singularities made manifest through the contraction function Δ. This is precisely Dyson's[77] technique for deriving Feynman diagrams and was used extensively in string theory. It is an example of what is now called an operator product expansion and, in the region $|z| > |\zeta|$, it reads by virtue of (5.1.7):

$$\begin{aligned} {}^\circ_\circ H^i(z)H^j(z){}^\circ_\circ\,{}^\circ_\circ H^k(\zeta)H^l(\zeta){}^\circ_\circ &= {}^\circ_\circ H^i(z)H^j(z)H^k(\zeta)H^l(\zeta){}^\circ_\circ + \Delta(z,\zeta){}^\circ_\circ\{H^i(z)H^l(\zeta)\delta^{jk} \\ &\quad + H^j(z)H^k(\zeta)\delta^{il} - H^i(z)H^k(\zeta)\delta^{jl} - H^j(z)H^l(\zeta)\delta^{ik}\}{}^\circ_\circ \\ &\quad + \Delta(z,\zeta)^2(\delta^{jk}\delta^{il} - \delta^{jl}\delta^{ik}) \qquad (5.2.1\text{a}) \\ &\equiv h^{ijkl}(z,\zeta). \qquad (5.2.1\text{b}) \end{aligned}$$

We can use the right-hand side of (5.2.1a) to define $h^{ijkl}(z,\zeta)$ more generally. Then it has the symmetry property

$$h^{ijkl}(z,\zeta) = h^{klij}(\zeta,z), \tag{5.2.2}$$

and so both $T^a_m T^b_n$ and $T^b_n T^a_m$ can be written as double integrals of the same integrand

$$(-1/4)\oint \frac{d\zeta\zeta^n}{2\pi i\zeta}\oint\frac{dzz^m}{2\pi iz} M^a_{ij}M^b_{kl}h^{ijkl}(z,\zeta) \tag{5.2.3}$$

differing only in the orientation of the contours of integration. In the first case $|z| > |\zeta|$ and in the second case this inequality is reversed. The only singularities of the integrand are at $z, \zeta = 0$ and ∞ and at $z = \zeta$. Thus the commutator $[T_m^a, T_n^b]$ is the same integrand integrated over the difference of the two contours which by Cauchy's theorem can be taken to be a z contour enclosing $z = \zeta$ once followed by a ζ integration contour enclosing the origin once:

$$[T_m^a, T_n^b] = -(1/4) \oint_0 \frac{d\zeta \zeta^n}{2\pi i \zeta} \oint_\zeta \frac{dz z^m}{2\pi i z} M_{ij}^a M_{kl}^b h^{ijkl}(z, \zeta). \tag{5.2.4}$$

The term in (5.2.1a) regular at $z = \zeta$ does not contribute to (5.2.4), the pole term (single contraction, Δ) yields $if^{abc}T_{m+n}^c$ while the double pole (double contraction or loop diagram) yields the Schwinger term as follows. It is

$$\frac{\mathrm{Tr}(M^a M^b)}{2} \oint_0 \frac{d\zeta \zeta^n}{2\pi i \zeta} \oint_\zeta \frac{dz z^m}{2\pi i z} \Delta(z, \zeta)^2. \tag{5.2.5}$$

By (5.1.8) we have the identity

$$\Delta(z, \zeta)^2 = \frac{z\zeta}{(z - \zeta)^2} + 4\varepsilon; \qquad \varepsilon = 0 \quad (\mathrm{NS}), \qquad 1/16 \quad (\mathrm{R}). \tag{5.2.6}$$

The ε term drops out of (5.2.5) as it is regular at $z = \zeta$ (so that there is no distinction between the Neveu-Schwarz and Ramond cases as far as the commutation relations are concerned), leaving $-m\delta_{m+n,0}\mathrm{Tr}(M^a M^b)/2$ for (5.2.5).

Since g is simple we have, as in (2.1.19) and (4.3.3),

$$-\mathrm{Tr}(M^a M^b) = y_M \delta^{ab} = x_M \psi^2 \delta^{ab}, \tag{5.2.7}$$

where ψ denotes the highest root of g (see Subsec. 3.1) and x_M is positive as the M^a are anti-Hermitian. Thus (1.1.2) is obtained with the central term k equaling $y_M/2 = \psi^2 x_M/2$, so that the level, given by (3.4.2), is

$$x = 2k/(\psi^2) = x_M. \tag{5.2.8}$$

Thus, by the general theorem of Subsec. 3.4, whereby the levels of highest weight representations are quantized, x_M must be an integer and indeed it is, for representations M which are real as we have supposed. x_M is called the Dynkin index of the representation.[78,79] M may describe a reducible representation which decomposes into two real components M_1 and M_2. Then

$$x_M = x_{M_1} + x_{M_2}. \tag{5.2.9}$$

The Dynkin index x_M can be evaluated using the methods of Subsec. 4.3. Thus by

(4.3.5) and (5.2.7), we have in terms of the weights μ of the representation M,

$$x_M = \Sigma\mu^2/(\psi^2 \operatorname{rank} g). \tag{5.2.10}$$

For example, the defining, n dimensional representation of SO(n) has weights $\pm e_i$ ($i = 1, 2, \ldots, [n/2]$) (and in addition 0 if n is odd) where the e_i are orthogonal unit vectors and the highest root $\psi = e_1 + e_2$. Thus, since SO(n) has rank $[n/2]$, (5.2.10) shows that $x_M = 1$. Thus we can construct a level 1 representation of $\hat{so}(n)$ by choosing fermions in the defining representation of SO(n).

Taking fermions in the adjoint representation of g (assumed simple) yields representations of $\hat{g}$ of level $\tilde{h}(g)$, the dual Coxeter number of g discussed in Subsec. 4.3 and listed at the end of that subsection.

Further Dynkin indices can be calculated as above or read off tables.[80,81] It is clear that irreducible representations of g carried by the fermions cannot yield all the possible Kac-Moody levels, and even if reducible representations are allowed, the lowest levels may be missing. This is so for E_8. All representations are real, and the smallest Dynkin index is 30, corresponding to the adjoint representation. Therefore there is no way of obtaining Kac-Moody representations of $\hat{E}_8$ of level less than 30 by the construction (5.1.1). On the other hand level 1 can certainly be obtained by vertex operators since E_8 is simply-laced.

5.3. *Quarks in complex representations of g*

In the original quark model the quarks were not in a real representation but in a complex one, the **3** of SU(3).[82] We shall now treat such representations by building on our results of the previous two sections for real representations and see that new phenomena occur. Let N^a be a complex antihermitian matrix satisfying the Lie algebra g (4.3.2) and split it into real and imaginary pieces A^a and B^a:

$$N^a = A^a + iB^a. \tag{5.3.1}$$

A^a is real, anti-Hermitian and hence antisymmetric. B^a is real, Hermitian and hence symmetric. We shall construct a real representation of twice the dimension of N^i by replacing i by the $2\dim N \times 2\dim N$ real matrix:

$$i \to J = \begin{pmatrix} 0 & I \\ -I & 0 \end{pmatrix}. \tag{5.3.2}$$

Obviously J is antisymmetric; its square is

$$J^2 = -I_{(2\dim N)}, \tag{5.3.3}$$

that is, minus the identity matrix of dimension $2\dim M$. Correspondingly

$$N^a \to M^a = \begin{pmatrix} A^a & B^a \\ -B^a & A^a \end{pmatrix} = I \otimes A^a + J \otimes B^a,$$

if we think of the 2 dim N space as a product of a 2-dimensional space times a dim N-dimensional space. Remembering that $\mathrm{Tr}(A^a B^b)$ vanishes since A^a and B^b are, respectively, antisymmetric and symmetric, we find

$$\mathrm{Tr}(M^a M^b) = 2\,\mathrm{Tr}(A^a A^b - B^a B^b) = 2\,\mathrm{Tr}(N^a N^b). \tag{5.3.4}$$

Now the real matrices M^a satisfy all the conditions of our construction (5.1.1) and we see from (5.3.4) that the resultant level is given by

$$x_M = 2x_N. \tag{5.3.5}$$

Since the M representation decomposes into two irreducible components generated by N^a and N^{a*} this is in accord with (5.2.9) which holds even if M_1 and M_2 are complex.

If we denote the matrix $J = M^0$ we see that it fulfills the conditions of our Kac-Moody construction (5.1.1). The matrices M^a, $a = 0, 1, \ldots, \dim g$, generate the algebra $u(1) \oplus g$ and hence we find the corresponding Kac-Moody algebra. This means that the generators T^0_m commute with the T^a_n, $a \neq 0$ and satisfy

$$[T^0_m, T^0_n] = (\dim N) m \delta_{m+n,0} \tag{5.3.6}$$

by our previous result that $2k = -\mathrm{Tr}(M^{0^2}) = \mathrm{Tr}\, I = 2 \dim N$ by (5.3.3).

Thus, when our fermion representation is complex we automatically get an additional $\hat{u}(1)$ Kac-Moody algebra commuting with $\hat{g}$ without adding any new fermions. When the $\hat{u}(1)$ generators are included in the Sugawara construction (4.1.1) of the Virasoro algebra the c-number is increased by unity since $u(1)$ has rank one (see (2.2.3)).[16] All we have said is illustrated by the defining, n-dimensional representation of su(n) which is complex, and which yields level one representations of su(n) as x_N equals 1/2 then so that x_M equals unity by (5.3.5).

When the complex representation is, in addition, pseudoreal, i.e. equivalent to its complex conjugate but not real, the above construction can be extended. It can be shown then that dim N is even and that the representation is actually quaternionic in that it can be written in terms of dim $N/2 \times$ dim $N/2$ matrices with real quaternionic entries. These quaternions can be represented by real 4×4 matrices, but commuting with these there exists a second set of 4×4 matrices representing quaternions (as so(4) = so(3) $\times$ so(3)). Thus it is possible to construct a $g \oplus$ su(2) algebra of real 2 dim $N \times$ 2 dim N matrices and hence the corresponding Kac-Moody algebra. Thus the $\hat{g}$ Kac-Moody algebra can automatically be extended by a s$\hat{u}$(2) Kac-Moody algebra commuting with it without adding any new fermions.[17] This phenomena is illustrated by the defining $2m$-dimensional representations of the symplectic sp(m)

algebra of rank m, which is pseudoreal and indeed yields level one representations of $\hat{sp}(m)$. In the case of sp(1) this is related to what used to be called the Pauli-Gürsey symmetry in particle physics.

5.4. *Two Virasoro algebras*

It is well known in fermionic string theory that it is possible to construct a Virasoro algebra bilinear in dim M fermions,

$$L(z) = \sum_{n\in\mathbb{Z}} z^{-n}L_n = {\circ \atop \circ}\frac{1}{2}\sum_{i=1}^{\dim M} z\frac{dH^i}{dz}(z)H^i(z){\circ \atop \circ} + \varepsilon \dim M\,, \tag{5.4.1}$$

where, as in (5.2.6), ε equals 0 or 1/16 according as Neveu-Schwarz or Ramond fields are considered. The c-number for this L_n Virasoro algebra is

$$c = \dim M/2\,. \tag{5.4.2}$$

This could be proved directly using the Wick expansion (5.2.1) but we shall obtain an indirect proof by relating (5.4.1) to the Sugawara construction (4.1.1) and using the results of Sec. 4. In terms of conformally invariant quantum field theories in two dimensions (5.4.1) relates to the energy momentum tensor for dim M real free massless fermions of definite helicity.

Note that the addition of $\varepsilon \dim M$ affects L_0 only and implies that whereas L_0 vanishes on the Neveu-Schwarz vacuum it yields dim $M/16$ on the Ramond vacuum (which is degenerate). If dim M equals unity, $c = 1/2$, (5.4.2), and the weight 1/16 is as predicted by the FQS formula (1.4.11) and is relevant to the critical exponents of the Ising model as described in Subsec. 2.5.

When we constructed a g algebra out of the same fermions by the quark model (5.1.1) Sugawara's construction (4.1.1) yields a second Virasoro algebra $\mathscr{L}_m$, apparently quadrilinear in fermions. When its c-number, (4.1.4), differs from dim $M/2$, (5.4.2), we can be sure that L_m, (5.4.1) and $\mathscr{L}_m$, (4.1.1), differ. Nevertheless we shall find a relation using the Wick identity (5.2.1) under the supposition that M now describes a real, irreducible representation of g. By the normal ordering of T^i_m's denoted by crosses (see (4.1.3)),

$$\sum_{a=1}^{\dim g} T^a(z)T^a(\zeta) = {\times \atop \times}\sum_{a=1}^{\dim g} T^a(z)T^a(\zeta){\times \atop \times} + \frac{k \dim g z\zeta}{(z-\zeta)^2}\,,$$

where $|z| > |\zeta|$. But according to (5.2.1) involving the fermionic normal ordering denoted by open dots the same expression equals (for $|z| > |\zeta|$)

$${\circ \atop \circ}\sum_{a=1}^{\dim g} T^a(z)T^a(\zeta){\circ \atop \circ} + Q_M {\circ \atop \circ}\sum_{i=1}^{\dim M} H^i(z)H^i(\zeta){\circ \atop \circ}\Delta(z,\zeta) + \frac{1}{2}\dim g\, y_M \Delta(z,\zeta)^2\,,$$

where Q_M is the quadratic Casimir operator $(-\Sigma M^{a^2})$ and is proportional to the unit matrix, using Schur's lemma and the irreducibility of M. The quantities which are normal ordered have finite limits as z and ξ coalesce. It follows that the singular quantities must cancel out of the above equality leaving something regular. The double pole indeed cancels if k equals $y_M/2$, thereby confirming our previous result (5.2.8). In fact, using (5.2.6) and (4.3.4) we have

$$\frac{1}{2}\dim g\, y_M \Delta(z,\zeta)^2 - \frac{kz\zeta \dim g}{(z-\zeta)^2} = 2\varepsilon \dim g\, y_M = 2\varepsilon Q_M \dim M\,.$$

The term with the single pole is regular as its residue, the normal ordered product of $H^i(z)$ and $H^i(\zeta)$ summed, vanishes at $z = \zeta$ by the antisymmetry property. By l'Hôpital's rule, the result is

$$Q_M {}^{\circ}_{\circ} \sum_{i=1}^{\dim M} z\frac{dH^i}{dz} H^i {}^{\circ}_{\circ}\,.$$

Putting these results together, we have, in the limit $z = \zeta$, a relation between the two types of normal ordering:

$$ {}^{\times}_{\times}\sum_{a=1}^{\dim g} T^a(z)T^a(z){}^{\times}_{\times} = {}^{\circ}_{\circ}\sum_{a=1}^{\dim g} T^a(z)T^a(z){}^{\circ}_{\circ} + 2Q_M L(z)\,, \tag{5.4.3}$$

with $L(z)$ given by (5.4.1). Notice that (5.4.3) holds for Neveu-Schwarz or Ramond fields.

According to the definition of fermionic normal ordering in Subsec. 2.1 the term with the open dots in (5.4.3) must be totally antisymmetric in the fermion fields and hence can be written,

$$\frac{1}{3}{}^{\circ}_{\circ}H^i(z)H^j(z)H^k(z)H^l(z){}^{\circ}_{\circ}\left\{\sum_{i=1}^{\dim g}(M^a_{ij}M^a_{kl} + M^a_{jk}M^a_{il} + M^a_{ki}M^a_{jl})\right\}.$$

It follows that this term vanishes if, and only if,

$$\sum_{a=1}^{\dim g}(M^a_{ij}M^a_{kl} + M^a_{jk}M^a_{il} + M^a_{ki}M^a_{jl}) = 0\,. \tag{5.4.4}$$

Equation (5.4.4) certainly holds for fermions in the adjoint representation of g since the M^i are the structure constants of g and Eq. (5.4.4) is simply the Jacobi identity for them. Then the identity (5.4.3) reads

$$\frac{1}{2Q_\psi}{}^{\times}_{\times}\sum_{a=1}^{\dim g} T^a(z)T^a(z){}^{\times}_{\times} = L(z)\,.$$

The left-hand side is the Sugawara construction of the Virasoro algebra (4.1.1) for level

$\tilde{h}(g)$ and is seen to equal to free fermion construction (5.4.1). By (4.1.4) the c number is $\dim g/2$. Thus we have established that the free fermion construction (5.4.1) does satisfy the Virasoro algebra with c-number (5.4.2) at least when $\dim M$ equals the dimension of a simple Lie algebra g, but it is not difficult to extend the result to $\dim M$ any integer.

Expression (5.4.4) also vanishes for fermions in the defining representation of $g = \mathrm{so}(n)$. It is convenient to label the generators as an antisymmetric tensor so that

$$(M^{\alpha\beta})_{ij} = \delta_i^\alpha \delta_j^\beta - \delta_j^\alpha \delta_i^\beta ,$$

and, by (5.1.1),

$$T^{\alpha\beta}(z) = i\,{}^\circ_\circ H^\alpha(z) H^\beta(z) {}^\circ_\circ\,.$$

Evidently ${}^\circ_\circ T^{\alpha\beta}(z) T^{\alpha\beta}(z) {}^\circ_\circ$ vanishes by the antisymmetry property as $H^\alpha(z)$ is repeated within the fermionic normal ordering. It follows again that the free fermion construction (5.4.1) with $n = \dim M$ satisfies the Virasoro algebra with c-number (5.4.2).

Whenever we have a real representation M of g with dimension $\dim M$ it follows automatically that

$$g \subset \mathrm{so}(\dim M).$$

Therefore we have precisely the subalgebra situation discussed in Subsec. 4.4 with g replaced by $\mathrm{so}(\dim M)$ and h by g. Hence, by that analysis (Eq. (4.4.3)),

$$L(z) = \mathscr{L}^g(z) + K(z), \tag{5.4.5}$$

where we use the fact, just shown, that the Sugarawa construction for $\mathrm{so}(\dim M)$ with $\dim M$ fermions equals the free fermion construction (5.4.1). Thus K_m is a Virasoro algebra commuting with the $\hat{g}$ Kac-Moody generators and hence with $\mathscr{L}^g$.[16] By the "quantum equivalence" theorem of Subsec. 4.5, K vanishes and $\mathscr{L}^g$ and L are equal if and only if c_K vanishes, i.e.,

$$c_g \equiv \frac{x_M \dim g}{x_M + \tilde{h}(g)} = (\dim M)/2. \tag{5.4.6}$$

By (4.3.3) this can be written as

$$2Q_M = \psi^2(x_M + \tilde{h}) = 2k + Q_\psi \tag{5.4.7}$$

which, by the identity (5.4.3), implies (5.4.4). Thus equations (5.4.6) and (5.4.4) are equivalent conditions for the equality of the Sugawara and free fermion constructions of the Virasoro algebra generators.

Let us illustrate condition (5.4.6) in the two cases in which we know K vanishes. If the fermions lie in the adjoint representation of g, x_M equals $\tilde{h}$, $\dim M = \dim g$ and hence (5.4.6) is indeed satisfied. If the fermions lie in the defining representation of so(n) ($n \geq 5$) we saw that $x_M = 1$, $\dim M = n$. As $\tilde{h}(\mathrm{so}(n)) = n - 2$, $\dim \mathrm{so}(n) = n(n-1)/2$, (5.4.6) is again satisfied.

Another interesting example concerns fermions in the n-dimensional representation of su(n). As explained in the previous subsection we must consider $2n$ real fermions lying in a real, reducible representation with generators M. Then $x_M = 1$ and as $\tilde{h}(\mathrm{su}(n)) = n$, $\dim(\mathrm{su}(n)) = n^2 - 1$, the left hand side of (5.4.6) equals $(n^2 - 1)/(1 + n) = n - 1$, one less than $\dim M/2 = 2n/2 = n$, the right-hand side of (5.4.6).

But we saw in the previous subsection that we can construct u(1) currents commuting with the su(n) currents. Furthermore, as we saw in Subsec. 4.2, we can construct a Virasoro algebra $\mathscr{L}^{\mathrm{u}(1)}$ with these u(1) currents. It commutes with $\mathscr{L}^{\mathrm{su}(n)}$ and has unit c-number. Hence, adding these two Virasoro generators $\mathscr{L}^{\mathrm{u}(1)}$ and $\mathscr{L}^{\mathrm{su}(n)}$ we obtain a Virasoro algebra with c-number n now equal to precisely half the number of free real fermions. We conclude that the Sugawara construction for $\hat{\mathrm{u}}(n)$ equals the free fermion construction (5.4.1). This example underlines the importance of recognizing the u(1) currents as in the previous section. Actually this example has been known for a long time in the physics literature although it was derived quite differently.[59,61]

Many more interesting solutions to (5.4.6) can be found by trial and error but we shall now see how to find the most general solution.

5.5. *Symmetric spaces and a no-interaction theorem*

We have seen that, at least if g is simple, the alternative conditions (5.4.4) or (5.4.6) are necessary and sufficient for the quantum equality of the two constructions of the Virasoro algebra out of fermions, namely the Sugawara form (4.1.1) using fermion currents (5.1.1) and the free form (5.4.1). Since the Virasoro generators correspond to either the $(++)$ or $(--)$ components of the energy momentum tensor in 2-dimensional conformally invariant quantum field theories, we see that we have a numerical condition that an apparently interacting theory of fermions (with Sugawara energy-momentum tensor) reduces to a free theory with no interactions. The mathematical significance of this result will be discussed in Subsec. 5.6.

We shall now find the complete solution to the alternative conditions (5.4.4) or (5.4.6) (and their generalizations when g is not simple). The key observation is that (5.4.4) constitutes the cyclic identity for the Riemann tensor of a symmetric space. The theorem, due to Goddard, Nahm and Olive,[45] states that the necessary and sufficient condition for the Sugawara $\mathscr{L}^g$ to equal the free field (5.4.1) is that there exists a group G' containing G, such that G'/G is a symmetric space whose tangent space generators transform under G in the same way as the fermions.

Thus, if we decompose g' into even and odd parts,

$$g' = g + p\,.$$

The even generators, being g generators $T^a \equiv T_0^a$, satisfy

$$[T^a, T^b] = if^{abc}T^c, \qquad (5.5.1)$$

while the odd generators p^i, being orthogonal to the even generators and transforming according to M, like the fermions, satisfy

$$[T^a, p^i] = iM^a_{ij}p^j. \qquad (5.5.2)$$

Finally because we have a symmetric space the odd generators must close on the even generators,

$$[p^i, p^j] = X^{ij}_a T^a.$$

If g is simple we can choose

$$\mathrm{Tr}(T^a T^b) = y\delta^{ab}, \quad \mathrm{Tr}(p^i p^j) = y\delta^{ij}.$$

Then we find

$$iyX^{ij}_a = \mathrm{Tr}(T^a[p^i, p^j]) = \mathrm{Tr}([T^a, p^i]p^j) = iyM^a_{ij}.$$

Thus

$$[p^i, p^j] = iM^a_{ij}T^a$$

so that by (5.5.2) and (5.5.3)

$$[[p^i, p^j], p^k] = \sum_{a=1}^{\dim g} M^a_{ij}M^a_{kl}p^l.$$

This means that $\sum_{a=1}^{\dim g} M^a_{ij}M^a_{kl}$ is the Riemann tensor of the symmetric space G'/G. The Jacobi identity for the p generators is the cyclic identity for the Riemann tensor and takes the form of our condition (5.4.4). Thus the existence of the symmetric space, as described, guarantees condition (5.4.4) and hence the equality of $\mathcal{L}^g$ and L as claimed. Conversely (5.4.4) guarantees that given (5.5.1) and (5.5.2) we can construct the symmetric space G'/G.

The virtue of the above result is that symmetric spaces have been classified by mathematicians, so that all possible cases of our "no interaction theorem" for fermions, $\mathcal{L}^g = L$, are listed in this classification, which can be found in Helgason's[83] book (p. 518 for type I) or in Ref. 45. In this list g need not be simple nor even semi-simple, but the above arguments can easily be modified to include these possibilities. We have already mentioned the case $g = \mathrm{u}(n)$ when g is not semi-simple. The corresponding symmetric space here is the complex projective space $\mathrm{CP}(n) = \mathrm{SU}(n+1)/\mathrm{U}(n)$. The example with n fermions transforming with respect to $\mathrm{SO}(n)$ corresponds to the sphere

$S^n = SO(n+1)/SO(n)$. These are the simplest examples of what are known as type I symmetric spaces. In our other example fermions transformed according to the adjoint representation of a simple group G. This corresponds to the type II symmetric space $G \times G/G$. More details can be found in Ref. 45.

The above "no interaction theorem" was known fifteen years ago in the cases that we now see correspond to the sphere S^n and the complex projective space CP(n), but the proof was "by hand."

Physicists seem to have concluded that all fermionic Sugawara models are therefore free and hence uninteresting but, as we have seen, this is false. Even the result that certain fermionic Sugawara theories are free is now seen to be interesting in view of Witten's observation, discussed in Subsecs. 2.3 and 2.4, that a totally different and apparently highly nonlinear theory, namely the Wess-Zumino model (2.3.12) with bosonic fields constrained to the manifold of a Lie group G, also exhibits the same Kac-Moody Virasoro structure with energy-momentum tensor of Sugawara form. As we have seen the same algebraic structure can be realized by the quark currents of free fermions but for dynamical consistency the Sugawara energy-momentum tensor must equal the free fermion energy-momentum tensor. The theorem of Goddard, Nahm and Olive above enumerates precisely the cases when this identity occurs in terms of the possible combinations of the choice of algebra g and level x. Thus we know precisely when the Wess-Zumino model can be quantum equivalent to a free fermion theory and have complemented the results of Sec. (3.5) concerning the conditions for quantum equivalence to a free boson theory (on the maximal torus, in fact). The only overlap between the two results is for level 1 of so($2n$) and this is a way of understanding the fermion-boson equivalence of Skyrme,[3] Coleman,[84] Mandelstam[85] and others (see Sec. 7). Let us recall that, in general, the Wess-Zumino model (2.3.12) cannot be quantum equivalent to a free field theory as its Virasoro c-number (4.1.5) is neither integer nor half-integer (see Subsecs. 2.2, 2.3 and 2.4).

5.6. *Finite reducibility with an* so(n) *illustration*

The representations of the Kac-Moody algebra $\hat{g}$ of physical interest, namely the highest weight, unitary ones (such that L_0 is positive), have been classified in Sec. 3.4 and there exists a finite number of inequivalent irreducible representations with a given level x.

The quark model construction (5.1.1) yields a specific level, $x = x_M$ and it is natural to enquire how many irreducible components it contains (these being necessarily of the same level). The answer could be finite or infinite and unfortunately the latter case is the generic case. Thus the irreducible components usually occur with infinite multiplicity but there are special cases where this mathematical horror is evaded and these are furnished by the theorem of the previous section stating that $K = L - \mathscr{L}^g$ vanishes when there is a symmetric space G'/G.

To see this, first note that if K_m does not vanish, it commutes with the $\hat{g}$ generators (see (4.4.1)) so that the $\hat{g}$ highest weight states form a representation of the K_m Virasoro algebra. Since all nontrivial (i.e. $K \neq 0$) representations are infinite dimensional the infinite multiplicity follows.[45] Conversely (supposing g to be simple), if K_n vanishes,

$\mathscr{L}_0$ and L_0 are equal, in particular. Now from (4.1.7) we see that on a highest weight state of $\hat{g}$, $\mathscr{L}_0$ has the eigenvalue

$$Q_\lambda/(2k + Q_\psi), \tag{5.6.1}$$

where Q_λ is the value of the quadratic Casimir operator in the vacuum representation whose highest weight is λ. Expression (5.6.1) is bounded as λ varies over the finite range of possibilities given by (3.4.2b) as $x = 2k/\psi^2$ is fixed. But $\mathscr{L}_0$ equals L_0, (5.4.1), whose eigenvalues are $\varepsilon \dim M$ plus the total mode number (Σn_i) of the Fock space element considered, $[b^{\alpha_1}_{-n_1} b^{\alpha_2}_{-n_2} \ldots |0\rangle, n_i > 0]$. Clearly there are only a finite number of states with given L_0 eigenvalue and the finite multiplicity therefore follows. These statements can be extended to the case where g is semi-simple but modification is needed otherwise. In the special case that g is simple, is of the same rank as g', and the Ramond fields only are used, the result has been already found by Kac and Peterson.[86] They used the method of characters and also gained information about the decomposition into irreducible representations in their special situation.

Actually if K_m vanishes (by virtue of the symmetric space theorem) the expression (5.6.1) for the $\mathscr{L}_0$ eigenvalue simplifies using (5.4.7) to

$$Q_\lambda/(2Q_M). \tag{5.6.2}$$

But the L_0 eigenvalue has to be $p + (\dim M)/16$ or $p/2$ (where p is an integer ≥ 0) according as Ramond or Neveu-Schwarz field are used. Hence

$$Q_\lambda/(2Q_M) = p + (\dim M)/16 \qquad \text{(R)} \tag{5.6.3a}$$

$$= p/2 \qquad \text{(NS)}. \tag{5.6.3b}$$

This is highly surprising but we shall illustrate its validity by considering fermions in the defining representation of $g = \mathrm{so}(n)$, corresponding to the symmetric space $S^n = SO(n + 1)/SO(n)$. As we have seen (after equation (5.2.10)) this yields level 1 and hence, by (3.4.2b) and the theory of representations of so(n), the possibilities $\lambda = 0$, $\lambda = \lambda_v \equiv e_1$ (the defining, n-dimensional representation of so(n)) and $\lambda = \lambda_s(\lambda_{\bar{s}})$, the spinor representation of so(n) which splits into two pieces when n is even. These weights are all fundamental weights and so correspond to specific points of the extended Dynkin diagrams in the usual way,

n even:

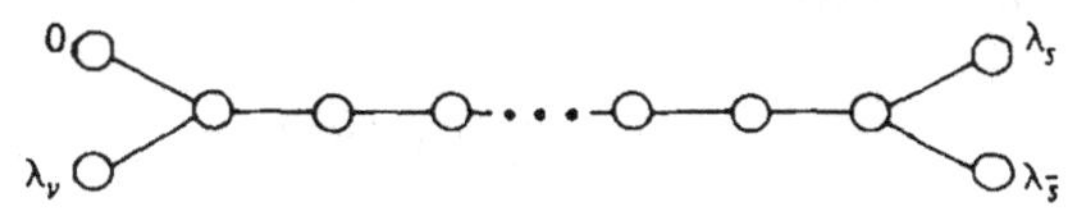

n odd:

Fig. 7. Extended Dynkin diagram for so(n).

If $\lambda = 0$, Q_λ vanishes, so by (5.6.2) $\mathscr{L}_0$ vanishes. There is only one state with zero L_0 eigenvalue and that is $|0_{NS}\rangle$, the vacuum in the Neveu-Schwarz sector. If $\lambda = \lambda_v$, we have, by (4.3.3) and (4.3.4) that Q_{λ_v} equals $\psi^2 \dim \mathrm{so}(n)/n = (n-1)\psi^2/2$. Hence $\mathscr{L}_0$ simply has eigenvalue $\frac{1}{2}$ by (5.6.2) and there are precisely n states with this L_0 eigenvalue, namely $b^{\alpha}_{-1/2}|0_{NS}\rangle$, $\alpha = 1, 2, \ldots, n$. These indeed form an n-dimensional representation of $g = \mathrm{so}(n)$.

For the spinor weight $\lambda = \lambda_s$ (or $\lambda_{\bar{s}}$) the quadratic Casimir equals $Q_{\lambda_s} = y_{\lambda_s} \dim g/\dim M = (\Sigma\mu^2)\dim g/(\dim M \operatorname{rank} g)$ by Eq. (4.3.5). But any weight of the spinor representation has components $\pm\sqrt{(\psi^2/8)}$ so that its length squared is always $\psi^2 \operatorname{rank} g/8$. Thus

$$Q_{\lambda_s} = (\dim \mathrm{so}(n)\psi^2)/8 = n(n-1)\psi^2/16.$$

Hence the $\mathscr{L}_0$ eigenvalue (3.6.3) equals

$$Q_{\lambda_s}/2Q_M = n/16 = (\dim M)/16. \tag{5.6.4}$$

The only possible states are the degenerate Ramond vacua which indeed fall into the irreducible spinor representations of so(n) with unit multiplicity. We see that the minimal L_0 eigenvalues indeed agree with (5.6.3), and that each possible level 1 irreducible representation occurs just once (with no degeneracy) and that both Ramond and Neveu-Schwarz fields are necessary to obtain this complete and neat picture.

These results concerning so(n) were first obtained, differently, by Frenkel.[66,87] As we have mentioned before, when n is even, so(n) is simply-laced and its level 1 representations can therefore be obtained, equivalently, by the vertex operator construction. This equivalence is the basis of Skyrme's fermion-boson equivalence as explained in Sec. 7.[3]

Fermions in the adjoint representation of a simple Lie group G satisfy the criterion of Subsec. 5.5 by virtue of the type II symmetric space $G \times G/G$, and lead to two interesting features.[16] The first is that it is possible to take the square root of $\mathscr{L}^g$ (which equals (5.4.2)) and obtain a super-Virasoro generator and hence a supersymmetric extension of the Kac-Moody Virasoro algebra structure discussed above retaining only fermion fields with no elementary bosons. The second is that the vacuum states in the Ramond sector have highest weights λ equal to ρ i.e. half the sum of positive roots of g (also given by (3.1.17)), so that $Q_\lambda = 3\rho^2$ by (4.3.9) and (5.6.3a) reduces to

$$\rho^2 = Q_\psi \dim g/24. \tag{5.6.5}$$

This is known as Freudenthal-de Vries strange formula[88] and is important in more advanced considerations.[5]

6. Vertex Operators and Affine Kac-Moody Algebras

6.1. *The vertex operators*

The quantum equivalence theorem of Subsec. 4.5 showed that in principle it should be possible to construct a level 1 representation of $\hat{g}$, whenever g is simply laced i.e. has roots of equal length and is hence of A, D or E type, from the level 1 representation of $\hat{t}$ where t is the Cartan subalgebra of g. This construction can be achieved using the vertex operator of string theory (or what is nearly the same thing, Skyrme's[3] soliton field operator construction) as was first realized for su(N) by Halpern[95] and for simply-laced groups by Frenkel and Kac,[72] and by Segal.[65]

Since g is simply-laced, choosing $\psi^2 = 2$ ensures that all roots have length $\sqrt{2}$. Then, in the Cartan-Weyl basis of g, the commutation relation of $\hat{g}$, given in Eqs. (3.2.4–6), read for level 1:

$$[H^i_m, H^j_n] = \delta^{ij} m\delta_{m+n,0}\,, \tag{6.1.1}$$

$$[H^i_m, E^\alpha_n] = \alpha^i E^\alpha_{m+n}\,, \tag{6.1.2}$$

$$[E^\alpha_m, E^\beta_n] = \begin{cases} \varepsilon(\alpha,\beta)E^{\alpha+\beta}_{m+n}\,, & \text{if } \alpha+\beta \text{ is a root of } g \\ \alpha\cdot H_{m+n} + m\delta_{m+n,0}\,, & \text{if } \alpha+\beta = 0 \\ 0\,, & \text{otherwise,} \end{cases} \tag{6.1.3}$$

where $\varepsilon(\alpha, \beta)$ equals ± 1, and satisfies certain consistency conditions to be discussed later in Subsec. 6.5.

We see from (6.1.1) that H^i_m, H^i_{-m} where $i = 1 \ldots \operatorname{rank} g$, $m = 1, \ldots, \infty$, form independent harmonic oscillators. In fact such a collection of oscillators is of fundamental importance in string theory (corresponding to the successive harmonics of a string vibrating in rank g dimensions). The generating function of these oscillators is called a Fubini-Veneziano momentum field[71] and we have, using the notation of (2.1.1),

$$H^i(z) = P^i(z) \equiv p^i + \sum_{n=1}^{\infty} (\alpha^i_n z^{-n} + \alpha^i_{-n} z^n), \tag{6.1.4}$$

where

$$\alpha^{i\dagger}_n = \alpha^i_{-n}\,, \quad p^{i\dagger} = p^i\,, \tag{6.1.5}$$

and

$$[\alpha^i_m, \alpha^j_n] = \delta^{ij} m\delta_{m+n,0}\,. \tag{6.1.6}$$

In string theory, there is also a Fubini-Veneziano coordinate field, $Q^i(z)$ obtained from $P^i(z)$ by integrating the relation

$$P^i(z) = iz\frac{dQ^i}{dz}, \tag{6.1.7}$$

so that

$$Q^i(z) = q^i - ip^i \ln z + i\sum_{n\neq 0} \alpha_n^i z^{-n}/n, \tag{6.1.8}$$

where q^i is a "constant" of integration, taken to be canonically conjugate to p^i, i.e.,

$$[q^i, p^j] = i\delta^{ij}. \tag{6.1.9}$$

The state space considered is a Fock space built up from the simultaneous vacua of all these oscillators, $|0\rangle$, which also carries zero momentum p

$$\alpha_n^i|0\rangle = 0, \quad n > 0, \qquad p^i|0\rangle = 0. \tag{6.1.10}$$

Momentum λ can be added to these states by acting with the "plane wave" $\exp(i\lambda \cdot q)$. We denote

$$|\lambda\rangle = e^{i\lambda\cdot q}|0\rangle. \tag{6.1.11}$$

Later we shall see that the values of λ must be restricted. Because of (6.1.10) we indeed have a highest weight representation of $\hat{t}$. It is natural to define a "normal ordering" whereby α_n^i, $n \geq 1$, is moved to the right of α_m^j with $m \leq -1$ and p is moved to the right of q. If this normal ordering is denoted by double dots it follows from (6.1.6) and (6.1.9) that we have the contraction identity

$$P^i(z)Q^j(\zeta) = :P^i(z)Q^j(\zeta): - i\delta^{ij}z/(z-\zeta), \qquad |z| > |\zeta|. \tag{6.1.12}$$

Differentiating with respect to ζ yields, using (6.1.7)

$$P^i(z)P^j(\zeta) = :P^i(z)P^j(\zeta): + \delta^{ij}z\zeta/(z-\zeta)^2, \qquad |z| > |\zeta|, \tag{6.1.13}$$

which is, by (6.1.4), a special case of the identity (4.1.3). We also find

$$Q^j(\zeta)P^i(z) = :Q^j(\zeta)P^i(z): - i\delta^{ij}z/(z-\zeta), \qquad |\zeta| > |z|. \tag{6.1.14}$$

Notice that the right-hand sides of (6.1.12) and (6.1.14) are the same apart from the range of validity specified by the inequality.

These relations make clear why the normal ordering has to be introduced: matrix elements of products of Fubini-Veneziano fields, taken between states of the Fock space, exhibit singularities unless the products are first normal ordered.

Now let us introduce the vertex operator[71,89,90,91]

$$U^\alpha(z) = z^{\alpha^2/2} :e^{i\alpha\cdot Q(z)}:. \tag{6.1.15}$$

Since, by (6.1.9),

$$[p^i, U^\alpha(z)] = \alpha^i U^\alpha(z), \tag{6.1.16}$$

we see that $U^\alpha(z)$ creates momentum α, and that we have taken a step towards representing (6.1.2), if α is taken to be a root of g. The normal ordering in the definition of the vertex operator ensures that it will have finite matrix elements in the Fock space. It will turn out that the quantities E^α_m will be proportional to the coefficients in the Laurent expansion of $U^\alpha(z)$, and so we must first ensure that $U^\alpha(z)$ is indeed single valued when the complex variable z encircles the origin once. This is not obvious since $Q^i(z)$ itself has a logarithmic branch point at the origin. In fact the factors in $U^\alpha(z)$ involving the oscillators are single valued and the only difficulty comes from the remaining factors which are

$$z^{\alpha^2/2} e^{i\alpha\cdot q} z^{\alpha\cdot p} = e^{i\alpha(q - ip\ln z)}. \tag{6.1.17}$$

Evidently this is single valued if

$$p\cdot\alpha + \alpha^2/2 \in \mathbb{Z}. \tag{6.1.18}$$

Let us imagine that we consider states built up from $|\bar{p}\rangle$ (see (6.1.11)) by the action of $U^\alpha(z)$ (or its Laurent coefficients) and the α^i_{-n}, $(n \geq 1)$. These states will have momenta p given by

$$p \in \Lambda_\Sigma + \bar{p}, \tag{6.1.19}$$

where Λ_Σ is the lattice generated by the set Σ of points α considered. If Σ is the set of roots of g then Λ_Σ is the root lattice of g, $\Lambda_R(g)$. Hence two possible values of p differ by an element of Λ_Σ and (6.1.18) then implies that

$$\alpha\cdot\lambda \in \mathbb{Z} \quad \text{if} \quad \lambda \in \Lambda_\Sigma \quad \text{and} \quad \alpha \in \Sigma, \tag{6.1.20}$$

i.e. Λ_Σ is what is called an integral lattice, the scalar product of any two points of Λ_Σ is an integer. If in addition, Λ_Σ is 'even', i.e. the square of any point of Λ_Σ is not just an integer but an even integer, then (6.1.18) and (6.1.19) imply that

$$\bar{p} \in \Lambda^*_\Sigma, \tag{6.1.21}$$

where Λ^* is the lattice dual to Λ, i.e., the lattice of points whose scalar product with any point of Λ is an integer,

$$\Lambda^* = \{x; x\cdot y \in \mathbb{Z}, y \in \Lambda\}. \tag{6.1.22}$$

We have made this statement in slightly greater generality than is necessary for the immediate context. If, as we have supposed, all $\alpha \in \Sigma$ have length $\sqrt{2}$, Λ_Σ is certainly even. Thus, if Σ is the set of roots of g, Λ_Σ is $\Lambda_R(g)$, the root lattice of g, and is even, while $\Lambda_\Sigma^* = \Lambda_W(g)$, the weight lattice of g (see Eq. (3.1.17)). Thus, by (6.1.18), p must be a weight of g. Having understood how both the root and weight lattices of g arise naturally, we can now proceed to verify Eqs. (6.1.2) and (6.1.3).

6.2. *The almost commutation relations of $\hat{g}$*

Given that condition (6.1.18) is satisfied as explained so that $U^\alpha(z)$ is single valued and has the Laurent expansion,

$$U^\alpha(z) = \sum_{m\in \mathbb{Z}} z^{-m} A_m^\alpha , \tag{6.2.1}$$

we can now investigate the properties of the A_m^α. First note that by (6.1.5), (6.1.6) and (6.1.15) we have the hermiticity property of the vertex operator,

$$U^\alpha(z)^\dagger = U^{-\alpha}(1/z^*), \tag{6.2.2}$$

whence, by (6.2.1),

$$A_m^{\alpha\dagger} = A_{-m}^{-\alpha}. \tag{6.2.3}$$

Now in order to check (6.1.2) we note that according to (6.1.12) and Wick's theorem,[76]

$$\begin{aligned} P^i(z)U^\alpha(\zeta) &= :P^i(z)U^\alpha(\zeta): + \alpha^i z U^\alpha(\zeta)/(z-\zeta), \qquad |z| > |\zeta|, \\ &\equiv h^{i\alpha}(z,\zeta). \end{aligned} \tag{6.2.4}$$

Notice that by (6.1.14) and Wick's theorem we have also

$$U^\alpha(\zeta)P^i(z) = h^{i\alpha}(z,\zeta); \qquad |\zeta| > |z|.$$

It follows that both $H_m^i A_n^\alpha$ and $A_n^\alpha H_m^i$ can be written as double integrals of the same integrand

$$\oint \frac{d\zeta\, \zeta^n}{2\pi i \zeta} \oint \frac{dz\, z^m}{2\pi i z} h^{i\alpha}(z,\zeta)$$

differing only in the orientation of the contours of integration. In both cases the z and

ζ contours encircle the origin once in a positive sense but in the first case $|z| > |\zeta|$ and in the second case the inequality is reversed. The only singularities of the integrand occur at $z, \zeta = 0$ and ∞, and at $z = \zeta$. Thus the commutator $[H^i_m, A^\alpha_n]$ equals the same integrand integrated over the difference of the two contours which, by Cauchy's theorem, can be taken to be a z contour enclosing the point $z = \zeta$ once, followed by a ζ contour enclosing the origin once:

$$[H^i_m, A^\alpha_n] = \oint_0 \frac{d\zeta\,\zeta^n}{2\pi i\zeta} \oint_\zeta \frac{dz\,z^m}{2\pi i z} h^{i\alpha}(z,\zeta).$$

The first term in $h^{i\alpha}(z,\zeta)$, the normal ordered product, is regular at $z = \zeta$ and hence contributes nothing by Cauchy's theorem. The second term has a simple pole at $z = \zeta$ and hence, by the residue theorem, contributes

$$[H^i_m, A^\alpha_n] = \alpha^i \oint \frac{d\zeta\,\zeta^{m+n}}{2\pi i\zeta} U^\alpha(\zeta) = \alpha^i A^\alpha_{m+n}\,. \tag{6.2.5}$$

Thus we have verified (6.1.2) and now proceed to check (6.1.3). It will turn out that A^α_m will "almost" satisfy (6.1.3), the discrepancy being some awkward signs. The first step is to normal order $U^\alpha(z)U^\beta(\zeta)$, an exercise of basic importance in evaluating the string scattering amplitudes. We find

$$:e^{i\alpha\cdot Q(z)}::e^{i\beta\cdot Q(\zeta)}: = (z-\zeta)^{\alpha\cdot\beta} :e^{i\{\alpha\cdot Q(z)+\beta(\zeta)\}}:, \quad |z| > |\zeta|, \tag{6.2.6}$$

because

$$e^{i\alpha\cdot Q_>(z)} e^{i\beta\cdot Q_<(\zeta)} = e^{[i\alpha\cdot Q_>(z),\, i\beta\cdot Q_<(\zeta)]} e^{i\beta\cdot Q_<(\zeta)} e^{i\alpha\cdot Q_>(z)}$$

$$= (1-\zeta/z)^{\alpha\cdot\beta} e^{i\beta\cdot Q_<(\zeta)} e^{i\alpha\cdot Q_>(z)}, \quad |z| > |\zeta|,$$

where

$$Q^i_>(z) = i\sum_{n=1}^{\infty} \frac{1}{n}\alpha^i_n z^{-n}, \quad Q^i_<(z) = i\sum_{n=-1}^{-\infty} \frac{1}{n}\alpha^i_n z^{-n}, \tag{6.2.7}$$

and

$$z^{\alpha\cdot p} e^{i\beta\cdot q} = e^{i\beta\cdot q} z^{\alpha\cdot p} z^{\alpha\cdot\beta}$$

Identity (6.2.6) is crucial in exhibiting the factorization and duality properties of the string scattering amplitudes. Indeed the vertex operators were originally constructed as a solution to this equation.[89,90] Equation (6.2.6) can be generalized to the product of any number of vertex operators. The dual scattering amplitudes were the vacuum expectation values of such products, suitably integrated.

Notice that because of our comment in the previous section that α and β had to lie on an integral lattice Λ_Σ, the singularity $(z-\zeta)^{\alpha\cdot\beta}$ exhibited in (6.2.6) is always an isolated singularity and never a branch point. There is no singularity if $\alpha\cdot\beta \geq 0$.

The simultaneous interchange of α with β and z with ζ, changes the right-hand side of (6.2.6) by a sign $(-1)^{\alpha\cdot\beta}$. Hence, by a now familiar argument,

$$A_m^\alpha A_n^\beta - (-1)^{\alpha\cdot\beta} A_n^\beta A_m^\alpha = \oint_0 \frac{d\zeta\,\zeta^n}{2\pi i\zeta} \oint_\zeta \frac{dz\,z^m}{2\pi i z}(z-\zeta)^{\alpha\cdot\beta} z^{\alpha^2/2}\zeta^{\beta^2/2} :e^{i\{\alpha\cdot Q(z)+\beta\cdot Q(\zeta)\}}:\,, \quad (6.2.8)$$

since we have the difference of two terms expressed in terms of the same integrand but with contours satisfying $|z| > |\zeta|$ and $|\zeta| > |z|$ respectively. The difference of the two contours can be arranged, by Cauchy's theorem, as a z contour enclosing ζ once followed by a ζ contour enclosing the origin once.

If $\alpha\cdot\beta \geq 0$ the right-hand side of (6.2.8) vanishes since there is no singularity inside the z contour. If $\alpha\cdot\beta = -1$, the z contour encloses a simple pole at ζ and is evaluated by the residue theorem as $A_{m+n}^{\alpha+\beta}$. When α and β are vectors in Euclidean space, as now, the only remaining possibility is that $\alpha\cdot\beta = -2$, which implies that $\alpha+\beta = 0$. Then the z contour encloses a double pole at ζ, so that (6.2.8) yields:

$$\oint \frac{d\zeta\,\zeta^n}{2\pi i}\frac{d}{dz}(z^m :e^{i\{\alpha\cdot Q(z)+\beta\cdot Q(\zeta)\}}:)|_{z=\zeta} = \oint \frac{d\zeta}{2\pi i}\zeta^{m+n-1}(m+\alpha\cdot P(\zeta))$$

$$= \alpha\cdot H_{m+n} + m\delta_{m+n,0}\,.$$

To recapitulate these results, with (6.1.1) and (6.2.5), we have

$$[H_m^i, H_n^j] = m\delta^{ij}\delta_{m+n,0}\,, \quad (6.2.9)$$

$$[H_m^i, A_n^\alpha] = \alpha^i A_{m+n}^\alpha\,, \quad (6.2.10)$$

$$A_m^\alpha A_n^\beta - (-1)^{\alpha\cdot\beta}A_n^\beta A_m^\alpha = \begin{cases} 0, & \alpha\cdot\beta \geq 0, \\ A_{m+n}^{\alpha+\beta}, & \alpha\cdot\beta = -1, \\ \alpha\cdot H_{m+n} + m\delta_{m+n,0}\,, & \alpha\cdot\beta = -2. \end{cases} \quad (6.2.11)$$

To compare with Eqs. (6.1.1), (6.1.2) and (6.1.3) note that if $\alpha\cdot\beta = -1$, $(\alpha+\beta)^2 = 2$ and that the Weyl reflection of β in α, namely $\sigma_\alpha(\beta)$, equals $\alpha+\beta$ so that $\alpha+\beta$ must be a root of g. Conversely if $\alpha+\beta$ is a root of g with the same length, $\sqrt{2}$, as α and β, then $\alpha\cdot\beta = -1$. Hence the two sets of equations coincide apart from the sign $(-1)^{\alpha\cdot\beta}$ appearing on the left-hand side of (6.2.11) and the sign $\varepsilon(\alpha,\beta)$ appearing on the right-hand side of (6.1.3).

We shall show that the discrepancy between the sets of commutation relations satisfied by the E_m^α and A_m^α can be rectified by constructing quantities c_α such that

$$E_m^\alpha = A_m^\alpha c_\alpha\,. \quad (6.2.12)$$

The most economical way of constructing the c_α, that is without introducing any more degrees of freedom, was proposed by Frenkel and Kac[72] and consisted of constructing the c_α out of the momenta p^i. Thus the c_α will not commute with the $e^{iq\cdot\alpha}$ factor in A^α_m but instead will have the property

$$E^\alpha_m = A^\alpha_m c_\alpha = c_{-\alpha} A^\alpha_m . \tag{6.2.13}$$

Furthermore c_α is Hermitian and has unit square, i.e.,

$$c^\dagger_\alpha = c_\alpha, \quad c^2_\alpha = 1 . \tag{6.2.14}$$

These properties guarantee that E^α_m indeed has the same hermiticity property as A^α_m, (6.2.3), i.e.,

$$E^{\alpha\dagger}_m = E^{-\alpha}_{-m} . \tag{6.2.15}$$

For the time being we assume this construction has been done and defer its treatment to Subsec. 6.5. It will be seen that several new ideas have to be introduced which link up naturally with ideas in Sec. 7 concerning fermions and in Ref. 5 concerning Lorentzian lattices.

6.3. *Equivalence of the Virasoro and Sugawara constructions*

In the Cartan-Weyl basis for g, the Sugawara construction (4.1.1) for the Virasoro algebra reads

$$\mathscr{L}^g(z) = (2k + Q_\psi)^{-1} {}^\times_\times\left\{\sum_{i=1}^{\text{rank } g} H^i(z)H^i(z) + \sum_{\alpha>0} (E^\alpha(z)E^{-\alpha}(z) + E^{-\alpha}(z)E^\alpha(z))\right\}{}^\times_\times . \tag{6.3.1}$$

The quantum equivalence theorem (4.5.1) showed that if g was simply-laced and the level was unity, then (6.3.1) must equal the Virasoro construction

$$L(z) = (1/2){}^\times_\times \sum_{i=1}^{\text{rank } g} H^i(z)H^i(z){}^\times_\times , \tag{6.3.2}$$

when the roots are all chosen to have length $\sqrt{2}$. Now that we have by (6.1.15) and (6.2.13)

$$E^\alpha(z) = z :e^{i\alpha\cdot Q(z)}: c_\alpha = c_{-\alpha} z :e^{i\alpha\cdot Q(z)}: , \tag{6.3.3}$$

we can explicitly check the equality between (6.3.1) and (6.3.2).

First note that as $\psi^2 = 2$, $x = 1$, therefore $k = 1$. Further $Q_\psi = 2\tilde{h} = 2h$ where h denotes the ordinary Coxeter number of g since the ordinary and dual Coxeter numbers coincide when g is simply-laced. We see from (4.3.6) that when g is simply-laced

$$h = (\text{number of roots of } g)/\operatorname{rank} g\,. \tag{6.3.4}$$

Actually this is the definition of h when g is not simply-laced. It follows that the prefactor in (6.3.1) reduces, in the situation considered, to

$$(2k + Q_\psi)^{-1} = \frac{1}{2(1+h)}\,. \tag{6.3.5}$$

It is clear that the result would follow from the following two identities

$${}^\times_\times E^\alpha(z)E^{-\alpha}(z) + E^{-\alpha}(z)E^\alpha(z){}^\times_\times = {}^\times_\times(\alpha\cdot H(z))^2{}^\times_\times \tag{6.3.6}$$

and

$$\sum_{\alpha>0} \alpha\alpha^T = hI_{\operatorname{rank} g} \tag{6.3.7}$$

because summing (6.3.6) over the positive roots α of g, using (6.3.7) would establish that the second term in the curly brackets of (6.3.1) equals h times the first term, so that by (6.3.5) the result follows.

To establish (6.3.7) note that the left-hand side is indeed proportional to the unit matrix $I_{\operatorname{rank} g}$ in the space of rank g dimensions and that the constant of proportionality is determined by taking the trace, using (6.3.4) and remembering that each α^2 equals 2.

To establish the identity (6.3.6), point split with $|z| > |\zeta|$ using (4.1.3):

$${}^\times_\times E^\alpha(z)E^{-\alpha}(z) + E^{-\alpha}(z)E^\alpha(z){}^\times_\times = \lim_{z\to\zeta}\left(E^\alpha(z)E^{-\alpha}(\zeta) + E^{-\alpha}(z)E^\alpha(\zeta) - 2z\zeta/(z-\zeta)^2\right).$$

Now introduce (6.3.3), remembering (6.2.13) and (6.2.14), and use the fundamental identity (6.2.6),

$$= \lim_{z\to\zeta}\frac{z\zeta}{(z-\zeta)^2}\{:e^{i\alpha\cdot(Q(z)-Q(\zeta))}: + :e^{-i\alpha\cdot(Q(z)-Q(\zeta))}: - 2\}\,.$$

This appears highly singular because of the denominator $(z-\zeta)^2$ but in fact the expression in the curly brackets possesses a double zero at $z = \zeta$ and so the limit can be evaluated by l'Hôpital's rule. Differentiating numerator and denominator once yields,

$$= \lim_{z\to\zeta}\frac{\zeta}{2(z-\zeta)}\{:\alpha\cdot P(z)[e^{i\alpha\cdot(Q(z)-Q(\zeta))} - e^{-i\alpha\cdot(Q(z)-Q(\zeta))}]:\}\,.$$

Differentiating yet again and taking the limit gives the desired result (6.3.6).

Notice that in the course of this argument we use in an essential way two distinct notions of normal ordering, one denoted by crosses which shifts Kac-Moody generators with positive suffixes to the right of ones with negative suffixes and ones denoted by small dots referring to the Bose oscillators (6.1.6) and the zero modes p^i, q^i. These coincide for the $H^i(z)$ but not for the $E^\alpha(z)$.

The above analysis was due to Frenkel,[66] although it is somewhat similar to an earlier analysis in string theory.[92]

Given the fact, proven above, that the Sugawara construction (6.3.1) collapses to the Virasoro construction (6.3.2) when vertex operators are used, it is easy to work out the commutation relations with the Kac-Moody generators H_n^i and E_n^α. It is well known in string theory,[27] and follows by our methods from (6.1.12) and (6.1.14), that

$$[L_n, Q^i(z)] = z^{n+1}\frac{dQ^i}{dz}. \tag{6.3.8}$$

Hence differentiating and remembering (6.1.7)

$$[L_n, P^i(z)] = z^{n+1}\frac{dP^i}{dz} + nz^n P^i(z). \tag{6.3.9}$$

The Laurent coefficients of $P^i(z)$, namely H_n^i do indeed satisfy (1.1.2). The corresponding calculation for the vertex operator is slightly more complicated. By Wick's theorem and Eq. (6.2.4)

$$(1/2){:}P^2(z){:}\, U^\alpha(\zeta) = (1/2){:}P^2(z)U^\alpha(\zeta){:} + z{:}\alpha\cdot P(z)U^\alpha(\zeta){:}/(z-\zeta) + \alpha^2 z^2 U^\alpha(z)/2(z-\zeta)^2$$

for $|z| > |\zeta|$ with a similar result on the right-hand side with the inequality between $|z|$ and $|\zeta|$ reversed when the factors on the left-hand side are reversed in order. By the usual contour argument we find[27]

$$[L_n, U^\alpha(z)] = z^{n+1}\frac{dU^\alpha}{dz} + (\alpha^2/2)nz^n U^\alpha(z). \tag{6.3.10}$$

We say that $U^\alpha(z)$ has conformal weight $\alpha^2/2$. Of course if $\alpha^2 = 2$ the conformal weight of $U^\alpha(z)$ is unity, the same as that of $P^i(z)$, (6.3.9), and is indeed appropriate to a Kac-Moody field which always has unit conformal weight, by Eq. (1.1.2).

Notice that by (6.3.8), $Q^i(z)$ has conformal weight 0, and so is suitable for exponentiation to obtain $U^\alpha(z)$, (6.1.15). Without normal ordering $U^\alpha(z)$ formally has zero conformal weight, but is divergent. Its conformal weight $\alpha^2/2$ is, in a sense, acquired through the normal ordering needed to make it finite.

6.4. *Simple reducibility of the vertex operator representation*

As we have seen, the space in which the vertex operators act consists of the oscillator Fock spaces equipped with a momentum in Λ_Σ^*, the weight lattice of g, the simply-laced algebra whose roots are α. This space must decompose into invariant subspaces with respect to $\hat{g}$ because of the vertex operator construction. Because L_0 is manifestly positive, these invariant subspaces must carry highest weight representations of $\hat{g}$ with level 1 as discussed in Subsec. 3.4. How many of these invariant subspaces are there? We shall show that the number precisely equals the number of elements of the centre of the simply connected Lie group obtained by exponentiating the algebra g.

First recall the results of Subsec. 3.4, obtained by general arguments. It was explained that the g weight of a highest weight state in a level-1 representation of $\hat{g}$ (with g simply-laced) has to be either 0, or a fundamental weight, λ, of g corresponding to a point of the Dynkin diagram of $\hat{g}$, $D(\hat{g})$, related to the point 0 added in constructing $D(\hat{g})$ from $D(g)$, by a symmetry of $D(\hat{g})$. These fundamental weights have the additional property that

$$\lambda \cdot \psi = 1, \quad \alpha \cdot \lambda \geq 0 \quad (\alpha \text{ a positive root of } g), \tag{6.4.1}$$

remembering $\psi^2 = 2$. It follows that $2\lambda \cdot \alpha/\alpha^2$ equals 0 or $\pm 1$where α is any root of g, i.e., λ is what is called a "minimal weight."

Since it is integral, the root lattice of g, Λ_R, is contained within the weight lattice $\Lambda_W = \Lambda_R^*$. The points of these lattices form a group which is Abelian since the multiplication law is simply addition. It follows that Λ_R is an invariant subgroup of Λ_W and that we can split Λ_W into disjoint cosets Λ_W/Λ_R. These cosets themselves constitute a group which is known to be isomorphic to $\mathbb{Z}(g)$ the centre of the simply connected group obtained by exponentiating g, and hence finite (if g is semi-simple). The points of each coset nearest the origin consist of a minimal weight and the weights conjugate to it under the action of the Weyl group. (For the root lattice itself the minimal weight is replaced by the zero weight.) Thus we write

$$\Lambda_W \equiv \Lambda_R^* = \Lambda_R \cup (\lambda_1 + \Lambda_R) \cup (\lambda_2 + \Lambda_R) \ldots, \tag{6.4.2}$$

where $\lambda_1, \lambda_2, \ldots$ are minimal weights and $A \cup B$ means the union of the sets of points A and B. Thus there is a correspondence between the elements of $\mathbb{Z}(g)$, the minimal weights of g (including 0) and the cosets Λ_W/Λ_R. This structure is of physical importance in understanding the theory of stable magnetic monopoles in non-Abelian gauge theories, and holds even if g is not simply-laced.[93,94]

Thus it follows that the inequivalent level-1 representations of $\hat{g}$ when g is simply laced correspond to the elements of $\mathbb{Z}(g)$. It remains to show that the vertex operator construction yields each of these level-1 representations once and once only.

Let $|\lambda\rangle$ be a highest weight state of g. It satisfies

$$H_n^i |\lambda\rangle = 0, \quad n \geq 1, \tag{6.4.3}$$

and hence is a Fock-space vacuum which may carry a momentum, λ say, in Λ_W. We also have

$$E_n^\alpha |\lambda\rangle = 0, \quad n \geq 1 \quad \text{or} \quad n = 0, \quad \alpha > 0. \tag{6.4.4}$$

It is sufficient to consider simple roots of g and hence by the discussion in Sec. 3.2, E_0^α, α a simple root of g, and $E_1^{-\psi}$. (6.4.4) is equivalent to

$$A_n^\alpha|\lambda\rangle = 0$$

which, by (6.2.1), (6.1.15), (6.1.16) means

$$\oint \frac{dz}{2\pi i z} z^{n+\alpha^2/2+\lambda\cdot\alpha} e^{i\alpha\cdot Q_<(z)}|\lambda+\alpha\rangle = 0.$$

Since $Q_<(z)$ contains positive powers of z only, this vanishes if and only if

$$n + \alpha^2/2 + \lambda\cdot\alpha \geq 1.$$

As $\alpha^2 = 2$ this yields $\alpha\cdot\lambda \geq 0$ applied to E_0^α (α simple). Applied to $E_1^{-\psi}$ this yields

$$\psi\cdot\lambda \leq 1.$$

The solutions λ are either $\lambda = 0$ or the minimal weights defined by (6.4.1).

In this way we obtain the important result that each possible level 1 highest weight irreducible representation occurs once, and only once; that is the vertex operator representation of g is "simply reducible." The corresponding highest weight states are

$$|\lambda\rangle = e^{i\lambda\cdot q}|0\rangle, \tag{6.4.5}$$

where λ is either the zero g weight or a minimal dominant weight of g.

If we define the character of an irreducible representation as

$$\chi^{\lambda,x}(\theta; z) = \mathrm{Tr}(e^{i\theta\cdot p} z^{L_0}), \tag{6.4.6}$$

where x is the level and λ the g weight of the highest weight state, we can easily evaluate this expression for any vertex operator representation:

$$\chi^{\lambda,x}(\theta; z) = \left\{ \sum_{\mu\in\lambda+\Lambda_R(g)} z^{\mu^2/2} e^{i\mu\cdot\theta} \right\} \prod_{n=1}^{\infty} (1-z^n)^{-\mathrm{rank}\, g}. \tag{6.4.7}$$

This is because L_0 in a vertex operator representation is given by (6.3.2) as

$$L_0 = p^2/2 + \sum_{n=1}^{\infty} \alpha_{-n}^i \alpha_n^i \tag{6.4.8}$$

and, for fixed i and n

$$\mathrm{Tr}(z^{\alpha_{-n}^i \alpha_n^i}) = (1-z^n)^{-1}.$$

Notice that despite the fact that the representation is infinite dimensional, so that the trace in the definition of the character pertains to an infinite sum, nevertheless it converges if $|z| < 1$.

6.5. *Sign compensation and a generalized Klein transformation*

Initially the sign $\varepsilon(\alpha, \beta)$ occurring in (6.1.3) is defined when α, β and $\alpha + \beta$ are all in Σ, but the key to constructing the correction factor c_α converting from (6.2.11) to (6.1.3) via (6.2.12) is to extend its definition to any α, β in Λ_Σ (the lattice generated by Σ). When this is done as indicated below,

$$c_\alpha = \sum_{\beta \in \Lambda_\Sigma} \varepsilon(\alpha, \beta) |\beta + \bar{p}\rangle\langle\beta + \bar{p}|, \tag{6.5.1}$$

if our ground state has momentum $\bar{p}$. Thus, for the time being, we are considering the action of c_α on states of the same momenta modulo Λ_Σ. In the situation explained above $\bar{p}$ can be taken to be a minimal weight or 0.

We shall call an integral lattice Λ an ε-lattice if it is furnished with $\varepsilon(\alpha, \beta) = \pm 1$ $(\alpha, \beta \in \Lambda)$ satisfying the properties:

$$\varepsilon(\alpha, \beta) = (-1)^{\alpha \cdot \beta + \alpha^2 \beta^2} \varepsilon(\beta, \alpha), \tag{6.5.2}$$

$$\varepsilon(\alpha, \beta)\varepsilon(\alpha + \beta, \gamma) = \varepsilon(\alpha, \beta + \gamma)\varepsilon(\beta, \gamma). \tag{6.5.3}$$

If Λ is even (like $\Lambda_R(g)$, g simply-laced), $(-1)^{\alpha^2 \beta^2} = 1$, but we shall see in Sec. 7 that the incorporation of fermions requires consideration of odd integral lattices as well. Furthermore our proof below that lattices such as $\Lambda_R(g)$ are ε-lattices will follow from considering odd lattices in a natural way.

Once given, $\varepsilon(\alpha, \beta)$ is not unique, for the quantity

$$\varepsilon'(\alpha, \beta) = \eta_\alpha \eta_\beta \eta_{\alpha+\beta} \varepsilon(\alpha, \beta), \qquad \eta_\alpha = \pm 1, \tag{6.5.4}$$

satisfies (6.5.2) and (6.5.3) if $\varepsilon(\alpha, \beta)$ does. (6.5.4) can be thought of as a $\mathbb{Z}_2$ gauge freedom at each lattice point. Judicious gauge fixing will impose extra, useful properties. Putting $\beta = 0$ in (6.5.3):

$$[\varepsilon(\alpha, 0) - \varepsilon(0, \gamma)]\varepsilon(\alpha, \gamma) = 0.$$

Thus $\varepsilon(\alpha, 0)$, $\varepsilon(0, \gamma)$ are all equal and can be $\mathbb{Z}_2$ gauged to equal unity by choosing η_0 in (6.5.4) to equal their common value. So then

$$\varepsilon(\alpha, 0) = \varepsilon(0, \gamma) = 1. \tag{6.5.5}$$

By choosing $\eta_\alpha \eta_{-\alpha} = \varepsilon(\alpha, -\alpha)$, we can arrange that

$$\varepsilon(\alpha, -\alpha) = 1, \tag{6.5.6}$$

and (6.5.5) and (6.5.6) will be preserved if, in future "gauge transformations" $\eta_\alpha \eta_{-\alpha} = 1$, $\eta_0 = 1$.

Choosing $\alpha + \beta + \gamma = 0$, (6.5.3) yields

$$\varepsilon(\alpha, \beta)\varepsilon(-\gamma, \gamma) = \varepsilon(\alpha, -\alpha)\varepsilon(\beta, -\alpha - \beta),$$

which, in view of (6.5.6), yields

$$\varepsilon(\alpha, \beta) = \varepsilon(\beta, -\alpha - \beta) = \varepsilon(-\alpha - \beta, \alpha), \tag{6.5.7}$$

where the second equality follows by repetition of the first. Choosing $\alpha + \beta = 0$, (6.5.3) yields

$$\varepsilon(\alpha, -\alpha)\varepsilon(0, \gamma) = \varepsilon(\alpha, \gamma - \alpha)\varepsilon(-\alpha, \gamma)$$

which, with the choices (6.5.5) and (6.5.6), reads, after relabelling

$$\varepsilon(\alpha, \beta) = \varepsilon(-\alpha, \alpha + \beta). \tag{6.5.8}$$

Combining this with (6.5.7) yields

$$\varepsilon(\alpha, \beta) = \varepsilon(-\beta, -\alpha). \tag{6.5.9}$$

We can now see the relevance of these conditions. The Jacobi identities for g imply that ε must satisfy (6.5.3) when α, β, γ, $\alpha + \beta$ and $\beta + \gamma$ are roots of g. The hermiticity property (6.2.15), combined with (6.1.1)–(6.1.3) requires (6.5.7) and (6.5.9) when α, β and $\alpha + \beta$ are roots of g.

It is straightforward to check that expression (6.5.1) for c_α is Hermitian and of unit square (6.2.14), if we understand that the momentum eigenstates are normalized

$$\langle \mu | \nu \rangle = \delta_{\mu\nu} \tag{6.5.10}$$

(which is possible because the spectrum is discrete). Checking that (6.1.1)–(6.1.3) follow from (6.2.9)–(6.2.12) is facilitated by defining

$$\hat{c}_\alpha = e^{iq\cdot\alpha} c_\alpha, \tag{6.5.11}$$

thereby absorbing the plane wave factor of A^α_m. The desired results follow, providing

$$\hat{c}_\alpha \hat{c}_\beta = (-1)^{\alpha\cdot\beta + \alpha^2\beta^2} \hat{c}_\beta \hat{c}_\alpha = \varepsilon(\alpha, \beta)\hat{c}_{\alpha+\beta}, \tag{6.5.12}$$

$$\hat{c}^\dagger_\alpha = \hat{c}_{-\alpha}. \tag{6.5.13}$$

By (6.5.1) and (6.5.11)

$$\hat{c}_\alpha \hat{c}_\beta = \sum_{\gamma, \delta \in \Lambda_I} \varepsilon(\alpha, \gamma)\varepsilon(\beta, \delta)|\alpha + \gamma + \bar{p}\rangle\langle\gamma + \bar{p}|\beta + \delta + \bar{p}\rangle\langle\delta + \bar{p}|$$

which, by (6.5.10) equals

$$\sum_{\gamma \in \Lambda_I} \varepsilon(\alpha, \gamma)\varepsilon(\beta, \gamma - \beta)|\alpha + \gamma + \bar{p}\rangle\langle\gamma - \beta + \bar{p}|$$

$$= \sum_{\gamma \in \Lambda_I} \varepsilon(\alpha, \beta + \gamma)\varepsilon(\beta, \gamma)|\alpha + \beta + \gamma + \bar{p}\rangle\langle\gamma + \bar{p}|$$

on relabelling γ. By (6.5.3), (6.5.11) and (6.5.1) the above equals

$$\varepsilon(\alpha, \beta)\hat{c}_{\alpha+\beta}\,.$$

Interchange of α and β and use of (6.5.2) yields the rest of Eq. (6.5.12). Equation (6.5.13) follows similarly.

Consider a cubic lattice Λ of dimension d,

$$\Lambda = (\Sigma n_j \mathbf{e}_i, \quad n_i \in \mathbb{Z})$$

where $\mathbf{e}_1, \ldots, \mathbf{e}_d$ are orthogonal unit vectors,

$$\mathbf{e}_i \cdot \mathbf{e}_j = g_{ij}\,,$$

and g_{ij} is diagonal with entries ± 1. If all diagonal entries equal $+1$, Λ is Euclidean; if only one is negative, Λ is "Lorentzian." We define the Clifford algebra

$$\gamma_i \gamma_j + \gamma_j \gamma_i = 2g_{ij} \tag{6.5.14}$$

and associate γ^i with the point e^i. More generally we associate γ^u with the point $\mathbf{u} = \sum_{i=1}^d u_i \mathbf{e}_i$ of the cubic lattice Λ by

$$\gamma^u = \xi_u (\gamma_1)^{u_1} (\gamma_2)^{u_2} \ldots (\gamma_d)^{u_d}, \tag{6.5.15}$$

where ξ_u is a sign ± 1 chosen at will. If we commute γ^u with γ^v we acquire a phase $(-1)^{\sum_{i \neq j} u_i v_j}$. But

$$\sum_{i \neq j} u_i v_j = (\Sigma u_i)(\Sigma v_j) - \Sigma u_i v_i \quad \text{and} \quad (\Sigma u_i)(\Sigma v_j) = u^2 v^2$$

modulo 2 as the u_i, v_j are integers. Hence

$$\gamma^u \gamma^v = (-1)^{u \cdot v + u^2 v^2} \gamma^v \gamma^u = \varepsilon(u, v) \gamma^{u+v}, \tag{6.5.16}$$

and this equation defines the sign $\varepsilon(u, v)$ for all u, v in the cubic lattice. It satisfies (6.5.2) as is seen by interchanging u and v in (6.5.16). Furthermore it satisfies (6.5.3) by virtue of the associativity of γ matrix multiplication. We conclude that the cubic lattice is an ε-lattice, whatever the signature of its metric.

We may wish to consider c_α acting on a space spanned by states with ground state momenta in different cosets with respect to Λ_Σ corresponding to taking $\bar{p} = \lambda_1, \lambda_2, \ldots$ where λ_i does not belong to $\lambda_j + \Lambda_\Sigma$ $(j \neq i)$. Defining $c_{\alpha,i}$ by replacing $\bar{p}$ by λ_i in (6.5.1) we have

$$\hat{c}_{\alpha,i} \hat{c}_{\beta,j} = 0 \quad (i \neq j),$$

so that we may define

$$\hat{c}_\alpha = \hat{c}_{\alpha,1} + \hat{c}_{\alpha,2} + \cdots \tag{6.5.17}$$

which will also satisfy (6.5.12) and (6.5.13).

In Subsec. 7.4 we shall see that this construction applied to the cubic lattice is essentially the "Klein transformation" familiar in quantum field theory. Thus the Frenkel-Kac construction of c_α is a generalization of this.

The final step, that the lattices of interest, including $\Lambda_R(g)$, g simply-laced, are ε-lattices, follows, as is now explained, from two observations: (i) that the cubic lattices $\mathbb{Z}^d$ in Euclidean or Lorentzian space are ε-lattices and (ii) that so are any sublattices.

As explained in the next Subsec. 7.1, when $D_n = \mathrm{so}(2n)$ has roots of length $\sqrt{2}$ the following part of the weight lattice

$$\Lambda_R(D_n) \cup (e_1 + \Lambda_R(D_n)) \tag{6.5.18}$$

is cubic. It follows that it is an ε-lattice and that so is $\Lambda_R(D_n)$, which is a sublattice. So also is $\Lambda_R(A_{n-1})$ as it is a sublattice of $\Lambda_R(D_n)$ (of one less dimension).

To see that the root lattices of the remaining simple simply-laced Lie algebras, E_6, E_7 and E_8 are also ε-lattices, we need some results concerning self-dual lattices, i.e., lattices for which $\Lambda = \Lambda^*$, (6.1.22).

If Λ is self-dual and either Euclidean or Lorentzian we can "add" to it an orthogonal unit lattice $\mathbb{Z}$ so as to obtain an odd self-dual lattice with one more dimension than Λ and a Lorentzian metric. There is a uniqueness theorem for such lattices which implies that this has to be a cubic Lorentzian lattice in $d + 1$ dimensions and hence an ε-lattice by the preceding discussion. Thus being sublattices of ε-lattices, all self-dual Euclidean or Lorentzian lattices are themselves ε-lattices.

$\Lambda_R(E_8)$ is self-dual (and even) and hence therefore an ε-lattice. $\Lambda_R(E_6)$ and $\Lambda_R(E_7)$ are sublattices (of lower dimension) and hence also ε-lattices.

It is very interesting in view of the later developments that the concept of self-dual lattice arises so naturally. For more information consult Ref. 5 whose treatment is the basis for the above discussion.

7. Vertex Operators and Fermi Fields

7.1. *Vectorial Fermi fields and lattice points of unit length*

In the previous section we assigned vertex operators to a set of points Σ corresponding to the roots of a simply-laced Lie algebra g. It was convenient to choose these points all to have length $\sqrt{2}$ since then the weight lattice of g was precisely dual to the root lattice. The condition that it was possible to form a sort of Fock space of these operators by letting a product of them act on a state of given momentum in a consistent way required (amongst other things) that the lattice Λ_Σ generated by the points Σ should be integral (as indeed it was in the cases considered). It is possible to satisfy this condition in another way, by starting with Σ consisting of points of unit length, again requiring Λ_Σ to be integral. Actually this is only possible if Λ_Σ is a cubic lattice. We now investigate this new possibility, showing that with some minor modifications, it works equally well, leading to quantities which tend to anticommute rather than commute, as the vertex operators associated with length $\sqrt{2}$ points did. Just as the Kac-Moody algebra $\hat{g}$ is an affinization of the finite dimensional Lie algebra g, so these new vertex operators lead to an affinization of the Clifford algebra. To a physicist they are simply Fermi fields, and arise in two versions, the Ramond and Neveu-Schwarz fields of Secs. 2 and 5. The only simply-laced, simple Lie algebras for which the cubic lattice occurs as a sublattice of the weight lattice are D_r (i.e. so$(2r)$), $r \geq 3$, and, like the Dirac γ-matrices (Clifford algebra elements) the Fermi fields will transform under D_r according to the $2r$-dimensional vectorial representation. In fermionic string theory this is precisely how the Neveu-Schwarz and Ramond fields transform with respect to the Lorentz group. In Subsec. 7.4 we shall see that there is one exceptional possibility, of great physical interest, which consists of constructing spinorial fermion fields with respect to D_4 (i.e. so(8)).

We now discuss the weight lattice of D_r, and how it relates to a cubic lattice. In general, as explained in Subsec. 6.4 the cosets of the weight lattice of a simple Lie algebra with respect to its root lattice form a finite group, $\mathbb{Z}(g)$, isomorphic to the centre of the simply connected Lie group obtained by exponentiating g. $\mathbb{Z}(\text{so}(2r))$ has four elements and the structure,

$$\mathbb{Z}(\text{so}(2r)) = \mathbb{Z}_2 \times \mathbb{Z}_2, \quad r \in 2\mathbb{Z}, \tag{7.1.1a}$$

$$= \mathbb{Z}_4, \qquad r \in 2\mathbb{Z}+1. \tag{7.1.1b}$$

Correspondingly, the four cosets of the weight lattices of so$(2r)$ are, by (6.4.2),

$$\Lambda_W(\text{so}(2r)) = \Lambda_R \cup (\lambda_v + \Lambda_R) \cup (\lambda_s + \Lambda_R) \cup (\lambda_{\bar{s}} + \Lambda_R), \tag{7.1.2}$$

where we take

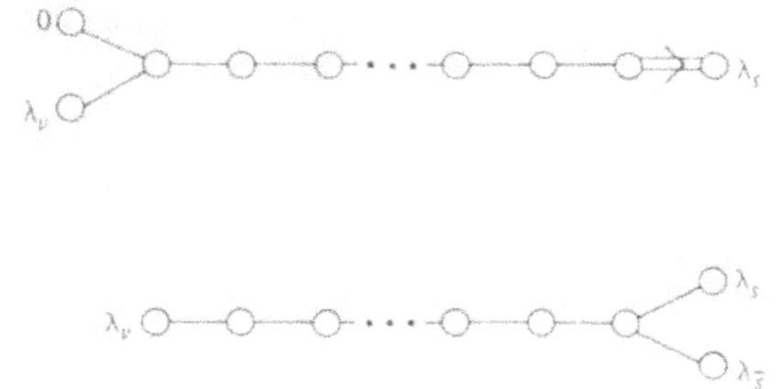

Fig. 8. Dynkin diagram of so(2r) showing the minimal weights.

$$\lambda_v = \mathbf{e}_1 , \tag{7.1.3a}$$

$$\lambda_s = \left(\sum_{i=1}^{r} e_i\right)\Big/2, \tag{7.1.3b}$$

$$\lambda_{\bar{s}} = \lambda_s - e_r , \tag{7.1.3c}$$

and $e_1, \ldots, e_r$ denote r orthonormal unit vectors. The roots have the form $(\pm e_i \pm e_j)$ $(i \neq j)$ and hence length $\sqrt{2}$. λ_v, λ_s and $\lambda_{\bar{s}}$ are the three minimal weights of so(2r) and define (by constituting the highest weights thereof) the vector, spinor and conjugate spinor representations respectively, i.e., the three "minimal representations" of so(2r). They correspond to the three points of the Dynkin diagram of so(2r) indicated in Fig. 8.

The Dirac gamma matrices γ^i are defined by the Clifford algebra relation

$$\{\gamma^i, \gamma^j\} = 2\delta^{ij}, \quad i, j = 1, 2, \ldots, 2r, \tag{7.1.4}$$

and it is a standard result, used by Dirac in his proof of the Lorentz covariance of his relativistic electron equation, that a representation of the so(2r) algebra is given by $i[\gamma^i, \gamma^j]/4$. This representation decomposes into two irreducible components, with highest weights λ_s and $\lambda_{\bar{s}}$. Therefore the fishtail of the so(2r) Dynkin diagram can be thought of as corresponding to the Dirac gamma matrices.

The minimal fundamental weights (7.1.3) and their Weyl group conjugate points are the closest points to the origin of their respective cosets (7.1.2). We see that

$$\lambda_v^2 = 1, \quad \lambda_s^2 = \lambda_{\bar{s}}^2 = r/4. \tag{7.1.5}$$

Thus generically the only points on the weight lattice of so(2r) of unit length are the points $\pm e_i$, $(i = 1, \ldots, r)$, which are conjugate to λ_v under the Weyl group. (The exceptional case $r = 4$ is treated in Subsec. 7.4.) These points constitute the set Σ to be assigned vertex operators and the lattice generated by them, Λ_Σ, is the cubic lattice:

$$\Lambda_\Sigma = Z^r = \Lambda_R \cup (\lambda_v + \Lambda_R). \tag{7.1.6}$$

If $e \in \Sigma$ (so that $e^2 = 1$) we see that according to (6.1.17) and (6.1.18), the vertex operator $U^e(z)$ is single-valued (periodic) when z encircles the origin once, i.e.,

$$U^e(ze^{2i\pi}) = U^e(z), \quad \text{if} \quad e\cdot p + 1/2 \in \mathbb{Z}, \tag{7.1.7a}$$

and double-valued (antiperiodic), i.e.,

$$U^e(ze^{2i\pi}) = -U^e(z), \quad \text{if} \quad e\cdot p \in \mathbb{Z}. \tag{7.1.7b}$$

We did not consider the possibility that $U^\alpha(z)$, be antiperiodic when $\alpha^2 = 2$ since we knew that the Hermitian Kac-Moody field was periodic (see Eq. (2.1.1)). However we saw in Secs. 2 and 5 that fermion fields were naturally periodic or antiperiodic and indeed $U^e(z)$, $e^2 = 1$ will turn out to be Fermi fields.

Since we imagine $U^e(z)$ acting on a product of vertex operators $U^{e'}(z')$, $e' \in \Sigma$, themselves acting on a ground state with momentum $\bar{p}$, we have $p \in \Lambda_\Sigma + \bar{p}$. Subtracting conditions (7.1.7) for two different values of p (for fixed $\bar{p}$) shows that Λ_Σ must be integral as indeed it is. Further we see that $U^e(z)$ is double-valued if and only if $p \in \Lambda_\Sigma$, and single-valued if and only if $p \in \Lambda_\Sigma + \lambda_s$, i.e., for $e^2 = 1$

$$U^e(z) \text{ is } \begin{cases} \text{of R type if } p \in \Lambda_\Sigma + \lambda_s \equiv (\Lambda_R + \lambda_s) \cup (\Lambda_R + \lambda_{\bar{s}}), & (7.1.8a) \\ \text{of NS type if } p \in \Lambda_\Sigma \equiv \Lambda_R \cup (\Lambda_R + \lambda_v), & (7.1.8b) \end{cases}$$

in terms of the Ramond-Neveu-Schwarz classification of (5.1.2). These vertex operators ($U^e, e^2 = 1$) have the same domain of definition as those of Sec. 6 ($U^\alpha, \alpha^2 = 2$) but fall into two sectors, which we call the Ramond and Neveu-Schwarz sectors respectively. Note that because Λ_W has two cosets with respect to Λ_Σ, i.e.,

$$\Lambda_W = \Lambda_\Sigma \cup (\Lambda_\Sigma + \lambda_s), \tag{7.1.9}$$

the choice of sector is determined solely by the ground state momentum $\bar{p}$. According as $U^e(z)$ is of R or NS type it can be expanded in integral or half integral powers of z so that we can define

$$A_r^e = \oint \frac{dz}{2\pi i z} z^r U^e(z) \begin{cases} r \in \mathbb{Z}, & \text{R sector}, \quad (7.1.10a) \\ r \in \mathbb{Z} + 1/2, & \text{NS sector}. \quad (7.1.10b) \end{cases}$$

The previous calculations of almost commutation relations hold good with minor modifications and with similar provisos about signs.

We have

$$[H_m^i, A_r^e] = e^i A_{m+r}^e \tag{7.1.11}$$

showing that $U^e(z)$ is indeed a vectorial field. If $e, f \in \Sigma$ so that $e^2 = f^2 = 1$, it follows that, as Λ_Σ is integral, $e \cdot f = 0, \pm 1$ only. We have from (6.2.8)

$$A_r^e A_s^f - (-1)^{e \cdot f} A_s^f A_r^e = \begin{cases} 0, & e \cdot f = 0, 1, \\ \delta_{r+s,0}, & e \cdot f = -1. \end{cases} \qquad \begin{matrix}(7.1.12a)\\ (7.1.12b)\end{matrix}$$

If $e \cdot f = 1$, then $e = f$ and we see that A_r^e, A_s^e anticommute, thereby establishing their fermionic nature (which will not be affected when we postmultiply by a factor c_e as in Sec. 6). Likewise, if $e \cdot f = -1$, then $e = -f$ and we again have anticommutation.

If $\alpha \in \Lambda_R$ and $\alpha^2 = 2$, $e \in \Lambda_\Sigma$ and $e^2 = 1$, we have $\alpha \cdot e = 0, \pm 1$ as the only possibilities again. Then (6.2.8) yields

$$A_r^e A_s^\alpha - (-1)^{\alpha \cdot e} A_s^\alpha A_r^e = \begin{cases} 0, & e \cdot \alpha = 0, 1, \\ A_{r+s}^{\alpha+e}, & e \cdot \alpha = -1. \end{cases} \qquad \begin{matrix}(7.1.13a)\\ (7.1.13b)\end{matrix}$$

Note that when $\alpha \cdot e = -1$, $(\alpha + e)^2 = 1$ so $\alpha + e \in \Sigma$.

The signs can be corrected by the same construction as in Sec. 6, defining

$$H_r^e = A_r^e c_e = c_{-e} A_r^e, \qquad (7.1.14)$$

since we saw that the cubic lattice was an ε-lattice. Then

$$\{H_r^e, H_s^f\} = \delta_{r+s,0}\delta_{e+f,0} \qquad (7.1.15a)$$

$$[E_m^\alpha, H_r^e] = \begin{cases} \varepsilon(\alpha, e) H_{m+r}^{e+\alpha}, & \alpha \cdot e = -1, \\ 0, & \alpha \cdot e = 0, 1. \end{cases} \qquad \begin{matrix}(7.1.15b)\\ (7.1.15c)\end{matrix}$$

Notice that we are using the fact that $(-1)^{e^2 f^2} = -1$ if and only if U^e and U^f are both fermion fields.

We can of course define $2r$ "real fields" $H^{(k)}(z)$ by

$$H^{e_k}(z) = (H^{(2k-1)}(z) - iH^{(2k)}(z))/\sqrt{2}, \qquad (7.1.16a)$$

so that, by the hermiticity property,

$$H^{-e_k}(z) = (H^{(2k-1)}(z) + iH^{(2k)}(z))/\sqrt{2}. \qquad (7.1.16b)$$

These fields satisfy the properties of the fields in Sec. 5, namely (5.1.4), (5.1.7), etc. The labelling in (7.1.16) has been chosen to fit standard conventions in the description of the so($2r$) algebra and the signs are chosen to mimic $[i\gamma^1\gamma^2/2, \gamma^1 \mp i\gamma^2] = \pm(\gamma^1 \mp i\gamma^2)$. We shall study this correspondence further in the next subsection.

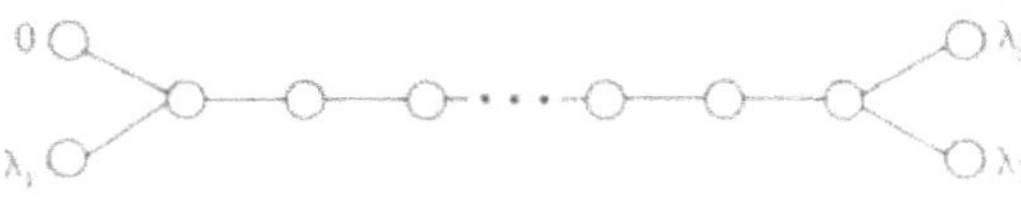

Fig. 9. Dynkin diagram for the affine algebra $\hat{so}(2r)$.

7.2. *Comparison of vertex operator and quark model representations of* $\hat{so}(2r)$

We have seen that the four level 1 representations of $\hat{so}(2r)$ can be constructed in two apparently different ways: by means of the vertex operator construction of Sec. 6, and by means of the quark model construction of Sec. 5 as bilinear in $2r$ real fermion fields. In the previous section we saw that such fermion fields can themselves be constructed out of r real boson fields by assigning vertex operators to the $2r$ points of unit length in the so($2r$) weight lattice. A detailed comparison of these two constructions will now be made and leads to the construction of the r bose fields from the $2r$ Fermi fields thereby completing a "quantum equivalence" theorem between fermions and bosons, originally due to Skyrme[3] and later developed by others.

In Subsec. 6.4 we described how the complete space of the vertex operator representation of $\hat{g}$, for g simple and simply-laced, namely a bosonic Fock space carrying momenta in the weight lattice of g, decomposed into irreducible invariant subspaces under $\hat{g}$. When $g = \text{so}(2r)$, there are four such subspaces with highest weight states carrying so($2r$) weights 0, λ_v, λ_s and $\lambda_{\bar{s}}$ respectively, as defined in Eq. (7.1.3). By Eq. (7.1.8) the fermion fields constructed in the previous section were of Neveu-Schwarz type in the subspace defined by the weights 0 and λ_v, and of Ramond type in the subspace defined by λ_s and $\lambda_{\bar{s}}$. Thus can be illustrated by the Dynkin diagram of $\hat{so}(2r)$, constructed in Subsec. 3.3, and shown in Fig. 9.

This differs from the so($2r$) Dynkin diagram, drawn in the previous section, by the addition of the point associated with the so($2r$) weight zero. The effect of this is to create a second "fishtail." The NS and R representations of the affinized Clifford algebra are associated with these two fishtails in the same way that the unique so($2r$) Clifford algebra representation is associated with the single so($2r$) fishtail. This provides a sort of "group theoretical" explanation of the two kinds of Fermi field (5.1.12) in string theory.

As we remarked in Subsec. 7.1 the representation $m^{ij} = i[\gamma^i, \gamma^j]/4$ of the so($2r$) algebra, constructed in terms of the Clifford algebra (7.1.4) decomposes into the two so($2r$) irreducible spinor representations with highest weights λ_s and $\lambda_{\bar{s}}$ respectively. The usual choice of basis for the Cartan subalgebra, $\{m^{12}, m^{34}, \ldots, m^{2r-1\,2r}\}$ indeed possesses as eigenvalues $\{\pm 1, \pm 1, \ldots, \pm 1\}/2$, namely the set of weights conjugate to λ_s and $\lambda_{\bar{s}}$ under the Weyl group, confirming that we have the normalization appropriate to root length $\sqrt{2}$.

We have already met in Subsecs. 5.1 and 5.6 the analogous result for the Kac-Moody algebra $\hat{so}(2r)$, namely that if $H^i(z)$ is a $2r$ component real Fermi field, the Laurent coefficients of $i[H^i(z), H^j(z)]/2$ yield a level 1 representation. This representation is simply reducible, decomposing into just two inequivalent components in the Ramond and Neveu-Schwarz sectors respectively. Thus it has precisely the same representation content as the vertex operator representation. In the Neveu-Schwarz sector one can check the equality of the two highest weight states in the two formulations:

$$|0\rangle = |0\rangle_{NS}, \quad e^{ie_1\cdot q}|0\rangle \equiv e^{i\lambda_v\cdot q}|0\rangle = A^{e_1}_{-\frac{1}{2}}|0\rangle_{NS}. \tag{7.2.1}$$

In the Ramond sector the 2^r-fold degenerate Ramond vacuum has the form

$$|0\rangle_R = e^{i\lambda\cdot q}|0\rangle, \tag{7.2.2}$$

where λ is any one of the 2^r points $\{\pm 1, \pm 1, \ldots, \pm 1\}/2$, Weyl group conjugate to λ_s or $\lambda_{\bar{s}}$; see (7.1.3).

Now let us compare the Kac-Moody generators in the two constructions. In the vertex operator construction, the step operator field for a root $e + f$ of so($2r$) is, by Eqs. (6.2.1) and (6.2.12)

$$E^{e+f}(z) \equiv \sum_{n\in Z} z^{-n}E^{e+f}_n = U^{e+f}(z)c_{e+f}. \tag{7.2.3}$$

In the quark model construction the same quantity is proportional to

$$U^e(z)c_eU^f(z)c_f = \varepsilon(e,f)U^{e+f}(z)c_{e+f}, \tag{7.2.4}$$

using the correction factor identities (6.2.14) and (6.2.15) and the normal ordering identity (6.2.6), remembering that the unit vectors e and f are orthogonal. Thus there is agreement to within a sign, which is as much as we can expect since, as we saw in Subsec. 6.5 there is a sign ambiguity in the step operators of g.

There is no such sign ambiguity for the Cartan subalgebra generators, and the comparison of the two constructions gives a more interesting result. Corresponding to the first Cartan subalgebra generator of so($2r$), $i\gamma^1\gamma^2/2$, we have, in the vertex operator construction

$$e_1\cdot P(z) = ize_1\cdot\frac{dQ}{dz}, \tag{7.2.5a}$$

whereas in the quark model construction the same quantity equals

$$iH^1(z)H^2(z). \tag{7.2.5b}$$

Now let us introduce the complex conjugate pair of Fermi fields $H^{\pm e_1}(z)$ according to Eqs. (7.1.14) and (7.1.16):

$$(H^1(z) - iH^2(z))/\sqrt{2} = H^{e_1}(z) = U^{e_1}(z)c_{e_1} = c_{-e_1}U^{e_1}(z), \tag{7.2.6a}$$

$$(H^1(z) + iH^2(z))/\sqrt{2} = H^{-e_1}(z) = U^{-e_1}(z)c_{-e_1} = c_{e_1}U^{-e_1}(z). \tag{7.2.6b}$$

It follows from this and Eq. (6.2.13) that

$$U^{e_1}(z)U^{-e_1}(\zeta) = H^{e_1}(z)H^{-e_1}(\zeta)$$

$$= (H^1(z)H^1(\zeta) + H^2(z)H^2(\zeta) + iH^1(z)H^2(\zeta) - iH^2(z)H^1(\zeta))/2, \quad (7.2.7)$$

at least in the region where the right-hand side makes sense. By the Wick contraction Eqs. (5.1.7), this is when $|z| > |\zeta|$, as a pole singularity occurs as z and ζ approach. We define the normal ordered product of the left-hand side in terms of that of the right-hand side, using fermionic normal ordering denoted by open dots. Since $\mathring{\circ} H^1(z)H^1(z) \mathring{\circ}$ vanishes, we have, in the limit z equals ζ,

$$\mathring{\circ} U^{e_1}(z)U^{-e_1}(z) \mathring{\circ} = \mathring{\circ} H^{e_1}(z)H^{-e_1}(z) \mathring{\circ} = iH^1(z)H^2(z). \quad (7.2.8)$$

Notice that normal ordering is unnecessary on the right-hand side and that the result is expression (7.2.5). Similar arguments can be repeated for any of the weights of the vector representation of so(2r). Thus

$$e \cdot P(z) \equiv ize \cdot \frac{dQ}{dz} = \mathring{\circ} H^e(z)H^{-e}(z) \mathring{\circ} = \mathring{\circ} U^e(z)U^{-e}(z) \mathring{\circ}, \quad (7.2.9)$$

whenever $e^2 = 1$.

This equality expresses the r bose fields $P^i(z)$ directly in terms of the r complex Fermi fields, thereby completing the boson-fermion quantum equivalence discussed further in the next section. The above argument has been put as it occurred in the mathematics literature and is due to Frenkel,[66] but some of the ideas involved had already appeared in the physics literature.[95,96]

7.3. *The boson-fermion quantum equivalence*

We have derived the following relations between a complex Fermi field $U^e(z)$, and one real Bose field:

$$U^e(z) = \sqrt{z} : e^{ie \cdot Q(z)} :, \quad (7.3.1)$$

$$e \cdot P(z) \equiv ize \cdot \frac{dQ}{dz} = \mathring{\circ} U^e(z)U^{-e}(z) \mathring{\circ} \quad (7.3.2)$$

by comparing the two ways of constructing the level 1 representations of the Kac-Moody alegbra $\hat{so}(2r)$, following Frenkel.[66,87] Two copies of the above results can be regarded as relating a free quantum scalar field to a complex Fermi field in two space-time dimensions. Such a relation was originally due to Skyrme[3] and has been subsequently developed by Streater and Wilde,[97] Coleman,[84] Mandelstam[85] and others.

The rather indirect derivation of (7.3.2) had the virtue of illustrating the unifying power of the representation theory of affine Kac-Moody algebras. We now present a direct proof emphasizing that it is equally valid for Fermi fields of Ramond or Neveu-Schwarz type.

The first step is to compare two different ways of exhibiting the singularity of $U^e(z)U^{-e}(\zeta)$ at $z = \zeta$. According to the fermionic normal ordering of Eq. (5.1.7), denoted by open dots, we have, remembering (2.2.7),

$$U^e(z)U^{-e}(\zeta) = {\circ \atop \circ} U^e(z)U^{-e}(\zeta) {\circ \atop \circ} + \Delta(z,\zeta), \quad |z| > |\zeta|.$$

On the other hand, using the vertex operator expression (6.1.15) and the identity (6.2.6), valid for the bosonic oscillator normal ordering, denoted by small dots:

$$U^e(z)U^{-e}(\zeta) = (:e^{i \cdot e(Q(z)-Q(\zeta))}:)\sqrt{z\zeta}/(z-\zeta), \quad |z| > |\zeta|.$$

It follows from these two expressions that, for $|z| > |\zeta|$,

$${\circ \atop \circ} U^e(z)U^{-e}(\zeta) {\circ \atop \circ} = \frac{\sqrt{z\zeta}}{z-\zeta}(:e^{ie\cdot(Q(z)-Q(\zeta))}: - 1) + \frac{\sqrt{z\zeta}}{z-\zeta} - \Delta(z,\zeta).$$

The left-hand side has to be regular at $z = \zeta$ and so indeed is the right-hand side as the residue of the apparent pole at $z = \zeta$ vanishes. By (3.1.10), the second term vanishes identically in the Neveu-Schwarz case while in the Ramond case it equals

$$\frac{\sqrt{z\zeta} - (z+\zeta)/2}{z-\zeta} = \frac{1}{2}\frac{\sqrt{z} - \sqrt{\zeta}}{\sqrt{z} + \sqrt{\zeta}}, \tag{7.3.3}$$

which certainly vanishes as z approaches ζ. By l'Hôpital's rule the first term yields the desired result (7.3.2) in the limit of z equal to ζ.

Notice the similarity of the argument to that leading to (6.3.6), the analogous result valid when $\alpha^2 = 2$ instead of 1. Notice also the equality of the Virasoro generator (6.3.2) constructed out of a single Bose field with (5.4.1) constructed out of two real Fermi fields. This follows from the theorem of Goddard, Nahm and Olive,[45] explained in Subsec. 5.5 when the symmetric space is SU(2)/U(1). Then Eq. (6.3.10) tells us that $U^e(z)$, or $H^e(z)$, $e^2 = 1$, has conformal weight $\frac{1}{2}$, as is indeed appropriate for a fermion.

In the Neveu-Schwarz sector, the zero mode of the identity (7.3.1) reads

$$e \cdot p = N_e, \tag{7.3.4}$$

that is, the e component of momentum equals the number operator for the complex fermions created by $H^e(z)$. As $(-1)^{N_e} = (-1)^{e \cdot p}$ anticommutes with $U^e(z)$, $(-1)^{N_e}U^f(z)$ likewise anticommutes with $U^e(z)$, providing that f is a unit vector orthogonal to e, instead of commuting, as $U^f(z)$ did (Eq. (7.1.12)). The extension of this construction to all unit vectors in Σ is well known in quantum field theory as the Klein transformation.[98] We see that the Frenkel-Kac momentum dependent correction factor (6.5.1) can be regarded as a generalization of this Klein transformation.

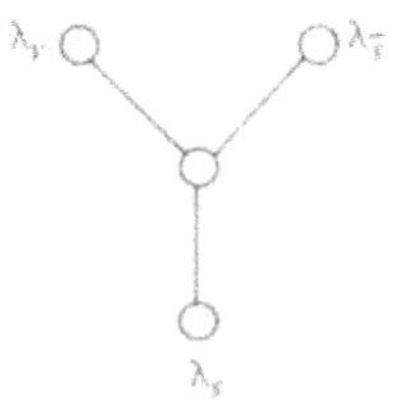

Fig. 10. The Dynkin diagram for so(8).

Related correction factors have appeared in the literature of dual string theory, (see Ref. 99 and Subsec. 3.1), where they were known as twisting operators. The correspondence was noted by Eguchi and Higashijima.[100]

We have make one proviso concerning the boson fermion quantum equivalence (7.3.1) and (7.3.2) and that is that the bosonic zero mode q cannot be reconstructed from the fermions (although $e^{ie\cdot q}$, $e \in \Sigma$, can). q is not a generator of the Kac-Moody algebra and only occurs in exponents. This relates to the observation of Segal[65] that it has to do with the global topological properties of the group into which the loops are mapped. This idea has its basis in Skyrme's soliton interpretation.[3]

Finally let us note that the boson-fermion quantum equivalence was not extensively used in the early formulation of string theory though early discussions of it occur in Refs. 59 and 95. Its role was further considered by Goddard and Olive,[5] and subsequently by Freund,[101] and Casher, Englert, Nicolai and Taormina.[102]

7.4. *The algebra* so(8) *and spinorial Fermi fields*

Fundamental physics often takes advantage of mathematical special cases. Equation (7.1.5) shows that there is an exception to our statement that the only weights of unit length of a simple, simply-laced Lie algebra are the vector weights of so($2r$) already exploited earlier in this section. When $r = 4$ so that so(8) is considered, the spinor weights also have unit length. There is a basic reason for this. The Dynkin diagram of so(8) has a greater symmetry than that of so($2r$) for other values of r, namely the six permutations of its three identical legs (see Fig. 10).

The extremities correspond to the fundamental weights λ_v, λ_s and $\lambda_{\bar{s}}$. It follows that the representations defined by these weights are related by an outer automorphism of so(8) and hence possess the same dimension, eight, and weights of the same length, unity. This symmetry is called triality symmetry; it plays a central role in superstring theory as explained in Subsec. 7.5; it and its generalizations are likely to be of great importance in mathematics and physics.

Let Σ_v, Σ_s and $\Sigma_{\bar{s}}$ denote respectively the sets of unit weights of the vector, spinor and conjugate spinor representations of so(8). These consist of the points conjugate under the action of the Weyl group of so(8) to the highest dominant weights of these representations, namely λ_v, λ_s and $\lambda_{\bar{s}}$, and therefore have the form:

$$\Sigma_v = \{\pm e_i ; i = 1, 2, 3, 4\}, \tag{7.4.1a}$$

$$\Sigma_s = \{(\pm e_1 \pm e_2 \pm e_3 \pm e_4)/2; \text{ with } n_- \text{ even}\}, \tag{7.4.1b}$$

$$\Sigma_{\bar{s}} = \{(\pm e_1 \pm e_2 \pm e_3 \pm e_4)/2; \text{ with } n_- \text{ odd}\}. \tag{7.4.1c}$$

n_- denotes the number of minus signs. Then the lattices Λ_{Σ_v}, Λ_{Σ_s} and $\Lambda_{\Sigma_{\bar{s}}}$ generated by Σ_v, Σ_s and $\Sigma_{\bar{s}}$, respectively, are each cubic and can be rotated into each other by "triality" rotations in the 4-dimensional space in which they lie. The union of all the points of these three lattices forms the weight lattice of so(8) which is not integral.

We can define three types of fermion field $U^e(z)$ by choosing the unit vector e in Σ_v, Σ_s or $\Sigma_{\bar{s}}$, respectively. Equation (6.1.16) then tells us that these fields transform under so(8) as vectors, spinors or conjugate spinors respectively. Thus the feature special to so(8) is that we can construct spinorial (and conjugate spinorial) fields as well as the vectorial ones of Subsec. 7.1. If Σ denotes one of Σ_v, Σ_s or $\Sigma_{\bar{s}}$, then Λ_Σ splits $\Lambda_W(\text{so}(8))$ into two cosets, Λ_Σ and M_Σ, say. By repetition of the analysis of Subsec. 7.1, $U^e(z)$, $e \in \Sigma$ is of Neveu-Schwarz type if it acts on a state with momentum $p \in \Lambda_\Sigma$ and it is of Ramond type if $p \in M_\Sigma$.

Just as bilinears in the vectorial fields (with correction factors) yield the four, unit level $\hat{so}(2r)$ representations, labelled by 0 and λ_v in the NS sector and λ_s and $\lambda_{\bar{s}}$ in the Ramond sector, so for $\hat{so}(8)$, bilinears in spinorial fields (with correction factors) yield the four level 1 representations labelled by 0 and λ_s in the NS sector, and λ_v and $\lambda_{\bar{s}}$ in the R sector. Similarly conjugate spinorial fields yield 0 and $\lambda_{\bar{s}}$ in the NS sector and λ_v and λ_s in the R sector.

The three types of SO(8) Fermi fields are not independent. This can be seen in several different ways.

It is evident from (7.4.1) that any root of so(8) can be split as the sum of two orthogonal unit weights in three different ways:

$$\alpha = e + e' = f + f' = g + g'; \quad e, e' \in \Sigma_v, \quad f, f' \in \Sigma_s, \quad g, g' \in \Sigma_{\bar{s}}. \tag{7.4.2}$$

By the identity (6.2.6)

$$U^\alpha(z) = U^e(z)U^{e'}(z) = U^f(z)U^{f'}(z) = U^g(z)U^{g'}(z). \tag{7.4.3}$$

Another interrelation is found by recalling that the Virasoro generators bilinear in the vectorial Fermi fields equal those bilinear in the bosonic fields, (6.3.2), because of the connection with the symmetric space SU(2)/U(1). Likewise (6.3.2) equals an expression bilinear in spinorial or conjugate spinorial fields. Hence the three fermionic Virasoro generators are equal and do not mutually commute as they would if they were composed out of independent fields.

Thirdly note that Eq. (7.3.2) expressing the boson field $e \cdot P(z)$ in terms of the fermionic field $H^e(z)$ is valid whether e lies in Σ_v, Σ_s or $\Sigma_{\bar{s}}$. Thus, starting with any of the three types of fermionic field, vectorial, spinoral or conjugate spinorial, the $P^i(z)$ can be reconstructed, and hence, by the vertex construction (6.1.15), so can any of the three fermionic fields which are therefore equivalent. Notice that, given the factors exponential in q for any one type of fermionic field, we can reconstruct the factors appropriate to the other types if we are permitted to take square roots.

Finally, if e and f are unit vectors taken from two distinct sets chosen from Σ_v, Σ_s and $\Sigma_{\bar{s}}$, then we see from (7.4.1) that their scalar product equals $\pm\frac{1}{2}$. If, and only if, it equals $-\frac{1}{2}$, $e+f$ equals a unit vector g, say, belonging to the third set. By the normal ordering identity (6.2.6) we than have, for $|z| > |\zeta|$ and $z-\zeta$ small,

$$U^e(z)U^f(\zeta) = (1-\zeta/z)^{-1/2}\big(U^g(z) + 0(z-\zeta)\big). \tag{7.4.4}$$

The above discussion is taken from Goddard, Olive and Schwimmer,[103] but the first observation of the relevance of so(8) triality to the boson-fermion quantum equivalence is due to Shankar.[104]

7.5. *Superstrings*

We have seen that rotations in eight dimensions produce a highly exceptional structure. It is precisely this which underlies the theory of superstrings as we now explain.[103,105]

The original fermionic string theory of Ramond[36] and Neveu and Schwarz[37] worked in any even space-time dimension, $2d$ say. Later it was realized that the theory simplified (and was unitary) when $2d = 10$ and when the ground state fermion had zero mass.[106] The theory then possessed 10-component vectorial bosonic and fermionic fields with elements of their combined Fock spaces corresponding to elementary particles which were fermions in the Ramond sector and bosons in the Neveu-Schwarz sector. A gauge invariance associated with the Virasoro algebra composed out of these fields implied that only a subspace of this Fock space contributed to physical processes. Furthermore this subspace could be taken to be the positive definite Fock space of 8-component oscillators transforming vectorially under the transverse SO(8) subgroup of the Lorentz group SO(9, 1). This generalized the fact that in Maxwell's electrodynamics, the photon is described by an SO(3, 1) 4-vector gauge potential yet has only two physical polarization states which are transverse to the direction of wave propagation owing to gauge invariance.

The "transverse" Neveu-Schwarz and Ramond Fock spaces each split into two invariant subspaces under the so(8) Kac-Moody algebra as we have already described. Physically these are distinguished as the "chiral projections" with $(1\pm\Gamma)/2$ where Γ is the G-parity:

$$\Gamma = -(-1)^{\sum_{i=1}^{8}\sum_{r=1/2}^{\infty} b^i_{-r}b^i_r} \quad : \mathrm{NS} \tag{7.5.1a}$$

$$= (\gamma^1\gamma^2\ldots\gamma^8)(-1)^{\sum_{i=1}^{8}\sum_{r=1}^{\infty} b^i_{-r}b^i_r} \quad : \mathrm{R}. \tag{7.5.1b}$$

Notice that Γ anticommutes with the Fermi fields and hence commutes with quantities bilinear in them such as the $\hat{so}(8)$ generators and the Virasoro generators (5.4.1).

Because of this, it is consistent to truncate the theory by retaining only the subspaces with $\Gamma = 1$. The troublesome "tachyon" is thereby eliminated and the resultant theory is potentially supersymmetric in 10 space-time dimensions.[35,107] This is the "superstring theory" as it is now known.

Its massless states consist of 8 transverse gauge particles transforming vectorially under SO(8) and 8 real fermions transforming spinorially. This pairing is obviously related to the triality symmetry of Sec. 7.4 and extends to higher mass levels as the number of states are related to the coefficients of the powers of z occurring in the partition function $\mathrm{Tr}(z^{L_0}(1 + \Gamma))/2 = \chi(z)$. In the NS and R sectors described these are easily evaluated as

$$\chi_{NS}(z) = \left\{ \prod_{n=1}^{\infty} (1 + z^{n-1/2})^8 - \prod_{n=1}^{\infty} (1 - z^{n-1/2})^8 \right\} \Big/ 2, \tag{7.5.2a}$$

$$\chi_R(z) = 8\sqrt{z} \prod_{n=1}^{\infty} (1 + z^n)^8. \tag{7.5.2b}$$

The equality of these two expressions, proven by Jacobi,[108] provided strong evidence for supersymmetry in space-time.[35,107] Mathematically, expressions (7.5.2) are the characters (6.4.6), (with $\theta = 0$) of the unit level representations of so(8) labelled by λ_v and λ_s respectively. By the boson-fermion quantum equivalence relations (7.3.1) and (7.3.2) these characters can equally well be evaluated in the vertex operator representation as (6.4.7). The two resultant expressions coincide apart from the μ summation which extend over the lattices $\lambda_v + \Lambda_R$ and $\lambda_s + \Lambda_R$ respectively. As we have seen both lattices are cubic (because of triality) and the sums therefore coincide, confirming Jacobi's identity directly, as suggested by Frenkel, Lepowsky and Meurman.[109]

Green and Schwarz[110] explicitly established this supersymmetry by constructing the supercharge out of the emission vertex for the 8 ground-state fermions. This vertex possesses a conventional factor U (Eq. (6.1.15)) times a factor $W_\lambda(z)$ constructed out of the fermionic fields,[111,112,113] the expression holding in any space-time dimension $2d$. The $W_\lambda(z)$ transform as a conjugate spinor with respect to SO($2d - 1, 1$) and have conformal weight $2d/16 = d/8$ (Eq. (6.3.20)). Furthermore

$$z^{-d/8} W_\lambda(z)|0\rangle|_{z=0} = |\lambda\rangle, \tag{7.5.3}$$

where λ is a weight of the conjugate spinor representation.

Such a quantity can immediately be written down:

$$W_\lambda(z) = z^{d/8} :e^{i\lambda\cdot Q(z)}:, \tag{7.5.4}$$

in terms of the d bosonic fields constructed out of the $2d$-component vectorial Fermi fields by the boson-fermion quantum equivalence (7.3.2). The requirements mentioned determine the vertex as previously constructed, so that (7.5.4) should coincide with it, although an explicit check has not yet been made.

Green and Schwarz[110] considered the transverse case in ten dimensions so that $2d = 8$, when $\lambda^2 = 1$ (Eq. (7.1.5)), so that the vertex (7.5.4) reduces to the spinorial fermion field of Sec. 7.4. The anticommutation relations (7.1.15a) are responsible for the supersymmetry algebra while the rotation and conformal properties follow by the work of this and the preceding section. The work of Green and Schwarz,[110] together with the earlier work, was unnecessarily complicated because of an unawareness of the simplification afforded by the boson-fermion quantum equivalence.

We see that the W_λ convert a state with momentum $p \in \lambda_v + \Lambda_R$ into one with momentum $p \in \lambda_s + \Lambda_R$ and vice versa (these being the positive G sectors of R and NS vectorial fields). It follows from the previous section what W_λ itself is of Ramond type, i.e., single-valued (in agreement with the construct of Green and Schwarz[110].)

It is still an outstanding problem to formulate the theory in a way which is simultaneously supersymmetric and SO(9, 1) Lorentz invariant. Very likely it is necessary to introduce Faddeev-Popov fields at the tree level as suggested by Friedan, Martinec and Shenker[114] and Knizhnik.[105] This corresponds to the old fact that time-component ghost states had to be explicitly projected out of fermion-tree amplitudes.[115]

The fact that the fermionic factor of all the vertices of the NSR theory (i.e., the NS and R fields and W_λ) can be written as a vertex operator in terms of d bose fields independent of the $2d$ original ones suggests that the fermionic theory may be a special case of a bosonic theory in higher dimensions ($3d$ on the fact of it). The truth of this evidently depends on a deeper understanding of the Faddeev-Popov ghost field structure.

8. Conclusions and Omissions

Our presentation has aimed to show that the theory of affine Kac-Moody algebras and their associated Virasoro algebras can fruitfully be regarded as a synthesis of the theory of roots and weights developed by mathematicians in classifying Lie algebras, and the theory of quantum fields created by physicists to describe the fundamental forces of nature, and taken to its most sophisticated extreme in the string theory of elementary particles. Such an unexpected and gratifying unification of ideas is bound to be capable of further development and to have implications for the physical theories beyond those we have already indicated.

Within this framework, we have pursued the theme that the Sugawara construction of the Virasoro algebra, originating in particle physics, furnishes a unifying thread to the representation theory of the Kac-Moody algebras provided by the diverse quantum field theory constructions.

Already an important lesson for the physicist can be perceived, namely that a certain simplicity can be attained only when quantum mechanics is treated exactly, that is to all orders in Planck's constant $\hbar$, (see Sec. 2). In fact many of the results have no classical analogue (that is when $\hbar$ vanishes) and cannot be treated perturbatively. Examples are provided by the normalization of the Sugawara construction (Sec. 5), the vertex operator constructions of Secs. 6 and 7, the boson-fermion quantum equivalence and the Virasoro algebra identities (4.5.2) and (5.4.3). Indeed the vertex operators originally arose in physics as quantum field operators for solitons arising when a scalar field wound around a circle,[3,65,85] and solitons are indeed a nonperturbative effect. Yet another example concerns a result mentioned at the end of Subsec. 5.6, but not explained in detail, the construction of supersymmetry algebras out of fermion fields without bosons.[16,17,116–118] It is also possible to construct an $N = 2$ supersymmetry algebra out of a Bose field with no Fermi fields.[119] As in the original work of Ramond[36] and Neveu and Schwarz,[37] the supercharges are explicitly constructed and their anticommutation relations verified using the algebraic structure of the constituent fields. The more recent treatments of supersymmetry have by contrast been essentially classical in that only transformation algebras are checked and then the elementary Bose and Fermi fields have to be paired just as the physical states are. Evidently a full quantum treatment opens new possibilities.

One natural future objective is to extend the power and the precision of the methods described to a wider class of physical theory including ones in four space-time dimensions. Nevertheless because a string world sheet is 2-dimensional the mathematical theory is already applicable to string theory which is currently the most favoured candidate for the ultimate unified theory of particle interactions. We saw in Sec. 7 that the mathematical formalism is particularly relevant to the fermionic version whose structure is much simplified by the use of the fermion-boson quantum equivalence. This result suggests further developments yet to be fully explored.

We close by mentioning some of the topics not treated in the preceding sections but nevertheless closely related.

In Subsec. 5.5 we stated that it was impossible to construct level-1 generators of $\hat{E}_8$ bilinear in fermions. Nevertheless it is possible to find such generators if we relax the condition of bilinearity and allow a trancendental construction utilising the triality properties of the $SO(8) \times SO(8)$ subgroup of E_8.[103]

Yet more generalizations of the vertex operator construction are possible. One interesting possibility, developed by Goddard and Olive[5] is to regard the Frenkel-Kac construction explained in Secs. 6 and 7 as a "transverse" or "light cone gauge" construction in the sense of string theory and consider instead a "covariant" version with contour integrals of the vertex operator associated with length squared 2 points of any integral lattice, with Euclidean, singular or Lorentzian metric, thereby obtaining simply-laced finite dimensional Lie algebras, affine untwisted Kac-Moody algebras or Lorentzian algebras respectively. Nestings of different sorts of algebra inside each other correspond to the nestings of the different types of lattice. Particularly interesting lattices to consider are the self-dual even ones because of their relation to electromagnetic duality symmetry conjectures[93,120] and their associated modularity properties. The no-ghost theorem of string theory[121,122] implies that only three such Lorentzian lattices are of interest, those in 10, 18 or 26 dimensions. The first two contain the weight lattices of E_8, $E_8 + E_8$ and Spin $(32)/Z_2$ in an interesting way as explained in Ref. 5. After the dramatic discovery by Green and Schwarz[123] that the latter two Lie algebras constituted the unique anomaly-free gauge algebras for supersymmetric gauge theories in ten dimensions, the above lattice ideas were further developed by

Gross, Harvey, Martinec and Rohm.[124] Another development is the construction of level 1 vertex operator representations for non-simply-laced algebras.[125]

A theory of non-unitary representations of Virasoro algebras has been developed by the Russian school[32] and applied to certain 2-dimensional statistical models.[126]

We end by repeating that although one of the remarkable and exciting features of the study of Kac-Moody and Virasoro algebras is the extent to which much of mathematics and physics becomes interrelated and unified, there is no doubt that much remains to be explored and discovered.

We wish to thank Adrian Kent, Werner Nahm, Adam Schwimmer and Neil Turok for discussions.

References

1. V. G. Kac. *Funct. Anal. App.* **1** (1967) 328; R. V. Moody, *Bull. Amer. Math. Soc.* **73** (1967) 217.
2. M. Jacob, ed., *Dual Theory* (North-Holland, Amsterdam, 1974).
3. T. H. R. Skyrme, *Proc. Roy, Soc.* **A262** (1961) 237.
4. M. Jimbo and T. Miwa, *Publ. RIMS*, Kyoto Univ. **19** (1983) 943; D. I. Olive and N. Turok, *Nucl. Phys.* **B257** [FS14] (1985) 277.
5. P. Goddard and D. Olive, *Vertex Operators in Mathematics and Physics*, MSRI Publication #3 (Springer, Heidelberg, 1984) p. 51.
6. P. Goddard, *Supplemento ai Rendiconti del Circolo Matematico di Palermo*. Serie II **9** (1985) 73 (Proceedings of 13th Winter School on Abstract Analysis at Srní).
7. D. I. Olive, *Supplemento ai Rendiconti del Circolo Matematico di Palermo*, Serie II **9** (1985) 177 (Proceedings of the 13th Winter School on Abstract Analysis, Srní).
8. D. I. Olive, "Kac-Moody and Virasoro algebras in local quantum physics," Imperial College preprint, TP/84-85/33.
9. L. Dolan, *Phys. Reports* **109** (1984) 1.
10. B. Julia, *Vertex operators in Mathematics and Physics*; MSRI publication #3 (Springer, Heidelberg, 1984) p. 393.
11. B. Julia, *Lectures in Applied Maths.* **21** (1985) 355 (Proceedings of Meeting on "Applications of Group Theory in Physics and Mathematical Physics," Chicago, 1982).
12. J. Lepowsky, *Vertex Operators in Mathematics and Physics*, MSRI publication #3 (Springer, Heidelberg 1984) p. 1.
13. I. MacDonald, *Lecture Notes in Maths.* **901** (1981) 258 (Springer, Heidelberg, 1981).
14. V. G. Kac, *Infinite-dimensional Lie Algebras—An Introduction* (Birkhauser, Boston, 1983; 2nd edition, Cambridge Univ. Press, Cambridge, 1985).
15. D. Friedan, Z. Qiu and S. Shenker, *Phys. Rev. Lett.* **52** (1984) 1575; *Vertex Operators in Mathematics and Physics*, MSRI publication #3 (Springer, Heidelberg, 1984) p. 491.
16. P. Goddard and D. Olive, *Nucl. Phys.* **B257** [FS14] (1985) 226.
17. P. Goddard, A. Kent and D. Olive, *Phys. Lett.* **152B** (1985) 88.
18. P. Goddard, A. Kent and D. Olive, *Commun. Math. Phys.* **103** (1986) 105.
19. V. G. Kac, Proceedings of the International Congress of Mathematicians, Helsinki, 1978.
20. V. G. Kac, *Lecture Notes in Physics* **94** (1979) 441.
21. B. L. Feigin and D. B. Fuchs, *Functs. Anal. Prilozhen.* **16** (1982) 47 [*Funct. Anal. App.* **16** (1982) 114].
22. C. B. Thorn, *Nucl. Phys.* **B248** (1984) 551.
23. V. G. Kac and M. Wakimoto, "Unitarizable highest weight representations of the Virasoro, Neveu-Schwarz and Ramond algebras," MIT preprint, 1985.

24. A. Kent, "Conformal invariance and current algebra," Enrico Fermi Inst. preprint 85-62.
25. J. F. Gomes, *Phys. Lett.* **171B** (1986) 75.
26. S. Adler and R. Dashen, *Current Algebras and Applications to Particle Physics* (Benjamin, New York, 1968).
27. S. Fubini and G. Veneziano, *Ann Phys.* **63** (1971) 12.
28. J.-L. Gervais and B. Sakita, *Nucl. Phys.* **B34** (1971) 477.
29. S. Ferrara, A. F. Grillo and R. Gatto, *Nuovo Cimento* **12A** (1972) 959.
30. S. Fubini, A. J. Hanson and R. Jackiw, *Phys. Rev.* **D7** (1973) 1732.
31. D. Friedan, in *Recent Advances in Field Theory and Statistical Mechanics*, 1982 Les Houches Summer School, eds. J. B. Zuber and R. Stora: Les Houches Session XXXIX, (North-Holland, Amsterdam, 1984).
32. A. A. Belavin, A. M. Polyakov and A. B. Zamolodchikov, *Nucl. Phys.* **B241** (1984) 333.
33. J. Schwinger, *Phys. Rev. Lett.* **3** (1959) 296 and Ref. 26, p. 235.
34. E. Witten, *Commun. Math. Phys.* **92** (1984) 455.
35. F. Gliozzi, D. Olive and J. Scherk, *Nucl. Phys.* **B122** (1977) 253.
36. P. Ramond, *Phys. Rev.* **D3** (1971) 2415.
37. A. Neveu and J. H. Schwarz, *Nucl. Phys.* **B31** (1971) 86; *Phys. Rev.* **D4** (1971) 1109.
38. P. A. M. Dirac, *Nuovo Cimento* **1** (1955) 16.
39. P. A. M. Dirac, *Rev. Mod. Phys.* **34** (1962) 592.
40. J. Schwinger, *Phys. Rev.* **127** (1962) 324.
41. J. Schwinger, *Phys. Rev.* **130** (1963) 406, **130** (1963) 800.
42. J. Wess and B. Zumino, *Phys. Lett.* **37B** (1971) 95.
43. H. Sugawara, *Phys. Rev.* **170** (1968) 1659.
44. C. Sommerfield, *Phys. Rev.* **176** (1968) 2019.
45. P. Goddard, W. Nahm and D. Olive, *Phys. Lett.* **160B** (1985) 111.
46. A. M. Polyakov, *Zh. Eksper. Teor. Fiz. Pis. Red.* **12** (1970) 538 [*JETP Lett.* **12** (1970) 381].
47. D. A. Huse, *Phys. Rev.* **B30** (1984) 3908.
48. G. E. Andrews, R. J. Baxter and P. J. Forrester, *J. Stat. Phys.* **35** (1984) 193.
49. J. L. Cardy, "Conformal Invariance," to appear in *Phase Transitions and Critical Phenomena*, Vol. 11, eds. C. Domb and J. L. Lebowitz (Academic Press, London, 1985).
50. D. I. Olive, "Lectures on gauge theories and Lie algebras with some applications to spontaneous symmetry breaking and integrable dynamical systems," Univ. of Virginia preprint, 1983.
51. J. E. Humphreys, *Introduction to Lie Algebras and Representation Theory*, (Springer, Heidelberg, 1972).
52. D. I. Olive and N. Turok, *Nucl. Phys.* **B215** [FS7] (1983) 470.
53. H. S. M. Coxeter, *Regular Polytopes* (MacMillan, New York, 1947; 3rd edition, Dover, New York, 1973) p. 194.
54. M. Gell-Mann, *Phys. Rev.* **125** (1962) 1067.
55. M. Gell-Mann and Y. Ne'eman, *The Eightfold Way* (Benjamin, New York, 1964).
56. V. G. Knizhnik and A. B. Zamolodchikov, *Nucl. Phys.* **B247** (1984) 83.
57. C. G. Callan, R. F. Dashen and D. H. Sharp, *Phys. Rev.* **165** (1968) 1883.
58. S. Coleman, D. Gross and R. Jackiw, *Phys. Rev.* **180** (1969) 1359.
59. K. Bardakci and M. Halpern, *Phys. Rev.* **D3** (1971), 2493.
60. G. F. Dell'Antonio, Y. Frishman and D. Zwanziger, *Phys. Rev.* **D6** (1972) 988.
61. R. Dashen and Y. Frishman, *Phys. Rev.* **D11** (1975) 278.
62. I. T. Todorov, *Phys. Lett.* **153B** (1985) 77.
63. D. Nemeschansky and S. Yankielowicz, *Phys. Rev. Lett.* **54** (1985) 620.
64. S. Jain, P. Shankar and S. Wadia, *Phys. Rev.* **D32** (1985) 2713.
65. G. Segal, *Comm. Math. Phys.* **80** (1981) 301.
66. I. B. Frenkel, *J. Funct. Anal.* **44** (1981) 259.

67. R. Goodman and N. R. Wallach, *J. Reine. Angew. Math.* **347** (1984) 69.
68. M. Virasoro, *Phys. Rev.* **D1** (1970), 2933.
69. J. Weis, see note in Ref. 27.
70. S. Mandelstam, *Phys. Rev.* **D7** (1973) 3763 and 3777.
71. S. Fubini and G. Veneziano, *Nuovo Cimento* **67A** (1970) 29.
72. I. B. Frenkel and V. G. Kac, *Inv. Math.* **62** (1980) 23.
73. Y. Nambu, Notes prepared for the Copenhagen Symposium, 1970 (unpublished).
74. P. G. Freund, P. Oh and J. T. Wheeler, *Nucl. Phys.* **B246** (1984) 371.
75. D. Altschuler, *Phys. Lett.* **163B** (1985) 193.
76. G. C. Wick, *Phys. Rev.* **80** (1950) 268.
77. F. J. Dyson, *Phys. Rev.* **75** (1949) 486.
78. E. B. Dynkin, *Transl. Amer. Math. Soc. (2)* **6** (1957) 111.
79. E. B. Dynkin, *Transl. Amer. Math. Soc. (1)* **9** (1962) 328.
80. W. Mackay and J. Patera, *Tables of Dimensions, Indices and Branching Ratios for Representations of Simple Groups*, (Dekker, New York, 1981).
81. R. Slansky, *Phys. Reports* **79** (1981) 1.
82. M. Gell-Mann, *Phys. Lett.* **8** (1964) 214.
83. S. Helgason, *Differential Geometry, Lie Groups and Symmetric Spaces*, (Academic Press, London, 1978).
84. S. Coleman, *Phys. Rev.* **D11** (1975) 2088.
85. S. Mandelstam, *Phys. Rev.* **D11** (1975) 3026.
86. V. G. Kac and D. H. Peterson, *Proc. Math. Acad. Sci.* U.S.A. **78** (1981) 3308.
87. I. B. Frenkel, *Proc. Natl. Acad. Sci.* U.S.A. **77** (1980) 6303.
88. H. Freudenthal and H. De Vries, *Linear Lie Groups* (Academic Press, New York, 1969) p. 243.
89. Y. Nambu, *Proceedings of the International Conference on Symmetries and Quark Models*, Wayne Univ., 1969 (Gordon and Breach, New York, 1970). p. 269.
90. S. Fubini, D. Gordon and G. Veneziano, *Phys. Lett.* **29B** (1969) 679.
91. J.-L. Gervais, *Nucl. Phys.* **B21** (1970) 192.
92. L. Brink and D. Olive, *Nucl. Phys.* **B56** (1973) 253.
93. P. Goddard, J. Nuyts and D. Olive, *Nucl. Phys.* **B125** (1977) 1.
94. P. Goddard and D. Olive, *Nucl. Phys.* **B191** (1981) 511 and 528.
95. M. Halpern, *Phys. Rev.* **D12** (1975) 1684.
96. T. Banks, D. Horn, and H. Neuberger, *Nucl. Phys.* **B108** (1976) 119.
97. R. Streater and I. F. Wilde, *Nucl. Phys.* **B24** (1970) 561.
98. O. Klein, *J. Phys. Radium* **9** (1938) 1.
99. V. Alessandrini, D. Amati, M. Le Bellac and D. Olive, *Phys. Reports.* **1C** (1971) 269 and in Ref. 1.
100. T. Eguchi and K. Higashijima, "Vertex operators and non-Abelian bosonization," Univ. of Tokyo preprint, presented at Niels Bohr Centennial, 1985.
101. P. G. Freund, *Phys. Lett.* **151B** (1985) 387.
102. A. Casher, F. Englert, H. Nicolai, and A. Taormina, *Phys. Lett.* **162B** 121.
103. P. Goddard, D. Olive and A. Schwimmer, *Phys. Lett.* **157B** (1985) 393.
104. R. Shankar, *Phys. Lett.* **92B** (1980) 333.
105. V. G. Knizhnik, *Phys. Lett.* **160B** (1985) 403.
106. L. Brink, D. Olive, C. Rebbi and J. Scherk, *Phys. Lett.* **45B** (1973) 379.
107. F. Gliozzi, D. Olive and J. Scherk, *Phys. Lett.* **65B** (1976) 282.
108. G. Jacobi, *Fundamenta* (Konigsberg, 1829).
109. I. B. Frenkel, J. Lepowsky and A. Meurman, *Vertex Operators in Mathematics and Physics*, MSRI Publication #3 (Springer, Heidelberg, 1984) p. 231.
110. M. Green and J. Schwarz, *Nucl. Phys.* **B181** (1981) 502.

111. C. B. Thorn, *Phys. Rev.* **D4** (1971) 1112.
112. E. Corrigan and D. Olive, *Nuovo Cimento* **11A** (1972) 749.
113. E. Corrigan and P. Goddard, *Nuovo Cimento* **18A** (1973) 339.
114. D. Friedan, E. Martinec and S. Shenker, *Phys. Lett.* **160B** (1985) 55.
115. D. I. Olive and J. Scherk, *Nucl. Phys.* **B69** (1974) 325.
116. D. Friedan and S. Shenker, *Proceedings of the Santa Fe Meeting*, eds. T. Goldman and M. N. Nieto (World Scientific, Singapore, 1985) p. 437.
117. V. G. Kac and I. T. Todorov, *Commun. Math. Phys.* **102** (1985) 175.
118. P. Di Vecchia, V. G. Knizhnik, J. L. Petersen and P. Rossi, *Nucl. Phys.* **B253** (1985) 701.
119. G. Waterson, *Phys. Lett.* **171B** (1986) 77.
120. C. Montonen and D. Olive, *Phys. Lett.* **72B** (1977) 117.
121. R. Brower, *Phys. Rev.* **D6** (1972) 1655.
122. P. Goddard and C. Thorn, *Phys. Lett.* **40B** (1972) 235.
123. M. Green and J. Schwarz, *Phys. Lett.* **149B** (1984) 117.
124. D. Gross, J. A. Harvey, E. Martinec and R. Rohm, *Phys. Rev. Lett.* **54** (1985) 502; *Nucl. Phys.* **B256** (1985) 253.
125. P. Goddard, D. Olive and A. Schwimmer, "Vertex operators for non-simply laced groups," DAMTP preprint 1986.
126. V. Dotsenko and Fateev, *Nucl. Phys.* **B240** [FS12] (1984) 312.
127. E. Del Giudice, P. Di Vecchia and S. Fubini, *Annals of Phys.* **70** (1972) 378.
128. V. G. Kac, *Functs. Anal. Prilozhen* **8** (1974) 77 [*Funct. Anal. App.* **8** (1974) 68].
129. L. Onsager, *Phys. Rev.* **65** (1944) 117.

CHAPTER 2

VERTEX OPERATORS

Reprinted Papers

3. I.B. Frenkel and V.G. Kac, "Basic Representations of Affine Lie Algebras and Dual Resonance Models", Invent. Math. **62** (1980) 23–66.

4. G. Segal, "Unitary Representations of some Infinite Dimensional Groups", Commun. Math. Phys. **80** (1981) 301–342.

5. P. Goddard and D. Olive, "Algebras, Lattices and Strings", in *Vertex Operators in Mathematics and Physics*, MSRI Publication #3 (Springer, Heidelberg, 1984) 51–96.

6. J. Lepowsky, "The Calculus of Twisted Vertex Operators", Proc. Natl. Acad. Sci. USA **82** (1985) 8295–8299.

Vertex operators have played a central role in the strong interplay between mathematics and physics in the development of the understanding of infinite-dimensional algebras. The history of vertex operators begins with T.H.R. Skryme [1] constructing a fermionic field operator for the soliton in the sine-Gordon model. The significance of this work was not properly appreciated and the operators were rediscovered when the scattering amplitudes of the dual resonance model, the precursor of string theory, were factorised using an operator formalism [2,3]. (The correspondence between the vertex operators of dual resonance models and the soliton operators of Skyrme only became fully apparent following the later work of S. Coleman [4] and S. Mandelstam [5].)

Vertex operators made their first appearance in the mathematical literature with an explicit construction by J. Lepowsky and R.L. Wilson [6] of a representation of the simplest example of an affine Kac-Moody algebra, which was generalised in [7]. The striking similarity of this work to the operators describing the interactions of strings was noticed but Lepowsky and Wilson had in fact, in the language now used in string theory, based their construction on a "twisted" boson field. Such fields and the corresponding vertices had already arisen in string theory [8,9], but the detailed connections between the areas of research became manifest following the construction by I.B. Frenkel and V.G. Kac [*Reprinted Paper* #3] and, independently, G. Segal [*Reprinted Paper* #4], which used the vertex operator on which the development of string theory had been based, that describing the basic interaction between strings.

These mathematical developments, motivated at least in part by string theory, have had an important back-reaction on our understanding of how internal symmetries, in particular gauge symmetries, can be incorporated in string theory. The Frenkel- Kac-Segal construction can be used to associate a finite- dimensional Lie group to an integral Euclidean lattice. This version of the construction is described in [10] and in *Reprinted Paper #5*, which also emphasises many aspects of the theory of lattices that have subsequently become of significance in string theory. In particular, considerations of monopole duality [11] motivated the study of self-dual lattices and this lead to speculation that the gauge groups associated to even self dual lattices, such as $E_8 \times E_8$ or $Spin(32)/Z_2$, might be of special significance. This speculation was later realised in precisely these cases with the discovery, by M.B. Green and J.H. Schwarz [12], of an anomaly cancellation in ten-dimensional superstring theories and the introduction of the "heterotic" string model by D.J. Gross *et al.* [13].

The twisted vertices used in the first vertex construction have found a recent and important application in string theory in the study of the motion of strings on orbifolds [14]. A general calculus for such operators is described in *Reprinted Paper #6*.

The vertex operator construction of Frenkel, Kac and Segal provides representations of the affine Kac-Moody algebras associated with the simply-laced finite-dimensional Lie algebras. Extensions of this construction to the case of non-simply-laced algebras are provided in [15] and [16].

References

[1] T.H.R. Skryme, "Particle states of a quantized meson field", Proc. Roy. Soc. **A262** (1961) 237–245.

[2] S. Fubini, D. Gordon and G. Veneziano, "A general treatment of factorization in dual resonance models", Phys. Lett. **29B** (1969) 679-682.

[3] Y. Nambu, "Quark model and the factorization of the Veneziano amplitude", in *Symmetries and quark models*, ed. R. Chand (Gordan and Breach, New York, 1970) 269–278.

[4] S. Coleman, "Quantum sine–Gordon equation as the massive Thirring model", Phys. Rev. **D11** (1975) 2088–2097.

[5] S. Mandelstam, "Soliton operators for the quantized sine-Gordon equation", Phys. Rev. **D11** (1975) 3026–3030.

[6] J. Lepowsky and R.L. Wilson, "Construction of the affine algebra $A_1^{(1)}$", Commun. Math. Phys. **62** (1978) 43–53.

[7] V.G. Kac, D.A. Kazdhan, J. Lepowsky and R.L. Wilson, "Realisation of the basic representations of the Euclidean Lie algebras", Adv. in Math. **42** (1981) 83–112.

[8] M.B. Halpern and C.B. Thorn, "Two faces of the dual pion-quark model. II Fermions and other things", Phys. Rev. **D4** (1974) 3084-3088.

[9] E.F. Corrigan and D.B. Fairlie, "Off-shell states in dual resonance theory", Nucl. Phys. **B91** (1975) 527–545.

[10] I.B. Frenkel, "Representations of Kac-Moody algebras and dual resonance models", *Lectures in Applied Mathematics*, Vol. **21** (American Mathematical Society, 1985) 325–353.

[11] F. Englert and P. Windey, "Quantization condition for 't Hooft monopoles in compact simple Lie groups", Phys. Rev. **D14** (1976) 2728–2731; P. Goddard, J. Nuyts and D. Olive, "Gauge theories and magnetic charge", Nucl. Phys. **B125** (1977) 1–28.

[12] M.B. Green and J.H. Schwarz, "Anomaly cancellations in supersymmetric $D = 10$ gauge theory and superstring theory", Phys. Lett. **149B** (1984) 117–122.

[13] D.J. Gross, J. Harvey, E. Martinec and R. Rohm, "Heterotic string", Phys. Rev. Lett. **54** (1985) 502–505; "Heterotic string theory (I). The free heterotic string", Nucl. Phys. **B256** (1985) 253–284; "Heterotic string theory (II). The interacting heterotic string", Nucl. Phys. **B267** (1986) 75–124.

[14] L. Dixon, J.A. Harvey, C. Vafa and E. Witten, "Strings on orbifolds", Nucl. Phys. **B261** (1985) 678–686.

[15] P.Goddard, W. Nahm, D. Olive and A. Schwimmer, "Vertex operators for non-simply-laced algebras", Commun. Math. Phys. **107** (1987) 179–212.

[16] D. Bernard and Jean Thierry-Mieg, "Level one representations of the simple affine Kac-Moody algebras in their homogeneous gradations", Commun. Math. Phys. **111** (1987) 181–246.

Inventiones math. 62, 23-66 (1980)

Inventiones mathematicae

Basic Representations of Affine Lie Algebras and Dual Resonance Models

I.B. Frenkel[1] and V.G. Kac[2]

[1] Department of Mathematics, Yale University New Haven, Conn. 06520, USA
[2] Department of Mathematics, MIT, Cambridge, MA 02139, USA

Introduction

0.1.

The remarkable link between the structures of mathematics and physics is exemplified by numerous instances in the development of these two disciplines. This link has remained ever mysterious. It comes up in the most unexpected situations and only further developments shed light on the meaning of that which is hidden behind the formal coincidence. The present article gives one more example of such a link between a mathematical and a physical theory.

Our original goal was to construct explicitly the simplest non-trivial highest weight representation (the so-called basic representation) of a Kac-Moody Lie algebra with the affine Cartan matrix. This seemed to be possible to do for two reasons. First, an explicit character formula is known [6] and second, there exists a construction of this representation [8], which is based on the specialization of the general character formula from [5]. Unfortunately the latter construction is very "inhomogeneous" (and has nothing to do with the explicit character formula). For a long time each of the authors had been trying unsuccessfully to attack the problem until they turned to physics where the answer had been found. It turned out that the basic representation can be described in terms of the vertex operators which play a crucial role in the dual resonance theory! H. Garland noticed the similarity between the operators in the "non-homogeneous" construction of the basic representation and the vertex operators. But that was only a similarity. In our construction the coincidence is complete, and this allows us to include the dual resonance models in the framework of the theory of affine Lie algebras. We hope that the mathematical instrument of affine Lie algebras will have useful physical applications, thereby repaying the debt.

Affine Lie algebras (which are alternatively known as Kac-Moody Lie algebras with an affine Cartan matrix, and also as loop algebras) form an important class of infinite dimensional Lie algebras. These algebras together

0020-9910/80/0062/0023/$08.80

with the Lie algebra of vector fields on the circle essentially exhaust all infinite-dimensional Lie algebras which admit a $\mathbb{Z}$-gradation by subspaces of bounded dimension and have no graded ideals [4]. The structure of affine Lie algebras is similar to that of simple finite dimensional Lie algebras, which permit to generalize many results of the classical theory [4], [11], [2]. On the other hand, the quotient of an affine Lie algebra by a 1-dimensional center is isomorphic to the Lie algebra of polynomial maps from the circle into a simple Lie algebra. This allows one to study the affine Lie algebras from a different point of view. In particular, they admit an extension by a Lie algebra of vector fields on the circle. As a result the Virasoro algebra, which is a central extension of the latter Lie algebra, operates on the space of a highest weight representation. This gives some information about representations of the Virasoro algebra [7], which are important in physical applications.

0.2.

It is interesting to follow up the history of the discoveries in the dual resonance theory, in particular the discovery of the vertex operators. The details can be found in surveys, e.g. [10], [12]. It is remarkable that most of the objects introduced in this theory naturally appear in the context of affine Lie algebras.

From the experimental data certain dependence between the characteristics of elementary particles was known. For instance, the dependence between the square mass and the spin turned out to be roughly linear in shape. The graph of the corresponding dependence is called the Regge trajectories. The dual resonance model starts from an approximation where all Regge trajectories are linear and all resonances infinitely narrow. In 1968 (i.e., roughly at the same time when investigations of Kac-Moody Lie algebras were begun) Veneziano discovered a four-point crossing-symmetric scattering amplitude with linear Regge trajectories. This initiated a great number of further investigations. In 1969 Koba and Nielsen and others generalized the Veneziano model to the case of any number of particles. Immediately after that several authors (Fubini-Gordon-Veneziano, Namby, Susskind) discovered the operator formalism which provides a powerful tool for studying dual-resonance models. The main object of this formalism is the so-called vertex operator $X(\gamma, z)$ and the most important observation was the fact that the Veneziano-Koba-Nielson amplitude can be expressed as an integral of the spherical function of the operators product $X(\gamma_1, z_1)\dots X(\gamma_N, z_N)$.

The duality property of the model follows immediately from the transformation law of the vertex operator $X(\gamma, z)$ under projective transformations of the complex variable z. In order to examine the effect of these transformations Gliozzi and Chin-Matsuda-Rebbi introduced the operators D_k, $k = -1, 0, 1$, which form a Lie algebra $sl_2(\mathbb{C})$. After that Virasoro in 1970 extended the definition of the Gliozzi operators to all integral values of k. These operators were used (Brower and Goddard-Thorn, 1972) to derive Ward-like identities which play a crucial role in showing that the physical states have positive norm (i.e., absence of "ghosts"). By 1974 the results of the dual resonance theory were summarized in several reviews ([10], [12] and others). We hope that further development of mathematical theory will clarify some physical points. One example of such a kind is a recent direct proof of Thorn of the "no-ghosts" theorem. He used the formula for the determinant of the contravariant form for the Virasoro algebra, which has been obtained in [7] in the framework of affine Lie algebras.

0.3.

We state now some main results of the paper. First of all, we recall the definition of the *affine Lie algebra* $\hat{\mathfrak{g}}$ associated with a complex finite dimensional simple Lie algebra $\mathfrak{g}$. Let $\mathfrak{h}$ be a Cartan subalgebra of $\mathfrak{g}$, Δ be the root system and Q be the lattice in $\mathfrak{h}^*$ generated by Δ. Let $\langle , \rangle$ denote an invariant bilinear form on $\mathfrak{g}$, normalized in such a way that the square length of a long root is 2. We identify $\mathfrak{h}^*$ and $\mathfrak{h}$ by the form $\langle , \rangle$. The Lie algebra $\hat{\mathfrak{g}}$ is the complex vector space

$$\hat{\mathfrak{g}}:=(\mathbb{C}[t,t^{-1}]\otimes\mathfrak{g})\oplus\mathbb{C}c$$

provided with the bracket

$$[x_1\oplus a_1 c, x_2\oplus a_2 c]=[x_1,x_2]_0\oplus \operatorname{Res}\left\langle\frac{dx_1}{dt}, x_2\right\rangle c,$$

where $x_1, x_2\in\mathbb{C}[t,t^{-1}]\otimes\mathfrak{g}$, $[,]_0$ denotes the bracket induced from $\mathfrak{g}$, $a_1, a_2\in\mathbb{C}$.

The *basic representation* π in a vector space V of an affine Lie algebra $\hat{\mathfrak{g}}$ is defined by the properties that it is irreducible and there exists a non-zero vector $v\in V$ such that

$$\pi(\mathbb{C}[t]\otimes_{\mathbb{C}}\mathfrak{g})v=0 \quad\text{and}\quad \pi(c)v=v.$$

This is the simplest among the irreducible highest weight representations, which can be exponentiated to the corresponding "loop group". The systematic study of the basic representations has been started in [6]. In this paper for $\mathfrak{g}$ of one of the types A_n, D_n, E_6, E_7, E_8 a character formula which determines explicitly the dimensions of the weight spaces was found. Starting from this formula we obtain an explicit construction of the basic representation.

Our construction is based on the notion of a *Heisenberg system*. An example of such a system is the pair $(\hat{\mathfrak{s}}, Q)$, where

$$\hat{\mathfrak{s}}=(\mathbb{C}[t,t^{-1}]\otimes_{\mathbb{C}}\mathfrak{h})\oplus\mathbb{C}c$$

is a subalgebra in $\hat{\mathfrak{g}}$ (isomorphic to a direct sum of $\mathfrak{h}$ and an infinite Heisenberg algebra) and Q is the root lattice of $\mathfrak{g}$.

We construct the *canonical representation* π_0 of the Heisenberg system $(\hat{\mathfrak{s}}, Q)$. We set $\mathfrak{s}_-=t^{-1}\mathbb{C}[t^{-1}]\otimes\mathfrak{h}\subset\hat{\mathfrak{s}}$; let $S(\mathfrak{s}_-)$ be the symmetric algebra of the space $\mathfrak{s}_-$ (this is the so-called Fock space) and let $\mathbb{C}(Q)$ be the group algebra of the lattice Q. We set $V=S(\mathfrak{s}_-)\otimes_{\mathbb{C}}\mathbb{C}(Q)$ and define the representation π_0 as follows. For an integer $k\neq 0$ and $h\in\mathfrak{h}$, $\pi_0(t^k\otimes h)$ acts trivially on $\mathbb{C}(Q)$, and acts as the operator of multiplication by $t^k\otimes h$ on $S(\mathfrak{s}_-)$ for $k<0$ (the creation operator), and as the derivation of the algebra $S(\mathfrak{s}_-)$ defined by:

$$\pi_0(t^k\otimes h)(t^{-k_1}\otimes h_1):=k\delta_{k,-k_1}\langle h,h_1\rangle,$$

when $k>0$ (the annihilation operator). On the other hand $\pi_0(1\otimes h)$ acts trivially on $S(\mathfrak{s}_-)$ and acts as the derivation ∂_h of the algebra $\mathbb{C}(Q)$ defined by

$$\partial_h e^\gamma := \gamma(h)\, e^\gamma, \qquad \gamma\in Q.$$

Finally, we define a projective representation of the group Q in $\mathbb{C}(Q)$ by:

$$\pi_0(\gamma)\, e^\beta := \varepsilon(\beta,\gamma)\, e^{\beta+\gamma},$$

where ε is a 2-cocycle of the group Q with values in $\{\pm 1\}$ satisfying two additional properties:

a) $\varepsilon(\alpha,\beta)\,\varepsilon(-\alpha,-\beta)=e^{\pi i\langle\alpha,\beta\rangle}$.

b) $\varepsilon(\alpha,-\alpha)=\varepsilon(\alpha,0)=1$.

Now, for any $\gamma\in Q$ and any non-zero complex number z we define a *vertex operator* $X(\gamma,z)$ by:

$$X(\gamma,z):=\exp\Big(\sum_{k\geq 1}\frac{z^k}{k}\gamma(-k)\Big)\exp(\gamma+(\ln z)\partial_\gamma)\exp\Big(-\sum_{k\geq 1}\frac{z^{-k}}{k}\gamma(k)\Big),$$

where $\gamma(k)$ denotes $\pi_0(t^k\otimes h_\gamma)$, $k\in\mathbb{Z}\setminus 0$. Exactly in this form the vertex operator appears in physics. The operator $X(\gamma,z)$ is a linear map from the space V into its completion $\bar V$. However, developing this operator by powers of z we obtain:

$$X(\gamma,z)=\sum_{k\in\mathbb{Z}} X_k(\gamma)\,z^k,$$

where each $X_k(\gamma)$ maps V into itself.

One can choose the Chevalley basis in $\mathfrak{g}$ (of ADE-type) such that (see Proposition 2.2):

$$[E_\gamma,E_{-\gamma}]=H_\gamma;\quad [H_\beta,E_\gamma]=\langle\beta,\gamma\rangle E_\gamma;\quad [H_\beta,H_\gamma]=0;$$

$$[E_\beta,E_\gamma]=0\quad\text{if }\beta+\gamma\notin\Delta\cup\{0\};\quad [E_\beta,E_\gamma]=\varepsilon(\beta,\gamma)\,E_{\beta+\gamma}\quad\text{if }\beta+\gamma\in\Delta.$$

Now we are in a position to state our main result.

Theorem A. *Let $\hat{\mathfrak{g}}$ be the affine Lie algebra associated with the Lie algebra $\mathfrak{g}$. Then the following formulas define a representation π of the Lie algebra $\hat{\mathfrak{g}}$ in the space V, which is equivalent to the basic representation:*

$$\pi(x)=\pi_0(x)\quad\text{for } x\in\tilde{\mathfrak{s}}\subset\hat{\mathfrak{g}};$$

$$\pi(E_\gamma\otimes t^k)=X_k(\gamma)\,c_\gamma,\qquad k\in\mathbb{Z},\ \gamma\in\Delta,$$

where $c_\gamma(e^\beta):=\varepsilon(\beta,\gamma)\,e^\beta$.

The proof of the theorem consists of two parts. First, we notice that the action of Q on $\hat{\mathfrak{g}}$ defined by $T_\gamma(t^k\otimes E_\beta)=t^{k-\langle\beta,\gamma\rangle}\otimes E_\beta$ gives rise to a projective representation of Q in the space of the basic representation. Next, the restriction of π to $\tilde{\mathfrak{s}}$ together with the latter representation of Q define a representation of the Heisenberg system $(\tilde{\mathfrak{s}}, Q)$, which turns out to be irreducible. Here we use the "explicit" character formula. This allows us to identify the spaces of the basic and the canonical representations.

The second step of the proof is to extend the canonical representation to the whole Lie algebra $\hat{\mathfrak{g}}$. For that we introduce the "Mellin transform" $\hat{E}_\gamma(z)$ of the element $E_\gamma \in \mathfrak{g}$, $z \in \mathbb{C} \setminus 0$, by

$$\hat{E}_\gamma(z) := \sum_{k \in \mathbb{Z}} z^{-k}(t^k \otimes E_\gamma).$$

$\hat{E}_\gamma(z)$ maps V into its completion $\bar{V}$. The commutation relations of $\hat{E}_\gamma(z)$ with the Heisenberg system $\{\tilde{\mathfrak{s}}, Q\}$ define the action of $\hat{E}_\gamma(z)$ uniquely, and it turns out that $\pi(\hat{E}_\gamma(z))$ coincides with the operator $X(\gamma, z)c_\gamma$!

Let ω be an antilinear involution of the Lie algebra $\mathfrak{g}$, the fixed point set of which is a compact form $\mathfrak{k}$ of $\mathfrak{g}$. We can define then an antilinear involution $\hat{\omega}$ of the affine Lie algebra $\hat{\mathfrak{g}}$ by:

$$\hat{\omega}(t^k \otimes x) = t^{-k} \otimes \omega(x), \qquad \hat{\omega}(c) = -c.$$

The fixed point set of $\hat{\omega}$ is called a *compact form* $\hat{\mathfrak{k}}$ of the Lie algebra $\hat{\mathfrak{g}}$. It follows from [2] that there exists a unique, up to a scalar multiple, positive definite Hermitian form in the space of the basic representation, which is invariant with respect to the action of $\hat{\mathfrak{k}}$. We construct such a form explicitly.

Let $u_1, \ldots, u_n$ be an orthonormal basis of $\mathfrak{h}_{\mathbb{R}}$. We order the basis of $\mathfrak{s}_-$:

$$k^{-\frac{1}{2}} t^k \otimes u_m, \qquad k = 1, 2, \ldots, m = 1, \ldots, n,$$

in some way: $v_1, v_2, \ldots$. We define a positive definite Hermitian form $\langle | \rangle$on V by the property that the monomials

$$\frac{v_1^{k_1}}{(k_1!)^{\frac{1}{2}}} \cdots \frac{v_s^{k_s}}{(k_s!)^{\frac{1}{2}}} e^\gamma, \qquad k_i = 1, 2, \ldots, \gamma \in Q,$$

form an orthonormal basis of V.

Our next result is

Theorem B. *The Hermitian form $\langle | \rangle$ on V is invarint with respect to the action of* $\hat{\mathfrak{k}}$.

Theorems A and B appear as statements a) and d) of Theorem 1 in the paper.

Now, as in the dual resonance theory, we introduce the Virasoro operators D_k, $k \in \mathbb{Z}$, in the space V by:

$$D_0 = -\sum_{k \geq 1} \sum_{i=1}^{n} u_i(-k) u_i(k) - \tfrac{1}{2} \sum_{i=1}^{n} u_i(0)^2,$$

$$D_m = -\tfrac{1}{2} \sum_{k \in \mathbb{Z}} \sum_{i=1}^{n} u_i(-k) u_i(k+m), \qquad m \in \mathbb{Z} \setminus 0.$$

where (as before) $u(k)$ denotes $\pi_0(t^k \otimes u)$.

In the framework of the basic representation the Virasoro operators receive a very simple interpretation. Let d_k denote the derivation of the Lie algebra $\hat{\mathfrak{g}}$ which is defined as $t^{k+1}\frac{d}{dt}$ on the subspace $\mathbb{C}[t,t^{-1}]\otimes\mathfrak{g}$ and sends c to 0. Notice that $\mathscr{D}=\sum_k \mathbb{C}d_k$ is the complexification of the Lie algebra of polynomial vector fields on the circle.

Theorem C. *The commutation with a Virasoro operator D_k preserves the subalgebra of operators $\pi(\hat{\mathfrak{g}})$ and induces on $\hat{\mathfrak{g}}$ the derivation d_k. Assigning $d_k \to D_k$ defines a projective representation of the Lie algebra $\mathscr{D}$ in the space V. More precisely, one has:*

$$[D_k, D_m]=(m-k)\,D_{m+k}+\frac{n}{12}(k^3-k)\,\delta_{k,-m}.$$

Theorem C appears as Proposition 2.8 and formula (2.30).

0.4.
Here is a Brief Account of the Contents of the Paper

In § 1 we describe in detail the structure of affine Lie algebras $\hat{\mathfrak{g}}$. Many of these results are now well-known and we represent them here for the convenience of the reader (having in mind also the physicists). We begin with the definition of an algebra $\tilde{\mathfrak{g}}$ which enlarges $\hat{\mathfrak{g}}$ by the derivation d_0 (see Section 1.1) and has some advantages compared with $\hat{\mathfrak{g}}$. In Sect 1.2 we construct explicitly the canonical invariant bilinear form $\langle\ ,\ \rangle$ on $\tilde{\mathfrak{g}}$. In Sect. 1.3 we describe the root system and the root space decomposition of $\tilde{\mathfrak{g}}$ and introduce some important subalgebras of $\tilde{\mathfrak{g}}$ such as the Cartan subalgebra, the Heisenberg subalgebra, the Borel subalgebra, etc. We explain why the affine Lie algebras are the Kac-Moody Lie algebras with an affine Cartan matrix in Sect. 1.4. In Sect. 1.5 we introduce an antilinear involution $\tilde{\omega}$, define the compact form $\tilde{\mathfrak{k}}$ of $\tilde{\mathfrak{g}}$ and establish the properties of the associated Hermitian form. In Sects. 1.6 and 1.7 we define and study the action of some extension of the Weyl group of $\tilde{\mathfrak{g}}$ in the $\tilde{\mathfrak{g}}$-modules from the category $\mathscr{K}$. This category is wide enough to include the heighest weight representations and the adjoint representation as well. In Sect. 1.8 we introduce the general notion of a Heisenberg system and in Sect. 1.9 define the "Mellin embedding" of $\mathfrak{g}$ in some completion of $\tilde{\mathfrak{g}}$.

We give an explicit construction of the basic representation in § 2. In Sect. 2.1 we study the weight system of the highest weight representations. In Sect. 2.2 we give an explicit description of the weight system of the basic representations. This is the point where we use the fact that the Cartan matrix is symmetric. It seems that in the cases of non-symmetric Cartan matrices the structure of the basic representation is more complicated (cf. [9], Example 2). We are planning to work on this matter in the future.

In Sect. 2.3 we describe in detail the action of the subgroup of translations of the Weyl group of $\tilde{\mathfrak{g}}$ in the basic representation (Proposition 2.3). Surprisingly, this gives a projective representation of the group Q with a nontrivial cocycle, and moreover, this cocycle, restricted to the root system, turns out to be nothing else but the structure constants of a Chevalley basis of $\mathfrak{g}$. As a side result (Proposition 2.2) we obtain a simple construction of the Chevalley basis of a simple Lie algebra with a symmetric Cartan matrix.

In Sect. 2.4 we introduce the canonical representation of a Heisenberg system and prove a version of the Stone-von Neumann theorem (Proposition 2.4). This allows us to make the first step of the proof of the main theorem (Proposition 2.5). The crucial point of the second step is Proposition 2.6, which is proven in Sect. 2.5. A special case of this proposition is treated in [8]. Proposition 2.6 leads to the definition of the vertex operators (Sect. 2.6), which had appeared before in physics literature.

In Sect. 2.7 we state our main theorem (Theorem 1) and complete its proof. We introduce the Virasoro operators in Sect. 2.8 and give their interpretation described above by Theorem C. In the last Sect. 2.9 we describe the action of the subalgebra $\mathfrak{g}$ and its universal enveloping algebra $U(\mathfrak{g})$ in the coherent states basis. It turns out that the Veneziano and the Koba-Nielson amplitudes are the matrix coefficients of the elements of $\mathfrak{g}$ and $U(\mathfrak{g})$, respectively, in our basis. So to some extent our exposition follows the reverse order of the history of the dual resonance theory!

In § 3 we transfer our constructions to the analytic language. This allows us to describe explicitly the action of several groups corresponding to subalgebras of $\tilde{\mathfrak{g}}$ and of $\mathscr{D}$ and give a direct proof of the main theorem. The space $\mathscr{V}$ of the basic representation turns out to be the subspace of $\mathscr{L}^2(\mathscr{H}, \mu)$, where $\mathscr{H}$ is a Hardy space of $\mathfrak{h}$-valued functions on the unit circle provided with a Gaussian measure μ (Sect. 3.1). In the next Sect. 3.2 we find the action of the Heisenberg group corresponding to the Heisenberg system $(\mathfrak{s}, Q)$. The formulas for this action (see Proposition 3.3) coincide with an example of the Vershik-Gelfand-Graev construction of the representations of the group G^X (cf. [16], p. 333). An interesting open problem is to extend these formulas to the whole group.

In Sect. 3.3 we describe the action of the vertex operators. As one can expect from the dual resonance theory the vertex operators create new poles (see formula (3.7)). In Sect. 3.4 we use simple facts about residues to prove explicit formulas for commutation relations between the vertex operators. This gives a new proof of our main theorem (Theorem 1').

In the next two Sects. 3.5 and 3.6 we describe the action of the Möbius transformations in the space $\mathscr{V}$ (Propositions 3.5 and 3.6). It is interesting that this representation arises from the important representation of the group $SU(1,1)$, which can be described as a limit of supplementary series twisted by a cocycle. The latter representation plays a crucial role in another Vershik-Gelfand-Graev construction related to the group $SU(1,1)^X$ [15]. At the end of our paper (Sect. 3.7) we sketch an alternative approach to our results using the language of current algebras, which is profoundly used in physics. We get here the projective properties of the Koba-Nielson amplitude and as a special case the Veneziano amplitude, which implies their duality properties. So, our last point is the starting point of the dual resonance theory which has inspired our construction.

Acknowledgements. We are grateful to J. Goldstone and C. Thorn for a number of consultations in dual resonance theory. These discussions gave us the idea to incorporate the vertex operators in our constructions. We are grateful to I. Bars for his computations of the commutation relations of the vertex operators based on our main theorem, which gives one more proof of our result. One of the authors (I.F.) wishes to thank A.M. Vershik for discussions of his papers [15] and [16] and for the debates concerning the construction of the projective representations of the group G^{S^1} isomorphic to the basic representations.

1. Affine Lie Algebras

1.1. Definition. Let $\mathbb{C}[t, t^{-1}]$ be the algebra of Laurent polynomials in the indeterminate t over the complex field. We recall that for a Laurent polynomial $P = \sum c_k t^k$ the *residue* is defined by $\operatorname{Res} P = c_{-1}$.

Let $\mathfrak{g}$ be a complex simple finite dimensional Lie algebra. Let $\langle\ ,\ \rangle$ denote an invariant bilinear symmetric form on $\mathfrak{g}$, normalized in such a way that the square length of a long root is equal to 2. We will call it the *standard bilinear form.*

The associated with $\mathfrak{g}$ *affine Lie algebra* $\tilde{\mathfrak{g}}$ is a complex infinite dimensional Lie algebra which is constructed as follows. First, consider a complex infinite dimensional Lie algeba $\bar{\mathfrak{g}} = \mathbb{C}(t, t^{-1}] \otimes_{\mathbb{C}} \mathfrak{g}$. This algebra can be naturally identified with the Lie algebra of polynomial maps of $\mathbb{C} \setminus 0$ into $\mathfrak{g}$. The invariant form on $\mathfrak{g}$ can be extended in a natural way to a bilinear $\mathbb{C}[t, t^{-1}]$-valued form on $\bar{\mathfrak{g}}$, which we denote by $\langle\ ,\ \rangle_t$. Also, any derivation D of $\mathbb{C}[t, t^{-1}]$ can be extended to a derivation of $\bar{\mathfrak{g}}$ by $D(P \otimes g) = D(P) \otimes g$.

We define a $\mathbb{C}$-valued bilinear form ψ on $\bar{\mathfrak{g}}$ by

$$\psi(x, y) = \operatorname{Res} \left\langle \frac{dx}{dt}, y \right\rangle_t .$$

A direct verification shows that

(i) $\psi(x, y) = -\psi(y, x)$ and

(ii) $\psi([x, y], z) + \psi([y, z], x) + \psi([z, x], y) = 0.$

In other words, ψ is a 2-cocycle, and therefore, we can define the corresponding 1-dimensional central extension of $\bar{\mathfrak{g}}$ which is denoted by $\hat{\mathfrak{g}}$. It is convenient to enlarge the algebra $\hat{\mathfrak{g}}$ by adding a derivation d which operates on $\bar{\mathfrak{g}}$ as $t\frac{d}{dt}$ and sends the center to 0. The complex Lie algebra thus obtained is the *affine algebra* $\tilde{\mathfrak{g}}$.

In more detail, $\tilde{\mathfrak{g}}$ is the complex space:

$$\tilde{\mathfrak{g}} = (\mathbb{C}[t, t^{-1}] \otimes_{\mathbb{C}} \mathfrak{g}) \oplus \mathbb{C}c \oplus \mathbb{C}d$$

with the following bracket:

$$[x \oplus \alpha c \oplus \beta d, x_1 \oplus \alpha_1 c \oplus \beta_1 d] = \left([x, x_1]_0 + \beta t \frac{dx_1}{dt} - \beta_1 t \frac{dx}{dt}\right) \oplus \left(\operatorname{Res} \left\langle \frac{dx}{dt}, x_1 \right\rangle_t\right) c. \tag{1.1}$$

Here $x, x_1 \in \bar{\mathfrak{g}}$, $[x, x_1]_0$ is the bracket in the Lie algebra $\bar{\mathfrak{g}}$ and $\alpha, \alpha_1, \beta, \beta_1 \in \mathbb{C}$.

It is clear from (1.1) that $1\otimes_{\mathbb{C}}\mathfrak{g}$ is a subalgebra in $\tilde{\mathfrak{g}}$; we identify $\mathfrak{g}$ with this subalgebra.

1.2. The Invariant Bilinear Form on $\tilde{\mathfrak{g}}$. We introduce a $\mathbb{C}$-valued bilinear form $\langle\ ,\ \rangle$ on $\tilde{\mathfrak{g}}$ by:

$$\langle x\oplus\alpha c\oplus\beta d,\ x_1\oplus\alpha_1 c\oplus\beta_1 d\rangle=\operatorname{Res}(t^{-1}\langle x, x_1\rangle_t)+\alpha\beta_1+\alpha_1\beta. \tag{1.2}$$

It is easy to see that this bilinear form is symmetric, non-degenerate and invarint. We check here that $\langle\ ,\ \rangle$ is invariant. Let

$$X_i=x_i\oplus\alpha_i c\oplus\beta_i d\in\tilde{\mathfrak{g}},\qquad i=1,2,3.$$

One has:

$$\langle[X_1,X_2],X_3\rangle=\left\langle\left([x_1,x_2]_0+\beta_1 t\frac{dx_2}{dt}-\beta_2 t\frac{dx_1}{dt}\right)\oplus\psi(x_1,x_2)c,\ x_3\oplus\alpha_3 c\oplus\beta_3 d\right\rangle$$

$$\begin{aligned}&=\langle[x_1,x_2]_0,x_3\rangle+\beta_1\operatorname{Res}\left\langle\frac{dx_2}{dt},x_3\right\rangle_t\\&\quad+\beta_2\operatorname{Res}\left\langle\frac{dx_3}{dt},x_1\right\rangle_t+\beta_3\operatorname{Res}\left\langle\frac{dx_1}{dt},x_2\right\rangle_t\\&=\langle X_1,[X_2,X_3]\rangle.\end{aligned}$$

We used the fact that $\operatorname{Res}\left\langle\frac{dx_1}{dt},x_3\right\rangle=-\operatorname{Res}\left\langle\frac{dx_3}{dt},x_1\right\rangle_t$ and the invariance of the form $\langle\ ,\ \rangle_t$.

Note also that the restriction of the form $\langle\ ,\ \rangle$ to the subalgebra $\mathfrak{g}\subset\tilde{\mathfrak{g}}$ induces the standard bilinear form on $\mathfrak{g}$. We also call the form $\langle\ ,\ \rangle$ on $\tilde{\mathfrak{g}}$ the ***standard bilinear form***. We can consider the elements from $\mathfrak{g}$ as linear functionals on $\tilde{\mathfrak{g}}$ with respect to the standard form. We will identify $\tilde{\mathfrak{g}}$ with a linear subspace of $\tilde{\mathfrak{g}}^*$.

1.3. The Root Decomposition of $\tilde{\mathfrak{g}}$ *and Some Important Subalgebras in* $\tilde{\mathfrak{g}}$. Let $\mathfrak{h}$ denote a Cartan subalgebra of $\tilde{\mathfrak{g}}$. Let $\mathfrak{g}=\mathfrak{h}\oplus\sum_{\alpha\in\Delta}\mathfrak{g}_\alpha$ be the root decomposition of $\mathfrak{g}$ with respect to $\mathfrak{h}$; here $\Delta\subset\mathfrak{h}^*$ is the system of roots. We fix a subsystem of positive roots $\Delta_+\subset\Delta$; let $\Pi=\{\alpha_1,\dots,\alpha_n\}$ be the system of simple roots and let $\tilde{\alpha}$ be the highest root.

Consider the following subalgebra in $\tilde{\mathfrak{g}}$:

$$\tilde{\mathfrak{h}}=\mathfrak{h}\oplus\mathbb{C}c\oplus\mathbb{C}d.$$

Clearly, this is a maximal commutative diagonalizable subalgebra in $\tilde{\mathfrak{g}}$. It is called a *Cartan subalgebra* of $\tilde{\mathfrak{g}}$. As usual for $\alpha\in\tilde{\mathfrak{h}}^*$ the attached *root space* is

$$\tilde{\mathfrak{g}}_\alpha=\{x\in\tilde{\mathfrak{g}}\,|\,[h,x]=\alpha(h)x,\ h\in\tilde{\mathfrak{h}}\},$$

and α is called a *root* if $\tilde{\mathfrak{g}}_\alpha\neq0$. We extend any linear function $\lambda\in\mathfrak{h}^*$ to a linear function on $\tilde{\mathfrak{h}}$, which we also denote by λ, setting $\lambda(c)=\lambda(d)=0$. Let δ be the linear function on $\tilde{\mathfrak{h}}$ defined by $\delta|_{\mathfrak{h}+\mathbb{C}c}=0$, $\delta(d)=1$.

The decomposition of $\tilde{\mathfrak{g}}$ into a sum of root spaces with respect to $\tilde{\mathfrak{h}}$ is

$$\tilde{\mathfrak{g}}=\tilde{\mathfrak{h}}\oplus\sum_{k,\gamma}(t^k\otimes_{\mathbb{C}}\mathfrak{g}_\gamma),$$

where (k,γ) ranges over $\mathbb{Z}\times(\Delta\cup 0)\setminus\{(0,0)\}$. Therefore the root system of $\tilde{\mathfrak{g}}$ with respect to $\tilde{\mathfrak{h}}$ is:

$$\tilde{\Delta}=\{k\delta+\gamma,\ \kappa\in\mathbb{Z},\ \gamma\in\Delta\cup 0\}\setminus\{0\}.$$

The root space $\tilde{\mathfrak{g}}_\alpha$, attached to a root $k\delta+\gamma$ is $\tilde{\mathfrak{g}}_{k\delta+\gamma}=t^k\otimes_{\mathbb{C}}\mathfrak{g}_\gamma$. So the multiplicity $\dim\tilde{\mathfrak{g}}_\alpha$ of a root $\alpha=k\delta+\gamma\in\tilde{\Delta}$ is 1 if $\gamma\neq 0$ and is n otherwise. A root $\alpha=k\delta+\gamma$ with $\gamma\in\Delta$ is called *real* and a root $\alpha=k\delta$, $k\in\mathbb{Z}\setminus 0$ is called *imaginary* (cf. [4]). The decomposition $\tilde{\mathfrak{g}}=\tilde{\mathfrak{h}}\oplus\sum_{\alpha\in\tilde{\Delta}}\tilde{\mathfrak{g}}_\alpha$ is called the *root space decomposition* of $\tilde{\mathfrak{g}}$.

From the properties of the standard form on $\mathfrak{g}$ one can immediately deduce the corresponding properties of the standard form on $\tilde{\mathfrak{g}}$:

$$\langle\ ,\ \rangle|_{\tilde{\mathfrak{h}}}\ \text{is non-degenerate;} \tag{1.3}$$

$$\langle\ ,\ \rangle|_{\tilde{\mathfrak{g}}_{-\alpha}\oplus\tilde{\mathfrak{g}}_\alpha}\ \text{is non-degenerate;} \tag{1.4}$$

$$\langle\tilde{\mathfrak{g}}_\alpha,\tilde{\mathfrak{g}}_\beta\rangle=0\quad \text{if } \alpha+\beta\neq 0. \tag{1.5}$$

Due to (1.3) we can identify $\tilde{\mathfrak{h}}$ and $\tilde{\mathfrak{h}}^*$. Note that $\langle k_1\delta+\gamma_1, k_2\delta+\gamma_2\rangle=\langle\gamma_1,\gamma_2\rangle$, and that a root $\alpha\in\tilde{\Delta}$ is real if and only if $\langle\alpha,\alpha\rangle\neq 0$ (in this case $\langle\alpha,\alpha\rangle>0$).

We define now a subsystem of *positive roots* $\tilde{\Delta}_+$ by

$$\tilde{\Delta}_+=\{k\delta+\gamma\,|\,\text{either } k>0,\text{ or } k=0,\ \gamma\in\Delta_+\}.$$

Then clearly, $\tilde{\Delta}=\tilde{\Delta}_+\cup(-\tilde{\Delta}_+)$. The corresponding system of *simple roots* is

$$\tilde{\Pi}=\{\alpha_0:=\delta-\tilde{\alpha}_1,\alpha_1,\ldots,\alpha_n\}.$$

In other words, the system $\alpha_0,\ldots,\alpha_n$ form a $\mathbb{Z}_+$-basis of $\tilde{\Delta}_+$.

The following subalgebras of $\tilde{\mathfrak{g}}$ can be regarded as straightforward generalizations of the maximal nilpotent and the Borel subalgebras of $\mathfrak{g}$:

$$\tilde{\mathfrak{n}}_-=\sum_{\alpha\in\tilde{\Delta}_+}\tilde{\mathfrak{g}}_{-\alpha},\quad \tilde{\mathfrak{n}}_+=\sum_{\alpha\in\tilde{\Delta}_+}\tilde{\mathfrak{g}}_\alpha,\quad \tilde{\mathfrak{b}}=\tilde{\mathfrak{h}}\oplus\tilde{\mathfrak{n}}_+.$$

We also introduce two important subalgebras which have no analog in the finite dimensional case. They play an important role in this paper:

$$\hat{\mathfrak{s}}=\mathbb{C}c\oplus\sum_{k\in\mathbb{Z}\setminus 0}(t^k\otimes_{\mathbb{C}}\mathfrak{h}),\quad \tilde{\mathfrak{s}}=\mathfrak{h}\oplus\hat{\mathfrak{s}}.$$

We call $\hat{\mathfrak{s}}$ the *homogeneous Heisenberg subalgebra* of $\tilde{\mathfrak{g}}$.

1.4. Connection with the Kac-Moody Lie Algebras. Set $a_{ij}=2\langle\alpha_i,\alpha_j\rangle/\langle\alpha_i,\alpha_i\rangle$, $i,j=0,1,\ldots,n$. The matrix $\tilde{A}=(a_{ij})_{i,j=0}^n$ is called the *Cartan matrix* of the Lie algebra $\tilde{\mathfrak{g}}$. This is at the same time the extended Cartan matrix of the Lie algebra $\mathfrak{g}$.

Let E_i, F_i, H_i, $i=1,2,\dots,n$, be the canonical generators[1] of the Lie algebra $\mathfrak{g}$. Choose $E_0\in\mathfrak{g}_{-\tilde{\alpha}}$ and $F_0\subset\mathfrak{g}_{\tilde{\alpha}}$ in such a way that $\tilde{\alpha}(H_0)=-2$, where $H_0=[E_0,F_0]$. Set

$$\begin{aligned} e_0&=t\otimes E_0, & e_i&=1\otimes E_i, & i&=1,\dots,n,\\ f_0&=t^{-1}\otimes F_0, & f_i&=1\otimes F_i, & i&=1,\dots,n,\\ h_0&=1\otimes H_0+c, & h_i&=1\otimes H_i, & i&=1,\dots,n. \end{aligned}$$

The elements e_i, f_i, h_i, $i=0,1,\dots,n$ are called the *canonical generators*. They generate the subalgebra $\hat{\mathfrak{g}}$ (of codimension 1) in $\tilde{\mathfrak{g}}$ and satisfy the following relations ($i,j=1,\dots,n$):

$$\begin{gathered} [e_i,f_j]=\delta_{ij}h_i,\quad [h_i,h_j]=0,\quad [h_i,e_j]=a_{ij},e_j,\quad [h_i f_j]=-a_{ij}f_j,\\ (ad\,e_j)^{1-a_{ij}}e_j=(ad f_i)^{1-a_{ij}}f_j=0\qquad i\neq j. \end{gathered}\tag{1.6}$$

Clearly the Lie algebra $\hat{\mathfrak{g}}$ does not contain graded (with respect to the root decomposition) ideals. Therefore $\hat{\mathfrak{g}}$ is the so-called Kac-Moody Lie algebra associated with the matrix $\tilde{A}$ (cf. [4], [11]). Relations (1.6) are defining relations for $\hat{\mathfrak{g}}$ (see [4], Proposition 13).

1.5. The Compact Form of $\tilde{\mathfrak{g}}$ (see also [2], [6]). Let $\tilde{\omega}$ be an involutive antilinear automorphism of the Lie algebra $\tilde{\mathfrak{g}}$ defined by: $\tilde{\omega}(e_i)=-f_i$, $\tilde{\omega}(f_i)=-e_i$, $\tilde{\omega}|\tilde{\mathfrak{h}}=-\mathrm{Id}$.

The subalgebra $\mathfrak{g}\subset\tilde{\mathfrak{g}}$ is invariant with respect to $\tilde{\omega}$ and it is well known that the real subalgebra $\mathfrak{g}_0\subset\mathfrak{g}$, which is a fixed point set of $\tilde{\omega}$ is a compact form of the Lie algebra $\mathfrak{g}$. Therefore, the involution $\tilde{\omega}$ can be defined in a more explicit way as follows. Let $\mathfrak{g}_0\subset\mathfrak{g}$ be a compact form of the Lie algebra $\mathfrak{g}$ and ω be the associated antilinear involution. Then $\tilde{\omega}$ is defined by:

$$\tilde{\omega}(\alpha c+\beta d+\sum_k t^k\otimes x_k)=-\bar{\alpha}c-\bar{\beta}d+\sum_k t^{-k}\otimes\omega(x_k).$$

The fixed point set $\tilde{\mathfrak{g}}_0$ of $\tilde{\omega}$ is a real subalgebra in $\tilde{\mathfrak{g}}$, and the complexification of $\tilde{\mathfrak{g}}_0$ is $\tilde{\mathfrak{g}}$. The real Lie algebra $\tilde{\mathfrak{g}}_0$ is called the *compact form* of the Lie algebra $\tilde{\mathfrak{g}}$.

We define a Hermitian form $\tilde{H}$ on $\tilde{\mathfrak{g}}$ by: $\tilde{H}(x,y)=-\langle x,\tilde{\omega}(y)\rangle$. Clearly one has: $\tilde{H}([u,x],y)=-\tilde{H}(x,[\tilde{\omega}(u),y])$, and hence $\mathrm{ad}\,u$ and $-\mathrm{ad}\,\tilde{\omega}(u)$ are adjoint to one another's operators on $\tilde{\mathfrak{g}}$ with respect to $\tilde{H}$. The restriction of $\tilde{H}$ to $\mathfrak{g}$, which we denote by H is also a Hermitian form satisfying $H([u,x],y)=-H(x,[\omega(u),y]$; in particular it is invariant on $\mathfrak{g}_0$.

Let $x=\alpha c+\beta d+\sum_k t^k\otimes x_k\in\tilde{\mathfrak{g}}$. Then one has

$$\tilde{H}(x,x)=\alpha\bar{\beta}+\bar{\alpha}\beta+\sum_k(-\langle x_k,\omega(x_k)\rangle).$$

All the summands in the second sum are positive since $\mathfrak{g}_0$ is a compact Lie algebra. Therefore the Hermitian form $\tilde{H}$ has in some basis the matrix diag$(-1,1,1,\dots)$, and its restriction to $\hat{\mathfrak{g}}\subset\tilde{\mathfrak{g}}$ is positive semidefinite.

---

[1] This means that $E_i\in\mathfrak{g}_{\alpha_i}$, $F_i\in\mathfrak{g}_{-\alpha_i}$, $H_i=[E_i,F_i]$ and $\alpha_i(H_i)=2$.

Consider now the Heisenberg subalgebra $\hat{\mathfrak{s}}$ and subalgebra $\tilde{\mathfrak{s}}=\mathfrak{h}\oplus\hat{\mathfrak{s}}$ of $\tilde{\mathfrak{g}}$. The involution $\tilde{\omega}$ leaves these subalgebras invariant. We denote by $\hat{\mathfrak{s}}_0$ and $\tilde{\mathfrak{s}}_0$, respectively, their intersections with $\mathfrak{g}_0$, and provide the subalgebras $\hat{\mathfrak{s}}$ and $\tilde{\mathfrak{s}}$ with a Hermitian form, defined as a restriction of the form $\tilde{H}$.

1.6. The Weyl Group of $\tilde{\mathfrak{g}}$. For a root $\alpha\in\tilde{\Delta}$ such that $\langle\alpha,\alpha\rangle\neq 0$ (real root) let $\alpha^\vee=2\alpha/\langle\alpha,\alpha\rangle$ denote the *dual* root. Let r_α be the reflection in the space $\tilde{\mathfrak{h}}^*$ with respect to α:

$$r_\alpha(\lambda)=\lambda-\langle\lambda,\alpha^\vee\rangle\alpha,\qquad \lambda\in\tilde{\mathfrak{h}}^*.$$

The group $\tilde{W}\subset GL(\tilde{\mathfrak{h}}^*)$ generated by all r_α such that $\alpha\in\tilde{\Delta}$, $\langle\alpha,\alpha\rangle\neq 0$ is called the *Weyl group* of the Lie algebra $\tilde{\mathfrak{g}}$.

Clearly the form $\langle\ ,\ \rangle|_{\tilde{\mathfrak{h}}^*}$ is $\tilde{W}$-invariant. Note also that any real root is a $\tilde{W}$-conjugate of a simple root and the line $\mathbb{C}\delta$ is the fixed point set for $\tilde{W}$. Usually we will write r_i instead of r_{α_i}, $i=0,1,\ldots,n$. The group $\tilde{W}$ is generated by r_i's, and the defining relations are $r_i^2=1$, $(r_ir_j)^{m_{ij}}=1$, where m_{ij} are given by the following table:

$a_{ij}a_{ji}$	0	1	2	3	4
m_{ij}	2	3	4	6	∞

To prove this we note that $\tilde{W}$ is the affine Weyl group of $\mathfrak{g}$. This becomes clear if we identify the linear span of $\Delta\cup\{\delta\}$ with space of affine linear functions on $\mathfrak{h}$, identifying δ with the function 1 (see e.g. [6], Proposition 3.3 for details).

Let W be the Weyl group of the Lie algebra $\mathfrak{g}$. We identify W with the subgroup in $\tilde{W}$ generated by reflections $r_1,\ldots,r_n$.

Let Q be the *root lattice* of $\mathfrak{g}$, i.e., the $\mathbb{Z}$-lattice generated by Δ, and let $Q^\vee$ denote the *dual root lattice*, i.e., lattice generated by $\alpha_i^\vee$, $i=1,2,\ldots,n$. It is well known that $Q^\vee$ contains all $\gamma^\vee$ for $\gamma\in\Delta$.

We set

$$T_i=r_{\delta-\alpha_i}r_i,\qquad i=1,\ldots,n.$$

For $\gamma=\sum k_i\alpha_i^\vee\in Q^\vee$ we set:

$$T_\gamma=T_1^{k_1}\ldots T_n^{k_n}.$$

One checks directly that:

$$T_\gamma(\lambda)=\lambda-(\langle\lambda,\gamma\rangle+\tfrac{1}{2}\langle\gamma,\gamma\rangle\langle\lambda,\delta\rangle)\delta+\langle\lambda,\delta\rangle\gamma \tag{1.7}$$

for any $\gamma\in Q^\vee$ and $\lambda\in\tilde{\mathfrak{h}}^*$. In particular,

$$T_\gamma(\alpha)=\alpha-\langle\alpha,\gamma\rangle\delta,\qquad \alpha\in\tilde{\Delta},\ \gamma\in Q^\vee. \tag{1.8}$$

The map $\gamma\to T_\gamma$ defines an embedding of the group $Q^\vee$ in $\tilde{W}$. This is a normal abelian subgroup in $\tilde{W}$ with free generators $T_1,\ldots,T_n$ and $\tilde{W}=W\times Q^\vee$ is a semidirect product of W and $Q^\vee$ (see e.g. [6], Proposition 3.4, for details).

1.7. The Category $\mathcal{K}$ *of* $\tilde{\mathfrak{g}}$*-Modules.* We denote by $\mathcal{K}$ the category of $\tilde{\mathfrak{g}}$-modules V satisfying the properties:

(i) $V=\sum_{\lambda\in\mathfrak{h}^*} V_\lambda$, where $V_\lambda=\{v\in V\,|\,h(v)=\lambda(h)\,v,\ h\in\tilde{\mathfrak{h}}^*\}$

are finite dimensional;

(ii) the elements e_i and f_i, $i=0,\ldots,n$, operate locally nilpotent on V.

Note that if V_1 is a submodule in a $\tilde{\mathfrak{g}}$-module V, then $V\in\mathcal{K}$ if and only if V_1 and $V/V_1\in\mathcal{K}$. Note also that the adjoint representation belongs to $\mathcal{K}$.

We fix in each $\mathfrak{g}_\gamma\subset\mathfrak{g}$, $\gamma\in\Delta$, an element E_γ such that $\gamma([E_\gamma, E_{-\gamma}])=2$ and $E_{\alpha_i}=E_i$, $i=1,\ldots,n$. For a real root $\alpha=k\delta+\gamma\in\Delta$, we set $E_\alpha=t^k\otimes E_\gamma\in\tilde{\mathfrak{g}}_\alpha$.

Let π be a representation from the category $\mathcal{K}$ of the Lie algebra $\tilde{\mathfrak{g}}$ in a space V. For a real root $\alpha\in\tilde{\Delta}$ we set:

$$r_\alpha^\pi=\exp-\pi(E_\alpha)\exp\pi(E_{-\alpha})\exp-\pi(E_\alpha).$$

Usually we write r_i^π instead of $r_{\alpha_i}^\pi$, $\alpha_i\in\tilde{\Pi}$.

Proposition 1.1. a) *The operator r_α^π is a well-defined automorphism of the space V such that*

(i) $r_\alpha^\pi(V_\lambda)=V_{r_\alpha(\lambda)}$;

(ii) $(r_\alpha^\pi)^2|_{V_\lambda}=\pm\mathrm{id}$;

(b) *Let W^π be the group generated by r_i^π, $i=0,\ldots,n$, and let A^π be the subgroup in W^π generated by $(r_i^\pi)^2$, $i=0,\ldots,n$. Then provided that* $\ker\pi\subset\mathbb{C}c$ *one has:*

(i) *A^π is a normal abelian subgroup of period 2;*

(ii) *the quotient group $\tilde{W}^\pi/A^\pi$ is isomorphic to $\tilde{W}$;*

(iii) *the map $r_i^\pi\mapsto r_i^{ad}$ induces an epimorphism ϕ_π: $\tilde{W}^\pi\to\tilde{W}^{ad}$ such that* $\ker\phi_\pi \subset\{\pm 1_V\}$ *if π is irreducible.*

(c) *For the adjoint representation one has:*

(i) *r_γ^{ad} is an automorphism of the Lie algebra $\tilde{\mathfrak{g}}$;*

(ii) *r_γ^{ad} preserves the standard bilinear form on $\tilde{\mathfrak{g}}$;*

(iii) *$\tilde{\mathfrak{h}}$ is r_α^{ad}-invariant and $r_\alpha^{ad}|_{\tilde{\mathfrak{h}}^*}=r_\alpha$;*

(iv) *$r_\alpha^{ad}(\tilde{\mathfrak{g}}_\beta)=\tilde{\mathfrak{g}}_{r_\alpha(\beta)}$, $\beta\in\tilde{\Delta}$.*

Proof. Set $\mathfrak{a}=\mathbb{C}e+\mathbb{C}h+\mathbb{C}f$. First, suppose that $\mathfrak{g}_\alpha$ and $\mathfrak{g}_{-\alpha}$ operate locally nilpotent on V. Then V as an $\mathfrak{a}$-module is a direct sum of finite-dimensional modules. Therefore, this module can be "integrated" to an $SL_2(\mathbb{C})$-module. We set $\sigma_\alpha=\left.\begin{pmatrix}0&1\\-1&0\end{pmatrix}\right|_V$ and check directly that $\sigma(v)=r_\alpha(v)$, $v\in V$. Now we apply a well known fact that $\sigma_\alpha(V_\lambda)=V_{r_\alpha(\lambda)}$ for a finite dimensional $SL_2(\mathbb{C})$-module. This proves a) under our assumption.

Now we prove that $\mathfrak{g}_\alpha$ and $\mathfrak{g}_{-\alpha}$ operate locally nilpotent on V. For a simple root α_i this is true since π is in the category $\mathcal{K}$. An arbitrary real root α can be transformed to a simple root α_i by an element $r_{i_1}\ldots r_{i_k}\in\tilde{W}$. Then $\tilde{\mathfrak{g}}_\alpha=\mathbb{C}r_{i_k}^\pi\ldots r_{i_1}^\pi(e_i)$, and therefore $\tilde{\mathfrak{g}}_\alpha$ operates locally nilpotent on V_1 which completes the proof of a). Since the adjoint representation belongs to the category $\mathcal{K}$, c) follows from a).

To prove b) note that under our hypothesis the conjugation by the operators from $\tilde{W}^\pi$ induces an epimorphism ϕ_π: $\tilde{W}^\pi\to\tilde{W}^{ad}$. Since the operators from A^π

are "diagonal" (see a)(ii)), we obtain that $\ker \phi_\pi|_{A^\pi} \subset \{\pm 1_V\}$. Due to c)(iii)the restriction of $\tilde{W}^{ad}$ to $\tilde{\mathfrak{h}} \cong \tilde{\mathfrak{h}}^*$ gives an epimorphism $\psi: \tilde{W}^{ad} \to \tilde{W}$ such that $\ker \psi = A^{ad}$. So $\psi \cdot \phi$ induces an epimorphism $\chi: \tilde{W}^\pi / A^\pi \to \tilde{W}$. To complete the proof of b) we have to show that χ is an isomorphism. For that it is sufficient to show that $(r_i^\pi r_j^\pi)^{m_{ij}} \in A^\pi$ if $m_{ij} < \infty$ (since $\tilde{W}$ is a Coxeter group). We employ the same argument as in a): the Lie algebra $\mathfrak{a}_1$, generated by e_i, e_j, f_i, f_j, is a finite dimensional semisimple Lie algebra of rank 2. The module V considered as an $\mathfrak{a}_1$-module is a direct sum of finite-dimensional modules, for which the relation in question is a well-known fact.

Proposition 1.2. a) *Let T_i^{ad} be an automorphism of the Lie algebra $\tilde{\mathfrak{g}}$ defined by:*

$$T_i^{ad} = r_{\delta-\alpha_i}^{ad} r_{\alpha_i}^{ad}, \qquad i = 1, \ldots, n.$$

Then the operators T_i^{ad} commute with each other.

b) *For $\gamma \in Q^\vee$, $\gamma = \sum k_i \alpha_i^\vee$, set*

$$T_\gamma^{ad} = (T_1^{ad})^{k_1} \ldots (T_n^{ad})^{k_n}.$$

Then

(i) *the map $\gamma \to T_\gamma^{ad}$ defines a representation of the group $Q^\vee$ in $\tilde{\mathfrak{g}}$;*

(ii) *T_γ^{ad} operates on $\tilde{\mathfrak{g}}$ as follows:*

$$T_\gamma^{ad}(t^k \otimes x_\beta) = t^{k-\langle \beta, \gamma \rangle} \otimes x_\beta, \qquad \beta \in \Delta \cup \{0\},\ x_\beta \in \mathfrak{g}_\beta,\ k\delta + \beta \neq 0, \tag{1.9}$$

$$T_\gamma^{ad}(h) = h - (\gamma(h) + \tfrac{1}{2}\langle \gamma, \gamma \rangle\, \delta(h))\, c + \delta(h)\, h_\gamma, \qquad h \in \tilde{\mathfrak{h}}. \tag{1.10}$$

in other words T_γ^{ad} induces on $\tilde{\mathfrak{h}}$ an operator adjoint to T_γ.
In particular

$$T_\gamma^{ad}(h) = h - \gamma(h)\, c, \qquad h \in \mathfrak{h}. \tag{1.11}$$

(iii) $T_{\gamma^\vee}^{ad} = r_{\delta-\gamma}^{ad} r_\gamma^{ad}$ *for $\gamma \in \Delta$.*

Proof. Formula (1.10) follows from (1.7) and Proposition 1.1c). It is sufficient to prove (1.9) only for $\gamma = \alpha_i^\vee$. Then a) also will be proven. We have:

$$T_\gamma^{ad}(t^k \otimes x_\beta) = c_{k,\beta}\, t^{k-\langle \beta, \gamma \rangle} \otimes x_\beta, \qquad c_{k,\beta} \in \mathbb{C} \setminus 0, \tag{1.12}$$

at least for $\beta \in \Delta$, $k \in \mathbb{Z}$. This follows from (1.8). Since the elements $t^k \otimes x_\beta$, $k \in \mathbb{Z}$, $\beta \in \Delta$, generate $\hat{\mathfrak{g}}$, (1.12) holds for any k, β such that $k\delta + \beta \neq 0$. Consider now a homomorphism $\phi: \tilde{\mathfrak{g}} \to \mathfrak{g}$ defined by:

$$\phi(t^k \otimes x) = x, \qquad k \in \mathbb{Z},\ x \in \mathfrak{g};\ \phi(c) = \phi(d) = 0.$$

Clearly, T_γ^{ad} preserves the kernel of ϕ and induces an identity automorphism on $\mathfrak{g}$ (see Proposition (1.1b)). This implies that in (1.12) all $c_{k,\beta}$ equal 1, which completes the proof of (1.9). The same argument shows that $r_{\delta-\gamma}^{ad} r_\gamma^{ad}$ operates on $\tilde{\mathfrak{g}}$ by formulas (1.9) and (1.0). This proves b)(iii).

1.8. Heisenberg Systems. As we shall see, some non-connected Heisenberg groups appear in our considerations. In order to deal with them it is convenient to introduce the notion of a Heisenberg system.

Definition. Let $\hat{\mathfrak{s}}$ be a complex Heisenberg Lie algebra, i.e., $\hat{\mathfrak{s}} = \mathbb{C}c \oplus \mathfrak{s}$, where $\mathbb{C}c$ is the 1-dimensional center of $\hat{\mathfrak{s}}$ and $\mathfrak{s}$ is a subspace (in general, infinite dimensional) provided with a non-degenerate alternate bilinear form ψ so that

$$[x, y] = \psi(x, y)\, c, \qquad x, y \in \mathfrak{s}.$$

Let $\mathfrak{h}$ be the n-dimensional complex vector space and $\Gamma \subset \mathfrak{h}^*$ be an n-dimensional lattice. Let $\tilde{\mathfrak{s}}$ be a direct sum of the Lie algebras $\hat{\mathfrak{s}}$ and $\mathfrak{h}$ (considered as a commutative Lie algebra). We define an action of the group Γ by automorphisms of the Lie algebra $\tilde{\mathfrak{s}}$ by:

$$T_\gamma(s \oplus h) = (s - \gamma(h)\, c) \oplus h, \qquad \gamma \in \Gamma,\ s \in \hat{\mathfrak{s}},\ h \in \mathfrak{h}. \tag{1.13}$$

The pair $(\hat{\mathfrak{s}}, \Gamma)$ is called a *Heisenberg system*.

Now we conisder an important example of a Heisenebrg system. Let Γ be an n-dimensional lattice with a real non-degenerate bilinear form $\langle\ ,\ \rangle$. We associate with this lattice a Heisenberg system $(\hat{\mathfrak{s}}, \Gamma)$ as follows. Set $\mathfrak{h} = (\Gamma \otimes_{\mathbb{Z}} \mathbb{C})^*$; the bilinear form $\langle\ ,\ \rangle$ on Γ induces a non-degenerate bilinear form on $\mathfrak{h}$ for which we keep the same notation. We set:

$$\tilde{\mathfrak{s}} = (\mathbb{C}[t, t^{-1}] \otimes_{\mathbb{C}} \mathfrak{h}) \oplus \mathbb{C}c.$$

We define a bilinear alternate form ψ on $\mathbb{C}[t, t^{-1}] \otimes_{\mathbb{C}} \mathfrak{h}$ by

$$\psi(t^{k_1} \otimes h_1, t^{k_2} \otimes h_2) = k_1 \delta_{k_1, -k_2} \langle h_1, h_2 \rangle.$$

Then $\tilde{\mathfrak{s}}$ becomes a Lie algebra with the bracket:

$$[x, y] = \psi(x, y)\, c, \qquad x, y \in \mathbb{C}[t, t^{-1}] \otimes_{\mathbb{C}} \mathfrak{h}; \qquad [c, \tilde{\mathfrak{s}}] = 0.$$

The Lie algebra $\tilde{\mathfrak{s}}$ is isomorphic to a direct sum of a commutative Lie algebra $\mathfrak{h}$ and a Heisenberg algebra $\hat{\mathfrak{s}} = \mathfrak{s} + \mathbb{C}c$, where $\mathfrak{s} = \sum_{k \neq 0} (t^k \otimes_{\mathbb{C}} \mathfrak{h})$. The space $\mathfrak{s}$ admits a canonical polarisation:

$$\mathfrak{s} = \mathfrak{s}_- \oplus \mathfrak{s}_+, \qquad \text{where } \mathfrak{s}_- = \sum_{k<0} t^k \otimes_{\mathbb{C}} \mathfrak{h},\ \mathfrak{s}_+ = \sum_{k>0} t^k \otimes_{\mathbb{C}} \mathfrak{h}.$$

The constructed pair $(\tilde{\mathfrak{s}}, \Gamma)$ is called a *Heisenberg system associated with the lattice* Γ.

One obtains immediately from Proposition 1.2 and the definition of the bracket (1.1) the following.

Proposition 1.3. *The subalgebra* $\tilde{\mathfrak{s}}$ *of* $\tilde{\mathfrak{g}}$ *is invariant with respect to the action* $\gamma \to T_\gamma^{ad}$ *of* $Q^\vee$ *on* $\tilde{\mathfrak{g}}$. *We restrict this action of* $Q^\vee$ *to* $\tilde{\mathfrak{s}}$. *Then the pair* $(\hat{\mathfrak{s}}, Q^\vee)$ *is a Heisenberg system associated with the the lattice* $Q^\vee$.

1.9. The Formal Mellin Transform. The dual space of the affine Lie algebra $\tilde{\mathfrak{g}}$ can be naturally identified with respect to the standard form $\langle\ ,\ \rangle$ with the following space:

$$\tilde{\tilde{\mathfrak{g}}} = (\mathbb{C}[[t, t^{-1}]] \otimes_{\mathbb{C}} \mathfrak{g}) \oplus \mathbb{C}c \oplus \mathbb{C}d$$

where $\mathbb{C}[[t, t^{-1}]]$ is the space of formal Laurent series in the indeterminate t. Of course $\bar{\bar{\mathfrak{g}}}$ is not a Lie algebra any more, but the action of $\tilde{\mathfrak{g}}$ on $\bar{\bar{\mathfrak{g}}}$ is well-defined. Also the action of r_γ^{ad} can be extended uniquely from $\tilde{\mathfrak{g}}$ to $\bar{\bar{\mathfrak{g}}}$. For $z \in \mathbb{C} \setminus 0$ and $x \in \mathfrak{g}$ we define the *formal Mellin transform* $\hat{x}(z)$ of x by:

$$\hat{x}(z) = \sum_{k \in \mathbb{Z}} z^{-k}(t^k \otimes x).$$

The following proposition is an immediate consequence of (1.1) and (1.9).

Proposition 1.4. *For $\gamma \in \Delta$ let $E_\gamma \in \mathfrak{g}$ be a root vector; let $z \in \mathbb{C} \setminus 0$ and let $\hat{E}_\gamma(z)$ be the formal Mellin transform of E_γ. Then:*

$$[t^k \otimes h, \hat{E}_\gamma(z)] = \gamma(h) z^{-k} \hat{E}_\gamma(z), \qquad h \in \mathfrak{h},\ k \in \mathbb{Z}.$$

$$T_\alpha^{ad} \hat{E}_\gamma(z) = z^{-\langle \alpha, \gamma \rangle} \hat{E}_\gamma(z), \qquad \alpha \in Q^\vee.$$

2. Basic Representations and Vertex Operators

2.1. Highest Weight Representations. Let $\tilde{\mathfrak{g}}$ be an affine Lie algebra, and let $\tilde{\mathfrak{h}}, \tilde{\mathfrak{n}}_+$ and $\tilde{\mathfrak{b}} = \tilde{\mathfrak{h}} \oplus \tilde{\mathfrak{n}}_+$ be its subalgebras introduced in Sect. 1.3. We recall that for $\Lambda \in \tilde{\mathfrak{h}}^*$ the *highest weight module* $V(\Lambda)$ is defined as an irreducible $\tilde{\mathfrak{g}}$-module for which there exists a non-zero vector $v_0 \in V(\Lambda)$ such that

$$\tilde{\mathfrak{n}}_+(v_0) = 0, \quad h(v_0) = \Lambda(h)\, v_0 \qquad \text{for } h \in \tilde{\mathfrak{h}}.$$

There exists a unique (up to isomorphism) such module for any $\Lambda \in \tilde{\mathfrak{h}}^*$ (see e.g. [6]); Λ is called the *highest weight* of the module $V(\Lambda)$. We denote the representation of $\tilde{\mathfrak{g}}$ in the space $V(\Lambda)$ by π_Λ.

The module $V(\Lambda)$ admits the weight decomposition with respect to $\tilde{\mathfrak{h}}$:

$$V(\Lambda) = \sum_{\lambda \in \tilde{\mathfrak{h}}^*} V(\Lambda)_\lambda.$$

Here λ ranges over the set $\left\{ \Lambda - \sum_{i=0}^{n} k_i \alpha_i,\ k_i \geq 0 \right\}$. The number $m_\lambda(\Lambda) = \dim V(\Lambda)_\lambda$ is called the *multiplicity* of λ; λ is called a *weight* of the module $V(\Lambda)$ if $m_\lambda(\Lambda) \neq 0$.

$\Lambda \in \tilde{\mathfrak{h}}^*$ is called *dominant* if $\Lambda(h_i)$ is a non-negative integer for all $i = 0, \ldots, n$. An important property of a $\tilde{\mathfrak{g}}$-module $V(\Lambda)$ with the dominant highest weight Λ is that $V(\Lambda)$ belongs to the category $\mathscr{K}$. Indeed, e_i is obviously locally nilpotent and for f_i this follows from the fact that $f_i^{\Lambda(h_i)+1}(v_0) = 0$. For a module $V(\Lambda)$ with a dominant highest weight Λ one knows a formula for the multiplicity of a weight[2]:

$$m_\lambda(\Lambda) = \sum_{w \in W} (\det w)\, K(w(\Lambda + \tilde{\rho}) - (\lambda + \tilde{\rho})). \tag{2.1}$$

[2] This formula is proven in [5] for all Kac-Moody Lie algebras with a symmetrizable Cartan matrix. We can apply this formula in our situation since by definition the weight decomposition for the $\tilde{\mathfrak{g}}$-module $V(\Lambda)$ is the same as for $V(\Lambda)$ considered as a $\hat{\mathfrak{g}}$-module; but $\hat{\mathfrak{g}}$ is a Kac-Moody Lie algebra with a symmetrizable Cartan matrix.

Here $\tilde{\rho} \in \mathfrak{h}^*$ is a function such that $\tilde{\rho}(h_i)=1$, $i=0, 1, \ldots, n$ and K is the (generalized) Kostant partition function.

Remark. Of course, the module $V(\Lambda)$ with a dominant Λ belongs to the category $\mathcal{K}$ for any Kac-Moody Lie algebra $\mathfrak{g}(A)$. An interesting open problem is to find a formula for the multiplicities of weights for any irreducible module from $\mathcal{K}$.

2.2. *Basic Representation: The Weight System.* We introduce the following subalgebra in $\tilde{\mathfrak{g}}$:

$$\mathfrak{m} = \mathbb{C}d + (\mathbb{C}[t] \otimes_{\mathbb{C}} \mathfrak{g}).$$

The *basic representation* of the Lie algebra $\tilde{\mathfrak{g}}$ is an irreducible $\tilde{\mathfrak{g}}$-module V for which there exists a non-zero vector $v_0 \in V$ such that

$$\mathfrak{m}(v_0)=0 \qquad c(v_0)=v_0.$$

It follows from Sect. 2.1 that the basic representation exists, is uniquely defined and is equivalent to a highest weight representation $V(\Lambda_0)$. To compute Λ_0 we note that $c = h_0 + \sum_{k=1}^{n} b_k h_k$. Therefore $\Lambda_0 \in \mathfrak{h}^*$ is defiend by:

$$\Lambda_0(h_0)=1, \qquad \Lambda_0(h_1)=\ldots=\Lambda_0(h_n)=\Lambda(d)=0.$$

Since Λ_0 is dominant, the multiplicities of the weights for the basic representation can be computed by formula (2.1). However, in the case when the Lie algebra $\mathfrak{g}$ is of one of the types A_n, D_n, E_6, E_7, E_8 (i.e., the Cartan matrix A of $\mathfrak{g}$ is symmetric) there exists a much simpler formula for these multiplicities proved in [6]. This formula plays a crucial role in our considerations. So from now on we assume (unless otherwise stated) that $\mathfrak{g}$ is a Lie algebra of one of the types A_n, D_n, E_6, E_7, E_8.

Let Q be the root lattice of $\mathfrak{g}$. Notice that $\langle \alpha, \alpha \rangle = 2$ for any root α and therefore the dual root lattice is again Q.

Let $p^{(n)}(k)$ denote the number of partitions of k into parts of n different colors; in other words:

$$\sum_{k \geq 0} p^{(n)}(k) q^k = \varphi(q)^{-n}, \qquad \text{where } \varphi(q) = \prod_{k \geq 1} (1-q^k).$$

Proposition 2.1. [6]. *The weight system P_0 of the basic representation consists of the elements of the form*

$$\Lambda_0 + \gamma - (\tfrac{1}{2}\langle \gamma, \gamma \rangle + k)\,\delta, \qquad \gamma \in Q,\ k \in \mathbb{Z}_+ = \{0, 1, \ldots\}, \tag{2.2}$$

the multiplicity of this weight being equal to $p^{(n)}(k)$. In other words:

$$ch\, V(\Lambda_0) = \sum_{\lambda} (\dim V(\Lambda_0)_\lambda)\, e^{\lambda} = e^{\Lambda_0} \sum_{\gamma \in Q} e^{\gamma - \frac{1}{2}\langle \gamma, \gamma \rangle \delta}\, \varphi(e^{-\delta})^n.$$

Proof. By formulas (1.7) and (1.8) one has:

$$T_\gamma(\Lambda_0) = \Lambda_0 + \gamma - \tfrac{1}{2}\langle \gamma, \gamma \rangle\, \delta,$$

$$T_\gamma(\gamma) = \gamma - \langle \gamma, \gamma \rangle\, \delta, \qquad \gamma \in Q^\vee (=Q).$$

On the other hand any $\mu \in P_0$ has clearly the form $\mu = \Lambda_0 - \gamma - k\delta$, $\gamma \in Q$. Therefore: $T_\gamma(\mu) = \Lambda_0 - (k - \frac{1}{2}\langle\gamma, \gamma\rangle)\delta$.

Since P_0 is $\tilde{W}$-invariant (by Proposition 1.1a), $T_\gamma(\mu) \in P_0$. So

$$P_0 = \{T_\gamma(\Lambda_0) - k\delta,\ \gamma \in Q,\ k \in \mathbb{Z}_+\}.$$

This proves (2.2).

For the proof of the multiplicity formula see [6], p. 130. A simpler proof based on the theory of modular forms can be found in [9]. Formula (2.1) is the base fact for both proofs.

Remark. (2.2) is equivalent to the fact that Λ_0 is the only maximal weight in P_0 [6].

2.3. Basic Representation: The Group of Translations. Let $\mathfrak{g}$ be a simple finite dimensional Lie algebra of type A_n, D_n or E_n. Then the root lattice Q of $\mathfrak{g}$ has a central extension T by the group $\mathbb{Z}/2\mathbb{Z} \cong \{\pm 1\}$:

$$1 \to \{\pm 1\} \to T \xrightarrow{\phi} Q \to 0,$$

which is uniquely defined by the property:

$$aba^{-1}b^{-1} = e^{\pi i\langle\alpha, \beta\rangle}, \quad \alpha, \beta \in Q,\ a, b \in T,\ \ \phi(a) = \alpha,\ \ \phi(b) = \beta.$$

Let ε be a 2-cocycle of Q with values in $\{\pm 1\}$ which defines this central extension. The following conditions are characteristic for ε: $Q \times Q \to \{\pm 1\}$:

$$\varepsilon(\alpha, \beta)\,\varepsilon(\alpha + \beta, \gamma) = \varepsilon(\beta, \gamma)\,\varepsilon(\alpha, \beta + \gamma); \tag{2.3}$$

$$\varepsilon(\alpha, \beta)\,\varepsilon(\beta, \alpha) = e^{\pi i\langle\alpha, \beta\rangle}; \tag{2.4}$$

$$\varepsilon(\alpha, 0) = 1. \tag{2.5}$$

One can also add the normalizing condition:

$$\varepsilon(\alpha, -\alpha) = 1. \tag{2.5$'$}$$

Remark. One can prove the existence of the cocycle ε satisfying (2.3)-(2.5)′ by several ways. The simplest and most elegant method was suggested by the referee of this paper[3]. He proposed to construct ε satisfying (2.3)-(2.5) explicitly with the additonal bilinearity condition

$$\varepsilon(\alpha + \beta, \gamma) = \varepsilon(\alpha, \gamma)\,\varepsilon(\beta, \gamma)$$

$$\varepsilon(\alpha, \beta + \gamma) = \varepsilon(\alpha, \beta)\,\varepsilon(\alpha, \gamma).$$

It is clear that this condition implies (2.3) and is agreeable with (2.4) and (2.5). Then ε is completely defined by the values $\varepsilon(\alpha_i, \alpha_j) = \varepsilon_{ij}$, where $1 \leq i \leq j \leq n$, $\varepsilon_{ij} = \pm 1$, and $\{\alpha_i\}_{i=1}^n$ a basis of Q. Now if we set $\varepsilon'(\alpha, \beta) = \varepsilon(\alpha, \beta)\,c(\alpha)\,c(\beta)\,c(\alpha + \beta)^{-1}$, where c: $Q \to \{\pm 1\}$, $c(0) = 1$, and $c(\alpha)\,c(-\alpha) = \varepsilon(\alpha, -\alpha)$, then ε' satisfies (2.3)-(2.5)′.

[3] We are grateful to the referee for this comment.

Proposition 2.2. *One can choose $E_\gamma \in \mathfrak{g}_\gamma$, $\gamma \in \Delta$, in such a way that:*

$$
\begin{aligned}
&[E_\gamma, E_{-\gamma}] = H_\gamma; \quad [E_\beta, E_\gamma] = 0 \quad \text{if } \beta+\gamma \notin \Delta \cup \{0\};\\
&[E_\beta, E_\gamma] = \varepsilon(\gamma, \beta)\, E_{\beta+\gamma} \quad \text{if } \beta+\gamma \in \Delta; \qquad (2.6)\\
&[H_\beta, E_\gamma] = \langle \beta, \gamma \rangle\, E_\gamma; \quad [H_\beta, H_\gamma] = 0.
\end{aligned}
$$

Proof. Consider a vector space $\mathfrak{g}_1$ with a basis E'_γ, $\gamma \in \Delta$; H'_{α_i}, $\alpha_i \in \Pi$; for $\gamma = \sum_i k_i \alpha_i \in Q$ set $H'_\gamma = \sum_i k_i H'_{\alpha_i}$. Define an anticommutative operation on $\mathfrak{g}_1$ by formulas (2.6). One checks immediately that the Jacoby identity follows from (2.3)-(2.5). Therefore $\mathfrak{g}_1$ is a Lie algebra. By a standard Lie theory argument, $\mathfrak{g}_1$ is isomorphic to $\mathfrak{g}$.

Remark. Proposition 2.2 gives in particular the existence of the simple Lie algebras E_6, E_7, E_8 (cf. [14]).

From now on we fix a cocycle ε, satisfying (2.3)-(2.5)′ and the basis E_γ, $\gamma \in \Delta$, H_{α_i}, $\alpha_i \in \Pi$, of $\mathfrak{g}$, satisfying (2.6). This basis is called the Chevalley basis.

Let $\tilde{\mathfrak{g}}$ be the associated with $\mathfrak{g}$ affine Lie algebra and let $\pi = \pi_{\Lambda_0}$ be the basic representation of $\tilde{\mathfrak{g}}$ in the space $V = V(\Lambda_0)$. For $\gamma \in \Delta$ we set

$$
\begin{aligned}
T^\pi_\gamma = &-\exp -\pi(t \otimes E_{-\gamma}) \exp \pi(t^{-1} \otimes E_\gamma) \exp -\pi(t \otimes E_{-\gamma})\\
&\cdot \exp -\pi(E_\gamma) \exp \pi(E_{-\gamma}) \exp -\pi(E_\gamma).
\end{aligned}
$$

By Proposition 1.1 b)(iii), the conjugation by T^π_γ induces the automorphism T^{ad}_γ of $\tilde{\mathfrak{g}}$, described by formulas (1.9) and (1.10). (Recall also that T^{ad}_γ induces on $\tilde{\mathfrak{h}}$ the translation $T_\gamma \in \tilde{W}$). We call the group $T \subset \operatorname{End} V$ generated by T^π_γ, $\gamma \in \Delta$, *the group of translations.*

The structure of the group T is described by the following.

Proposition 2.3. a) *Let $v = v_{\Lambda_0} \in V$ be a highest weight vector. Then one has:*

(i) $T^\pi_\gamma(v) = (t^{-1} \otimes E_\gamma)(v)$, $\gamma \in \Delta$;
(ii) $T^\pi_\beta T^\pi_\gamma(v) = (t^{-1} \otimes [E_\gamma, E_\beta])(v)$ *if* $\beta, \gamma, \beta+\gamma \in \Delta$, *or equivalently*
$T^\pi_\beta T^\pi_\gamma = \varepsilon(\beta, \gamma)\, T^\pi_{\beta+\gamma}$ *if* $\beta, \gamma, \beta+\gamma \in \Delta$;
$T^\pi_\beta T^\pi_\gamma(v) = (t^{-1} \otimes E_\beta)(t^{-1} \otimes E_\gamma)(v)$, *if* $\beta, \gamma \in \Delta$, $\beta+\gamma \notin \Delta \cup \{0\}$;
$T^\pi_\gamma T^\pi_{-\gamma} = 1_V$, $\gamma \in \Delta$;
(iii) $T^\pi_\beta T^\pi_\gamma T^\pi_{-\beta} T^\pi_{-\gamma} = e^{\pi i \langle \beta, \gamma \rangle} 1_V$, $\beta, \gamma \in \Delta$.

b) *The map ϕ: $T^\pi_\gamma \to \gamma$ induces an isomorphism $T/\{\pm 1\} \cong Q$.*

c) *The group T is isomorphic to the group (denoted by the same letter) introduced in the beginning of the section.*

Proof. One has:

$$
\begin{aligned}
T^\pi_\gamma(v) &= \exp -\pi(t \otimes E_{-\gamma}) \exp \pi(t^{-1} \otimes E_\gamma)(v)\\
&= (1 - \pi(t \otimes E_\gamma))(1 + \pi(t^{-1} \otimes E_\gamma))(v)
\end{aligned}
$$

since $\Lambda_0 - 2\delta + 2\gamma$ is not a weight (by (2.2)). This gives formula a)(i). Therefore

$$
\begin{aligned}
T^\pi_\beta T^\pi_\gamma(v) &= T^\pi_\beta(t^{-1} \otimes E_\gamma) = T^\pi_\beta(t^{-1} \otimes E_\gamma)\, T^\pi_{-\beta} T^\pi_\beta(v)\\
&= (t^{-1-\langle \beta, \gamma \rangle} \otimes E_\gamma)(t^{-1} \otimes E_\beta)(v).
\end{aligned}
$$

This gives formulas a)(ii) and (iii). b) follows from a) and Lemma 1.1b)(iii), and c) follows from b).

Proposition 2.3 allows us to choose in each fiber of the map $\phi: T \to Q$ an operator T_γ^π, $\gamma \in Q$, such that it is the same as before for $\gamma \in \Delta$ and

$$T_\beta^\pi T_\gamma^\pi = \varepsilon(\beta, \gamma) T_{\beta+\gamma}^\pi, \quad \beta, \gamma \in Q. \tag{2.7}$$

Computation similar to that in the proof of Proposition 2.3 gives the following relations between T_γ^π, $\gamma \in \Delta$, and $r_\beta^\pi = \exp - \pi(E_\beta) \exp \pi(E_{-\beta}) \exp - \pi(E_\beta)$, $\beta \in \Delta$:

$$r_\gamma^\pi T_\beta^\pi (r_\gamma^\pi)^{-1} = \begin{cases} \varepsilon(\beta - \gamma, \gamma) T_{\beta-\gamma}^\pi & \text{if } \langle \beta, \gamma \rangle = 1, \\ -\varepsilon(\beta + \gamma, -\gamma) T_{\beta+\gamma}^\pi & \text{if } \langle \beta, \gamma \rangle = -1, \\ -T_{-\beta} & \text{if } \beta = \pm\gamma, \\ T_\beta & \text{if } \langle \beta, \gamma \rangle = 0. \end{cases} \tag{2.8}$$

$$(r_\gamma^\pi)^2 T_\beta^\pi (r_\gamma^\pi)^{-2} = e^{\pi i \langle \beta, \gamma \rangle} T_\beta^\pi. \tag{2.9}$$

2.4. The Canonical Representation of a Heisenebrg System. Let $(\tilde{\mathfrak{s}}, \Gamma)$ be a Heisenberg system and let ε be a 2-cocycle of Γ with values in $\{\pm 1\}$, i.e., $\varepsilon: \Gamma \times \Gamma \to \{\pm 1\}$ is a map, which satisfies properties (2.3) and (2.5). By a ***representation associated with the cocycle*** ε of this system in a vector space V we mean a pair $\pi = \{\pi_1, \pi_2\}$ of a representation π_1 of the Lie algebra $\tilde{\mathfrak{s}}$ in the space V and a projective representation π_2 of the group Γ in V such that

$$\pi_2(\beta + \gamma) = \varepsilon(\beta, \gamma) \pi_2(\beta) \pi_2(\gamma), \quad \beta, \gamma \in \Gamma,$$

for which the following consistency condition is satisfied:

$$\pi_2(\gamma) \pi_1(x) \pi_2(\gamma)^{-1} = \pi_1(T_\gamma(x)), \quad \gamma \in \Gamma, \ x \in \tilde{\mathfrak{s}}. \tag{2.10}$$

Suppose now that one has a polarisation of the subspace $\mathfrak{s}$ of the Heisenberg algebra $\hat{\mathfrak{s}}$:

$$\mathfrak{s} = \mathfrak{s}_- \oplus \mathfrak{s}_+,$$

i.e., decomposition into direct sum of subspaces isotropic with respect to the bilinear form ψ. Then one can construct a representation of the Heisenberg system $(\tilde{\mathfrak{s}}, \Gamma)$ as follows.

Consider the symmetric algebra $S(\mathfrak{s}_-)$ over $\mathfrak{s}_-$. For an element $p \in \mathfrak{s}_+$ we define a derivation ∂_p of the algebra $S(\mathfrak{s}_-)$ by:

$$\partial_p(q) = \psi(p, q), \quad q \in \mathfrak{s}_-.$$

Denote by L_x the multiplication of $S(\mathfrak{s}_-)$ by an element $x \in S(\mathfrak{s}_-)$. Let $\mathbb{C}(\Gamma)$ denote the group algebra of the group Γ, i.e., a $\mathbb{C}$-algebra with a basis e^γ, $\gamma \in \Gamma$, and multiplication $e^{\gamma_1} e^{\gamma_2} = e^{\gamma_1 + \gamma_2}$. For $h \in \mathfrak{h}$ let ∂_h denote a derivation of the algebra $\mathbb{C}(\Gamma)$ defined by:

$$\partial_h(e^\gamma) = \gamma(h) e^\gamma.$$

Similarly we define ∂_β for $\beta\in\Gamma$ by: $\partial_\beta(e^\gamma)=\langle\beta,\gamma\rangle e^\gamma$. For $\gamma\in\Gamma$ we define an operator $\tilde{T}_\gamma$ on $\mathbb{C}(\Gamma)$ by:

$$\tilde{T}_\gamma(e^\beta)=\varepsilon(\gamma,\beta)\,e^{\beta+\gamma}.$$

We set

$$V=S(\mathfrak{s}_-)\otimes_{\mathbb{C}}\mathbb{C}(\Gamma)$$

and define a representation $\pi=\{\pi_1,\pi_2\}$ of the Heisenberg system $(\hat{\mathfrak{s}},\Gamma)$ in the space V as follows:

$$\begin{aligned}&\pi_1(p)=\partial_p\otimes 1,\ p\in\mathfrak{s}_+;\quad \pi_1(q)=L_q\otimes 1,\ q\in\mathfrak{s}_-\\ &\pi_1(c)=1_V;\quad \pi_1(h)=1\otimes\partial_h,\ h\in\mathfrak{h};\quad \pi_2(\gamma)=\tilde{T}_\gamma,\ \gamma\in Q.\end{aligned}\tag{2.11}$$

One checks immediately that consistency condition (2.10) is equivalent to $\varepsilon(-\gamma,\beta)\varepsilon(\gamma,\beta-\gamma)=1$; the latter formula follows from (2.3) and (2.5) (if one replaces γ by $-\alpha-\beta$ in (2.3)).

We call the constructed representation π of the Heisenberg system $(\hat{\mathfrak{s}},\Gamma)$ the *canonical representation associated with the cocycle* ε. Note that this representation is irreducible in the sense that there are no non-trivial subspaces in V invariant with respect to both representation π_1 and π_2. Note also that the canonical representations associated with equivalent cocycles are equivalent. The following proposition might be called the Stone-von Neuman theorem for Heisenberg systems.

Proposition 2.4. *Let $(\hat{\mathfrak{s}},\Gamma)$ be a Heisenberg system with a fixed polarisation $\mathfrak{s}=\mathfrak{s}_-\oplus\mathfrak{s}_+$, and let π' be an irreducible representation of $(\hat{\mathfrak{s}},\Gamma)$ associated with a cocycle ε in a vector space V'. Suppose that there exists a vacuum vector $v_0\in V'$, i.e., v_0 is a non-zero vector such that:*

$$\pi'(p)(v_0)=0,\quad p\in\mathfrak{s}_+,\quad \pi'(h)(v_0)=0,\quad h\in\mathfrak{h};\quad \pi'(c)(v_0)=v_0.$$

Then the respresentation π' is equivalent to the canonical representation π.

Proof. We set $v_\gamma=\pi'_2(\gamma)(v_0)$, $\gamma\in\Gamma$, and $M_\gamma=U(\hat{\mathfrak{s}})(v_\gamma)$, where $U(\hat{\mathfrak{s}})$ is the universal enveloping algebra of the Lie algebra $\hat{\mathfrak{s}}$. Clearly one has:

$$\pi'_1(p)v_\gamma=0,\ p\in\mathfrak{s}_+,\quad\text{and}\quad \pi'_1(c)v_\gamma=v_\gamma.$$

Therefore the restriction of the representation π to $\hat{\mathfrak{s}}$ in M_γ is equivalent to canonical commutation relations (due to an "ordinary" Stone-von Neuman theorem, see e.g. [8]); in particular, the $\hat{\mathfrak{s}}$-module M_γ is irreducible. Due to (2.10) one has:

$$\begin{aligned}\pi'_1(h)v_\gamma&=\pi'_1(h)\pi'_2(\gamma)(v_0)=\pi'_2(\gamma)\pi'_1(-T_\gamma(h))(v_0)\\ &=\pi'_2(\gamma)\pi'_1(\gamma(h)c+h)(v_0).\end{aligned}$$

So we obtain:

$$\pi'_1(h)(v_\gamma)=\gamma(h)\,v_\gamma.$$

In particular, since V' is irreducible, we obtain that $V'=\sum_{\gamma\in\Gamma}M_\gamma$. In order to prove the proposition one has only to show that this is a direct sum. Suppose the contrary. Then we have a non-trivial linear dependence: $\sum_k c_{\gamma_k} v_{\gamma_k} = 0$. We can assume that m is the minimal possible positive integer, $m \geqq 2$. But then, taking $h \in \mathfrak{h}$ such that $\gamma_1(h) = 1$, $\gamma_2(h) = 0$ and applying $\pi_1'(h)$ we obtain a "shorter" linear dependence: $c_{\gamma_1} v_{\gamma_1} + \gamma_3(h) c_{\gamma_3} v_{\gamma_3} + \ldots = 0$. This contradiction completes the proof of the proposition.

Let now $(\tilde{\mathfrak{s}}, \Gamma)$ be the Heisenberg system associated with an even integral lattice Γ and ε be a 2-cocycle of Γ with values in $\{\pm 1\}$. We define a Hermitian form $\langle | \rangle$ on the space $V = S(\mathfrak{s}_-) \otimes \mathbb{C}(\Gamma)$ as follows. First we define $\langle | \rangle$ on $\mathfrak{h}^* := \Gamma \otimes_{\mathbb{Z}} \mathbb{C}$ extending $\langle , \rangle$ from Γ by antilinearity, and on $\mathfrak{h}$, identifying $\mathfrak{h} \cong \mathfrak{h}^*$. Then we define $\langle | \rangle$ on $\mathfrak{s}$ by:

$$\langle t^k \otimes h_1 | t^l \otimes h_2 \rangle = k \delta_{k,l} \langle h_1, h_2 \rangle, \; h_1, h_2 \in \mathfrak{h}, \tag{2.12}$$

and extend this form to $S(\mathfrak{s}_-)$ in a usual way:

$$\langle x_1 \ldots x_m | y_1 \ldots y_n \rangle = \delta_{m,n} \sum_{\sigma \in S_m} \prod_{i=1}^{m} \langle x_i | y_{\sigma(i)} \rangle. \tag{2.13}$$

Secondly we define $\langle | \rangle$ on $\mathbb{C}(\Gamma)$ by:

$$\langle e^\beta | e^\gamma \rangle = \delta_{\beta,\gamma}. \tag{2.14}$$

This gives a Hermitian form $\langle | \rangle$ on the whole space $V = S(\mathfrak{s}_-) \otimes \mathbb{C}(\Gamma)$, which is positive definite, provided that the form $\langle , \rangle$ is positive definite on Γ.

One can check immediately that this Hermitian form is invariant with respect to the action of the lattice Γ in V and contravariant with respect to the action of $\tilde{\mathfrak{s}}$ in V. We recall that a Hermitian form $\langle | \rangle$ on the space V is *contravariant* with respect to a Lie algebra $\mathfrak{l}$ with an antilinear involution $\tilde{\omega}$, operating in V, if

$$\langle x(u) | v \rangle = -\langle u | \tilde{\omega}(x)(v) \rangle, \qquad u, v \in V, \; x \in \mathfrak{l}.$$

The antilinear involution $\tilde{\omega}$ on $\tilde{\mathfrak{s}}$ is defined by:

$$\tilde{\omega}(c) = -c, \qquad \tilde{\omega}(t^k \otimes h) = -t^{-k} \otimes \bar{h}, \qquad h \in \mathfrak{h}.$$

The irreducibility of the canonical representation implies that the invariantness properties define the Hermitian form $\langle | \rangle$ up to a constant factor.

Remark. The space $S(\mathfrak{s}_-)$ is usually called the Fock space, the unity 1 being called the vacuum vector.

Now we are in a position to make the first step of the construction of the basic representation $\pi = \pi_{\Lambda_0}$. Again, $\mathfrak{g}$ is a simple Lie algebra of type A_n, D_n or E_n. Then $Q^\vee = Q$ and taking for $\tilde{\mathfrak{s}}$ the subalgebra of $\tilde{\mathfrak{g}}$ introduced in Sect. 1.3, we obtain (by Proposition 1.3) that $(\tilde{\mathfrak{s}}, Q)$ is a Heisenberg system with a given polarisation described above.

Proposition 2.5. a) *Let π_1 denote the restriction of the basic representation to the subalgebra $\tilde{\mathfrak{s}}$, let π_2 denote the representation of Q defined by $\pi_2(\gamma) = T_\gamma^*$, $\gamma \in Q$. Then the representation (π_1, π_2) of the Heisenberg system $(\tilde{\mathfrak{s}}, Q)$ in the space $V(\Lambda_0)$*

is equivalent to the canonical representation of this system, associated with the cocycle ε, *satisfying* (2.3)–(2.5)′.

b) *Any contravariant Hermitian form on* $V(\Lambda_0)$ *with respect to the Lie algebra* $\tilde{\mathfrak{g}}$ *with involution* $\tilde{\omega}$ *is proportional to the form* $\langle\,|\,\rangle$.

Proof. π_2 is a representation of the group Q by Proposition 1.2. Let $V(\Lambda_0)=\oplus V_\lambda$ be the weight space decomposition for the basic representation. By Proposition 1.1 a) (i) one has:

$$T_\gamma^\pi(V_\lambda)=V_{T_\gamma(\lambda)},\qquad \gamma\in Q.$$

Set $v_\gamma=T_\gamma^\pi(v_0)$. By (1.7) we obtain:

$$V_{\Lambda_0-\frac{1}{2}\langle\gamma,\gamma\rangle\delta+\gamma}=\mathbb{C}v_\gamma,\qquad \gamma\in Q. \tag{2.15}$$

Each vector v_γ is a vacuum vector for the Heisenberg subalgebra $\hat{\mathfrak{s}}$. Therefore by the Stone-von Neuman theorem one has

$$\dim U(\hat{\mathfrak{s}})v_\gamma\cap V_{\Lambda_0-\frac{1}{2}\langle\gamma,\gamma\rangle\delta+\gamma-k\delta}=p^{(n)}(k). \tag{2.16}$$

Now Proposition 2.2 gives that the left-hand side of (2.16) is the whole space $V_{\Lambda_0-(\frac{1}{2}\langle\gamma,\gamma\rangle+k)\delta+\gamma}$. Formulas (2.15) and (2.16) show now that the representation (π_1,π_2) of the Heisenberg system $(\bar{\mathfrak{s}},Q)$ is irreducible. Now the statement a) follows from Proposition 2.4.

The statement b) follows from the fact that any contravariant Hermitian form, say H, on $V(\Lambda_0)$ is even $\tilde{W}^\pi$-invariant and in particular it is T-invariant. The form H is (in particular) contravariant with respect to $\bar{\mathfrak{s}}$ and T-invariant. Since $V(\Lambda_0)$ is irreducible with respect to the action of tne pair $\{\bar{\mathfrak{s}},T\}$, these properties define the form H up to a constant factor.

2.5. *A Statement About Differential Operators.* Let again $(\bar{\mathfrak{s}},\Gamma)$ be a Heisenberg system with a fixed polarisation $\mathfrak{s}=\mathfrak{s}_-\oplus\mathfrak{s}_+$ and let $\pi=\{\pi_1,\pi_2\}$ be the canonical representation associated with a cocycle ε of this system in the space $V=S(\mathfrak{s}_-)\otimes\mathbb{C}(\Gamma)$. We assume that the lattice Γ is integral even. The spaces $\bar{\mathfrak{s}}_-=\mathfrak{s}_+^*$ and $\bar{\mathfrak{s}}_+=\mathfrak{s}_-^*$ are clearly the completions of $\mathfrak{s}_-$ and $\mathfrak{s}_+$, respectively. We extend the bilinear form ψ from $\mathfrak{s}_-\oplus\mathfrak{s}_+$ to $\bar{\mathfrak{s}}_-\oplus\mathfrak{s}_+$ and to $\mathfrak{s}_-\oplus\bar{\mathfrak{s}}_+$. We set $\bar{V}=\bar{S}(\bar{\mathfrak{s}}_-)\otimes_{\mathbb{C}}\mathbb{C}(\Gamma)$, where $\bar{S}(\bar{\mathfrak{s}}_-)$ denotes the completion of the symmetric algebra $S(\bar{\mathfrak{s}})$. One has the canonical embedding $V\subset\bar{V}$.

We call a *differential operator* on V a linear map $V\to\bar{V}$. The left multiplication L_R by $R\in V$ is an example of a differential operator. Other examples are: $\partial_p\otimes 1$, $p\in\bar{\mathfrak{s}}_+$, $1\otimes\partial_h$, $h\in\mathfrak{h}$, $1\otimes\tilde{T}_\gamma$, $1\otimes\partial_\gamma$, $\gamma\in\Gamma$, which we shall denote by $\partial_p,\partial_h,T_\gamma$ and ∂_γ for short. We also write for short q instead of $q\otimes 1$ and e^γ instead of $1\otimes e^\gamma$. For $p\in\bar{\mathfrak{s}}_+$, $z\in\mathbb{C}\setminus 0$, and $\gamma\in\Gamma$ we define a differential operator $M=\exp(\partial_p+(\ln z)\partial_\gamma)$. By the Taylor formula one has:

$$M(q)=q+\psi(p,q),\qquad M(e^\alpha)=z^{\langle\alpha,\gamma\rangle}e^\alpha,\qquad q\in\mathfrak{s}_-,\ \alpha\in\Gamma.$$

Proposition 2.6. *Let* $A\colon V\to V$ *be a differential operator for which there exist* $p_0\in\bar{\mathfrak{s}}_+$, $q_0\in\bar{\mathfrak{s}}_-$, $\beta,\gamma\in\Gamma$ *and* $z\in\mathbb{C}\setminus 0$ *such that*

$$[\partial_p, A] = \psi(p, q_0) A, \quad p \in \mathfrak{s}_+,$$
$$[L_q, A] = \psi(p_0, q) A, \quad q \in \mathfrak{s}_-,$$
$$[\partial_\alpha, A] = \langle \beta, \alpha \rangle A \quad \alpha \in \Gamma,$$
$$\tilde{T}_\alpha A \tilde{T}_{-\alpha} = z^{\langle \gamma, \alpha \rangle} A, \quad \alpha \in \Gamma.$$

Then $A = a\tilde{T}_\beta(\exp q_0)(\exp - ((\ln z)\partial_\gamma + \partial_{p_0}))$, *where* $a \in \mathbb{C}$.

Proof. (cf. [8]). We replace A by $A_1 = \tilde{T}_{-\beta}\exp(-q_0)\cdot A \cdot \exp((\ln z)\partial_\gamma + \partial_{p_0})$. Then our statement is equivalent to the following: if $[\partial_p, A] = [L_q, A] = [\partial_\alpha, A] = [\tilde{T}_\alpha, A] = 0$ for any $p \in \mathfrak{s}_+$, $q \in \mathfrak{s}_-$, $\alpha \in \Gamma$, then $A = \text{const}$. The latter statement is obvious.

A special case of Proposition 2.6 is treated in [8] (Proposition 3.1).

2.6. The Vertex and Some Other Operators. Now we consider the special case of a Heisenberg system $(\mathfrak{s}, Q)$ associated with an even integral lattice Q, and the canonical representation associated with a cocycle ε of this system in the vector space $V = S(\mathfrak{s}_-) \otimes_{\mathbb{C}} \mathbb{C}(Q)$.

We introduce the following operators in V. For $h \in \mathfrak{h}$ and $k > 0$ the *creation operator* is

$$h(-k) := \text{multiplication by } t^{-k} \otimes h; \tag{2.17}$$

and *the annihilation operator* is

$$h(k) := \partial_{(t^k \otimes h)}, \tag{2.18}$$

where

$$\partial_{(t^k \otimes h)}(t^{-m} \otimes h') := k\delta_{k,m}\langle h, h' \rangle.$$

We will often write $\gamma(k)$ instead of[4] $h_\gamma(k)$, $\gamma \in Q$.

We define also for $h \in \mathfrak{h}$ the operator

$$h(0) := 1 \otimes \partial_h, \quad \text{where } \partial_h(e^\beta) := \beta(h) e^\beta, \quad \gamma \in Q; \tag{2.19}$$

and for $\gamma \in Q$ the operators.

$$\partial_\gamma := 1 \otimes \partial_\gamma, \quad \text{where } \partial_\gamma(e^\beta) = \langle \beta, \gamma \rangle e^\beta; \tag{2.20}$$
$$c_\gamma := 1 \otimes c_\gamma, \quad \text{where } c_\gamma(e^\beta) = \varepsilon(\gamma, \beta) e^\beta. \tag{2.21}$$

We define a $\mathbb{Z}_+$-gradation $V = \bigoplus_{k \geq 0} V_{-k}$ by:

$$\deg t^{-k} \otimes \mathfrak{h} = -k, \quad k = 1, 2, \ldots; \quad \deg e^\gamma = -\tfrac{1}{2}\langle \gamma, \gamma \rangle, \quad \gamma \in Q,$$

and the *energy operator* D_0 by:

$$D_0(v) = -kv, \quad \text{for } v \in V_{-k}. \tag{2.22}$$

The following differential operator, which is called the *vertex operator*, plays an important role in the dual resonance theory (see e.g. [10], [12]). This

[4] Here as usual $h_\gamma \in \mathfrak{h}$ is defined by $\langle h, h_\gamma \rangle = \gamma(h)$, $h \in \mathfrak{h}$.

operator is defined for any $\gamma \in \Gamma$ and a non-zero complex number z:

$$X(\gamma, z) := \exp\left(\sum_{k \geq 1} \frac{z^k}{k} \gamma(-k)\right) \exp(\gamma + (\ln z)\partial_\gamma) \cdot \exp\left(-\sum_{k \geq 1} \frac{z^{-k}}{k} \gamma(k)\right). \tag{2.23}$$

Notice that the middle factor can be written as

$$z^{\frac{1}{2}\langle\gamma,\gamma\rangle} e^\gamma \exp(\ln z)\partial_\gamma. \tag{2.24}$$

Notice also that

$$\exp(\ln z)\partial_\gamma = \sum_{k \in \mathbb{Z}} z^k P_k, \quad \text{where } P_k(e^\beta) := \delta_{k,\langle\beta,\gamma\rangle} e^\beta.$$

The operator $X(\gamma, z)$ is a map from V to $\bar{V}$. However, developing this operator by powers of z we obtain:

$$X(\gamma, z) = \sum_{k \in \mathbb{Z}} X_k(\gamma) z^k. \tag{2.25}$$

where each $X_k(\gamma)$ is a differential operator, which maps V into itself.

2.7. The Main Theorem. Now we are in a position to prove the main result of the paper. Let $\mathfrak{g}$ be a complex simple finite dimensional Lie algebra of one of the types A_n, D_n, E_6 E_7, E_8. Let $\mathfrak{h}$ be a Cartan subalgebra of $\mathfrak{g}$, Δ be the associated root system and Q be the lattice in $\mathfrak{h}^*$ generated by Δ. We denote by $\langle\,,\rangle$ the standard form, i.e., the invariant form, such that $\langle\gamma,\gamma\rangle = 2$ for and $\gamma \in \Delta$. Let E_γ, $\gamma \in \Delta$, H_{α_i}, $\alpha_i \in \Pi$, be a Chevalley basis of $\mathfrak{g}$ and ε be an associated cocycle (see Sect. 2.3).

Let $\tilde{\mathfrak{g}}$ be the affine Lie algebra associated with $\mathfrak{g}$. We set $\tilde{\mathfrak{s}}_k = \sum_{k \in \mathbb{Z}} \mathfrak{s}_k$, $\mathfrak{s}_- = \sum_{k<0} \mathfrak{s}_k$, where $\mathfrak{s}_k = t^k \otimes \mathfrak{h}$, $k \neq 0$ and $\mathfrak{s}_0 = \mathbb{C}c + \mathfrak{h}$.

We set $V = S(\mathfrak{s}_-) \otimes_{\mathbb{C}} \mathbb{C}(Q)$ and introduce the operators $h(k)$, ∂_γ, c_γ, D_0, $X_k(\gamma)$ by formulas (2.17)-(2.25). We introduce also the Hermitian form $\langle\,|\,\rangle$ on V by formulas (2.12)-(2.14).

Theorem 1. a) *The map* $\pi: \tilde{\mathfrak{g}} \to \operatorname{End} V$ *defined by:*

$$\pi(c) = 1_V, \quad \pi(d) = D_0, \quad \pi(t^k \otimes h) = h(k), \quad k \in \mathbb{Z},$$
$$\pi(t^k \otimes E_\gamma) = X_k(\gamma) c_\gamma, \quad \gamma \in \Delta.$$

is a representation of the Lie algebra $\tilde{\mathfrak{g}}$, *which is equivalent to the basic representation.*

b) *Under the identification of* V *with* $V(\Lambda_0)$ *one has:*

$$T_\gamma^{\pi_0} = e^\gamma \cdot c_\gamma, \quad \gamma \in Q.$$

c) *The weight decomposition of* V *(with respect to* $\tilde{\mathfrak{h}}$*) has the following form:*

$$V = \sum_{k,\gamma} V_{\Lambda_0 - (k + \frac{1}{2}\langle\gamma,\gamma\rangle)\delta + \gamma}, \quad \textit{where } (k,\gamma) \textit{ ranges over } \mathbb{Z}_+ \times Q,$$

and is defined by $\deg(t^{-k} \otimes h) = -k\delta$, $\deg e^\gamma = \gamma - \frac{1}{2}\langle\gamma,\gamma\rangle\delta$.

d) *The Hermitian form* $\langle\,|\,\rangle$ *on* V *is positive definite and contravariant. In particular, it is invariant with respect to the compact form* $\tilde{\mathfrak{g}}_0$.

Proof. By Proposition 1.3, the pair $(\tilde{\mathfrak{s}}, Q)$ is the Heisenberg system, associated with the root lattice $Q(=Q^\vee)$. By Proposition 2.5 we can identify the space $V(\Lambda_0)$ with the space V in such a way that: $v_{\Lambda_0}=1$; $\pi_{\Lambda_0}=\pi$ on $\tilde{\mathfrak{s}}$; b) holds; c) holds and in particular, $\pi_{\Lambda_0}(d)=\pi(d)$.

In order to show that then $\pi_{\Lambda_0}(t^k \otimes E_\gamma)=X_k(\gamma)c_\gamma$, $\gamma\in\Delta$, we consider the Mellin transform $\hat{E}_\gamma(z)$ of E_γ (see Sect. 1.8). For each $z\in\mathbb{C}\setminus 0$ we obtain an operator $\pi_{\Lambda_0}(\hat{E}_\gamma(z))$: $V\to\bar{V}$. By Proposition 1.4 this operator satisfies all the relations of Proposition 2.6. This implies that $\pi_{\Lambda_0}(\hat{E}_\gamma(z))=c(z)X(\gamma,z)c_\gamma$, where $X(\gamma,z)$ is the vertex operator associated with the root $\gamma\in\Delta$, and $c(z)$ is a scalar function depending only on z. To complete the proof of a) one has to show that $c(z)=1$. For that we compare the coefficients by $e^\gamma\in V$ in the developments with respect to the weight decomposition of the elements $\pi(\hat{E}_\gamma(z))(1)$ and $c(z)X(\gamma,z)(1)$; denote this coefficients by a and b, respectively. Since $\langle\gamma,\gamma\rangle=2$ we obtain from (2.24):

$$b=zc(z).$$

Using Proposition 2.3a) (i) one has:

$$e^\gamma=T_\gamma^{*\Lambda_0}(1)=\tilde{T}_\gamma(1)=\pi(t^{-1}\otimes E_\gamma)(1). \tag{2.26}$$

But $\pi(\hat{E}_\gamma(z))(1)=1+z\pi(t^{-1}\otimes E_\gamma)(1)+\dots$.

Therefore from (2.26) we obtain:

$$a=z.$$

Now the equality $a=b$ implies $c(z)=1$. This completes the proof of a) and also of b) and c). To prove d) notice that the existence of a non-degenerate contravariant Hermitian form, say $\langle\,|\,\rangle_1$ on $V(\Lambda_0)$ is a simple general fact, which holds for any highest weight module (see e.g. [1]). In particular, the form $\langle\,|\,\rangle_1$ is invariant with respect to T_γ^*, $\gamma\in Q$. Therefore, by Proposition 2.5b), the forms $\langle\,|\,\rangle$ and $\langle\,|\,\rangle_1$ are proportional. This proves that the form $\langle\,|\,\rangle$ is contravariant with respect to the whole Lie algebra $\tilde{\mathfrak{g}}$. The form $\langle\,|\,\rangle$ is clearly positive definite. This completes the proof of d) and of the theorem.

Remark. One can determine the action of the whole group $\tilde{W}^*$ (not only of the subgroup of translations). Indeed, the action of the operators r_γ^*, $\gamma\in\Delta$, can be read off from formulas (2.8) and (2.9) and the condition $r_\gamma^*(1)=1$.

2.8. *The Virasoro Operators.* We introduce now the so-called *Virasoro operators*, which are also of great importance in the dual resonance theory (see e.g. [10], [12]). We consider again a Heisenberg system, associated with an even integral lattice Q and the canonical representation $\pi=(\pi_1,\pi_2)$ associated with a cocycle ε of this system in the space $V=S(\mathfrak{s}_-)\otimes_{\mathbb{C}}\mathbb{C}(Q)$. We choose an orthonormal basic $u_1,\dots,u_n$ of $\mathfrak{h}$. The Virasoro operator D_0 is the energy operator, which has already appeared in Sect. 2.6:

$$D_0=-\sum_{k\geq 1}\sum_{i=1}^{n}u_i(-k)u_i(k)-\tfrac{1}{2}\sum_{i=1}^{n}u_i(0)^2.$$

All the other Virasoro operators are defined y:

$$D_m = -\tfrac{1}{2}\sum_{k\in\mathbb{Z}}\sum_{i=1}^{n} u_i(-k)u_i(k+m), \qquad m\in\mathbb{Z}\setminus 0.$$

Notice that D_m maps V into itself.

The following relations between the Virasoro operator and the operators introduced earlier hold:

Proposition 2.7

$$[D_m, \pi_1(h)] = \pi_1\left(t^{m+1}\frac{d}{dt}(h)\right), \qquad h\in\mathfrak{h}. \tag{2.27}$$

$$\tilde{T}_\gamma D_m \tilde{T}_\gamma^{-1} = D_m + \gamma(m) - \tfrac{1}{2}\delta_{m,0}\langle\gamma,\gamma\rangle, \tag{2.28}$$

$$[D_m, X(\gamma, z)] = -z^m\left(\frac{m}{2}\langle\gamma,\gamma\rangle + z\frac{z}{dz}\right)X(\gamma, z); \tag{2.29}$$

$$[D_m, D_l] = (l-m)D_{m+l} + \frac{n}{12}(m^3-m)\delta_{m,-l} \tag{2.30}$$

Proof. Relations (2.27) and (2.28) can be verified immediately from the definition of D_m and the obvious formula $[AB, C] = A[B, C] + [A, C]B$.

Relations (2.30) for $m \neq -l$ or for $m = -l$ up to a constant summand are deduced from (2.27). In order to compute the constant summand in (2.30) we compute the constant term in $[D_m, D_{-m}](1)$, $m > 0$. We have:

$$[D_m, D_{-m}](1) = D_m D_{-m}(1) = \tfrac{1}{4}\left(\sum_{k=1}^{m-1}\sum_{i=1}^{n} u_i(k)u_i(m-k)\right).$$

$$\left(\sum_{k=1}^{m-1}\sum_{i=1}^{n} u_i(-k)u_i(-m+k)\right)(1) = \frac{n}{2}\sum_{k=1}^{m-1} k(m-k) = \frac{n}{12}(m^3-m).$$

In order to prove the relation (2.29) we compare the commutation relations

$$\left[D_m, \exp\frac{z^k}{k}\gamma(-k)\right] = -z^k\gamma(m-k)\exp\frac{z^k}{k}\gamma(-k);$$

$$[D_m, \exp(\gamma + (\ln z)\partial_\gamma)] = -\gamma(m)\exp(\gamma + (\ln z)\partial_\gamma;$$

and the differentiation

$$-z^{m+1}\frac{d}{dz}\exp\frac{z^k}{k}\gamma(-k) = -z^{k+m}\gamma(-k)\exp\frac{z^k}{k}\gamma(-k);$$

$$-z^{m+1}\frac{d}{dz}\exp(\gamma + (\ln z)\partial_\gamma) = -z^m\gamma(0)\exp(\gamma + (\ln z)\partial_\gamma).$$

Therefore $[D_m, X(\gamma, z)]$ and $-z^{m+1}\frac{d}{dz}X(\gamma, z)$ differ only by m permutations of the factors. Each permutation adds a scalar factor $\frac{1}{2}z^{m\langle\gamma,\gamma\rangle}$ and we get (2.29).

In the framework of the basic representation the Virasoro operators have a very simple interpretation, which explains all the relations (2.27)-(2.30). Let $\tilde{\mathfrak{g}}$ be an affine Lie algebra. Let d_k denotes the derivation of the subalgebra $\hat{\mathfrak{g}} = (\mathbb{C}[t,t^{-1}]\otimes_{\mathbb{C}}\mathfrak{g})\oplus\mathbb{C}c$ of $\tilde{\mathfrak{g}}$ which on the subspace $\mathbb{C}[t,t^{-1}]\otimes_{\mathbb{C}}\mathfrak{g}$ is defined as $t^{k+1}\frac{d}{dt}$ and $d_k(c)=0$; note that $d_0=d$.

Proposition 2.8. *Let $\pi: \tilde{\mathfrak{g}} \to \operatorname{End} V$ be the representation of the Lie algebra $\tilde{\mathfrak{g}}$ defined in the statement of Theorem 1. Then one has:*

$$[D_m, \pi(x)] = \pi(d_m(x)) \quad \textit{for any } x\in\hat{\mathfrak{g}}\subset\tilde{\mathfrak{g}},$$

in other words, the commutation with D_m induces on $\hat{\mathfrak{g}}$ the derivation d_m.

Proof follows from Theorem 1, formulas (2.27) and (2.29) and the obvious relation:

$$d_m(\hat{E}_\gamma(z)) = -z^m\left(m+z\frac{d}{dz}\right)\hat{E}_\gamma(z), \qquad \gamma\in\Delta.$$

We will denote now by $\mathscr{D}$ the complex infinite-dimensional algebra spanned by d_m, $m\in\mathbb{Z}$. Of course, we have the following commutation relations:

$$[d_m, d_l] = (l-m)d_{m+l}.$$

This algebra is known as the Lie algebra of vector fields on the circle and the representation by operators $e^{2\pi i n\varphi}\frac{d}{d\varphi}$ in the space of $\mathfrak{g}$-valued functions on the circle is its natural representation. Let $\hat{\mathscr{D}}$ be the central extension of $\mathscr{D}$ defined by formula (2.30). We have obtained above the representation of $\hat{\mathscr{D}}$ by the Virasoro operators in the space V. The decomposition of the space V into a direct sum (see footnote 5) of irreducible representations of $\hat{\mathscr{D}}$ gives very useful information about the structure of $\hat{\mathscr{D}}$-modules [7]. Note that both algebras $\mathscr{D}$ and $\hat{\mathscr{D}}$ have a 3-dimensional subalgebra generated by d_{-1}, d_0, d_1 and D_{-1}, D_0, D_1 which is isomorphic to $SL(2,\mathbb{C})$ and plays an essential role in the theory (see § 3). One can extend the antilinear involution $\tilde{\omega}$ of $\tilde{\mathfrak{g}}$ to $\hat{\mathscr{D}}$ by

$$\tilde{\omega}(D_m) = -D_{-m}$$

and consider the fixed point subalgebra $\hat{\mathscr{D}}_0$ of $\hat{\mathscr{D}}$. Then it is easy to see that the Hermitian form $\langle\,|\,\rangle$ on the space V is contravariant with respect to $\hat{\mathscr{D}}$.[5]

Note finally that the intersection $\hat{\mathscr{D}}\cap sl(2,\mathbb{C})$ is a subalgebra of $\hat{\mathscr{D}}_0$ isomorphic to $su(1,1)$, which will become important in § 3.

2.9. Scalar Subalgebra and Veneziano Amplitude. In our realization of the basic representation the action of the homogeneous Heisenberg subalgebra $\hat{\mathfrak{s}}$ and of the Cartan subalgebra $\tilde{\mathfrak{h}}$ look particularly simple. However, for some other important subalgebras the formulas are very complicated. A striking example is the "scalar" subalgebra $\mathfrak{g}\subset\tilde{\mathfrak{g}}$. Recently one of the authors obtained a simple

[5] This implies that V is completely reducible with respect to $\hat{\mathscr{D}}$.

formula for the multiplicities of irreducible constitutents for the action of g in the eigenspaces of the energy operator D_0 [18].

Unfortunately we do not know a more explicit description of this decomposition. In this section we make some steps in this direction and show a connection between the problem and the scattering amplitudes of the scalar particles.

First, we remark that the completion $\bar{V}$ of V can be also viewed as the completion defined by the $\mathbb{Z}_+$-gradation $V=\oplus V_{-k}$ by the eigenspaces of the energy operator D_0. We extend the Hermitian form $\langle | \rangle$ to any two elements $u=\sum u_k$, $v=\sum v_k$, u_k, $v_k \in V_{-k}$, from $\bar{V}$ by

$$\langle u | v \rangle := \sum_{k \geq 0} \langle u_k | v_k \rangle,$$

provided that this series converges absolutely. In particular, we obtain: $V \subset V_H \subset \bar{V}$, where V_H is the Hilbert completion of V. In what follows it is easy to check that the form $\langle | \rangle$, each time when it appears, is well-defined.

Now we introduce the following "overcomplete basis" of $\bar{V}$, which is called the *coherent states basis:*

$$v(\gamma_1, z_1; \gamma_2, z_2; \ldots; \gamma_N, z_N) := X(\gamma_1, z_1) \ldots X(\gamma_N, z_N) v_0,$$

where $|z_1|>|z_2|>\ldots>|z_N|$ (due to this condition the expression above has sense). One has the following well-known fact in the dual-resonance theory.

Proposition 2.9.

a) $\langle v(\gamma_1, z_1; \ldots; \gamma_N, z_N) | v_0 \rangle = \prod_{i=1}^{N} z_i \prod_{i<j} (z_i - z_j)^{\langle \gamma_i, \gamma_j \rangle}$,

if $\sum_{i=1}^{N} \gamma_i = 0$, *and* 0 *otherwise* ($|z_1|>\ldots>|z_N|$).

b) $\langle v(\beta_1, w_1; \ldots; \beta_M, w_M) | v(\alpha_1, z_1; \ldots; \alpha_N, z_N) \rangle$

$$= \prod_{i=1}^{M} w_i \prod_{j=1}^{N} \bar{z}_j \prod_{i<j}^{M} (w_i - w_j)^{\langle \beta_i, \beta_j \rangle}$$

$$\prod_{i>j}^{N} (\bar{z}_i - \bar{z}_j)^{\langle \alpha_i, \alpha_j \rangle} \prod_{i=1}^{M} \prod_{j=1}^{N} (1 - \bar{z}_j w_i)^{-\langle \alpha_j, \beta_i \rangle}$$

if $\sum_{i=1}^{M} \beta_i = \sum_{j=1}^{N} \alpha_j$, *and* 0 *otherwise* ($|w_M|^{-1}>\ldots>|w_1|^{-1}>|z_1|>\ldots>|z_N|$).

c) $v(\gamma_1, z_1; \ldots \gamma_N, z_N) \in V_H$ *provided that* $|z_k|<1$, $k=1, \ldots, N$.

Proof of a) is based on the formula $e^A e^B = e^{[A,B]} e^B e^A$, when $[A, B]$ commutes with A and B. We have

$$e^A e^B = \left(1 - \frac{z_2}{z_1}\right)^{\langle \gamma_1, \gamma_2 \rangle} e^B e^A,$$

when

$$A=-\sum_{n>0}\frac{z_1^{-n}}{n}\gamma_1(n),\qquad B=\sum_{n\geq 0}\frac{z_2^n}{n}\gamma_2(-n),\qquad |z_1|>|z_2|,$$

and

$$e^Ae^B=z_1^{\langle\gamma_1,\gamma_2\rangle}e^Be^A,$$

when

$$A=(\ln z_1)\partial_{\gamma_1},\qquad B=\gamma_2.$$

b) follows from a) and from the contravariance of the Hermitian form $\langle\,|\,\rangle$.

c) follows from b).

Now we can find the matrix elements of the operators of $\mathfrak{g}$ and of the universal enveloping algebra $U(\mathfrak{g})$ as well. We have

$$X_0(\gamma)v=\int_{C_R}X(\gamma,z)v\frac{dz}{z},$$

were C_R is a circle of radius R with the center at the origin. Let $v_1=v(\beta_1,w_1;\ldots;\beta_M,w_M)$ and $v_2=v(\alpha_1,z_1;\ldots;\alpha_N,z_N)$ be two coherent states. Then the matrix element

$$\langle X_0(\gamma)v_1|v_2\rangle=\left(\int_{C_R}\prod_{j=1}^{N}(z-z_j)^{\langle\gamma,\alpha\rangle}\prod_{i=1}^{M}(z-w_i)^{\langle\gamma,\beta_i\rangle}dz\right)\langle v_1|v_2\rangle,$$

where

$$|w_1|^{-1}>R>|z_1|,$$

can be calculated explicitly by making use of the residue formula.

We can also calculate the matrix elements for the operators of $U(\mathfrak{g})$, i.e.,

$$\langle X_0(\gamma_1)X_0(\gamma_2)\ldots X_0(\gamma_k)v_1|v_2\rangle.$$

This is equivalent to the calculation of the following important function:

$$A=\int_{C_{R_r}}\ldots\int_{C_{R_s}}\langle X(\gamma_1,z_1)\ldots X(\gamma_r,z_r)\ldots X(\gamma_s,z_s)\ldots X(\gamma_N,z_N)v_0|v_0\rangle \cdot\frac{dz_r}{z_r}\frac{dz_{r+1}}{z_{r+1}}\cdots\frac{dz_s}{z_s} \tag{2.31}$$

where $|z_1|>\ldots>R_r>\ldots>R_s>\ldots>|z_N|$. This calculation also can be done explicitly by the residue formula.

We mention here that the function A (defined by (2.31)) plays an important role in physics. If we take z_k on the real line and the domain of integration is

$$z_{r-1}<z_r<\ldots<z_s<z_{s+1},$$

then A, up to a scalar multiple, becomes the scattering amplitude of the high energy processes. For example, if $r=2$ and $s=N-2$, then we get the so-called Koba-Nielson amplitude. As we will show in the next section, A has interesting projective properties. Due to these properties we can choose three parameters arbitrarily, say, $z_1=0$, $z_{N-1}=1$, $z_N=\infty$. Then A receives the standard form of the

Koba-Nielson amplitude:

$$A=\int_U \prod_{\substack{i,j=1\\ i>j}}^{N-1} (z_i-z_j)^{\langle\gamma_i,\gamma_j\rangle}dz_2\ldots dz_{N-2},$$
$$U=\{0=z_1<z_2\ldots<z_{N-1}=1\}. \tag{2.32}$$

If, finally, $N=4$, $z_2=x$, then we have

$$A=\int_0^1 (1-x)^{\langle\gamma_3,\gamma_2\rangle}x^{\langle\gamma_2,\gamma_1\rangle}dx=B(\langle\gamma_2,\gamma_3\rangle+1,\langle\gamma_2,\gamma_1\rangle+1)$$
$$=\frac{\Gamma(\langle\gamma_3,\gamma_2\rangle+1)\Gamma(\langle\gamma_2,\gamma_1\rangle+1)}{\Gamma(\langle\gamma_3,\gamma_2\rangle+\langle\gamma_2,\gamma_1\rangle+2)} \tag{2.33}$$

where B and Γ are the classical beta and gamma functions. The last formula for the scattering amplitude was discovered by Veneziano. This was the starting point for the dual resonance models.

3. Functional Realization of Basic Representations

In this section we give a functional realization of the basic representation. In this realization the vertex and the Virasoro operators receive a very simple form. This allows us to construct explicitly the action of the group corresponding to some subalgebras in the affine Lie algebra and in the Virasoro algebra. Finding explicit formulas for the action of the whole affine Lie group in the space of the basic representation is a very interesting open problem.

3.1. Holomorphic Functionals on a Hardy Space. First, we construct a realization of the canonical representation of a general Heisenberg system. We consider again the infinite-dimensional Heisenberg algebra $\hat{\mathfrak{s}}=\sum_{k\neq 0}\mathfrak{s}_k\oplus\mathbb{C}c$. For each $k>0$ we consider the $(2n+1)$-dimensional subalgebra $\mathfrak{a}_k=\mathfrak{s}_{-k}\oplus\mathbb{C}c\oplus\mathfrak{s}_k$ of $\hat{\mathfrak{s}}$; recall that the commutation relations in $\mathfrak{a}_k$ are

$$[t^k\otimes h_1,t^{-k}\otimes h_2]=k\langle h_1,h_2\rangle c, h_1,h_2\in\mathfrak{h}.$$

It is well known that an irreducible representation π of $\mathfrak{a}_k$ in a Hilbert space, for which

$$\pi(c)=\mathrm{Id},\qquad \pi(t^k\otimes h)^*=\pi(t^{-k}\otimes\bar{h}),\qquad h\in\mathfrak{h}, \tag{3.1}$$

(* denotes the Hermitian conjugation) is equivalent to the representation π_k in the space $\mathscr{H}_k$ of complex analytic functions on $\mathfrak{h}$ with the scalar product (ν_k is the normalized Lebesgue measure on $\mathfrak{h}$):

$$\langle x|y\rangle_k=\int_{\mathfrak{h}} x(h)\overline{y(h)}\, e^{-\frac{1}{k}H(h,h)}d\nu_k(h),$$

which is defined by:

$$\left.\begin{aligned}\pi_k(t^{-k}\otimes h_1)x(h)&=\langle h,h_1\rangle x(h)\\ \pi_k(t^k\otimes h_1)x(h)&=k\frac{\partial}{\partial h_1}x(h)\\ \pi_k(c)&=\mathrm{Id}.\end{aligned}\right\}\qquad(3.2)$$

It follows that an irreducible representation of an algebra $\mathfrak{s}^N=\sum_{\substack{k=-N\\k\neq 0}}^{N}\mathfrak{s}_k\oplus\mathbb{C}c$, which satisfies (3.1), is equivalent to the representation π^N in the space $\mathscr{H}^N$ of complex analytic functions on $\mathfrak{s}_+^N=\sum_{k=1}^{N}\mathfrak{s}_k$ with the scalar product

$$\langle x|y\rangle^{(N)}=\int_{\mathfrak{s}_+^N}x(h)\overline{y(h)}e^{-\left(\sum_{k=1}^{N}\frac{1}{k}H(h_k,h_k)\right)}d\nu^N,$$

where ν^N is the product of the measures ν_k and $h=\sum_{k=1}^{N}h_k\otimes t^k\in\mathfrak{s}_+^N$. Let μ^N denote the measure such that

$$d\mu^N(h)=\exp\left(-\sum_{k=1}^{N}\frac{1}{k}H(h_k,h_k)\right)d\nu^N.$$

We tend N to ∞. It is well-known that then $\mathfrak{s}_+$ has a zero total measure and therefore it is necessary to consider a completion of $\mathfrak{s}_+$.

From now on we shall consider the space $\mathbb{C}[t,t^{-1}]\otimes_{\mathbb{C}}\mathfrak{g}$ as the space of polynomial mappings from $\mathbb{C}\setminus 0$ to $\mathfrak{g}$. From this point of view it is more natural to write ht^k instead of $t^k\otimes h$. The space $\mathfrak{s}_+$ can be viewed then as the space of $\mathfrak{h}$-valued polynomials on $\mathbb{C}$, vanishing at $0\in\mathbb{C}$.

We introduce on $\mathfrak{s}_+$ a new scalar product $(\,,)$ which is defined by $\left(t\frac{dh}{dt},g\right)=\tilde{H}(h,g)$. More explicitly:

$$\left(\sum_{k\geq 1}h_kt^k,\sum_{k\geq 1}g_kt^k\right)=\sum_{k\geq 1}\frac{1}{k}H(h_k,g_k).$$

The space $\mathfrak{s}_+$ provided with the scalar product $(\,,)$ is a pre-Hilbert space. Let $\mathfrak{s}_+^H$ be the completion of $\mathfrak{s}_+$.

We recall now the well-known fact (which is easy to check) that the Hilbert space of sequences of complex numbers

$$\left\{a=(a_1,a_2,\ldots)\mid \|a\|^2=\sum_{k=1}^{\infty}\frac{1}{k}a_k\bar{a}_k<\infty\right\}$$

is isomorphic under the identification $a=a(t):=\sum_{k=1}^{\infty}a_kt^k$ to the Hilbert space of analytic functions on the unit disc D vanishing at the origin and such that

$$\|a(t)\|^2=\int_D|a(t)|^2\frac{d\nu(t)}{t\bar{t}}<\infty,$$

where $\nu(t)$ is the normalized Lebesgue measure on D.

As a consequence we obtain a realization of the Hilbert space $\mathfrak{s}_+^H$ as the space of $\mathfrak{h}$-valued analytic functions on D vanishing at 0 with the scalar product

$$(h, g) = \int_D H(h(t), g(t)) \frac{d\nu(t)}{t\bar{t}}.$$

Now we are in a position to construct the Gaussian measure associated with the Hilbert structure of $\mathfrak{s}_+^H$ (see [3]). Let $\mathfrak{s}_+^\infty \subset \mathfrak{s}_+^H$ be the topological space of $\mathfrak{h}$-valued C^∞-functions which can be extended inside the circle to an analytic function, vanishing at the origin. Let $\mathfrak{s}_+^D \supset \mathfrak{s}_+^H$ be the space dual to $\mathfrak{s}_+^\infty$. The Gaussian measure μ_+ on $\mathfrak{s}_+^D$ is defined by the property that the measures μ^N are projections of μ_+ on the finite dimensional subspaces $\mathfrak{s}_+^N$. Let now $\mathscr{L}^2(\mathfrak{s}_+^D, \mu_+)$ be the space of the square integrable functions on $\mathfrak{s}_+^D$ and $\mathscr{L}^2_{\text{anal}}(\mathfrak{s}_+^D, \mu_+)$ be the subspace of square integrable analytic functions on $\mathfrak{s}_+^D$ with respect to the natural complex structure on $\mathfrak{s}_+^D$. We denote by μ_0 the discrete measure on $\mathfrak{h}$ such that $\mu_0(\gamma) = 1$ for $\gamma \in \Gamma$ and $\mu_0(h) = 0$ otherwise.

Let $\mathscr{H}(\mathfrak{h}) = \mathfrak{s}_+^H \oplus \mathfrak{h}$ be a Hardy space of $\mathfrak{h}$-valued analytic functions in D with the semidefinite scalar product which is a trivial extension from $\mathfrak{s}_+^H$. Set $\mathscr{H}^D(\mathfrak{h}) = \mathfrak{s}_+^D \oplus \mathfrak{h}$ and let μ denote the measure $\mu_+ \otimes \mu_0$ on $\mathscr{H}^D(\mathfrak{h})$.

We set

$$\mathscr{V}^H = \mathscr{L}^2_{\text{anal}}(\mathfrak{s}_+^D, \mu_+) \otimes \mathscr{L}^2(\mathfrak{h}, \mu_0) = \mathscr{L}^2_{\text{anal}}(\mathscr{H}^D(\mathfrak{h}), \mu).$$

There is a natural imbedding of the space $V = S(\mathfrak{s}_-) \otimes \mathbb{C}(\Gamma)$ into $\mathscr{V}^H$ with a dense image. We denote this image by $\mathscr{V}$.

We denote $h'(t) := t \dfrac{dh(t)}{dt}$ for $h(t) \in \mathfrak{s}_+$. We have now the following:

Proposition 3.1. *The canonical representation of the Heisenberg system $(\bar{\mathfrak{s}}, \Gamma)$ associated with the cocycle ε in the Hilbert space $\mathscr{V}^H$ is given by the following formulas:*

$$\left.\begin{aligned} \pi(h_1(t)) F(h(t)) &= \langle h_1(t), h(t) \rangle F(h(t)), \; h_1(t) \in \mathfrak{s}_-; \\ \pi(h_1(t)) F(h(t)) &= \frac{\partial}{\partial h_1'(t)} F(h(t)), \; h_1(t) \in \mathfrak{s}_+; \\ \pi(h_1) F(h(t)) &= \langle h_1, h(0) \rangle F(h(t)); \; \pi(c) = \mathrm{Id}. \end{aligned}\right\} \tag{3.3}$$

$$\pi(\gamma) F(h(t)) = \tilde{\varepsilon}(h(0), \gamma) F(h(t) + \gamma), \quad \gamma \in \Gamma, \tag{3.4}$$

where $\tilde{\varepsilon}$ is any function on $\mathfrak{h} \times \mathfrak{h}$ such that $\tilde{\varepsilon}(\beta, \gamma) = \varepsilon(\beta, \gamma)$, for $\beta, \gamma \in \Gamma$.

Let now $\mathfrak{g}$ be a Lie algebra of one of the types A_n, D_n, E_n, and $(\bar{\mathfrak{s}}_1, Q)$ be a Heisenberg system associated with the root lattice Q. We can construct, using Proposition 3.1, the canonical representation of the Heisenberg system associated with the cocycle ε satisfying (2.3)–(2.5)′ in the Hilbert space $\mathscr{V}^H$. Our main result (Theorem 1) asserts that we can extend this representation to the representation of the whole algebra $\tilde{\mathfrak{g}}$ in the space $\mathscr{V}^H$, which is equivalent to the basic representation. Later on we will reprove this result directly. Now we only mention that the representation of the derivation $d \in \tilde{\mathfrak{g}}$ is given by

$$\pi(d)F(h(t)) = \left(-t\frac{d}{dt} - \tfrac{1}{2}\langle h(0), h(0)\rangle\right)F(h(t)), \tag{3.5}$$

which easily follows from the definition of the energy operator (2.22).

3.2. The Action of the Heisenberg Group. Now we consider the representation of the group corresponding to the Lie algebra $\hat{\mathfrak{s}}$ in the space $\mathcal{V}^H$. For $h(t) = \sum_k h_k t^k \in \hat{\mathfrak{s}}$ we denote by $h_+(t)$ (respectively, $h_-(t)$) the projection on $\mathfrak{s}_+ \oplus \mathfrak{h}$ (respectively $\mathfrak{s}_- \oplus \mathfrak{h}$).

Proposition 3.2. *One has:*

a) $\exp\pi(h_1(t))F(h(t)) = \exp(\tfrac{1}{2}\langle h_1'(t)_+, h_1(t)_-\rangle$
$\qquad + \langle h_1(t)_-, h(t)\rangle)F(h(t) + h_1'(t)_+),$

where $h_1(t) \in \hat{\mathfrak{s}}$.

b) $\exp\pi(h_1(t))\exp\pi(h_2(t))$
$\qquad = \exp\tfrac{1}{2}(\langle h_1'(t)_+, h_2(t)_-\rangle - \langle h_2'(t)_+, h_1(t)_-\rangle \exp\pi(h_1(t) + h_2(t)),$

where $h_1(t), h_2(t) \in \hat{\mathfrak{s}}$.

Proof. We use that $e^{A+B} = e^A e^B e^{-\frac{1}{2}[A,B]}$ if $[A, B]$ commutes with A and B.

a) We have:

$$\exp\pi(h(t)) = \exp\left(\tfrac{1}{2}\pi(c)\sum_{k=1}^{\infty} k\langle h_k, h_{-k}\rangle\right)$$
$$\cdot\exp\pi(h(t)_-)\exp\pi(h(t)_+ - h(0)).$$

Using Proposition 3.1 we obtain:

$$\exp\pi(h_1(t))F(h(t)) = \exp(\tfrac{1}{2}\langle h_1'(t)_+, h_1(t)_-\rangle)\exp\langle h_1(t)_-, h(t)\rangle$$
$$\cdot\exp\frac{\partial}{\partial h_1'(t)_+}F(h(t)).$$

The proof of b) is similar.

Let $\hat{\mathfrak{s}}_0$ be the real subalgebra of $\hat{\mathfrak{s}}$ fixed by the antilinear involution $\tilde{\omega}$ (see Sect. 1.5). Clearly, this subalgebra is represented in $\mathcal{V}^H$ by Hermitian operators. Denote by $\tilde{S}_0$ the group of (unitary) operators generated by $\exp\pi(h)$, $h \in \hat{\mathfrak{s}}_0$.

Finally, let $\tilde{S}_T$ denote the group generated by $\tilde{S}_0$ and the group of translations T (which also acts by unitary operators in $\mathcal{V}^H$, see Theorem 1). The following proposition describes the action of the group $\tilde{S}_T$ in the space $\mathcal{V}^H$ (cf. [16], p. 333).

Proposition 3.3. *The representation of the group $\tilde{S}_T$ in the space $\mathcal{V}^H$ is an irreducible unitary representation which is given by the following formulas:*

$$\pi((g(t), a))F(h(t)) = a e^{-\frac{1}{2}(\eta(g), \eta(g)) - (h, \eta(g))} F(h(t) + \eta(g))$$
$$\pi((g, \gamma, a))F(h(t)) = a\tilde{\varepsilon}(h(0), \gamma)e^{\langle x(t), h(0)\rangle}F(h(t) + \gamma).$$

Here $g(t) = \exp x(t)$, *where*

$$x(t) = \sum_{k \neq 0} x_k t^k \in \hat{\mathfrak{s}}_0, \quad \eta(g) := (g^{-1}(t)g'(t))_+, \quad \alpha \in \mathbb{C}, \quad |a| = 1, \ \gamma \in Q.$$

3.3. The Realization of the Vertex Operators. Now we write down the action of the vertex operators $X(\gamma, z)$, $\gamma\in\Gamma$, in the space $\mathscr{V}\subset\mathscr{V}^H$. From Proposition 3.1 and the definition of the vertex operator (see formula (2.23)) we obtain:

$$X(\gamma, z)\,F(h(t))=\exp\left(\sum_{k\geq 1}\frac{z^k}{k}\langle\gamma, h_k\rangle\right)\exp\left((\ln z)\langle\gamma, h_0\rangle-\frac{\partial}{\partial\gamma}\right)$$
$$\exp\left(-\sum_{k\geq 1}\frac{\partial}{\partial(\gamma t^k)}z^{-k}\right)F(h(t)),$$

where $h(t)=\sum_{k\geq 0}h_k t^k$.

We can extend the function $F\in\mathscr{V}$ from the space $\mathfrak{s}_+^D\oplus\mathfrak{h}$ to the space $\bar{\mathfrak{s}}_+\oplus\mathfrak{h}$. Since

$$\exp\left(-\frac{\partial}{\partial(\gamma t^k)}z^k\right)F(h(t))=F(h(t)-\gamma t^k z^{-k}),$$

we obtain (see (2.24)):

$$X(\gamma, z)\,F(h(t))$$
$$=z^{-\frac{1}{2}\langle\gamma,\gamma\rangle}\exp\left\langle\gamma, h_0\ln z+\sum_{k=1}^{\infty}\frac{1}{k}h_k z^k\right\rangle F\left(h(t)-\sum_{k=0}^{\infty}\gamma t^k z^{-k}\right). \tag{3.6}$$

One notices immediately that $h_0\ln z+\sum_{k=1}^{\infty}\frac{1}{k}h_k z^k$ is the antiderivative

$$\tilde{h}(z):=\int h(z)\frac{dz}{z}.$$

The series $\sum_{k=0}^{\infty}\gamma t^k z^{-k}$ converges inside the circle $|t|<|z|$ to $\frac{\gamma}{1-tz^{-1}}$, and we can consider the analytic continuation inside the unit circle. Note that $h(t)-\frac{\gamma}{1-tz^{-1}}$ does not lie in $\mathscr{H}^D(\mathfrak{h})$; however, $F\in\mathscr{V}$ can be extended by the formula:

$$F\left(h(t)-\frac{\gamma}{1-tz^{-1}}\right):=F\left(h(t)-\frac{\gamma}{1-tz^{-1}}+\frac{\gamma(tz^{-1})^N}{1-tz^{-1}}\right),$$

where N is sufficiently large.

So we obtain a simpler form for the vertex operator.

Proposition 3.4. *For $F\in\mathscr{V}$, $|z|=R<1$, one has:*

$X(\gamma, z)\,F(h(t))$

$$X(\gamma, z)\,F(h(t))=z^{-\frac{1}{2}\langle\gamma,\gamma\rangle}\exp\langle\gamma, \tilde{h}(z)\rangle\, F\left(h(t)-\frac{\gamma}{1-tz^{-1}}\right); \tag{3.7}$$

$$X_k(\gamma)\,F(h(t))=\int_{C_R}\exp\langle\gamma, \tilde{h}(z)\rangle\, F\left(h(t)-\frac{\gamma}{1-tz^{-1}}\right)z^{-k}dz, \tag{3.8}$$

where $C_R=\{z\in\mathbb{C}\,|\,|z|=R\}$ and $\langle\gamma,\gamma\rangle=2$.

Note that if $\bar{h}(z)=\int h(z)\frac{dz}{z}$ has an analytic continuation inside the circle C_R of radius $R>1$, then (3.7), (3.8) can be defined inside this circle.

Similarly, we can determine the sequential action of two vertex operators:

$$\begin{aligned} &X(\beta, z_0)\,X(\gamma, z)\,F(h(t))= \\ &=(z^{-1}-z_0^{-1})^{\langle\beta,\gamma\rangle} z_0^{-\frac{1}{2}\langle\beta,\beta\rangle} z^{-\frac{1}{2}\langle\gamma,\gamma\rangle} e^{\langle\beta,\bar{h}(z_0)\rangle} e^{\langle\gamma,\bar{h}(z)\rangle} \\ &\quad F\left(h(t)-\frac{\beta}{1-tz_0^{-1}}-\frac{\gamma}{1-tz^{-1}}\right), \end{aligned} \tag{3.9}$$

provided that $|z|<|z_0|$. We used here (3.6) and two formulas:

$$\left[\exp-\frac{\partial}{\partial\beta},\ \exp(\ln z)\langle\gamma,\cdot\rangle\right]=z^{-\langle\beta,\gamma\rangle}$$

and

$$\left[\exp\left(-\sum_{k\geq 1}\frac{\partial}{\partial(\beta t^k)}z_0^{-k}\right),\ \exp\sum_{k\geq 1}\frac{z^k}{k}\langle\gamma t^{-k},\cdot\rangle\right]=(1-zz_0^{-1})^{\langle\beta,\gamma\rangle}.$$

Finally, we write down the formula for the sequential action of N vertex operators

$$\begin{aligned} &X(\gamma_1, z_1)\ldots X(\gamma_N, z_N)\,F(h(t))= \\ &=\sum_{k=1}^{N} z_k^{-\frac{1}{2}\langle\gamma_k,\gamma_k\rangle}\prod_{i>j}(z_i^{-1}-z_j^{-1})^{\langle\gamma_i,\gamma_j\rangle} e^{\sum_{n=1}^{N}\langle\gamma_k,\bar{h}(z_k)\rangle} F\left(h(t)-\sum_{k=1}^{N}\frac{\gamma_k}{1-tz_k^{-1}}\right) \end{aligned} \tag{3.10}$$

provided that

$$|z_1|>|z_2|>\ldots>|z_N|. \tag{3.11}$$

If $F=\delta_{0,h(0)}$ is an element in $\mathscr{V}$ corresponding to the vacuum vector $v_0\in V$, then we get easily from (3.10) the result of Proposition 2.9.

Notice that the right-hand side of (3.10) has sense even when (3.11) is not satisfied. We denote the operator defined by the right-hand side of (3.10) for any $(z_1, \ldots, z_N)$, such that $z_i\neq z_j$ for $i\neq j$, by

$$:X(\gamma_1, z_1)\ldots X(\gamma_N, z_N):.$$

For example, we have

$$:X(\beta, z_1)\,X(\gamma, z_2):=\begin{cases} X(\beta, z_1)\,X(\gamma, z_2) & \text{if } |z_1|<|z_2| \\ (-1)^{\langle\beta,\gamma\rangle}X(\gamma, z_2)\,X(\beta, z_1) & \text{if } |z_2|>|z_2|. \end{cases}$$

3.4. The Main Theorem: A Direct Proof. In this section we want to obtain the commutation relations between the homogeneous parts $X_k(\gamma)$ of the vertex operators $X(\gamma, z)$ by direct computations and give another proof of Theorem 1.

From now on we assume that our affine Lie algebra $\tilde{\mathfrak{g}}$ is associated with a simple Lie algebra of type A_n, D_n or E_n. In this case we have $\Gamma=Q$, where Q is the root lattice of $\mathfrak{g}$, and the cocycle ε of Q defined in Sect. 2.3. First, we prove one important proposition which contains the essential information about the commutation relations.

Proposition 3.5. *Let $\beta, \gamma \in \Delta \subset Q$. Then*

$$\operatorname*{Res}_{z=z_0} \{z^m(:X(\beta, z_0)\, X(\gamma, z):)\, F(h(t))\}$$

$$= \begin{cases} 0 & \text{if } \beta+\gamma \notin \Delta \cup \{0\} \\ -z_0^{m+1} X(\beta+\gamma, z_0)\, F(h(t)) & \text{if } \beta+\gamma \in \Delta \\ z_0^{m+1}(\pi(\gamma(z_0)) + (m+1))\, F(h(t)) & \text{if } \beta = -\gamma. \end{cases}$$

Proof. We denote $z^m(:X(\beta, z_0)\, X(\gamma, z):)\, F(h(t))$ by $f(z)$. It is clear from (3.9) that $f(z)$ has no poles in $z = z_0$ when $\langle \beta, \gamma \rangle \geqq 0$, or equivalently, when $\beta+\gamma \notin \Delta \cup \{0\}$. This proves the formula in the first case.

If $\langle \beta, \gamma \rangle = -1$, or equivalently $\beta+\gamma \in \Delta$, we have a pole of order 1 in $z = z_0$. Therefore the residue is equal to $z_0^{m+1} X(\beta+\gamma, z_0)\, F(h(t))$ by formula (3.9) and Proposition 3.4.

Finally, if $\langle \beta, \gamma \rangle = -2$, or equivalently, $\beta + \gamma = 0$, we have a pole of order 2 in $z = z_0$. Therefore, the residue is equal to

$$\frac{\partial}{\partial z}(z - z_0)^2 f(z)\Big|_{z=z_0}.$$

Using the calculation of the derivative

$$\frac{\partial}{\partial z} F\left(h(t) - \frac{\gamma}{1 - tz^{-1}} + \frac{\gamma}{1 - tz_0^{-1}}\right)\Bigg|_{z=z_0}$$

$$= \sum_{k=0}^{\infty} \sum_{i=1}^{n} \frac{\partial}{\partial z}\left((\langle u_i, \gamma \rangle z^{-k}) \frac{\partial}{\partial(\gamma t^k)} F(h(t))\right)\Bigg|_{z=z_0}$$

$$= z_0^{-1} \sum_{k \geqq 1} k z_0^{-k} \frac{\partial}{\partial(\gamma t^k)} F(h(t)).$$

we obtain the formula in the third case:

$$\operatorname*{Res}_{z=z_0} f(z) = (m+1)\, z_0^{m+1} F(h(t)) + z_0^{m+1} \langle \gamma, h(z_0) \rangle\, F(h(t))$$
$$+ z_0^{m+1} \sum_{k \geqq 1} k z_0^{-k} \frac{\partial}{\partial(\gamma t^k)} F(h(t)).$$

We define an operator c_γ, $\gamma \in Q$, on $\mathscr{V}$ by:

$$c_\gamma F(h(t)) = \bar{\varepsilon}(\gamma_1, h(0))\, F(h(t))$$

and an operator T_γ by

$$T_\gamma F(h(t)) = F(h(t) + \gamma).$$

We set:

$$X^\varepsilon(\gamma, z) = X(\gamma, z)\, c_\gamma, \quad X_k^\varepsilon(\gamma) = X_k(\gamma)\, c_\gamma. \tag{3.12}$$

Now we can give the second proof of Theorem 1.

Theorem 1'. a) *The map* π: $\tilde{\mathfrak{g}} \to \mathrm{End}\,\mathscr{V}$ *such that:* $\pi(c) = 1_{\mathscr{V}}$; $\pi(d)$ *is defined by* (3.5); $\pi(t^k \otimes h)$, $k \in \mathbb{Z}$, *is defined by* (3.3), $\pi(t^k \otimes E_\gamma) = X_k^\varepsilon(\gamma)$, $\gamma \in \Delta$, $k \in \mathbb{Z}$, *is a representation of the Lie algebra* $\tilde{\mathfrak{g}}$, *which is equivalent to the basic representation* $\pi_0 (= \pi_{\Lambda_0})$.

b) *Under the identification of* $\mathscr{V}$ *with* $V(\Lambda_0)$ *one has:* $T_\gamma^{\pi_0} = T_\gamma \cdot c_\gamma$, $\gamma \in Q$.

c) *The weight decomposition of* $\mathscr{V}$ *(with respect to* $\tilde{\mathfrak{h}}$) *has the following form:*

$$\mathscr{V} = \sum_{k,\gamma} \mathscr{V}_{\Lambda_0 - (k + \frac{1}{2}\langle\gamma,\gamma\rangle)\delta + \gamma},$$

where (k, γ) *ranges over* $\mathbb{Z}_+ \times Q$), *and*

$$\mathscr{V}_{\Lambda_0 - (k + \frac{1}{2}\langle\gamma,\gamma\rangle)\delta + \gamma} = \{F(\cdot) \in \mathscr{V} \mid F = \delta_{\gamma, h(0)} F_0,$$

where $\left(t\dfrac{d}{dt} - k\right) F_0(h(t)) = 0$ *and* $F_0(h(t))$ *does not depend on* $h(0)\}$.

d) *The Hermitian form* $\langle\ |\ \rangle$ *on* $\mathscr{V} \subset \mathscr{L}^2_{\mathrm{anal}}(\mathscr{H}(\mathfrak{h}), \mu)$ *is positive definite and contravariant.*

Proof. The central point we have to check is the commutation relations between $\pi(t^k \otimes E_\beta)$ and $\pi(t^m \otimes E_\gamma)$. We do it as follows:

$$\begin{aligned}
&[\sum_{k \in \mathbb{Z}} \pi(t^k \otimes E_\beta) z_0^{-k}, \pi(t^m \otimes E_\gamma)]\, F(h(t)) \\
&\quad = [X^\varepsilon(\beta, z_0), X_m^\varepsilon(\gamma)]\, F(h(t)) \\
&\quad = \int_{C_{R_1}} \{X^\varepsilon(\beta, z_0) X^\varepsilon(\gamma, z) F(h(t))\} z^{-m-1} dz \\
&\quad\quad - \int_{C_{R_2}} \{X^\varepsilon(\gamma, z) X^\varepsilon(\beta, z_0) F(h(t))\} z^{-m-1} dz,
\end{aligned}$$

where $R_1 < |z_0| < R_2$. Using the property of the cocycle ε we can rewrite this expression as

$$\varepsilon(\beta, \gamma) \int {:}X(\beta, z_0) X(\gamma, z){:}\, F(h(t)) z^{-m-1} dz$$

where we integrate along two circles with radii R_1 and R_2 in opposite directions. This integral can be calculated by the residues of the integrand in singular pojnts inisde the ring $R_1 < |z| < R_2$. From (3.7) we obtain that the integrand has only one singular point z_0 and

$$\varepsilon(\beta, \gamma) \operatorname*{Res}_{z = z_0} \{z^{-m-1} ({:}X(\beta, z_0) X(\gamma, z){:})\, F(h(t))\}$$

calculated in Proposition 3.5 gives us the right answer.

3.5. On the Vershik-Gelfand-Graev Representation τ [15]. Let τ_m be the representation of the group $SL_2(\mathbb{C})$ defined by:

$$\tau_m \begin{pmatrix} \alpha & \beta \\ \gamma & \delta \end{pmatrix} f(t) = (\alpha + \gamma t^{-1})^{-m} (\delta + \beta t)^{-m} f\left(\frac{\alpha t + \gamma}{\beta t + \delta}\right), \quad m \in \mathbb{Z}, \tag{3.13}$$

in some space of functions, which will be specified later. Set

$$a_0(s)=\begin{pmatrix} e^{-\frac{1}{2}s} & 0 \\ 0 & e^{\frac{1}{2}s}\end{pmatrix},\quad a_1(s)=\begin{pmatrix} 1 & s \\ 0 & 1\end{pmatrix},\quad a_{-1}(s)=\begin{pmatrix} 1 & 0 \\ -s & 1\end{pmatrix} \tag{3.14}$$

and let d_0, d_1, d_{-1}, be the corresponding elements of the Lie algebra. Then it is easy to check that

$$\tau_m(d_0)=-t\frac{d}{dt};\quad \tau_m(d_1)=-t^2\frac{d}{dt}-mt;\quad \tau_m(d_{-1})=-\frac{d}{dt}+mt^{-1}. \tag{3.15}$$

Consider now the subgroup $SU(1,1)\subset SL(2,\mathbb{C})$ $\left(\text{consisting of the matrices } \begin{pmatrix}\alpha & \beta \\ \bar{\beta} & \bar{\alpha}\end{pmatrix} \text{ with } |\alpha|^2-|\beta|^2=1\right)$. The representations τ_m, $m=1,2,\ldots$, of $SU(1,1)$ form the so-called discrete series in the space of analytic or antianalytic functions on the unit disc, which have zero of multiplicity at least m at the origin. (These functions are also determined by their boundary values on the circle). If m is any real number, formula (3.13) still has sense for the group $SU(1,1)$ operating in the space of C^∞-function on the circle. One can introduce a Hermitian structure such that the representations τ_m for $0<m<1$ become unitary representations of the so-called complementary series (notice that τ_m and τ_{1-m} are equivalent representations).

Proposition 3.6 [15]. *As $m\to 1$, the representation of the complementary series τ_m of $SU(1,1)$ becomes reducible and has three irreducible subquotients τ_1^0, τ_1^+ and τ_1^-. Here τ_1^0 is the 1-dimensional representation and $\tau_1^\pm$ are unitary representations in the Hardy space of analytic or antianalytic functions on the unit disc with the scalar product*

$$(f,g)=\sum_{k=1}^{\infty}\frac{1}{k}f_k\bar{g}_k,$$

where

$$f(z)=\sum_{k=1}^{\infty}f_kz^k\quad\left(\text{or } f(z)=\sum_{k=1}^{\infty}f_k\bar{z}^k\right).$$

Following [15], define a 1-cocycle on $SU(1,1)$ by $\beta(g)f(t)=f(0)-\tau_1(g)f(0)$, $g\in SU(1,1)$. The representation $\tau(g)=\tau_1(g)+\beta(g)$ turns out to be completely reducible. This representation plays a crucial role in the representation theory of the group $SL(2,\mathbb{R})^X$ [15].

We obtain from (3.15) that:

$$\left.\begin{aligned}&\tau(d_0)f(t)=-t\frac{d}{dt}f(t);\quad \tau(d_1)f(t)=-t^2\frac{d}{dt}f(t)-tf(t)+tf(0);\\ &\tau(d_{-1})f(t)=-\frac{d}{dt}f(t)+t^{-1}f(t)-t^{-1}f(0).\end{aligned}\right\} \tag{3.16}$$

Notice that the representation τ can be extended to a (non-unitary) representation of $SL(2,\mathbb{C})$.

Finally, we note that we can split off the antianalytic part of the space of representation τ. Then in a Hardy space $\mathscr{H}=\left\{f(t)=\sum_{k=0}^{\infty} f_k t^k\right\}$ with the scalar product degenerate on constants we can define a representation $\tau^+(g)=\tau_1(g)+\beta_+(g)$:

$$\tau^+\begin{pmatrix}\alpha & \beta\\ \gamma & \delta\end{pmatrix} f(t)=(\alpha+\gamma t^{-1})^{-1}(\delta+\beta t)^{-1} f\left(\frac{\alpha t+\gamma}{\beta t+\delta}\right)+\left[1-\left(\frac{\alpha}{\alpha+\gamma t^{-1}}\right)\right] f(0) \tag{3.17}$$

and we have

$$\left.\begin{aligned}&\tau^+(d_0) f(t)=-t\frac{d}{dt} f(t); \quad \tau^+(d_1) f(t)=-t^2\frac{d}{dt} f(t)-t f(t);\\ &\tau^+(d_{-1}) f(t)=-\frac{d}{dt} f(t)+t^{-1} f(t)-t^{-1} f(0).\end{aligned}\right\} \tag{3.18}$$

In the same way we can define $\tau^-(g)$, splitting off the analytic part.

3.6. The Action of the Group SL(2, ℂ) Under the Basic Representation. Consider the 3-dimensional subalgebra in the Virasoro algebra spanned by the operators D_0, D_1 and D_{-1}. Recall that (see Sect. 2.8):

$$D_0=-\frac{1}{2}\sum_{i=1}^{n} u_i(0)^2-\sum_{k\geq 1}\sum_{i=1}^{n} u_i(-k)\,u_i(k)$$

$$D_1=-\sum_{i=1}^{n} u_i(0)\,u_i(1)-\sum_{k\geq 1}\sum_{i=1}^{n} u_i(-k)\,u_i(k+1)$$

$$D_{-1}=-\sum_{i=1}^{n} u_i(-1)\,u_i(0)-\sum_{k\geq 1}\sum_{i=1}^{n} u_i(-k-1)\,u_i(k).$$

We denote the two summands in D_k, $k=0, \pm 1$, by D'_k and D''_k, respectively. Proposition 2.7 implies that the operators D_k, $k=0, \pm 1$, satisfy the same commutation relations as the elements d_k, $k=0, \pm 1$. The same is true for the operators D''_k, $k=0, \pm 1$. Moreover, we have:

Proposition 3.7. *The operators D''_k act in the space $\mathscr{V}$ by:*

$$\pi(D''_k)\,F(h(t))=F(\tau(d_k)\,h(t))$$

and therefore the action of the corresponding group is

$$\pi(g)\,F(h(t))=F(\tau(g)\,h(t)), \qquad g\in SL(2, \mathbb{C}).$$

Proof is a straightforward computation based on (3.16) and the following formula:

$$\frac{d}{ds}\pi(g(s))\,F(h(t))\Big|_{s=0}=\sum_{k=0}^{\infty}\sum_{i=1}^{n}\frac{\partial\langle\tau(g(s))\,h(t), u_i t^{-k}\rangle}{\partial s}\Big|_{s=0}\frac{\partial}{\partial(u_i t^k)} F(h(t)).$$

Remark. Notice that the restriction of the representation τ to the subspace of analytic function vanishing at 0 (which is isomorphic to s_+^0) is the representation τ_1^+. It is unitary with respect to the scalar product on s_+^H under the action of the subgroup $SU(1,1)$.

Proposition 3.8. *The action of the group $SL(2,\mathbb{C})$ in the space $\mathscr{V}$ corresponding to the Lie algebra $\mathbb{C}D_{-1}+\mathbb{C}D_0+\mathbb{C}D_1$,* is:

$$\pi(g)F(h(t))=\gamma^{-\langle h(0),h(0)\rangle}(\exp\langle h(0),\tilde{h}(\gamma\delta^{-1})\rangle)F(\tau^+(g)h(t))$$

where $g=\begin{pmatrix}\alpha & \beta\\ \gamma & \delta\end{pmatrix}\in SL(2,\mathbb{C})$, $\delta\neq 0$, $\tau^+(g)$ *is defined by* (3.17), *and* $\tilde{h}(z)=\int h(z)\frac{dz}{z}$.

Proof. From Proposition 3.7 and Definition of D_k we can find easily the following formulas for the operators D_k and the corresponding one-parameter subgroups:

$$e^{sD_0}F(h(t))=e^{-\frac{s}{2}\langle h(0),h(0)\rangle}F(\tau^+(a_0(s))h(t)),$$

$$e^{sD_1}F(h(t))=F(\tau^+(a_1(s))h(t)),$$

$$e^{-sD_{-1}}F(h(t))=\xi(h(\cdot),s)F(\tau^+(a_{-1}(-s))h(t)),$$

where

$$\xi(h(\cdot),s):=\exp\int\left\langle h(0),\frac{h(s)-h(0)}{s}\right\rangle ds.$$

It is easy to see that $\xi(h(\cdot),s)$ is a 1-cocycle and that

$$\frac{d}{ds}\xi(h(\cdot),s)\Big|_{s=0}=\langle h(0),h'(0)\rangle.$$

To finish the proof we use the decomposition:

$$\begin{pmatrix}\alpha & \beta\\ \gamma & \delta\end{pmatrix}=\begin{pmatrix}1 & \beta\delta^{-1}\\ 0 & 1\end{pmatrix}\begin{pmatrix}\delta^{-1} & 0\\ 0 & \delta\end{pmatrix}\begin{pmatrix}1 & 0\\ \gamma\delta^{-1} & 1\end{pmatrix},\qquad \delta\neq 0.$$

Remark. There is a striking similarity between the action of the subgroup generated by D_{-1}:

$$\pi\begin{pmatrix}1 & 0\\ z & 1\end{pmatrix}F(h(t))=z^{-\langle h(0),h(0)\rangle}(\exp\langle h(0),\tilde{h}(z)\rangle)F\left(\frac{t}{t+z}h(t+z)+\frac{h(0)}{1+tz^{-1}}\right)$$

and the action of the vertex operator:

$$X(h(0),z)F(h(t))=z^{-\frac{\langle h(0),h(0)\rangle}{2}}(\exp\langle h(0),\tilde{h}(z)\rangle)F\left(h(t)-\frac{h(0)}{1-tz^{-1}}\right).$$

Using the action of $SL(2,\mathbb{C})$ on $\tilde{\mathfrak{g}}$ we can get also the invariance of the matrix elements of $U(g)$ (see Sect. 2.9). Due to (2.29) for $m=0,\ \pm 1$, we have

$$[D_0, X(\gamma, z)] = -z\frac{d}{dz}X(\gamma, z),$$

$$[D_1, X(\gamma, z)] = \left(-z^2\frac{d}{dz} - z\right)X(\gamma, z),$$

$$[D_{-1}, X(\gamma, z)] = \left(-\frac{d}{dz} + z^{-1}\right)X(\gamma, z), \quad \gamma \in \Delta.$$

Hence, the action of the group $SL(2, \mathbb{C})$ defined by (3.13) is:

$$\pi(g)X(\gamma, z)\pi(g^{-1}) = (\alpha + \gamma z^{-1})^{-1}(\delta + \beta z)^{-1}X\left(\gamma, \frac{\alpha z + \gamma}{\beta z + \delta}\right) \tag{3.19}$$

and

$$\pi(g)\left(X(\gamma, z)\frac{dz}{z}\right)\pi(g^{-1}) = X(\gamma, w)\frac{dw}{w}, \quad \text{where } w = \frac{\alpha z + \gamma}{\beta z + \delta}, \tag{3.20}$$

since $dw = (\delta + \beta z)^{-2}dz$.

We will use (3.19) and (3.20) in the next section to prove the projective properties of the scattering amplitude.

3.7. Current Algebras and Scattering Amplitudes. In the latter subsection we have realized the space of the basic representation as the space of analytic functions. There is an alternative language of generalized functions which is profoundly used in physics. We sketch here several important points of this approach. We begin with the definition of a current algebra.

Let again $\mathfrak{g}$ be a simple Lie algebra with the standard form $\langle\ ,\ \rangle$. The *current algebra* corresponding to $\mathfrak{g}$ is defined by (cf. [13]):

$$[x(\theta), y(\theta')] = \delta(\theta - \theta')z(\theta) + \delta'(\theta - \theta')\langle x, y\rangle c, \tag{3.21}$$

where $x, y, z \in \mathfrak{g}$, $[x, y] = z$, c is a central element, $\theta, \theta' \in [0, 2\pi]$, δ and δ' are the delta function and the derivative of the delta function. The Lie algebra generated by $x(\theta)$, $x \in \mathfrak{g}$, and c is nothing else but the affine Lie algebra $\hat{\mathfrak{g}}$ and $x(\theta)$ is exactly the Mellin transform of $x \in \mathfrak{g}$. Therefore for the basic representation π constructed in Sect. 3.4 we get

$$\pi(E_\gamma(\theta)) = X^\epsilon(\gamma, e^{2\pi i\theta}), \quad \gamma \in \Delta.$$

It is possible to check directly the commutation relations for the vertex operators, using the theory of generalized functions[6]. We can define also the coherent states

$$X(\gamma_1, e^{2\pi i\theta_1}) \ldots X(\gamma_N, e^{2\pi i\theta_N})v_0, \quad \theta_i \neq \theta_j,$$

which lie in the extension of the Hilbert space of the basic representation. The formula (2.31) for the matrix element A also has sense if one takes the main value of the integral. It is convenient to split the domain of integration into the parts where the integrand is continuous.

[6] This was done recently [17] on the basis of our main theorem.

From (3.19) and (3.20) we get the transformation properties of the split matrix elements of $U(g)$ under the action of the group $SU(1,1)$.

Proposition 3.9. a) *The following corrected amplitudes*

$$A'=\int\frac{(z_1-z_2)}{z_1}\frac{(z_2-z_3)}{z_2}\cdots\frac{dz_r}{z_r}\cdots\frac{dz_s}{z_s}\cdots\frac{(z_N-z_1)}{z_N}\langle X(\alpha_1,z_1)\dots X(\alpha_N,z_N)v_0|v_0\rangle$$

are independent of the action of the group $SU(1,1)$, *i.e., under the change of variables and parameters* $z'_k=\dfrac{\alpha z_k+\beta}{\bar{\beta}z_k+\bar{\alpha}}$, *where* $z_k=e^{2\pi i\theta_k}$, $k=1,2,\dots,N$.

b) *If the number of variables is equal to* $N-3$, *i.e.,* $s-r+1=N-3$ *then* A' *does not depend on the choice of parameters absolutely.*

c) [10] *If we change the circle by the real line and the group* $SU(1,1)$ *by*

$$SL(2,\mathbb{R})=\left\{\begin{pmatrix}a & b\\ c & d\end{pmatrix}\middle|\, ad-bc=1,\ a,b,c,d\in\mathbb{R}\right\}$$

then a) *and* b) *hold as well.*

Proof. Notice first that

$$z'_1-z'_2=\frac{\alpha z_1+\gamma}{\beta z_1+\delta}-\frac{\alpha z_2+\gamma}{\beta z_2+\delta}=(\beta z_1+\delta)^{-1}(\beta z_2+\delta)^{-1}(z_1-z_2),$$

then if we combine it with (3.19) and (3.20) we get a).

b) follows from a) and well-known properties of projective transformations.

Finally, we deduce c) from a) and b), applying the Cayley transformation $g=\frac{1}{2}\begin{pmatrix}1 & i\\ -i & 1\end{pmatrix}\in SL(2,\mathbb{C})$ to the variables and the parameters z_k, $k=1,2,\dots,N$. This transformation transfers the unit circle into the real line and the group $SU(1,1)$ into the group $SL(2,\mathbb{R})$ and by (3.19) and (3.20) preserves the form of A'.

Proposition 3.9 together with Proposition 2.9 shows that the Koba-Nielsen amplitude defined by

$$A'=\int\{dz_1\dots[dz_a]\dots[dz_b]\dots[dz_c]\dots dz(z_b-z_a)\cdot(z_c-z_b)(z_i-z_a)\prod_{i>j}(z_i-z_j)^{\langle\gamma_i,\gamma_j\rangle},$$

where N variables and parameters z_i are ordered cyclically along the real line and the square brackets indicate omitting, is independent of the choice of the three z's which are kept fixed. If we take for example $z_0=0$, $z_{N-1}=1$, $z_N=\infty$, we get (2.32). If $N=4$, then we get $A'=B(\langle\gamma_2,\gamma_3\rangle+1,\langle\gamma_2,\gamma_1\rangle+1)$ (see [10]). This is the Veneziano amplitude which is the foundation stone of dual resonance models.

References

1. Dixmier, J.: Algèbres enveloppantes, Paris: Gauthier-Villars 1974
2. Garalnd, H.: The arithemtic theory of loop algebras, J. Algebra **53**, 480–551 (1978)
3. Gelfand, I.M., Vilenkin, N.Y.: Generalized Functions, Vol. 4, Application of Harmonic Analysis, New York: Academic Press 1964
4. Kac, V.G.: Simple irreducible graded Lie algebras of finite growth, Math. USSR-Izv. **2**, 1271–1311 (1968)
5. Kac, V.G.: Infinite-dimensional Lie algebras and Dedekind's η-function, Functional Anal. Appl. **8**, 68–70 (1974)
6. Kac, V.G.: Infinite dimensional algebras, Dedekind's η-function, classical Möbius function and the very strange formula, Advances in Math. **30**, 85–136 (1978)
7. Kac, V.G.: Contravariant form for Lie algebras and superalgebras, Lecture Notes in Physics **94**, 441–445 (1979)
8. Kac, V.G., Kazhdan, D.A., Lepowsky, J., Wilson, R.L.: Realization of the basic representations of the Euclidean Lie algebras, in press (1980)
9. Kac, V.G., Peterson, D.H.: Affine Lie algebras and Hecke modular forms. Bull. Amer. Math. Soc. **3** (1980)
10. Mandelstam, S.: Dual-resonance models, Physics Rep. **13**, 259–353 (1974)
11. Moody, R.V.: A new class of Lie algebras, J. Algebra **10**, 211–230 (1968)
12. Schwartz, J.H.: Dual-resonance theory, Physics Rep. **8**, 269–335 (1973)
13. Sugawara, H.: A field theory of currents, Physical Reviews **170**, 1659 (1968)
14. Tits, J.: Sur les constantes de structure et le théorème d'existence des algèbres de Lie semi-simple, Inst. Hautes Études Sci. Publ. Math., **31**, 21–58 (1966)
15. Vershik, A.M., Gelfand, I.M., Graev, M.I.: Representations of the group $SL(2, R)$ where R is a ring of functions, Russian Math. Surveys, **28**, 83–128 (1973)
16. Vershik, A.M., Gelfand, I.M., Graev, M.I.: Representations of the group of smooth mappings of a manifold X into a compact Lie group, Compositio Math. **35**, 299–339 (1977)
17. Bars, I., Garland, H.: Private communication
18. Kac, V.G.: An elucidation of "Infinite-dimensional algebras ... and the very strange formula." $E_8^{(1)}$ and the cube root of the modular invariant j. Advances in Math. **35**, 264–273 (1980)

Received March 24 / Revised June 30, 1980

Commun. Math. Phys. 80, 301–342 (1981)

Communications in
Mathematical
Physics

Unitary Representations of some Infinite Dimensional Groups

Graeme Segal
St. Catherine's College, Oxford University, Oxford OX1 3UJ, England

Abstract. We construct projective unitary representations of (a) $\mathrm{Map}(S^1;G)$, the group of smooth maps from the circle into a compact Lie group G, and (b) the group of diffeomorphisms of the circle. We show that a class of representations of $\mathrm{Map}(S^1;T)$, where T is a maximal torus of G, can be extended to representations of $\mathrm{Map}(S^1;G)$,

Introduction

One object of this paper is to describe a series of projective unitary representations of the group of (orientation preserving) diffeomorphisms of the circle. They are characterized, and distinguished from other known representations ([8], [13]), by the property of having "positive energy", which means that the rotation of the circle through an angle α is represented by $e^{-i\alpha K}$ where K is a positive operator.

In their infinitesimal form, i.e. as representations of $\mathrm{Vect}(S^1)$, the Lie algebra of smooth vector fields on the circle, the representations have been known for some time to physicists ([5], [3]) in connection with the quantization of strings moving relativistically. ($\mathrm{Vect}(S^1)$ is called by physicists the Virasoro algebra.) I have tried to explain briefly in an appendix to this paper how the representations are relevant to the theory of strings; but as a crude oversimplification one can say that one wants to describe unparametrized strings but finds it more convenient to describe parametrized strings: the group of diffeomorphisms acts on the Hilbert space of states of a parametrized string by changing parametrization.

The infinitesimal version of the representations has also been described by Kač([7][7a]).

My approach to the construction of the representations involves constructing irreducible representations of another family of groups. For any Lie group G the group $\mathrm{Diff}(S^1)$ of orientation preserving diffeomorphisms of the circle S^1 is a group of automorphisms of the group $\mathrm{Map}(S^1;G)$ of smooth maps from S^1 to G (under pointwise composition). Taking first $G = \mathbb{T}$, the circle group, I shall construct an irreducible projective unitary representation of $\mathrm{Map}(S^1;\mathbb{T})$ on a Hilbert space H. Then I shall show, what seems to me rather surprising, that any representation

0010-3616/81/0080/0301/$08.40

of Map(S^1; $\mathbb{T}$) belonging to a certain class extends canonically to a representation of Map(S^1; SU_2) on the same Hilbert space H, when $\mathbb{T}$ is identified with the diagonal matrices in SU_2. The next step is to show that Diff(S^1) acts projectively on H, intertwining with the action of Map(S^1; SU_2), so that we have a projective action of the semidirect product Diff(S^1) $\tilde{\times}$ Map(S^1; SU_2). This last group contains the product Diff(S^1) $\times SU_2$, where SU_2 is identified with the constant maps $S^1 \to SU_2$. We shall see that under the action of Diff(S^1) $\times SU_2$ the space H can be decomposed

$$H = \bigoplus_{q=0,1,2,\ldots} P_{q^2} \otimes D_q,$$

where the P_{q^2} are distinct irreducible representations of Diff(S^1), and D_q is the $(2q+1)$-dimensional irreducible representation of SU_2.[1]

Defining the action of Diff(S^1) on H involves constructing the metaplectic representation of a certain infinite dimensional symplectic group. This construction is due in essence to Shale [10], but has been developed more explicitly by Vergne [12]. My method is superficially, but not fundamentally, different from hers.

The theorem that a suitable projective representation of Map(S^1; $\mathbb{T}$) can be extended to one of Map(S^1; SU_2) can be generalized in the following way. If G is a simply connected and simply laced (cf. Sect. 4) compact Lie group with a maximal torus T then a class of projective representations of Map(S^1; T) can be extended to Map(S^1; G). Among other things this gives one a new and interesting explicit construction of the fundamental irreducible projective representations of Map(S^1; G). On these representations too the group Diff(S^1) acts, and again they can be decomposed under Diff(S^1) $\times G$.

Since writing the present work I have learnt that the extension theorem has been proved independently by Frenkel and Kač [0], who have also observed that the essential ingredient is the "Veneziano vertex" of the theory of strings [5].

All the representations we shall be concerned with are projective, i.e. they are really representations of central extensions of the groups in question by the circle $\mathbb{T}$. The extensions are of some interest in their own right: they are described from various points of view in Sect. 7.

To get some idea of the position of the representations constructed here among the totality of representations of Map(S^1; G) and Diff(S^1) one can consider the "orbits" in the coadjoint representation in the manner of Kirillov and Kostant. That is done in Sect. 8. In the case of Map(S^1; G) the results are very satisfactory, in that for projective representations with the cocycle we are considering the orbit method suggests that the unitary representations constructed in this paper are the only ones which exist. These representations are precisely the ones found by Kač, whose method, like mine, constructs only representations of positive energy (i.e. "with a lowest weight"). I should perhaps mention at this point that Gel'fand and others have constructed non-projective representations of Map(S^1; G) of a completely different type, not of positive energy (cf. [14].)

In the case of Diff(S^1) the predictions of the orbit method are not so clear. The

1 To avoid misunderstanding I should emphasize that in this paper the irreducibility of P_{q^2} is proved only when $q = 0$ (cf. Sect. 6). The irreducibility has been proved in general by Kač

coadjoint representation of $\mathrm{Diff}(S^1)$ has a natural interpretation in terms of second-order differential equations on the circle—so-called "Hill's equations"—and the orbits are classified by their monodromy. There seems to be at least a rough correspondence between the orbits and the unitary representations.

My interest in the subject of this paper was aroused by Goldstone, and all the results about the decomposition of the irreducible representations of $\mathrm{Map}(S^1;\mathbb{T})$ under the action of $\mathrm{Diff}(S^1)$ were told to me by him. I am most grateful for the stimulus and instruction he has given me.

The decomposition of the representation of $\mathrm{Map}(S^1;G)$ in the general case was explained to me by Macdonald. I have also been greatly helped in connection with the metaplectic representation by Kazhdan and Vergne, and in understanding the action of $\mathrm{Diff}(S^1)$ on Hill's equations by Hitchin.

It will be obvious that the results of this paper have mostly already been obtained, at least on the infinitesimal level, by Kač. I hope nevertheless that my methods and point of view are sufficiently different to justify their publication. My methods, on the other hand, are in some sense familiar among physicists: in connection with Sect. 5 I should mention the work of Goddard and Horsley [3], and for the construction of Sect. 4 the volume [5], passim.

2. The Central Extension of Map $(S^1;T)$

The group $M = \mathrm{Map}(S^1;\mathbb{T})$ is disconnected, with its connected components indexed by the winding number. There is an exact sequence

$$0 \to \mathbb{Z} \xrightarrow{\times 2\pi} \mathrm{Map}(S^1;\mathbb{R}) \to \mathrm{Map}(S^1;\mathbb{T}) \xrightarrow{w} \mathbb{Z} \to 0,$$

where the middle map is $f \mapsto e^{if}$, and w is the winding number. Our first task is to define a central extension $\hat{M}$ of M by $\mathbb{T}$.

Let F denote the vector space of smooth functions $f:\mathbb{R} \to \mathbb{R}$ such that

$$\Delta_f = \frac{1}{2\pi}(f(\theta + 2\pi) - f(\theta))$$

is constant. The subgroup of F consisting of f such that $\Delta_f \in \mathbb{Z}$ will be denoted by $F_{\mathbb{Z}}$. Any element of M can be written e^{if}, with $f \in F_{\mathbb{Z}}$, when S^1 is thought of as $\mathbb{R}/2\pi\mathbb{Z}$.

There is a skew bilinear form $S:F \times F \to \mathbb{R}$ defined by

$$S(f,g) = \frac{1}{4\pi}\int_0^{2\pi}(f'(\theta)g(\theta) - f(\theta)g'(\theta))d\theta + \frac{1}{4\pi}(f(2\pi)g(0) - f(0)g(2\pi)).$$

(Observe that $S(f,1) = \Delta_f$.) We define the group $\hat{M}$ as the space $\mathbb{T} \times M$ with the composition law

$$(\lambda, e^{if})\cdot(\mu, e^{ig}) = (\lambda\mu e^{-iS(f,g)}, e^{i(f+g)}).$$

It is fundamental for our purposes that the group $\mathrm{Diff}(S^1)$ acts on $\hat{M}$ as a group of automorphisms. That is true because the cocycle $c:M \times M \to \mathbb{T}$ given by $c(e^{if}, e^{ig}) = e^{-iS(f,g)}$ which defines $\hat{M}$ is invariant under $\mathrm{Diff}(S^1)$.

It is not hard to determine *all* the central extensions of M by $\mathbb{T}$ which are invariant under $\mathrm{Diff}(S^1)$. Indeed any such extension $\tilde{M}$ can be identified with $\mathbb{T} \times M$ as a space. Let C denote the circle of constant function in M, and let ε denote any element of M of winding number 1. Then $c \mapsto \varepsilon c \varepsilon^{-1} c^{-1}$, where the multiplication is in $\hat{M}$, is a map $C \to \mathbb{T}$.

Proposition (2.1) *The central extensions* $\mathbb{T} \to \tilde{M} \to M$ *which are invariant under* $\mathit{Diff}(S^1)$ *are completely classified by the winding number of the map* $C \to \mathbb{T}$ *just described. Any integral winding number can occur.*

I shall omit the proof. The extension we are denoting by $\hat{M}$ has winding number 2. To obtain an extension with winding number 1 one can replace the skew form S used to define M with the (non-skew) bilinear form $s: F \times F \to \mathbb{R}$ defined by

$$s(f,g) = \frac{1}{4\pi}\int_0^{2\pi} f'(\theta)g(\theta)d\theta + \tfrac{1}{2}\Delta_f g(0).$$

(Notice that $s(f,g) - s(g,f) = S(f,g)\cdot$) The form s is not invariant under $\mathrm{Diff}(S^1)$, but nevertheless a double covering of $\mathrm{Diff}(S^1)$ acts on the associated extension $\tilde{M}$ by

$$\phi^*(\lambda,\gamma) = (\lambda\{\gamma(\phi(0))\gamma(0)^{-1}\}^{-(1/2)w(\gamma)}, \phi^*\gamma).$$

3. The Projective Representation of Map $(S^1; T)$

We shall construct a unitary representation of $\hat{M}$ on a Hilbert space H. It is more convenient notationally, however, to describe it as a projective representation of M, i.e. to associate to each γ in M an operator $T(\gamma): H \to H$ such that

$$T(\gamma)T(\gamma') = c(\gamma,\gamma')T(\gamma\gamma'),$$

where $c: M \times M \to \mathbb{T}$ is the cocycle described in the previous section.

We begin with the identity component M_0 of M. If $C \subset M_0$ is the constants, then M_0/C can be identified with the vector space $V = \mathrm{Map}(S^1; \mathbb{R})/(\text{constants})$. We shall construct a projective unitary representation of V on a Hilbert space H_0; then, regarding H_0 as a representation of M_0, we shall define H as the representation of M induced from H_0.

The skew form S induces a skew form $S: V \times V \to \mathbb{R}$. To represent the Lie algebra of the desired central extension of V is to associate linearly to each $f \in V$ an operator $A(f)$ so that

$$[A(f), A(g)] = 2i\, S(f,g).$$

(Then the group element e^{if} in M_0 will be represented by $e^{iA(f)}$.) That is, we must represent the "canonical commutation relations" associated to V and S. We do this using the standard representation on Fock space.

A *complex polarization* of V for the form S means a decomposition $V_{\mathbb{C}} = W \oplus \bar{W}$, where $V_{\mathbb{C}}$ is the complexification of V, such that S is identically zero on W. ($\bar{W}$ denotes the complex conjugate of W.) Then

$$(w_1, w_2) \mapsto \langle w_1, w_2 \rangle = 2i\, S(\bar{w}_1, w_2)$$

is a hermitian form on W. If this is positive definite, making W a pre-Hilbert-space, we shall call W a *positive* polarization.

In our case, where $V = \mathrm{Map}(S^1;\mathbb{R})/(\text{constants})$, there is a canonical positive polarization, in which W is the space of smooth maps $f: S^1 \to \mathbb{C}$ which extend to holomorphic functions on the disk $D = \{z \in \mathbb{C} : |z| \leq 1\}$ (modulo constants). The spaces W and $\bar{W}$ are isotropic for S by Cauchy's theorem, and the form $\langle\ ,\ \rangle$ is positive-definite. In fact if $f(z) = \sum_{n>0} a_n z^n$ and $g(z) = \sum_{n>0} b_n z^n$ then

$$\begin{aligned}\langle f, g\rangle &= \sum 2n\bar{a}_n b_n \\ &= \frac{2}{\pi}\int_D \overline{f'(z)}g'(z)d\mu(z) \\ &= \frac{1}{\pi i}\int_D \overline{df} \wedge dg\end{aligned}$$

where μ is Lebesgue measure on D.

Let us consider the symmetric algebra $S(W)$ of W. The elements of W act on this by multiplication: we write $A(w): S^k(W) \to S^{k+1}(W)$ for multiplication by w. For each $\bar{w}$ in $\bar{W}$ there is a unique derivation $A(\bar{w}): S^k(W) \to S^{k-1}(W)$ of the algebra $S(W)$ whose action on $S^1(W) = W$ is given by

$$A(\bar{w})\cdot u = \langle w, u\rangle.$$

For any f in V we define $A(f): S(W) \to S(W)$ by

$$A(f) = A(f_+) + A(f_-),$$

where $f = f_+ + f_-$ with $f_+ \in W$ and $f_- \in \bar{W}$. The relation

$$[A(f), A(g)] = 2iS(f, g)$$

follows at once from the fact that $A(f_-)$ and $A(g_-)$ are derivations.

The inner product $\langle\ ,\ \rangle$ on W extends to $S(W)$ by

$$\langle w_1 w_2 \ldots w_k, w_1 w_2 \ldots w_k\rangle = \sum \langle w_{i_1}, w_1\rangle\langle w_{i_2}, w_2\rangle \ldots \langle w_{i_k}, w_k\rangle,$$

where the sum is over all permutations $(i_1, \ldots, i_k)$ of $(1, \ldots, k)$. With respect to this inner product the operators $A(w)$ and $A(\bar{w})$ are adjoint for any w in W, and so $A(f)$ is self-adjoint for real f in V.

We define H_0 as the Hilbert space completion $\hat{S}(W)$ of $S(W)$. The $A(f)$ for f in V can be thought of as unbounded operators in H_0; we have to show that they are self-adjoint in the appropriate sense so that they define unitary operators $e^{iA(f)}$ in H_0.

It is easy to check that the following estimates hold:

$$\|A(w)\cdot\xi\| \leq \sqrt{k+1}.\|w\|\ \|\xi\|,$$

$$\|A(\bar{w})\cdot\xi\| \leq \sqrt{k}.\|w\|\ \|\xi\|,$$

when $w \in W$ and $\xi \in S^k(W)$. From these we deduce

$$\|A(f)\cdot\xi\| \leq 2\sqrt{k+1}.\|f\|\ \|\xi\|$$

when $f \in V$ and $\xi \in S^0(W) \oplus \ldots \oplus S^k(W)$, and $\| f \|$ is defined as $\| f_+ \|$. It follows that the series

$$\sum \frac{1}{k!}(iA(f))^k \cdot \xi$$

converges in H_0 for each $\xi \in S(W)$, and defines

$$e^{iA(f)} : S(W) \to H_0.$$

But if $\xi, \eta \in S(W)$ then

$$\begin{aligned}\langle e^{iA(f)}\xi, e^{iA(f)}\eta \rangle &= \sum_{k,m} \left\langle \xi, \frac{(-iA(f))^k}{k!}\frac{(iA(f))^m}{m!}\eta \right\rangle \\ &= \langle \xi, \eta \rangle,\end{aligned}$$

because $\sum_{k+m=n} (-1)^k/k!m! = 0$ when $n > 0$.

So $e^{iA(f)}$ extends to a unitary operator $H_0 \to H_0$. In view of the identity $e^P e^Q = e^{(1/2)[P,Q]} e^{P+Q}$, which holds whenever [P, Q] commutes with P and Q, we find

$$e^{iA(f)} e^{iA(g)} = e^{-iS(f,g)} e^{iA(f+g)},$$

which shows that $T(e^{if}) = e^{iA(f)}$ defines the desired projective unitary representation of the group $V = M_0/C$ on H_0.

Let us regard M_0 as $V \times C$ by identifying V with the subspace $\{f : \int f = 0\}$ of $\mathrm{Map}(S^1; \mathbb{R})$. Then for each integer k there is a representation H_k of M_0 which coincides with H_0 as a representation of V, but in which $\lambda \in C$ acts as λ^k. The projective representation of M induced from H_0 can be best be described as the Hilbert space direct sum $H = \bigoplus_{k \in \mathbb{Z}} H_{2k}$, on which the canonical element $e^{in\theta}$ of M of winding number n acts as the "identity" map $H_{2k} \to H_{2(k+n)}$. The "vacuum vector" in H_{2k} will be denoted by Ω_{2k}, and H_{2k} will be thought of as $\hat{S}(W).\Omega_{2k}$.

To conclude this section I shall show that the representation of $\hat{M}$ we have constructed is irreducible, and is essentially the only representation of its kind.

We shall see in Sect. 5 that $\mathrm{Diff}(S^1)$ acts (projectively) on H intertwining with the representation of $\hat{M}$. But it is meanwhile easy to see that the subgroup R of rigid rotations of the circle does so. For R acts naturally on W, and hence on $\hat{S}(W) = H_0$. To fix its action on the other H_{2k} it is enough to see how it acts on the vacuum vectors Ω_{2k}. We have no choice in prescribing this, for if r_α is the rotation through α, and ε_k is the function $e^{ik\theta}$ in M, then we must have

$$\begin{aligned} r_\alpha \Omega_{2k} &= r_\alpha T(\varepsilon_k)\Omega_0 = r_\alpha T(\varepsilon_k) r_\alpha^{-1}\Omega_0 \\ &= T(r_\alpha \cdot \varepsilon_k)\Omega_0 = T(r_\alpha \cdot \varepsilon_k)T(\varepsilon_{-k})\Omega_{2k} \\ &= c(r_\alpha \cdot \varepsilon_k, \varepsilon_{-k})T(e^{ik\alpha})\Omega_{2k} = e^{ik^2\alpha} \cdot e^{-2ik^2\alpha} \cdot \Omega_{2k} \\ &= e^{-ik^2\alpha}\Omega_{2k}. \end{aligned}$$

On the other hand there is no difficulty in seeing that this action on the Ω_{2k} does define an action of R on H which intertwines with M.

Let $H(q)$ denote the part of H where r_α acts as $e^{-iq\alpha}$. $H(q)$ is finite dimensional and vanishes when $q<0$. $\left(\text{Its dimension is the coefficient of } x^q \text{ in } \prod_{n>0}(1-x^n)^{-1}\cdot\sum_{m\in Z}x^{m^2}\right)$

I shall call a representation of M *symmetric* if its isomorphism class does not change when twisted by a rotation of the circle. If the intertwining operator corresponding to the rotation r_α is $e^{-i\alpha K}$ where K is a positive operator I shall say the representation has *positive energy*. (K is determined only up to the addition of an integral of 2π, so it would be better to require the spectrum of K to be bounded below. But the ambiguity is removed when one extends the action of R to $\mathrm{Diff}(S^1)$).

Proposition (3.1) *The projective representation of M on H is irreducible, symmetric, and of positive energy. It is one of precisely two such representations of M with the given cocycle c.*

Proof. First consider the action of M_0 on $H_0=\hat{S}(W)$. Clearly the vacuum vector Ω_0 is cyclic. Any possible decomposition $H_0=H_0'\oplus H_0''$ must respect the grading by the eigenspaces $H_0(q)=H_0\cap H(q)$, so Ω_0 must belong to either H_0' or H_0''. Thus H_0 is irreducible under M_0. It follows that all H_k are irreducible under M_0, and hence that H is irreducible under M.

But the H_k are the only irreducible representations of M_0 of positive energy. To see that, it is enough (since $M_0=C\times V$) to show that $\hat{S}(W)$ is the only irreducible representation of V of positive energy. But any such representation would have to contain a vector Ω_0 annihilated by $A(\bar{w})$ for $\bar{w}\in\bar{W}$, for $A(\bar{w})$ lowers energy. On the other hand $\hat{S}(W)$ is freely generated by the action of the $A(f)$, for $f\in V$, subject to this constraint; so it is the only possible representation.

Finally, the group $\hat{M}$ is a semidirect product $\mathbb{Z}\tilde{\times}M_0$, and the representations H_k and H_m of $\hat{M}_0$ are conjugate under $\hat{M}$ if and only if $k\equiv m \pmod 2$. So by Mackey's theorem there are two irreducible positive-energy representations of $\hat{M}$ which are faithful on the centre, restricting respectively to $\bigoplus H_{2k}$ and $\bigoplus H_{2k+1}$ as representations of $\hat{M}_0$.

4. The action of Map(S^1 ; SU_2)

The group $M=\mathrm{Map}(S^1;\mathbb{T})$ can be thought of as a subgroup of $G=\mathrm{Map}(S^1;SU_2)$ by identifying $\mathbb{T}$ with the diagonal matrices in SU_2. In this section I shall prove

Proposition (4.1) *Let H be a projective unitary representation of M with the multiplier c. Suppose that H is symmetric with positive energy, and that each eigenspace $H(q)$ of the rotations of the circle is finite-demensional. Then the action of M on H extends canonically to a projective unitary representation of G.*

To begin with we shall define the representation on the Lie algebra of G. In fact we shall consider first the smaller Lie algebra $\mathrm{Map}_{\mathrm{alg}}(S^1;su_2)$ of all algebraic maps $f:S^1\to su_2$, i.e. those of the form

$$f(\theta)=\sum_{k=-N}^{N}L_k e^{ik\theta},$$

where the L_k are complex 2×2 matrices.

As we shall be considering unbounded operators it is useful to introduce a "rigging" (cf. [1] (Chap. 1, Sect. 4)) of the Hilbert space H. Let $\check{H}$ denote the direct sum and $\hat{H}$ the direct product of the $H(q)$. Then $\check{H} \subset H \subset \hat{H}$, and $\check{H}$ and $\hat{H}$ are dual to each other.

For any $\alpha \in S^1$ and $0 \leqq \lambda < 1$ let us consider the element $\gamma_{\alpha,\lambda} : S^1 \to \mathbb{T}$ of M defined by

$$\gamma_{\alpha,\lambda}(\theta) = \frac{\lambda - e^{i(\theta-\alpha)}}{1 - \lambda e^{i(\theta-\alpha)}}.$$

This should be thought of as a "blip" situated at the point α of the circle, becoming sharper as $\lambda \to 1$. It has winding number 1, and if λ is close to 1 then $\gamma_{\alpha,\lambda}(\theta)$ is close to 1 except in a small neighbourhood of $\theta = \alpha$, of length roughly $1 - \lambda$, where it winds once around $\mathbb{T}$. In fact $\gamma_{\alpha,\lambda}(\theta) = e^{if_{\alpha,\lambda}(\theta)}$, where $f_{\alpha,\lambda}(\theta)$ is the angle APA' in the diagram, where $P = e^{i\theta}$, $A = \lambda e^{i\alpha}$, and $A' = \lambda^{-1} e^{i\alpha}$.

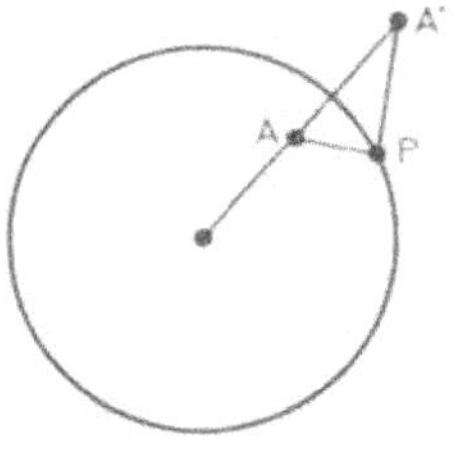

Fig. 1

To $\gamma_{\alpha,\lambda} \in M$ corresponds a unitary operator $T(\gamma_{\alpha,\lambda})$ on H. I shall prove that as $\lambda \to 1$ the operator $(1-\lambda^2)^{-1} T(\gamma_{\alpha,\lambda})$ tends in a certain sense to a definite but highly singular operator B_α.[2]

We can write $f_{\alpha,\lambda} = q_\alpha + f^+_{\alpha,\lambda} + f^-_{\alpha,\lambda}$, where

$$f^+_{\alpha,\lambda}(\theta) = i\log(1 - \lambda e^{i(\theta-\alpha)}) = -i \sum_{n>0} \frac{\lambda^n}{n} e^{in(\theta-\alpha)},$$

$$f^-_{\alpha,\lambda}(\theta) = -i\log(1 - \lambda e^{-i(\theta-\alpha)}), \text{ and}$$

$$q_\alpha(\theta) = \pi + \theta - \alpha.$$

Then $f^+_{\alpha,\lambda} \in W$ and $f^-_{\alpha,\lambda} \in \bar{W}$. Furthermore

$$\begin{aligned} S(f^+_{\alpha,\lambda}, f^-_{\alpha,\lambda}) &= -\frac{1}{2\pi} \int f^+_{\alpha,\lambda}\, df^-_{\alpha,\lambda} \\ &= -\frac{1}{2\pi} \int \frac{\log(1-\lambda z)}{1-\lambda z^{-1}} . \lambda z^{-2}\, dz \\ &= -i\log(1-\lambda^2); \end{aligned}$$

so that

$$T(\gamma_{\alpha,\lambda}) = (1-\lambda^2)\, T(e^{iq_\alpha}) e^{iA^+_{\alpha,\lambda}} e^{iA^-_{\alpha,\lambda}},$$

where $A^\pm_{\alpha,\lambda} = A(f^\pm_{\alpha,\lambda})$.

2 In the language of quantum field theory B_α is, as a function of α, an "operator-valued distribution"

The operators $e^{iA^+_{\alpha,\lambda}}$ and $e^{iA^-_{\alpha,\lambda}}$ can be expanded

$$e^{iA^+_{\alpha,\lambda}} = \sum_{n\geqq 0} \lambda^n e^{-in\alpha} C_n,$$
$$e^{iA^-_{\alpha,\lambda}} = \sum_{n\geqq 0} \lambda^n e^{in\alpha} D_n.$$

Here C_n and D_n are operators independent of λ and α which map $H(q)$ into $H(q+n)$ and $H(q-n)$ respectively.

As $H(q-n)$ is zero when $n > q$ it follows that

$$e^{iA^+_{\alpha,\lambda}} e^{iA^-_{\alpha,\lambda}} = \sum_{m,n\geqq 0} \lambda^{n+m} e^{i(n-m)\alpha} C_m D_n$$

tends to a well-defined operator $\check{H} \to \hat{H}$ as $\lambda \to 1$. Now $T(e^{iq_\alpha})$ maps H_{2k} into H_{2k+2}, and takes $H_{2k}(q)$ into $H_{2k+2}(q+2k+1)$. On H_{2k} we have $T(e^{iq_\alpha}) = e^{-(2k+1)i\alpha} T_0$, where $T_0 = T(e^{iq_0})$. So

$$B_\alpha = \lim_{\lambda\to 1} (1-\lambda^2)^{-1} T(\gamma_{\alpha,\lambda}) = \sum_{n,m\geqq 0} e^{i(n-m-2k-1)\alpha} T_0 C_m D_n$$

on H_{2k}.

If $\phi : S^1 \to \mathbb{C}$ is a trigonometric polynomial, i.e. $\phi(\theta) = \sum_{r=-N}^{N} \phi_r e^{ir\theta}$, let

$$B(\phi) = \frac{1}{2\pi} \int_0^{2\pi} \phi(\alpha) B_\alpha \, d\alpha.$$

Then $B(\phi) = \Sigma \phi_{m-n+2k+1} T_0 C_m D_n$ on $\check{H}_{2k}$, and so it maps $\check{H}$ into itself.

The adjoint operator B^*_α of B_α is given by

$$B^*_\alpha = \lim_{\lambda\to 1} (1-\lambda^2)^{-1} T(\gamma^{-1}_{\alpha,\lambda}) = \sum e^{-i(n-m-2k+1)} T_0^{-1} \tilde{C}_m \tilde{D}_n$$

on $\check{H}_{2k}$, where $\tilde{C}_m = D^*_m$ and $\tilde{D}_n = C^*_n$. We define

$$B^*(\phi) = \frac{1}{2\pi} \int_0^{2\pi} \phi(\alpha) B^*_\alpha \, d\alpha,$$

so that $B(\phi)^* = B^*(\bar{\phi})$.

For a trigonometric polynomial ϕ the operator $A(\phi)$ can be written $A(\phi) = \Sigma \phi_k a_k$, where $a_k = A(e^{ik\theta})$ maps $H(q)$ into $H(q+k)$ and hence $\check{H}$ into itself. Furthermore for any *distribution* ϕ on S^1 one has an operator $A(\phi) : \check{H} \to \hat{H}$.

We shall now show that the operators $A(\phi)$, $B(\phi)$, $B^*(\phi)$ define a projective representation of the complex Lie algebra of algebraic maps $S^1 \to sl_2(\mathbb{C})$ on the vector space $\check{H}$. More precisely, we have

Proposition (4.2) (i) $[A(\phi), B(\psi)] = 2B(\phi\psi)$,

(ii) $[A(\phi), B^*(\psi)] = -2B^*(\phi\psi)$,

(iii) $[B(\phi), B^*(\psi)] = A(\phi\psi) + iS(\phi, \psi)$,

(iv) $[B(\phi), B(\psi)] = [B^*(\phi), B^*(\psi)] = 0$.

These are the commutation relations for a projective representation of $\mathrm{Map}_{\mathrm{alg}}(S^1; sl_2(\mathbb{C}))$ under the correspondence

$$A(\phi) \leftrightarrow \begin{pmatrix} \phi & 0 \\ 0 & -\phi \end{pmatrix} \quad B(\phi) \leftrightarrow \begin{pmatrix} 0 & \phi \\ 0 & 0 \end{pmatrix}, \quad B^*(\phi) \leftrightarrow \begin{pmatrix} 0 & 0 \\ \phi & 0 \end{pmatrix}.$$

The cocycle giving the projective multiplier is

$$(f, g) \mapsto \omega(f, g) = \frac{1}{2\pi} \int_0^{2\pi} \langle f'(\theta), g(\theta) \rangle \, d\theta,$$

where $f, g: S^1 \to sl_2(\mathbb{C})$, and $\langle \ , \ \rangle$ is the invariant form $(A, B) \mapsto -\operatorname{trace}(AB)$ on $sl_2(\mathbb{C})$.

It is easy to check that in the representation the elements of $\mathrm{Map}_{\mathrm{alg}}(S^1; su_2)$ act on $\check{H}$ by skew operators.

Proof of Proposition (4.2). (i) It suffices to show that $[A(\phi), B_\alpha] = 2\phi(\alpha)B_\alpha$. Now $B_\alpha = \lim B_\alpha^\lambda$, where $B_\alpha^\lambda = (1 - \lambda^2)^{-1} e^{iA(f_{\alpha,\lambda})}$. As the commutator of $A(\phi)$ and $A(f_{\alpha,\lambda})$ is scalar we have

$$[A(\phi), B_\alpha^\lambda] = i[A(\phi), A(f_{\alpha,\lambda})]B_\alpha^\lambda = -2S(\phi, f_{\alpha,\lambda})B_\alpha^\lambda.$$

But, regarded as a distribution, $f_{\alpha,\lambda} \to 2\pi H_\alpha$ as $\lambda \to 1$, where H_α is the Heaviside function defined by $H_\alpha(\theta) = 1$ if $\theta \geqq \alpha$, and $H_\alpha(\theta) = 0$ if $\theta < \alpha$. So $S(\phi, f_{\alpha,\lambda}) \to 2\pi S(\phi, H_\alpha) = -\phi(\alpha)$, as desired.

(ii) This is simply the adjoint of (i).

(iii) We first calculate

$$[B_\alpha^\lambda, (B_\beta^\lambda)^*] = e^{iA_{\alpha,\lambda}^+} e^{iA_{\alpha,\lambda}^-} e^{-iA_{\beta,\lambda}^+} e^{iA_{\beta,\lambda}^-} T(e^{iq_\alpha}) T(e^{-iq_\beta})$$
$$- \text{(similar expression)}$$
$$= \{ e^{[A_{\alpha,\lambda}^-, A_{\beta,\lambda}^+] + i(\alpha-\beta)} - e^{[A_{\beta,\lambda}^-, A_{\alpha,\lambda}^+] - i(\alpha-\beta)} \} e^{i(A_{\alpha,\lambda}^+ - A_{\lambda,\lambda}^+)} e^{i(A_{\alpha,\lambda}^- - A_{\beta,\lambda}^-)} T(e^{-i(\alpha-\beta)}).$$

The expression in braces is

$$\frac{e^{i(\alpha-\beta)}}{(1 - \lambda^2 e^{i(\alpha-\beta)})^2} - \frac{e^{-i(\alpha-\beta)}}{(1 - \lambda^2 e^{-i(\alpha-\beta)})^2}$$
$$= \frac{2i(1 - \lambda^4)\sin(\alpha - \beta)}{(1 - 2\lambda^2\cos(\alpha - \beta) + \lambda^4)^2}$$
$$= \frac{i}{\lambda 2} \frac{d}{d\alpha} \frac{1 - \lambda^4}{1 - 2\lambda^2\cos(\alpha - \beta) + \lambda^4}$$

This tends to $-2\pi i\delta'(\alpha - \beta)$ as $\lambda \to 1$.

Furthermore as $\lambda \to 1$ we have

$$A_{\alpha,\lambda}^+ \to A_\alpha^+ = A(-i \sum_{n>0} e^{in(\theta-\alpha)}), \textit{ and}$$
$$A_{\alpha,\lambda}^- \to A_\alpha^- = A(+i \sum_{n>0} e^{-in(\theta-\alpha)}),$$

and the expression

$$e^{i(A_\alpha^+ - A_\beta^+)} e^{i(A_\alpha^- - A_\beta^-)} = \sum_{p,q,r,s} e^{i(r-p)\alpha + i(s-q)\beta} C_p \tilde{C}_q D_r \tilde{D}_s$$

is a well-defined operator $\check{H} \to \hat{H}$.

Thus we find

$$\lim_{\lambda\to 1}[B_\alpha^\lambda,(B_\beta^\lambda)^*] = -\pi i\delta'(\alpha-\beta)e^{i(A_\alpha^+ - A_\beta^+)}e^{i(A_\alpha^- - A_\beta^-)}T(e^{-i(\alpha-\beta)})$$

$$= -2\pi i\delta'(\alpha-\beta) = 2\pi\delta(\alpha-\beta)\frac{d}{d\alpha}(A_\alpha^+ + A_\alpha^- - \alpha a_0),$$

where we have used the relation

$$\delta'(x)F(x) = \delta'(x)F(0) - \delta(x)F'(0),$$

and also that $T(e^{-i(\alpha-\beta)}) = e^{-i(\alpha-\beta)a_0}$.

But $A_\alpha^+ + A_\alpha^- - \alpha a_0 = A(2\pi H_\alpha - \theta - \theta)$, so $\frac{d}{d\alpha}(A_\alpha^+ + A_\alpha^- - \alpha a_0) = -2\pi A(\delta_\alpha)$, where δ_α is the delta-function at α. Hence

$$[B_\alpha, B_\beta^*] = -2\pi i\delta'(a-\beta) + 4\pi^2\delta(\alpha-\beta)A(\delta_\alpha),$$

and accordingly

$$[B(\phi), B^*(\psi)] = iS(\phi,\psi) + A(\phi\psi),$$

as we want.

(iv) The proof of this is precisely similar to that of (iii), but easier, because when we calculate $[B_\alpha^\lambda, B_\beta^\lambda]$ the analogue of the expression in braces above is

$$(1-\lambda^2 e^{i(\alpha-\beta)})^2 e^{-i(\alpha-\beta)} - (1-\lambda^2 e^{-i(\alpha-\beta)})^2 e^{i(\alpha-\beta)}$$
$$= -2i(1-\lambda^4)\sin(\alpha-\beta),$$

which tends to zero as $\lambda \to 1$.

That completes the proof of (4.2).

Let us now write $P(\phi) = B(\phi) + B^*(\phi)$ and $Q(\phi) = i(B(\phi) - B^*(\phi))$, so that $\{iA(\phi), iP(\phi), iQ(\phi)\}$ correspond respectively to the standard generators

$$\begin{pmatrix} i\phi & 0 \\ 0 & -i\phi \end{pmatrix}, \begin{pmatrix} 0 & i\phi \\ i\phi & 0 \end{pmatrix}, \begin{pmatrix} 0 & -\phi \\ \phi & 0 \end{pmatrix}$$

of Map$(S^1; su_2)$. The operators $\{iA(1), iP(1), iQ(1)\}$ define an action of the algebra su_2 on $\check{H}$ which preserves each finite dimensional subspace $H(q)$. This action can be exponentiated to give a continuous action of the group SU_2 on H. We shall derive a number of consequences of this.

Let $\check{H}$ denote the subspace of H consisting of sums $\sum\xi_q$, with $\xi_q \in H(q)$, such that $\{\xi_q\}$ is rapidly decreasing, i.e. such that $(1+q)^n\|\xi_q\|$ is bounded as $q\to\infty$ for each n. $\check{H}$ is precisely the space of vectors $\xi\in H$ such that $\theta\mapsto r_\theta\xi$ is a smooth function $S^1 \to H$. The dual of $\check{H}$ is the subspace $\hat{H}$ of $\check{H} = \prod H(q)$ consisting of series $\sum\xi_q$ such that $\{\xi_q\}$ has polynomial growth, i.e. such that $(1+q)^{-n}\|\xi_q\|$ is bounded as $q\to\infty$ for some n. In other words $\xi\in\hat{H}$ if $\theta\mapsto r_\theta\xi$ is a distribution on S^1.

When $\phi = \sum\phi_k e^{ik\theta}$ is a smooth function on S^1 the operator $A(\phi) = \sum\phi_k a_k$ maps $\check{H}$ into $\check{H}$, and when ϕ is a distribution we find $A(\phi): \check{H} \to \hat{H}$. But the group SU_2 acts on H, and by conjugating with appropriate elements of it, one can transform the operators $P(\phi)$ and $Q(\phi)$, which are defined when ϕ is a trigonometric polynomial, into $A(\phi)$. It follows not only that $P(\phi)$ and $Q(\phi)$ map $\check{H}$ into itself but also that they can be defined for any smooth function ϕ on S^1. That proves

Proposition (4.3) *The Lie algebra of smooth maps $S^1 \to su_2$ acts projectively on $\check{H}$.*

We can now complete the proof of (4.1). We have seen that the operators $A(\phi)$ can be exponentiated to give unitary transformations of H. The same must therefore hold for $P(\phi)$ and $Q(\phi)$. This gives us projective unitary representations of three subgroups of $\mathrm{Map}(S^1;SU_2)$. The three subgroups clearly generate the whole group. To see that they fit together to give a projective unitary representation of $\mathrm{Map}(S^1;SU_2)$ one need only observe that if $f \in \mathrm{Map}(S^1;SU_2)$ is expressed in two different ways in terms of elements of the three subgroups, leading to two operators U_f and U'_f representing it, then $U_f^{-1}U'_f$ commutes with the action of the Lie algebra $\mathrm{Map}(S^1;su_2)$. As this action is irreducible U_f and U'_f can differ only by a scalar.

Generalization to other groups

Suppose that G is a compact Lie group with a maximal torus T. The discussion in this section can be generalized to prove that in certain circumstances a projective representation of $\mathrm{Map}(S^1;T)$ can be extended canonically to one of $\mathrm{Map}(S^1;G)$.

The method seems to apply only to groups G which are *simply laced*. Let us recall that G is simply laced if there is an invariant inner product on its Lie algebra $\mathfrak{g}$ in terms of which all the roots have the same length. (Equivalently one can say: if the Weyl group of G acts transitively on the roots.) This happens if and only if G is a product of circles and simple groups of types A, D and E, i.e. if it has no factors of type, B, C, F or G. I shall also assume that G is simply connected, although that is not essential.

Proposition (4.4) *Suppose that G is simply laced and simply connected with maximal torus T. Let H be a projective unitary representation of $M = \mathrm{Map}(S^1;T)$ with the cocycle c described below. Suppose that H is symmetric with positive energy, and that each eigenspace $H(q)$ of the rotations of the circle is finite dimensional. Then the action of M on H extends canonically to an action of $\mathrm{Map}(S^1;G)$.*

To define the cocycle c I need to recall some facts about simply laced groups.

We write $T = t/2\pi L$, where t is the Lie algebra of T, and L is a lattice in t. There is an exact sequence

$$0 \to L \xrightarrow{\times 2\pi} \mathrm{Map}(S^1;t) \to \mathrm{Map}(S^1;T) \to L \to 0.$$

Let F denote the smooth function $f: \mathbb{R} \to t$ such that

$$\Delta_f = \frac{1}{2\pi}(f(\theta + 2\pi) - f(\theta))$$

is constant; and let $F_L = \{f \in F: \Delta_f \in L\}$. Then $\mathrm{Map}(S^1;\mathrm{T}) = F_L/2\pi L$.

The roots of G are certain linear maps $\alpha: t \to \mathbb{R}$. For each α we define $h_\alpha \in t$ so that $\alpha(\xi) = \langle h_\alpha, \xi \rangle$ for all $\xi \in t$. For a simply laced group the inner product $\langle , \rangle$ on $\mathfrak{g}$ can be normalized so that $\langle h_\alpha, h_\alpha \rangle = 2$ for each root α. (In the case of SU_n this amounts to defining $\langle \xi, \eta \rangle = -\,\mathrm{trace}\,(\xi\eta)$, and in the case of SO_{2n} it corresponds to $\langle \xi, \eta \rangle = -\frac{1}{2}\mathrm{trace}(\xi\eta)$.) It is well known that each h_α belongs to the lattice L, and for a simply connected group the h_α even generate L and are precisely the set of all $\lambda \in L$ such that $\langle \lambda, \lambda \rangle = 2$. In that case, $\lambda \mapsto \frac{1}{2}\langle \lambda, \lambda \rangle$ is an integer-valued quadratic form on L. Let us choose a bilinear form

$$\sigma : L \times L \to \mathbb{Z}/2 \text{ such that } \sigma(\lambda, \lambda) \equiv \tfrac{1}{2}\langle \lambda, \lambda \rangle \pmod 2.$$

Then

$$\sigma(\lambda, \mu) + \sigma(\mu, \lambda) \equiv \langle \lambda, \lambda \rangle \pmod 2$$

for all $\lambda, \mu \in L$.

The complexified Lie algebra $\mathfrak{g}_{\mathbb{C}}$ can be decomposed

$$\mathfrak{g}_{\mathbb{C}} = t_{\mathbb{C}} \oplus \bigoplus_{\alpha \in R} \mathbb{C}.e_\alpha,$$

where R is the set of roots. The relations are

$$\begin{aligned} [\xi, e_\alpha] &= \alpha(\xi)e_\alpha \text{ for } \xi \in t_{\mathbb{C}} \\ [e_\alpha, e_{-\alpha}] &= h_\alpha \\ [e_\alpha, e_\beta] &= 0 \text{ if } \alpha \neq -\beta \text{ and } \langle \alpha, \beta \rangle \neq -1, \\ [e_\alpha, e_\beta] &= (-1)^{\sigma(\alpha,\beta)} e_{\alpha+\beta} \text{ if } \langle \alpha, \beta \rangle = 1. \end{aligned}$$

(Here $\sigma(\alpha, \beta)$ means $\sigma(h_\alpha, h_\beta)$. These are not quite the usual form of the generators and relations for $\mathfrak{g}$, but are easily checked to be equivalent to them. Notice that the choice of σ is immaterial, for if σ' is another choice then $\sigma'(\lambda, \mu) - \sigma(\lambda, \mu) = \zeta(\lambda + \mu) - \zeta(\lambda) - \zeta(\mu)$ for some $\zeta : L \to \mathbb{Z}/2$.)

We define a blinear from $s : F \times F \to \mathbb{R}$ by

$$s(f, g) = \frac{1}{4\pi}\int_0^{2\pi} \langle f'(\theta), g(\theta) \rangle d\theta + \tfrac{1}{2}\langle \Delta_f, g(0) \rangle.$$

If g is constant then $s(f, g) = \langle \Delta_f, g \rangle$, and if f is constant then $s(f, g) = 0$, so $s((2\pi L \times F_L) + (F_L \times 2\pi L)) \subset 2\pi\mathbb{Z}$, and we can define a cocycle c on $M = \mathrm{Map}(S^1; T)$ by

$$c(e^{if}, e^{ig}) = (-1)^{\sigma(\Delta_f, \Delta_g)} e^{-is(f,g)}$$

Suppose now that $e^{if} \mapsto T(e^{if})$ is a projective unitary representation of M on a Hilbert space H, associated to the cocycle c. Passing to the Lie algebra of M gives us unbounded operators $A(f)$ on H for each $f : S^1 \to t$, and we have

$$[A(f), A(g)] = 2is(f, g).$$

As before, we assume that the group of rotations of the circle acts on H with positive energy, compatibly with the action of M.

Let us identify the roots of G with the vectors ξ of length 2 in the lattice L. To each ξ, and $\theta \in S^1$, we associate a "blip" $\gamma^\xi_{\theta,\lambda} = \exp(\xi \cdot f_{\theta,\lambda}) \in M$, where $0 \leqq \lambda < 1$ and $f_{\theta,\lambda} : \mathbb{R} \to \mathbb{R}$ is as on page 8. We find just as before that $(-\lambda^2)^{-1} T(\gamma^\xi_{\theta,\lambda})$ tends to an operator $B^\xi_\theta : \check{H} \to \hat{H}$ as $\lambda \to 1$, where

$$B^\xi_\theta = T(e^{i\xi q_\theta}) e^{iA(\xi f^+_\theta)} e^{iA(\xi_\theta^-)}$$

For a trigonometrical polynomial $\phi: S^1 \to \mathbb{C}$ we define

$$B^\xi(\phi) = \frac{1}{2\pi}\int_0^{2\pi} \phi(\theta) B^\xi_\theta d\theta : \check{H} \to \check{H}.$$

Proposition (4.2) generalizes to

Proposition (4.5) (i) $[A(f), B^\xi(\phi)] = B^\xi(\langle f, S\phi\rangle)$,

(ii) $[B^\xi(\phi), B^{-\xi}(\psi)] = A(\xi\phi\psi) + iS(\phi,\psi)$,

(iii) $[B^\xi(\phi), B^\eta(\psi)] = 0$ if $\xi \neq -\eta$ and $\langle \xi, \eta\rangle \neq -1$,

(iv) $[B^\xi(\phi), B^\eta(\psi)] = (-1)^{\sigma(\xi,\eta)} B^{\xi+\eta}(\phi\psi)$ if $\langle \xi,\eta\rangle = -1$.

These are the commutation relations for a projective representation of the Lie algebra $\mathrm{Map}_{\mathrm{alg}}(S^1; \mathfrak{g}_{\mathbb{C}})$ corresponding to the cocycle

$$(f, g) \mapsto \omega(f, g) = \frac{1}{2\pi}\int_0^{2\pi} \langle f'(\theta), g(\theta)\rangle d\theta.$$

The only comment worth making about the proof of (4.5) is that in proving (iv) one proceeds exactly as in the proof of (4.2) (iii), and obtains a formula for $[B^{\lambda,\xi}_\theta, B^{\lambda,\eta}_\phi]$ where the expression in braces in the earlier proof in replaced by

$$(-1)^{\sigma(\xi,\eta)}\left\{\frac{e^{(1/2)i(\theta-\phi)}}{1-\lambda^2 e^{i(\theta-\phi)}} + \frac{e^{-(1/2)i(\theta-\phi)}}{1-\lambda^2 e^{-i(\theta-\phi)}}\right\}$$

$$= (-)^{\sigma(\xi,\eta)}\frac{2(1-\lambda^2)\cos\frac{1}{2}(\theta-\phi)}{1-2\lambda^2\cos(\theta-\phi)+\lambda^4}.$$

This tends to $(-1)^{\sigma(\xi,\eta)} 2\pi\delta(\theta-\phi)$ as $\lambda \to 1$, which gives the formula we want.

Extending the representation of the Lie algebra $\mathrm{Map}_{\mathrm{alg}}(S^1; \mathfrak{g}_{\mathbb{C}})$ to one of the group $\mathrm{Map}(S^1; G)$ is done just as when $G = SU_2$, and presents nothing new.

It is worth considering how many representations of $\Gamma = \mathrm{Map}(S^1; G)$ – or, more precisely, of its central extension $\tilde{\Gamma}$—we obtain by this method. They correspond to representations of $\hat{M}$ which are symmetric and of positive energy. These are constructed and classified exactly as in Sect. 3. $\hat{M}$ is a semidirect product $L \times \hat{M}_0$, where $\hat{M}_0$ is the identity component. As before $M_0 \cong T \times V$, where $V = \mathrm{Map}(S^1; \mathfrak{t})/(\text{constants})$ has a nondegenerate skew form induced by s. The arguments of Sect. 3. show that $\hat{M}_0$ has a unique irreducible representation H_χ of the appropriate kind for each character χ of T, in which T, which is in the centre of $\hat{M}_0$, acts scalarly by χ. The character group of T is $L^* = \mathrm{Hom}(L; \mathbb{Z})$. Conjugation by $\lambda \in L$ transforms H_χ to $H_{\chi+\lambda}$, where L is embedded in L^* by the inner product $\langle\ ,\ \rangle$. Thus we obtain an irreducible representation $\bigoplus_{\lambda\in L} H_{\chi+\lambda}$ of $\hat{M}$ for each coset $\chi + L$ of L in L^*. Now L^*/L is precisely the centre Z of G. The common centre of $\hat{M}$ and $\tilde{\Gamma}$ is $Z \times \mathbb{T}$, and the representations we have obtained are classified by the action of Z. (By using the inner product we have identified $Z = L^*/L$ with its character group.)

These conclusions agree with Kač's theory of the representation of $\mathrm{Map}(S^1; G)$, because L^*/L is precisely the set of orbits of the lattice of weights L^* under the affine Weyl group of G.

5. The Metaplectic Representation

We have seen that the group $\mathrm{Diff}(S^1)$ is a group of automorphisms of $\hat{M}$. Our next task is to show that it acts projectively by unitary transformations on H, intertwining with the action of $\hat{M}$. As the representation of $\hat{M}$ on H is irreducible this would follow from Schur's Lemma if one knew that the isomorphism class of the representation of $\hat{M}$ on H did not change when twisted by a diffeomorphism: but unfortunately that is not clear, as it is not obvious that the property of having positive energy is preserved.

The action of $\mathrm{Diff}(S^1)$ on $\hat{M}$ leaves fixed the constants $C \subset M$, so $\mathrm{Diff}(S^1)$ ought to act independently in each eigenspace H_{2k} of C. We shall show that it acts on $H_0 = \hat{S}(W)$. It is then automatic that it acts on all of H because H is the representation of $\hat{M}$ induced from the representation H_0 of $\hat{M}_0$, and $\hat{M}/\hat{M}_0 \cong (\mathrm{Diff}(S^1) \tilde{\times} \hat{M})/(\mathrm{Diff}(S^1) \tilde{\times} \hat{M}_0.)$

$\mathrm{Diff}(S^1)$ acts on $V = \mathrm{Map}(S^1;\mathbb{R})/\mathbb{R}$ preserving the skew form S, so it can be thought of as a subgroup of the symplectic group $Sp(V)$. If V were finite-dimensional then a double converging of $Sp(V)$ would act, by the metaplectic representation, on $\hat{S}(W)$. In the infinite-dimensional case one must replace $Sp(V)$ by a subgroup $Sp_0(V)$ consisting of maps which are not too far from preserving the polarization $V_{\mathbb{C}} = W \oplus \bar{W}$. The definition, which is due to Shale [10], is as follows.

Let us suppose that we have a real vector space V with a skew form $S: V \times V \to \mathbb{R}$ and a positive polarization $V_{\mathbb{C}} = W \oplus \bar{W}$. As usual we regard W as a pre-Hilbert-space with $\langle w_1, w_2 \rangle = 2iS(\bar{w}_1, w_2)$. With respect to the decomposition $V_{\mathbb{C}} = W \oplus \bar{W}$ an endomorphism A of V can be expressed as a matrix

$$A = \begin{pmatrix} a & b \\ \bar{b} & \bar{a} \end{pmatrix}.$$

We find that A belongs to $Sp(V)$ if and only if

(i) $\bar{a}^t a - b^t \bar{b} = 1$, and

(ii) $\bar{a}^t b$ is symmetric.

(These equations should be thought of as shorthand: they do not really presuppose the existence of the transposed operators.)

Definition (5.1) $Sp_0(V)$ is the subset of $Sp(V)$ consisting of elements A as above such that $b: \bar{W} \to W$ is a Hilbert–Schmidt operator.

("*Hilbert–Schmidt*" means that $\sum \|b\bar{\varepsilon}_k\|^2$ converges for any orthonormal family $\{\bar{\varepsilon}_k\}$ in $\bar{W}$.)

If $A \in Sp_0(V)$ it follows from (i) above that $a: W \to W$ is a bounded operator, and hence that $Sp_0(V)$ is a subgroup of $Sp(V)$, for the Hilbert–Schmidt operators are closed under composition with bounded operators.

Proposition (5.2) (*Shale*) *$Sp_0(V)$ acts projectively by unitary transformations on $\hat{S}(W)$.*

Before proving this I shall show that $Sp_0(V)$ does contain $\mathrm{Diff}(S^1)$, using a simple argument pointed out to me by Kazhdan.

Proposition (5.3) $\mathrm{Diff}(S^1) \subset Sp_0(V)$.

Proof. Let us write $\{\varepsilon_n\}$ for the standard basis $\{e^{in\theta}\}_{n \neq 0}$ of $V_{\mathbb{C}} = \mathrm{Map}(S^1;\mathbb{C})/\mathbb{C}$.

If $\phi \in \mathrm{Diff}(S^1)$ we write $\{\lambda_{nm}\}$ for its matrix elements with respect to this basis, i.e. $\phi^*(\varepsilon_n) = \sum \lambda_{nm}\varepsilon_m$. We have to show that the matrix $\{\lambda_{n,-m}\}_{n,m>0}$ defines a Hilbert-Schmidt operator. Clearly it is enough to show that $\{\lambda_{n,-m}\}$ is rapidly decreasing, in the sense that for each integer k we have $|\lambda_{n,-m}| \leqq C(n+m)^{-k}$ for some constant C (depending on k).

Now

$$\lambda_{n,-m} = \frac{1}{2\pi}\int_0^{2\pi} e^{in\phi(\theta)+im\theta}d\theta.$$

(Here ϕ is regarded as a map $\phi: \mathbb{R} \to \mathbb{R}$ such that $\phi(\theta+2\pi) = \phi(\theta)+2\pi$.) For any $t \in [0,1]$ the function ϕ_t defined by $\phi_t(\theta) = t\phi(\theta)+(1-t)\theta$ is also a diffeomorphism, and when $t = n/(n+m)$ we have

$$\lambda_{n,-m} = \frac{1}{2\pi}\int_0^{2\pi} e^{i(n+m)\phi_t(\theta)}d\theta$$
$$= \frac{1}{2\pi}\int_0^{2\pi} e^{i(n+m)\theta}\psi_t'(\theta)d\theta,$$

where ψ_t is the inverse function to ϕ_t. So

$$\lambda_{n,-m} = \frac{i^k}{2\pi}.(n+m)^{-k}\int_0^{2\pi} e^{i(n+m)\theta}\psi_t^{(k+1)}(\theta)d\theta$$

on integrating by parts k times, and

$$|\lambda_{n,-m}| \leqq (n+m)^{-k}\sup\{|\psi_t^{(k+1)}(\theta)| : 0 \leqq \theta \leqq 2\pi, 0 \leqq t \leqq 1\},$$

as we want.

To construct the metaplectic representation of $Sp_0(V)$ I shall introduce an infinite-dimensional analogue of the Siegel bounded complex domain $Sp_{2n}(R)/U_n$. Let us recall that $Sp_{2n}(R)/U_n$ can be realized as the space of complex symmetric $n \times n$ matrices Z such that $\bar{Z}Z < 1$.

Let X denote the space of symmetric Hilbert-Schmidt operators $Z: \bar{W} \to W$ such that $\bar{Z}Z < 1$, i.e. those such that

(i) $S(\bar{w}_1, Z\bar{w}_2) = S(\bar{w}_2, Z\bar{w}_1)$, and

(ii) $1 - \bar{Z}Z$ is positive-definite.

The group $Sp_0(V)$ acts transitively on X by the formula

$$\begin{pmatrix} a & b \\ \bar{b} & \bar{a} \end{pmatrix}.Z = (aZ+b)(\bar{b}Z+\bar{a})^{-1}:$$

the proof is the same as in the finite-dimensional case. The stabiliser of $0 \in X$ is the unitary group $U(W)$, so $X \cong Sp_0(V)/U(W)$.

In the finite-dimensional case $Sp_{2n}(R)/U_n$ can be identified with the set of all positive polarizations of V. In our case X is to the thought of as the set of positive polarizations of V which are not too far away from W. (One gets a polarization $V_{\mathbb{C}} = U \oplus \bar{U}$ from $Z: \bar{W} \to W$ by taking $\bar{U}$ to be the graph of Z.)

The symmetric Hilbert-Schmidt operators $Z: \bar{W} \to W$ can be regarded as elements of $\hat{S}^2(W)$ by identifying Z with $\sum \varepsilon_k \cdot Z\bar{\varepsilon}_k$, where $\{\varepsilon_k\}$ is an orthonormal basis for W. For each $Z \in X$ we consider the element $e^{(1/2)z}$ of $\hat{S}(W)$

Lemma (5.4) $\langle e^{(1/2)Z_1}, e^{(1/2)Z_2} \rangle = \det(1 - \bar{Z}_1 Z_2)^{-1/2}$
(Here one chooses the branch of the square root which is $+1$ when $\hat{Z}_1 Z_2 = 0$. Notice also that the determinant is defined, because $\bar{Z}_1 Z_2$, being the product of two Hilbert–Schmidt operators, is of trace class.)

It is enough, by continuity, to prove (5.4) when Z_1 and Z_2 have finite rank, in which case I shall leave it as an exercise for the reader.

In view of (5.4) the map $Z \mapsto \varepsilon_Z = \det(1 - \bar{Z}Z)^{1/4} e^{(1/2)Z}$ is an embedding of the domain X into the space of unit vectors in $S(W)$, and

$$\langle \varepsilon_{Z_1}, \varepsilon_{Z_2} \rangle = \det(1 - \bar{Z}_1 Z_1)^{1/4} \det(1 - \bar{Z}_2 Z_2)^{1/4} \det(1 - \bar{Z}_1 Z_2)^{-1/2}. \dots (*)$$

Now let F_X denote the free vector space generated by the symbols $\{\varepsilon_Z\}_{Z \in X}$, and let H_X be the Hilbert space obtained by completing F_X using the inner product defined by the formula (*). (This inner product on F_X is positive because it is induced from that of $\hat{S}(W)$.) Clearly H_X is a closed subspace of $\hat{S}(W)$.

Proposition (5.5) $H_X = \hat{S}^{\text{even}}(W)$

I shall postpone the proof of this for the moment.

For each $A \in Sp_0(V)$ one can define a unitary operator $T_A : H_X \to H_X$ by

$$T_A \cdot \varepsilon_Z = \mu \det(1 + \bar{a}^{-1} \bar{b} Z)^{1/2} \cdot \varepsilon_{AZ}.$$

where $\mu : \mathbb{C}^\times \to \mathbb{T}$ is radial projection. To see that T_A is well-defined one must check that

$$\langle \varepsilon_{Z_1}, \varepsilon_{Z_2} \rangle = \mu \det(1 + \bar{a}^{-1} \bar{b} Z_1)^{-1/2} \mu \det(1 + \bar{a}^{-1} \bar{b} Z_2)^{1/2} \langle \varepsilon_{A Z_1}, \varepsilon_{A Z_2} \rangle,$$

which, however, is a simple calculation. It is also straightforward to check that

$$T_{A_1} T_{A_2} = c(A_1, A_2) T_{A_1 A_2},$$

where $c(A_1, A_2) = \det(a_1^{-1} a_3 a_2^{-1})^{-1/2}$. Here

$$A_i = \begin{pmatrix} a_i & b_i \\ \bar{b}_i & \bar{a}_i \end{pmatrix}$$

for $i = 1, 2, 3$, and $A_3 = A_1 A_2$. The determinant is well-defined because $a_1^{-1} a_3 a_2^{-1} = 1 + a_1^{-1} b_1 \bar{b}_2 a_2^{-1}$ is of the form $1 +$ (trace class), and $\| a_1^{-1} b_1 \bar{b}_2 a_2^{-1} \| < 1$.

The cocycle c evidently measures the extent by which $A \mapsto a$ fails to be a homomorphism, i.e. by which A fails to preserve the polarization $V_{\mathbb{C}} = W \oplus \bar{W}$. I shall return to this point in Sect. 7.

To understand what is going on in the preceding formulae one should adopt the following point of view. There is a natural holomorphic line bundle L on $X = \{\text{polarizations of } V\}$ whose fibre at U (where $V_{\mathbb{C}} = U \oplus \bar{U}$) is $\det(U)^{-1/2}$. We have chosen a trivialization of L by identifying U with W by $U \subset V_{\mathbb{C}} \xrightarrow{pr} W$. In the finite dimensional case the action of $Sp_{2n}(\mathbb{R})$ on X is covered by an action of a double covering $\tilde{S}p_{2n}(\mathbb{R})$ on L, which in terms of our trivialization is given by

$$A \cdot (Z, \lambda) = (A \cdot Z, \mu \det(\bar{b} Z + \bar{a})^{1/2} \cdot \lambda).$$

An element of $\tilde{S}p_{2n}(\mathbb{R})$ is an element of $Sp_{2n}(\mathbb{R})$ together with a choice of a square root of det (a). But in our case we cannot define det (a), let alone its square root; so we are forced to make an extension $\tilde{S}p_0(V)$ of $Sp_0(V)$ by $\mathbb{T}$ act on L by

$$A\cdot(Z,\lambda)=(A\cdot Z,\mu\det(1+\bar{a}^{-1}\bar{b}Z)^{1/2}\cdot\lambda).$$

Instead of embedding X in $\hat{S}(W)$ above it would have been more natural to embed the line bundle L linearly in $\hat{S}(W)$. The action of $\tilde{S}p_0(V)$ is then extended by linearity to $\hat{S}^{\text{even}}(W)$.

We have now made $Sp_0(V)$ act projectively on $\hat{S}^{\text{even}}(W)$. We can extended the action to $\hat{S}^{\text{odd}}(W)$ by the following device. Consider $Sp_0(V\oplus\mathbb{R}^2)$, where $\mathbb{R}^2$ has the obvious skew form. This contains $Sp_0(V)\times\mathbb{T}$, where $\mathbb{T}=SO_2\subset Sp(\mathbb{R}^2)$. There is a polarization $(V\oplus\mathbb{R}^2)_{\mathbb{C}}=(W\oplus\mathbb{C})\oplus(\bar{W}\oplus\bar{\mathbb{C}})$ where $\mathbb{T}$ acts in the natural way on $\mathbb{C}$. Then $Sp_0(V)\times\mathbb{T}$ acts on

$$\hat{S}^{\text{even}}(W\oplus\mathbb{C})=\hat{S}^{\text{even}}(W)\hat{\otimes}\hat{S}^{\text{even}}(\mathbb{C})\ \oplus\ \hat{S}^{\text{odd}}(W)\hat{\otimes}\hat{S}^{\text{odd}}(\mathbb{C}),$$

and $Sp_0(V)$ acts in each isotypical component of the action of $\mathbb{T}$. The action on $\hat{S}^{\text{odd}}(W)\otimes S^k(\mathbb{C})\cong\hat{S}^{\text{odd}}(W)$, for any odd k, gives us what we want.

It remains to give the proof of (5.5).

Clearly H_X contains e^{tww} for each $w\in W$ and $t\in\mathbb{R}$. By differentiating this repeatedly with respect to t at $t=0$ we find, because H_X is a closed subspace, that it contains w^{2k} for each k. Then from the identity

$$m!w_1w_2\ldots w_m=\sum_{\sigma}(-1)^{m-|\sigma|}\Big(\sum_{i\in\sigma}w_i\Big)^m,$$

where σ runs through the subsets of $\{1,2,\ldots,m\}$, and $|\sigma|$ denotes the number of elements in σ, we find that H_X contains $S^m(W)$ whenever m is even.

A possible disadvantage of the method adopted here for constructing the metaplectic representation is that it does not make manifest that the action of $Sp_0(V)$ intertwines correctly with the unbounded operators $A(v)$ for $v\in V$, and hence with the group M of Sect. 2. The rest of this section is devoted to establishing that it does so.

We begin by considering the action of the Lie algebra $sp_0(V)$ of $Sp_0(V)$ on $\hat{S}(W)$. An element of $sp_0(V)$ is a matrix

$$\begin{pmatrix}\alpha & \beta\\ \bar{\beta} & \bar{\alpha}\end{pmatrix},$$

where α is bounded and skew-Hermitian and β is symmetric and Hilbert–Schmidt. The complexification $sp_0^{\mathbb{C}}(V)$ consists of operators

$$\begin{pmatrix}\alpha & \beta\\ \bar{\gamma} & -\alpha^t\end{pmatrix},$$

where α is an arbitrary bounded operator and β and γ are symmetric Hilbert–Schmidt operators.

Proposition (5.6) *The element*

$$\begin{pmatrix}\alpha & \beta\\ \bar{\gamma} & -\alpha^t\end{pmatrix}$$

of $sp_0^{\mathbf{C}}(V)$ *acts on* $\hat{S}(W)$ *as* $D_\alpha + \frac{1}{2}M_\beta + \frac{1}{2}M_\gamma^*$, *where* D_α *is the derivation of* $\hat{S}(W)$ *induced by* $\alpha: W \to W$, M_β *is multiplication by* $\beta \in \hat{S}^2(W)$, *and* M_γ^* *is the adjoint of* M_γ.

Proof. One can consider separately elements of the forms

$$\begin{pmatrix} \alpha & 0 \\ 0 & -\alpha^t \end{pmatrix}, \quad \begin{pmatrix} 0 & \beta \\ 0 & 0 \end{pmatrix}, \text{ and } \begin{pmatrix} 0 & 0 \\ \bar{\gamma} & 0 \end{pmatrix}.$$

For the first kind, which preserve the polarization $V_{\mathbf{C}} = W \oplus \bar{W}$, the result is clear.

By direct calculation one finds that the effect of

$$\begin{pmatrix} 0 & \beta \\ \bar{\beta} & 0 \end{pmatrix} \in sp_0(V)$$

on $e^{(1/2)Z} \in \hat{S}(W)$ is multiplication by $\frac{1}{2}(\beta - Z\bar{\beta}Z - \operatorname{tr}(\bar{\beta}Z))$. It follows that the action of

$$\begin{pmatrix} 0 & \beta \\ 0 & 0 \end{pmatrix}$$

on $e^{(1/2)Z}$, and hence on all of $\hat{S}(W)$, is multiplication by $\frac{1}{2}\beta$. Finally, in any unitary representation

$$\begin{pmatrix} 0 & 0 \\ \bar{\gamma} & 0 \end{pmatrix}$$

has to be represented by the adjoint of the representative of

$$\begin{pmatrix} 0 & \gamma \\ 0 & 0 \end{pmatrix}.$$

Returning now to the intertwining of $sp_0(V)$ with the operators $A(v)$ for $v \in V$, what we want to show is that

$$[X, A(v)] = A(Xv)$$

for $X \in sp_0^{\mathbf{C}}(V)$ and $v \in V_{\mathbf{C}}$. Because $A(\bar{v}) = A(v)^*$ it is enough to consider $[X, A(w)]$ for $w \in W$. When X is of the form

$$\begin{pmatrix} \alpha & 0 \\ 0 & -\alpha^t \end{pmatrix}$$

the result is obvious. When

$$X = \begin{pmatrix} 0 & \beta \\ 0 & 0 \end{pmatrix}$$

then $Xw = 0$; but on the left we have two multiplication operators, so the commutator vanishes too. When

$$X = \begin{pmatrix} 0 & 0 \\ \bar{\gamma} & 0 \end{pmatrix}$$

then $Xw = \bar{\gamma}w \in \bar{W}$, and we have to show that $[X, A(w)] = A(\gamma\bar{w})^*$. By adjunction this is equivalent to $[A(w)^*, X^*] = A(\gamma\bar{w})$. Now X^* is multiplication by $\frac{1}{2}\gamma$ and

$A(w)^*$ is a derivation. So the commutator is multiplication by $\frac{1}{2}A(w)^*\gamma$, which is precisely $\bar{w}$. That completes the proof.

Finally we come to the group $Sp_0(V)$. It is of course not true that the elements of this group can be obtained by exponentiating bounded operators in the Lie algebra $sp_0(V)$. On the other hand $Sp_0(V)$ has a normal subgroup $Sp_{cpt}(V)$ consisting of elements which differ from the identity by a compact operator. The exponentials of bounded operators clearly generate $Sp_{cpt}(V)$, so for this group the desired intertwining with the $A(v)$ follows from what we have proved. $Sp_0(V)$ also contains the unitary group $U(W)$ of W, the transformations which preserve the polarization $V_{\mathbf{C}} = W \oplus \bar{W}$. For these the intertwining relation is obvious. But that is all we need, in view of

Proposition (5.7) $Sp_0(V) = Sp_{cpt}(V) \cdot U(W)$.

Proof. We have already remarked that $Sp_0(V)/U(W)$ is the space of symmetric Hilbert-Schmidt operators $\bar{W} \to W$, so it is enough to see that $Sp_{cpt}(V)$ acts transitively on X. But that is obvious.

6. The Action of Diff (S^1) on H

In the last section we showed that $\mathrm{Diff}(S^1)$ acts projectively on H, intertwining with the action of $\mathrm{Map}(S^1; SU_2)$. It follows that the product $\mathrm{Diff}(S^1) \times SU_2$ acts on H. In this section I shall discuss the decomposition of H under this action.

I shall write A, B, B^* for the operators $A(1)$, $B(1)$, $B(1)^*$ of Sect. 4. They form a basis for the complexification of the Lie algebra of SU_2, and satisfy

$$[A, B]^* = 2B,$$
$$[A, B^*] = -2B^*,$$
$$[B, B^*] = A.$$

Recall that $H = \bigoplus_{k \in Z} H_{2k}$, where H_{2k} is the $2k$-eigenspace of A. Thus $B(H_{2k}) \subset H_{2k+2}$ and $B^*(H_{2k}) \subset H_{2k-2}$.

From the elementary representation theory of SU_2 we know that we can write

$$H = \bigoplus_{q=0,1,2,\ldots} P_{q^2} \otimes D_q,$$

where D_q is the irreducible representation of SU_2 of dimension $2q+1$, and $P_{q^2} = \{\xi \in H_{2q} : B\xi = 0\}$ is a representation of $\mathrm{Diff}(S^1)$. Furthermore the operator $B^* : H_{2k+2} \to H_{2k}$ is injective when $k \geqq 0$, and

$$H_{2k} = P_{k^2} \oplus B^*(H_{2k+2}). \qquad (*)$$

It follows that $H_{2k} \cong \bigoplus_{q \geqq k} P_{q^2}$ as a representation of $\mathrm{Diff}(S^1)$.

Each space H_{2k} is graded by the action of the group R of rotations of $S^1 : H_{2k} = \bigoplus_{q \geqq 0} H_{2k}(q)$. Let $\pi_{2k}(t) = \sum \dim H_{2k}(q) \cdot t^q$ be the Poincaré series of H_{2k}. We know that $\pi_{2k}(t) = t^{k^2} \pi_0(t)$, and that $\pi_0 = \pi$ is the partition function:

$$\pi(t) = \prod_{n>0} (1 - t^n)^{-1}.$$

From (*) we find

Proposition (6.1) *The Poincaré series of* P_{q^2} *is* $t^{q^2}(1 - t^{2q+1})\pi(t)$.

Kač has proved the conjecture of Goldstone that the representations P_{q^2} of Diff(S^1) are irreducible. Here I shall prove the irreducibility only for P_0. (But cf. Prop. (6.7).) Before doing so it seems appropriate to make some elementary general remarks about the representation theory of the Lie algebra of Diff(S^1), i.e. the Lie algebra $\mathscr{V} = \text{Vect}(S^1)$ of smooth vector fields on the circle.

The complexification $\mathscr{V}_{\mathbf{C}}$ has an obvious basis $\{v_k\}_{k\in\mathbf{Z}}$, where $v_k = ie^{-ik\theta}\dfrac{d}{d\theta}$; and the relations are

$$[v_k, v_m] = (m-k)v_{k+m}.$$

Suppose that we have a projective representation of $\mathscr{V}$. If V_k is the operator representing v_k then

$$[V_k, V_m] = (m-k)V_{k+m} + c(v_k, v_m), \qquad \text{.......} (\mathscr{I})$$

where $c: \mathscr{V}_{\mathbf{C}} \times \mathscr{V}_{\mathbf{C}} \to \mathbf{C}$ is a cocycle of the Lie algebra. It is easy to calculate c for our representation of $\mathscr{V}$ on H, but it is worth noticing that there are few possibilities, in view of the following well-known proposition [2].

Proposition (6.2) $H^2(\mathscr{V};\mathbf{C}) \cong \mathbf{C}$, *with generator c, where*

$$c(\xi,\eta) = \frac{1}{24\pi i}\int_0^{2\pi} \xi'(\theta)(\eta''(\theta) + \eta(\theta))d\theta,$$

so that

$$c(v_k, v_m) = \tfrac{1}{12}m(m^2-1) \text{ if } k+m=0$$
$$= 0 \text{ if } k+m \neq 0.$$

Proof. If $c: \mathscr{V} \times \mathscr{V} \to \mathbf{C}$ represents an element of $H^2(\mathscr{V};\mathbf{C})$ we can assume it is invariant under the group R of rotations of the circle, for averaging c over a compact group of inner automorphisms of $\mathscr{V}$ does not change its cohomology class. But if c is R-invariant then $c(v_k, v_m) = 0$ when $k+m \neq 0$, and we have only to determine $c_k = c(v_{-k}, v_k)$. From the cocycle condition

$$c([\xi,\eta],\zeta) + c([\eta,\zeta],\xi) + c([\zeta,\xi],\eta) = 0$$

we find (taking $\xi = v_k, \eta = v_m, \zeta = v_{-k-m}$, and noticing that $c_{-n} = -c_n$)

$$(k-m)c_{k+m} = (k+2m)c_k - (2k+m)c_m.$$

The most general solution of this is $c_k = \lambda k^3 + \mu k$, with $\lambda, \mu \in \mathbf{C}$. But c can be altered by adding to it a coboundary, i.e. a cocycle of the form $(\xi,\eta) \mapsto f([\xi,\eta])$, where $f: \mathscr{V} \to \mathbf{C}$ is an R-invariant linear map. The only possible f is $f(v_k) = \nu\delta_{k0}$, for some $\nu \in \mathbf{C}$; and this changes c_k by $2\nu k$. The value of μ is therefore irrelevant. If we normalize c by requiring it to vanish on the Lie algebra of $PSL_2(\mathbf{R})$, which is spanned by $\{v_1, v_0, v_{-1}\}$ then $c_k = \lambda k(k^2-1)$, and we have the result of (6.2).

The cocycle which arises in the representation of Diff(S^1) on H will be discussed further in Sect 7. It is precisely the cocycle c given in Proposition (6.2), corresponding to $\lambda = \frac{1}{12}$.

Suppose now that we have a projective unitary representation of Diff(S^1) on a Hilbert space K, corresponding to the cocycle we have been considering. Then $\mathscr{V}_{\mathbb{C}}$ acts on K by the unbounded operators $\{V_k\}$. The group R of rotations has the generator iV_0, so $e^{2\pi i V_0} = 1$, and V_0 is self-adjoint with integral spectrum. Let $K = \bigoplus K(q)$, where $K(q)$ is the q-eigenspace of V_0. Then $V_k \cdot K(q) \subset K(q+k)$. Elements of K which are annihilated by V_k for all $k < 0$ will be called *lowest weight vectors*. If the spectrum of V_0 is bounded below by q_0 then $K(q_0)$ consists of lowest weight vectors. Using the relations (‡) we see that if $\Omega \in K(q_0)$ then the cyclic subrepresentations K_Ω of Diff(S^1) generated by Ω is the closed subspace of K spanned by the vectors of the form $V_{k_1} V_{k_2} \ldots V_{k_r} \Omega$, where $k_1 \geqq k_2 \geqq \ldots \geqq k_r > 0$. Because $K_\Omega(q_0) = \mathbb{C} \cdot \Omega$ it follows that if K is irreducible then $K(q_0) = \mathbb{C} \cdot \Omega$. In fact K is irreducible if and only if it contains no lowest weight vectors other than multiples of Ω. For on the one hand any other lowest weight vector Ω' would be orthogonal to K_Ω, and on the other, if K were reducible each piece would contain at least one lowest vector. This means that to decompose H under Diff(S^1) we have to find all the lowest weight vectors in it. This was done, at least conjecturally, by Goldstone, and his conclusions have been verified by Kač [7].

The lowest weight vectors we have found so far (and in fact there are no others) form the SU_2-invariant subspace generated by the vacuum vectors $\{\Omega_{2k}\}_{k \in Z}$. In other words, there is a sequence of lowest weight vectors $\Omega_{2k}^{(m)}$ in H_{2k}, where $\Omega_{2k}^{(m)}$ has weight $(k+m)^2$. (The weight is the eigenvalue of V_0.) Up to a scalar multiple $\Omega_{2k}^{(m)}$ is $(B^*)^m \Omega_{2k+2m}$ if $k \geqq 0$, and $B^m \Omega_{2k-2m}$ if $k \leqq 0$. Goldstone has found elegant explicit expressions for the vectors $\Omega_{2k}^{(m)}$. To explain them let us recall from Sect. 4 the definition of the operators $a_k = A(e^{ik\theta})$ on H. The a_k for $k \geqq 1$ commute, and H_{2k} is a completion of the free cyclic module $\mathbb{C}[a_1, a_2, \ldots] \cdot \Omega_{2k}$ for the polynomial algebra $\mathbb{C}[a_1, a_2, \ldots]$. We also introduced operators $c_1, c_2, \ldots$ related to $a_1, a_2, \ldots$ by Newton's formulae

$$\exp \sum_{n>0} \frac{a_n}{n} t^n = \sum_{n>0} c_n t^n,$$

i.e.

$$c_1 = a_1,$$
$$2c_2 = a_1^2 + a_2,$$
$$6c_3 = a_1^3 + 3a_1 a_2 + 2a_3, \text{etc.}$$

(The signs, however, are not the same as in the usual version.) Then $\mathbb{C}[a_1, a_2, \ldots] = \mathbb{C}[c_1, c_2, \ldots]$; and Goldstone's formulae are

Proposition (6.4) *If $k \geqq 0$ then $\Omega_{\pm 2k}^{(m)} = f_{2k}^{(m)} \Omega_{\pm 2k}$, up to a scalar multiplier, where*

$$f_{2k}^{(m)} = \begin{vmatrix} c_{2k+1} & c_{2k+2} & c_{2k+3} & \cdots & c_{2k+m} \\ c_{2k+2} & c_{2k+3} & c_{2k+4} & \cdots & c_{2k+m+1} \\ \vdots & \vdots & & & \\ c_{2k+m} & c_{2k+m+1} & \cdots & \cdots & c_{2k+2m-1} \end{vmatrix}.$$

Proof. We know $\Omega_{-2k}^{(m)} = B^m \Omega_{-2k-2m}$. In the notation of Sect. 4 this is

$$(2\pi)^{-m} \int \ldots \int B_{\theta_1} B_{\theta_2} \ldots B_{\theta_m} \Omega_{-2k-2m} \, d\theta_1 \ldots d\theta_m.$$

Now $B_\theta = e^{iA_\theta^+} e^{iA_\theta^-} T(e^{iq\theta})$. Because

$$e^{iA_\theta^-} e^{iA_\phi^+} = (1 - e^{i(\theta-\phi)})^2 e^{iA_\phi^+} e^{iA_\theta^-}$$

and

$$T(e^{iq\theta})\Omega_{-2n} = e^{(2n-1)i\theta}\Omega_{-2n+1}$$

we find that the integrand in the above expression is

$$\Delta(\theta_1, \ldots, \theta_m)^2 e^{i(2k+1)(\theta_1+\cdots+\theta_m)} e^{iA_{\theta_1}^+} \ldots e^{iA_{\theta_m}^+}\Omega_{-2k},$$

where

$$\Delta(\theta_1, \ldots, \theta_m) = \prod_{p<q} (e^{i\theta_p} - e^{i\theta_q}).$$

Now let us write $z_k = e^{i\theta_k}$, and observe that $\Delta = \Delta(\theta_1, \ldots, \theta_m)$ is the Vandermonde determinant

$$\begin{vmatrix} 1 & z_1 & z_1^2 & \ldots & z_1^{m-1} \\ 1 & z_2 & z_2^2 & \ldots & z_2^{m-1} \\ \vdots & & & & \\ 1 & z_m & z_m^2 & \ldots & z_m^{m-1} \end{vmatrix}$$

Accordingly Δ^2 is obtained by summing

$$\text{sign}\,(f)\Delta z_{f_1}^{m-1} z_{f_2}^{m-2} \ldots z_{f_m}^0 = \begin{vmatrix} 1 & z_{f_1} & z_{f_1}^2 & \ldots & z_{f_1}^{m-1} \\ z_{f_2} & z_{f_2}^2 & z_{f_2}^3 & \ldots & z_{f_2}^m \\ \vdots & & & & \\ z_{f_m}^{m-1} & z_{f_m}^m & & \ldots & z_{f_m}^{2m-2} \end{vmatrix}$$

over all permutations $f = (f_1, f_2, \ldots, f_m)$ of $(1, 2, \ldots, m)$. So, because the integrand is symmetric in $(\theta_1, \ldots, \theta_m)$ we have

$$B^m\Omega_{-2k-2m} = m!(2\pi)^{-m}\int \ldots \int J(z_1, \ldots, z_m) e^{iA_{\theta_1}^+} \ldots e^{iA_{\theta_m}^+}\Omega_{-2k}d\theta_1 \ldots d\theta_m,$$

where

$$J(z_1, \ldots, z_m) = \begin{vmatrix} z_1^{2k+1} & z_1^{2k+2} & \ldots & z_1^{2k+m} \\ z_2^{2k+2} & z_2^{2k+3} & \ldots & z_2^{2k+m+1} \\ \vdots & & & \\ z_m^{2k+m} & z_m^{2k+m+1} & \ldots & z_m^{2k+2m-1} \end{vmatrix}.$$

This yield the desired formula, because

$$e^{iA_{\theta_1}^+} \ldots e^{iA_{\theta_m}^+}\Omega_{-2k} = \sum_{k_1, \ldots, k_m} e^{-i(k_1\theta_1+\ldots+k_m\theta_m)} c_{k_1} c_{k_2} \ldots c_{k_m}\Omega_{-2k}.$$

The case of $\Omega_{2k}^{(m)}$ is precisely similar.

The last result to be proved in this section is

Proposition (6.5) *The representation P_0 is irreducible.*

Proof. It is enough to show that P_0 is cyclic with respect to its vacuum vector Ω_0, for we have seen that if P_0 contains another lowest weight vector Ω' then Ω' is perpendicular to the cyclic representation generated by Ω_0.

Recall from Sect. 5 that an element $T: V \to V$ of the complexified Lie algebra of $Sp_0(V)$ can be written

$$T = \begin{pmatrix} \alpha & \beta \\ \bar{\gamma} & -\alpha' \end{pmatrix}$$

with respect to the decomposition $V_{\mathbb{C}} = W \oplus \bar{W}$. Here α is an endomorphism of W, and β and γ are Hilbert–Schmidt operators $\bar{W} \to W$ which can be regarded as elements of $\hat{S}^2(W)$. The action of T on $\hat{S}(W)$ is $\frac{1}{2}M_\beta + D_\alpha - \frac{1}{2}M_\gamma^*$ where M_β is multiplication by $\beta \in s^2(W)$, and D_α is the derivation of $S(W)$ induced by $\alpha: W \to W$. If T is the action of the vector field v_k, with $k > 0$, then it is easy to check that the corresponding $\beta \in \hat{S}^2(W)$ is $\beta_k = \frac{1}{4}\sum_{i=1}^{k-1} a_i a_{k-i}$ in terms of the basis $\{a_m = e^{im\theta}\}$ of W.

Now the Poincaré series of P_0 is $\prod_{k>1}(1-t^k)^{-1}$. So for dimensional reasons it will certainly be cyclic if the vectors $V_{k_1} V_{k_2} \cdots V_{k_r} \Omega_0$ in $H_0 = \hat{S}(W)$ are all linearly independent when $k_1 \geqq k_2 \geqq \ldots \geqq k_r > 1$. But

$$V_{k_1} V_{k_2} \cdots V_{k_r} \Omega_0 = \beta_{k_1} \beta_{k_2} \cdots \beta_{k_r} \Omega_0 + (\text{terms of lower degree}).$$

These vectors will be linearly independent if the elements $\beta_2, \beta_3, \beta_4, \ldots$ of the polynomial algebra $S(W)$ are algebraically independent. But that is obvious, for if β_k were algebraic cover $\mathbb{C}[\beta_2, \ldots, \beta_{k-1}]$ then it would be algebraic over $\mathbb{C}[a_1, \ldots, a_{k-2}]$, and so a_{k-1} would be algebraic over $\mathbb{C}[a_1, \ldots, a_{k-2}]$, which is absurd.

To conclude this section I should remark that it is easy to see that for each $\lambda \in \mathbb{R}$ the Lie algebra $\mathscr{V}$ has a unique irreducible projective representation P_λ (with the correct cocycle) generated by a lowest weight vector ω_λ such that $V_0 \omega_\lambda = \lambda \omega_\lambda$. If this is to come from the extension of Diff(S^1) we are considering then λ must be an integer. If the representation is to be unitary then λ must not be negative, for otherwise

$$\langle V_1 \omega_\lambda, V_1 \omega_\lambda \rangle = \langle \omega_\lambda, V_{-1} V_1 \omega_\lambda \rangle = 2\langle \omega_\lambda, V_0 \omega_\lambda \rangle = 2\lambda \langle \omega_\lambda, \omega_\lambda \rangle$$

would be negative. (But of course if $\lambda \geqq 0$ then $\bar{P}_\lambda$ is a representation of negative energy with a highest weight $-\lambda$.) If λ is not a square Kač has shown that the Poincaré series of P_λ is $\chi_\lambda(t) = t^\lambda \pi(t)$. But when $\lambda = q^2$ the representation P_λ suddenly becomes smaller, and its Poincaré series is

$$\chi_{q^2}(t) = t^{q^2}(1 - t^{2q+1})\pi(t).$$

Notice that $t^{q^2}\pi(t) = \sum_{r \geqq q} \chi_{r^2}(t)$.

The Decomposition of the Representation of Map$(S^1; G)$ *for other G*

Let us now turn to the more general situation of the last part of Sect. 4. We considered there some irreducible projective representations of $M = \mathrm{Map}(S^1 : T)$, where T was a torus. The representations are of the form

$$H = \bigoplus_{\lambda \in X} H_\lambda,$$

where H_λ is an irreducible representation of the identity component M_0 of M on which $T \subset M$ acts by the character λ. The set X of characters is a coset of $L = \pi_1(T)$ in $L^* = \mathrm{Hom}(T; \mathbb{T})$.

The group $\mathrm{Diff}(S^1)$ is a group of automorphisms of the central extension $\hat{M}$ of M: it does not preserve the cocycle c defining $\hat{M}$, but acts by

$$\phi^*(z, e^{if}) = ze^{-(1/2)i\langle \Delta_f, f(\phi(0)) - f(0)\rangle}, \phi^*(e^{if}).$$

This means that $\mathrm{Diff}(S^1)$ acts projectively on H, as before, preserving each subspace H_λ. To find the cocycle which arises it is enough to consider H_0, which is an irreducible representation of the vector space $V = M_0/T$. This vector space, with its skew form, is simply the product of d copies of $V^{(1)} = \mathrm{Map}(S^1; \mathbb{R})/\mathbb{R}$, where d is the dimension of T. H_0 is therefore the tensor product of d copies of $H_0^{(1)}$, the irreducible representation of $V^{(1)}$. This gives us

Proposition (6.6) $\mathrm{Diff}(S^1)$ *acts projectively with the cocycle d.c on the faithful irreducible representations of* $\mathrm{Map}(S^1; T)$, *where* $d = \dim(T)$, *and c is cocycle of Proposition (6.2).*

By considering the action of the Lie algebra $\mathscr{V}$ on H_0 we know a priori that under $\mathrm{Diff}(S^1)$ the decomposition of H_0 must be of the form

$$H_0 = \bigoplus_q (P_q^d)^{m_q},$$

where P_q^d is an irreducible representation of $\mathscr{V}$ with the cocycle $d.c$ generated by a lowest weight vector with positive integral lowest weight q. (It is clear from general considerations that up to isomorphism there can be only one such representation P_q^d.) The case $d > 1$ is slightly easier than the one we have already studied, and we have

Proposition (6.7) *The Poincaré series of* P_q^d is $t^q\pi(t)$ *if* $q > 0$, *and is* $(1 - t)\pi(t)$ *if* $q = 0$.
Proof. When $q = 0$ the proof is essentially the same as that of (6.5). P_q^d can be identified with the cyclic $\mathscr{V}$-module generated by the vacuum vector in $H_0 = \hat{S}(W)$. Choose an orthonormal basis $\{\xi_1, \dots, \xi_d\}$ of the Lie algebra t of T. This gives one a basis $\{a_M^j = \xi^j e^{im\theta}\}$ $(j = 1, \dots, d; m = 1, 2, 3, \dots)$ of W. The elements β_k of (6.5) are replaced by

$$\beta_k^{(d)} = \frac{1}{2} \sum_{i=1}^{k-1} \sum_{j=1}^{d} a_i^j a_{k-j}^j,$$

and $\beta_2^{(d)}, \beta_3^{(d)}, \dots$ are algebraically independent as before.

To treat P_q^d for $q > 0$ we consider the action of $\mathscr{V}$ on H_λ. The vacuum vector Ω_λ of H_λ is obtained from Ω_0 by the action of $e^{i\lambda} \in M$, so the argument we have used earlier when $d = 1$ shows that the group of rotations R acts on Ω_λ with weight $\frac{1}{2}\langle \lambda, \lambda \rangle$. We choose λ so that $\frac{1}{2}\langle \lambda, \lambda \rangle = q$ and let P_q^d be the $\mathscr{V}$-module generated by Ω_λ. We can calculate the action $\tilde{V}_k$ of $v_k \in V$ on H_λ in terms of its action V_k on H_0: when H_0 and H_λ are both identified with $\hat{S}(W)$ we have

$$\tilde{V}_k = V_k + \sum_j \lambda_j a_k^j,$$

when $k \neq 0$, where $\{\lambda_j\}$ are the components of λ with respect to the basis of t. In other words, when $k > 0$ the action of $\tilde{V}_k$ on $\hat{S}^r(W)$ is multiplication by $\tilde{\beta}_k^{(d)} = \beta_k^{(d)} + \sum \lambda_j a_k^j$ modulo $\hat{S}^r(W)$. The elements $\tilde{\beta}_k^{(d)}$ for $k = 1, 2, 3, \ldots$ are algebraically independent, and that establishes our result.

In view of (6.7) we can write down at once the decomposition of H_0 under $\mathrm{Diff}(S^1)$: the result is that P_q^d occurs with multiplicity m_q equal to the coefficient of t^q in $t + \pi(t)^{d-1}$.

It is more interesting, however, to decompose the space $H = \bigoplus_{\lambda \in L} H_\lambda$ under the group $G \times \mathrm{Diff}(S^1)$. To do so we write down the character of H as a representation $T \times R$. It is

$$F = \pi(t)^d \sum_{\lambda \in L} e^{i\lambda} t^{1/2\langle \lambda, \lambda \rangle},$$

where t is now regarded as the "identity" character $R \to \mathbb{T}$. We should like to write this in the form $\sum_{\mu \in P} f_\mu(t) \chi_\mu$, where P is the set of dominant weights in L, and χ_μ is the character of the irreducible representation of G associated to $\mu \in P$. The following result was shown to me by Macdonald.

Proposition (6.8) *The character of H as a representation of $G \times R$ is $\sum_{\mu \in P} f_\mu(t) \chi_\mu$, where*

$$f_\mu(t) = \pi(t)^d t^{(1/2)\langle \mu, \mu \rangle} \prod_{\alpha > 0} (1 - t^{\langle \alpha, \mu + \rho \rangle}).$$

(Here the product is over the positive roots of G, and ρ is half the sum of the positive roots.)

Proof. By the Weyl character formula $f_\mu(t)$ is the coefficient of $e^{i(\mu + \rho)}$ in $\Delta \cdot F$, where $\Delta = \sum_{w \in W} (-1)^w e^{iw\rho}$ is the Weyl denominator, W being the Weyl group of G. This gives

$$f_\mu(t) = \pi(t)^d \sum_{w \in W} (-1)^w t^{(1/2)\langle \mu + \rho - w\rho, \mu + \rho - w\rho \rangle}.$$

But we know from the Weyl denominator identity that

$$\sum_{w \in W} (-1)^w t^{\langle w\rho, \xi \rangle} = \prod_{\alpha > 0} (t^{(1/2)\langle \alpha, \xi \rangle} - t^{-(1/2)\langle \alpha, \xi \rangle})$$

for any $\xi \in L$; and applying that when $\xi = -\mu - \rho$ gives us the desired formula.

The character of H can also be calculated in a completely different way by using the generalized Weyl character formula of Kač[6]. I shall not give the details here, but the result is

$$F = \Pi^{-1} \sum_{\lambda \in L} \chi_{n\lambda} t^{(1/2)n\langle \lambda, \lambda \rangle + \langle \lambda, \rho \rangle},$$

where $\Pi = \pi(t)^d \prod_{k=1}^{\infty} \prod_{\alpha} (1 - t^k e^{i\alpha})$, α runs over all the roots of G, and n is the Coxeter number of G. Equating this to our earlier expression gives the interesting identity

$$\sum_{\lambda \in L} \chi_{n\lambda} t^{(1/2)n\langle \lambda, \lambda \rangle + \langle \lambda, \rho \rangle} = \sum_{\lambda \in L} e^{i\lambda} t^{(1/2)\langle \lambda, \lambda \rangle} \prod_{k=1}^{\infty} \prod_{\alpha} (1 - t^k e^{i\alpha}),$$

which has also been found by Kač from another point of view ([6](3.38)).

7. Group Extensions

(*a*) Map(S^1;G)

In Sect. 4 we constructed some projective unitary representations of the group $\Gamma = \mathrm{Map}(S^1;G)$, where G was a compact, simply connected, simply laced Lie group, but we did not describe explicitly the central extension $\mathbb{T} \to \hat{\Gamma} \to \Gamma$ involved. We shall now discuss it further.

The first point to notice is that as a topological space $\hat{\Gamma}$ is not the product of Γ and $\mathbb{T}$, so it cannot be described by a continuous cocycle $\Gamma \times \Gamma \to \mathbb{T}$ as in Sect. 2. (This is closely related to the fact that $\hat{\Gamma}$ does not come from an extension of Γ by $\mathbb{R}$. It is easy to see that if an extension of a group Γ by $\mathbb{T}$ lifts to $\mathbb{R}$ then it is topologically a product. The converse is true if Γ is connected, but not otherwise: the extension $\hat{M} \to M$ of Sect. 2 is topologically a product but does not lift to $\mathbb{R}$.)

The topological type of the circle bundle $\hat{\Gamma} \to \Gamma$ can be determined from the extension of Lie algebras

$$\mathbb{R} \to \mathrm{Lie}(\hat{\Gamma}) \to \mathrm{Lie}(\Gamma)$$

in view of the following simple result, which I think is well-known.

Proposition (7.1) *If a group extension $\mathbb{T} \to \hat{\Gamma} \to \Gamma$ corresponds to the Lie algebra cohomology class $\omega \in H^2(\mathrm{Lie}(\Gamma);\mathbb{R})$, then the image of $\frac{\omega}{2\pi}$ under the map*

$$H^2(\mathrm{Lie}(\Gamma);\mathbb{R}) \to H^2(\Gamma;\mathbb{R})$$

is the first Chern class of the circle-bundle $\hat{\Gamma} \to \Gamma$ (with real coefficients).

The map in (7.1) is the one which interprets a skew multilinear form on Lie (Γ) as a left-invariant differential form on the manifold Γ. (The 2π comes in because we are identifying $\mathbb{T}$ with $\mathbb{R}/2\pi\mathbb{Z}$.)

Let us recall from Sect. 4 that in our case the extension of Lie algebras is defined by the cocycle ω, where

$$\omega(\phi, \psi) = \frac{1}{2\pi}\int_0^{2\pi} \langle \phi'(\theta), \psi(\theta)\rangle d\theta,$$

and $\langle , \rangle$ is the invariant inner product on $\mathfrak{g} = \mathrm{Lie}(G)$ described in Sect. 4.

For any compact Lie group G we can define elements λ in $H^2(\Gamma;\mathbb{Z})$, and hence natural circle-bundles on Γ, by the transgression of elements $\alpha \in H^3(G;\mathbb{Z})$: one pulls α back to $\varepsilon^*(\alpha) \in H^3(S^1 \times \Gamma)$ by the evaluation map $\varepsilon: S^1 \times \Gamma \to G$, and integrates over S^1 to get $\lambda \in H^2(\Gamma)$.

With real coefficients α can be represented by a left-invariant form, again denoted α, given by

$$\alpha(\xi, \eta, \zeta) = \frac{1}{8\pi^2}\langle [\xi, \eta], \zeta \rangle,$$

where $\xi, \eta, \zeta \in \mathfrak{g}$ are thought of as tangent vectors to G at some point, and $\langle , \rangle$ is an invariant inner product on $\mathfrak{g}$. We have

Proposition (7.2) *If G is simply connected then α is an integral class if and only if $\langle \lambda, \lambda \rangle \in 2\mathbb{Z}$ for each co-root λ in the Lie algebra of the maximal torus of G.*

Let us recall that the definition of co-roots does not involve choosing an inner product on g: in particular, if λ_α is the co-root associated to a root α then $\alpha(\lambda_\alpha) = 2$. The condition of (7.2) is satisfied for the inner product of Sect. 4.

Now suppose $\phi, \psi : S^1 \to \mathfrak{g}$ are two tangent vectors to Γ at $f : S^1 \to G$. If ϕ, ψ and $\frac{d}{d\theta}$ are regarded as tangent vectors to $S^1 \times \Gamma$ at (θ, f) then

$$\varepsilon^* \alpha_{(\theta, f)}\left(\phi, \psi, \frac{d}{d\theta}\right) = \langle [\phi(\theta), \psi(\theta)], f(\theta)^{-1} f'(\theta) \rangle,$$

and accordingly λ is given at $f \in \Gamma$ by

$$\lambda_f(\phi, \psi) = \frac{1}{8\pi^2} \int_0^{2\pi} \langle [\phi(\theta), \psi(\theta)], f(\theta)^{-1} f'(\theta) \rangle d\theta.$$

This is not an invariant differential form on Γ, but it is cohomologous to $\frac{\omega}{2\pi}$, where ω is the invariant form defined above. In fact $\lambda = \frac{\omega}{2\pi} + d\beta$, where

$$\beta_f(\phi) = \frac{1}{8\pi^2} \int_0^{2\pi} \langle \phi(\theta), f(\theta)^{-1} f'(\theta) \rangle d\theta.$$

If G is connected and simply connected then so is Γ, because $\pi_2(G) = 0$. In that case $H^2(\Gamma; \mathbb{Z}) \to H^2(\Gamma; \mathbb{R})$ is injective, and we have

Proposition (7.3) *If $\mathbb{T} \to \hat{\Gamma} \to \Gamma$ is a group extension corresponding to the Lie algebra cocycle ω then topologically $\hat{\Gamma}$ is the circle-bundle on Γ with Chern class $\lambda \in H^2(\Gamma; \mathbb{Z})$.*

We have still not proved the existence of any such extension $\hat{\Gamma}$, except by the indirect method of Sect. 4. One way to remedy this is by analogy with the discussion in Sect. 5.

Let M be a finite dimensional real representation of G with an inner product $\langle , \rangle$, and let V denote the vector space of smooth maps $S^1 \to M$ modulo constants. V has an inner product given by

$$\langle \phi, \psi \rangle = \frac{1}{2\pi} \int_0^{2\pi} \langle \xi(\theta), \eta(\theta) \rangle d\theta.$$

The group Γ acts orthogonally on V. As in Sect. 5 we decompose the complexification of V as $V_{\mathbb{C}} = W \oplus \bar{W}$, where W is the holomorphic maps of the unit disk into M (modulo constants). If f is an orthogonal transformation of V we write it as a matrix

$$\begin{pmatrix} a(f) & b(f) \\ \overline{b(f)} & \overline{a(f)} \end{pmatrix}$$

with respect to the decomposition $V_{\mathbb{C}} = W \oplus \bar{W}$.

In the symplectic case the operator $a(f) : W \to W$ was necessarily invertible. This is no longer true in the present situation, but nevertheless for any $f \in \Gamma$ it is easy to see that $a(f)$ is a Fredholm operator, and $b(f)$ is Hilbert–Schmidt. It follows that $f \mapsto a(f)$ defines a homomorphism $\Gamma \to \mathrm{Aut}(W)/\mathrm{Aut}_1(W)$, where $\mathrm{Aut}_1(W)$ are the automorphisms of W of the form (identity) + (operator of trace class). Thus the extension

$$\mathrm{Aut}_1(W) \to \mathrm{Aut}(W) \to \mathrm{Aut}(W)/\mathrm{Aut}_1(W)$$

pulls back to define an extension of Γ by $\mathrm{Aut}_1(W)$. But on $\mathrm{Aut}_1(W)$ we have the determinant, a homomorphism

$$\det : \mathrm{Aut}_1(W) \to \mathbb{C}^\times,$$

and this gives us an extension $\tilde{\Gamma}$ of Γ by $\mathbb{C}^\times$.

Proposition (7.4) *The extension $\mathbb{C}^\times \to \tilde{\Gamma} \to \Gamma$ corresponds to the Lie algebra cocycle $\tilde{\omega}$ defined by*

$$\tilde{\omega}(\phi, \psi) = \frac{1}{2\pi} \int_0^{2\pi} \langle \phi'(\theta), \psi(\theta) \rangle_M d\theta,$$

where $\langle\, ,\rangle_M$ is the trace-form of M (i.e. $\langle \xi, \eta \rangle_M = \mathrm{trace}(\xi_M \eta_M)$, where ξ_M is the action of $\xi \in \mathfrak{g}$ on M).

Proof. Let the derivative of $f \mapsto a(f)$ at the identity be denoted by $\phi \mapsto A(\phi)$. We have to show that

$$\mathrm{trace}\,\{[A(\phi), A(\psi)] - A([\phi, \psi])\} = -4i\tilde{\omega}(\phi, \psi) \qquad \ldots (*)$$

for $\phi, \psi : S^1 \to \mathfrak{g}$.

We can write

$$\mathrm{Lie}(\Gamma) = \bigoplus_{k=-\infty}^{\infty} L_k \text{ and } W = \bigoplus_{k=1}^{\infty} W_k.$$

where $L_k = \mathfrak{g} \cdot z^k$ and $W_k = M_{\mathbb{C}} \cdot z^k$. Obviously $[L_k, L_m] \subset L_{k+m}$, and if $\phi \in L_k$ then $A(\phi) \cdot W_k \subset W_{k+m}$, where $W_n = 0$ if $n \leqq 0$. It is enough to prove (*) for homogeneous elements $\phi \in L_k$ and $\psi \in L_m$. If $k + m \neq 0$ then the left-hand side is zero because the matrices of $[A(\phi), A(\psi)]$ and $A([\phi, \psi])$ have no diagonal entries; and $\tilde{\omega}(\phi, \psi) = 0$ also. If $\phi = \xi z^k \in L_k$ with $\xi \in \mathfrak{g}$ and $k \geqq 0$, and $\psi = \eta z^{-k} \in L_{-k}$ then $[A(\phi), A(\psi)]$ and $A([\phi, \psi])$ both preserve each subspace W_m and coincide on W_m if $m > k$. If $m \leqq k$ then

$$\begin{aligned}\{[A(\phi), A(\psi)] - A([\phi, \psi])\} \cdot \zeta z^m &= -(\eta_M \xi_M \zeta + [\xi, \eta]_M \zeta) z^m \\ &= \xi_M \eta_M \zeta z^m.\end{aligned}$$

so the trace on W_m is $\langle \xi, \eta \rangle_M$. The trace on all of W is accordingly

$$\begin{aligned} k\langle \xi, \eta \rangle_M &= \frac{1}{2\pi i} \int \langle \xi z^k \cdot d(\eta z^{-k}) \rangle \\ &= -4i\tilde{\omega}(\phi, \psi), \end{aligned}$$

in accordance with (*).

The extension $\tilde{\Gamma}$ which we want can now be constructed. Choose a real orthogonal representation M of G such that $\langle \xi, \eta \rangle_M = n\langle \xi, \eta \rangle$, where n is an integer.

Then $\hat{\Gamma}$ is an n-fold covering of $\tilde{\Gamma}$: in fact $\pi_1(\tilde{\Gamma}) = \mathbb{Z}/n$, and $\hat{\Gamma}$ is the universal covering of $\tilde{\Gamma}$. In the case of simple simply laced groups one can take for M the adjoint representation of G: then n is the Coxeter number of G.

Remark. The Infinite Dimensional Spin Group

One could carry the preceding discussion further in precise analogy with Sect. 5. One can define a subgroup $O_0(V)$ of the orthogonal group $O(V)$, and a central extension

$$\mathbb{T} \to \mathrm{Spin}_0^c(V) \to O_0(V).$$

$\mathrm{Spin}_0^c(V)$ acts unitarily on the completion of the exterior algebra $\Lambda(W)$: this is the "spin" representation [11], analogous to the metaplectic representation of $Sp_0(V)$. It can be constructed by starting with a space X of polarizations of V such that $X \cong O_0(V)/U(W)$, defining a holomorphic line bundle L on X whose fibre at U (where $V_{\mathbb{C}} = U \oplus \bar{U}$) is $\det(U)^{1/2}$, observing that $\mathrm{Spin}_0^c(V)$ acts on L, and associating to each fibre of L a ray in $\hat{\Lambda}(W)$. The only significant difference between this and the symplectic case is that L is now not a trivial bundle. One might have guessed that $\pi_0(O_0(V)) = \mathbb{Z}$ because any $f \in O_0(V)$ defines a Fredholm operator $a(f)\colon W \to W$ which has an index in $\mathbb{Z}$; but it turns out that $a(f)$ must have index zero, because $w \mapsto b(f)\bar{w}$ is an isomorphism $\ker a(f) \to \ker a(f)^*$; and $\pi_0(O_0(V)) = \mathbb{Z}/2$, with the components distinguished by the parity of the dimension of $\ker a(f)$.

(*b*) $\mathrm{Diff}(S^1)$

We have already seen how a central extension of $\mathrm{Diff}(S^1)$ by $\mathbb{T}$ arises from the embedding $\mathrm{Diff}(S^1) \to Sp_0(V)$, where $V = \mathrm{Map}(S^1; \mathbb{R})/(\text{constants})$. This extension is topologically a product, so it can be defined by a cocycle $\omega\colon \mathrm{Diff}(S^1) \times \mathrm{Diff}(S^1) \to \mathbb{T}$. Indeed we know that

$$\omega(\phi, \psi) = \mu \det(a(\phi)^{-1} a(\phi\psi) a(\psi)^{-1})^{-1/2},$$

where $a(\phi)\colon W \to W$ is the component of $\phi\colon V \to V$ in the decomposition $V_{\mathbb{C}} = W \oplus \bar{W}$, and $\mu\colon \mathbb{C}^\times \to \mathbb{T}$ is the projection. We should like, however, to have a more explicit formula for the cocycle.

Such a formula has been found by Bott. If $\phi\colon S^1 \to S^1$ is a diffeomorphism (where S^1 is regarded as $\mathbb{R}/2\pi\mathbb{Z}$) define $\hat{\phi}\colon \mathbb{T} \to \mathbb{T}$ (where $\mathbb{T} = \{z \in C : |z| = 1\}$) by $\phi(e^{i\theta}) = e^{i\phi(\theta)}$. Then Bott's cocycle $\tilde{\omega}$ is given by

$$\tilde{\omega}(\phi, \psi) = e^{iw(\phi, \psi)},$$

where

$$w(\phi, \psi) = \mathrm{Re} \int_{S^1} \log \phi' \cdot d \log \chi',$$

and $\chi = \psi\phi$. (Here $\phi'(e^{i\theta})$ is to be interpreted as $\phi'(\theta) e^{i(\phi(\theta) - \theta)}$.)

I shall show that the cocycles ω and $\tilde{\omega}$ are cohomologous, i.e. that they define the same extension of $\mathrm{Diff}(S^1)$ by $\mathbb{T}$. I do not know whether they are actually equal: it seems to me an interesting question.

Before comparing ω and $\tilde{\omega}$ we again need some general remarks. A group extension $\mathbb{T} \to \hat{G} \to G$ of the kind we are considering is a smooth principal $\mathbb{T}$-bundle on the manifold G. If one chooses a vector space splitting of the extension of Lie algebras $\mathbb{R} \to \hat{\mathfrak{g}} \to \mathfrak{g}$ then the elements of $\mathfrak{g}$ define left-invariant vector fields on $\hat{G}$, and hence a left-invariant connection in the $\mathbb{T}$-bundle. The curvature of this connection is a left-invariant 2-form α on G. It is easy to check that α represents the Lie algebra cohomology class of the extension $\hat{\mathfrak{g}}$: it is the representative defined by the chosen splitting $\hat{\mathfrak{g}} \cong \mathfrak{g} \oplus \mathbb{R}$.

The class $\alpha \in H^2(\mathfrak{g};\mathbb{R})$ nearly determines the extension $\hat{G}$. To see that, let us recall that the extensions of G by $\mathbb{T}$ form an abelian group Ext($G;\mathbb{T}$). One way to obtain an extension is to take a homomorphism $\theta : \pi_1(G) \to \mathbb{T}$ and to define $\hat{G}$ as the extension $(\tilde{G} \times \mathbb{T})/\pi_1(G)$ induced by θ from the universal covering $\tilde{G} \to G$. This gives a homomorphism

$$\mathrm{Hom}(\pi_1(G);\mathbb{T}) \to \mathrm{Ext}(G;\mathbb{T}).$$

the extensions so obtained are flat vector bundles, and we have

Proposition (7.4) *The sequence*

$$\mathrm{Hom}(\pi_1(G);\mathbb{T}) \to \mathrm{Ext}(G;\mathbb{T}) \to H^2(\mathfrak{g};\mathbb{R}) \to H^2(G;\mathbb{T})$$

is exact. (The right-hand map is the composite of $H^2(\mathfrak{g};\mathbb{R}) \to H^2(G;\mathbb{R})$, which regards a left-invariant form as a de Rham cohomology class, and $H^2(G;\mathbb{R}) \to H^2(G;\mathbb{T})$.)

I shall not give a proof of this here: but exactness at Ext($G;\mathbb{T}$), which is what we shall use, holds because if $\alpha \in H^2(\mathfrak{g};\mathbb{R})$ is zero then the splitting $\hat{\mathfrak{g}} \cong \mathfrak{g} \oplus \mathbb{R}$ can be chosen so that the connection in the circle-bundle $\hat{G} \to G$ is flat, and so defines a homomorphism $\pi_1(G) \to \mathbb{T}$. The result is of course very well-known when G is finite-dimensional, and I state it here to emphasize that finite-dimensionality is not required. The consequence that is relevant for us is

Corollary (7.5) *An extension of* Diff(S^1) *by $\mathbb{T}$ is determined by its infinitesimal class in H^2 (Vect(S^1):$\mathbb{R}$) together with its restriction to* $\mathrm{PSL}_2(\mathbb{R}) \subset \mathrm{Diff}(S^1)$. *In fact* $\mathrm{Ext}(\mathrm{Diff}(S^1);\mathbb{T}) \cong \mathbb{T} \times \mathbb{R}$.

The corollary follows from (7.4) because $\pi_1(\mathrm{Diff}(S^1)) = \pi_1(\mathrm{PGL}_2(\mathbb{R})) = \mathbb{Z}$, and $\mathrm{Hom}(\pi_1(\mathrm{PGL}_2(\mathbb{R}));\mathbb{T}) \cong \mathrm{Ext}(\mathrm{PGL}_2(\mathbb{R});T)$ because $\mathrm{PGL}_2(\mathbb{R})$ is semisimple. We saw in Sect. 6 that $H^2(\mathrm{Vect}(S^1);\mathbb{R}) \cong \mathbb{R}$.

Another way to formulate the result (7.5) is to say that the group $G = \mathrm{Diff}(S^1)$ has a *universal* central extension $A \to G$ with $A = \mathbb{Z} \oplus \mathbb{R}$. This E is an extension $\mathbb{R} \to E \to \tilde{G}$, where $\tilde{G}$ is the simply connected covering of G, the group of diffeomorphisms $\phi : \mathbb{R} \to \mathbb{R}$ satisfying $\phi(\theta + 2\pi) = \phi(\theta) + 2\pi$.

Returning to the cocycles ω and $\tilde{\omega}$, we observe first that they both vanish identically on $\mathrm{PSL}_2(\mathbb{R})$, which is the group of holomorphic automorphisms of the unit disk in $\mathbb{C}$, and is the subgroup of Diff(S^1) which preserves the polarization $V_{\mathbb{C}} = W \oplus \bar{W}$. In the case of ω we have only to notice that the restriction of $a : \mathrm{Diff}(S^1) \to \mathrm{Aut}(W)$ to $\mathrm{PSL}_2(\mathbb{R})$ is a homomorphism. In the case of $\tilde{\omega}$ the integral defining $w(\phi, \psi)$ vanishes for $\phi, \psi \in \mathrm{PSL}_2(\mathbb{R})$ by Cauchy's theorem, as then $\hat{\phi}$ and $\hat{\chi}$ extend to holomorphic functions in the disk.

Now let us consider the Lie algebra cocycles induced by ω and $\tilde{\omega}$. The general formula for the Lie algebra cocycle associated to a cocycle $c : G \times G \to \mathbb{R}$ is

$$(\xi, \eta) \mapsto D^2c(\xi, n) - D^2c(n, \xi),$$

where $D^2c : \mathfrak{g} \times \mathfrak{g} \to \mathbb{R}$ is the mixed second derivative of c at $(1, 1)$.

In the case of $\tilde{\omega}$ an elementary calculation gives

$$(\xi,\eta)\mapsto\frac{1}{24\pi}\int_0^{2\pi}(\xi'''(\theta)+\xi'(\theta))\eta(\theta)d\theta,$$

whereas from ω we get

$$(\xi,\eta)\mapsto\tfrac{1}{2}i\,\mathrm{trace}([\xi,\eta]_W-[\xi_W,\eta_W]),$$

where $\xi_W : W\to W$ denotes the $W-W$ component of the action of ξ on V. Each of these formulae is invariant under the group R of rotations, so to prove they coincide it is enough to evaluate them on $\xi=v_{-k}$ and $\eta=v_k$. The first formula gives $\frac{i}{12}k(k^2-1)$, so we must show

$$\mathrm{trace}(2kv_{0,W}-[v_{-k,W},v_{k,W}])=\tfrac{1}{6}k(k^2-1).$$

The calculation is just like that in the proof of (7.3). The operators $2kv_{0,W}$ and $[v_{-k,W},v_{k,W}]$ are both diagonal with respect to the basis elements $\{a_m\}_{m\geqq 1}$ of W. They coincide on a_m if $m>k$. If $m\leqq k$ then $2kv_{0,W}$ multiplies a_m by $2km$. But

$$\begin{aligned}[v_{-k,W},v_{k,W}]a_m&=v_{-k,W}v_{k,W}a_m\\&=m(m+k)a_m,\end{aligned}$$

and so the trace is $\sum_{k=1}^{m}m(k-m)=\frac{1}{6}k(k^2-1)$, as we want.

Before leaving the central extension of $\mathrm{Diff}(S^1)$ it would be pointed that the extension of Lie algebras

$$\mathbb{R}\to\mathrm{Vect}(S^1)^{\wedge}\to\mathrm{Vect}(S^1)$$

is closely related to the "Schwarzian derivative". One ordinarily thanks of this as a third-order non-linear differential operator S defined on functions on the circle, characterized by the property that

$$S(a\phi+b/c\phi+d)=S(\phi)$$

for any constants a,b,c,d. The formula is

$$S(\phi)=\frac{1}{2}\frac{\phi'''}{\phi'}-\frac{3}{4}\left(\frac{\phi''}{\phi'}\right)^2.$$

I shall think of it, however, as defined for diffeomorphisms ϕ of $S^1=R/2\pi Z$ by the formula

$$\sigma(\phi)=\tfrac{1}{6}S(\phi)+\tfrac{1}{24}((\phi')^2-1).$$

Then it has the properties

(a) $\sigma(\phi\psi)=\sigma(\psi)$ if $\phi\in\mathrm{PSL}_2(R)$, and

(b) $\sigma(\phi\,\psi)=\psi^*\sigma(\phi)\cdot(\psi')^2+\sigma(\psi)$.

Here ψ^* is the operation defined by $(\psi^*f)(\theta)=f(\psi(\theta))$. The second property can be expressed by saying that σ is a crossed homomorphism from $\mathrm{Diff}(S^1)$ to Q, the space of quadratic differentials on S^1.

Now Q is naturally dual to Vect (S^1). To describe an extension of the Lie algebra Vect(S^1) by $\mathbf{R}$ it certainly suffices to give the adjoint action of Diff(S^1) on $\mathbf{R} \oplus$ Vect(S^1), or dually on $\mathbf{R} \oplus Q$. The latter action preserves the affine hyperplane $1 \oplus Q$, and is obviously determined by its restriction to that. This restriction can be thought of as an affine action of Diff(S^1) on Q.

Proposition (7.6) *For the extension we have been studying the affine action of* Diff(S^1) *on Q is given by*

$$(\phi^{-1}, q) \mapsto \phi^* q + \sigma(\phi),$$

where if $q = a(\theta)d\theta^2$ then $\phi^* q = a(\phi(\theta))\phi'(\theta)^2 d\theta^2$.

The proof of this is obvious from what has preceded. For an affine action is the same thing as a crossed homomorphism, and the crossed homomorphism σ : Diff$(S^1) \to Q$ is determined by its derivative $D\sigma$ at the identity, which is given by

$$D\sigma(\xi) = \tfrac{1}{12}(\xi''' + \xi'),$$

in agreement with the formula we found earlier for the cocycle defining the extension.

8. Orbits

According to Kirillov and Kostant's theory of "orbits" the irreducible unitary representations of a group G are roughly in correspondence with a class of orbits of the action of G on $\mathfrak{g}^*$, the dual of the Lie algebra of G. More precisely, if $\alpha \in \mathfrak{g}^*$ then the orbit X_α of α is naturally a symplectic manifold with a closed 2-form ω_α. The orbit is called *integral* if ω_α defines an integral cohomology class in $H^2(X_\alpha; \mathbf{R})$. If G_α is the isotropy group of α, and $\mathfrak{g}_\alpha$ is its Lie algebra, then $\alpha|\mathfrak{g}_\alpha$ is a homomorphism of Lie algebras $\mathfrak{g}_\alpha \to \mathbf{R}$, and one can ask whether it lifts to a character $\chi: G_\alpha \to \mathbf{T}$. If one is lucky then representations of G will correspond to pairs (X_α, χ), where X_α is an integral orbit and $\chi: G_\alpha \to \mathbf{T}$ is a lift of $\alpha|\mathfrak{g}_\alpha$.

It would be very optimistic to hope for such a correspondence to exist in the case of the infinite dimensional groups considered in this paper. Nevertheless it is interesting to inspect the orbits to see whether any of them show signs of corresponding to the representations we have found.

The groups we have considered are central extensions by $\mathbf{T}$ of $\Gamma = $ Map$(S^1; G)$ and Diff(S^1), where G is a compact group. Their Lie algebras can accordingly be identified with $\mathbf{R} \oplus$ Map$(S^1; \mathfrak{g})$ and $\mathbf{R} \oplus$ Vect(S^1). The duals of Map$(S^1; \mathfrak{g})$ and Vect(S^1) are spaces of distributions on the circle; but we shall consider here only the "smooth" part of the duals, identified respectively with $L = $ Map$(S^1; \mathfrak{g})$ (using an invariant inner product on $\mathfrak{g}$) and Q, the space of quadratic differentials on S^1. Thus we have to consider the action of Γ on $\mathbf{R} \oplus L$ and the action of Diff(S^1) on $\mathbf{R} \oplus Q$. As we are interested only in representations which are faithful on the centre $\mathbf{T}$ we need only consider the orbits in $1 \oplus L$ and $1 \oplus Q$. It turns out in both cases that these orbits are very easy to classify (unlike the orbits in $0 \oplus L$ and $0 \oplus Q$).

(*a*) *The Case of $\Gamma = $* Map$(S^1; G)$, *where G is Compact*

The action of Γ on $1 \oplus L$ will be thought of as an affine action on L:

$$(f, \xi) \mapsto f * \xi = f \cdot \xi + a_f,$$

where $f \in \Gamma$ and $\xi \in L$, and $f \cdot \xi$ denotes the obvious adjoint action of f on ξ. Because $f \mapsto a_f$ is a crossed homomorphism it is determined by its derivative at the identity element, which is $\eta \mapsto \omega(\eta, \cdot)$, where $\omega : L \times L \to \mathbb{R}$ is the Lie algebra cocycle defining the extension $\tilde{\Gamma}$ of Γ. Recalling that

$$\omega(\xi, \eta) = \frac{\pi}{2\pi} \int_0^{2\pi} \langle \xi'(\theta), \eta(\theta) \rangle d\theta$$

we find

Proposition (8.1) $a_f(\theta) = f'(\theta) f(\theta)^{-1}$

Note. Here, and from now on, I shall employ notation as if G were a matrix group, and $f : S^1 \to G$ a matrix-valued function. Thus the adjoint action of f on ξ above is given by $(f \cdot \xi)(\theta) = f(\theta)\xi(\theta)f(\theta)^{-1}$.

For any $\xi \in L$ there is a unique function $F_\xi : [0, 2\pi] \to G$ such that $F_\xi(0) = 1$ and $F'_\xi(\theta) = \xi(\theta) F_\xi(\theta)$ for $0 \leqq \theta \leqq 2\pi$. Let us define $g : L \to G$ by $g(\xi) = F_\xi(2\pi)$.

Proposition (8.2) (i) $g(f * \xi) = f(0) g(\xi) f(0)^{-1}$,

(ii) *the orbits of the affine action of Γ on L are the inverse-images by $g : L \to G$ of the conjugacy classes of G,*

(iii) *each orbit contains a constant map $\xi : S^1 \to \mathfrak{g}$, and*

(iv) *the isotropy group of any $\xi \in L$ is isomorphic (via $f \mapsto f(0)$) to the centralizer of $g(\xi)$ in G.*

Proof. (i) Because the solutions of ordinary differential equations are unique one has $F_{f*\xi}(\theta) = f(\theta) F_\xi(\theta) f(0)^{-1}$.

(ii) If $g(\xi) = g(\eta)$ then define $f \in \Gamma$ by $f(\theta) = F_\eta(\theta) F_\xi(\theta)^{-1}$. (This does belong to Γ, being smooth at $\theta = 0 = 2\pi$.) By calculation it appears that $f * \xi = \eta$.

(iii) Given any $\xi \in L$ choose $\xi_0 \in \mathfrak{g}$ so that $\exp(2\pi\xi_0) = g(\xi)$. Regarding ξ_0 as a constant element of L we have $F_{\xi_0}(\theta) = \exp(\theta\xi_0)$ and so $g(\xi_0) = g(\xi)$.

(iv) We have $f * \xi = \xi$ if and only if

$$f'(\theta) = [\xi(\theta), f(\theta)].$$

This equation is uniquely soluble for f given $f(0)$ providing $f(0)$ commutes with $g(\xi)$. (It is enough to consider the case when ξ is constant: then the solution is $f(\theta) = e^{\theta\xi} f(0) e^{-\theta\xi}$.)

Now let us consider some constant $\xi \in \mathfrak{g} \subset L$ and ask whether $\langle \xi, \ \rangle$ lifts to a character of its isotropy group, the centralizer Z of $\exp(2\pi\xi)$ in G. Let T be a maximal torus of G whose Lie algebra contains ξ. It is well-known that if $\langle \xi, \ \rangle$ is a weight of T then $\exp(2\pi\xi)$ is in the centre of G, so that $Z = G$. On the other hand if $\exp(2\pi\xi)$ is in the centre of G then it is easy to see that $\langle \xi, \ \rangle$ is liftable. The orbit of $0 \in \mathfrak{g}$ is $\Gamma/G \cong \Omega G$, the loop-space of G, and one can identify it with a subgroup of Γ. It is easy to check that its symplectic form is the 2-form discussed in Sect. 7, the Chern class of the central extension regarded as a circle-bundle on ΩG. The other orbits can be identified with the connected components of $\Omega G'$, where G' is the quotient of G by its centre.

We have seen in Sect. 4 that the representations of Γ of positive energy are indexed by the centre of G, and Kač's method leads to the same conclusion. So the suggestion of the orbit method would be that all the projective unitary representations with the cocycle we are considering are of positive energy.

(b) Diff(S^1) *and Hill's Equations*

The smooth part of the dual of the Lie algebra of the central extension of Diff(S^1) that we are concerned with can be identified with $R \oplus Q$, where Q is the space of quadratic differentials on S^1, and the projective representations of Diff(S^1) with the given cocycle ought to correspond to the orbits of Diff(S^1) on $1 \oplus Q \cong Q$. We have seen in Sect. 7 that the action is the affine action

$$(\phi^{-1}, q) \mapsto \phi^* q + \sigma(\phi),$$

where $\sigma(\phi)$ is the Schwarzian derivative.

The space $1 \oplus Q$ has an interesting interpretation in terms of Hill's equations. A Hill's operator is a second-order linear differential operator on the circle of the form

$$D_q = \left(\frac{d}{d\theta}\right)^2 + q(\theta),$$

where q is a real-valued function on S^1. If the parameter θ is changed by a diffeomorphism $\theta \mapsto \phi(\theta)$ so that $d/d\theta$ becomes $(\phi')^{-1} d/d\theta$ then D_q becomes

$$\frac{1}{(\phi')^2}\left(\frac{d}{d\theta}\right)^2 - \frac{\phi''}{(\phi')^3}\frac{d}{d\theta} + q(\phi(\theta)).$$

It is very easy to check that this is

$$M((\phi')^{3/2}) \cdot D_{\tilde{q}} \cdot M((\phi')^{-1/2}),$$

where $\tilde{q}(\theta) = q(\phi(\theta)) \cdot \phi'(\theta)^2 + S(\phi)$, and $M(\psi)$ is the operation of multiplication by ψ. This means that if D_q is regarded not as an operator on functions on S^1 but as an operator taking of weight $\frac{1}{2}$ to densities of weight $\frac{3}{2}$ then the natural transform of D_q by ϕ is $D_{\tilde{q}}$. We have, therefore, a natural action of Diff(S^1) on the space of Hill's operators, and can identify the Hill's operators with the affine space $1 \oplus Q$. To make the fit precise we shall redefine D_q as $(d/d\theta)^2 + \hat{q}$, where $\hat{q} = 6q + \frac{1}{4}$.

The orbits of this action have been studied in [9]. I shall recall the result here, as I wish to state it differently (and the result in [9] seems not quite correct). Consider the equation $D_q f = 0$ as an equation on $\mathbb{R}$ rather than S^1. Let V_q be its two-dimensional space of solutions. Evaluation at a point $\theta \in \mathbb{R}$ defines an element of the dual space V_q^*, or, more accurately, of the projective line $P(V_q^*)$, as V_q consists of densities rather than functions. Thus to D_q is associated a natural map $F_q : \mathbb{R} \to P(V_q^*)$, which is easily seen to be a local homeomorphism. Because D_q has periodic coefficients the operation of translation by 2π induces a map $M_q : V_q \to V_q$ called the *monodromy of* D_q, such that the following diagram commutes:

$$\begin{array}{ccc} \mathbb{R} & \xrightarrow{F_q} & P(V_q^*) \\ {\scriptstyle T}\downarrow & & \downarrow{\scriptstyle M_q} \\ \mathbb{R} & \xrightarrow{F_q} & P(V_q^*) \end{array}$$

where T is $\theta \mapsto \theta + 2\pi$. Conversely if one is given a smooth immersion $F: \mathbb{R} \to P^1_{\mathbb{R}}$ and a projective automorphism M of $P^1_{\mathbb{R}}$ such that $MF = FT$ then it is clear that the pair (F, M) arises from a unique Hill's equation. So Hill's equations can be identified with pairs (F, M), where two such pairs are regarded as the same if they differ by an automorphism of $P^1_{\mathbb{R}}$. The action of a diffeomorphism ϕ of S^1 on (F, M) simply replaces it by $(F\tilde{\phi}, M)$, were $\tilde{\phi}: \mathbb{R} \to \mathbb{R}$ is a lift of ϕ. Now for a given operator D_q the monodromy M_q determines a well-defined conjugacy class in $\mathrm{PGL}_2(\mathbb{R})$; but (because $\pi_1\ \mathrm{PGL}_2(\mathbb{R}) \cong \pi_1\ P^1_{\mathbb{R}}$) the path F_q determines a definite lift of this to a conjugacy class in $\widetilde{SL}_2(\mathbb{R})$, the simply-connected covering group of $\mathrm{PGL}_2(\mathbb{R})$. The following result is now obvious:

Proposition (8.3) *The orbits of* $\mathrm{Diff}(S^1)$ *on* $1 \oplus Q$ *correspond precisely (by assigning to* D_q *its lifted monodromy) to the conjugacy classes of* $\widetilde{SL}_2(\mathbb{R})$.

Note. $\widetilde{SL}_2(\mathbb{R})$ has an outer involution, and the preceding statement really refers to conjugacy under the full automorphism group.

The conjugacy classes in $SL_2(\mathbb{R})$ are determined by the trace, and are of three types, elliptic, parabolic and hyperbolic. The conjugacy classes in $\widetilde{SL}_2(\mathbb{R})$ are correspondingly of three types. The elliptic ones are all on one-parameter subgroups, and are determined by giving, up to sign, a number $\theta \in \mathbb{R}$ such that $2\cos\theta =$ (trace). The parabolic and hyperbolic classes from a disconnected space with components corresponding to $\mathbb{Z}$: they are determined by their trace and the number of their component—evidently only those in the 0-component lie on one-parameter subgroups.

When the lifted monodromy lies on a one-parameter subgroup (and not otherwise) the corresponding orbit contains a representative with q constant. (The other orbits can presumably be represented by Mathieu equations, but the results of [9] seem wrong at this point.) For these orbits the isotropy group always contains the group R of rotations of the circle. It is possible to lift q to a character of R only if q is an integer. The monodromy M_q is

$$\begin{pmatrix} \cos 2\pi\sqrt{q} & -\sin 2\pi\sqrt{q} \\ \sin 2\pi\sqrt{q} & \cos 2\pi\sqrt{q} \end{pmatrix}$$

in $\mathrm{PGL}_2(\mathbb{R})$ if $q \geqq 0$ (and an analogous hyperbolic element if $q < 0$). This is the identity if $\hat{q} = \frac{1}{4}m^2$, i.e. $q = \frac{1}{24}(m^2 - 1)$, where m is an integer. The isotropy group of q is R if $\hat{q} \neq \frac{1}{4}m^2$, but if $\hat{q} = \frac{1}{4}m^2$ it jumps in dimension from 1 to 3, and becomes the m-fold covering of $\mathrm{PGL}_2(\mathbb{R})$ consisting of the elements of $\mathrm{Diff}(S^1)$ which are periodic with period $2\pi/m$ and are m-fold coverings of diffeomorphisms in $\mathrm{PGL}_2(\mathbb{R})$.

All this means that the elliptic orbits correspond rather well in a qualitative way with the results of Sect. 6, and with Kač's results, but the values of q predicted are not quite right, as often happens with the orbit method.

Appendix

A string is the image of a smooth map $x : [0, 1] \to \mathbb{R}^3$. When it moves it sweeps out a "world-surface" $x : [0, 1] \times \mathbb{R} \to \mathbb{R}^4$ in Minkowski space-time (I shall take the metric in Minkowski space to be $-dt^2 + dx^2 + dy^2 + dz^2$). The particles of which the string is made up are regarded as indistinguishable, so the parametrization has no physical significance. Nevertheless I shall suppose that the string can be parametrized in such a way that its individual particles move no faster than light, i.e. $\left\langle \frac{\partial x}{\partial t}, \frac{\partial x}{\partial t} \right\rangle \leq 0$. Then the world-surface between times t_0 and t_1 has an area

$$\int_{t_0}^{t_1} dt \int_0^1 \left\{ \left\langle \frac{\partial x}{\partial s}, \frac{\partial x}{\partial t} \right\rangle^2 - \left\langle \frac{\partial x}{\partial s}, \frac{\partial x}{\partial s} \right\rangle \left\langle \frac{\partial x}{\partial t}, \frac{\partial x}{\partial t} \right\rangle \right\}^{1/2} ds.$$

The motion of the string is supposed to be governed by an action-principle which requires this area to be minimized, or at any rate to be stationary.

If the world-surface lay in Euclidean rather than Minkowski space this would not make sense, as we should have a soap-film without a wire, and the string would collapse completely. But in Minkowski space the problem is well-posed, and we see that the ends of the string must move with the velocity of light at right angles to the string, for then and only then will the area of the surface not change to first order if one simply shrinks the string in on itself.

I shall make two more assumptions. The first is that the motion is such that if one sends a light signal along the string from either end then it will get to the other end in a finite time. That seems reasonable physically, as the part of the string reachable by light signals from one end will presumably behave independently of the rest. Granting this assumption there is a natural parametrization of the world-surface which assigns to a point P the pair (t, t'), where t is the time at which one must send a light-signal from the end along the string to arrive at P, at t' is the time at which a signal from P will reach the end. (I am choosing a preferred end of the string.)

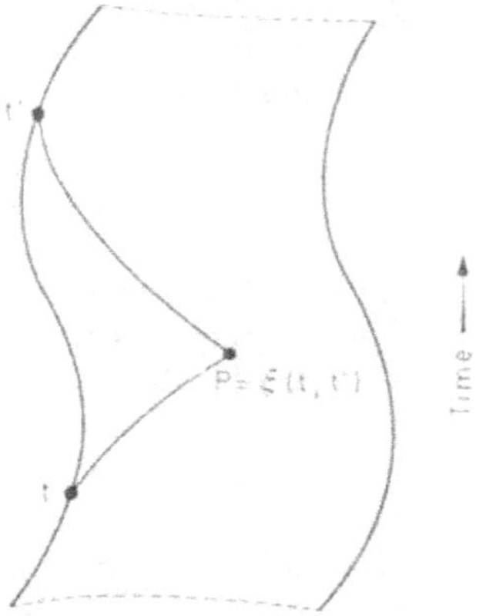

Fig. 2

This parametrization is a map

$$\xi : \{(t, t') \in \mathbb{R}^2 : t \leq t' \leq t + 2\ell(t)\} \to \mathbb{R}^4,$$

where $2\ell(t)$ is the time taken for a signal to travel from the preferred end of the string to the other end and back, beginning at time t.

The second assumption is more technical. Suppose that for a point x near the edge of the world-surface the times taken for forwards and backwards light signals from x to reach the edge are $t_+(x)$ and $t_-(x)$. I shall suppose that the ratio $t_+(x)/t_-(x)$ tends to 1 as x approaches the edge of the surface. (I think that this is equivalent to assuming that the string does not curl up at its ends, i.e. that its curvature is bounded at any time.) Then we have

Proposition (A.1) (a) $\ell(t) = \ell$ *is independent of* t, *and*

(b) $\xi(t, t') = \frac{1}{2}(f(t) + f(t'))$,

where $f: \mathbb{R} \to \mathbb{R}^4$ *satisfies*

(i) $f(t + 2\ell) - f(t) = 2\pi p$ *is independent of* t, *and*

(ii) $\langle f'(t), f'(t) \rangle = 0$.

It is easy to check that p is the total momentum of the string.

The map $f: \mathbb{R} \to \mathbb{R}^4$ is the trajectory of the end of the string parametrized by the time coordinate in the particular Lorentz frame. It is more natural to reparametrize it invariantly by a parameter θ so that

$$\left.\begin{array}{l} \langle f'(\theta), p \rangle = \langle p, p \rangle \text{ and} \\ f(\theta + 2\pi) = f(\theta) + 2\pi p. \end{array}\right\} \qquad (*)$$

These requirements fix the parametrization up to an additive constant.

The significance of Proposition (A.1) is that the motions of the string are completely equivalent to those of a point particle which moves with the velocity of light along a trajectory $f: \mathbb{R} \to \mathbb{R}^4$ satisfying (*). The particle has a well-defined rest-frame in which it describes a periodic closed orbit with the speed of light.

Proof of Proposition (A.1)

If the world-surface is parametrized arbitrarily in terms of parameters (u, v) then the Euler–Larange equations for the variational problem are

$$\frac{\partial}{\partial u}\left(\frac{1}{\sqrt{F}}(\langle x_u, x_v \rangle x_v - \langle x_v, x_v \rangle x_u)\right) + \frac{\partial}{\partial v}\left(\frac{1}{\sqrt{F}}(\langle x_u, x_v \rangle x_u - \langle x_u, x_u \rangle x_v)\right) = 0,$$

where $F = \langle x_u, x_v \rangle^2 - \langle x_u, x_u \rangle \langle x_v, x_v \rangle$, and $x_u = \partial x/\partial u$, $x_v = \partial x/\partial v$. If the parametrization is such that x_u and x_v are light-like then the equations simplify to

$$\frac{\partial^2 x}{\partial u \partial v} = 0.$$

That is the case for the preferred parametrization introduced above, so we find

$$\xi(t, t') = f_1(t) + f_2(t'),$$

for some functions $f_1, f_2: \mathbb{R} \to \mathbb{R}^4$ such that $f_1'(t)$ and $f_2'(t')$ are light-like. The trajectory of one end of the string is $t \mapsto f_1(t) + f_2(t)$, and so $f_1'(t) + f_2'(t)$ must be light-like too. This means that $f_1'(t)$ is parallel to $f_2'(t)$.

Now we use the second assumption. Let f_i^0 denote the time component of f_i. The time-coordinate of $\xi(t, t')$ is $f_i^0(t) + f_2^0(t')$, and the times taken for forward and backward signals from $\xi(t, t')$ to the end are $t' - f_1^0(t) - f_2^0(t)$ and $f_1^0(t) + f_2^0(t) - t$, i.e. $f_1^0(t') - f_1^0(t)$ and $f_2^0(t') - f_2^0(t)$. If the ratio of these tends to 1 as $t' \to t$ then $f_1^{0'}(t) = f_2^{0'}(t)$, and so $f_1' = f_2'$, and we can suppose $f_1 = f_2 = \frac{1}{2}f$.

The trajectory of the other end of the string is $t \mapsto \frac{1}{2}(f(t - \ell(t)) + f(t + \ell(t)))$. For this to be light-like $f'(t - \ell(t))$ and $f'(t + \ell(t))$ must be parallel. But the time-component of each is 1, so $f'(t - \ell(t)) = f'(t + \ell(t))$.

Finally we apply the second assumption at this end. Suppose that t' is slightly greater than t. The point $\xi(t', t + 2\ell))$ has time-coordinate $\frac{1}{2}(t + t') + \ell(t)$, and signals from it research the end at $t + \ell(t)$ and $t' + \ell(t')$. From this we find that $(\frac{1}{2}(t' - t) + \ell(t') - \ell(t))/\frac{1}{2}(t' - t) \to 1$ as $t' \to t$, and so that $\ell'(t) = 0$. Thus $\ell(t)$ is constant, and $f'(t + 2\ell) = f'(t)$. This gives the desired result $f(t + 2\ell) = f(t) + 2\pi p$ for some vector p.

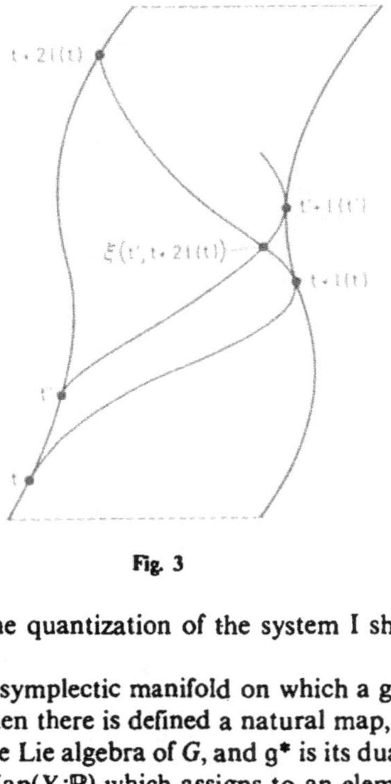

Fig. 3

Before turning to the quantization of the system I shall make a few general remarks.

Suppose that Y is a symplectic manifold on which a group G acts, preserving the symplectic form. Then there is defined a natural map, the "momentum" map, $P: Y \to \mathfrak{g}^*$, where $\mathfrak{g}$ is the Lie algebra of G, and $\mathfrak{g}^*$ is its dual. P is simply the transpose of the map $\mathfrak{g} \to \mathrm{Map}(Y; \mathbb{R})$ which assigns to an element of $\mathfrak{g}$ a Hamiltonian function for the corresponding flow on Y. (The Hamiltonian are determined only up to an additive scalar, and P is more properly a map $Y \to \hat{\mathfrak{g}}^*$, where $\hat{\mathfrak{g}}$ is the central extension of $\mathfrak{g}$ by $\mathbb{R}$ determined by Y.) Clearly P is equivariant with respect to G. If one chooses an orbit ω of the coadjoint action of G on * then $Y_\omega = P^{-1}(\omega)$ is G-invariant, and $X_\omega = Y_\omega/G$ has, if it is a manifold, a natural symplectic structure. X_ω is called the *quotient system* of Y by G with momentum ω. If ω_0 is a point of the orbit ω notice that X_ω can be identified with $X_{\omega_0} = Y_{\omega_0}/G_0$, where $Y_{\omega_0} = P^{-1}(\omega_0)$, and G_0 is the stabilizer of ω_0.

The state space X of a string is a quotient system of this form. Suppose that Y is the space of all smooth maps $f : \mathbb{R} \to \mathbb{R}^4$ such that

$$p_f = \frac{1}{2\pi}(f(\theta + 2\pi) - f(\theta))$$

is constant. Y is a vector space, and has a symplectic structure given by the skew form $S : Y \times Y \to \mathbb{R}$ defined by

$$S(f_1, f_2) = \frac{1}{4\pi}\int_0^{2\pi} (\langle f_1, f_2' \rangle - \langle f_1', f_2 \rangle) d\theta + \tfrac{1}{2}(\langle f_1(0), p_{f_2} \rangle - \langle p_{f_1}, f_2(0) \rangle).$$

Let G denote the universal covering group of $\mathrm{Diff}(S^1)$, the space of diffeomorphisms $\phi : \mathbb{R} \to \mathbb{R}$ such that $\phi(\theta + 2\pi) = \phi(\theta) + 2\pi$. G acts on Y preserving S, and the Hamiltonian function H_ξ inducing the action of the vector field $\xi(\theta) d/d\theta$ is given by

$$H_\xi(f) = S(\xi f', f) = \frac{1}{2\pi}\int_0^{2\pi} \xi(\theta)\langle f'(\theta), f'(\theta) \rangle d\theta.$$

In other words, the momentum map $P : Y \to \mathfrak{g}^*$ is $f \mapsto \langle f', f' \rangle$.

This means that $Y_0 = P^{-1}(0)$ consists of the trajectories $f \in Y$ which travel with the speed of light. Each such has a unique preferred parametrization, and so $X_0 = Y_0/G$ is precisely the state-space X of a string.

In favorable cases quotient systems of the form X_ω can be quantized as follows. One first chooses a quantization of Y, a certain Hilbert space H. The action of G on Y corresponds to a unitary action of G on H. This can be decomposed as a direct integral

$$H = \int H_\omega \otimes P_\omega,$$

where P_ω runs through the irreducible unitary representations of G. If the irreducible representations P_ω can be indexed by the orbits ω of G in $\mathfrak{g}^*$ then the Hilbert space H_ω ought to be the quantization of the quotient system X_ω.

When one attempts to apply this ideal procedure to the states of a string at least two difficulties arise. The first appears when one quantizes any infinite dimensional linear system Y on which a group G acts: the extension of $\mathfrak{g}$ by $\mathbb{R}$ defined by Y is trivial, as each $\xi \in \mathfrak{g}$ has a canonical Hamiltonian which is a homogeneous quadratic function on Y; on the other hand when Y is quantized in the standard "metaplectic" way a non-trivial central extension $\hat{G}$ of G acts on H, and the representations of $\hat{G}$ which occur in H correspond, if they correspond to orbits at all, to orbits in a certain *affine* action of G on $\mathfrak{g}^*$ (cf. Sect. 8). It is thus not clear which H_ω to associate to which values of the classical momentum.

The second difficulty is that one cannot quantize the linear system Y satisfactorily. For to do so we should presumably begin by observing that the $\mathbb{R}^4$-valued function $f \mapsto p_f$ on Y corresponds to the total momentum of the system, and the corresponding operator (actually four commuting operators) should break up H as

$$\int_{p \in \mathbb{R}^4} H_p,$$

where H_p is got by quantizing the linear system $V_p = \{f \in Y : p_f = p\}/\mathbb{R}^4$. (Thus

V_p is a quotient system of Y by the group $\mathbb{R}^4$ of translations.) To quantize V_p one would choose a positive polarization $V_{p,\mathbb{C}} = W \oplus \bar{W}$ and define $H_p = \hat{S}(W)$, as the body of this paper. But unfortunately, because the inner product in Minkowski space is indefinite one cannot choose the polarization both positive and invariant under the Lorentz group. The standard procedure is to choose an invariant polarization W which is not positive (it is the part of $V_{p,\mathbb{C}}$ of "positive energy" in the sense of this paper), and to form a pseudo-Hilbert-space $H_p = \hat{S}(W)$ which has an indefinite inner product. The group G acts projectively on H_p, and it has a discrete decomposition

$$H_p = \bigoplus_{n \geq 0} H_p^{(\lambda_n)} \oplus P_{\lambda_n},$$

where P_{λ_n} is the irreducible representation of G with lowest weight $\lambda_n = n + \frac{1}{2}\langle p, p\rangle$ which need not be integral here as G is now not $\mathrm{Diff}(S^1)$ but its covering group.

One wants to pick out in H_p the isotypical piece $H_p^{(\alpha)}$ associated to a certain irreducible representation P_α of G taken to correspond to the momentum condition $\langle f', f'\rangle = 0$. This will be zero unless $\frac{1}{2}\langle p, p\rangle - \alpha$ is a negative integer. Physicists customarily take $\alpha = 1$, for a reason I shall explain in a moment, though this has the great disadvantage that the Hilbert space then contains state vectors with $\langle p, p\rangle = 2$, i.e. particles moving faster than light with imaginary mass, so called "tachyons". The general nature of the model, on the other hand, leads one to suppose that the lowest state of a string ought to be when it collapses to a particle moving with the speed of light: that would correspond to $\alpha = 0$.

Goddard and Thorn [4] have shown that when $\alpha = 1$ the metric of $H_p^{(\alpha)}$ is positive semi-definite. By dividing $H_p^{(\alpha)}$ by its radical one obtains a genuine Hilbert space $\tilde{H}_p^{(\alpha)}$, and physicists take

$$\tilde{H} = \int \tilde{H}_p^{(\alpha)},$$

where p runs over all momentum vectors in $\mathbb{R}^4$ such that $\frac{1}{2}\langle p, p\rangle + \alpha$ is a negative integer, as the Hilbert space of quantum states of the string. (I should perhaps mention that physicists define $H_p^{(\alpha)}$ as the subspace $\{\xi \in H_p : V_0\xi = \alpha\xi,\ V_{-k}\xi = 0$ for $k > 0\}$ of H_p (where the V_k are the basis for $\mathscr{V}$), exploiting the fact that each irreducible representation of G contains a unique lowest-weight vector.)

To produce a space of states of a free string is not of interest unless one can describe how strings interact. I am certainly not competent to do that, but I shall simply point out, as it was the motivation for Sect. 4 of this paper, that one can define for certain vectors $v \in \mathbb{R}^4$ an operator $B^v : H_p \to H_{p+v}$ which commutes with the action of G and is supposed to correspond to the process of absorption of a particle of momentum v by a string of momentum p to form a string of momentum $p + v$.

One wants B_v to be defined only when $\frac{1}{2}\langle p+v, p+v\rangle - \frac{1}{2}\langle p, p\rangle$, i.e. $\langle p, v\rangle + \frac{1}{2}\langle v, v\rangle$, is integral. Now the elements f of the additive group Y described above act projectively on H by operators $A(f)$. If $\langle v, v\rangle = 2$, and we restrict ourselves to the H_p such that $\langle p, v\rangle$ is integral, then just as in the last part of Sect. 4 we can define for each $\theta \in \mathbb{R}$ a "blip" $B_\theta^v : H_p \to H_{p+v}$. The integral

$$B^v = \frac{1}{2\pi}\int_0^{2\pi} B_\theta^v d\theta$$

is an operator which commutes with G, and is what we want. But it makes sense only when $\langle v, v\rangle = 2$, and the particle absorbed is a tachyon. Physicists want the particles absorbed to correspond to unexcited states of a string, i.e. to those such that $\frac{1}{2}\langle p, p\rangle - \alpha = 0$, and they are therefore led to assume that $\alpha = 1$.

References

0. Frenkel, I. B., Kač, V. G. : Basic representations of affine Lie algebras and dual resonance models. Inventiones math. **62**, 23–66 (1980)
1. Gel' fand, I. M., Vilenkin, N. Ya. : Generalized Functions, Vol. 4. New York: Academic Press 1964
2. Gel'fand, I. M., Fuks, D. B. : Funkt. Anal. Jego Prilozh. **2**, 92–93 (1968)
3. Goddard, P., Horsley, R. : Nucl. Phys. **B111**, 272–296 (1976)
4. Goddard, P., Thorn, C. B. : Phys. Lett. **40B**, 235–238 (1972)
5. Jacob, M. (ed.): Dual theory. Amsterdam: North-Holland 1974
6. Kač, V. G. : Adv. Math. **30**, 85–136 (1978)
7. Kač, V. G. : Contravariant form for infinite- dimensional Lie algebras and superalgebras. (to apper)

7a. Kač, V. G. : Adv. Math. **35**, 264–273 (1980)

8. Kirillov, A. A. : Unitary representations of the group of diffeomorphisms of a manifold. (to appear)
9. Lazutkin, V. F., Pankratova, T. F. : Funkt. Anal. Jego Prilozh. **9**, 41–48 (1975) (Russian)
10. Shale, D. : Trans. Am. Math. Soc. **103** 149–167 (1962)
11. Shale, D. : Stinespring, W. F. : J. Math. Mech. **14**, 315–322 (1965)
12. Vergne, M. : Seconde quantification et groupe symplectique. C. R. Acad. Sci. (Paris), **285**, A 191–194 (1977)
13. Vershik, A. M., Gel'fand, I. M., Graev, M. I. : Usp. Mat. Nauk **30**, 1–50 (1975)
14. Vershik, A. M., Gel'fand, I. M., Graev, M. I. Dokl. Akad. Nauk. SSSR **232**, 745–748 (1977)

Communicated by A. Jaffe

Received June 17, 1980; in revised form October 14, 1980

ALGEBRAS, LATTICES AND STRINGS

P. Goddard[†] and D. Olive[†]

Abstract

A unified construction is given of various types of algebras, including finite dimensional Lie algebras, affine Kac-Moody algebras, Lorentzian algebras and extensions of these by Clifford algebras. This is done by considering integral lattices (i.e. ones such that the scalar product between any two points is an integer) and associating to the points of them the square of whose length is 1 or 2, the contour integral of the dual model vertex operator for emitting a "tachyon". If the scalar product is positive definite, the algebra of these quantities associated with the points of length 2 closes, when the momenta are included, to form a finite dimensional Lie algebra. If the scalar product is positive semi definite, this algebra closes to an affine Kac-Moody algebra when the vertex operators for emitting "photons" are added. If the scalar product is Lorentzian, the algebra closes if the vertex operators for all the emitted states in the dual model are added. Special lattices in 10, 18, and 26 dimensional Lorentzian space are discussed and implications of the dual model no ghost theorem for these algebras are mentioned. This framework links many physical ideas, including concepts in magnetic monopole theory and the fermion-boson equivalence as well as the dual model. (Knowledge of dual models is not assumed but familiarity with aspects of the theory of Lie algebras is presumed in the latter part of this paper.)

[†] Partially supported by the National Science Foundation through the Mathematical Sciences Research Institute.

Vertex Operators in Mathematics and Physics - Proceedings of a Conference November 10-17, 1983. Publications of the Mathematical Sciences Research Institute #3, Springer-Verlag, 1984.

1. INTRODUCTION

Many of the ideas occurring in the quantum theory of the relativistic string (or dual model), such as supersymmetry or dimensional reduction, have found applications outside its immediate context. (For reviews of the subject, see the collection of ref. [1], or ref [2], for example.) Although exciting and important developments are still being made [3], interest in the theory was most intense in the period 1968 to 1974. The objectives of those working on it progressively broadened from the phenomenological description of high energy scattering (resonances and Regge behavior in particular) to the construction of a completely consistent theory of the strong, and possibly other, interactions. The demand for consistency led to the realization that each dual model or string theory should be considered in a particular space-time dimension, 26 for the original model of Veneziano and others and 10 for the theory of Neveu, Schwarz and Ramond, which includes fermions and introduced supersymmetry. Although formulated in spaces of nonphysical dimensions, these theories possess a high level of consistency and contain very rich algebraic structures; for instance, Yang-Mills gauge theory and supergravity appear as "subtheories" by taking suitable limits. One explanation for why these theories moved from the center of interest is that the technical difficulties presented by handling amplitudes with many fermions proved insuperable on the time scale that theoretical physicists usually expect to solve their problems. Nor could help be found in the mathematical literature because mathematicians were only just discovering the sort of structures which have to be exploited in order to gain an economical understanding of string theories. Subsequently the trade has so far been mainly the other way, with mathematicians taking advantage of the constructions made by physicists.

One of the ideas to have proved useful in mathematics is that of vertex operators $U(r,z)$, which are analytic operators functions of momentum r and a complex variable, z. They occur naturally in dual models and they, or rather their moments

$$(1.1) \qquad A_n(r) = \frac{1}{2\pi i}\oint \frac{dz}{z} z^n\, V(r,z) \, ,$$

are important in constructing representations [4] of affine Kac-Moody Algebras [5]. Here the "momenta" r are Euclidean vectors and correspond to the roots of a simply-laced Lie algebra. In dual theory, the n = 0 operators are particularly interesting because they create the physical states of the model by application to some basic state. This is because they commute with the operators, L_n, of the Virasoro algebra [6], which define gauge conditions in the model. They were crucial in the work which led to the proof of the absence of ghosts (i.e. physically unacceptable negative norm states) from the model [7-10].

In this paper we shall show how these physical state creation operators can be made to associate to any integral lattice a Lie algebra. [This describes an approach we developed last winter; after completing this work we learned of unpublished work of I.B. Frenkel which adopted a very similar point of view and who obtained some further results [11]. Our approach is expressed in a formalism which might be familiar to physicists.] If the construction is applied to a Euclidean Lattice Λ, we obtain a finite dimensional compact Lie algebra, g_Λ, which will be semi-simple if and only if the points of squared length 2 in Λ, Λ_2, span a space of the same dimension as Λ. In this case, if we extend the lattice to $\Lambda' = \Lambda \oplus \mathbb{Z}$ where the new direction is taken to be null, the construction applied to Λ' yields the affine Lie algebra $\tilde{g}_\Lambda$ associated with g_Λ.

For a lattice Λ in a real vector space V with an inner product (i.e. symmetric bilinear form), which is not necessarily positive definite, or even non-singular, the construction produces a Lie algebra of rank equal to the dimension of V, with roots corresponding to points v, with $v^2 \leq 2$, on the even lattice Λ_R generated by the set Λ_2 of point r of Λ with $r^2 = 2$. The root spaces corresponding to $r \in \Lambda_2$ have dimension one and, if the inner product is non-singular, those corresponding to non-zero $v \in \Lambda_R$ with $v^2 = 0$ have dimension at most dim Λ_R-2.

A particularly interesting lattice to consider is the twenty-six dimensional even Lorentzian lattice $II^{25,1}$. Conway and Sloane [12] have shown that one can take as a set of simple roots for this lattice a set of points, ℓ, isometric to the Leech Lattice. (A set of simple roots is a minimal set of vectors having the property that the reflections in the hyperplanes perpendicular to them generate the Weyl group of the lattice, that is the group of all reflections which are automorphisms of the lattice.) It has been suggested [13] that these simple roots can be used to construct an infinite rank Lie algebra L_∞, which might be related to the Fischer-Griess Monster group. Application of the construction to $II^{25,1}$ yields a representation for L_∞ which has rank 26. However it seems that a more elaborate setting [14] is necessary, which is in fact related to the vertices which describe the interactions of dual models with electromagnetic currents [15], in order to accommodate the action of the group.

In this case the algebra obtained has a direct interpretation in terms of the relativistic string theory. The lattice $II^{25,1}$ can be interpreted as the momentum space for a (particular) toroidal version of 26-dimensional Minkowski space. This is especially intriguing because the dimension 26 is a critical dimension for the relativistic string model. In higher dimensions the spectrum of physical states has unacceptable states with negative norms, which give rise to nonphysical negative probabilities. (In lower dimensions, it seems, at least in any interacting version of the theory, that the spectrum has other features which are at best unnattractive.) Here the construction gives rise to physical creation operators for the relativistic string. Thus they can be represented in the Fock space of physical states. This space has a non-negative inner-product. It is tempting to suppose that the root spaces can be identified with the physical state creation operators for the corresponding momentum. In fact Frenkel [11] has shown that one can set up an isomorphism between those root spaces and those physical states generated by the physical state creation operators from the single particle states. (These are not necesarily all the physical states for number theoretic reasons.) This puts bounds on the dimensions of the root spaces.

Our paper is organized as follows. In Section 2 we review the background information we shall need about lattices and we establish some notation. In Section 3 we give our construction of a Lie algebra associated with an integral lattice. In Section 4 we discuss the rank and the roots of this algebra. Then, in Section 5, we consider the relationship of this construction to the Virasoro algebra and the way that previous representations of affine Lie algebras are subsumed in it. The following section applies the construction to even Lorentzian lattices and, in particular, a Lie algebra L_∞ to which Conway, Queen and Sloane draw attention. In Section 7 we extend the construction of Section 3 to include anticommuting operators and in Section 8 we apply the idea of associating an algebra with a lattice to enable us to obtain information about self-dual lattices.

2. LATTICES AND THEIR PROPERTIES

To set the scene for our subsequent discussions we shall survey briefly the main facts about lattices that will be relevant for us (see e.g. [16]). Suppose that V is a real N-dimensional vector space with an inner-product, that is a symmetric bilinear form, not necessarily non-singular, denoted $x \cdot y$ for $x, y \in V$. We shall be principally interested in the cases where V is Euclidean space, $\mathbb{R}^N$, or Minkowski space, $\mathbb{R}^{N-1,1}$. We can define a *lattice* in V as a set of points of the form

$$\Lambda = \{ \sum_{i=1}^{N} n_i \underline{e}_i : n_i \in \mathbb{Z} \} \tag{2.1}$$

where $\underline{e}_i$, $1 \leqslant i \leqslant N$, forms a basis for V; it is also called a basis for Λ. The lattice Λ will be *Euclidean* or *Lorentzian* in the cases that V is Euclidean or Minkowski space, respectively. It is said to be *unimodular* if $|\det(\underline{e}_i \cdot \underline{e}_j)| = 1$. This is equivalent to the condition that Λ should contain one point of V per unit volume. We can also consider the situation where Λ spans a subspace rather than the whole of V.

For any lattice $\Lambda \subset V$, we define the *dual* of Λ, denoted Λ^*, to be the set of points $y \in v$ for which $x \cdot y$ is integral for all $x \in \Lambda$. If Λ spans V, and the inner product is non-singular, Λ^* is a lattice, the *dual lattice* to Λ. In that case we can form a basis for Λ^* by taking the basis $\underline{e}_i^*$, $1 \leq i \leq N$, for V dual to $\underline{e}_i$, $1 \leq i \leq N$, so that $\underline{e}_i \cdot \underline{e}_j^* = \delta_{ij}$. The lattice Λ is *integral* if $x \cdot y$ is an integer for every $x,y \in \Lambda$. This is equivalent to the condition $\Lambda \subset \Lambda^*$. Since Λ and Λ^* have reciprocally related numbers of points per unit volume in V, the condition that Λ be both integral and unimodular is equivalent to the condition that it be self-dual, i.e. $\Lambda = \Lambda^*$.

A vector $x \in \Lambda$ is said to be *primitive* if x is not a multiple of any other point of the lattice, that $x \neq ny$ for any $y \in \Lambda$, $n \in \mathbb{Z}$ with $|n| \geq 2$. Using a basis it is easy to see that x is primitive if and only if $x \cdot x' = 1$ for some x' in the dual lattice Λ^*.

A lattice is the direct sum of lattices $\Lambda_i \subset \Lambda$, $1 \leq i \leq r$ if Λ_i is perpendicular to Λ_j, for $i \neq j$, and any $x \in \Lambda$ can be written in the form

$$x = \sum_{i=1}^{r} x_i, \quad x_i \in \Lambda_i \,. \tag{2.2}$$

A lattice is *indecomposable* if it can not be expressed as the direct sum of two non zero sublattices. A Euclidean lattice has a unique decomposition into a direct sum of indecomposable sublattices but, in spaces of indefinite signature, such decompositions are not unique, in general.

The simplest examples of self-dual lattices are the cubic lattices $\mathbb{Z}^{m,n}$, consisting of those points in $\mathbb{R}^{m,n}$ with integral coordinates (with respect to some orthonormal basis). Any integral Euclidean lattice is isomorphic to one of the form $\mathbb{Z}^m \oplus \Lambda$, for some m, where Λ contains no points of length one. So, in classifying integral Euclidean lattices, it is only necessary to consider those whose nonzero points have squared length at least 2. Amongst these are the <u>even</u> lattices, that is the integral lattices with $x^2 \in 2\mathbb{Z}$ for all x. Even lattices are sometimes called type II, whilst the remainder are called <u>odd</u> or type I. This terminology also applies to non-Euclidean lattices.

If we consider self-dual Euclidean lattices, the choice in low dimensions is limited (see, e.g. [17]). The first non-trivial example is in dimension 8, the root lattice of the group E_8, which is even.

The next is a weight lattice associated with D_{12}, the algebra of SO(24), and this is odd. Even self-dual Euclidean lattices only occur in dimensions which are multiples of 8. There are two in dimension 16 and 24 in dimension 24 where, for the first time there is a lattice whose minimum nonzero squared length is 4. This is called the Leech lattice.

The possibilities for self-dual Lorentzian lattices are even more limited. If Λ is an odd self-dual lattice in a space of indefinite signature, $\mathbb{R}^{m,n}$, it can be shown that it is isomorphic to $\mathbb{Z}^{m,n}$. Thus, in particular, there is just one odd self-dual Lorentzian lattice in each dimension. Even Lorentizan lattices exist only in dimensions of the form $N = 8n+2$, n an integer, and again such lattices are determined by their dimension. These lattices we denoted by $II^{N-1,1}$, and they can be defined as consisting of those x for which

(2.3a) (i) either $x \in \mathbb{Z}^{N-1,1}$ or $x - \ell \in \mathbb{Z}^{N-1,1}$

and

(2.3b) (ii) $x \cdot \ell \in \mathbb{Z}$

where $\ell = (\frac{1}{2},\frac{1}{2},\ldots,\frac{1}{2};\frac{1}{2})$.

We can understand the relationship between the constraints on the dimensions of even self-dual Euclidean and Lorentzian lattices by considering a light-like vector k in a Lorentzian lattice in Minkowski space V; so $k^2 = 0$ but $k \neq 0$. If V_k denotes the subspace orthogonal to k, consisting of $x \in V$ with $x \cdot k = 0$, V_k contains K_k the one-dimension of subspace of vectors parallel to k. The restriction to V_k of the inner product on V is singular but non-negative, with null space K_k. We can form a Euclidean space V_k/K_k of dimension two less than V, by identifying vectors of the form $x + \lambda k$ for different $\lambda \in \mathbb{R}$ and fixed $x \in V_k$. The length of such a vector $(x+\lambda k)^2$ equals x^2, independent of λ. We might as well take k to be primitive. Then $V_k \cap \Lambda$ defines a lattice, which, which becomes a Euclidean lattice Λ_k in V_k/K_k when we identify all points of the form $x + mk$

for different $m \in \mathbb{Z}$. If Λ is even and self-dual, then so is Λ_k, showing that dim $\Lambda = 8n+2$, given the result on the dimensionality of even self-dual Euclidean lattices.

Two vectors in a lattice λ are *equivalent* if they are related by an automorphism of the lattice, that is a (pseudo)-orthogonal transformation R of V such that $R(\Lambda) = \Lambda$. Clearly equivalent light-like vectors in a self-dual Lorentzian lattice Λ (i.e. $II^{8n+1,1}$ for some n) yield isomorphic self-dual Euclidean lattices Λ_k. Conversely, it is sufficient to check that if isomorphic Euclidean lattices are obtained from different light vectors in Λ, the isomorphism can be extended to an automorphism of the whole lattice Λ, under which the light-vectors are equivalent. Further, any self-dual Euclidean lattice can be obtained in this way since, if Λ' is such a lattice, $\Lambda = \Lambda' \oplus II^{1,1}$ is a self-dual Lorentzian lattice and, if k $= (0,\ell)$ where $\ell = (\frac{1}{2};\frac{1}{2}) \in II^{1,1}$, $\Lambda_k = \Lambda'$. Hence the problem of classifying self-dual Euclidean lattices is equivalent to that of classifying the inequivalent primitive light-like vectors in $II^{8n+1,1}$. Hence, it follows from the results we quoted on self-dual Euclidean lattices, that there is just one such vector for n = 1, two for n = 2 but 24 for n = 3. Conway and Sloane have shown that, taking k to be the particular light-like vector

$$w = (0,1,2,\ldots,24;70) \tag{2.4}$$

produces a Λ_k isomorphic to the Leech lattice.

3. THE LIE ALGEBRA ASSOCIATED WITH AN INTEGRAL LATTICE

We give a construction which associates to any integral lattice Λ, a Lie algebra g_Λ, of rank N = dim Λ, which is finite dimensional if Λ is Euclidean. This construction also works in the more general circumstance that $\Lambda \subset \Lambda_R^*$ where Λ_R is the sublattice of Λ generated by the points Λ_2 of squared length 2 in Λ. If the inner product, on the vector space V containing Λ, is singular, or if Λ_R

does not span all of V, the requirement that $\Lambda \subset \Lambda_R^*$ places no restriction on the components, of points of Λ, orthogonal to points of Λ_R.

We define the algebra g_Λ by means of a Fock space representation. To this end we choose a basis for V with respect to which the metric tensor is $g^{\mu\nu}$. We introduce an infinite set of annihilation and creation operators α_m^μ, $m \in \mathbb{Z}$, $1 \leq r \leq N$, satisfying the commutation relations

(3.1) $$[\alpha_m^\mu, \alpha_n^\nu] = m g^{\mu\nu} \delta_{m,-n}$$

and the hermiticity condition

(3.2) $$\alpha_m^{\mu\dagger} = \alpha_{-m}^\mu \ .$$

We interpret $p^\mu = \alpha_0^\mu$ as a momentum operator and introduce orthonormal vectors ψ_γ, $\gamma \in \Lambda$, of momentum γ,

(3.3) $$p^\mu \psi_\gamma = \gamma^\mu \psi_\gamma,$$

(3.4) $$\langle \psi_\gamma, \psi_{\gamma'} \rangle = \delta_{\gamma\gamma'},$$

(3.5) $$\alpha_n^\mu \psi_\gamma = 0, \qquad n > 0 \ .$$

The Fock space, $\mathcal{H}$, in which we work is that generated from the vectors ψ_γ by the operators α_m^μ $(m < 0)$. It is the direct sum of the momentum eigenspaces $\mathcal{H}_\gamma = \{\psi \in \mathcal{H}: p\psi = \gamma\psi\}$.

We can introduce operators $e^{i\gamma\cdot q}$, $\gamma \in \Lambda$, which generate momentum

(3.6) $$e^{i\gamma\cdot q}\psi_{\gamma'} = \psi_{\gamma+\gamma'}.$$

Finally, we have the commutation relations

(3.7) $$[q^\mu, p^\nu] = i g^{\mu\nu}.$$

The whole of $\mathcal{H}$ is generated from the vacuum vector Ψ_0 by α^{μ}_{m}, $m<0$, and $e^{i\gamma\cdot q}$, $\gamma \in \Lambda$.

Now introduce

(3.8) $$Q^{\mu}(z) = q^{\mu} - ip^{\mu}\log z + i \sum_{n\neq 0} \frac{1}{n} \alpha^{\mu}_{n} z^{-n}$$

so that

(3.9) $$p^{\mu}(z) \equiv iz\frac{dQ^{\mu}}{dz} = \sum_{n} \alpha^{\mu}_{n} z^{-n}$$

Let

(3.10a) $$Q^{\mu}_{>}(z) = i \sum_{n>0} \frac{1}{n} \alpha^{\mu}_{n} z^{-n}, \qquad Q^{\mu}_{<}(z) = i \sum_{n<0} \frac{1}{n} \alpha^{\mu}_{n} z^{-n}$$

and

(3.10b) $$Q^{\mu}_{0} = q^{\mu} - ip^{\mu} \log z.$$

Then the vertex operator of "momentum" r is defined by the normal ordered expression

$$U(r,z) = z^{r^2/2} : \exp(ir\cdot Q(z)):$$

(3.11) $$= z^{r^2/2} \exp(ir\cdot Q_{<}(z)) e^{ir\cdot q} z^{r\cdot p} \exp(ir\cdot Q_{>}(z))$$

$$= \exp(ir\cdot Q_{<}(z)) \exp(ir\cdot Q_{0}(z)) \exp(ir\cdot Q_{>}(z))$$

Note in the normal ordered expression $Q_{>}$ is moved to the right of $Q_{<}$ and p to the right of q. Defined in this way $U(r,z)$, $r^2 = 2$ has a well defined action on $\mathcal{H}$ as $r\cdot\gamma \in \mathbb{Z}$ for $r \in \Lambda_2$, $\gamma \in \Lambda$ if $\Lambda \subset \Lambda_k^*$ and so there is no branch cut at $z = 0$. Note the hermiticity property

$$U(r,z)^{\dagger} = U(-r, 1/z^{*})$$

The physical state creation operator corresponding to the

vertex U is defined by

(3.12) $$A_r = \frac{1}{2\pi i}\oint \frac{dz}{z} U(r,z)$$

with the integration contour positively encircling $z = 0$. It follows from the hermiticity property that

(3.13) $$A_r^\dagger = A_{-r}$$

To investigate the commutation relations of the A_r, $r \in \Lambda_2$ we need to normal order $U(r,z)U(s,\zeta)$. This is a standard dual model calculation followed in evaluating particle scattering amplitudes for example. We find

(3.14) $$:\exp[ir.Q(z)]::\exp[is.Q(\zeta)]: =$$

$$:\exp[ir.Q(z)+is\cdot Q(\zeta)]:(z-\zeta)^{r\cdot s} \text{ for } |\zeta| < |z|$$

because

(3.15) $$\exp[ir\cdot Q_>(z)]\exp[is\cdot Q_<(\zeta)] =$$

$$(1-\zeta/z)^{r\cdot s} \exp[is\cdot Q_<(\zeta)]\exp[ir\cdot Q_>(z)] \text{ if } |\zeta| < |z|$$

and

$$z^{r\cdot p}e^{is\cdot q} = e^{is\cdot q}z^{r\cdot p}\, z^{r\cdot s}$$

Note that the right hand side of equation (3.14) is defined by analytic continuation for all z,ζ ($z \neq 0$, $\zeta \neq 0$, $z \neq \zeta$) and that simultaneous interchange of r with s and z with ζ changes it by a factor $(-1)^{r\cdot s}$. Hence

(3.16) $$A_r A_s - (-1)^{r\cdot s} A_s A_r =$$

$$= \frac{1}{(2\pi i)^2}\left\{\oint_{|z|>|\zeta|}\frac{dz}{z}\oint\frac{d\zeta}{\zeta} - \oint_{|\zeta|>|z|}\frac{dz}{z}\oint\frac{d\zeta}{\zeta}\right\} z^{r^2/2}\zeta^{s^2/2}$$

$$:\exp\{ir\cdot Q(z)+is\cdot Q(\zeta)\}:(z-\zeta)^{r\cdot s}$$

$$= \frac{1}{(2\pi i)^2}\oint_0\frac{d\zeta}{\zeta}\oint_\zeta\frac{dz}{z}\,(z-\zeta)^{r\cdot s} z^{r^2/2}\zeta^{s^2/2} :\exp\{ir\cdot Q(z)+is\cdot Q(\zeta)\}:$$

where the z integral is taken on a contour positively encircling ζ, excluding $z = 0$ and the ζ contour is then taken positively about $\zeta = 0$. The integrand is non singular at $z = \zeta$ if $r\cdot s \geqslant 0$, it has a simple pole if $r\cdot s = -1$, in which case $r+s \in \Lambda_2$ and a double pole if $r\cdot s = -2$ which happens in particular if $r = -s$. Thus

(3.17) $A_r A_s - (-1)^{r\cdot s} A_s A_r = 0 \quad$ if $r\cdot s \geqslant 0$

(3.18) $\qquad = A_{r+s} \quad$ if $r\cdot s = -1$

(3.19) $\qquad = \frac{1}{2\pi i}\oint\frac{d\zeta}{\zeta} r\cdot P(\zeta) = r\cdot p \quad$ if $r\cdot s = -2$ and if $r+s = 0$

In addition we have

(3.20) $$[p^\mu, A_r] = r^\mu A_r$$

Now we wish to modify equations (3.17) - (3.19) so they become commutators. This is done by introducing quantities C_r, $r \in \Lambda_R$, commuting with the previously mentioned oscillators and satisfying

$$\left.\begin{aligned} C_u C_v &= (-1)^{u\cdot v} C_v C_u \\ C_u C_v &= \epsilon(u,v) C_{u+v} \end{aligned}\right\} \quad u, v \in \Lambda_R$$

where $\epsilon(u,v)$ takes values ± 1. Such quantities can be constructed for any even integral lattice. Generalizations to odd integral lattices and more discussion will be found in the appendix. The C_u can be taken to be functions of momentum, $C_u\psi_\gamma = \epsilon(u,\gamma)\psi_\gamma$, so that we are not increasing the size of the representation space, or to be some generalisation of Dirac γ matrices.

If we set

(3.21) $$e(r) = A_r C_r \qquad r \in \Lambda_2$$

we find

(3.22a) $$[e(r),e(s)] = 0 \qquad \text{if } r\cdot s \geq 0$$

(3.22b) $$= \epsilon(r,s)e(r+s) \qquad \text{if } r\cdot s = -1$$

(3.22c) $$= r\cdot p \qquad \text{if } r = -s$$

Further it follows from eqn. (3.20) that

(3.23) $$[p^\mu,e(r)] = r^\mu e(r)$$

The algebra g_Λ associated with the lattice Λ is defined to be the Lie algebra generated by $e(r)$, $r \in \Lambda_2$ p^μ, $1 \leq \mu \leq N$. In the given representation, the generators have the hermiticity property

(3.24) $$e(r)^\dagger = e(-r) \qquad p^{\mu\dagger} = p^\mu$$

If Λ_R is Euclidean, we always have $(r\pm s)^2 \geq 0$ with zero only if $r\pm s = 0$. Hence the cases covered by equation (3.22) exhaust all the possibilities. In this case $e(r)$, $r \in \Lambda_2$, p^μ, $1\leq\mu\leq N$ close under commutation and form a basis for g_Λ. It is semisimple if and only if dim Λ_R = dim Λ and simple if in addition Λ_R is indecomposable. The semisimple part of g_Λ is always simply laced (i.e. has roots of equal length).

From equations (3.6) and (3.11) we find

$$A_r\Psi_\lambda = \oint \frac{dz}{2\pi i z} z^{r^2/2 + r\cdot\lambda} \exp\{ir\cdot Q_<(z)\} \Psi_{\lambda+r} \tag{3.25}$$

As $Q_<(z)$ contains only positive powers of z we see, given $r^2 = 2$, that this expression vanishes if and only if

$$r\cdot\lambda \geqslant 0 \tag{3.26}$$

When g_Λ is a semisimple finite dimensional Lie algebra all its finite dimensional irreducible representations possess unique "highest weight states" annihilated by the step operators for positive roots. The corresponding weight, which is dominant, can be used to label the irreducible representation. We deduce in this case from equation (3.25) and (3.26) that our construction (3.11) furnishes a reducible representation which includes in its decomposition all irreducible representations of g_Λ whose highest weights occur in Λ.

More generally if Λ_R is Euclidean, g_Λ is the Lie algebra of a compact Lie group G. A specific global structure is obtained for G when the antihermitian linear combinations of the generators are exponentiated. This global form is the one for which Λ is the lattice of weights.

Finally let us complete the calculation of $A_rA_s - (-1)^{r\cdot s}A_sA_r$ when $(r+s)^2$ vanishes without $r+s$ necessarily vanishing. Of course this can only happen if Λ is not Euclidean. Because $r\cdot s = -2$ we have

$$A_rA_s - A_sA_r = \oint \frac{d\zeta}{2\pi i} \frac{d}{dz}: \exp\{ir\cdot Q(z) + is\cdot Q(\zeta)\}:\Big|_{z=\zeta}$$

$$= \oint \frac{d\zeta}{2\pi i\zeta} : r\cdot P(\zeta) \exp\{i(r+s)Q(\zeta)\}:$$

by (3.9). This does not appear to exhibit the required antisymmetry in r and s until we realize that

$$\frac{1}{\zeta}\left(\frac{r+s}{2}\right): P(\zeta)\exp\{i(r+s)Q(\zeta)\}:$$

is a total derivative and therefore integrates to zero. Hence

$$A_rA_s - A_sA_r = \oint \frac{dz}{2\pi i z} \left(\frac{r-s}{2}\right) \cdot P(z) \exp\left(i(r+s)Q(z)\right) \tag{3.27}$$

Note that normal ordering has been omitted. It is unnecessary as

$$(r+s)^2 = 0, \qquad (r-s)\cdot(r+s) = 0 \tag{3.28}$$

so that all quantities in (3.27) mutually commute. The operator on the right side of (3.27) is known as the DDF operator (7) and played an important role in dual theory (8,9,10). It is the integral of the "photon emission vertex." The polarization of the "photon" is $\frac{r-s}{2}$ and is automatically transverse to the photon momentum $(r+s)$ by (3.28) as it must be in a physical situation.

4. THE RANK AND ROOTS OF g_Λ

We have seen that, in the Euclidean case, the rank of g_Λ is $N = \dim \Lambda$ and its roots are just the points of Λ_2. We shall now establish that, in the non-Euclidean case, its rank is still N and its roots are points $v \in \Lambda_R$ with $v^2 \leqslant 2$.

The algebra g_Λ is spanned by elements of the form

$$e' = [e(r_1), [e(r_2), \ldots [e(r_{m-1}), e(r_m)] \ldots]] \tag{4.1}$$

where each $r_i \in \Lambda_2$ (together with the p^μ, which are necessary if $\dim \Lambda_R < \dim \Lambda$, which we shall call the non-semi-simple case). Then e' belongs to the simultaneous eigenspace of the p labelled by

$$v = \sum_{i=1}^{m} r_i. \tag{4.2}$$

We first show that $e' = 0$ if $v^2 > 2$. To this end we calculate

$$U_{r_1,r_2,\ldots r_m}(z_1,z_2,\ldots z_m) = U(r_1,z_1)U(r_2,z_2)\ldots U(r_m,z_m)$$

$$= \exp\left(i \sum_{j=1}^{m} r_j Q_<(z_j)\right) e^{ir\cdot q} \prod_{j=1}^{m} z_j^{r_j\cdot p+1} \exp\left(i \prod_{j=1}^{m} r_j Q_>(z_j)\right) \tag{4.3}$$

$$\prod_{i<j} (z_i - z_j)^{r_i \cdot r_j}$$

$$\text{if } |z_1| > |z_2| > \ldots > |z_m|$$

Arguments similar to those used in eq. (3.14), show that the multiple commutator (4.1) equals, up to a sign,

$$(4.4) \quad \frac{1}{(2\pi i)^m} \oint_0 \frac{dz_m}{z_m} \oint_{z_n} \frac{dz_{m-1}}{z_{m-1}} \cdots \oint_{z_2} \frac{dz_1}{z_1} U_{r_1, r_2, \ldots, r_m}(z_1, z_2, \ldots, z_m)$$

Here the integration contours are such that z_j encircles z_{j+1} positively excluding z_i for $j+1 < i \leqslant m+1$, where $z_{m+1} = 0$. Now since A_r takes a finite occupation number state (i.e. one of the form $\alpha_{-m_1}^{\mu_1} \alpha_{-m_2}^{\mu_2} \ldots \alpha_{-m_M}^{\mu_M} \psi_\gamma$) into a finite linear combination of such states, to establish that e' vanishes, we need only prove that all the matrix elements of (4.4) between such states vanish. These matrix elements have the form

$$(4.5) \quad \frac{1}{(2\pi i)^m} \oint_0 \frac{dz_m}{z_m} \oint_{z_m} \frac{dz_{m-1}}{z_{m-1}} \cdots \oint_{z_1} \frac{dz_1}{z_1} \prod_{i<j} (z_i - z_j)^{r_i \cdot r_j}$$

$$f(z_1, z_2, \ldots, z_m)$$

where f is analytic provided that all the z_i are nonzero and finite. It is easy to see that the integral over z_{m-1} (or one of the previous integrals over z_i, $1 \leqslant i \leqslant m-2$) will vanish unless

$$(4.6) \qquad \sum_{i<j} r_i \cdot r_j \leqslant -(m-1),$$

that is

$$(4.7) \qquad v^2 = 2m + 2 \sum_{i<j} r_i \cdot r_j \leqslant 2,$$

our desired result.

Any element of g_Λ commuting with p^μ, $1 \leqslant \mu \leqslant N$, would have to be a linear combination of them together with expressions of the form (4.1) with $v = 0$, or, equivalently of expressions of the form $[e(-v), e']$ where $v^2 = 2$. In this case the integral of eq. (4.5) is

$$\frac{\lambda}{2\pi i} \oint \frac{dz}{z} f(z, z, \ldots, z) \tag{4.8}$$

where λ is a constant (possibly zero) depending only on $r_1, r_2, \ldots, r_n$. We see this as follows. In evaluating the integral we take residues at the poles $(z_i - z_j)^{r_i r_j}$. If we have a multiple pole we either differentiate f or the remaining terms in the product $\Pi(z_i - z_j)^{r_i \cdot r_j}$. If we have differentiated f M times the integral over z_{m-1} has a factor $(z_{m-1} - z_m)^L$ where

$$\begin{aligned} L &= \sum_{i<j} r_i \cdot r_j + (m-2) + M \\ &= \tfrac{1}{2}v^2 + M - 2 = M - 1 \end{aligned} \tag{4.9}$$

so to have any pole left we must have $M = 0$; all derivatives must be applied to the product $\Pi(z_i - z_j)^{r_i \cdot r_j}$ giving the result (4.8). Thus, in this case, e' equals $\lambda e(v)$ up to a sign. From this we see that any element of g_Λ commuting with the p^μ is a linear combination of them. In consequence, the rank of g_Λ is dim Λ and the roots of g_Λ are vectors of Λ_R with $v^2 \leqslant 2$. If g_Λ^v is the corresponding root space, we have established that

$$\dim g_\Lambda^v = 1 \quad \text{if } v^2 = 2 \tag{4.10}$$

If we consider those e' with $v^2 = 0$, arguments of the sort we have just used show that e' is of the form

$$e' = \xi \cdot a(v) C_v \tag{4.11}$$

where

(4.12) $$a^{\mu}(k) = \frac{1}{2\pi i}\oint \frac{dz}{z} : P^{\mu}(z)\exp(ik\cdot Q(z)):$$

for some vector $\xi \in V$, or rather that subspace spanned by Λ_R. We shall show in the next section that

(4.13) $$\xi\cdot v = 0$$

(as in equation (3.27) and (3.28).) Further, since $v\cdot a(v) = 0$,

(4.14) $$\dim g_{\Lambda}^{v} \leq \dim \Lambda_R - 2 \quad \text{if } v^2 = 0$$

and, if Λ_R is irreducible, we would expect equality. (In terms of dual theory, $a^{\mu}(k)$ are physical creation operators for "photons" and eq. (4.13) the transversality condition on the "photon polarization" vector ξ.)

5. THE VIRASORO ALGEBRA AND AFFINE LIE ALGEBRAS

The Virasoro Algebra

If the inner product on V is non-singular, we can define a representation of the Virasoro algebra in $\mathcal{H}$ in the familiar way, by introducing the operators

(5.1) $$L_n = \tfrac{1}{2}: \sum \alpha_m\cdot\alpha_{n-m}:$$

It is the algebra with basis L_n, $n \in \mathbb{Z}$, together with the central element 1; they satisfy the commutation relations

(5.2) $$[L_m, L_n] = (m-n)L_{m+n} + \frac{N}{12}m(m^2-1)\delta_{m,-n}\ .$$

The vertex operators $U(r,z)$ were constructed specifically so that they have the property

(5.3) $$[L_n, U(r,z)] = \left(z^{n+1}\frac{d}{dz} + n\frac{r^2}{2}z^n\right)U(r,z)$$

and $r^2 = 2$ has been chosen because then

$$[L_n, e(r)] = \frac{1}{2\pi i}\oint \frac{d}{dz}\,(z^n U(r,z))C_r = 0 \tag{5.4}$$

Also L_n clearly commutes with p^μ so that the subspace $\mathcal{H}_\lambda$ of $\mathcal{H}$ defined by the equations

$$L_0\psi = \lambda\psi, \qquad L_n\psi = 0, \qquad n > 0\ , \tag{5.5}$$

is an invariant subspace for g_Λ. In particular cases $\mathcal{H}_\lambda$ may give nontrivial irreducible representations.

We can use the fact that e', as defined by eq. (4.11) commutes with L_n to establish eq. (4.13). Calculations [7] shows that

$$[L_n, a^\mu(k)] = \frac{n^2}{2}k^\mu \oint\frac{dz}{z}\, z^n \exp(ikQ(z)), \tag{5.6}$$

where normal ordering is actually unnecessary because $k^2 = 0$, so that $\xi\cdot a$ commutes with L_n if and only if eq. (4.13) holds.

<u>Affine Lie algebras</u>

In the Euclidean case, we can extend g_Λ to an affine Lie algebra $\tilde{g}_\Lambda$ by using the Frenkel-Kac construction [4], which amounts to taking other moments of the vertex operator U(k,z) and which we shall describe next. However such algebras can also be obtained directly from our construction by enlarging the lattice Λ to a lattice Λ' and we shall describe this afterwards.

We can construct $\tilde{g}_\Lambda$ by defining[3],

$$e_n(r) = \frac{1}{2\pi i}\oint\frac{dz}{z}\, z^n U(r,z)C_r \tag{5.7}$$

for $r \in \Lambda_2$. It has a basis consisting of $e_n(r)$, $r \in \Lambda_2$, $n \in \mathbb{Z}$; $p^\mu_n = \alpha^\mu_n$ $1 \leq \mu \leq N$, $n \in \mathbb{Z}$ and the central element 1. We then find

$$[p^\mu_n, p^\nu_n] = m g^{\mu\nu}\delta_{m,-n} \tag{5.8}$$

(5.9) $$[p^\mu_m, e_n(r)] = r^r e_{m+n}(r)$$

(5.10a) $$[e_m(r), e_n(s)] = 0 \qquad \text{if } r\cdot s \geqslant 0$$

(5.10b) $$= \epsilon(r,s)\, e_{m+n}(r+s) \qquad \text{if } r\cdot s = -1$$

(5.10c) $$= r\cdot p_{m+n} + m\delta_{m,-n} \qquad \text{if } r = -s$$

Since we have assumed that Λ_k is Euclidean, these exhaust the possibilities for $r\cdot s$ and the algebra closes.

We also have

(5.11) $$[L_m, e_n(r)] = -ne_{m+n}(r)$$

and

(5.12) $$e_n(r)\Psi_\gamma = 0 \text{ if and only if } n+\gamma\cdot r \geqslant 0$$

by extending the argument leading to (3.26). Making a particular choice of positive roots for g_Λ, the positive root spaces of $\tilde{g}_\Lambda$ are spanned by $e_0(r)$, $r > 0$; α^μ_n, $e_n(s)$, $n > 0$. We see from eq. (5.12) that the states in $\mathcal{H}$ with momentum of the form $\gamma+s$ will give a representation space for $\tilde{g}_\Lambda$ with highest weight vector Ψ_γ if and only if γ is a minimal fundamental weight [18] (or zero).

Eqs. (5.10) still hold even if Λ_k is not Euclidean, and then the $e_n(r)$, p^μ generate an algebra $\tilde{g}_\Lambda \supset g_\Lambda$.

To obtain $\tilde{g}_\Lambda$ from the construction of Section 3 as an algebra $g_{\Lambda'}$, take Λ' to be a lattice consisting of the points $x+nk$, $x \in \Lambda$, $n \in \mathbb{Z}$, in a vector space $V' = V\oplus K_k$ where K_k denotes a one-dimensional space spanned by a null vector k. Since $(x+nk)^2 = x^2$, Λ'_2 consists of all the points $r + nk$, $r \in \Lambda_2$, $n \in \mathbb{Z}$. The Lie algebra $g_{\Lambda'}$ is generated by elements $e(r+nk)$, $r \in \Lambda_2$, $n \in \mathbb{Z}$, and p^μ, $q \leqslant \mu \leqslant N+1$. Writing

(5.13) $$\hat{e}_n(r) = e(r+nk)$$

and

$$\hat{p}^{\mu}_{n} = \frac{1}{2\pi i}\oint P^{\mu}(z)\exp[ink\cdot Q(z)], \tag{5.14}$$

so that $k\cdot\hat{p}_n = k\cdot p\delta_{n0}$, we have the algebra

$$[\hat{p}^{\mu}_{m},\hat{p}^{\nu}_{n}] = mg^{\mu\nu}\delta_{m,-n}k\cdot p \tag{5.15}$$

$$[\hat{p}^{\mu}_{m},\hat{e}_{n}(r)] = r^{\mu}\hat{e}_{m+n}(r) \tag{5.16}$$

$$[\hat{e}_{m}(r),\hat{e}_{n}(s)] = 0 \qquad \text{if } r\cdot s \geq 0 \tag{5.17}$$

$$= \epsilon(r,s)\hat{e}_{m+n}(r+s) \qquad \text{if } r\cdot s = -1 \tag{5.18}$$

$$= r\cdot\hat{p}_{m+n} + mk\cdot p\delta_{m-n} \qquad \text{if } r = -s \tag{5.19}$$

(Note that $g^{\mu\nu}$ is now singular and $k_{\mu}g^{\mu\nu} = 0$.) Thus we see that we have regained the affine Lie algebra of eqs. (5.8–10) with $e_n(r)$ replaced by $\hat{e}_n(r)$, p_m by $\hat{p}_m$ and the central element 1 by $k\cdot p$.

In the expressions for $\hat{e}_n(r)$, $\hat{p}^{\mu}_{n}$, $k\cdot\alpha_n$ commutes with everything. Therefore we can take the quantities $k\cdot\alpha_n$ to be constant without altering the commutation relation (5.15–5.19). The substitution

$$k\cdot p \longrightarrow 1 \quad k\cdot\alpha_n \longrightarrow 0 \quad \text{if } n \neq 0 \tag{5.20}$$

explicitly yields the Kac-Frenkel construction (5.7).

Note that the derivation d does not occur in our construction unless we extend the lattice to a Lorentzian one containing a vector $\bar{k}$ satisfying $\bar{k}\cdot r = 0$, $\bar{k}\cdot k = 1$ and taking

$$d = \bar{k}\cdot p$$

Again we have equality between the rank of the algebra and the dimension of the lattice.

6. A LORENTZIAN ALGEBRA

If we take $\Lambda = II^{25,1}$, g_Λ is a representation of the algebra L_∞ introduced by Conway, Queen and Sloane [13], which they conjectured might be a natural setting for the Fischer-Griess monster group [19]. They defined L_∞ by means of a set of relations. For this purpose, let ℓ denote the points of Λ satisfying $r \cdot w = -1$ and $r^2 = 2$ where w is the light-like vector of eq. (2.4). These point are called Leech roots and the set ℓ is isometric to the Leech lattice. Then L_∞ has three generators e_r, e_{-r} and h_r for each $r \in \ell$, satisfying the relations

(6.1) $$[h_r, e_s] = r \cdot s e_s \qquad r \in \ell, \pm s \in \ell;$$

(6.2) $$[e_r, e_{-s}] = h_r \delta_{rs} \qquad r, s \in \ell;$$

(6.3) $$[h_r, h_s] = 0 \qquad r, s \in \ell;$$

(6.4) $$(\text{ad } e_r)^{1-r\cdot s}(e_s) = 0$$

for $r \neq s$ and either both $r, s \in \ell$ or both $-r, -s \in \ell$. (Here ad $e_r(x) = [e_r, x]$.)

Taking $e_r = e(r)$ as defined by eq. (3.14) and $h_r = r \cdot p$ clearly provides a representation of eqs. (6.1-3). Further, if $r \neq s$,

(6.5) $$(s + r(1 - r \cdot s))^2 = (r-s)^2 \geqslant 4$$

for $r, s \in \ell$, because the minimum squared distance between two points of the Leech Lattice is 4. Thus since the roots v of g_Λ have $v^2 \leqslant 2$ we see that eq. (6.5) also holds. Because the e_r, $\pm r \in \ell$, generate g_Λ, it provides a representation of L_∞, of rank 26. This reduction in rank corresponds to process, advocated by Conway <u>et. al.</u>, of setting $\sum n_i h_{r_i} = 0$ whenever $\sum n_i r_i = 0$.

Taking k to be any of the 23 other inequivalent light-like vectors in Λ, g_{Λ_k} provides us with a simply-laced finite dimensional subalgebra of g_Λ associated with the 24-dimensional Euclidean lattice Λ_k. Further taking the lattice $\Lambda_k' = \{x \in \Lambda: x\cdot k = 0\}$ gives us the corresponding affine Lie algebra $g_{\Lambda_k'} = \tilde{g}_{\Lambda_k}$ in Section 5 with $g_{\Lambda_k'} \subset g_{\Lambda_k} \subset g_\Lambda$.

More generally, consider g_Λ where Λ is the even Lorentzian lattice $II^{N-1,1}$. The representation in $\mathcal{H}$ involves negative norm states. If we wish to avoid these dual theory suggests we consider the invariant subspaces $\mathcal{H}$ defined by eqs. (5.5). In fact we have the following result [9,10]

(6.6) $\langle\psi,\psi\rangle \geq 0$ for all $\psi \in \mathcal{H}_\lambda$ if and only if $N = 26$ and $\lambda = 1$ or $N \leq 25$ and $\lambda \leq 1$.

So we can provide a non-negative representation of g_Λ for $\Lambda = II^{25,1}$ by using $\mathcal{H}_1$. (Thorn has given an alternative proof of the "no ghost" theorem (6.6) based on a formula of Kac [20] of which a proof is given in ref [21].)

Frenkel [11] has shown that the adjoint representation of g_Λ is isomorphic to its action on the subspace of $\mathcal{H}$, generated from the states Ψ_γ with $\gamma^2 = 2$ so that the root spaces of g_Λ with root vector v, E_v, have

(6.7) $$\dim E_v \leq d(-\tfrac{1}{2}v^2)$$

where

(6.8) $$q^{-1}\prod_n(1-q^n)^{-24} = \sum_{m=-1}^{\infty} d(m)q^m.$$

7. OPERATORS ASSIGNED TO LATTICE POINTS OF UNIT LENGTH.

So far we have assigned operators to the points Λ_2 of the lattice Λ (i.e. those with $r^2 = 2$). Now we shall see that this can also be done for the points of $\Lambda_1 = \{r \in \Lambda: r^2 = 1\}$, by virtually the same construction. The difference is that the new operators will tend to anticommute rather than commute and so resemble a generalized Clifford algebra. For ordinary orthogonal Lie groups the generators can be represented as bilinear in gamma matrices (This is how Dirac established the Lorentz covariance of his relativistic electron equation). The analogue of this holds for certain affine Kac-Moody algebras [22] and the basic identities so obtained provide the basis for Skyrme's fermion-boson equivalence theorem [23], obtained by him and others [24] in the context of the Sine-Gordon-massive Thirring model. A difference is that in the Kac-Moody case there are two sorts of Clifford field, recognized by physicists as the "Neveu-Schwarz" [25] and "Ramond" [26] fields of the spinor dual string model. This difference can be understood in terms of the structure of the Dynkin diagrams for the even dimensional orthogonal group and the affine Kac-Moody algebra based on it.

In fact all our previous equations (3.11) - (3.20) all hold good whether r^2 and s^2 equal 1 or 2, provided the action of A_r on $\mathcal{H}$ is well defined. The dangerous factor in U(r,z) which has to be single valued is

$$z^{r^2/2 + r \cdot p} \tag{7.1}$$

When $r^2 = 2$ this requires $p \in \Lambda_R^*$, as we said in section 3. When $r^2 = 1$ the discussion is more complicated. First let us suppose Λ is Euclidean. Then, if e and f $\in \Lambda_1$, $|e.f| \leq 1$ so that Λ_1 must consist of a set $\pm e_i$ where $e_i \cdot e_j = \delta_{ij}$. This is suggestive of the weights of the 2N dimensional representation of D_N and we shall suppose this is g_Λ (For D_4 we could also have the weights of one of the spinor representations). The weight lattice of D_N splits into four cosets with respect to Λ_R, namely

$$\Lambda_R, \quad (\lambda_v + \Lambda_R), \quad (\lambda_s + \Lambda_R), \quad (\lambda_{\bar{s}} + \Lambda_R)$$

where λ_v, λ_s, $\lambda_{\bar{s}}$ are the fundamental (minimal) weights defining the vector (2N dimensional), and the two inequivalent spinor representations respectively. They correspond to the points of the Dynkin diagram of D_N as indicated.

The requirement that (7.1) and hence $U(r,z)$ be single valued when $r \in \Lambda_1$ is that $r \cdot p \in \mathbb{Z} + \frac{1}{2}$, i.e.

(7.2) $$p \in (\lambda_s + \Lambda_R) \cup (\lambda_{\bar{s}} + \Lambda_R)$$

This is invariant with respect to addition by $\Lambda_R \cup (\lambda_v + \Lambda_R)$, that is, for D_N, the quotient group of Λ_w by $\Lambda_R \cup (\lambda_v + \Lambda_R$ is the $\mathbb{Z}_2$ group with generators

$$(\Lambda_R + \lambda_s) \cup (\Lambda_R + \lambda_{\bar{s}})$$

This guarantees that A_e $(e^2 = 1)$ is well defined on any state constructed from a product of A_r's, $(r^2 = 1$ or $2)$ operating on Ψ_γ with γ satisfying (7.2).

Now let us adopt the convention $e,f \in \Lambda_1$; $r,s, \in \Lambda_2$. Then we see from (3.17) and (3.18) that

(7.3) $$\begin{cases} A_e^2 = 0 & \\ A_e A_f - A_f A_e = 0 & e \cdot f = 0 \\ A_e A_{-e} + A_{-e} A_e = A_0 = 1 & \end{cases}$$

and

(7.4) $$\begin{cases} A_e A_r - (-1)^{e \cdot r} A_r A_e = 0 & r \cdot e = 0 \text{ or } 1 \\ A_e A_r + A_r A_e = A_{r+e} & r \cdot e = -1 \end{cases}$$

This covers all possibilities in Euclidean space. As explained in the appendix the left hand side of equations (7.3) can be converted into the form of anticommutators (and the left hand side of equations 7.4 into the form of commutators) by a similar construction to the previous one. Then (7.3) becomes the algebra of N fermionic oscillators and can be rewritten as a D_N Clifford algebra by defining

$$(7.5)\quad \gamma_i = A_{e_i} + A_{-e_i}, \quad \gamma_{i+N} = \frac{A_{e_i} - A_{-e_i}}{i} \quad : \; i = 1 \ldots N$$

Now let us discuss the singular lattice obtained by adding to the weight lattice of D_N a single, orthogonal, light-like vector k. Our construction (3.11) applied to Λ_2 yields the Frenkel-Kac constructions for the D_N Kac-Moody algebra on substituting (equation 5.20)

$$(7.6)\qquad k \cdot p \longrightarrow 1 \qquad k \cdot \alpha_n \longrightarrow 0 \quad n \neq 0$$

Λ_1 consists of the points $\pm e_i + nk$, $n \in \mathbb{Z}$, so that making the same substitution (7.6) in our construction (3.11) applied to Λ_1 yields

$$(7.7)\qquad A_{e+nk} \longrightarrow B_e^n = \oint \frac{dz}{2\pi i z} z^{n+\frac{1}{2}} : \exp(ie \cdot Q(z)) :$$

of course

$$(B_e^n)^\dagger = B_{-e}^{-n}$$

By equations (3.17) and (3.18) we obtain

$$(7.8)\qquad \left.\begin{aligned} B_e^n B_e^m + B_e^m B_e^n &= 0 \\ B_e^n B_f^m - B_f^m B_e^n &= 0 \qquad e \cdot f = 0 \\ B_e^n B_{-e}^{-m} + B_{-e}^{-m} B_e^n &= \delta_{nm} \end{aligned}\right\}$$

Let us discuss the single valuedness of the integrand in (7.7). One possibility generalizes the previous one (7.2)

$$(7.9)\qquad n \in \mathbb{Z}, \quad e_i \cdot p \in \mathbb{Z}+\tfrac{1}{2} \quad \text{so} \quad p \in (\Lambda_R+\lambda_s) \cup (\Lambda_R+\lambda_{\bar{s}})$$

Another is to alter what we said above and let n be a half integer:

$$(7.10) \qquad n \in \mathbb{Z}+\tfrac{1}{2}, \quad e_i p \in \mathbb{Z} \quad \text{so} \quad p \in \Lambda_R \cup (\Lambda_R + \lambda_v)$$

Equations (7.8) apply in both cases and when we convert them into anticommutators and make the substitutions generalizing (7.7) we obtain quantities known to physicists as respectively the Ramond (7.9) and Neveu-Schwarz (7.10) oscillators, [25,26].

These occur in the spinor dual string model in the context of fermionic and bosonic excitations of the string respectively. More precisely we define

$$(7.11) \qquad B_e(z) = \sum_{r \in \mathbb{Z}+1/2} z^{-r} B_e^r = \sqrt{z} :e^{ieQ(z)}: .$$

and the Neveu-Schwarz field is $H_i(z)$, $i = 1 \ldots 2N$ where

$$H_i(z) = \begin{cases} B_{e_i}(z) + B_{-e_j}(z) & i = 1 \ldots N \\ (B_{e_i}(z) - B_{-e_i}(z))/i & i = N+1, \ldots 2N \end{cases}$$

This field and the correspondingly defined Ramond fields are related by an "intertwining operator" called the "fermion emission vertex" which is much studied in the literature, [27].

We have seen that the generators of the D_N Kac-Moody algebra in the vertex operator representation can only act in highest weight representations whose highest D_N weight is either 0 or one of the D_N minimal weights λ_v, λ_s or $\lambda_{\bar{s}}$. These weights correspond to one of the four tips of the extended Dynkin diagram:

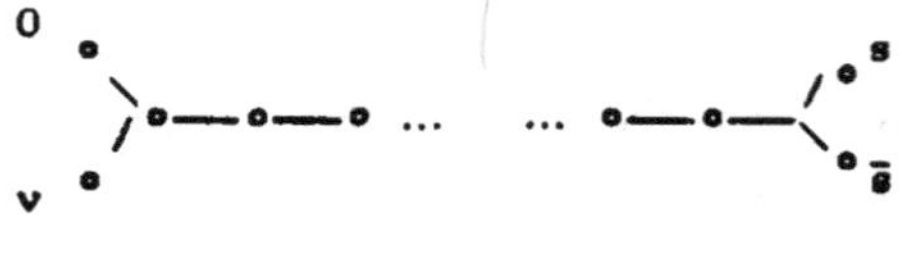

The Neveu-Schwarz oscillators act in the representations defined by 0 and λ_v while the Ramond oscillators act in the representations defined by λ_s and $\lambda_{\bar{s}}$. The fact that the affine Kac-Moody algebra has two "Clifford algebra" representations as compared to the single one for the D_N algebra is evidently due to the enhanced symmetry of the extended Dynkin diagram relative to the ordinary one, drawn above (7.2).

Let us concentrate on the Neveu-Schwarz oscillators. By (3.25) we find

$$B_e^r \Psi_0 = 0 \qquad r \geq 1/2\ , \quad \text{all } e^2 = 1.$$

Thus we can think of $B_e^r (r \geq 1/2)$ as destruction operators and B_e^r $(r \leq -1/2\ )$ as creation operators and hence define a new normal ordering operation denoted with open dots. From equation (7.8) we find

$$(7.12)\quad B_e(z)B_f(\zeta) = {}^\circ_\circ B_e(z)B_f(\zeta){}^\circ_\circ + \frac{\sqrt{z\zeta}}{z-\zeta}\,\delta_{e+f,0} \qquad |\zeta|<|z|$$

Hence using (7.8)

$$(7.13)\quad \begin{aligned} B_e(z)B_e(z) &= {}^\circ_\circ B_e(z)B_e(z){}^\circ_\circ = 0 \\ B_e(z)B_f(z) &= {}^\circ_\circ B_e(z)B_f(z){}^\circ_\circ = :\exp\{i(e+f)\cdot Q(z)\}: \quad \text{if } e\cdot f = 0 \end{aligned}$$

while if $e = -f$ we have by (7.11), (7.12) and (3.14)

$${}^\circ_\circ B_e(z)B_{-e}(\zeta){}^\circ_\circ = \sqrt{z\zeta}\ (:\exp\{ie(Q(z) - Q(\zeta))\}: - 1)/(z-\zeta)$$

Taking $z \to \zeta$ using l'Hôpital's rule we find

$$(7.14)\qquad {}^\circ_\circ B_e(z)B_{-e}(z){}^\circ_\circ = iz\frac{d}{dz}e\cdot Q(z) = e\cdot P(z)$$

by (3.9). Equations (7.13) and (7.14) show us how the D_N Kac-Moody algebra generators can be expressed as bilinear in the Neveu-Schwarz fields, at least in the representation with highest weights 0 and λ_v. Of course the Ramond fields must be used in the other two representations [22].

Identities (7.11) and (7.14) are also remarkable in relating anticommuting fields B to commuting fields Q. In fact they are the basis for the boson-fermion equivalence originally discovered by Skyrme [23]. In two dimensional field theories there are two identities of the above form one for each light like component.

We think it remarkable that the Kac-Moody theoretical framework enables us to unify such apparently disparate physical ideas.

8. SELF-DUAL LATTICES

Interactions between elementary particles are believed to be controlled by a gauge group, which is some compact Lie group. The fields describing particles transform according to some representation of the gauge group and the corresponding particles carry quantum numbers which are the weights of the group in that representation. Thus physicists are led to consider weight lattices and root lattices. It has been shown that the possible magnetic monopoles, that can occur in gauge theoreies, carry magnetic quantum numbers that correspond to points of a lattice dual to the weight lattice of the gauge group [28]. It has been conjectured that there may exist a gauge field theory of such monopoles, whose weight lattice is this dual lattice [29]. An interesting symmetry arises if the weight lattice and its dual are the same. The physical ideas are highly speculative but they provide some physical motivation for the study of self-dual lattices.

Euclidean integral lattices

Consider an integral Euclidean lattice Λ and the even lattice Λ_R spanned by the point of Λ with squared length 2, i.e. Λ_2. We shall say Λ is saturated if dim Λ_R = dim Λ. According to the construction of section 3, there is a simply laced Lie algebra g_Λ whose root lattice is Λ_R and whose root system is Λ_2. If Λ is saturated, g_Λ is semisimple and

(8.1) $$\Lambda_R \subset \Lambda \subset \Lambda^* \subset \Lambda_R^* = \Lambda_W$$

where Λ_W denotes the weight lattice of the simply-connected Lie group with Lie algebra g_Λ.

Semisimple Lie algebras g are classified; each one is the algebra of a finite number of compact Lie groups. The weight lattices Λ of such groups satisfy (8.1) where Λ_R is the root lattice of the Lie algebra. Thus saturated Euclidean integral lattices can be enumerated by Lie algebra theory. Now

(8.2) $$\Lambda_W/\Lambda_R = Z(g),$$

the center of the simply connected group G whose Lie algebra is g. Further there must exist subgroups K, K^V of $Z(g)$, with $K^V \subset K$, such that

(8.3) $$\Lambda = \Lambda(G/K), \qquad \Lambda^* = \Lambda(G/K^V),$$

where $\Lambda(H)$ is the weight lattice of the group H, and

(8.4) $$\Lambda(G)/\Lambda(G/K^V) = \Lambda_R^*/\Lambda^* \cong K^V,$$

(8.5) $$\Lambda(G/K)/\Lambda(G/Z) = \Lambda/\Lambda_R \cong Z/K$$

Since these are equal,

(8.6) $$Z/K \cong K^V$$

These results are of significance in the theory of magnetic monopoles [29]. In particular, if Λ is self-dual, $K \cong K^V$ and so

(8.7) $$|Z| = |K|^2$$

where $|Z|$ denotes the order of Z.

<u>Self-dual Euclidean lattices whose associated Lie algebras are simple</u>

It is instructive to use the previous results to construct some Euclidean self-dual saturated lattices. The easiest ones for us to find are those whose associated Lie algebras are simple. As it is also simply laced it must be one of E_6, E_7, E_8, D_n, A_n. We saw that in order for a lattice to be self-dual the center of the simply connected group with the associated Lie algebra must have an order which is a perfect square. This means we only need to consider E_8, D_n, A_{n^2-1}, whose centres have orders 1, 4 and n^2, respectively. We consider these in turn:

E_8. As the center is trivial $\Lambda_R(E_8) = \Lambda_W(E_8)$ and hence it is self-dual and even. Clearly it has dimension 8.

A_{n^2-1}. As the centre is $\mathbb{Z}_{n^2}$, the candidate subgroup for K is $\mathbb{Z}_n$ leading to the possibly self-dual lattices $\Lambda(SU(n^2)/\mathbb{Z}_n)$. This lattice is generated by $\lambda_n + \Lambda_R$ where λ_n is the nth fundamental weight and comprises n cosets with respect to Λ_R:

(8.8) $$m\lambda_n + \Lambda_R, \quad m = 1,2,\dots,n$$

Now if $x \in m\lambda_n + \Lambda_R$ and $x' \in m'\lambda_n + \Lambda_R$, $x \cdot x'$ differs from $mm'\lambda_n^2$ by an integer. Using an algorithm from magnetic monopole theory [30]

(8.9) $$2\lambda^2/\alpha^2 = |Z(G_\alpha)|/|Z(G)|$$

where λ is a fundamental weight, α the corresponding simple root of G and G_α the simply connected group whose Dynkin diagram is obtained by deleting the point corresponding to α from the Dynkin diagram for G. As $\alpha^2 = 2$ in the present context, we have

(8.10) $$\begin{aligned}\lambda_n^2 &= |Z(SU(n)\times SU(n^2-n))|/|Z(SU(n^2))| \\ &= n(n^2-n)/n^2 = n-1.\end{aligned}$$

Thus $\Lambda(SU(n^2)/\mathbb{Z}_n)$ is integral for all n and even or odd as n is odd or even. As $\mathbb{Z}_{n^2}/\mathbb{Z}_n \cong \mathbb{Z}_n$, it is self-dual. If the lattice is even, with n = 2s+1 say, then its dimension $n^2 - 1 = 4s(s+1)$ which is a multiple of 8. If the lattice is odd, with n = 2s say, it has dimension $4s^2 - 1$, s= 1,2,....

In the cases n = 2 and 3 we have $\lambda_2^2 = 1$ and $\lambda_3^2 = 2$ whilst otherwise $\lambda_n^2 \geq 3$. We now discuss the special cases of n = 2 and 3. If n = 2 the points of the lattice with $\lambda^2 = 1$ are the weights of the 6-dimensional (real) representation of SU(4) which is isomorphic to the covering group of SO(6); thus the lattice so obtained is

(8.11) $$\Lambda(SU(4)/\mathbb{Z}_2) = \mathbb{Z}^3.$$

If n = 3, the lattice $\Lambda(SU(9)/\mathbb{Z}_3)$ contains extra points of length squared 2 which when added to the roots of SU(9) yield the root system of a bigger algebra, E_8 in fact. Thus we obtain again the lattice already mentioned,

(8.12) $$\Lambda(SU(9)/\mathbb{Z}_3) = \Lambda_R(E_8)$$

D_n. The centre is $\mathbb{Z}_2 \times \mathbb{Z}_2$ if n is even and $\mathbb{Z}_4$ if n is odd. A spinor weight, λ_{sp} generates a subgroup $\mathbb{Z}_2$ if n is even, but $\mathbb{Z}_4$ if n is odd. The subgroup G_α of the algorithm (8.9) is SU(n) so that

(8.13) $$\lambda_{sp}^2 = n/4$$

and the corresponding lattices Λ are even and self-dual only if n is a multiple of 8; if n differs by 4 from a multiple of 8, Λ is odd and self-dual. If n is not a multiple of 4, the lattice is not integral. In the special cases of n = 4 and 8, the spinor weight λ_{sp} yields $\mathbb{Z}^4$ and $\Lambda_R(E_8)$ again respectively.

The vector fundamental weight, λ_v, generates a subgroup $\mathbb{Z}_2$ for every n. The corresponding lattices are $\mathbb{Z}^n$, as the algorithm (8.9) show that $\lambda_v^2 = 1$ for each n.

Note that all self-dual even Euclidean lattices found above

have dimensions which are multiples of 8, in agreement with the well-known result [16]. The same methods can be applied to semisimple Lie algebras g, but this is more tedious. It turns out that, at least for even self-dual lattices with dimensions less than 24, the various simple components of the semisimple algebra must have the same Coxeter number. We shall understand this result another way below.

Relation between Euclidean and Lorentzian even self-dual lattices

If Λ is an Euclidean even self-dual lattice, then as mentioned in section 2,

(8.14) $$\Lambda \oplus II^{1,1} = II^{8n+1,1} ,$$

where dim Λ = 8n, by the uniqueness theorem for even self-dual Lorentzian lattices.

If k denotes one of the two primitive light-like vectors in $II^{1,1}$, we can use it to define $\tilde{g}_\Lambda$, the affine Kac-Moody algebra associated with g_Λ. Its simple roots are those of g_Λ together with $k - \psi_i$, where ψ_i is the highest root of g_i, the i-th simple component of

(8.15) $$g_\Lambda = \bigoplus_i g_i .$$

Thus

(8.16) $$\Lambda_2 = \Phi(g_\Lambda) \subset \Phi(\tilde{g}_\Lambda) \subset II_2^{8n+1,1} ,$$

the set of points of squared length 2 on $II^{8n+1,1}$, where $\Phi(g)$ denotes the roots of the Lie algebra g.

Our aim is to deduce more about such lattices Λ and about $II^{8n+1,1}$, from such interrelationships. Our analysis will be incomplete in that we are going to make some plausible assumptions for which we do not yet have general proofs. These assumptions lead to results which we can check, thereby adding credence to the assumptions. We shall assume:

(a) that there exists $\Delta \in II^{8n+1,1}$ such that the simple roots α of $II^{8n+1,1}$ satisfy $\alpha \cdot \Delta = 1$, at least for $n = 1,2,3$;

b) that the simple roots of the affine Kac-Moody algebra $\tilde{g}_\Lambda$ can also be taken as simple roots of $II^{8n+1,1}$;

(c) $II^{8n+1,1}$ has a simple root orthogonal to all the simple roots of the Lie algebra g_Λ, and so in $II^{1,1}$.

Conway and Sloane [12] have proved (a) for $n = 1,2,3,$. It follows that if $\Delta^2 < 0$ it lies in the positive Weyl chamber defined on the "mass shell" hyperboloid, $x^2 = -1$, or, if $\Delta^2 = 0$, the light cone $x^2 = 0$.

Let $\bar{k}$ be the unique light-like lattice vector in $II^{1,1}$ satisfying

(8.17) $$\bar{k}^2 = 0,\ k \cdot \bar{k} = 1.$$

then we write

(8.18) $$\Delta = \sum_i \delta_i + nk + m\bar{k}$$

where δ_i is half the sum of the positive roots of g_i, since if α is a simple root of g_Λ it is also a simple root of $\tilde{g}_\Lambda$ and hence, by (b) and (a), $\alpha \cdot \Delta = 1$ which forces Δ to have the stated form. Now

(8.19) $$\sum_i \delta_i = \frac{1}{2} \sum \alpha$$

where the sum is over all the positive roots of g_Λ. The remaining simple roots of the Kac-Moody algebra $\tilde{g}_\Lambda$ are the $k-\psi_i$. So, using assumptions (a) and (b) again,

(8.20) $$1 = \Delta \cdot (k-\psi_i) = 1-h_i + m$$

where h_i is the Coxeter number of g_i. Hence each $h_i = m$, and the Coxeter numbers of each of the components of g have a common value which henceforth will simply be called h; this result was mentioned earlier. It remains to determine the integer n in eq. (8.18), and to do this we use assumption (c). The only roots of $II^{1,1}$ are $\pm(k+\bar{k})$ and the sign must be chosen so that the scalar products with the other simple roots are non positive. In fact

$$-(k+\bar{k})\cdot(k-\psi_i) = -1 \tag{8.21}$$

so the simple root in question must be

$$\alpha_{-1} = -(k+\bar{k}) \tag{8.22}$$

Now $\alpha_{-1}\cdot\Delta = -h-n$ and so, by (c), $n = -h-1$ giving

$$\Delta = \sum_i \delta_i + h\bar{k} - (h+1)k. \tag{8.23}$$

This shows that $\sum \delta_i \in \Lambda$, not just Λ_W. $II^{8n+1,1}$ may well have more simple roots, but the information obtained so far may be used to help construct its Dynkin diagram. Before discussing this we shall derive a "strange formula" for Δ^2. This will be the main result of this section. Clearly,

$$\Delta^2 = \sum_i \delta_i^2 - 2h(h+1)\ . \tag{8.24}$$

By the Freudenthal-de Vries "strange formula",

$$\delta_i^2 = \frac{1}{12}h_i \dim g_i = \frac{1}{12}h_i(h_i+1)\text{rank } g_i. \tag{8.25}$$

But, since $h_i = h$ for each component,

$$\sum \delta_i^2 = \frac{1}{12}h(h+1)\sum_i \text{rank } g_i = \frac{1}{12}h(h+1)\dim \Lambda \tag{8.26}$$

since Λ is assumed saturated. Hence the result

$$\Delta^2 = \frac{1}{12}h(h+1)(\dim \Lambda - 24) \tag{8.27}$$

indicating that 24 is a critical dimension for Euclidean even self-dual lattices.

Notice that Δ is intrinsic to the Lorentzian lattice whilst h is an attribute of the Euclidean lattice Λ imbedded transversely in it. It follows that for $II^{9,1}$ and $II^{17,1}$, i.e. for dim Λ = 8 and 16, h must be the same for all transverse Euclidean self-dual even Lattices of the same dimension; we shall verify this. Formally our formula works for $II^{1,1}$ as there h = 0 so that $\Delta = -k$, though in that case the Δ satisfying (a) is not unique.

Let us discuss $II^{9,1}$. We found the eight-dimensional self-dual Euclidean lattice

(8.28) $$\Lambda = \Lambda_R(E_8)$$

with Dynkin diagram

Adding to this the points corresponding to $\alpha_0 = k-\psi$ and α_{-1} yields

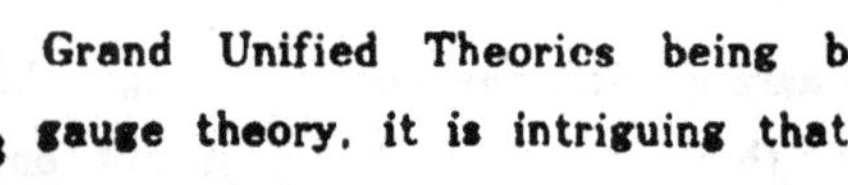

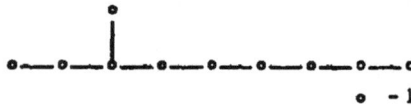

This is indeed the Dynkin diagram of $II^{9,1}$; it can be thought of as E_{10}, being a natural extension of the E series. In view of the recent speculation about Grand Unified Theories being based on N = 4 supersymmetric E_8 gauge theory, it is intriguing that E_8 is related to a lattice in a 10-dimensional Lorentzian space as that is a natural space for N = 4 supersymmetry. In the argument singling out E_8 as a possible gauge group, it was essential that its root lattice was self-dual [31].

Now let us turn to $II^{17,1}$. From our previous work, there are at least two possible 16-dimensional even saturated self-dual Euclidean lattices $\Lambda_R(E_8)\oplus\Lambda_R(E_8)$ and $\Lambda(\widetilde{SO}(32)/\mathbb{Z}_2)$, where $\widetilde{SO}(32)$ is the simply connected group with the Lie algebra of SO(32). Note that D_{16} and E_8 both have the same Coxeter number 30 as "predicted."

The Dynkin diagram of $E_8\oplus E_8$ with the extended roots $\alpha_{0i} = k-\psi_i$, i = 1,2, added is

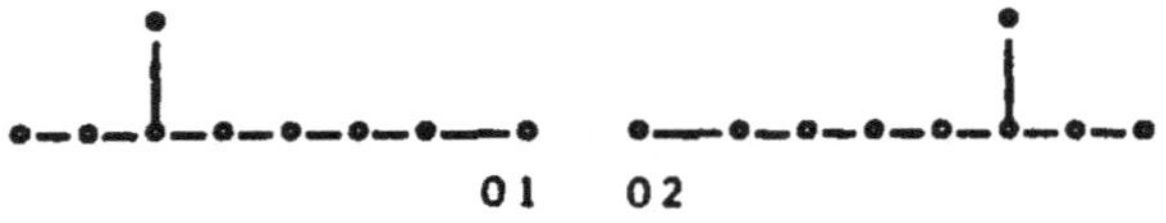

The over extended root α_{-1} is orthogonal to all the simple roots except α_{01} and α_{02} with which it has inner product -1. Hence we deduce that, granted our assumptions (a)-(c), the diagram

describes at least some of the simple roots of $II^{17,1}$. It is in fact the correct diagram.

Now consider starting from the Dynkin diagram for D_{16} instead. Adding the extended and overextended roots yields

i.e. the previous diagram with one point missing. In this case, unlike the previous cases, there exist points of Λ not on $\Lambda_R(D_{16})$. These are defined by a special minimal weight Λ_{sp} of D_{16} as discussed previously. Here λ_{sp} is the highest weight of one of the two spinor representations of D_{16}. Let us now try to construct a root of $II^{17,1}$ by adding to $-\lambda_{sp}$ an element of $II^{1,1}$,

(8.29) $$s = -\lambda_{sp} + nk + m\bar{k}$$

Then for the simple roots α_i of D_{16}, $s \cdot \alpha_i = 0$ unless α_i is the root corresponding to λ_{sp}, for which it equals -1; so if s is a simple root it is joined to the point of the Dynkin diagram corresponding to λ_{sp}. Now

(8.30) $$s^2 = \lambda_{sp}^2 + 2nm = 4 + 2nm,$$

by our previous calculation of λ_{sp}^2 for D_{16}. So if $s^2 = 2$, $nm = -1$. Now

$$s \cdot \alpha_0 = 1+m$$

as λ_{sp} is minimal (i.e. $\lambda_{sp} \cdot \psi = 1$). Thus to get $s \cdot \alpha_0 \leq 0$, consistently with $s^2 = 2$, we need $m = -1$, $n = 1$. Then s is automatically orthogonal to α_{-1} and so it furnishes the missing point.

Note that E_8 and D_{16} both have Coxeter number 30 as predicted above and this gives $\Delta^2 = -1240$ for $II^{9,1}$ and $\Delta^2 = -620$ for $II^{17,1}$, in agreement with Conway's calculation [17].

Finally we turn to $II^{25,1}$. The procedure described above can be repeated but does not terminate, indicating an infinite Dynkin diagram, as is indeed the case. Note that as $\Delta^2 = 0$ we can not deduce that the different 24-dimensional Euclidean self-dual even lattices have the same Coxeter number. It does follow however that Δ furnishes a light-like vector orthogonal to none of the roots of $II^{25,1}$ Thus the lattice transverse to Δ has no roots (and so can not be saturated). It is the Leech lattice. The associated Lie group is $U(1)^{24}$ with $h = 0$, proving that eq. (8.23) for Δ again holds.

We conclude that the formulae (8.23) and (8.24) for Δ and Δ^2 hold in 28 different cases, lending support to assumptions (a), (b) and (c) for dimensions 26 or less. In higher dimensions they must break down as it is easy to find Euclidean even self-dual saturated lattices of the same dimension with different Coxeter numbers.

9. CONCLUSION

Our main point has been to relate integral lattices directly to algebras of various kinds by associating to the points of length squared equal to 1 or 2 (the sets Λ_1 and Λ_2 respectively) the contour integral of the vertex operator for emitting the "tachyon" state of the dual string model. The advantage of this point of view is that we can understand a wide range of results in a unified way, as well as viewing the structure of the dual string model in a new light.

If the scalar product, defining the notion of integrality on the lattice, is positive definite, the Λ_2 operators generate a finite dimensional Lie algebra, requiring only the addition of the momenta p^μ for closure. If the scalar product is positive semi-definite with a unique null direction (a vector in this direction being included in the lattice), the algebra generated is an affine Kac-Moody algebra and requires the addition of "photon" emission vertices for closure. These are derived as commutators of "tachyon" operators, a circumstance which would not generally obtain in the physical theory. If the lattice is Lorentzian (i.e. the scalar product corresponds to a nonsingular matrix with a single negative eigenvalue), the algebra generated is yet larger, with additional roots of length squared equal to -2, -4, -6, ... , whose corresponding step operators are contour integrals of vertices for emitting the massive states of the dual string model. The same construction for the Λ_1 points extends the Λ_2 algebra by a Clifford algebra. Corresponding to the possibilities of nesting lattices in each other, there exist natural subalgèbras. The point of view we have advocated here is not entirely new, being similar to that of Frenkel [11], but we hope that our account will at least make more intelligible to physicists. The basic technique is an exercise in normal ordering of free quantum fields which was developed and exploited in the work on the dual model [1,2,3].

We think the picture developed should be of interest to physicists since it correlates in a more uniform way diverse results they already know. We have already mentioned the role of contour integrals of vertex operators for emitting physical states of the dual string model. In section 7 we explained how the Ramond and Neveu-Schwarz fields arose naturally in connection with the Λ_1

points of a positive semidefinite lattice, constituting a generalised Clifford algebra for the D_N Kac-Moody algebra. These "fermi" fields were constructed out of the "bose" fields, a point of view not envisaged by the dual string literature. Nevertheless this relationship is precisely the fermion-boson quantum equivalence established by Skyrme [23] ten years earlier, in relating the quantum Sine-Gordon and massive Thirring models, and subsequently much exploited in other two dimensional field theories. This same fermion-boson equivalence relation (and its generalizations) has been used to construct general solutions of an enormous class of soliton equations, whose classification has recently been related to that of the affine Kac-Moody algebras in what is possibly their most remarkable physical applications so far [32].

Particle physicists are more interested in spontaneously broken gauge theories in four space-time dimensions and possible soliton-like behavior occurring there in the guise of magnetic monopoles. Conjectures have been made concerning alternative electric and magnetic formulations of the theory based on the construction of a "dual group" from the lattice reciprocal to the weight lattice of the original gauge group [29]. The construction of section 3 achieves this in a more concrete way. A key question concerns the analogue of the fermion-boson equivalence theorem which would facilitate the construction of monopole quantum field operators, just as the known construction yields the Sine-Gordon soliton field operators. Subsequent work [33] indicates that if this can be done it is more likely to be in the N=4 supersymmetric theory which naturally lives in a ten dimensional Lorentzian space [34], as it is a limiting case of the spinor string theory which exists there.

In this context, Lie algebraic methods have been used when studying higher gauge groups and many of the concepts developed as useful in the study of magnetic monopoles, e.g. the importance of minimal weights [35,36] and the importance of the quantity δ, equal to one half the sum of positive roots, in grading the roots [17], find echoes in the work of section 8.

These comments show intriguing links between the study of gauge theories and affine Kac-Moody algebras, a point of view put forward by other speakers at this meeting from different standpoints. Possibly some larger algebra, perhaps Lorentzian, which might involve supersymmetry also, plays a role as well, but it is yet to be identified.

Particularly interesting are the Lorentzian algebras $II^{9,1}$, $II^{17,1}$, $II^{25,1}$, named after the lattices to which they correspond. The latter two are not hyperbolic in the sense that the deletion of any point of their Dynkin diagrams leaves an ordinary or extended Dynkin diagram. Nevertheless the discussions of section 6 and 8 indicate that these algebras possess simplifying features. Firstly, the no ghost theorem establishes the existence of an invariant subspace for the algebra, which is positive definite, despite the Lorentzian metric associated with the lattice. Secondly, the existence of the vector Δ, which has unit scalar product with all the simple roots, seems to be a very special feature which these particular Lorentzian algebras share with all finite dimensional semisimple algebras. Thus these algebras may conceivably be simpler in structure than the other Lorentzian algebras and the present tenous connection may be the hint of a more substantial connection.

These comments suggest that both mathematicians and physicists are in the happy state of seeing the glimmerings of a vast new structure, of considerable importance in their respective subjects, which it will be their task to illuminate in the future.

ACKNOWLEDGEMENT

We are grateful to I. Frenkel, V.G. Kac and J. McKay for helpful and stimulating conversations. Most of this work was done while visiting the Mathematics and Physics departments of the University of Virginia, and whilst one of us (DO) was a member of the Institute for Advanced study there. We are very grateful for their hospitality.

APPENDIX

The operator construction determines equations with quantities *of the following form on the left hand side:*

$$A_r A_s - (-1)^{r \cdot s} A_s A_r \tag{A1}$$

where r^2 and s^2 could equal 1 or 2. We should like to modify the quantities A_r such that instead of (A1) we have an anticommutator if both r^2 and s^2 equal 1 and a commutator otherwise. We now explain how to do this.

Suppose $\{e_i\}$ constitute a basis for the integral lattice in question, containing A_2 (and if relevant A_1). Suppose further that we can construct corresponding quantities γ_i satisfying

$$\gamma_i \gamma_j = (-1)^{[(e_i \cdot e_j)^2 - e_i^2 e_j^2]} \gamma_j \gamma_i \tag{A2}$$

and commuting with all the A_r's. Then for any point of the lattice

$$u = \sum n_i e_i$$

define

$$\gamma_u = (\gamma_1)^{n_1} (\gamma_2)^{n_2} \ldots$$

It is easy to check that (A2) is generalized:

$$\gamma_u \gamma_v = (-1)^{[(u \cdot v)^2 - u^2 v^2]} \gamma_v \gamma_u$$

Note that the square bracket in the exponent of (-1) is the square of the area of the parallelogram with sides u and v. Further

$$\gamma_u \gamma_v = \epsilon(u,v) \gamma_{u+v}$$

where $\epsilon(u,v)$ equals ± 1. Then we define

$$E_r = \gamma_r A_r$$

so that

$$E_r E_s - (-1)^{r^2 s^2} E_s E_r = \epsilon(r,s)\gamma_{r+s}(A_r A_s - (-1)^{r\cdot s} A_s A_r)$$

If for example $r \cdot s \geq -1$ we have by the equations of the text

$$= \begin{cases} 0 & r\cdot s \geq 0 \\ \epsilon(r,s)E_{r+s} & r\cdot s = -1 \end{cases}$$

irrespective of the metric. It is understood $E_0 = A_0 = 1$ effectively. This is the desired form of a $\mathbb{Z}_2$ graded algebra with elements odd or even as $r^2 = 1$ or 2. It may be necessary to redefine the γ_i's by a sign to get the conventional Lie algebra signs on the right hand side.

Thus it remains to show that solutions to (A2) exist. For example the lattice $\mathbb{Z}^n$ in Euclidean or Lorentzian space has a basis e_i satisfying $e_i e_j = \pm\delta_{ij}$. A solution to (A2) is given by the Clifford algebra (Dirac gamma matrices).

Let Λ be any Euclidean or Lorentzian self-dual lattice. Then $\Lambda\oplus\mathbb{Z}$ is odd self dual and Lorentzian if the added $\mathbb{Z}$ is respectively time like or space like. By the uniqueness theorem for odd self dual Lorentzian lattices

$$\Lambda\oplus\mathbb{Z} = I_{\dim\Lambda,1} \cdot$$

Since γ matrices (A2) exist for $I_{\dim \Lambda,1}$ they do also for Λ by this equation, that is for any self dual Euclidean or Lorentzian lattice. We suspect solutions to (A2) exist for any integral lattice but have not shown it.

Instead of increasing the dimension of the representation space by introducing γ matrices, we can achieve the same effect by considering a function of momentum u, C_u, defined by $C_u\Psi_\gamma = \epsilon(u,\gamma)\Psi_\gamma$ for $\gamma \in \Lambda_R$. Such a function can be extended to $\gamma \in \Lambda$

by choosing a γ_0 in each coset of Λ by Λ_R and setting $\epsilon(u,\gamma) = \epsilon(u,\gamma-\gamma_0)$ for γ in that coset.

REFERENCES

[1] M. Jacob (ed.) Dual Theory (North Holland, Amsterdam, 1975).

[2] J. Scherk, Rev. Mod. Phys. 47, 123 (1975).

[3] J.H. Schwarz, Phys. Report 89, 223 (1982); M.B. Green, Surveys in High Energy Physics 3, 127 (1983)

[4] I.B. Frenkel and V.G. Kac, Inv. Math. 62, 23 (1980).

[5] V.G. Kac, Math USSR-Izv. 2, 1271 (1968); R.V. Moody, J. Algebra 10, 211 (1968)

[6] M.A. Virasoro, Phys. Rev. D1, 2933 (1970).

[7] E. Del Giudice, P. Di. Vecchia and S. Fubini, Ann. Phys. (N.Y.) 70, 378 (1972).

[8] R.C. Brower and P. Goddard, Nuc. Phys. B40, 437 (1972).

[9] R.C. Brower, Phys. Rev. D6, 1655 (1972).

[10] P. Goddard and C.B. Thorn, Phy. Letters 40B, 235 (1972).

[11] I.B. Frenkel, Representations of Kac-Moody algebras and dual resonance models, preprint.

[12] J.H. Conway and N.J.A. Sloane, Lorentzian forms for the Leech lattice, preprint

[13] J.H. Conway, L. Queen and N.J.A. Sloane, A monster Lie algebra?, preprint.

[14] I.B. Frenkel, J. Lepowsky and A. Meurman, An E_8 approach to F_1, preprint.

[15] E. Corrigan and D.B. Fairlie, Nuc. Phys. B91, 527 (1975); M. Green Nucl. Phys. B103, 313 (1976); M. Green and J. Shapiro Phys. Lett. 64B, 454 (1976); R. Horsley, Nuc. Phys. B138, 474 (1978).

[16] J.-P. Serre, A course in arithmetic (Springer Verlag, New York, 1973).

[17] J.H. Conway, The automorphism group of the 26-dimensional even unimodular Lorentizian Lattice, preprint.

[18] J.E. Humphreys, Introduction to Lie Algebras and Representation Theory, (Springer-Verlag, 1972).

[19] R.L. Griess, Inv. Math. 69, 1 (1982).

[20] V.G. Kac in: W. Berginböck, A Böhm and E. Takasugi (eds.) Group Theoretical Methods in Physics, Lecture Notes in Physics, Vol. 94 (Springer-Verlag, New York, 1979).

[21] B.L. Feigin and D.B. Fuks, Funct. Anal. 16, 2, 47 (1982).

[22] I.B. Frenkel, Journal of Funtional Analysis, 44, 259 (1981).

[23] T.H.R. Skyrme, Proc. R. Soc. A247, 260(1958), A252, 236(1959), A260, 127(1961), A262, 237(1961).

[24] R.F. Streater and I.F. Wilde, Nucl. Phys. B24, 561 (1970); S. Coleman, Phys. rev. D11, 2088(1975); S. Mandelstam, Phys. Rev D11, 3026(1975).

[25] A. Neveu and J. Schwarz, Nucl. Phys. B31, 86(1971).

[26] P. Ramond, Phys. Rev. D3, 2415(1971).

[27] C.B. Thorn, Phys. Rev. D4, 1112(1971); E. Corrigan and D. Olive, Nuovo Dim. 11A, 749(1972).

[28] F. Englert and P. Windey, Phys. Rev. D14, 2728 (1976).

[29] P. Goddard, J. Nuyts and D. Olive, Nucl. Phys. B125, 1 (1977).

[30] P. Goddard and D. Olive, Nucl. Phys. B191, 511(1981).

[31] G. Chapline and R. Slansky, Nucl. Phys. B209, 461 (1982);
D. Olive and P. West, Nucl. Phys B217, 1 (1983).

[32] M. Jimbo: talk at the meeting.

[33] D. Olive, Magnetic monopoles and electromagnetic conjectures in Monopoles in Quantum Field theory (e.d N. Craigie et al., world Scientific, Singapore (1982).

[34] F. Gliozzi, D. Olive and J. Scherk, Nucl. Phys. B122, 253(1977).

[35] R. Brandt and F. Neri, Nucl. Phys. B161, 253 (1979);
S. Coleman in Proceedings of the 1981 School of Subnuclear Physics "Ettore Majorana".

[36] P, Goddard and D. Olive, Nucl. Phys. B191, 528 (1981).

[37] N Ganoulis, P. Goddard and D. Olive, Nucl. Phys. B205 [FS5], 601(1982).

P. Goddard
Department of Applied Mathematics and Theoretical Physics
University of Cambridge, England

D. Olive
Blackett Laboratory
Imperial College, London, England

Proc. Natl. Acad. Sci. USA
Vol. 82, pp. 8295–8299, December 1985
Mathematics

Calculus of twisted vertex operators

(affine Lie algebras/basic modules/string theory)

J. LEPOWSKY

Department of Mathematics, Rutgers University, New Brunswick, NJ 08903; and The Institute for Advanced Study, Princeton, NJ 08540

Communicated by G. D. Mostow, August 26, 1985

ABSTRACT **Starting from an arbitrary isometry of an arbitrary even lattice, twisted and shifted vertex operators are introduced. Under commutators, these operators provide realizations of twisted affine Lie algebras. This construction, generalizing a number of known ones, is based on a self-contained "calculus."**

Section 1. Introduction

The discovery that vertex operators can provide constructions of affine Lie algebras has stimulated much activity. In the first construction (1) and its generalization (2), affine algebras of types A, D, and E, twisted by the principal automorphism of the underlying finite-dimensional simple Lie algebra $\mathfrak{g}$, were realized by means of certain "twisted vertex operators." (The principal automorphism extends the Coxeter element of the Weyl group, acting on a Cartan subalgebra.) In the next construction (refs. 3 and 4), untwisted affine algebras of types $A^{(1)}$, $D^{(1)}$, and $E^{(1)}$ were represented using the "untwisted vertex operators" of dual-string theory. This was modified in ref. 5 to obtain twistings by certain outer involutions of $\mathfrak{g}$. In ref. 6, another twisted construction was found, this time based on an involution of $\mathfrak{g}$ that is -1 on a Cartan subalgebra, providing a generalization of ref. 1 different from that of ref. 2.

Certain aspects of these constructions were uniformly generalized in ref. 7, in which the role of an automorphism of the Cartan subalgebra was emphasized. Here I continue the approach of ref. 7 to give a common generalization of all the existing constructions, starting from an arbitrary isometry of an arbitrary even lattice and based entirely on elementary calculations with vertex operators. I construct an affine algebra as an explicit algebra of operators on a "generalized Fock space" and, in the process, obtain a formula for an extension to $\mathfrak{g}$ of the automorphism of the root system. The case of the principal automorphism is analyzed in detail in ref. 8.

I also give a "shifted" generalization of the construction. Over $\mathbb{C}$, I thus obtain continuous families of twisted realizations of affine Lie algebras of types A, D, and E. When the "shifting parameter" is rational, the resulting twisted affine algebra has a corresponding $\mathbb{Z}$-grading.

One of the motivations was to try to find an algebraic approach to reducing the critical space-time dimension in string theory from 26 or 10 to 4. In our construction, the dimension of the span of the lattice is "reduced" to the dimension of the span of the sublattice fixed by the automorphism. When applied to the E_8 root lattice, perhaps this can be used to break the E_8-symmetry of the heterotic string (9).

The details of this work will appear elsewhere.

The publication costs of this article were defrayed in part by page charge payment. This article must therefore be hereby marked "*advertisement*" in accordance with 18 U.S.C. §1734 solely to indicate this fact.

Section 2. Assumptions

Suppose the following.

2.1. L is a finitely generated free abelian group.

2.2. $\langle\cdot,\cdot\rangle$ is a nonsingular symmetric $\mathbb{Z}$-bilinear form on L such that

$$\langle\alpha,\alpha\rangle \in 2\mathbb{Z} \quad \text{for} \quad \alpha \in L.$$

2.3. ν is an automorphism of L such that $\langle\nu\alpha,\nu\beta\rangle = \langle\alpha,\beta\rangle$ for $\alpha, \beta \in L$.

2.4. m is a positive integer such that $\nu^m = 1$.

2.5. If m is even, $\langle\nu^{m/2}\alpha,\alpha\rangle \in 2\mathbb{Z}$ for $\alpha \in L$.

Remark: Given L, $\langle\cdot,\cdot\rangle$, ν and m satisfying assumptions 2.1–2.4, assumption 2.5 can always be arranged by doubling m if necessary.

Observe that

$$\Big\langle \sum_{p\in\mathbb{Z}/m\mathbb{Z}} \nu^p\alpha, \alpha\Big\rangle \in 2\mathbb{Z} \text{ for } \alpha \in L. \qquad [2.1]$$

Section 3. Notation

Let $\mathbb{F}$ be a field of characteristic 0 containing a primitive mth root of unity ω. (Later we shall assume that $\mathbb{F}$ contains additional roots of unity. We may take $\mathbb{F} = \mathbb{C}$, $\omega = e^{2\pi i/m}$.) Embed L canonically in the $\mathbb{F}$-vector space $\mathfrak{h}$, where $\mathfrak{h} = \mathbb{F}\otimes_{\mathbb{Z}} L$, and extend $\langle\cdot,\cdot\rangle$ and ν by $\mathbb{F}$-linearity to $\mathfrak{h}$.

We shall use some elementary results in the self-contained sections 2 and 3 of ref. 7.

For $n \in \mathbb{Z}$, set $\mathfrak{h}_{(n)} = \{x \in \mathfrak{h} \mid \nu x = \omega^n x\} \subset \mathfrak{h}$, so that $\mathfrak{h} = \coprod_{p\in\mathbb{Z}/m\mathbb{Z}} \mathfrak{h}_{(p)}$. (We identify $\mathfrak{h}_{(n \bmod m)}$ with $\mathfrak{h}_{(n)}$ for $n \in \mathbb{Z}$.) For $p \in \mathbb{Z}/m\mathbb{Z}$, denote by

$$P_p: \mathfrak{h} \to \mathfrak{h}_{(p)} \qquad [3.1]$$

the pth projection and, for $x \in \mathfrak{h}$ and $n \in \mathbb{Z}$, set $x_{(n)} = P_{(n \bmod m)}x$. Viewing $\mathfrak{h}$ as an abelian Lie algebra, consider the ν-twisted affine Lie algebra (in the terminology of ref. 10) $\hat{\mathfrak{h}}(\nu) = \coprod_{n\in\mathbb{Z}} \mathfrak{h}_{(n)}\otimes t^n \oplus \mathbb{F}c \oplus \mathbb{F}d$ with brackets determined by

$$[x\otimes t^i, y\otimes t^j] = m^{-1}\langle x,y\rangle i\delta_{i+j,0}c \qquad [3.2]$$
$$[c, \hat{\mathfrak{h}}(\nu)] = 0, \ [d, x\otimes t^i] = ix\otimes t^i$$

for $i, j \in \mathbb{Z}$, $x \in \mathfrak{h}_{(i)}$, $y \in \mathfrak{h}_{(j)}$. Consider also its commutator subalgebra $\hat{\mathfrak{h}}(\nu)' = \coprod_{n\neq 0} \mathfrak{h}_{(n)}\otimes t^n \oplus \mathbb{F}c$ and the subalgebras $\hat{\mathfrak{h}}(\nu)_\pm = \coprod_{\pm n>0} \mathfrak{h}_{(n)}\otimes t^n$, $\mathfrak{b}(\nu) = \hat{\mathfrak{h}}(\nu)_+ \oplus \mathfrak{h}_{(0)} \oplus \mathbb{F}c \oplus \mathbb{F}d$. The form $\langle\cdot,\cdot\rangle$ being nonsingular on $\mathfrak{h}$, we observe that $\hat{\mathfrak{h}}(\nu)'$ is a Heisenberg Lie algebra, in the sense that its commutator subalgebra equals its center and is one-dimensional.

Make $\mathbb{F}$ a (one-dimensional) $\mathfrak{b}$-module as follows: $\hat{\mathfrak{h}}(\nu)_+\cdot 1 = 0$, $\mathfrak{h}_{(0)}\cdot 1 = 0$, $c\cdot 1 = 1$, $d\cdot 1 = 0$. Form the following induced

$\hat{\mathfrak{h}}(\nu)$-module, which we denote S:

$$S = S(\hat{\mathfrak{h}}(\nu)^{-}) = U(\hat{\mathfrak{h}}(\nu)) \otimes_{U(\mathfrak{b})} \mathbb{F}. \qquad [3.3]$$

Here $U(\cdot)$ denotes the universal enveloping algebra, $S(\cdot)$ denotes the symmetric algebra, and we have used the Poincaré–Birkhoff–Witt theorem to make the identification. Then S is irreducible even under the Heisenberg subalgebra $\hat{\mathfrak{h}}(\nu)'$, and S provides the "canonical realization of the Heisenberg commutation relations." The action of d defines a $\mathbb{Z}$-grading on S: $S = \coprod_{n \in -(1/m)\mathbb{N}} S_n$, where S_n is the n-eigenspace of d and $\mathbb{N}$ denotes $\{0, 1, 2, \ldots\}$.

For $n \in \frac{1}{m}\mathbb{Z}$ and $x \in \mathfrak{h}_{(mn)}$, write $x(n)$ for the operator on S corresponding to $x \otimes t^n$.

We shall use the elementary calculus of formal variables ζ, ζ_1, ζ_2, . . . explained in sections 2 and 3 of ref. 7. For $\alpha \in \mathfrak{h}$, define $E^{\pm}(\alpha, \zeta) = \exp(\sum_{\pm n > 0} m\alpha_{(mn)}(n)\zeta^{n}/n)$, a pair of formal Laurent series in ζ with coefficients in End S. (Here exp denotes the formal exponential series.) By proposition 3.4 of ref. 7, we have

$$E^{+}(\alpha, \zeta_1)E^{-}(\beta, \zeta_2) = E^{-}(\beta, \zeta_2)E^{+}(\alpha, \zeta_1) \prod_{p \in \mathbb{Z}/m\mathbb{Z}} (1 - \omega^{-p}\zeta_1/\zeta_2)^{\langle \nu^p\alpha, \beta\rangle} \qquad [3.4]$$

for $\alpha, \beta \in \mathfrak{h}$.

Section 4. An Identity

Define the function

$$C: L \times L \to \mathbb{F}^{\times}$$
$$(\alpha, \beta) \mapsto (-1)^{\langle \sum \nu^p\alpha, \beta\rangle}\omega^{\langle \sum p\nu^p\alpha, \beta\rangle} = \prod(-\omega^{p})^{\langle \nu^p\alpha, \beta\rangle}, \qquad [4.1]$$

where the sums and product range over $p \in \mathbb{Z}/m\mathbb{Z}$. Then C is bilinear into the abelian group $\mathbb{F}^{\times}$; i.e.,

$$C(\alpha + \beta, \gamma) = C(\alpha, \gamma)C(\beta, \gamma)$$
$$C(\alpha, \beta + \gamma) = C(\alpha, \beta)C(\alpha, \gamma) \qquad [4.2]$$

for $\alpha, \beta, \gamma \in L$. Also,

$$C(\alpha, \alpha) = 1 \qquad [4.3]$$
$$C(\nu\alpha, \nu\beta) = C(\alpha, \beta). \qquad [4.4]$$

The verification of [4.3] uses [2.1]. Note that $C(\beta, \alpha) = C(\alpha, \beta)^{-1}$.

Recall from section 2 of ref. 7 the notation $\delta(\zeta) = \sum_{n \in \mathbb{Z}} \zeta^n$, $D\delta(\zeta) = \sum_{n \in \mathbb{Z}} n\zeta^n$, where $D = D_\zeta = \zeta(d/d\zeta)$. For $\alpha, \beta \in L$, set $\varepsilon_1(\alpha, \beta) = \prod_{0<p<m}(1 - \omega^{-p})^{\langle \nu^p\alpha, \beta\rangle}$,

$$F(\alpha, \beta) = \frac{1}{2} \sum_{0<p<m} \frac{\langle (\nu^p - \nu^{-p})\alpha, \beta\rangle}{1 - \omega^p}.$$

Note that ε_1 (resp. F) is a bilinear function from $L \times L$ to $\mathbb{F}^{\times}$ (resp. $\mathbb{F}$). Also,

$$\varepsilon_1(\nu\alpha, \nu\beta) = \varepsilon_1(\alpha, \beta),$$
$$\varepsilon_1(\alpha, \beta)/\varepsilon_1(\beta, \alpha) = (-1)^{\langle \alpha, \beta\rangle}C(\alpha, \beta)^{-1},$$
$$F(\alpha, \alpha) = 0. \qquad [4.5]$$

PROPOSITION 4.1. *Let* $\alpha, \beta \in L$. *Set*

$$I(n) = \{p \in \mathbb{Z}/m\mathbb{Z} \mid \langle \nu^p\alpha, \beta\rangle = n\}$$

for $n \in \mathbb{Z}$, *and suppose that* $\langle \nu^p\alpha, \beta\rangle \geq -2$ *for all* $p \in \mathbb{Z}/m\mathbb{Z}$. *Then*

$$\prod(1 - \omega^{-p}\zeta_1/\zeta_2)^{\langle \nu^p\alpha, \beta\rangle} - C(\beta, \alpha)(\zeta_1/\zeta_2)^{\langle \sum \nu^p\alpha, \beta\rangle}\prod(1 - \omega^{-p}\zeta_2/\zeta_1)^{\langle \nu^p\beta, \alpha\rangle}$$
$$= \sum_{r \in I(-1)} \varepsilon_1(\nu^r\alpha, \beta)\delta(\omega^{-r}\zeta_1/\zeta_2) + \sum_{r \in I(-2)} \varepsilon_1(\nu^r\alpha, \beta)\Big[D\delta(\omega^{-r}\zeta_1/\zeta_2) - \Big(\frac{1}{2}\Big\langle \sum \nu^p\alpha, \beta\Big\rangle + F(\nu^r\alpha, \beta)\Big)\delta(\omega^{-r}\zeta_1/\zeta_2)\Big],$$

where the unindexed sums and products range over $p \in \mathbb{Z}/m\mathbb{Z}$.

We now obtain another description of the function ε_1. For $\alpha, \beta \in L$, define $\varepsilon_2(\alpha, \beta) = (-1)^{\langle \sum \nu^p\alpha, \beta\rangle}\omega^{-\langle \sum p\nu^p\alpha, \beta\rangle} = \prod(-\omega^{-p})^{\langle \nu^p\alpha, \beta\rangle}$, where the sums and product range over $-m/2 < p < 0$. Then ε_2 is bilinear, $\varepsilon_2(\nu\alpha, \nu\beta) = \varepsilon_2(\alpha, \beta)$ and

$$\varepsilon_2(\alpha, \beta)/\varepsilon_2(\beta, \alpha) = (-1)^{\langle \alpha, \beta\rangle}C(\alpha, \beta)^{-1}. \qquad [4.6]$$

Write $\varepsilon'(\alpha, \beta) = \prod_{0<p<m/2}(1 - \omega^{-p})^{\langle (\nu^p + \nu^{-p})\alpha, \beta\rangle}$,

$$\varepsilon^{*}(\alpha, \beta) = \begin{cases} \varepsilon'(\alpha, \beta)2^{\langle \nu^{m/2}\alpha, \beta\rangle} & \text{if } m \in 2\mathbb{Z} \\ \varepsilon'(\alpha, \beta) & \text{if } m \in 2\mathbb{Z} + 1. \end{cases}$$

Then

$$\varepsilon_1(\alpha, \beta) = \varepsilon_2(\alpha, \beta)\varepsilon^{*}(\alpha, \beta). \qquad [4.7]$$

Recalling assumption 2.5, we define $\sigma'(\alpha) = \prod_{0<p<m/2}(1 - \omega^{-p})^{\langle \nu^p\alpha, \alpha\rangle}$

$$\sigma(\alpha) = \begin{cases} \sigma'(\alpha)2^{\langle \nu^{m/2}\alpha, \alpha\rangle/2} & \text{if } m \in 2\mathbb{Z} \\ \sigma'(\alpha) & \text{if } m \in 2\mathbb{Z} + 1. \end{cases} \qquad [4.8]$$

Then

$$\sigma(\nu\alpha) = \sigma(\alpha) \qquad [4.9]$$
$$\varepsilon^{*}(\alpha, \beta) = \sigma(\alpha + \beta)/\sigma(\alpha)\sigma(\beta). \qquad [4.10]$$

Section 5. $\hat{L}$ and $\hat{\nu}$

Set $\omega_0 = (-1)^m\omega$. Then ω_0 is a primitive $2m$th root of unity if m is odd and, for any m, -1 and ω are powers of ω_0. In view of formulas **4.2** and **4.3**, there is a unique (up to equivalence) central extension

$$1 \to \langle \omega_0\rangle \to \hat{L} \xrightarrow{\ -\ } L \to 1 \qquad [5.1]$$

of L by the cyclic group generated by ω_0 with commutator map C—i.e., such that

$$aba^{-1}b^{-1} = C(\bar{a}, \bar{b}) \quad \text{for} \quad a, b \in \hat{L}. \qquad [5.2]$$

If we replace the map $^{-}$ in [5.1] by $\nu \circ {}^{-}$, we obtain another central extension of L by $\langle \omega_0\rangle$ with commutator map C, by [4.4]. By the uniqueness, there is an automorphism $\hat{\nu}$ of $\hat{L}$ (fixing ω_0) such that $\hat{\nu}$ covers ν—i.e., such that

$$\overline{(\hat{\nu}a)} = \nu\bar{a} \quad \text{for} \quad a \in \hat{L}. \qquad [5.3]$$

It is easy to see that the automorphisms of $\hat{L}$ covering the identity automorphism of L are precisely the maps $\rho^{*}: a \mapsto a\rho(\bar{a})$ for a homomorphism $\rho: L \to \langle \omega_0\rangle$. Similarly, there is a homomorphism ρ_0: $L \cap \mathfrak{h}_{(0)} \to \langle\omega_0\rangle$ such that $\hat{\nu}a = a\rho_0(\bar{a})$ if $\nu\bar{a} = \bar{a}$. Now ρ_0 can be extended to a homomorphism $\rho: L \to \langle\omega_0\rangle$ since the map $1 - P_0$ induces an isomorphism from $L/L \cap \mathfrak{h}_0$ to the free abelian group $(1 - P_0)L$ (the projection of L to the orthogonal complement of $\mathfrak{h}_{(0)}$). Multiplying $\hat{\nu}$ by the inverse of ρ^* gives us an automorphism $\hat{\nu}$ of $\hat{L}$ satisfying [5.3] and

$$\hat{\nu}a = a \quad \text{if} \quad \nu\bar{a} = \bar{a}. \qquad [5.4]$$

Section 6. Irreducible $\hat{N}$-Modules

Let $N = (1 - P_0)\mathfrak{h} \cap L = \{\alpha \in L | \langle\alpha, \mathfrak{h}_{(0)}\rangle = 0\}$, $M = (1 - \nu)L \subset N$. On N, the commutator map C (see [4.1]) simplifies to

$$C_N(\alpha, \beta) = \omega^{\langle\sum p\nu^p\alpha, \beta\rangle}. \qquad [6.1]$$

Let $R = \{\alpha \in N | C_N(\alpha, N) = 1\}$, the radical of the alternating bilinear form C_N. It is easy to see that $M \subset R$. Denoting by $\hat{Q}$ the subgroup of $\hat{L}$ obtained by pulling back a subgroup Q of L (see [5.1]), we observe that $\hat{R}$ is the center of $\hat{N}$ and that $\hat{M} \subset \hat{R}$. Observe that $a\hat{\nu}a^{-1} \in \hat{M}$ for all $a \in \hat{L}$. Using [5.4], we can prove the following.

PROPOSITION 6.1. *There is a unique homomorphism* τ: $\hat{M} \to F^{\times}$ *such that* $\tau(\omega_0) = \omega_0$ *and* $\tau(a\hat{\nu}a^{-1}) = \omega^{-\langle\sum\nu^p\bar{a}, \bar{a}\rangle/2}$ *for* $a \in \hat{L}$.

Now N/M is a finite group, and hence so are $\hat{N}/\hat{M}$ and $\hat{N}/\text{Ker } \tau$. Choose $m_0 \in \mathbb{N}$ such that $x^{m_0} = 1$ for all $x \in \hat{N}/\text{Ker } \tau$.

PROPOSITION 6.2. *Assume that* F *contains a primitive* m_0*th root of unity. There are exactly* $|R/M|$ *extensions of* τ *to a homomorphism* $\chi: \hat{R} \to F^{\times}$. *For each such* χ, *there is a unique (up to equivalence) irreducible* $\hat{N}$*-module on which* $\hat{R}$ *acts according to* χ, *and every irreducible* $\hat{N}$*-module on which* $\hat{M}$ *acts according to* τ *is equivalent to one of these. Every such module has dimension* $|N/R|^{1/2}$. *To construct the module corresponding to* χ, *let* A *be any subgroup of* N *(necessarily containing* R*) that is maximal such that the alternating bilinear form* 6.1 *is trivial on* A. *Then* $\hat{A}$ *is a maximal abelian subgroup of* $\hat{N}$. *Let* $\psi: \hat{A} \to F^{\times}$ *be any homomorphism extending* χ *and denote by* F_ψ *the* $\hat{A}$*-module* F *with character* ψ. *Then* T *is isomorphic to the induced* $\hat{N}$*-module* $F[\hat{N}] \otimes_{F[\hat{A}]} F_\psi \simeq F[N/A]$ *(*$F[\cdot]$ *denoting group algebra).*

Section 7. The Vertex Operators

Let T be any $\hat{N}$-module on which $\hat{M}$ acts as multiplication by the character τ (see *Proposition 6.1*). Form the induced $\hat{L}$-module $U_T = F[\hat{L}] \otimes_{F[\hat{N}]} T$. Since T may be viewed as a module for the finite group $\hat{N}/\text{Ker } \tau$, T is completely reducible. In case F contains enough roots of unity, the structure of T follows from *Proposition 6.2* and, in the irreducible case, $U_T = F[\hat{L}] \otimes_{F[\hat{A}]} F_\psi \simeq F[L/A]$.

In general, $\hat{L}$ and $\mathfrak{h}_{(0)}$ act on U_T as follows:

$$a \cdot b \otimes t = ab \otimes t,$$

$$\alpha \cdot b \otimes t = \langle\alpha, \bar{b}\rangle b \otimes t \qquad [7.1]$$

for $a, b \in \hat{L}$, $t \in T$, $\alpha \in \mathfrak{h}_{(0)}$, and we have $[\alpha, a] = \langle\alpha, \bar{a}\rangle a$. In case $\langle\alpha, L\rangle \subset \mathbb{Z}$, define the End U_T-valued formal Laurent series ζ^{α} as follows:

$$\zeta^{\alpha} \cdot b \otimes t = \zeta^{\langle\alpha, \bar{b}\rangle} b \otimes t \qquad [7.2]$$

and the operator ω^{α} on U_T by $\omega^{\alpha} \cdot b \otimes t = \omega^{\langle\alpha, \bar{b}\rangle} b \otimes t$. Then $\zeta^{\alpha}a = a\zeta^{\alpha + \langle\alpha, \bar{a}\rangle}$, $\omega^{\alpha}a = a\omega^{\alpha + \langle\alpha, \bar{a}\rangle}$. Moreover,

$$\hat{\nu}a = a\omega^{-\sum\nu^p\bar{a} - \langle\sum\nu^p\bar{a}, \bar{a}\rangle/2} \qquad [7.3]$$

on U_T. It follows that $\hat{\nu}^m = 1$ on U_T and hence on $\hat{L}$.

Define a $\mathbb{Z}$-grading on U_T by

$$\deg(b \otimes t) = -\frac{1}{2}\left\langle\sum \nu^p\bar{b}, \bar{b}\right\rangle \qquad [7.4]$$

and define an operator d on $U_T = \coprod_{n \in \frac{1}{m}\mathbb{N}} (U_T)_n$ accordingly. Then U_T becomes an $\hat{\mathfrak{h}}(\nu)$-module by making $\hat{\mathfrak{h}}(\nu)'$ act trivially.

Since the projection operator P_0 (see [3.1]) induces an isomorphism from L/N to P_0L, we have a natural isomorphism

$$U_T \simeq F[P_0L] \otimes_F T \qquad [7.5]$$

of $\hat{\mathfrak{h}}(\nu)$-modules, $\hat{\mathfrak{h}}(\nu)$ acting in the obvious way on $F[P_0L] \otimes T$.

Recalling [3.3], set

$$V_T = S \otimes_F U_T = \coprod_{n \in \frac{1}{m}\mathbb{N}} (V_T)_n, \qquad [7.6]$$

a tensor product $\hat{\mathfrak{h}}(\nu)$-module on which $\hat{L}$ acts by its action on the second factor and whose $\mathbb{Z}$-grading is determined by the action of d. For $a \in \hat{L}$, we define the corresponding vertex operator $X(a, \zeta)$ and its coefficients $x_a(n) \in \text{End } V_T$ as follows:

$$X(a, \zeta) = E^-(-\bar{a}, \zeta)E^+(-\bar{a}, \zeta)a\zeta^{-\sum\nu^p\bar{a} - \langle\sum\nu^p\bar{a}, \bar{a}\rangle/2} \qquad [7.7]$$

$$= \sum_{n \in \mathbb{Z}} x_a(n)\zeta^n,$$

an End V_T-valued formal Laurent series.

Using [7.3] we obtain

$$X(\hat{\nu}a, \zeta) = X(a, \omega\zeta) \qquad [7.8]$$

$$DX(a, \zeta) = [d, X(a, \zeta)], \qquad [7.9]$$

the last formula showing that each operator $x_a(n)$ has degree n [i.e., takes $(V_T)_m$ to $(V_T)_{m+n}$ for all $m \in \mathbb{Z}$].

It will be convenient to renormalize the vertex operators as follows: Recalling [4.8], define $Y(a, \zeta) = m^{-\langle\bar{a}, \bar{a}\rangle/2}\sigma(\bar{a})X(a, \zeta) = \sum_{n \in \mathbb{Z}} y_a(n)\zeta^n$. Then formulas 7.8 and 7.9 also hold with Y in place of X (recall [4.9]).

Section 8. Commutators

For $\alpha \in \mathfrak{h}$, write $\alpha(\zeta) = \sum_{n \in \mathbb{Z}} \alpha_{(n)}(n)\zeta^n$, $\alpha^{\pm}(\zeta) = \sum_{n \in \mathbb{Z}, \pm n > 0} \alpha_{(n)}(n)\zeta^n + (1/2)\alpha_{(0)}(0)$, where $\alpha_{(n)}(n)$ is understood as an operator on V_T in the obvious way. Then, $\alpha(\zeta) = \alpha^+(\zeta) + \alpha^-(\zeta)$. For $a \in \hat{L}$ and $x \in F$, set $:(\alpha(\zeta) + x)Y(a, \zeta): = \alpha^-(\zeta)Y(a, \zeta) + Y(a, \zeta)\alpha^+(\zeta) + xY(a, \zeta)$.

Using [3.4], *Proposition 4.1*, [4.7], [4.10], [5.2], and the rules for multiplying suitable expressions by δ and $D\delta$ (see, e.g., proposition 3.9 of ref. 7), we can prove the following:

THEOREM 8.1. *Let* $a, b \in \hat{L}$ *and suppose that* $\langle\nu^p\bar{a}, \bar{b}\rangle \geq -2$ *for all* $p \in \mathbb{Z}/m\mathbb{Z}$. *For* $n \in \mathbb{Z}$, *set* $J(n) = \{p \in \mathbb{Z}/m\mathbb{Z} | \langle\nu^p\bar{a}, \bar{b}\rangle = n\}$. *Then*

$$[Y(a, \zeta_1), Y(b, \zeta_2)] =$$

$$m^{-1}\sum_{p \in J(-1)} \varepsilon_2(\nu^p\bar{a}, \bar{b})Y((\hat{\nu}^pa)b, \zeta_2)\delta(\omega^{-p}\zeta_1/\zeta_2)$$

$$+ m^{-2}\sum_{p \in J(-2)} \varepsilon_2(\nu^p\bar{a}, \bar{b})[Y((\hat{\nu}^pa)b, \zeta_2)D\delta(\omega^{-p}\zeta_1/\zeta_2)$$

$$+ :(m\nu^p\bar{a}(\zeta_2) - F(\nu^p\bar{a}, \bar{b}))Y((\hat{\nu}^pa)b, \zeta_2):\delta(\omega^{-p}\zeta_1/\zeta_2)].$$

By equating the coefficients of $\zeta_1^{n_1}\zeta_2^{n_2}$ on the two sides, we obtain a formula for $[y_a(n_1), y_b(n_2)]$.

Sometimes it is useful to parametrize the vertex operators by means of elements of L rather than $\hat{L}$. To do this, choose a normalized section of $\hat{L}$—i.e., a map $\alpha \mapsto e_\alpha$ from L to $\hat{L}$ such that $e_0 = 1$ and $\overline{e_\alpha} = \alpha$ for all $\alpha \in L$. As is well known, the function ε_C: $L \times L \to \langle\omega_0\rangle$ defined by $e_\alpha e_\beta = \varepsilon_C(\alpha, \beta)e_{\alpha+\beta}$ for $\alpha, \beta \in L$ satisfies the conditions

$$\varepsilon_C(\alpha, \beta)\varepsilon_C(\alpha + \beta, \gamma) = \varepsilon_C(\beta, \gamma)\varepsilon_C(\alpha, \beta + \gamma) \quad [8.1]$$

$$\varepsilon_C(0, 0) = 1 \quad [8.2]$$

$$\varepsilon_C(\alpha, \beta)/\varepsilon_C(\beta, \alpha) = C(\alpha, \beta); \quad [8.3]$$

that is, ε_C is a normalized 2-cocycle associated with C. Conversely, any such cocycle comes from a normalized section of $\hat{L}$. Recall also that it is easy to construct such cocycles by means of bilinear functions ε_C (which automatically satisfy [8.1] and [8.2]).

Define the function η: $\mathbf{Z}/m\mathbf{Z} \times L \to \langle\omega_0\rangle$ by the condition $\hat{\nu}^p e_\alpha = \eta(p, \alpha)e_{\nu^p\alpha}$.

Also define ε: $L \times L \to \langle\omega_0\rangle$ by $\varepsilon(\alpha, \beta) = \varepsilon_2(\alpha, \beta)\varepsilon_C(\alpha, \beta)$. By [4.6] and [8.3], ε is a normalized cocycle associated with the bilinear map $(-1)^{\langle\alpha,\beta\rangle}$.

It is easy to reformulate *Theorem 8.1* in terms of the vertex operators defined as follows for $\alpha \in L$: $Y(\alpha, \zeta) = Y(e_\alpha, \zeta) = \sum_{n\in\mathbf{Z}} y_\alpha(n)\zeta^n$. Formula 7.8 is now replaced by the following:

$$Y(\alpha, \omega^p\zeta) = \eta(p, \alpha)Y(\nu^p\alpha, \zeta) \quad [8.4]$$

for $\alpha \in L$, $p \in \mathbf{Z}/m\mathbf{Z}$.

For convenience, we assume from now on that L is *positive definite*—that is, the natural $\mathbf{R}$-linear extension of $\langle\cdot,\cdot\rangle$ to the real vector space $\mathbf{R} \otimes_{\mathbf{Z}} L$ is positive definite. Setting $L_n = \{\alpha \in L \mid \langle\alpha, \alpha\rangle = n\}$ for $n \in \mathbf{Z}$, we have that $\langle\alpha, \beta\rangle = 0, \pm1, \pm2$ for $\alpha, \beta \in L_2$.

THEOREM 8.2. *Let $\alpha, \beta \in L_2$. Recall the notation $I(n)$ from* Proposition 4.1. *We have*

$$[Y(\alpha, \zeta_1), Y(\beta, \zeta_2)] =$$

$$m^{-1} \sum_{p\in I(-1)} \eta(p, \alpha)\varepsilon(\nu^p\alpha, \beta)Y(\nu^p\alpha + \beta, \zeta_2)\delta(\omega^{-p}\zeta_1/\zeta_2)$$

$$+ m^{-2}\varepsilon(-\beta, \beta) \sum_{p\in I(-2)} \eta(p, \alpha)[D\delta(\omega^{-p}\zeta_1/\zeta_2)$$

$$- m\beta(\zeta_2)\delta(\omega^{-p}\zeta_1/\zeta_2)].$$

(We use that $\nu^p\alpha = -\beta$ for $p \in I(-2)$ and that in this case $F(\nu^p\alpha, \beta) = 0$ by [4.5].)

Section 9. The Twisted Affine Algebra

Motivated by *Theorem 8.2*, we define a nonassociative algebra $(\mathfrak{g}, [\cdot,\cdot])$ over $\mathbf{F}$ as follows: $\mathfrak{g} = \mathfrak{h} \oplus \coprod_{\alpha\in L_2} \mathbf{F}x_\alpha$ ($\{x_\alpha\}_{\alpha\in L_2}$ a set of symbols), with $[\mathfrak{h}, \mathfrak{h}] = 0$, $[h, x_\alpha] = \langle h, \alpha\rangle x_\alpha = -[x_\alpha, h]$,

$$[x_\alpha, x_\beta] = \begin{cases} \varepsilon(\alpha, -\alpha)\alpha & \text{if } \alpha + \beta = 0 \\ \varepsilon(\alpha, \beta)x_{\alpha+\beta} & \text{if } \langle\alpha, \beta\rangle = -1 \\ 0 & \text{if } \langle\alpha, \beta\rangle \geq 0 \end{cases}$$

for $h \in \mathfrak{h}$, $\alpha, \beta \in L_2$. The fact that ε is a normalized cocycle associated with the bilinear map $(-1)^{\langle\alpha,\beta\rangle}$ implies that $\mathfrak{g}$ is a Lie algebra (refs. 3 and 4). We extend the form $\langle\cdot,\cdot\rangle$ from $\mathfrak{h}$ to a nonsingular symmetric form on $\mathfrak{g}$ as follows: $\langle h, x_\alpha\rangle = 0 = \langle x_\alpha, h\rangle$,

$$\langle x_\alpha, x_\beta\rangle = \begin{cases} \varepsilon(\alpha, -\alpha) & \text{if } \alpha + \beta = 0 \\ 0 & \text{if } \alpha + \beta \neq 0. \end{cases}$$

Then $\langle\cdot,\cdot\rangle$ is $\mathfrak{g}$-invariant; i.e., $\langle[x, y], z\rangle + \langle y, [x, z]\rangle = 0$ for $x, y, z \in \mathfrak{g}$. We also extend the automorphism ν of $\mathfrak{h}$ to a linear automorphism ν of $\mathfrak{g}$ by the following: $\nu x_\alpha = \eta(1, \alpha)x_{\nu\alpha}$. Then $\nu^m = 1$ on $\mathfrak{g}$.

$$\nu^p x_\alpha = \eta(p, \alpha)x_{\nu^p\alpha} \quad [9.1]$$

for $p \in \mathbf{Z}/m\mathbf{Z}$, and ν preserves $[\cdot,\cdot]$ and $\langle\cdot,\cdot\rangle$.

Assume for convenience that L_2 spans $\mathfrak{h}$. Then $\mathfrak{g}$ is a semisimple Lie algebra. All the conditions of section 2 of ref. 7 hold, and we can form the twisted affine Lie algebra $\hat{\mathfrak{g}}(\nu) = \coprod_{n\in\mathbf{Z}} \mathfrak{g}_{(n)} \otimes t^n \oplus \mathbf{F}c \oplus \mathbf{F}d$ ($\mathfrak{g}_{(n)}$ denoting the ω^n-eigenspace of ν in $\mathfrak{g}$), with brackets normalized so that $[x \otimes t^i, y \otimes t^j] = [x, y] \otimes t^{i+j} + m^{-1}\langle x, y\rangle i\delta_{i+j,0}\, c$, $i, j \in \mathbf{Z}$, $x \in \mathfrak{g}_{(i)}$, $y \in \mathfrak{g}_{(j)}$ (cf. [3.2]). All the hypotheses of theorem 2.6 of ref. 7 hold (see [8.4] and *Theorem 8.2*), where in the notation of that theorem, we take $E = \operatorname{End} V_T$, $X'(\alpha, \zeta) = Y(\alpha, \zeta)$ for $\alpha \in L_2$, $c \mapsto 1$ and the scalar $e = m^{-1}$, and so we have the following.

THEOREM 9.1. *The representation of $\hat{\mathfrak{h}}(\nu)$ on V_T extends uniquely to a Lie algebra representation of $\hat{\mathfrak{g}}(\nu)$ on V_T such that $\sum_{n\in\mathbf{Z}}((x_\alpha)_{(n)} \otimes t^n)\zeta^n \mapsto Y(\alpha, \zeta)$ for all $\alpha \in L_2$—i.e., $(x_\alpha)_{(n)} \otimes t^n \mapsto y_\alpha(n)$ for all $\alpha \in L_2$, $n \in \mathbf{Z}$.*

Thus we have constructed the twisted affine Lie algebra by means of twisted vertex operators.

PROPOSITION 9.2. *The $\hat{\mathfrak{g}}(\nu)$-module V_T is irreducible if and only if the $\hat{N}$-module T is irreducible* (*cf.* Sections 6 and 7).

Section 10. Shifted Operators and Their Commutators

Fix an element $\gamma \in \mathfrak{h}_{(0)}$. We shall construct "$\gamma$-shifted" analogues of the vertex operators already considered, and we shall compute certain brackets: The shifted operators should be thought of as adapted to the coset $L + \gamma$ of L.

Define a γ-shifted action of $\mathfrak{h}_{(0)}$ on U_T as follows: $a^\gamma \cdot b \otimes t = \langle a, \bar{b} + \gamma\rangle b \otimes t$ for $a \in \mathfrak{h}_{(0)}$, $b \in \hat{L}$, $t \in T$ (cf. [7.1]). Correspondingly, define

$$\zeta^{a^\gamma} \cdot b \otimes t = \zeta^{\langle a, \bar{b}+\gamma\rangle} b \otimes t$$

(cf. [7.2]). Here and throughout this section, we allow arbitrary (not necessarily integral) values of $\mathbf{F}$ as exponents of formal variables such as ζ. The expected algebraic rules shall hold for such symbols.

We also define a γ-shifted $\mathbf{F}$-grading on U_T as follows: $\deg^\gamma(b \otimes t) = -(1/2)\langle\Sigma\nu^p(\bar{b} + \gamma), \bar{b} + \gamma\rangle$ (cf. [7.4]). Correspondingly, we have $U_T = \coprod_{n\in\mathbf{F}}(U_T)_n$, and we define a γ-shifted action d^γ of the operator d on U_T so as to act as multiplication by $\deg^\gamma(b \otimes t)$ on $b \otimes t$.

Retaining the original (trivial) action of $\hat{\mathfrak{h}}(\nu)'$ on U_T, we obtain a new $\hat{\mathfrak{h}}(\nu)$-module structure on U_T, and hence on V_T, by the tensor product action. The isomorphism 7.5 still holds, with $\hat{\mathfrak{h}}(\nu)$ acting in the γ-shifted way on $\mathbf{F}[P_0L]$. We have the corresponding analogue of [7.6], where the grading is now an $\mathbf{F}$-grading.

For $a \in \hat{L}$, we use formula 7.7, with $\Sigma\nu^p\bar{a}$ replaced by $(\Sigma\nu^p\bar{a})^\gamma$, to define the γ-shifted vertex operator $X^\gamma(a, \zeta)$, and we define its components as follows: $X^\gamma(a, \zeta) = \sum_{n\in\mathbf{F}} x_a^\gamma(n)\zeta^n$; similarly, for $Y^\gamma(\alpha, \zeta)$ ($\alpha \in L$). Then $DX^\gamma(a, \zeta) = [d^\gamma, X^\gamma(a, \zeta)]$ (cf. [7.9]). Also, $x_a^\gamma(n) \neq 0$ only if $n \in -m\langle\bar{a}, \gamma\rangle + \mathbf{Z}$. Since $X^\gamma(a, \zeta) = X(a, \zeta)\zeta^{m\langle\bar{a},\gamma\rangle}$, it is easy to derive commutator formulas for γ-shifted vertex operators from *Theorems 8.1 and 8.2*.

Suppose that γ lies in the rational span of L. Choose $M \in \mathbf{N}$, $M > 0$ such that $M\langle\alpha, \gamma\rangle \in \mathbf{Z}$ for $\alpha \in L$, and let $\omega_\gamma \in \mathbf{F}$ be a primitive mMth root of unity such that $\omega_\gamma^M = \omega$. For $a \in \hat{L}$ define $Y^{\gamma,M}(\alpha, \zeta) = Y^\gamma(\alpha, \zeta^M)$. Define a linear automorphism ν_γ of $\mathfrak{g}$ by asserting that $\nu_\gamma = \nu$ on $\mathfrak{h}$, $\nu_\gamma = \nu\omega_\gamma^{-mM\langle\alpha,\gamma\rangle}$ on x_α for $\alpha \in L_2$. Then ν_γ is a Lie algebra automorphism of $\mathfrak{g}$ preserving $\langle\cdot,\cdot\rangle$, and $\nu_\gamma^{mM} = 1$ on $\mathfrak{g}$. For $p \in \mathbf{Z}/mM\mathbf{Z}$ and $\alpha \in L$, define $\eta_\gamma(p, \alpha) = \omega_\gamma^{-pmM\langle\alpha,\gamma\rangle}\eta(p, \alpha)$. Then the obvious analogue of [9.1] holds for η_γ. Under the assumptions of *Theorem 8.2*, we find that $[Y^{\gamma,M}(\alpha, \zeta_1), Y^{\gamma,M}(\beta, \zeta_2)]$ equals the expression in the right-hand side in that theorem, with m replaced by mM; $I(n)$, by its analogue for $\mathbf{Z}/mM\mathbf{Z}$; η, by η_γ; Y, by $Y^{\gamma,M}$; ω, by ω_γ; and $\beta(\zeta_2)$, by $\beta(\zeta_2^M)_- + \langle\beta, \gamma\rangle$.

Consider the twisted affine algebra $\hat{\mathfrak{h}}(\nu_\gamma)$, where we take the scalar e of section 2 of ref. 7 to be $(mM)^{-1}$. Then the $\hat{\mathfrak{h}}(\nu)$-module V_T becomes an $\hat{\mathfrak{h}}(\nu_\gamma)$-module with $c \mapsto 1$, when we use the Lie algebra isomorphism $\tilde{\mathfrak{h}}(\nu_\gamma) \to \tilde{\mathfrak{h}}(\nu)$ taking $x \otimes t^{Mn} \mapsto x \otimes t^n$ ($n \in \mathbf{Z}$, $x \in \mathfrak{h}_{(Mn)}$), $c \mapsto c$ and $d \mapsto Md$. Using theorem 2.6 of ref. 7 with $E = \mathrm{End}\, V_T$, $X'(\alpha, \zeta) = Y^{\gamma,M}(\alpha, \zeta)$ for $\alpha \in L_2$, $c \mapsto 1$ and $e = (mM)^{-1}$, we have the following generalization of the results of *Section 9*.

THEOREM 10.1. *The representation of* $\hat{\mathfrak{h}}(\nu_\gamma)$ *on* V_T *extends uniquely to a representation of the twisted affine algebra* $\tilde{\mathfrak{g}}(\nu_\gamma)$ *on* V_T *such that* $\sum_{n\in\mathbf{Z}}((x_\alpha)_{(n)} \otimes t^n)\zeta^n \mapsto Y^{\gamma,M}(\alpha, \zeta)$ *for* $\alpha \in \hat{L}_2$ ($(x_\alpha)_{(n)}$ *defined with respect to* ν_γ), *and* V_T *is* $\tilde{\mathfrak{g}}(\nu_\gamma)$-*irreducible if and only if* T *is* $\hat{N}$-*irreducible.*

Remark: The $\tilde{\mathfrak{g}}(\nu_\gamma)$-modules V_T in the irreducible case are basic modules (level 1 standard modules for affine algebras of types A, D, and E). For distinct pairs ν, γ, these algebras and modules can be isomorphic, and it can be very interesting to examine such isomorphisms, as in refs. 11 and 12. In particular, one gets nontrivial character identities, using [7.5]. In this connection, it is often convenient to replace t by $t^{1/mM}$ in the definition of the twisted affine algebra and ζ by $\zeta^{1/mM}$ in the definition of the vertex operator, as in formula 4 and theorem 4 of ref. 11, for example.

Note Added in Proof. *Theorem 8.1* has the following complete and surprisingly concise generalization. In the field of formal Laurent series $\sum_{n>N} a_n z^n$ ($a_n \in F$) in a new variable z, set $y = 1 - (1 - z)^{1/m}$. Then for all $a, b \in \hat{L}$, $[Y(a, \zeta_1), Y(b, \zeta_2)]$ is the coefficient of z^{-1} in

$$m^{-1} \sum_{r\in\mathbf{Z}/m\mathbf{Z}} Y(\hat{\nu}^r a, \zeta_2(1 - y))Y(b, \zeta_2)(1 - y)^{-D-m}\delta(\omega^{-r}\zeta_1/\zeta_2).$$

Here the product of vertex operators is to be put into "normal ordered" form using [3.4] and then expanded in powers of y and hence z. Lie algebras generated by the components of given vertex operators can be studied by iterating this formula.

This paper is dedicated to the memory of my friend and collaborator Gerald Wilbur McCollum (April 4, 1944–August 18, 1984). I thank George Chapline, Igor Frenkel, Arne Meurman, Cumrun Vafa, and Edward Witten for stimulating conversations. This research was partially supported by the National Science Foundation through the Mathematical Sciences Research Institute, the Institute for Advanced Study, and Grant MCS83-01664 and by Rutgers University through the Faculty Academic Study Program.

1. Lepowsky, J. & Wilson, R. L. (1978) *Commun. Math. Phys.* **62**, 43–53.
2. Kac, V. G., Kazhdan, D. A., Lepowsky, J. & Wilson, R. L. (1981) *Adv. Math.* **42**, 83–112.
3. Frenkel, I. B. & Kac, V. G. (1980) *Invent. Math.* **62**, 23–66.
4. Segal, G. (1981) *Commun. Math. Phys.* **80**, 301–342.
5. Frenkel, I. B. (1981) *J. Funct. Anal.* **44**, 259–327.
6. Frenkel, I. B., Lepowsky, J. & Meurman, A. (1985) *Contemp. Math.* **45**, 99–120.
7. Lepowsky, J. & Wilson, R. L. (1984) *Invent. Math.* **77**, 199–290.
8. Figueiredo, L. (1985) Dissertation (Rutgers Univ., New Brunswick, NJ).
9. Gross, D. J., Harvey, J. A., Martinec, E. & Rohm, R. (1985) *Nucl. Phys. B* **256**, 253–284.
10. Lepowsky, J. (1985) in *Vertex Operators in Mathematics and Physics*, Mathematical Sciences Research Institute Publ. No. 3, eds. Lepowsky, J., Mandelstam, S. & Singer, I. M. (Springer, New York), pp. 1–13.
11. Frenkel, I. B., Lepowsky, J. & Meurman, A. (1984) *Proc. Natl. Acad. Sci. USA* **81**, 3256–3260.
12. Frenkel, I. B., Lepowsky, J. & Meurman, A. (1985) in *Vertex Operators in Mathematics and Physics*, Mathematical Sciences Research Institute Publ. No. 3, eds. Lepowsky, J., Mandelstam, S. & Singer, I. M. (Springer, New York), pp. 231–273.

CHAPTER 3

FERMIONS

Reprinted Papers

7. V.G. Kac and D.H. Peterson, "Spin and Wedge Representations of Infinite-Dimensional Lie Algebras and Groups", Proc. Natl. Acad. Sci. USA **78** (1981) 3308- 3312.

8. I.B. Frenkel, "Two Constructions of Affine Lie Representations and Boson-Fermion Correspondence in Quantum Field Theory", J. Funct. Anal. **44** (1981) 259–327.

The simplest way to construct an explicit representation of an affine Kac-Moody algebra in terms of operators is to use bilinears in fermion fields rather than bosonic vertex operators. This is the way in which they first (unwittingly) appeared in the physics literature. Such an algebra results if we consider a two-dimensional current algebra with space compactified to a circle. (For a review of current algebra see Ref.[1].) Such periodic fermion fields were introduced into string theory, or rather the dual resonance models that preceded it, by K. Bardacki and M.B. Halpern [2], who considered the current algebra construction, and by P. Ramond [3].

The use in the mathematical literature of fermion fields to construct Kac-Moody algebras begins with the papers of V.G. Kac and D.H. Peterson [*Reprinted Paper* #7] and I.B. Frenkel [*Reprinted Paper* #8]. The equivalence of the fermionic current algebra and bosonic vertex operator constructions, in suitable circumstances, is provided by the fermion-boson equivalence, which originated in the work of T.H.R. Skyrme [4] and was subsequently developed by R. Streater and I.F. Wilde [5], S. Coleman [6], S. Mandelstam [7] and others.

The vertex operator construction of representations of non-simply-laced algebras, referred to in chapter 2, leads to the introduction of "interacting" fermi fields with properties related to the structure of the division algebras [8]. Another generalisation of the idea of free fermion fields are the parafermion fields introduced by V.A. Fateev and A.B. Zamolodchikov. (See [9] and references therein.)

References

1. S. Adler and R. Dashen, *Current Algebras and Applications to Particle Physics* (Benjamin, New York, 1968).

2. K. Bardacki and M.B. Halpern, "New dual quark models", Phys. Rev. **D3** (1971) 2493–2506.

3. P. Ramond, "Dual theory for free fermions", Phys. Rev. **D3** (1971) 2415–2418.

4. T.H.R. Skyrme, "Particle states of a quantized meson field", Proc. Roy. Soc. **A262** (1961) 237–245.

5. R. Streater and I.F. Wilde, "Fermion states of a boson field", Nucl. Phys. **B24** (1970) 561–575.

6. S. Coleman, "Quantum sine-Gordon equation as the massive Thirring model", Phys. Rev. **D11** (1975) 2088–2097.

7. S. Mandelstam, "Soliton operators for the quantized sine-Gordon equation", Phys. Rev. **D11** (1975) 3026–3030.

8. P. Goddard, W. Nahm, D. Olive, H. Ruegg and A. Schwimmer, "Fermions and octonions", *to appear in* Commun. Math. Phys. (1987).

9. V.A. Fateev and A.B. Zamoldchikov, "Non-local (parafermion) currents in two-dimensional conformal quantum field theory and self-dual critical points in Z_N-symmetric statistical systems", Sov. Phys. JETP **62** (1985) 215–225; "Disorder fields in two-dimensional quantum field theory and $N = 2$ extended supersymmetry", Sov. Phys. JETP **63** (1986) 913–919; 'Conformal quantum field theory models in two dimensions having Z_3 symmetry", Nucl. Phys. **B280 [FS18]** (1987) 644–660.

Proc. Natl. Acad. Sci. USA
Vol. 78, No. 6, pp. 3308–3312, June 1981
Mathematics

Spin and wedge representations of infinite-dimensional Lie algebras and groups

(Clifford algebra/spin representation/affine Kac-Moody Lie algebra/highest weight representation/line bundle)

VICTOR G. KAC* AND DALE H. PETERSON†

*Massachusetts Institute of Technology, Cambridge, Massachusetts 02139; and †University of Michigan, Ann Arbor, Michigan 48109

Communicated by Bertram Kostant, November 24, 1980

ABSTRACT We suggest a purely algebraic construction of the spin representation of an infinite-dimensional orthogonal Lie algebra (sections 1 and 2) and a corresponding group (section 4). From this we deduce a construction of all level-one highest-weight representations of orthogonal affine Lie algebras in terms of creation and annihilation operators on an infinite-dimensional Grassmann algebra (section 3). We also give a similar construction of the level-one representations of the general linear affine Lie algebra in an infinite-dimensional "wedge space." Along these lines we construct the corresponding representations of the universal central extension of the group $SL_\infty(k[t,t^{-1}])$ in spaces of sections of line bundles over infinite-dimensional homogeneous spaces (section 5).

1. Let V be a vector space over a field k with $2 \neq 0$ (we do *not* assume that $\dim V < \infty$) and ϕ be a nondegenerate k-valued symmetric bilinear form on V. Define Lie algebras (with the usual bracket):

$$o(V;\phi) = \{a \in \mathrm{End}_k V \mid \phi(ax,y) + \phi(x,ay) = 0,\ x,y \in V\},$$

$$o_{\mathrm{fin}}(V;\phi) = \{a \in o(V;\phi) \mid \dim a(V) < \infty\}.$$

Suppose U is an *isotropic subspace of* V satisfying

$$U^{\perp\perp} = U;\ \dim U^{\perp}/U \le 1. \qquad [1]$$

Introduce a Lie subalgebra of $o(V;\phi)$:

$$o(V,U;\phi) = \{a \in o(V;\phi) \mid \dim\,[U + a(U)]/U < \infty\}.$$

We shall construct the projective spin representation $\sigma_{V,U}$ of the Lie algebra $o(V,U;\phi)$.

Recall (ref. 2) that the *Clifford algebra* $C\ell V$ associated to (V,ϕ) is an associative k-algebra with unit 1, defined as the quotient of the tensor algebra over V by the two-sided ideal generated by elements of the form $x \otimes y + y \otimes x - \phi(x,y)1$, $x,y \in V$. We identify V with a subspace of $C\ell V$. If $a \in o(V;\phi)$, the action of a on V extends uniquely to a derivation $\pi(a)$ of $C\ell V$, and π is a representation of $o(V;\phi)$ on $C\ell V$.

Let $C\ell_2 V$ be the linear span in $C\ell V$ of elements of the form $[x,y]: = xy - yx$ for $x,y \in V$. Then $C\ell_2 V$ is a Lie subalgebra of $C\ell V$. If $a \in C\ell_2 V$, $x \in V$, then $[a,x] \in V$, and $x \mapsto [a,x]$ lies in $o_{\mathrm{fin}}(V;\phi)$, allowing us to identify $C\ell_2 V$ and $o_{\mathrm{fin}}(V)$. For $a \in C\ell_2 V$, $x \in C\ell V$, one has $\pi(a)x = [a,x]$. Let $(C\ell V)U$ be the left ideal of $C\ell V$ generated by U, and set $s(V,U) = C\ell V/(C\ell V)U$. For $a \in o(V,U;\phi)$, there exists $\hat{a} \in C\ell_2 V$ such that

$$\pi(a - \hat{a})U \subset U. \qquad [2]$$

Furthermore, $\hat{a}$ is defined modulo $(C\ell V)U + k$ by Eq. 2. This allows us to define the projective *spin representation* $\sigma_{V,U}$ of $o(V,U;\phi)$ on the space $s(V,U)$ by

$$\sigma_{V,U}(a)[x + (C\ell V)U] = \pi(a)x + x\hat{a} + (C\ell V)U.$$

Note that if $a \in o_{\mathrm{fin}}(V) = C\ell_2 V$, then $\pi(a)x = ax - xa$, and we may take $\hat{a} = a$, so that we obtain the usual definition (ref. 2): $\sigma_{V,U}(a)[x + (C\ell V)U] = ax + (C\ell V)U$. Hence, on $o_{\mathrm{fin}}(V)$, $\sigma_{V,U}$ is the usual spin representation.

Let α be the involutive automorphism of $C\ell V$ defined by $\alpha(x) = -x$, $x \in V$ and $C\ell V = C\ell^+ V \oplus C\ell^- V$ be the corresponding eigenspace decomposition. This induces the decomposition $s(V,U) = s^+(V,U) \oplus s^-(V,U)$ into a direct sum of irreducible *half-spin representations*, $\sigma^+_{V,U}$ and $\sigma^-_{V,U}$.

As soon as the choice $a \mapsto \hat{a}$ is made, we obtain a k-valued two-cocycle γ of the Lie algebra $o(V,U;\phi)$:

$$\gamma(a,b)I_{s(V,U)} = [\sigma_{V,U}(a), \sigma_{V,U}(b)] - \sigma_{V,U}([a,b]).$$

To make this choice, suppose that U' is a subspace of U with $\dim U/U' < \infty$ and that U'' is a subspace of V such that we have (nonorthogonal) direct sum decompositions $V = U' \oplus U''$, $V = U'^{\perp} \oplus U''^{\perp}$. Let $p': V \to U'$, $p'': V \to U''$, $p'^{\perp}: V \to U'^{\perp}$, $p''^{\perp}: V \to U''^{\perp}$ be the associated projections. Then for $a \in o(V,U;\phi)$, we choose

$$\hat{a} = p''ap'^{\perp}. \qquad [3]$$

Then we have

$$\gamma(a,b) = \frac{1}{2}\,\mathrm{trace}\ p'(ap''b - bp''a)p'^{\perp}. \qquad [4]$$

Now suppose in addition that $U''^{\perp}$ is isotropic and that there exists a (indexed) subset $\{u_i : i \in I\} \subset U'$ and a basis $\{u^i : i \in I\}$ of $U''^{\perp}$ such that $\phi(u_i, u^j) = \delta_{ij}$. Let p_0 be a projection of V along $U' + U''^{\perp}$ onto the finite-dimensional subspace $U'^{\perp} \cap U''$ of V. Then for any $a \in o(V,U;\phi)$, there exists a set $\{a_j : j \in J\} \subset C\ell V$ such that for any $x \in C\ell V$, there exists a finite set $J(x) \subset J$ such that $a_j x \in (C\ell V)U$ for $j \notin J(x)$, and $\sigma(a)[x + (C\ell V)U] = \Sigma_{j \in J(x)} a_j x + (C\ell V)U$. In this case, we write $\sigma(a) = \Sigma_{j \in J} a_j$. Then for $a \in o(V,U;\phi)$, we have

$$\sigma_{V,U}(a) = \frac{1}{2}(ap_0 + p_0 a) + \frac{1}{2}\sum_{i \in I} (au^i)u_i - u^i(au_i). \qquad [5]$$

2. Let $k[t,t^{-1}]$ be the ring of finite Laurent series over k. For $P = \Sigma_{i \in \mathbf{Z}} c_i t^i \in k[t,t^{-1}]$, set Res $P = c_{-1}$. Let V be a finite dimensional vector space over k, and let $\tilde{V} = k[t,t^{-1}] \otimes_k V = \oplus_{s \in \mathbf{Z}}(t^s \otimes V)$ be the associated *loop space*, regarded as a vector space over k.

Call $A \in \mathrm{End}_k \tilde{V}$ homogeneous of degree m if $A(t^s \otimes V) \subset t^{s+m} \otimes V$ for $s \in \mathbf{Z}$. In this case, we assign to A the sequence $A^{(i)} \in \mathrm{End}_k V$, $i \in \mathbf{Z}$, defined by $A(t^i \otimes x) = t^{i+m} \otimes A^{(i)}(x)$, $x \in V$. Let $g\ell_*(V)$ be the Lie subalgebra of $\mathrm{End}_k \tilde{V}$ spanned by the homogeneous $A \in \mathrm{End}_k \tilde{V}$.

The publication costs of this article were defrayed in part by page charge payment. This article must therefore be hereby marked "*advertisement*" in accordance with 18 U. S. C. §1734 solely to indicate this fact.

Fix $r \in \mathbf{Z}$, and set $r' = [r/2]$. Let ϕ be a nondegenerate k-valued symmetric bilinear form on V. Define a nondegenerate k-valued symmetric bilinear form $\tilde{\phi}_r$ on $\tilde{V}$ by

$$\tilde{\phi}_r(P \otimes x, Q \otimes y) = \operatorname{Res} t^r PQ\phi(x,y),\ P,Q \in k[t,t^{-1}],\ x,y \in V.$$

Set $o_*(V;r) = gl_*(V) \cap o(\tilde{V};\tilde{\phi}_r)$. For $s \in \mathbf{Z}$, define a decomposition

$$\tilde{V} = \tilde{V}_s \oplus \tilde{V}'_s,\ \tilde{V}_s = \bigoplus_{i \geq s} (t^i \otimes V),\ \tilde{V}'_s = \bigoplus_{i < s} (t^i \otimes V). \qquad [6]$$

If r is even, set $\tilde{U}_r = \tilde{V}_{-r'}$; if r is odd, suppose U is an isotropic subspace of V satisfying Eq. 1 and set $\tilde{U}_r = \tilde{V}_{-r'} \oplus (t_{-r'-1} \otimes U)$. Then $\tilde{U}_r$ is a $\tilde{\phi}_r$-isotropic subspace of $\tilde{V}$ satisfying Eq. 1, and $o_*(V;r) \subset o(\tilde{V},\tilde{U}_r;\tilde{\phi}_r)$. Therefore, we may define the projective *spin representation* σ_r of $o_*(V;r)$ on $s(\tilde{V},\tilde{U}_r)$ to be the restriction of $\sigma_{\tilde{V},\tilde{U}_r}$ to $o_*(V;r)$.

Set $\tilde{U}'_r = \tilde{V}_{-r'}$, $\tilde{U}''_r = \tilde{V}'_{-r'}$, and define $\tilde{a}$ by Eq. 3 for $a \in o(\tilde{V},\tilde{U}_r;\tilde{\phi}_r)$. Then the cocycle γ is given on $o_*(V;r)$ by Eq. 4 as follows. If A and B are homogeneous of degrees m and s, respectively, then $\gamma(A,B) = 0$ unless $m + s = 0$, and if $m \geq 0$,

$$\gamma(A,B) = \frac{1}{2}\delta_{m,-s}\operatorname{trace}_V \sum_{j=0}^{m-1} A^{(j-r'-m)}B^{(j-r')}. \qquad [7]$$

Fix a basis $u_1,\ldots,u_n$ of V, and let $u^1,\ldots,u^n$ be the dual basis of V with respect to ϕ. For $a \in \operatorname{End}_k V$, let (a_{ij}), $c_{ij} = \phi(u^i, au_j)$, be the matrix of a. Then, if $A \in o_*(V;r)$ is homogeneous of degree m, Eq. 5 gives the following expression for $\sigma_r(A)$ as a pointwise convergent series of left multiplication operators on $s(\tilde{V},\tilde{U}_r)$:

$$\sigma_r(A) = -\frac{1}{2}\sum_{s \in \mathbf{Z}}\sum_{i,j=1}^{n} A_{ji}^{(s)}(t^{-s-r-1} \otimes u^i)(t^{s-m} \otimes u_j), \quad \text{if } m \neq 0;$$

$$\sigma_r(A) = -\sum_{s \geq -r'}\sum_{i,j=1}^{n} A_{ji}^{(s)}(t^{-s-r-1} \otimes u^i)(t^s \otimes u_j)$$

$$-\frac{1}{2}\delta_{r,-2r'-1}\sum_{i,j=1}^{n} A_{ji}^{(-r'-1)}(t^{-r'-1} \otimes u^i)(t^{-r'-1} \otimes u_j), \qquad [8]$$

if $m = 0$.

3a. We assume in section 3 that k is algebraically closed of characteristic 0.

Given a Lie algebra $\mathfrak{g}$ over k, we define the associated *loop algebra* $\tilde{\mathfrak{g}} = k[t,t^{-1}] \otimes_k \mathfrak{g}$ with the obvious bracket, regarded as a Lie algebra over k. Given a finite-dimensional representation ν: $\mathfrak{g} \to \operatorname{End}_k V$, we define in the obvious way a representation $\tilde{\nu}$: $\tilde{\mathfrak{g}} \to \operatorname{End}_k \tilde{V}$. In particular, the orthogonal loop algebra $\tilde{o}(V;\phi) = \oplus_{s \in \mathbf{Z}}(t^s \otimes o(V;\phi))$ acts on $\tilde{V}$, and because $\tilde{o}(V;\phi) \subset o_*(V,r)$, we may restrict σ_r to $\tilde{o}(V;\phi)$. We obtain a projective representation that becomes a linear representation of the central extension $\hat{o}(V;\phi)$ of $\tilde{o}(V;\phi)$ defined below.

Let $\mathfrak{g}$ be a finite-dimensional simple Lie algebra over k, $\mathfrak{h}$ a Cartan subalgebra of $\mathfrak{g}$, Δ the set of roots of $\mathfrak{g}$ in $\mathfrak{h}$, W the Weyl group of Δ, Δ_+ a set of positive roots, $\mathfrak{n}_+$ the corresponding maximal nilpotent subalgebra of $\mathfrak{g}$, $\Pi = \{\alpha_1,\ldots,\alpha_\ell\}$ the set of simple roots, θ the highest root, and $(\ ,\ )$ a nondegenerate invariant symmetric bilinear form on $\mathfrak{g}$ (and, hence, on $\mathfrak{g}^*$ and $\mathfrak{h}^*$). For $\alpha \in \mathfrak{h}^*$ with $(\alpha,\alpha) \neq 0$, define $H_\alpha \in \mathfrak{h}$ by $\beta(H_\alpha) = 2(\beta,\alpha)/(\alpha,\alpha)$ for $\beta \in \mathfrak{h}^*$. The *affine Lie algebra* $\hat{\mathfrak{g}}$ associated to $\mathfrak{g}$ is an extension $\hat{\mathfrak{g}} = \tilde{\mathfrak{g}} \oplus kc$ of the loop algebra $\tilde{\mathfrak{g}}$ by a one-dimensional center kc, with bracket

$$[A,B]_{\hat{\mathfrak{g}}} = [A,B]_{\tilde{\mathfrak{g}}} + \frac{1}{2}(\theta,\theta)\operatorname{Res}\left(\frac{dA}{dt},B\right)c, \quad \text{for } A,B \in \tilde{\mathfrak{g}}. \qquad [9]$$

Let $\hat{\mathfrak{h}} = \mathfrak{h} \oplus kc$, with basis $\{h_0 := c - H_\theta, h_i := H_{\alpha_i}, 1 \leq i \leq \ell\}$, called the set of dual simple roots. Define reflections $r_i \in GL(\hat{\mathfrak{h}}^*)$, $0 \leq i \leq \ell$, by $r_i(\lambda) = \lambda - \lambda(h_i)\alpha_i$, and let $\hat{W} \subset GL(\hat{\mathfrak{h}}^*)$ be the group generated by the r_i. We regard W as the subgroup of $\hat{W}$ generated by $r_1,\ldots,r_\ell$. Set $\hat{\mathfrak{n}}_+ = \mathfrak{n}_+ \oplus (tk[t] \otimes \mathfrak{g})$. Finally, define derivations d_s, $s \in \mathbf{Z}$, by $[d_s,c] = 0$, $[d_s,A] = t^{s+1}\,dA/dt$ for $A \in \tilde{\mathfrak{g}}$.

Define fundamental weights $\Lambda_i \in \hat{\mathfrak{h}}^*$ by $\Lambda_i(h_j) = \delta_{ij}$, and set $\rho = \Sigma_{i=0}^{\ell}\Lambda_i$. Let $\hat{P}_+$ be the semigroup generated by the Λ_i. For $\Lambda \in \hat{P}_+$, there exists a unique irreducible $\hat{\mathfrak{g}}$-module $L(\Lambda)$, called the *highest weight module* (ref. 3), admitting a nonzero $v \in L(\Lambda)$ such that $\hat{\mathfrak{n}}_+(v) = 0$ and $h(v) = \Lambda(h)v$ for $h \in \hat{\mathfrak{h}}$. $\Lambda(c)$ is called the *level* of the module $L(\Lambda)$; it is a nonnegative integer and is 0 if and only if $L(\Lambda)$ is the trivial one-dimensional module.

Let V be a vector space over k with basis $u_1,\ldots,u_n$, $n \geq 3$, and let ϕ be the symmetric bilinear form on V defined by $\phi(u_i,u_{n-j+1}) = \delta_{ij}$. Then $\mathfrak{g} = o(V;\phi)$ is identified with the Lie algebra of $n \times n$ matrices skew symmetric with respect to the side diagonal. Let $\mathfrak{h}$ be the set of diagonal matrices in $\mathfrak{g}$ and $\mathfrak{n}_+$ be the set of strictly upper triangular matrices in $\mathfrak{g}$. Set $\ell = [n/2]$, and let U be the $\mathfrak{n}_+$-stable maximal isotropic subspace $ku_1 \oplus \ldots \oplus ku_\ell$ of V.

Let E_{ij} be the standard basis elements of the space of $n \times n$ matrices. Set $(A,B) = 1/2$ trace AB. Then setting $s_i = E_{ii} - E_{n-i+1,n-i+1}$ for $1 \leq i \leq \ell$, the dual simple roots of $\hat{o}(V;\phi)$ are $h_0 := c - s_1 - s_2$; $h_i = s_i - s_{i+1}, 1 \leq i \leq \ell-1$; $h_\ell = s_\ell + s_{n-\ell-1}$. [If $n = 4$, we define $\hat{o}(V;\phi)$ by Eq. 9, where $(\theta,\theta) = 2$, even though $o(V;\phi)$ is not simple.] Then the highest weight modules of $\hat{o}(V;\phi)$ of level one are the $L(\Lambda_i)$, where $i = 0,1,\ell-1,\ell$ if n is even, and $i = 0,1,\ell$ if n is odd. These will be realized in section 3b as half-spin modules.

3b. Let $n \in \mathbf{Z}$, $n \geq 3$, and let $r \in \mathbf{Z}$, $r' = [r/2]$. Set $J_- = \{(i,-s) \mid 1 \leq i \leq n,\ s \in (r+1)/2 + \mathbf{Z};\ s > 0 \text{ or } s = 0, \text{ and } i \geq n - \ell + 1\}$, $J_+ = \{(n-i+1,s) \mid (i,-s) \in J_-\}$, $J = J_- \cup J_+$.

For $(i,s) \in J$, set $\xi_{i,s} = t^{s-\frac{r+1}{2}} \otimes u_i$ if n is even; $\xi_{i,s} = \sqrt{-2}\,(t^{s-\frac{r+1}{2}} \otimes u_i)(t^{-\frac{r+1}{2}} \otimes u_{\ell+1})$ if n is odd. Then $\{\xi_{i,s}, \xi_{i',s'}\} = \delta_{i+i',n+1}\delta_{s,-s'}1$ where $\{a,b\} = ab + ba$. Let $X_{r,n}$ be the (Grassmann) subalgebra of $C\ell\tilde{V}$ generated by the anticommuting elements $\xi_{i,-s}$, $(i,-s) \in J_-$. Set $X^{\pm}_{r,n} = X_{r,n} \cap C\ell^{\pm}\tilde{V}$. For $(i,-s) \in J_-$, define the *creation operator* $\xi^+_{i,-s}$, and the *annihilation operator* $\xi^-_{i,-s}$ by: $\xi^+_{i,-s}(x) = \xi_{i,-s}\,x$, $\xi^-_{i,-s}(1) = 0$, $\xi^-_{i,-s}(\xi_{i',-s'}x) = \delta_{i,i'}\delta_{s,s'}\,x - \xi_{i',-s'}\xi^-_{i,-s}(x)$ for $x \in X_{r,n}$, $(i',-s') \in J_-$.

If nr is even (respectively nr is odd), then $C\ell\tilde{V} = (C\ell\tilde{V})\tilde{U}_r \oplus X_{r,n}$ (respectively $C\ell^+\tilde{V} = (C\ell^-\tilde{V})\tilde{U}_r \oplus X_{r,n}$). Therefore, we may identify $X^{\pm}_{r,n}$ with $s^{\pm}_{\tilde{V},\tilde{U}_r}$ (respectively $X_{r,n} = X^+_{r,n}$ with $s^+_{\tilde{V},\tilde{U}_r}$). Then for $(i,s) \in J$, the action of left multiplication by $\xi_{i,s}$ on $s_{\tilde{V},\tilde{U}_r}$ (respectively $s^+_{\tilde{V},\tilde{U}_r}$) is given on $X_{r,n}$ by the operator $\xi_i(s)$, where for $(i,-s) \in J_-$, $\xi_i(-s) = \xi^+_{i,-s}$, $\xi_{n-i+1}(s) = \xi^-_{i,-s}$.

The following theorems describe the action of $\hat{o}(V;\phi)$ on the half-spin modules $X^{\pm}_{r,n}$.

THEOREM 1. *Let* n *and* r *be integers with* nr *even and* $n \geq 3$. *Then (i) the following formulas define a linear representation* $\hat{\sigma}_r$ *of* $\hat{o}(V;\phi)$ *on* $X_{r,n}$, *which is a linearization of the projective spin representation* σ_r:

$$\hat{\sigma}_r(c) = I; \quad \textit{for } m \neq 0,$$

$$\hat{\sigma}_r(t^m \otimes a) = \frac{1}{2}\sum_{s \in \mathbf{Z}}\sum_{i,j} a_{ij}\xi_i\left(-s - \frac{r+1}{2}\right)\xi_{n-j+1}\left(s + m + \frac{r+1}{2}\right);$$

$$\hat{\sigma}_r(1 \otimes a) = \sum_{s > -r'}\sum_{i,j} a_{ij}\xi_i\left(-s - \frac{r+1}{2}\right)\xi_{n-j+1}\left(s + \frac{r+1}{2}\right)$$

$$+ \frac{1}{2}\delta_{r,2r'+1}\sum_{i,j} a_{ij}\xi_i(0)\xi_{n-j+1}(0).$$

(ii) *For* $m \in \mathbf{Z}$, *define operators* D_m *on* $X_{r,n}$ *by*

$$D_m = -\frac{1}{4}\sum_{s\in\mathbf{Z}}\sum_{i}(2s+m+r+1)\xi_i\left(-s-\frac{r+1}{2}\right) \times \xi_{n-i+1}\left(s+m+\frac{r+1}{2}\right) \quad \textit{for } m \neq 0;$$

$$D_0 = -\frac{1}{2}\sum_{s\geq -r'}\sum_{i}(2s+r+1) \times \xi_i\left(-s-\frac{r+1}{2}\right)\xi_{n-i+1}\left(s+\frac{r+1}{2}\right).$$

Then

$$[D_m, D_{m'}] = (m'-m)\,D_{m+m'} + \frac{n}{24}\delta_{m,-m'}\, m(m^2 - 1 + 3\delta_{r,2r'+1})I;$$

$$[D_m, \hat{\sigma}_r(A)] = \hat{\sigma}_r(d_m(A)) \quad \textit{for } A \in \hat{o}(V;\phi).$$

(iii) *As* $\hat{o}(V;\phi)$*-modules*, $X^{+}_{r,n} \cong L(\Lambda_0)$, $X^{-}_{r,n} \cong L(\Lambda_1)$ *for r even, and* $X^{+}_{r,n} \cong L(\Lambda_\ell)$, $X^{-}_{r,n} \cong L(\Lambda_{\ell-1})$ *for r odd.*

In the statement of *Theorem* 2, Σ' denotes a sum to be extended only over summands not involving $\xi_{\ell+1}(0)$, which is undefined.

THEOREM 2. *Let* n *and* r *be odd integers with* $n \geq 3$. *Then* (a) *The following formulas define a linear representation* $\hat{\sigma}^{+}_r$ *of* $\hat{o}(V;\phi)$ *on* $X_{r,n}$, *which is a linearization of the projective half-spin representation* σ^{+}_r:

$$\hat{\sigma}^{+}_r(c) = 1, \quad \textit{for } m \neq 0,$$

$$\hat{\sigma}^{+}_r(t^m \otimes a) = \frac{1}{2}\sum_{s\in\mathbf{Z}}\sum_{i,j}{}' a_{ij}\xi_i\left(-s-\frac{r+1}{2}\right) \times \xi_{n-j+1}\left(s+m+\frac{r+1}{2}\right) + \frac{1}{\sqrt{-2}}\sum_{i} a_{i,\ell+1}\,\xi_i(m);$$

$$\hat{\sigma}^{+}_r(1 \otimes a) = \sum_{s\geq -r'}\sum_{i,j} a_{ij}\xi_i\left(-s-\frac{r+1}{2}\right)\xi_{n-j+1}\left(s+\frac{r+1}{2}\right) + \frac{1}{2}\sum_{i,j}{}' a_{ij}\xi_i(0)\xi_{n-j+1}(0) + \frac{1}{\sqrt{-2}}\sum_{i}{}' a_{i,\ell+1}\xi_i(0).$$

(b) *For* $m \in \mathbf{Z}$, *define an operator* D_m *on* $X_{r,n}$ *by*:

$$D_m = -\frac{1}{4}\sum_{s\in\mathbf{Z}}\sum_{i}{}'(2s+m+r+1)\xi_i\left(-s-\frac{r+1}{2}\right) \times \xi_{n-i+1}\left(s+m+\frac{r+1}{2}\right) + \frac{m}{2\sqrt{-2}}\xi_{\ell+1}(m)$$

for $m \neq 0$.

$$D_0 = -\frac{1}{2}\sum_{s\geq r'}\sum_{i}(2s+r+1) \times \xi_i\left(-s-\frac{r+1}{2}\right)\xi_{n-i+1}\left(s+\frac{r+1}{2}\right).$$

Then:

$$[D_m, D_{m'}] = (m'-m)D_{m+m'} + \frac{n}{24}\delta_{m,-m'}\, m(m^2+2)I;$$

$$[D_m, \hat{\sigma}^{+}_r(A)] = \hat{\sigma}^{+}_r(d_m(A)) \quad \textit{for } A \in \hat{o}(V;\phi).$$

(c) *As* $\hat{o}(V;\phi)$*-modules*, $s\zeta^{-}_{.(r)} \cong s\zeta^{+}_{.(r)} \cong X_{r,n} \cong L(\Lambda_\ell)$.

These theorems are proved as follows.

Parts *a* follow immediately from Eqs. 7, 8, and 9.

To prove part *b*, consider the Lie algebra $\mathscr{A} := k[t,t^{-1},d/dt] \otimes_k gl(V)$ of differential operators on $\dot{V}$. Let $\mathscr{A}_r = \mathscr{A} \cap o(\dot{V};\dot{\phi}_r)$ be the subalgebra of $\dot{\phi}_r$-skew self-adjoint elements of $\mathscr{A}$, spanned by elements of the form

$$\dot{d}_{m,\ell,a} = \frac{1}{2}t^{\ell+m}\left(\left(\frac{d}{dt}\right)^{\ell} + t^{-(r+\ell+m)}\left(\frac{d}{dt}\right)^{\ell} t^{r-\ell+m}\right) \otimes a,$$

where $\phi(ax,y) + (-1)^{\ell}\phi(x,ay) = 0$ for $x,y \in V$. In particular, $\dot{d}_m := \dot{d}_{m,1,I} = t^{m+1}\,d/dt + 1/2(r+m+1)t^m$ lie in $\mathscr{A}_r$, and $D_m = \sigma_r(\dot{d}_m)$. Define a two-cocycle γ_0 on $\mathscr{A}$ by:

$$\gamma_0\left(t^{\ell+m}\left(\frac{d}{dt}\right)^{\ell} \otimes a,\ t^{\ell'+m'}\left(\frac{d}{dt}\right)^{\ell'} \otimes a'\right) = \frac{1}{2}\delta_{m,-m'}(\mathrm{trace}_V\, aa')(-1)^{\ell}\ell!\ell'!\binom{m+\ell}{\ell+\ell'+1}$$

Then if $r = 0$, the cocycle γ defined by Eq. 7 is the restriction of γ_0 to $\mathscr{A}_0$, and is given by

$$\gamma_0(\dot{d}_{m,\ell,a}, \dot{d}_{m',\ell',a'}) = \frac{1}{4}\delta_{m,-m'}(\mathrm{trace}_V\, aa') \times [(\ell-\ell')! + (-1)^{\ell}\ell!\ell'!]\binom{m+\ell}{\ell+\ell'+1}.$$

This and Eq. 8 suffice to verify part *b* for $r = 0$; the general case is similar.

Finally, part *c* is proved by using the so-called Weyl–Kac character formula for $L(\Lambda)$. One checks the equality of the characters by applying the "principal specialization" Φ to both sides and using formula 3.29 from ref. 3.

3c. For $\mathfrak{m}$, a semisimple Lie algebra, we define the affine Lie algebra $\hat{\mathfrak{m}}$ to be the direct sum of the affine Lie algebras associated to the simple summands of $\mathfrak{m}$. Then the notions of section 3a generalize in an obvious way to $\mathfrak{m}$ and $\hat{\mathfrak{m}}$. Given a homomorphism $\tau: \mathfrak{m} \to o(V;\phi)$, we obtain an induced homomorphism $\hat{\tau}: \hat{\mathfrak{m}} \to \hat{o}(V;\phi)$. Then, using the composition $\hat{\sigma}_{-1} \circ \hat{\tau}$ and some choice of maximal isotropic subspace U of V, we regard $X_V := X_{-1,\dim V}$ as an $\hat{\mathfrak{m}}$-module, called the *spin module of* $\hat{\mathfrak{m}}$ *associated to* τ.

Let $\mathfrak{g}$ be a simple Lie algebra, $\mathfrak{g} = \mathfrak{k} \oplus \mathfrak{p}$ a Cartan decomposition of $\mathfrak{g}$ such that $\mathfrak{k}$ is semisimple and contains a Cartan subalgebra $\mathfrak{h}$ of $\mathfrak{g}$. Then, because $\mathfrak{k}$ preserves the Killing form on $\mathfrak{p}$, we have an inclusion $\mathfrak{k} \subset o(\mathfrak{p})$. Let Δ (respectively $\Delta_{\mathfrak{k}}$) be the sets of roots of $\mathfrak{g}$ (respectively $\mathfrak{k}$) in $\mathfrak{h}$, Δ_+ a set of positive roots, $\Delta_{\mathfrak{k}+} = \Delta_+ \cap \Delta_{\mathfrak{k}}$, $\hat{\rho}$ (respectively $\hat{\rho}_{\mathfrak{k}}$) the sum of the fundamental weights of $\hat{\mathfrak{g}}$ (respectively $\hat{\mathfrak{k}}$), W the Weyl group of $\mathfrak{g}$ in $\mathfrak{h}$, $W^1 = \{w \in W: w(\Delta_+) \supset \Delta_{\mathfrak{k}+}\}$. Choose the maximal isotropic subspace $U = \mathfrak{p}_+ := \mathfrak{p} \cap \mathfrak{n}$, of $\mathfrak{p}$.

PROPOSITION 1. *The spin module* $X_{\mathfrak{p}}$ *of* $\hat{\mathfrak{k}}$ *associated to the inclusion* $\mathfrak{k} \subset o(\mathfrak{p})$ *has the decomposition*

$$X_{\mathfrak{p}} \cong \bigoplus_{w\in W^1} L(w(\hat{\rho}) - \hat{\rho}_{\mathfrak{k}}).$$

The highest weight vectors are pure spinors lying in $C\ell\mathfrak{p}$.

The proof is essentially the same as that of the finite-dimensional analogue in ref. 4.

We shall describe in detail the most beautiful case

$$\mathfrak{g} = sp(V \oplus V') \supset sp(V) \oplus sp(V') = \mathfrak{k},$$

where $\mathfrak{p} \cong V \otimes V'$ as a $\mathfrak{k}$-module. Let $\dim V = 2n$, $\dim V' = 2m$. Recall that a *composition* of n into $m+1$ parts is an $(m+1)$-tuple $(k_0,\ldots,k_m)$ of nonnegative integers with $k_0 + \ldots + k_m = n$; denote by $P_{n,m}$ the set of such compositions. Then, there is a natural bijection between $P_{n,m}$ and the set of all m-element subsets of $\{1,2,\ldots,m+n\}$, so that taking the complementary subset induces a bijection $\pi \mapsto \hat{\pi}$ from $P_{n,m}$ to $P_{m,n}$. We label the fundamental weights of $\hat{sp}(U)$, $\dim U = 2\ell$, according to the diagram $\overset{0}{\circ} \Longrightarrow \overset{1}{\circ} — \dots — \overset{\ell-1}{\circ} \Longleftarrow \overset{\ell}{\circ}$. For $\pi = (k_0,\dots,k_m) \in P_{n,m}$, so that $\hat{\pi} = (k_0',\dots,k_n') \in P_{m,n}$, set

$$\Lambda(\pi) = k_0\Lambda_0 + \dots + k_m\Lambda_m, \Lambda'(\hat{\pi}) = k_0'\Lambda_0' + \dots + k_n'\Lambda_n'.$$

PROPOSITION 2. *The spin module* $X_{V\otimes V'}$ *associated to* $sp(V) \oplus sp(V') \subset o(V \otimes V')$ *has the following* $\hat{sp}(V) \oplus \hat{sp}(V')$*-module decomposition*:

$$X_{V\otimes V'} \cong \bigoplus_{\pi \in P_{n,m}} L(\Lambda(\pi)) \otimes L(\Lambda'(\hat{\pi})).$$

We note that as a special case of *Proposition 2* we obtain the decomposition of the $\hat{sl}_2 \oplus \hat{sp}_{2n}$-module:

$$X_{-1,4n} \cong \bigoplus_{s=0}^{n} L(s\Lambda_0 + (n-s)\Lambda_1) \otimes L(\Lambda_s').$$

Remarks. (*a*) One gets similar decompositions for the spin module associated to $o(V) + o(V') \subset o(V \otimes V')$; (*b*) *Proposition 1* can be generalized to the case of $\mathfrak{f}$ and $\mathfrak{g}$ of equal rank, for instance

$$\mathfrak{g} = sl(V \oplus V') \supset sl(V \oplus V') \cap (gl(V) \oplus gl(V')) = \mathfrak{f}.$$

(*c*) The spin representation associated to the adjoint representation of a simple Lie algebra $\mathfrak{g}$ is decomposed into a direct sum of several copies of the $\hat{\mathfrak{g}}$-module $L(\hat{\rho})$. This holds for an arbitrary Kac–Moody Lie algebra and follows from the obvious product decomposition for the character of $L(\rho)$.

4a. Let the assumptions on k, V, ϕ, U be as in section 1. Set $O(V;\phi) = \{g \in GL(V) | \phi(gx,gy) = \phi(x,y) \text{ for } x,y \in V\}$. For $g \in O(V;\phi)$, extend the action of g on V to an automorphism $\beta \mapsto g \cdot \beta$ of $C\ell V$. Set

$$O(V,U;\phi) = \{g \in O(V;\phi) \mid \dim (U + g(U))/U < \infty\}.$$

Then, one can show that $O(V,U;\phi)$ is a subgroup of $O(V,\phi)$. Define a map det: $O(V,U;\phi) \to \{\pm 1\}$ by

$$\det g = \exp \pi i \dim [(U + (I-g)U^{\perp})/U].$$

One can show that this is a homomorphism. Let $SO(V,U;\phi)$ be the kernel of det. We shall define the projective spin representation of the group $O(V,U;\phi)$.

Let $\mathfrak{M}$ be the set of all isotropic subspaces A of V with $A^{\perp\perp} = A$ and $\dim(A+U)/A = \dim(A+U)/U < \infty$. For $A,B \in \mathfrak{M}$ set

$$\langle A|B\rangle_{\pm 1} = \{\bar{\beta} = \beta + (C\ell V)B \mid \beta \in C\ell^{\pm}V,\ A\beta \subset (C\ell V)B\}.$$

Then one can show (cf. ref. 2) that $\langle A|B\rangle_{\pm 1}$ is a subspace of $C\ell V/(C\ell V)B$ of dimension at most 1. Set

$$\hat{O}(V,U;\phi) = \{(g,\bar{\beta}) \mid g \in O(V,U;\phi),\ \bar{\beta} \in \langle gU|U\rangle_{\det g},\ \bar{\beta} \neq 0\}.$$

This is a group, with multiplication $(g,\bar{\beta})\,(g',\bar{\beta}') = (gg',\overline{(g\cdot\beta')\beta})$; moreover, it is a central extension of $O(V,U;\phi)$ by k^*. Define the *spin representation* $\Sigma_{V,U}$ of the group $\hat{O}(V,U;\phi)$ on the space $s_{V,U}$ by

$$\Sigma_{V,U}((g,\bar{\beta}))[x + (C\ell V)U] = (g \cdot x)\beta + (C\ell V)U.$$

Set $\hat{SO}(V,U;\phi) = \{(g,\bar{\beta}) \in \hat{O}(V,U;\phi) | \det g = 1\}$. Then, it is easy to see that if $U^{\perp} = U$, $s_{V,U}$ is an irreducible $\hat{O}(V,U;\phi)$-module, which decomposes under $\hat{SO}(V,U;\phi)$ into a direct sum $s^+_{V,U} \oplus s^-_{V,U}$ of inequivalent irreducible half-spin modules. If $U^{\perp} \neq U$, then $O(V,U;\phi) = \{\pm I\} \times SO(V,U;\phi)$, and $s_{V,U}$ decomposes under $\hat{SO}(V,U;\phi)$ into a direct sum $s^+_{V,U} \oplus s^-_{V,U}$ of equivalent irreducible half-spin modules.

Remarks. (*a*) Suppose for simplicity that $U^{\perp} = U$. Then for $A,B \in \mathfrak{M}$, the map $u_1 \wedge \dots \wedge u_n \mapsto u_1 \dots u_n + (C\ell V)B$ from $\Lambda^{\max}(A/A \cap B)$ to $\langle A|B\rangle$ is an isomorphism. Let $A_i \in \mathfrak{M}$, $i \in \mathbb{Z}/3\mathbb{Z}$. Then by considering the alternating two-form $\bar{\phi}$ on $A_1 \oplus A_2 \oplus A_3$ given by $\bar{\phi}(x_1 \oplus x_2 \oplus x_3, y_1 \oplus y_2 \oplus y_3) = \Sigma_i \phi(x_i,y_{i+1}) - \phi(y_i,x_{i+1})$, one obtains a canonical element of $\mathrm{Hom}_k(\langle A_1|A_2\rangle \otimes \langle A_2|A_3\rangle, \langle A_1|A_3\rangle)$, which is the product $\bar{\beta} \otimes \bar{\gamma} \mapsto \bar{\beta}\bar{\gamma}$. (*b*) $\hat{O}(V,U;\phi)$ is a double cover of a corresponding subgroup of $GL(V)$, which is constructed in section 5c. (*c*) Results similar to those of section 5 hold for the spin representations of $o(V,U;\phi)$ and $\hat{O}(V,U;\phi)$. (*d*) One can write down the representation $\Sigma_{V,U}$ and the corresponding cocycle in terms of the Fermi integral (cf. ref. 1). (*e*) One can show that $\Sigma_{V,U} \otimes \Sigma_{V,U}$ is isomorphic to $\Lambda_{V,U}$ (see section 5a) restricted to $O(V,U;\phi)$, provided that $U^{\perp} = U$.

4b. Recall the notation of section 2. Let G be an algebraic group over k. The group $\tilde{G} := G(k[t,t^{-1}])$ is called the associated *loop group*. Given a representation $\nu: G \to \mathrm{End}_k V$, one associates the representation $\tilde{\nu}: \tilde{G} \to \mathrm{End}_k \tilde{V}$. In particular, we have the general linear and the orthogonal loop groups $\tilde{GL}(V)$ and $\tilde{O}(V)$. Fix $r \in \mathbb{Z}$. Then $\tilde{O}(V) \in O(\tilde{V},\tilde{U}_r;\tilde{\phi}_r)$, so that we may restrict $\Sigma_{\tilde{V},\tilde{U}_r}$ to $\tilde{O}(V)$, obtaining a projective representation Σ_r of $\tilde{O}(V)$ with differential σ_r.

From *Theorems 1* and *2c* we deduce:

PROPOSITION 3. *The projective representations* $\sigma_r^{\pm}$ *of the group* $SO_n(k[t,t^{-1}])$ *in* $s^{\pm}(\tilde{V},\tilde{U}_r)$ *are irreducible, provided that* char $k = 0$ *and* $n \geq 3$.

5a. Let V be a vector space over a field k (we do not assume that $\dim V < \infty$). Let $\Lambda(V) = \oplus_{k=0}^{\infty} \Lambda^k(V)$ be the exterior algebra over V. If $\dim V = n < \infty$, set $\Lambda^{\max}(V) = \Lambda^n(V)$. For subspaces $A \subset B$ of V with $\dim A < \infty$, we have a canonical inclusion $\Lambda(B/A) \otimes \Lambda^{\max}(A) \subset \Lambda(B)$; if also $\dim B < \infty$, this gives a canonical isomorphism $\Lambda^{\max}(B/A) \otimes \Lambda^{\max}(A) \cong \Lambda^{\max}(B)$.

Two subspaces A,B of V are called commensurable if $\dim(A+B)/(A \cap B) < \infty$; in this case, set $(A|B) = \mathrm{Hom}_k(\Lambda^{\max}(A/A \cap B), \Lambda^{\max}(B/A \cap B))$. For A,B,C, commensurable, $\alpha \in (A|B)$, $\beta \in (B|C)$, define $\alpha\beta \in (A|C)$ by using the canonical isomorphism $(A|B) \otimes (B|C) \cong (A|C)$. Then $(\alpha\beta)\gamma = \alpha(\beta\gamma)$. For A,B, commensurable and $g \in GL(V)$, we have an obvious isomorphism $\alpha \mapsto g \cdot \alpha$ from $(A|B)$ onto $(g(A)|g(B))$. Then $g \cdot (\alpha\beta) = (g \cdot \alpha)(g \cdot \beta)$, $(gg') \cdot \alpha = g \cdot (g' \cdot \alpha)$.

Fix a nonempty family $\mathfrak{M}$ of pairwise commensurable subspaces of V, such that if $A,B \in \mathfrak{M}$, then $A \cap B \in \mathfrak{M}$, and if $A \in \mathfrak{M}$ and $B \supset A$ is a subspace of V with $\dim B/A < \infty$, then $B \in \mathfrak{M}$. Fix $U \in \mathfrak{M}$. Then for $A,B \in \mathfrak{M}$ with $A \subset B$, we have a canonical inclusion $\Lambda(V/B) \otimes (U|B) \subset \Lambda(V/A) \otimes (U|A)$, which allows us to define as an inductive limit the vector space

$$\Lambda(V,U;\mathfrak{M}) = \bigcup_{A \in \mathfrak{M}} \Lambda(V/A) \otimes (U|A).$$

Define a grading $\Lambda(V,U;\mathfrak{M}) = \oplus_{k \in \mathbb{Z}} \Lambda_{(k)}(V,U;\mathfrak{M})$ by the following: if $k \in \mathbb{Z}_+$, $A \in \mathfrak{M}$, $k' = k + \dim A/(A \cap U) - \dim U/(A \cap U)$; then $\Lambda^k(V/A) \otimes (U|A) \subset \Lambda_{(k')}(V,U;\mathfrak{M})$. Then $\Lambda(V,U;\mathfrak{M})$ is a graded $\Lambda(V)$-module.

Let $GL(V;\mathfrak{M})$ be the group of all $g \in GL(V)$ so that g,g^{-1} preserve the family $\mathfrak{M}$. For $U \in \mathfrak{M}$, set $\hat{GL}(V,U;\mathfrak{M}) = \{(\alpha,g) | g \in GL(V;\mathfrak{M}), \alpha \in (U|gU), \alpha \neq 0\}$, which is a group with product $(\alpha,g)(\alpha',g') = (\alpha(g \cdot \alpha'), gg')$. Then $\hat{GL}(V,U;\mathfrak{M})$ is a central extension of $GL(V;\mathfrak{M})$ by k^*.

If $U' \in \mathfrak{M}$, the isomorphism $(U'|U) \otimes \Lambda(V,U;\mathfrak{M}) \cong \Lambda(V,U';\mathfrak{M})$ induces a map $(U'|U) \to \mathrm{Hom}_k(\Lambda(V,U;\mathfrak{M}), \Lambda(V,U';\mathfrak{M}))$, denoted by $\alpha \mapsto \alpha_*$. In particular, this gives a canonical isomorphism $\phi_{U,U'}: \hat{GL}(V,U;\mathfrak{M}) \to \hat{GL}(V,U';\mathfrak{M})$. Finally, for $g \in GL(V;\mathfrak{M})$ we have an obvious isomorphism $g_*: \Lambda(V,U;\mathfrak{M}) \to \Lambda(V,gU;\mathfrak{M})$. Now we can define the *wedge representation* $\Lambda_{V,U}$ of the group $\hat{GL}(V,U;\mathfrak{M})$ on the space $\Lambda(V,U;\mathfrak{M})$ by $\Lambda_{V,U}(\alpha,g) = \alpha_* \circ g_*$.

For $\alpha \in (U'|U)$, $\hat{g} \in \hat{GL}(V,U;\mathfrak{M})$, $\Lambda_{V,U'}(\phi_{U,U'}(\hat{g})) \circ \alpha_* = \alpha_* \circ \Lambda_{V,U}(\hat{g})$. Therefore, the projective wedge representation of $GL(V;\mathfrak{M})$ on $\Lambda(V,U;\mathfrak{M})$ depends only on V and $\mathfrak{M}$.

Define a subgroup of $GL(V;\mathfrak{M})$ (which is independent of the choice of $U \in \mathfrak{M}$) by

$$GL_0(V;\mathfrak{M}) = \{g \in GL(V;\mathfrak{M}) \mid \dim (U + g(U))/U = \dim (U + g^{-1}(U))/U\}.$$

Then the corresponding subgroup $\hat{GL}_0(V,U;\mathfrak{M}) \subset \hat{GL}(V,U;\mathfrak{M})$ consists of the $\hat{g}$ such that $\Lambda_{V,U}(\hat{g})$ is degree-preserving on $\Lambda(V,U;\mathfrak{M})$.

Set $P = \{U' \in \mathfrak{M} | \dim U/(U \cap U') = \dim U'/(U \cap U')\}$, and define a line bundle $\mathfrak{L}$ on P by $\mathfrak{L} = \{(\alpha',U') | U' \in P, \alpha' \in (U|U')\}$. Then $GL_0(V;\mathfrak{M})$ acts transitively on P, and $\hat{GL}_0(V,U;\mathfrak{M})$ acts on $\mathfrak{L}$ by $(\alpha,g) \cdot (\alpha',U') = (\alpha(g \cdot \alpha'), g(U'))$. Let $\mathfrak{L}^* = \{(\alpha',U') | U' \in P, \alpha' \in (U'|U)\}$ be the dual line bundle.

For $A,B \in \mathfrak{M}$, $A \subset B$, set $P_{A,B} = \{U' \in P | A \subset U' \subset B\}$. Then we regard $P_{A,B}$ as a Grassmanian, thus as a smooth algebraic variety. Then the restrictions of $\mathfrak{L},\mathfrak{L}^*$ to $P_{A,B}$ are algebraic varieties. We call a section of $\mathfrak{L}^*$ over P *regular* if its restriction to each $P_{A,B}$ is a regular map, and denote by $H^0(P,\mathfrak{L}^*)$ the space of regular sections of $\mathfrak{L}^*$ over P. Then the obvious map $\mathfrak{L} \to \Lambda_{(0)}(V,U;\mathfrak{M})$ induces a linear map $\Lambda_{(0)}(V,U;\mathfrak{M})^* \to H^0(P,\mathfrak{L}^*)$. It is easy to see that this map is an isomorphism of $\hat{GL}_0(V,U;\mathfrak{M})$-modules.

5b. Let $d\Lambda$ denote the natural representation of the Lie algebra $gl(V)$ in $\Lambda(V)$. Define the Lie algebra $gl(V;\mathfrak{M}) = \{a \in gl(V) |$ for any $A \in \mathfrak{M}$ there exists $B \in \mathfrak{M}$ such that $A \supset B + a(B)\}$. Let $a \in gl(V;\mathfrak{M})$, $A \in \mathfrak{M}$. Choose $B \in \mathfrak{M}$, with $A \supset B + a(B)$, and a decomposition $V = B \oplus B'$. Consider the sequence of maps: $\Lambda(V/A) \otimes (U|A) \subset \Lambda(V/B) \otimes (U|B) \simeq \Lambda(B') \otimes (U|B) \xrightarrow{d\Lambda(a) \otimes 1} \Lambda(V) \otimes (U|B) \to \Lambda(V/B) \otimes (U|B)$. Their composition defines a map $a_*^A : \Lambda(V/A) \otimes (U|A) \to \Lambda(V,U;\mathfrak{M})$. Up to addition of scalars times the inclusion, a_*^A is independent of the choices of B,B'. From this we may define a degree-preserving projective representation $a \mapsto a_* = \lim_{A \in \mathfrak{M}} a_*^A$ of $gl(V;\mathfrak{M})$ on $\Lambda(V,U;\mathfrak{M})$.

To make a definite choice of a_*, choose a direct sum decomposition $V = U \oplus U'$, with associated projections $p: V \to U$, $p' : V \to U'$. Then we require $a_*(\Lambda^0(V/U) \otimes (U|U)) \subset U' \Lambda(V,U;\mathfrak{M})$, where the right-hand side is defined by using the action of $\Lambda(V)$ on $\Lambda(V,U;\mathfrak{M})$. We define the projective wedge representation $d\Lambda_{V,U}$ of $gl(V;\mathfrak{M})$ on $\Lambda(V,U;\mathfrak{M})$ by $d\Lambda_{V,U}(a) = a_*$. One can show that its two-cocycle is given by (cf. Eq. 4): $\gamma(a,b) = \text{trace } p(ap'b - bp'a)p$.

The representations $\Lambda_{V,U}$ and $d\Lambda_{V,U}$ satisfy the following relations. Let $a \in gl(V;\mathfrak{M})$, $g \in GL(V;\mathfrak{M})$, $\hat{g} = (\alpha,g) \in \hat{GL}(V,U;\mathfrak{M})$, and set $\tilde{p} = g^{-1}pg$. Then

$$\Lambda_{V,U}(\hat{g}) d\Lambda_{V,U}(a) \Lambda_{V,U}(\hat{g}^{-1}) = d\Lambda_{V,U}(gag^{-1}) + [\text{trace } (\tilde{p} - p)(a\tilde{p} + pa)]I.$$

5c. Let k be a field of characteristic 0, $V = ku_1 \oplus \ldots \oplus ku_n$, a vector space over k. Recall the notation of section 2.

Regard $\tilde{k}$ as a commutative Lie algebra, so that $\tilde{k} = k[t,t^{-1}]$. Define the affine Lie algebra $\hat{k} = \tilde{k} \oplus kc'$, where c' centralizes $\tilde{k}$ and $[p,q] = (\text{Res } dp/dt\, q)c'$ for $p,q \in \tilde{k}$. Writing $gl(V) = sl(V) \oplus kI$, we set: $\hat{gl}(V) = \hat{sl}(V) \oplus \hat{k}$.

Let $\mathfrak{h}$ be a Cartan subalgebra of $sl(V)$, and set $\hat{\mathfrak{h}} = \mathfrak{h} \oplus kc \oplus kI \oplus kc' \subset \hat{gl}(V)$. Define $\Lambda_0', \Lambda_0'' \in \hat{\mathfrak{h}}^*$ by $\Lambda_0'(\mathfrak{h} \oplus kc) = 0 = \Lambda_0''(\mathfrak{h} \oplus kc)$, $\Lambda_0'(I) = 0 = \Lambda_0''(c')$, $\Lambda_0'(c') = 1 = \Lambda_0''(I)$. Then we have a theory of irreducible highest weight modules $L(\Lambda)$ of $\hat{gl}(V)$ (cf. section 3a).

Let $U = k[t] \otimes_k V \subset \tilde{V}$, and let $\mathfrak{M}_0$ be the set of subspaces A of $\tilde{V}$ such that for some $k \in \mathbf{Z}$, $t^{-k}U \supset A \supset t^kU$. Then it is easy to see that $\hat{gl}(V) \subset gl(V;\mathfrak{M}_0)$.

THEOREM 3. *For* $0 \le i \le n-1$, *the* $\hat{gl}(V)$*-modules* $\Lambda_{(i)}(\tilde{V},U;\mathfrak{M}_0)$ *and* $L(\Lambda_i + n\Lambda_0' + i\Lambda_0'')$ *are isomorphic*.

5d. Now we again turn to the corresponding group. Set $U_0 = k[t^{-1}] \otimes_k V$, and let $\mathfrak{M}_0'$ be the set of subspaces A of $\tilde{V}$ such that for some $k \in \mathbf{Z}$, $t^kU_0 \supset A \supset t^{-k}U_0$. Then $\tilde{SL}(V) \subset GL_0(\tilde{V};\mathfrak{M}_0')$, so that we obtain a central extension $\hat{SL}(V) \subset \hat{GL}_0(V,U_0;\mathfrak{M}_0')$ of $\tilde{SL}(V)$. Then for $j > 0$, there is a unique inclusion $H_j := \{g \in \tilde{SL}(V) | g \equiv I \bmod t^jk[t]\} \subset \hat{SL}(V)$ such that $\Lambda_{\tilde{V},U_0}(a)$ is locally unipotent for $a \in H_j$, $j > 0$; a subgroup of $\hat{SL}(V)$ is called *open* if it contains H_j for some $j > 0$.

Fix $i \in \mathbf{Z}$, $0 \le i \le n-1$. Set

$$U_i = (t^{-1}k[t^{-1}] \otimes_k V) \oplus \left(\bigoplus_{s=i+1}^{n} ku_s \right).$$

Let P_i be the set of subspaces A of $\tilde{V}$ such that $A \supset t^{-1}A$, A and U_i are commensurable, and $\dim A/(A \cap U_i) = \dim U_i/(A \cap U_i)$. Define a filtration $P_i^{(0)} \subset P_i^{(1)} \subset \ldots$ of P_i by

$$P_i^{(s)} = \{A \in P_i | t^{s-1}k[t^{-1}] \otimes V \supset A \supset t^{-s+1}k[t^{-1}] \otimes V\}.$$

We regard the $P_i^{(s)}$ as projective varieties.

Consider the line bundle $\mathfrak{L}_i^* = \{(\alpha,A) | A \in P_i, \alpha \in (A|U_i)\}$ over P_i. For $s \in \mathbf{Z}_+$, we regard $\mathfrak{L}_i^*$ restricted to $P_i^{(s)}$ as an algebraic variety, and call a section of $\mathfrak{L}_i^*$ regular if its restriction to each $P_i^{(s)}$ is a regular map. Then the group $\hat{SL}(V)$ acts on the space $H^0(P_i,\mathfrak{L}_i^*)$ of regular sections of $\mathfrak{L}_i^*$.

THEOREM 4. *The* $\hat{SL}(V)$*-module* $H^0(P_i,\mathfrak{L}_i^*)^0$ *of regular sections with open stabilizer of the line bundle* $\mathfrak{L}_i^*$ *is irreducible. As* $\hat{sl}(V)$*-modules, we have* $H^0(P_i,\mathfrak{L}_i^*)^0 \cong L(\Lambda_i)$.

Note Added in Proof. Recently, Frenkel (5) independently obtained, by a different method, some of the results of section 3b of the present paper. He referred to a paper by Bardakci and Halpern (6) in which they actually construct the restriction of the spin representation of $o(6)$ to $gl(3)$.

We owe to G. Lusztig the idea of the construction of the group $SL_n(k[t,t^{-1}])$, which inspired our work on the wedge representation. We would like to thank him for discussions on this matter. The research has been partially supported by National Science Foundation Grants MCS-8005813 and MCS-800286.

1. Berezin, F. A. (1966) *The Method of Second Quantization* (Academic, New York).
2. Chevalley, C. C. (1954) *The Algebraic Theory of Spinors* (Columbia Univ. Press, New York).
3. Kac, V. G. (1978) *Adv. Math* **30**, 85–136.
4. Parthasarathy, R. (1972) *Ann. Math.* **96**, 1–30.
5. Frenkel, I. B. (1980) *Proc. Natl. Acad. Sci. USA* **77**, 6303–6306.
6. Bardakci, K. & Halpern, M. B. (1971) *Phys. Rev. D* **3**, 2493–2506.

JOURNAL OF FUNCTIONAL ANALYSIS 44, 259–327 (1981)

Two Constructions of Affine Lie Algebra Representations and Boson–Fermion Correspondence in Quantum Field Theory

I. B. FRENKEL

Department of Mathematics, Yale University, New Haven, Connecticut 06520

Communicated by the Editor

We establish an isomorphism between the vertex and spinor representations of affine Lie algebras for types $D_l^{(1)}$ and $D_{l+1}^{(2)}$. We also study decomposition of spinor representations using the infinite family of Casimir operators and prove that they are either irreducible or have two irreducible components. We show that the vertex and spinor constructions of the representations can be reformulated in the language of two-dimensional quantum field theory. In this physical context, the two constructions yield the generalized sine-Gordon and Thirring models, respectively, already in renormalized form. The isomorphism of representations implies an equivalence of these two models which is known in quantum field theory as the boson–fermion correspondence

INTRODUCTION

I. The theory of affine Lie algebras is in many features analogous to the theory of simple finite-dimensional Lie algebras. The classification, the structure and the representations of these two classes of Lie algebras are quite similar, (see e.g. [17]). The role of the finite-dimensional irreducible representations of the simple Lie algebras belongs, in the affine theory, to the so-called standard representations introduced by Kac [16]. These representations, as in the classical theory, can be classified by the sets of nonnegative integral numbers corresponding to the points of the Dynkin diagrams. The representations corresponding to the simplest nontrivial sets (0,..., 0, 1, 0,..., 0)

0022-1236/81/150259-69$02.00/0

are called fundamental representations and play an important role in the theory.

The fundamental representations of simple Lie algebras admit a number of constructions, which depend on the type of the algebra and the point of the Dynkin diagram which determine the representation. Important examples are the spinor representation of the classical orthogonal Lie algebras $\mathfrak{o}(n)$ (see e.g. [5]). The first construction of the simplest affine Lie algebra of the type $A_1^{(1)}$ has been obtained by Lepowsky and Wilson [20], and generalized in [18] for a majority of affine Lie algebras. Two other constructions discovered in [9][1] and [10] (see also [27]), and the relation between them comprise the first part of this paper.

Our construction of affine Lie algebra representations, which we call vertex representations, is rather general and has no analogues in the classical theory. We will consider affine Lie algebras of the types $A_l^{(1)}$, $D_l^{(1)}$, $E_l^{(1)}$ (see [10]) and of the types $A_{2l-1}^{(2)}$, $D_{l+1}^{(2)}$, $E_6^{(2)}$ (see [11]). The vertex representation is constructed in the space $V = S \otimes \mathbb{C}[\Gamma]$, in which S is the space of the canonical representation of the infinite-dimensional Heisenberg subalgebra

$$[h_i(n), h_j(m)] = n\,\delta_{ij}\,\delta_{n,-m}, \qquad n, m \in \mathbb{Z}\backslash 0,\ i, j = 1, \ldots, l, \tag{1}$$

and $\mathbb{C}[\Gamma] = \sum_{\gamma \in \Gamma} \mathbb{C}e^{\gamma}$ is a group algebra of a lattice $\Gamma \subset \mathbb{R}^l$. We define $h_i(0)\,e = \gamma_i e^{\gamma}$, $\gamma \in \Gamma$, $i = 1, \ldots, l$. The representation of the whole affine Lie algebra can be obtained with the help of one universal operator, which is called the vertex operator

$$X(\alpha, z) = \exp\left(\sum_{n=1}^{\infty} \frac{z^n}{n}\,\alpha(-n)\right) \exp(\ln z\alpha(0) + \alpha) \exp\left(-\sum_{n=1}^{\infty} \frac{z^{-n}}{n}\,\alpha(n)\right), \tag{2}$$

$\|\alpha\|^2 = 2$ and α runs through the classical root system $\Delta \subset \mathbb{R}^l$; $\alpha(n) = \sum_{i=1}^{l} \alpha_i h_i(n)$, where α_i are coordinates of α in the orthonormal basis $\{h_i\}_{i=1}^{l}$ of $\mathbb{R}^l$.

The second construction is the generalization of the classical spinor representation and is applicable for the orthogonal affine Lie algebras, i.e., algebras of the types $D_l^{(1)}$, $B_l^{(1)}$, $D_{l+1}^{(2)}$ (see [9]). Following the analogy with the classical case we call these representations spinor representations. Thus the spinor representation for the type $D_l^{(1)}$ is constructed in the space of the canonical representation of the infinite-dimensional Clifford algebra generated by the elements $b_i(n)$, $i = \pm 1, \ldots, \pm l$, $n \in Z$ ($Z = \mathbb{Z}$ or $Z = \mathbb{Z} + \frac{1}{2}$) with the anticommutation relations

$$\{b_i(n), b_j(m)\} = -\delta_{i,-j}\,\delta_{n,-m}. \tag{3}$$

[1] Independent construction of the spinor representation has been obtained by V. G. Kac and D. Peterson [28]

The action of the elements of affine Lie algebra is defined by the quadratic expressions in the elements $b_i(n)$. We prove in this paper that the spinor representations of affine Lie algebras are either irreducible or have two irreducible components, which depends on the form of the Dynkin diagram of the algebra. Our method uses an analysis of the following Lie algebra

$$[D(m), D(n)] = (n-m)\,D(m+n) + (l/12)(m^3 - m)\delta_{m,-n}, \quad m, n \in \mathbb{Z}. \tag{4}$$

This algebra is in fact the central extension of the algebra of vector fields on the circle. Segal noted that every highest-weight representation of an affine Lie algebra can be extended to the semidirect product with the algebra (4). In Section I.1 we introduce the Casimir operators $C(m)$, $m \in \mathbb{Z}$, defined for the invariant form $\langle\ ,\ \rangle_m$ on the affine Lie algebra and we show that the Segal operators constitute the main parts of the Casimir operators. The Casimir operator $C(0)$ had been considered first by Kac in [16]. In the vertex and spinor representations Segal's operator admit another realization which we call Virasoro operators. Thus

$$D(m) = -\tfrac{1}{2} \sum_{i=1}^{l} \sum_{k \in \mathbb{Z}} :h_i(k)\,h_i(m-k): \tag{5}$$

in the vertex representation (see [10]), and

$$D(m) = -\sum_{i=1}^{l} \sum_{k \in \mathbb{Z}} (k - m/2)\,:b_i(k)\,b_{-i}(m-k): \tag{6}$$

in the spinor representation, where dots denote normal ordering (see (I.1.9) and (I.3.7)). We will obain easily that the Virasoro operators together with the operators of the affine Lie algebra generate the whole operator algebra or the direct sum of two operator algebras. It proves that the spinor representation is either irreducible or has two irreducible components and each irreducible component is isomorphic to the fundamental representation. This provides us with an explicit construction of certain fundamental representations: four for the type $D_l^{(1)}$, three for the type $B_l^{(1)}$, two for the type $D_{l+1}^{(2)}$.

In the case of affine orthogonal Lie algebras of the types $D_l^{(1)}$ and $D_{l+1}^{(2)}$, we obtain therefore two constructions of the same fundamental representations corresponding to the endpoints of their Dynkin diagrams.

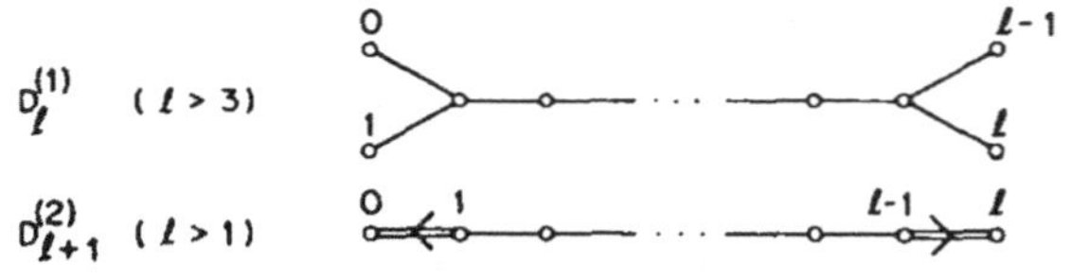

We ascertain the isomorphism between these two constructions in

Section I.4. It turns out that the vertex operators $X(\alpha, z)$, $\alpha \in \mathbb{R}^l$, $z \in \mathbb{C}\backslash 0$ again help to solve the problem. In this case $\|\alpha\|^2 = 1$ and α runs through the orthonormal vectors $\pm h_i$, $i = 1,\dots, l$, of $\mathbb{R}^l$, i.e., through fundamental weights of the natural $2l$-dimensional representation of the Lie algebra $\mathfrak{o}(2l)$. Let us define the generating function of $b_i(n)$, $n \in Z$, by $\mathcal{B}_i(z) = \sum_{n\in Z} b_i(n)\, z^{-n}$, $z \in \mathbb{C}\backslash 0$. Then the isomorphism between the vertex and spinor representations of the affine Lie algebra $\hat{\mathfrak{o}}(2l)$ follows from the correspondence for $i = 1,\dots, l$

$$
\begin{aligned}
X(h_i, z)\,\varepsilon_i &\leftrightarrow \mathcal{B}_i(z),\\
X(-h_i, z)\,\varepsilon_i &\leftrightarrow \mathcal{B}_{-i}(z),
\end{aligned}
\tag{7}
$$

where $\varepsilon_i v \otimes e^\gamma = (-1)^{N_i} v \otimes e^\gamma$, $N_i = \langle h_i, \gamma\rangle$, $v \in S$, $\gamma \in \Gamma$. Under this isomorphism the Virasoro operators, $D(m)$, defined in both constructions coinside for every $m \in Z$. For the generating functions we have the correspondence

$$
-\frac{1}{2}\sum_{i=1}^{l} :(\mathcal{h}_i(z))^2: \leftrightarrow -\sum_{i=1}^{l} :\mathcal{B}_i(z)\left(z\frac{d}{dz}\right)\mathcal{B}_{-i}(z):. \tag{8}
$$

Comparing the character formulas in the two constructions of the representations we obtain Jacobi identities for elliptic θ-functions.

$$
\begin{aligned}
&\prod_{i=1}^{l}\left(\prod_{n=1}^{\infty}(1-q^{2n})^{-1}\sum_{m\in Z} e^{mh_i}q^{m^2}\right)\\
&\quad=\prod_{i=1}^{l}\left(\prod_{n=1}^{\infty}(1+q^{2m-1}e^{h_i})(1+q^{2m-1}e^{-h_i})\right),\\
&\prod_{i=1}^{l}\left(\prod_{n=1}^{\infty}(1-q^{2n})^{-1}\sum_{n\in Z} e^{(m+1/2)h_i}q^{m^2+m}\right)\\
&\quad=\prod_{i=1}^{l}\left(e^{h_i/2}\prod_{m=1}^{\infty}(1+q^{2m-2}e^{h_i})(1+q^{2m}e^{-h_i})\right).
\end{aligned}
\tag{9}
$$

When we pass from the affine Lie algebra of type $D_l^{(1)}$ to $D_{l+1}^{(2)}$, we have to enlarge the spaces of the representations multiplying them by the same space V_0. This space is interesting by itself, since one can construct there irreducible representations of two quite different algebras, namely, the Heisenberg algebra,

$$
[E(m), E(n)] = -2m\,\delta_{m,-n}, \qquad m, n \in 2\mathbb{Z}+1, \tag{10}
$$

and the algebra of the anticommutation relations

$$
\{e(m), e(n)\} = -2\delta_{m,-n}, \qquad m, n \in \mathbb{Z}. \tag{11}
$$

Comparison of the two corresponding character formulas gives the simple combinatorial identity

$$\prod_{n=1}^{\infty} (1-q^{2n-1})^{-1} = \prod_{n=1}^{\infty} (1+q^n). \tag{12}$$

In the representation, the operators $E(m)$ can be obtained as quadratic expressions of operators $e(n)$, and, on the other hand, the operators $e(n)$ can be expressed in terms of Lepowsky–Wilson operators ([20, formula in 5.6]), if one will substitutes the coefficient 2 for 4.

In the classical theory one knows a number of special isomorphisms of semisimple Lie algebras ([5]). One of them, namely, $\mathfrak{o}(4) \cong \mathfrak{sl}(2) \oplus \mathfrak{sl}(2)$, implies the isomorphism of the corresponding affine Lie algebras, i.e. $D_2^{(1)} \cong A_1^{(1)} + A_1^{(1)}$. Another isomorphism of affine Lie algebras, namely, $D_1^{(2)} \cong A_1^{(1)}$, has no analogue in the classical theory. Thanks to the last isomorphism we obtain the spinor constructions of the simplest affine Lie algebra. Using, in addition, the realization of V_0 as the space of the Heisenberg algebra (10) representation we can realize the basis representation of $A_1^{(1)}$ in the space

$$S[x_1^0, x_3^0, \ldots] \otimes A[x_1^+, x_3^+, \ldots] \otimes A[x_1^-, x_3^-, \ldots], \tag{13}$$

where S (resp. A) denotes symmetric (resp. antisymmetric) algebra of polynomials in infinitely many variables. One of the applications of the last construction will become clear when we pass to the second part of the paper.

II. One can see from our exposition that the subject and methods of Part I are purely algebraic. Our main prolem was to describe two constructions of the fundamental representations of affine Lie algebras and to find relations between them. However, we did not notice that we get into another region, where everybody speaks a different language. In fact it is sufficient to change the notation and the terminology and we find ourselves among the immense variety of models and insights of particle physics. One of the main goals of Part II is to provide both parts with a (brief) dictionary which will help to communicate results and ideas of two important areas.

In a joint paper with V. G. Kac [10] we already discussed the discovery of the vertex operator in the dual-resonance model (see e.g. [21]), which plays a crucial role in the construction of the vertex representation. Note that the physicists never guessed that the vertex operator can help to construct the representation of the infinite-dimensional Lie algebra. The important part of the spinor representations was found in the study of some quark models [2], although the whole affine Lie algebra of type $D_l^{(1)}$, the irreducibility of the representations, and generalizations to the type $D_{l+1}^{(2)}$ had not been considered. In both models the Virasoro algebra has been defined and studied, e.g., the identity (I.3.18) has been known before (see Footnote 12 in [2]).

But the most important and deep connection of affine Lie algebra representations wth particle physics concern the quantum field theory in one space and one time dimension. The main objects of the quantum field theory are Bose fields $\varphi_i(u)$ and Fermi fields $\psi_i(u)$, $\psi_i^+(u)$, $u \in \mathbb{R}$, $i = 1,\dots, l$, which satisfy the following relations

$$[\varphi_i(u_1), \varphi_j(u_2)] = -\frac{i}{4}\delta_{ij}\,\mathrm{sign}(u_1 - u_2), \tag{14}$$

$$\begin{aligned} \{\psi_i(u_1), \psi_j^+(u_2)\} &= \delta_{ij}\,\delta(u_1 - u_2), \\ \{\psi_i(u_1), \psi_j(u_2)\} &= \{\psi_i^+(u_1), \psi_j^+(u_2)\} = 0. \end{aligned} \tag{15}$$

Usually physicists consider two copies of these fields defined in two independent spaces $u \in \mathbb{R}$, $v \in \mathbb{R}$. Then they introduce physical time and space coordinates by $u = t + x$, $v = t - x$. The variables u and v are called light cone coordinates and (14), (15) are called light cone quantization. The use of the light cone coordinates is quite convenient in quantum electrodynamics at infinite momentum, in parton models and other questions. We will use it systematically, and we will deal mainly with one of the two light cone coordinates, i.e., with one-half of physical fields. The canonical representation spaces of Bose and Fermi fields are called Fock spaces and denoted by $\mathscr{H}_b$ and $\mathscr{H}_f$, respectively. Using the fundamental Bose and Fermi fields one can construct in Fock spaces a special type of operator, which are called currents on the real line with values in a finite dimensional algebra $\mathfrak{g}$. They satisfy the commutation relations

$$[J_k(u_1), J_m(u_2)] = 2if^{kmn}J_n(u_1)\,\delta(u_1 - u_2) + 2i\frac{c}{2\pi}\delta_{km}\,\delta'(u_1 - u_2), \tag{16}$$

with f^{kmn} the structure constants of $\mathfrak{g}$. The Hamiltonians and Lagrangians which define field models can be expressed in terms of the current algebra. We will consider first light cone models, which comprise one-half of the physical picture. We will also introduce a canonical two-dimensional model by the assignment of the Lagrangian density

$$\mathscr{L}(u, v) = J^{(1)k}(u)\,J_k^{(1)}(u) + J^{(2)k}(v)\,J_k^{(2)}(v) + gJ^{(1)k}(u)\,J_k^{(2)}(v), \tag{17}$$

where $J^{(1)}$, $J^{(2)}$ denote the first and the second copy of the current algebra (16), g an arbitrary constant.

The construction of the current algebra and a great many calculations in quantum field theory lead to divergences which physicists skillfully bypass using the following two main ideas. In the first approach they work in the spatial box of length L with periodic (antiperiodic) boundary conditions for the fields and they take the limit $L \to +\infty$ at the end of all calculations (see e.g. [1]). The representation theory of the affine Lie algebras described in Part I allows us to obain this approach if we set

$$\partial_u \varphi_{L,j}(u) = \frac{\sqrt{\pi}}{L} \tilde{h}_j(e^{2\pi i u/L}), \tag{18}$$

$$\psi^{\pm}_{L,j}(u) = \frac{\pm 1}{\sqrt{L}} \tilde{b}_{\pm j}(e^{2\pi i u/L}). \tag{19}$$

In particular, the exponents of Bose fields can be defined in terms of the vertex operator (2).

$$\frac{1}{L} :e^{2i\sqrt{\pi}(\alpha, \varphi_L(u))}: = \frac{1}{2\pi} X(\alpha, e^{2\pi i u/L}). \tag{20}$$

The current algebra in the spartial box which is, in fact, a Fourier transform of the affine Lie algebra consists of the operators (18) and (20) in Bose representation and of the quadratic expressions of the operators (19) in Fermi representation. Returning to the discovery of the vertex operator *we get the formal connection between dual resonance models (which have physical dimension essentially bigger than* 2) *with the two-dimensional field models: the underlying mathematical structures are basically the same in both theories, though this connection has no physical sense.*

The second appraoch is based on the renormalization procedure of all the fields operators by means of introducing a large cutoff constant M. In a cutoff field theory all mathematical expressions and manipulations are meaningful, at the end we can again take the limit $M \to +\infty$. It turns out that the renormalization pocedure can be also obtained from the results of Part I if we use the Cayley transform from the circle to the upper half plane, i.e., define Bose and Fermi fields by the formulas

$$\partial_u \varphi_j(u) = \frac{1}{2\sqrt{\pi}} \frac{2M}{M^2 + u^2} \tilde{h}_j \left(\frac{Mi - u}{Mi + u}\right), \tag{21}$$

$$\psi^{\pm}_j(u) = \frac{\pm 1}{\sqrt{2\pi}} \left(\frac{2M}{M^2 + u^2}\right)^{1/2} \tilde{b}_{\pm j} \left(\frac{Mi - u}{Mi + u}\right). \tag{22}$$

Note that the cutoff constant M multiplied by i is the image of the origin under the Cayley transform. We note that the operators $b_{\pm j}(m)$, which form spinor representations, appeared first in quark models [2]. Thus the quark field $\tilde{b}_{-j}(z)$ can be transformed into the one-half of Dirac–Fermi field, i.e., into the light cone Fermi field $\psi_j(u)$, which again has no physical sense, but exhibits the coincidence of mathematical structures. This formal correspondence was first noticed by Halpern (Appendix B in [15]).

To obtain the explicit renormalized expressions for currents we develop the theory of continuous analogues of the affine Lie algebras, which is quite similar to the original theory of affine Lie algebras. Thus one has a continuous analogue of the Heisenberg subalgebra (1)

$$[h_i\langle p\rangle, h_i\langle q\rangle] = p\,\delta_{ij}\,\delta(p+q), \tag{1$'$}$$

the analogue of the vertex operator, which we call the Mandelstam operator

$$X\langle\alpha, w\rangle = (2M)^{-\langle\alpha,\alpha\rangle/2}\exp\left(\int_0^\infty \alpha\langle -p\rangle(e^{ipw} - e^{-Mp})\frac{dp}{p}\right)$$
$$\times \exp\alpha \exp\left(-\int_0^\infty \alpha\langle p\rangle(e^{-ipw} - e^{-Mp})\right)\frac{dp}{p}, \tag{2$'$}$$

the continuous algebra of anticommutation relations

$$\{b_i\langle p\rangle, b_j\langle q\rangle\} = -\delta_{i,-j}\,\delta(p+q). \tag{3$'$}$$

Discrete and continuous algebras (1) and (1)$'$, (3) and (3)$'$ are related among themselves by means of the Laguerre transform. Thanks to this transform commutation relations for Mandelstam operators can be obtained from those of vertex operators or verified directly (see (23)).

The above considerations allows us to write down explicit renormalized expressions for currents in terms of Mandelstam operators,

$$\frac{1}{2M}N_M e^{2i\sqrt{\pi}\langle\alpha,\varphi(u)\rangle}: = \frac{1}{2\pi}X\langle\alpha, u\rangle = \frac{1}{2\pi}\frac{2M}{M^2+u^2}X\left(\alpha, \frac{Mi-u}{Mi+u}\right), \tag{23}$$

where $\partial_u\varphi_j(u) = (1/2\sqrt{\pi})\int_{-\infty}^{\infty} h_j\langle p\rangle e^{-ipu}\,dp$, and N_M denotes ordering and renormalization determined by a cutoff constant M. (One should understand "ordering" and "renormalization" in the sense of the formula (2)$'$, N_M is a standard notation accepted in physical literature.) Therefore both approaches to infrared regularization in quantum field theory can be formulated in the framework of affine Lie algebra representations. Bose and Fermi Fock spaces $\mathscr{H}_b$ and $\mathscr{H}_f$ from this point of view become completions of the spaces of affine Lie algebras representations in the canonical Hermitian structure defined in [13]. The fact that $\mathscr{H}_b$ and $\mathscr{H}_f$ are essentially the same allows us to apply different results of Part I to quantum field theory and this gives physical meanings to the mathematical statements: Thus the question of diagonalization of elements from $\mathfrak{g} \oplus \mathbb{C}\,d(0)$ is interpreted now as the solution of the generalized light cone sine-Gordon and Thirring models.

Two-dimensional canonical models (17) can be written in much simpler form due to the connection between the Casimir operators and the Virasoro operators; e.g., for the root system of type A_{l-1} one has

$$\mathcal{L}(u, v) = \left[(\partial_u \varphi^{(1)})^2 + (\partial_v \varphi^{(2)}) + \frac{g_v}{2\pi} \partial_u \varphi^{(1)k} \partial_v \varphi_k^{(2)}\right] + \frac{2g_v}{(2M)^2} N_M \sum_{j<k} \cos 2\sqrt{\pi}\,(\varphi_j - \varphi_k), \tag{24}$$

where $\varphi_k = \varphi_k^{(1)} - \varphi_k^{(2)}$, $g_v = 2\pi g/(l+1)$, for the boson model, which is called the generalized since-Gordon model and

$$\mathcal{L}(u, v) = |\psi^{(1)+} i\overleftrightarrow{\partial}_u \psi^{(1)} + \psi^{(2)+} i\overleftrightarrow{\partial}_v \psi^{(2)}| + g_v \psi^{(1)+} \psi^{(1)} \psi^{(2)+} \psi^{(2)} \tag{25}$$

for the fermion model, which is called the generalized Thirring model. But the most remarkable application of the affine Lie algebra theory is a result of the isomorphism between two construction of representations. It is interesting to recall here how the correspondece between the two field models (24) and (25) had been discovered in physics.

In 1975 Coleman [6] established the surprising result that the quantum sine-Gordon model is equivalent to the massive Thirring model. Then the exact correspondences between sine-Gordon operators and bilinear functions of the Fermi operators were listed in the paper [19] by Kogut and Susskind. Soon afterward, Mandelstam [22] gave the bosonic expression for the Fermi field operator itself and obtained therefore a new simple proof of Coleman's equivalence. Halpern in [15] extended this correspondence to the $SU(N)$-Thirring model in he cutoff approach. Banks, *et al.* [1] again undertook the study of the $SU(N)$-Thirring model by considering field operators in the spatial box with periodic boundary conditions. They found in particular that the $SU(2)$-Thirring model is equivalent to the theory of a free massless scalar field and a sine-Gordon field. Thus the correspondence between canonical two-dimensional boson theory (24) and fermion theory (25) was established in both approaches. A couple of years later Witten [26] reformulated these models using $2N$-component Majorana fermions instead of usual N-component Dirac fermions. In the new form $SU(N)$-Thirring model is naturally extended to the $O(2N)$-model. In particular he found out that the $O(4)$-model is equivalent to two decoupled sine-Gordon system. He continued to study the $O(2N)$-model with Shankar. Recently Shankar [24] using the symmetry of the Dynkin diagram for $O(8)$ has studied a trility symmetry of Majorana particle and two kinds of kinks. To date a great number of papers have been written about the boson–fermion correspondence, we have mentioned only a few of them.

We give here Halpern's generalization of the Mandelstam correspondence, which becomes now one of the basic results in the two-dimensional quantum field theory

$$\begin{aligned} \psi_j(u) &\leftrightarrow -(2M)^{-1/2} N_M e^{-2i\sqrt{\pi}\varphi_j(u)} \varepsilon_j, \\ \psi_j^+(u) &\leftrightarrow (2M)^{-1/2} N_M e^{2i\sqrt{\pi}\varphi_j(u)} \varepsilon_j. \end{aligned} \tag{26}$$

We summarize the above exposition in the following statements:

(L) *The isomorphism of two constructions of the representation of the affine Lie algebra* $\hat{D}(2l)$ *given by* (7) *is one half of the Banks–Hon–Neuberger equivalence. Vertex and spinor constructions of the affine Lie algebra representation are the boson and fermion representations of one-half of the current algebra in momentum realization, defined in a spatial box of length L with periodic boundary conditions.*

(M) *The isomorphism of two constructions of the representation of the affine Lie algebra* $\hat{D}(2l)$ *given by* (7) *is a Cayley transform of the one half the Mandelstam–Halpern equivalence* (26). *Vertex and spinor constructions of the affine Lie algebra representation is the Laguerre transform of the boson and fermion representation of one-half of the current algebra in momentum realization.*

(L–M) *The generalized sine-Gordon model of type* A_{l-1} *(resp.* D_l*) is isomorphic to the generalized* $SU(l)$ *(resp.* $D(2l)$*). Thirring model for every finite L or M (e.g.,* (24) *is isomorphic to* (25)*). Two approaches to the infrared regularization described above are equivalent in the limits* $L \to +\infty$, $M \to +\infty$.

Finally, we consider in the Banks *et al.* approach one particular model containing three fields ψ, ψ^+, ϕ defined in the spatial box with antiperiodic boundary conditions. The special isomorphism of affine Lie algebras of types $A_1^{(1)}$ and $D_2^{(2)}$ allows us to construct an irreducible representation of $\mathfrak{sl}(2)$-current algebra from fermion–antifermion fields ψ, ψ^+ and boson field ϕ. The space of representation is the completion of the space (13). We establish the isomorphism of this model with the sine-Gordon model containing one Bose field φ defined in the spatial box of double length. The canonical model associated with the three-particle representation of the $\mathfrak{sl}(2)$-current algebra is quite similar with the two-dimensional quantum electrodynamic or Schwinger model [19].

The plan of the paper is quite simple. It consists of two parts: mathematical and physical. Each part contains four sections which are in "one-to-one" correspondence. In Section I.1 we describe structural theory of affine Lie agebras, in Section II.1 we introduce current algebras and define the connection between them by means of Laguerre transformation. In

Section I.2 we consider vertex representation of affine Lie algebras, in Section II.2 we describe boson field representation of the current algebras and sine-Gordon model. Section I.3 is devoted to the spinor representation and Section II.3 to its physical analogue, fermion field representation. In Section I.4 we consider isomorphism between two constructions of representations which is known in physics, see Section II.4, as boson–fermion correspondence. Each of the eight sections is divided into two subsections, in the first we study simply laced algebras, i.e., of types $A_l^{(1)}$, $D_l^{(1)}$, $E_l^{(1)}$, in the second, twisted algebras, i.e., of types $A_{2l-1}^{(2)}$, $D_{l+1}^{(2)}$, $E_6^{(2)}$. Thus the special isomorphism for $\widehat{\mathfrak{sl}}(2)$ considered in Section I.4.2 turns into the equivalence between QED-like model and sine-Gordon model in Section II.4.2. We preface the mathematical part with the basic results for simple Lie algebras and conclude the physical part with the Appendix about the Laguerre polynomials and Bateman k-functions.

We tried to expose the material in the form understandable for both mathematicians and physicists in order to draw their attention to instruments of the adjacent science which can be fruitful in their own work. We hope that mathematicians will appreciate the variety of different models and constructions of two-dimensional field theory and physicists will comprehend the unifying picture of some of their discoveries, which is drawn on the tested material of the representation theory.

Simple Lie Algebras: Basic Results

We recall some facts about complex finite-dimensional simple Lie algebras and fix the necessary notations. All the details can be found in [5].

Let $\mathfrak{g}$ be a simple Lie algebra over complex field $\mathbb{C}$ of one of the types A_l ($l \geqslant 1$), D_l ($l \geqslant 4$), E_l ($l = 6, 7, 8$). Let $\mathfrak{h}$ be a Cartan subalgebra of $\mathfrak{g}$, Δ the set of roots with respect to $\mathfrak{h}$ in $\mathfrak{g}$, Δ_+ a set of positive roots, $\Pi = \{\alpha_1, \ldots, \alpha_l\}$ the corresponding set of simple roots, $\tilde{\alpha}$ the highest root, Q the root lattice spanned by Π. Let $\langle\ ,\ \rangle$ be an invariant symmetric bilinear form on $\mathfrak{g}$ normalized by $\langle \alpha, \alpha \rangle = 2$, $\alpha \in \Delta$. We will identify $\mathfrak{g}$ and $\mathfrak{g}^*$ by means of this form.

Let $\varepsilon\colon Q \times Q \to \{\pm 1\}$ be a bilinear function satisfying the conditions

$$\varepsilon(\alpha, \beta)\, \varepsilon(\beta, \alpha) = (-1)^{\langle \alpha, \beta \rangle}, \qquad \alpha, \beta \in Q; \tag{S.1}$$

$$\varepsilon(\alpha, \alpha) = -1, \qquad \alpha \in \Delta. \tag{S.2}$$

One can easily check that arbitrary assignment of values $\varepsilon(\alpha_i, \alpha_j)$, $i < j$, uniquely determines the function ε with such properties.

We choose a Chevalley basis $\Pi \cup \{x_\alpha^\epsilon, \alpha \in \Delta\}$ in $\mathfrak{g}$ satisfying the following commutation relations ([10]):

$$\begin{aligned}
&[h, x_\alpha^\epsilon] = \langle h, \alpha \rangle, && h \in \mathfrak{h}, \\
&[x_\alpha^\epsilon, x_\alpha^\epsilon] = 0, && \alpha + \beta \notin \Delta \cup \{0\}, \\
&[x_\alpha^\epsilon, x_\beta^\epsilon] = \varepsilon(a, \beta)\, x_{\alpha+\beta}^\epsilon, && \alpha + \beta \in \Delta, \\
&[x_\alpha^\epsilon, x_{-\alpha}^\epsilon] = -\alpha.
\end{aligned} \tag{S.3}$$

We define the canonical antilinear involutive automorphism θ of $\mathfrak{g}$ by the conditions

$$\begin{aligned}
&\theta(\alpha_i) = -\alpha_i, && \alpha_i \in \Pi, \\
&\theta(x_\alpha^\epsilon) = x_{-\alpha}^\epsilon, && \alpha \in \Delta,
\end{aligned} \tag{S.4}$$

The invariant subalgebra $\mathfrak{g}_c$ of $\mathfrak{g}$ with respect to θ is a compact simple Lie algebra. It is clear that $\langle\ ,\ \rangle$ is invariant with respect to this automorphism.

We denote by C he Casimir element in the center of the universal envelopping algebra of $\mathfrak{g}$. Let h_i $(i = 1,\ldots, l)$ compose an orthonormal basis of $\mathfrak{h}$. Then we have

$$C = \sum_{i=1}^{l} h_i^2 - 2 \sum_{\alpha \in \Delta} x_{-\alpha}^\epsilon x_\alpha^\epsilon + 2\rho = \sum_{i=1}^{l} h_i^2 - \sum_{\alpha \in \Delta} x_{-\alpha}^\epsilon x_\alpha^\epsilon, \tag{S.5}$$

where $\rho = \frac{1}{2} \sum_{\alpha \in \Delta} \alpha$.

Let P be a lattice spanned by fundamental weights $\Omega = \{\omega_1,\ldots, \omega_l\}$, i.e., $2\langle \omega_i, \alpha_j \rangle / \langle \alpha_j, \alpha_j \rangle = \delta_{ij}$, $P_{++} \subset P$ contains elements with nonnegative coefficient with respect to Ω. A one-to-one correspondence between elements $\lambda \in P_{++}$ and irreducible finite dimensional $\mathfrak{g}$-modules V_λ is classical (see e.g. [5]). It is defined by the condition: there is a $v_0 \in V_\lambda$ such that $hv_0 = \langle \lambda, h \rangle v_0$, $h \in \mathfrak{h}$, and $x_\alpha^\epsilon v_0 = 0$, $\alpha \in \Delta_+$.

Let $\mathfrak{g}$ be a simple Lie algebra of one of the types (i) A_{2l-1} $(l \geq 2)$, (ii) D_{l+1} $(l \geq 3)$, (iii) E_6. Let σ be the involutive antilinear automorphism of $\mathfrak{g}$ corresponding to the involution of the Dynkin diagram. One has the decomposition, with respect to σ;

$$\mathfrak{g} = \mathfrak{g}^0 + \mathfrak{g}^1, \tag{S.6}$$

where $x \in \mathfrak{g}^0$ iff $\sigma x = x$ and $x \in \mathfrak{g}^1$ iff $\sigma x = -x$. It is a classical fact (see e.g. [5]) that $\mathfrak{g}^0$ is a simple Lie algebra of the types C_l $(l \geq 2)$, $B_l (l \geq 3)$, F_4 corresponding to (i), (ii), (iii), respectively. Let $\mathfrak{h} = \mathfrak{h}^0 + \mathfrak{h}^1$ be a decomposition of a Cartan subalgebra of $\dot{\mathfrak{g}}$ with respect to σ and let $l_\sigma = \dim \mathfrak{h}$. We denote by Δ^0 the root system of $\mathfrak{g}^0$ with respect to $\mathfrak{h}^0$, by Q^0 the root lattice spanned by Δ^0, and by $\Pi^0 = \{\beta_1, \beta_2, ..., \beta_l\}$ the set of simple roots. Let $\varepsilon: Q \times Q \to \{\pm 1\}$ be a bilinear function satisfying the condition (S.1), (S.2) and an additional condition

$$\varepsilon(\sigma\alpha, \beta) = \varepsilon(\alpha, \sigma\beta) = \varepsilon(\alpha, \beta), \qquad \alpha, \beta \in \Delta. \tag{S.7}$$

One can check that arbitrary assignment of values $\varepsilon(\alpha_i, \alpha_j)$, $i < j$, satisfying (S.7) uniquely determines the function σ with such properties. We can define now a bilinear function $\varepsilon^0: Q^0 \times Q^0 \to \{\pm 1\}$ setting

$$\varepsilon^0(\alpha_i^0, \alpha_j^0) = \varepsilon(\alpha_i, \alpha_j). \tag{S.8}$$

We will describe now one special construction of a simple Lie algebra $\mathfrak{o}(2l)$ (type D_l), a simple Lie algebra $\mathfrak{o}(2l+1)$ (type B_l) and a simple Lie algebra $\mathfrak{o}(2l+2)$ with an involution σ. Let $C(n)$ denote a Clifford algebra of dimension 2^n over $\mathbb{C}$, i.e. $C(n)$ is an associative algebra with unit 1 generated by the elements e_i, $i = 1, ..., n$, satisfying the relations

$$\{e_i, e_j\} = e_i e_j + e_j e_i = -2\delta_{ij}, \qquad i, j = 1, ..., n. \tag{S.9}$$

We denote by $\mathfrak{e}$ the complex vector space spanned by $\{e_i\}_{i=1}^n$. The Clifford algebra has a natural $\mathbb{Z}_2$–grading

$$C(n) = C^0(n) \oplus C^1(n), \tag{S.10}$$

in which $C^0(n)$ is a subalgebra of $C(n)$ spanned by the products of an even number of elements e_i, $i = 1, ..., n$. We define a bilinear symmetric form in $C(n)$ by specifying that the basis $\{e_{i_1}, ..., e_{i_k}, \ 1 \leqslant i_1 < \cdots < i_k \leqslant n\}$ be orthogonal and that

$$\langle e_{i_1} \cdots e_{i_k}, e_{i_1} \cdots e_{i_k} \rangle = 4(-1)^{k(k+1)/2} \tag{S.11}$$

Thus we obtain the decomposition $C(n) = \sum_{k=0}^n C_k(n)$, where $C_k(n)$ spanned by products of k different element e_i, $i = 1, ..., n$. One can verify that the form $\langle \ , \ \rangle$ is invariant, i.e.,

$$\langle x_1 x_2, x_3 \rangle = \langle x_1, x_2 x_3 \rangle, \qquad x_1, x_2, x_3 \in C(n). \tag{S.12}$$

We define also a Lie bracket in $C(n)$ as usual

$$[x_1, x_2] = x_1 x_2 - x_2 x_1, \qquad x_1, x_2 \in C(n). \tag{S.13}$$

Now we let $n = 2l$, then $C_2(2l)$ with Lie bracket (S.13) is isomorphic to a simple orthogonal Lie algebra $\mathfrak{o}(2l)$, and the restriction of the form $\langle \ , \ \rangle$ to $C_2(2l)$ is a scalar multiple of Killing form. Let us denote

$$b_j = \tfrac{1}{2}(e_j + ie_{j+l}), \ b_{-j} = \tfrac{1}{2}(e_j - ie_{j+l}), \qquad j = 1, ..., l. \tag{S.14}$$

We choose a Cartan subalgebra $\mathfrak{h}$ of $\mathfrak{o}(2l)$, spanned by the orthonormal basis

$$h_j = \tfrac{1}{2}(b_{-j}b_j - b_j b_{-j}). \tag{S.15}$$

One can check easily that the form $\langle\ ,\ \rangle$ normed in such a way that $\langle \alpha, \alpha \rangle = 2$, $\alpha \in \Delta$, has roots of $\mathfrak{o}(2l)$ with respect to $\mathfrak{h}$.

Let $n = 2l + 1$; then the space $C_2(2l + 1) \oplus C_1(2l + 1)$ with the Lie bracket (S.13) is isomorphic to a simple orthogonal Lie algebra $\mathfrak{o}(2l + 2)$, $C_2(2l + 1)$ is isomorphic to subalgebra $\mathfrak{o}(2l + 1)$. $\mathbb{Z}_2$-grading (S.10) of the Clifford algebra provides $\mathfrak{o}(2l + 2)$ with the outer involution

$$\mathfrak{o}(2l + 2) = \mathfrak{o}(2l + 1) \oplus \mathfrak{e}. \tag{S.16}$$

Note that the Cartan subalgebra $\mathfrak{h}$ of $\mathfrak{o}(2l)$ will be also a Cartan subalgebra of $\mathfrak{o}(2l + 1)$.

Let θ be a complex conjugation in the space $\mathfrak{e}$. It extends to a antilinear involutive automorphism of the Clifford algebra and therefore of the simple orthogonal algebras constructed above.

We fix bases in the root systems of types D_l and B_l as follows:

$$\begin{aligned} \Pi(D_l) &= \{\alpha_1 = h_1 - h_2, \ldots, \alpha_{l-1} - h_l, \alpha_l = h_{l1} + h_l\}, \\ \Pi(B_l) &= \{\alpha_1 = h_1 - h_2, \ldots, \alpha_{l-1} = h_{l-1} - h_l, \alpha_l = h_l\}. \end{aligned} \tag{S.17}$$

Then the highest weights of standard representations are

$$\omega_1(D_l) = h_1, \qquad \omega_1(B_l) = h_1, \tag{S.18}$$

the weights of semispinor representations of $\mathfrak{o}(2l)$ and spinor representation of $\mathfrak{o}(2l + 1)$ are

$$\begin{aligned} \omega_{l-1}(D_l) &= \tfrac{1}{2}(h_1 + \cdots + h_{l-1} - h_l), \\ \omega_l(D_l) &= \tfrac{1}{2}(h_1 + \cdots + h_{l-1} + h_l), \\ \omega_l(B_l) &= \tfrac{1}{2}(h_1 + \cdots + h_{l-1} + h_l). \end{aligned} \tag{S.19}$$

PART I. TWO CONSTRUCTIONS OF AFFINE LIE ALGEBRAS REPRESENTATIONS

1. *Structural Theory*

1.1. Let $\mathbb{C}[t, t^{-1}]$ be the algebra of Laurent polynomials in the indeterminates t and t^{-1} over the complex field. We define the *affine Lie* *algebra* $\hat{\mathfrak{G}}$ associated with a simple Lie algebra $\mathfrak{g}$ as the finite-dimensional vector pace $C[t, t^{-1}] \otimes_C \mathfrak{g} \oplus Cc$ with the Lie bracket

$$[x(n), y(m)] = [x, y](n + m) + n\delta_{n,-m}\langle x, y\rangle c, \qquad (.1.1)$$

where $x, y \in \mathfrak{g}$, $n, m \in Z$, $x(n)$ denotes the element $t^n \otimes x \in \hat{\mathfrak{G}}$, and c is a central element of $\hat{\mathfrak{G}}$. The subalgebra $\hat{\mathfrak{H}} = C[t, t^{-1}] \otimes_C \mathfrak{h} \oplus Cc$ is called the *Heisenberg subalgebra of* $\hat{\mathfrak{G}}$. We will identify $\mathfrak{g} \cong 1 \otimes \mathfrak{g}$ and call it the *scalar subalgebra* of $\hat{\mathfrak{G}}$. The subalgebra $\hat{\mathfrak{h}} = \mathfrak{h} \oplus Cc$ is called the *Cartan subalgebra* of $\hat{\mathfrak{G}}$.

We can extend the canonical automorphism θ of $\mathfrak{g}$ to the affine Lie algebra $\hat{\mathfrak{G}}$ by the following formulas

$$\theta(x(n)) = (\theta x)(-n), \qquad x \in \mathfrak{g}, \quad n \in Z, \qquad \theta(c) = -c. \qquad (I.1.2)$$

The invariant subalgebra $\hat{\mathfrak{G}}_c$ of $\hat{\mathfrak{G}}$ with respect to θ is called the compact form of $\hat{\mathfrak{G}}$.

Let $d(n)$, $n \in Z$, be the derivation of the algebra of Laurent polynomials $C[t, t^{-1}]$ defined as $t^{n+1}(d/dt)$. We can extend it to the derivation of affine Lie algeba $\hat{\mathfrak{G}}$ assuming that it acts trivially on the central element. One has

$$[d(n), x(m)] = mx(n + m), \qquad n, m \in Z, \quad x \in \mathfrak{g}. \qquad (I.1.3)$$

We define a bilinear symmetric invariant form $\langle\ ,\ \rangle_n$ on the semidirect product $\hat{\mathfrak{G}} \oplus C\, d(n)$ by the formula

$$\langle x(m) + x_0 c + x_1 d(n), y(k) + y_0 c + y_1 d(n)\rangle_n = \delta_{m+k,n}\langle x, y\rangle + x_0 y_1 + x_1 y_0, \qquad (I.1.4)$$

where $x, y \in \mathfrak{g}$, $x_0, x_1, y_0, y_1 \in C$. Let $\mathfrak{D}$ denote the Lie algebra spanned by the elements $d(n)$, $n \in Z$, with Lie bracket

$$[d(n), d(m)] = (m - n)\, d(n + m), \qquad (I.1.5)$$

which is induced by the commutations between $t^{n+1}(d/dt)$ and $t^{m+1}(d/dt)$. For every $d = \sum_k a_k d(k) \in \mathfrak{D}$, $a_k \in C$, we can introduce an invariant form $\langle\ ,\ \rangle_d = \sum_k a_k \langle\ ,\ \rangle_k$ in $\hat{\mathfrak{G}} \oplus Cd$.

The derivation $d(0)$ plays a special role in the structural theory of $\hat{\mathfrak{G}}$. One has a root decomposition with respect to $\hat{\mathfrak{h}} \oplus C\, d(0)$:

$$\hat{\mathfrak{G}} = \hat{\mathfrak{h}} \oplus \sum_{a \in \hat{\Delta}} \hat{\mathfrak{G}}^a, \quad \hat{\mathfrak{G}}^a = \{x \in \hat{\mathfrak{G}} : [h, x] = \langle a, h\rangle x, h \in \hat{\mathfrak{h}} \oplus C\, d(0)\}. \qquad (I.1.6)$$

$\hat{\Delta}$ is an affine root system and consists of the real roots $\hat{\Delta}_R = \{nc + \alpha, n \in Z,$

$\alpha \in \Delta\}$ and the imaginary roots $\hat{\Delta}_I = \{nc, n \in \mathbb{Z}\backslash 0\}$. The subset $\hat{\Pi} = \{\alpha_0 = c - \tilde{\alpha}, \alpha_1, ..., \alpha_l\}$ of $\hat{\Delta}$ is a basis of the affine root system.

Let us consider now $\hat{\mathfrak{G}} \oplus \mathbb{C}\, d(m)$-module V with the property that any $v \in V$ is annihilated by all $x(n)$, $x \in \mathfrak{g}$, when n is suficiently large. We choose dual bases in $\mathfrak{G}^a$ and $\mathfrak{G}^{mc-a}$ with respect to the form $\langle\ ,\ \rangle_m$: $x_{a,i}$, $x_{mc-a,i}$, $1 \leqslant i \leqslant \dim \mathfrak{G}^a$. The following endomorphism of V is called the Casimir element

$$C(m) = 2c\, d(m) + \sum_{i=1}^{l} h_i(0)\, h_i(m) + 2 \sum_{a \in \hat{\Delta}_+} \sum_i x_{mc-a,i} x_{a,i} + 2\tilde{\rho}(m), \qquad (\text{I.1.7})$$

where $\tilde{\rho}(m) = h \cdot d(m) + \rho(m)$ ($\rho(m)$ as in (I.1.1) denotes $t^m \otimes \rho$, ρ as in (S.5)), h is the Coxeter number. The element $C(0)$ first was considered by Kac [16]. It is convenient to introduce also the operator

$$L(m) = C(m) - d(m)(2c + 2h), \qquad (\text{I.1.8})$$

which is contained in the algebra generated by $\hat{\mathfrak{G}}$. This operator has a simple form in the Chevalley basis (S.3). Let us define the ordering of two elements by

$$\begin{aligned} :x(k)\, y(n): &= x(k)\, y(n), && n > k, \\ &= \tfrac{1}{2}(x(k)\, y(n) + y(n)\, x(k)), && n = k, \qquad (\text{I.1.9}) \\ &= y(n)\, x(k), && n < k. \end{aligned}$$

Then we have

$$L(m) = \sum_{n \in \mathbb{Z}} \left(\sum_{i=1}^{l} :h_i(n)\, h_i(m-n): - \sum_{\alpha \in \Delta} :x^{\epsilon}_{-\alpha}(n)\, x^{\epsilon}_{\alpha}(m-n): \right) (\text{I.1.10})$$

Remark. Segal was the first to consider operators $L(m)$ for arbitrary $m \in \mathbb{Z}$ (unpublished results).

Using the same arguments as in [17] we obtain the following result:

PROPOSITION I.1.11. (i) *$C(m)$ commutes with the action of $\hat{\mathfrak{G}} \oplus \mathbb{C}\, d(m)$ in V.*

(ii) *Let c be represented by c_0 Id, $c_0 \in \mathbb{C}$, in V. Then*

$$[L(m), x(n)] = -2(c_0 + h)\, n x(m+n), \qquad m, n \in \mathbb{Z}.$$

Further on we will consider only the $\hat{\mathfrak{G}}$-modules in which c is a scalar multiple of the identity element. v is called a highest weight vector of V if

$$\begin{aligned} x_a v &= 0, && a \in \hat{\Delta}_+,\ \ x_a \in \mathfrak{G}^a, \\ hv &= \langle \lambda, h \rangle v, && h \in \mathfrak{h} \oplus \mathbb{C}\, d(0). \end{aligned} \qquad (\text{I.1.12})$$

λ is called a highest weight of V. A highest weight is called a dominant integral if for every $a \in \Delta_+$, $2\langle\lambda, a\rangle/\langle a, a\rangle \in \mathbb{Z}_+$. Simple $\mathfrak{G}$-modules with dominant integral highest weight have been classified in [16]. We let

$$\hat{\Omega} = \{\hat{\omega}_0, \hat{\omega}_1, ..., \hat{\omega}_l\}, \tag{I.1.13}$$

where $2\langle\hat{\omega}_1, a_j\rangle/\langle a_j, a_j\rangle = \delta_{ij}$, $i,j = 0, 1,..., l$, then $\hat{\omega} = d + \text{Const} \cdot c$, $\hat{\omega}_l = n_l d + \omega_l + \text{Const} \cdot c$, n_l a positive integer. For example, if $\mathfrak{G} = \hat{\mathfrak{D}}(2l)$, one has from (S.18), (S.19)

$$\begin{aligned} \hat{\omega}_1 &= d + h_1 + \text{Const} \cdot c, \\ \hat{\omega}_{l-1} &= d + (h_1 + \cdots + h_{l-1} - h_l)/2 + \text{Const} \cdot c, \\ \hat{\omega}_l &= d + (h_1 + \cdots + h_{l-1} + h_l)/2 + \text{Const} \cdot c. \end{aligned} \tag{I.1.14}$$

A simple highest weight module V with $\lambda = \hat{\omega}_i$ is called the fundamental module corresponding to the ith point of the Dynkin diagram of $\mathfrak{G}$. We will denote it by V_i. V_0 is also called basic module. Fundamental modules with $n_l = 1$ admit direct constructions which we will consider in the next sections.

1.2. Let $\mathfrak{G}$ be the affine Lie algebra associated with a simple Lie algebra $\mathfrak{g}$; i.e., $\mathfrak{G} \cong \mathbb{C}[\tau, \tau^{-1}] \otimes_{\mathbb{C}} \mathfrak{g} \oplus \mathbb{C}(c/2)$ with a Lie bracket defined by (I.1.1). The subalgebra $\mathfrak{G}_\sigma$ of $\mathfrak{G}$ consisting of the elements $x \otimes \tau^n$, $n \in \mathbb{Z}$, $x \in \mathfrak{g}^{\bar{n}}$ $x \in \mathfrak{g}^{\bar{n}}$, where $\bar{n} = 0$ for n even and $\bar{n} = 1$ for n odd, is called *the affine Lie algebra associated with a simple Lie algebra and involution* σ. For any $x \in \mathfrak{g}$, $h \in \mathbb{Z}$, we denote by $x(n/2)$ the element $x^{\bar{n}} \otimes t^n$ of $\mathfrak{G}_\sigma$. The subalgebra $\mathfrak{H}_\sigma = \mathfrak{H} \cap \mathfrak{G}_\sigma$ is called the *Heisenberg subalgebra of* $\mathfrak{G}$, $\mathfrak{g}^0 \cong 1 \otimes \mathfrak{g}^0$ is called the *scalar subalgebra* and $\mathfrak{h}_\sigma = \mathfrak{h}^0 \oplus \mathbb{C}c$ is called the *Cartan subalgebra* of $\mathfrak{G}_\sigma$. We denote by $\mathfrak{G}^0 \cong \mathbb{C}[\tau^2, \tau^{-2}] \otimes_{\mathbb{C}} \mathfrak{g}^0 \oplus \mathbb{C}(c/2)$ the affine Lie algebra associated with $\mathfrak{g}^0$.

We introduce the derivatives of the affine Lie algebra $d(n)$, $n \in \mathbb{Z}$, by the formula (I.1.3) and the root decomposition with respect to $\mathfrak{h}_\sigma \oplus \mathbb{C}\, d(0)$ by the formula (I.1.6). Affine root system $\hat{\Delta}_\sigma$ of $\mathfrak{G}_\sigma$ consists now of the elements $\{nc,\ n \in \frac{1}{2}\mathbb{Z}\backslash 0\} \cup \{nc + \alpha,\ n \in \mathbb{Z},\ \alpha \in \Delta^0$, or $n \in \mathbb{Z} + \frac{1}{2},\ \alpha \in \Delta^0,\ \alpha\text{-short}\}$. Then $\hat{\Pi}_\sigma = \{\beta_0 = c/2 - \bar{\beta}^\vee, \beta_1, ..., \beta_l\}$ is a basis of $\hat{\Delta}_\sigma$, $\hat{\Pi}^0 = \{\beta_0 = c - \bar{\beta}, \beta_1, ..., \beta_l\}$ is a basis of $\hat{\Delta}^0$, where $\bar{\beta}$, $\bar{\beta}^\vee$ are, respectively, long and short dominant roots of Δ^0.

The Casimir elements $C(m)$ and operators $L(m)$ for $\mathfrak{G}_\sigma$ are defined as in Section 1.1 by the formulas (I.1.7) and (I.1.8), respectively, in which h is a Coxeter member of $\mathfrak{g}$ or equivalently of $\mathfrak{g}^0$. The analogue of (I.1.10) has the following form

$$L(m) = \sum_{n \in \mathbb{Z}/2} \left(\sum_{i=1}^{l_0} :h_i(n)\, h_i(m-n): - \sum_{\alpha \in \Delta} :x^{\epsilon}_{-\alpha}(n)\, x^{\epsilon}_{\alpha}(m-n): \right). \tag{I.1.15}$$

One can verify (I.1.14) using the simple equality

$$-\sum_{\alpha\in\Delta} x^{\epsilon}_{-\alpha}x^{\epsilon}_{\alpha} = \sum_{\alpha\in\Delta^0}\frac{(x^{\epsilon}_{-\alpha})^0(x^{\epsilon}_{\alpha})^0}{\langle(x^{\epsilon}_{-\alpha})^0,(x^{\epsilon}_{\alpha})^0\rangle} + \sum_{\substack{\alpha\in\Delta^0\\ \alpha\text{-short}}}\frac{(x^{\epsilon}_{-\alpha})^1(x^{\epsilon}_{\alpha})^1}{\langle(x^{\epsilon}_{-\alpha})^1,(x^{\epsilon}_{\alpha})^1\rangle}. \tag{I.1.16}$$

The generalization of Proposition I.1.11 for affine Lie algebras $\hat{\mathfrak{G}}_\sigma$ is certainly valid as the proof does not depend of the type of affine root system.

Dominant highest weight modules for affine Lie algebras $\hat{\mathfrak{G}}_\sigma$ and $\hat{\mathfrak{G}}^0$ are defined as in Section 1.1. We denote by $\hat{\Omega}_\sigma$ a basis dual for $\hat{\Pi}_\sigma$ and by $\hat{\Omega}^0$ a basis dual to $\hat{\Pi}^0$ in both cases $\hat{\omega}_0 = d + \text{Const}\cdot c$. For example, if $\hat{\mathfrak{G}}_\sigma = \hat{\mathfrak{O}}^{(2)}(2l+2)$, one has from (S.18), (S.19)

$$\hat{\omega}_l = d + (h_1 + \cdots + h_l)/2 + \text{Const}\cdot c \tag{I.1.17}$$

for $\hat{\mathfrak{G}}^0 = \hat{\mathfrak{O}}(2l+1)$;

$$\begin{aligned}\hat{\omega}_1 &= d + h_1 + \text{Const}\cdot c,\\ \hat{\omega}_l &= d + (h_1 + \cdots + h_l)/2 + \text{Const}\cdot c.\end{aligned} \tag{I.1.18}$$

2. *Vertex Representations of Affine Lie Algebras*

2.1. We recall first the main construction of [10]. Let P be a weight lattice in $\mathfrak{h}$. The pair $(\hat{\mathfrak{G}}, P)$ with the action of P on $\hat{\mathfrak{G}}$ defined below is called the *Heisenberg system* associated with a lattice P; i.e., one has

$$\begin{aligned}[h_1(m), h_2(n)] &= m\,\delta_{m,-n}\langle h_1, h_2\rangle c, && h_1, h_2\in\mathfrak{h},\\ \lambda(h) &= h - \langle h,\lambda\rangle c, && \lambda\in P.\end{aligned} \tag{I.2.1}$$

Let $\varepsilon: P\times P\to T$, $T=\{z\in\mathbb{C}: |z|=1\}$ be a bilinear function, then

$$\begin{aligned}&\varepsilon(\lambda,0)=\varepsilon(0,\mu)=1,\\ &\varepsilon(\lambda,\mu)=\varepsilon(-\lambda,\mu)^{-1}=\varepsilon(\lambda,-\mu)^{-1}=\varepsilon(-\lambda,-\mu), \qquad \lambda,\mu\in P\end{aligned} \tag{I.2.2}$$

Let $S(\mathfrak{H}^-)$ denotes a symmetric algebra of $\mathfrak{H}^-$ and $\mathbb{C}[P]$ be a group algebra of P. We construct an irreducible projective representation with the cocycle ε of the Heisenberg system $(\hat{\mathfrak{H}}, P)$ in the space $V = S(\mathfrak{H}^-)\otimes\mathbb{C}[P]$ in a standard way:

$$\begin{aligned}h(m)\cdot v\otimes e^{\lambda} &= m(\partial_{h(-m)}v)\otimes e^{\lambda}, && m>0;\\ h(-m)\cdot v\otimes e^{\lambda} &= (h(-m)\,v)\otimes e^{\lambda}, && m>0;\\ h(0)\cdot v\otimes e^{\lambda} &= v\otimes(\partial_h e^{\lambda}) = \langle h,\lambda\rangle\, v\otimes e^{\lambda};\\ \mu\cdot v\otimes e^{\lambda} &= \varepsilon(\mu,\lambda)\, v\otimes e^{\lambda+\mu},\\ c\cdot v\otimes e^{\lambda} &= v\otimes e^{\lambda},\end{aligned} \tag{I.2.3}$$

where $\lambda, \mu \in P$, $v \in S(\mathfrak{H}^-)$. We obtain immediately that

$$\begin{aligned} \mu \cdot (\lambda \cdot v) &= \varepsilon(\mu, \lambda)(\mu + \lambda) \cdot v \\ \mu \cdot (h(0) \cdot (\mu^{-1} \cdot v)) &= (\mu(h(0))) \cdot v, \qquad v \in V, \end{aligned} \tag{I.2.4}$$

and also

$$\mu \cdot (\lambda \cdot (\mu^{-1} \cdot (\lambda^{-1} \cdot v))) = \varepsilon(\mu, \lambda)\, \varepsilon(\lambda, \mu)^{-1} v, \qquad v \in V. \tag{I.2.5}$$

We define now a positive Hermitian structure (,) in V. First note that $S(\mathfrak{H}^-)$ has a unique Hermitian structure so that

$$\begin{aligned} h^* &= -\theta h, \qquad h \in \hat{\mathfrak{H}} \backslash \mathfrak{h}, \\ (1, 1) &= 1, \end{aligned} \tag{I.2.6}$$

where * denotes Hermitian conjugation. We set

$$(v \otimes e^\lambda, u \otimes e^\mu) = (v, u) \cdot \delta_{\lambda\mu}, \qquad v, u \in S(\mathfrak{H}^-), \quad \lambda, \mu \in P. \tag{I.2.7}$$

We denote by V' the space of formal infinite sums of elements from V. Then the Hermitian scalar product (,) can be extended to the pairing of V and V'.

Let Γ be a lattice satisfying the condition

$$Q \subset \Gamma \subset P. \tag{I.2.8}$$

We will consider the decomposition $P = P_1 \cup P_2 \cup \cdots$ into the orbits of Γ, $P_1 = \Gamma$. Let us consider the restriction of our representation V to the Heisenberg system $(\hat{\mathfrak{H}}, \Gamma)$. The space V is decomposed into the direct sum of the representations $V = V_1 \oplus V_2 \oplus \cdots$ according to the orbit decomposition of P; i.e., $V_k = S(\mathfrak{H}^-) \otimes \mathbf{C}[P_k]$, where $\mathbf{C}[P_k] = \sum_{\lambda \in P_k} \mathbf{C} e^\lambda$.

In this paper we will consider only the lattice Γ satisfying the properties

$$\begin{aligned} &\text{(i)} \quad \langle \lambda, \mu \rangle \in \mathbf{Z}, \qquad \lambda, \mu \in \Gamma \\ &\text{(ii)} \quad \text{for every orbit } P_k \text{ of } \Gamma, \text{ either} \\ &\qquad \langle \lambda, \mu \rangle \in \mathbf{Z}, \quad \lambda \in \Gamma, \quad \mu \in P_k, \qquad \text{or} \\ &\qquad \langle \lambda, \mu \rangle \in \mathbf{Z} + \tfrac{1}{2}, \quad \lambda \in \Gamma, \quad \mu \in P_k. \end{aligned} \tag{I.2.9}$$

Definition. The operator $X(\mu, z)\colon V_k \to V'_k$, $\mu \in \Gamma$, $z \in \mathbf{C} \backslash 0$, defined by the formula

$$X(\mu, z) = \exp\left(\sum_{n=1}^{\infty} \frac{z^n}{n}\mu(-n)\right)\exp(\ln z \cdot \mu(0) + \mu)$$

$$\times \exp\left(-\sum_{n=1}^{\infty} \frac{z^{-n}}{n}\mu(n)\right) \tag{I.2.10}$$

is called the *vertex operator*.

Vertex operators play an important role in the dual resonance theory (see [21]). Note that the first and the last factor of (I.2.8) act only in $S(\mathfrak{H}^-)$ and the middle factor only in $\mathbb{C}[P_k]$ by

$$\exp(\ln z \cdot \mu(0) + \mu)\, e^{\lambda} = z^{\langle\mu,\lambda\rangle + \langle\mu,\mu\rangle/2} e^{\lambda+\mu}. \tag{I.2.11}$$

We define now the operators

$$X_n(\mu)\, v = \operatorname*{Res}_{z=0}\, (X(\mu, z)\, v \cdot z^{n-1}), \qquad v \in V_k, \tag{I.2.12}$$

where $n \in \mathbb{Z}$ (resp. $\mathbb{Z} + \frac{1}{2}$) if $\langle \Gamma, P_k \rangle \subset \mathbb{Z}$ (resp. $\mathbb{Z} + \frac{1}{2}$). The formulas (I.2.10), (I.2.11) imply that $X(\mu, z)\, v \cdot z^{n-1}$ can have only a finite pole at $z = 0$, therefore $X_n(\mu)$ is a well-defined operator in V_k. We introduce also the operators depending on the cocycle ε,

$$X^{\varepsilon}(\mu, z) = X(\mu, z)\, \varepsilon_{\mu}, \qquad X^{\varepsilon}_n(\mu) = X_n(\mu)\, \varepsilon_{\mu}. \tag{I.2.13}$$

Now we will find the commutation and anticommutation relations between these operators. We let

$$X^{\varepsilon}(\lambda, \mu, z, z_0) = {:}X(\lambda, z)\, X(\mu, z_0){:}\; \varepsilon_{\lambda+\mu} \tag{I.2.14}$$

Proposition I.2.15. *Let $\lambda, \mu \in \Gamma$ and $v \in vV_k$. One has*

(i) $X^{\varepsilon}(\lambda, z)\, X^{\varepsilon}(\mu, z) = \varepsilon(\lambda, \mu)(z - z_0)^{\langle\lambda,\mu\rangle}(zz_0)^{-\langle\lambda,\mu\rangle/2}\, X^{\varepsilon}(\lambda, \mu, z, z_0)$, *where* $|z| > |z_0|$.

(ii) *If* $\varepsilon(\lambda, \mu)\, \varepsilon(\mu, \lambda)^{-1} = \pm(-1)^{\langle\lambda,\mu\rangle}$ *then*

$$[X^{\varepsilon}_n(\lambda), X^{\varepsilon}(\mu, z^0]_{\mp}\, v = \varepsilon(\lambda, \mu) \operatorname*{Res}_{z=z_0}\, ((z - z_0)^{\langle\lambda,\mu\rangle}(zz_0)^{-\langle\lambda,\mu\rangle/2}$$

$$\times X^{\varepsilon}(\lambda, \mu, z, z_0)\, v \cdot z^{n-1}),$$

where $[\ ,\ ]_- \equiv [\ ,\ ]$ *denotes commutator,* $[\ ,\ ]_+ \equiv \{\ ,\ \}$ *denotes anticommutator.*

Proof. Part (i) follows from the equalities

$$\exp\left(-\sum_{n=1}^{\infty}\frac{z^{-n}}{n}\lambda(n)\right)\exp\left(\sum_{n=1}^{\infty}z_0^n\mu(-n)\right)$$

$$=(1-z_0z^{-1})^{\langle\lambda,\mu\rangle}\exp\left(\sum_{n=1}^{\infty}\frac{z_0^n}{n}\mu(-n)\right)\exp\left(-\sum_{n=1}^{\infty}\frac{z^{-n}}{n}\lambda(n)\right)$$

$$\exp(\ln z\cdot\lambda(0)+\lambda)\exp(\ln z_0\cdot\mu(0)+\mu)$$

$$=(zz_0^{-1})^{\langle\lambda,\mu\rangle/2}\exp(\ln z\cdot\lambda(0)+\ln z_0\cdot\mu(0)+\lambda+\mu).$$

(ii) $[X_n(\lambda),X^\varepsilon(\mu,z_0)]_\mp v=\int_{R_1}X^\varepsilon(\lambda,z)\,X^\varepsilon(\mu,z_0)\,v\cdot z^{n-1}\,dz$
$-\int_{R_2}\pm X^\varepsilon(\mu,z_0)\,X^\varepsilon(\lambda,z)\,v\cdot z^{n-1}\,dz$, where $R_1>|z_0|>R_2$.

Now (i) implies that the integrands are equal and its analytical continuation inside the ring has only one finite pole at $z=z_0$.

Let us choose $\Gamma=P$, and let restriction ε to $Q\times Q$ satisfy the properies (S.1), (S.2), then we have [10]:

THEOREM I.2.16. *Let the action of the Heisenberg algebra $\hat{\mathfrak{H}}$ is defined by the formulas* (I.2.3) *in V_k. We define the action of $x_\alpha^\varepsilon(n)\in\mathfrak{G}$, $\alpha\in\Delta$, $n\in\mathbb{Z}$ as the action of the components of the vertex operator $X_n^\varepsilon(\alpha)$. Then V_k is a fundamental module of $\mathfrak{G}$, V_1 is a basic module, and we have*

$$x^*=-\theta x,\qquad x\in\mathfrak{G}.$$

Proof. Follows immediately from Proposition I.2.15 (ii) and calculations of Res in $z=z_0$. See [10] for details.

We note that the vertex operator $X^\varepsilon(\alpha,z)$ defines the action of the formal sum $\sum_{n\in Z}x_\alpha^\varepsilon(n)\,z^{-n}\colon V_k\to V_k'$. Many formulas in the vertex representation admit a simpler version using the formal sums. We denote

$$\tilde{x}(z)=\sum_{n\in Z}x(n)\,z^{-n},\qquad x\in\mathfrak{g},\quad z\in\mathbb{C}\backslash 0. \tag{I.2.17}$$

Now we will introduce the Virasoro algebra which was considered first in the dual resonance theory (see e.g. [21]).

$$D(m)=-\tfrac{1}{2}\sum_{k\in Z}\sum_{l=1}^{l}:h_l(k)\,h_l(m-k):,\qquad m\in\mathbb{Z}, \tag{I.2.18}$$

in which $\{h_i\}_{i=1}^l$ is an orthonormal basis of $\mathfrak{h}$. We can also write (I.2.18) using the notation of (I.2.18).

$$\tilde{D}(z)=-\tfrac{1}{2}\sum_{l=1}^{l}:\tilde{h}_l^2(z):. \tag{I.2.19}$$

One can notice from the definition of the Virasoro operators that they consist only of a part of the Segal operators (I.1.10). However, in the vertex representation they are proportional.

PROPOSITION I.2.20. (i) $D(m) = -[1/2(h+1)]\, L(m)$, *in which h is a Coxeter number of* $\mathfrak{g}$.

(ii) $[D(m), x(k)] = kx(m+k), \qquad x \in \mathfrak{g},$

(iii) $[D(m), D(k)] = (k-m)\, D(m+k) + (l/12)(m^3 - m)\, \delta_{m,-k}$

Proof. We will prove (i) by the direct verification, (ii) follows from (i) and (I.1.11), the proof of (iii) can be found in [10].

The proof of (i) is based on the following equality for every $\alpha \in \Delta$

$$:X^{\epsilon}(\alpha, z)\, X^{\epsilon}(-\alpha, z) + X(-\alpha, z)\, X^{\epsilon}(\alpha, z): = -:\tilde{\alpha}^2(z):. \qquad (I.2.21)$$

We then obtain statement (i) from the definitions of $D(m)$ and $L(m)$, and the classical equation $\sum_{\alpha \in \Delta} \alpha^2 = 2h \sum_{i=1}^{l} h_i^2$.

Let $|z| > |z_0|$ then the left side of (I.2.21) can be transformed as follows

$$\begin{aligned}
&\lim_{z \to z_0} \Big(X^{\epsilon}(\alpha, z)\, X^{\epsilon}(-\alpha, z_0) + X^{\epsilon}(-\alpha, z)\, X^{\epsilon}(\alpha, z_0) + 2 \sum_{n=1}^{\infty} n z^{-n} z_0^n\Big) \\
&= \lim_{z \to z_0} \frac{z z_0}{(z - z_0)^2} (-X(\alpha, -\alpha, z, z_0) - X(-\alpha, \alpha, z, z_0) + 2) \\
&= -\frac{1}{2} \left(z \frac{d}{dz}\right)^2 (X(\alpha, -\alpha, z, z_0) + X(-\alpha, \alpha, z, z_0))\big|_{z=z_0} \\
&= -\frac{1}{2} \left(z \frac{d}{dz}\right) (\tilde{\alpha}^-(z)\, X(\alpha, -\alpha, z, z_0) + X(\alpha, -\alpha, z, z_0)\, \tilde{\alpha}^+(z) \\
&\qquad - \tilde{\alpha}^-(z)\, X(-\alpha, \alpha, z, z_0) - X(-\alpha, \alpha, z, z_0)\, \tilde{\alpha}^+(z))\big|_{z=z_0} \\
&= -(\tilde{\alpha}^-(z_0)\, \tilde{\alpha}(z_0) + \tilde{\alpha}(z_0)\, \tilde{\alpha}^+(z_0)),
\end{aligned}$$

in which $\tilde{\alpha}(z) = \tilde{\alpha}^-(z) + \tilde{\alpha}^+(z)$, $\tilde{\alpha}^-(z) \in \mathfrak{H}^-$, $\tilde{\alpha}(z) \in \mathfrak{H}^+ \oplus \mathfrak{h}$.

Theorem I.2.16 and Proposition I.2.20 imply in particular that any element H of the scalar subalgebra $\mathfrak{g} \oplus \mathbb{C}\, d(0)$ acts in the vertex representation as the operator

$$H = \int_{C_R} \left\{ a \sum_{j=1}^{l} :(h_j(z))^2: + \sum_{j=1}^{l} a_j h_j(z) + \sum_{\alpha \in \Delta} a_\alpha X^{\epsilon}(\alpha, z) \right\} \frac{dz}{z}, \qquad (I.2.22)$$

where a, a_j, $a_\alpha \in \mathbb{C}$. If the projection of H to $\mathfrak{g}$ is semisimple, then there exists $g \in G$ and $H' \in \mathfrak{h} \oplus \mathbb{C}\, d(0)$ that $H = \operatorname{Ad} g \cdot H'$. This fact and simple spectral decomposition of H' allows one to find the spectrum of an arbitrary operator of the form (I.2.22).

2.2. Now we recall the construction of [11]. Let P^0 be a weight lattice of $\mathfrak{g}^0$ in $\mathfrak{h}_R$. Let $\varepsilon^0: P^0 \times P^0 \to T$, be a bilinear function. The representation of the Heisenberg system $(\hat{\mathfrak{H}}_\sigma, P^0)$ is defined by the formulas (I.2.3) with $m \in \frac{1}{2}\mathbb{Z}$, in the space $V^\sigma = S(\mathfrak{H}_\sigma^-) \otimes \mathbb{C}[P^0]$, provided with Hermitian structure analogues to (I.2.6).

Let Γ be a lattice and $Q^0 \subset \Gamma^0 \subset P^0$. Let $P^0 = P_1^0 \cup P_2^0 \cup \cdots$ be a decomposition into the orbits of Γ, $V^\sigma = V_1^\sigma \oplus V_2^\sigma \oplus \cdots$ be a corresponding decomposition into the irreducible representations of $(\hat{\mathfrak{H}}_\sigma, \Gamma^0)$. We will assume that Γ^0 satisfies (I.2.9). Let Γ be a lattice, $Q \subset \Gamma \subset P$, such that its projection on $\mathfrak{h}^0$ is Γ^0.

DEFINITION. The vertex operator $X^\sigma(\mu, \zeta): V_k^\sigma \to V_{k'}^\sigma$, $\mu \in \Gamma$, $\zeta \in \mathbb{C}\backslash 0$, defined by the formula

$$X^\sigma(\mu, \zeta) = \exp\left(2 \sum_{n=1}^{\infty} \frac{\zeta^n}{n} \mu\left(-\frac{n}{2}\right)\right) \exp(2 \ln \zeta \cdot \mu(0) + \mu^0)$$
$$\times \exp\left(-2 \sum_{n=1}^{\infty} \frac{\zeta^{-n}}{n} \mu\left(\frac{n}{2}\right)\right). \tag{I.2.23}$$

Note that if $\mu \in \Gamma^0$, then $\mu(n/2) = 0$ for $n \in 2\mathbb{Z} + 1$ and we obtain $X^\sigma(\mu, \zeta) = X(\mu, \zeta^2)$. We also define

$$X^{\sigma\varepsilon}(\mu, \zeta) = X^\sigma(\mu, \zeta)\varepsilon^0_{\mu^0} \quad \text{and} \quad X_n^\sigma(\mu), X_n^{\sigma\varepsilon}(\mu), X^{\sigma\varepsilon}(\lambda, n, z, z_0)$$

as above (see (I.2.12)–(I.2.14)). One has the generalization of Proposition I.2.15 (i) (see [11]).

$$X^{\sigma\varepsilon}(\lambda, \zeta)\, X^{\sigma\varepsilon}(\mu, \zeta) = \varepsilon^0(\lambda^0, \mu^0)(\zeta - \zeta_0)^{\langle\lambda, \mu\rangle}(\zeta + \zeta_0)^{\langle\lambda, \sigma(\mu)\rangle}$$
$$\times (\zeta\zeta_0)^{-\langle\lambda^0, \mu^0\rangle} X^{\sigma\varepsilon}(\lambda, \mu, \zeta, \zeta_0), \tag{I.2.24}$$

where $|\zeta| > |\zeta_0|$. It implies an analogue of (I.2.15) (ii).

Let us choose $\Gamma^0 = P^0$ and let restriction ε^0 to $Q^0 \times Q^0$ be a projection of the cocycle ε on $Q \times Q$ satisfying (S.7). Then we have [11]:

THEOREM I.2.25. *Let the action of the Heisenberg algebra $\hat{\mathfrak{H}}_\sigma$ be defined by the formulas* (I.2.3) *in V_k^σ. We define the action of $x_\alpha^\varepsilon(n) \in \hat{\mathfrak{G}}_\sigma$, $\alpha \in \Delta$, $n \in \frac{1}{2}\mathbb{Z}$ by the components of the vertex operator $X_n^{\sigma\varepsilon}(\alpha)$. Then V_k is a fundamental module of $\hat{\mathfrak{G}}_\sigma$, V_1^σ is a basic module, and we have*

$$x^* = -\theta x, \qquad x \in \hat{\mathfrak{G}}_\sigma.$$

We denote

$$\tilde{x}^{\sigma}(\zeta)=\sum_{n\in\mathbb{Z}} x(n/2)\zeta^{-n}, \qquad x\in\mathfrak{g},\quad \zeta\in\mathbb{C}\backslash 0; \tag{I.2.26}$$

thus for $\tilde{x}\in\mathfrak{g}^0$ one has $\tilde{x}(\zeta^2)=\tilde{x}^{\sigma}(\zeta)$. We introduce the Virasoro operators $D(m)$, $m\in\mathbb{Z}$, by the formula

$$\tilde{D}(\zeta^2)=-\tfrac{1}{2}\sum_{i=1}^{l} :(\tilde{h}_i^{\sigma}(\zeta))^2:. \tag{I.2.27}$$

Using the same arguments as above we obtain exact generalization of Proposition I.2.20. Note that the Coxeter numbers for $\mathfrak{g}$ and $\mathfrak{g}^0$ coincide.

3. *Spinor Representations of Orthogonal Affine Lie Algebras*

3.1. We will describe in this section another construction of the vertex representations of orthogonal affine Lie algebras given in [9]. This construction is analogous to the construction of the spinor representations of the simple Lie algebras [5]. We will begin with the type $D_l^{(1)}$ (Note that $D_3\simeq A_3$, $D_2\simeq A_1+A_1$. In the last case the Lie algebra $so(4,\mathbb{C})$ is not simple, nevertheless all the results are still valid in this case as well.)

Let $C(Z^n)$ denote an infinite-dimensional Clifford algebra generated by elements $e_i(m)$, $i=1,\dots,n$; $m\in Z$, in which Z denotes either $\mathbb{Z}$ or $\mathbb{Z}+\frac{1}{2}$, satisfying the relations

$$\{e_i(k), e_j(m)\}=e_i(k)e_j(m)+e_j(m)e_i(k)$$
$$=-2\delta_{ij}\delta_{k,-m}, \qquad i,j=1,\dots,n;\ k,m\in Z \tag{I.3.1}$$

We define a $\mathbb{Z}_2$-grading $C(Z^n)=C^0(Z^n)\oplus C^1(Z^n)$ as usual (see S.10). We denote by $\mathfrak{E}$ the complex vector space spanned by $e_i(m)$, $i=1,\dots,$; $m\in Z$. Let θ be an antilinear involutive automorphism of $\mathfrak{e}$ defined by

$$\theta(e_i(m))=e_i(-m). \tag{I.3.2}$$

It extends naturally to the Clifford algebra $C(Z^n)$.

Let $n=2l$. We define the simple $C(Z^{2l})$-module, which we denote by $V(Z^l)$, as follows. Let

$$b_j(m)=(e_j(m)+ie_{j+l}(m))/2,$$
$$b_{-j}(m)=(e_j(m)-ie_{j+l}(m))/2, \qquad j=1,\dots,l;\quad m\in Z. \tag{I.3.3}$$

Then for these elements the only nonzero anticommutators are

$$\{b_j(m), b_{-j}(-m)\}=-1, \qquad j=\pm 1,\dots,\pm l;\quad m\in Z. \tag{I.3.4}$$

We call $b_j(m)$, $m<0$, $b_j(0)$, $j<0$, creation operators and $b_j(m)$, $m>0$, $b_j(0)$,

$j > 0$ annihilation operators, in whch $j = 1, \ldots, l$; $m \in Z$. We define $V(Z^l)$ to be the free module generated by the creation operators acting on the vacuum vector v_0. The action of $C(Z^{2l})$ is defined by the condition

$$b_j(m)\, v_0 = 0, \qquad b_j(m) \text{ an annihilation operator.} \tag{I.3.5}$$

$V(Z^l)$ has a natural $\mathbb{Z}_2$-grading compartible with the $\mathbb{Z}_2$-grading of $C(Z^{2l})$; i.e., $V(Z^l) = V^0(Z^l) \oplus V^1(Z^l)$, $V^0(Z^l) = C^0(Z^{2l})\, v_0$.

We define now a positive Hermitian structure (,) in $V(Z^l)$ by the characteristic conditions

$$\begin{gathered} b_j^*(m) = -\theta(b_j(m)) = -b_{-j}(-m), \qquad j = \pm 1, \ldots, \pm l, \quad m \in Z, \\ (v_0, v_0) = 1. \end{gathered} \tag{I.3.6}$$

Let $r = \sum_{i=1}^{2l} a_i e_i \in \mathfrak{e}$, $a_i \in \mathbb{C}$. We denote by $r(n)$ the element $\sum_{i=1}^{2l} a_i e_i(n)$. We introduce the normal ordering of two elements of $\mathfrak{E}$ similar to (I.1.9)

$$\begin{aligned} :r(k)\, s(n): &= r(k)\, s(n), && n > k, \\ &= (r(k)\, s(n) - s(n)\, r(k))/2, && n = k, \\ &= -s(n)\, r(k), && n < k. \end{aligned} \tag{I.3.7}$$

We wll consider $\mathfrak{o}(2l)$ realized as a linear subspace of the finite-dimensional Clifford algebra $C(2l)$ with the Lie bracket (S.13). Let $r_1, r_2 \in \mathfrak{e}$, then $x = :r_1 r_2:$ belongs to $\mathfrak{o}(2l)$, where $:r_1 r_2: = (r_1 r_2 - r_2 r_1)/2$. We define

$$x(m) = \sum_{k \in Z} :r_1(k)\, r_2(m-k):, \qquad m \in \mathbb{Z}. \tag{I.3.8}$$

This operator is well defined in $V(Z^l)$, because for every $v \in V(Z^l)$ only a finite number of terms in (I.3.8) does not annul v. We write, as in (I.2.17),

$$\tilde{r}(z) = \sum_{n \in Z} r(n)\, z^{-n}, \qquad r \in \mathfrak{e}. \tag{I.3.9}$$

Then (I.3.8) can be written in the form

$$\tilde{x}(z) = :\tilde{r}_1(z)\, \tilde{r}_2(z):. \tag{I.3.10}$$

The formal operators $\tilde{r}(z)$, $\tilde{x}(z)$ acts from $V(Z^l)$ to its dual $V(Z^l)'$ as the operator in (I.2.17).

PROPOSITION I.3.11. *One has a representation of* $\hat{\mathfrak{O}}(2l)$ *in* $V(Z^l)$ *defined by* (I.3.8), c *acts as the identity operator, and also*

$$x^* = -\theta x, \qquad x \in \hat{\mathfrak{O}}(2l).$$

Proof. To prove the first statement it is sufficient to verify the commutation relations (I.1.1) for monomials $x = r_1 r_2 \in \mathfrak{o}(2l)$, $y = s_1 s_2 \in \mathfrak{o}(2l)$, $m, n \in Z$. This verification is similar to the proof of the commutation relations of the Virasoro algebra (cf. [10]). For $s \in \mathfrak{e}$, $n \in Z$ one has

$$[x(m), s(n)] = \sum_{k \in Z} r_1(k)\{r_2(m-k), s(n)\} - \{r_1(k), s(n)\}\, r_2(m-k)$$
$$= r_1(m+n)\{r_2, s\} - \{r_1, s\}\, r_2(m+n) = [x, s](m+n). \quad (I.3.12)$$

Then we obtain

$$[x(m), y(n)] = \sum_{k \in Z} :([x(m), s_1(k)]\, s_2(n-k)$$
$$+ s_1(k)[x(m), s_2(n-k)]): + \text{Const}\, \delta_{m,-n}$$
$$= \sum_{k \in Z} :([x, s_1](m+k)\, s_2(n-k)$$
$$+ s_1(k)[x, s_2](m+n-k)): + \text{Const}\, \delta_{m,-n}$$
$$= [x, y](m+n) + \text{Const}\, \delta_{m,-n}. \quad (I.3.13)$$

Finally using definitions (S.9), (S.11) we calculate the constant in (I.3.13)

$$(v_0, x(m)\, y(-m)\, v_0) = \Big(v_0, \sum_{0 < k < m} r_1(k)\, r_2(m-k)$$
$$\times \sum_{0 < k < m} s_1(-m+k)\, s_2(-k)\, v_0\Big) = m(\{r_1 s_1\}\{r_2 s_2\}$$
$$= \{r_1 s_2\}\{r_2 s_1\}) = m\langle x, y\rangle, \qquad m > 0. \quad (I.3.14)$$

The last statement follows from (I.3.6).

In the spinor representation we again can extend the action of affine Lie algebra $\hat{\mathfrak{O}}(2l)$ to the semidirect product with the derivation algebra $\mathfrak{D}$ using the Segal operators (I.1.10). However, as in the case of vertex representations we will obtain a simpler form of these operators. This reduction will allow us to decompose $V(Z^l)$ into irreducible components.

We call the Virasoro operators the following endomorphisms of $V(Z^l)$:

$$D_i(m) = -\sum_{k \in Z} (k - m/2) :b_i(k)\, b_{-i}(m-k):, \qquad m \in \mathbb{Z},$$
$$D(m) = \sum_{i=1}^{l} D_i(m), \quad (I.3.15)$$

or as the formal series

$$\tilde{D}_i(z) = - :\tilde{b}_i(z)\left(z \frac{\vec{d}}{dz}\right) \tilde{b}_{-i}(z):$$
$$= -\frac{1}{2} :\tilde{b}_i(z)\left(z \frac{d}{dz} \tilde{b}_{-i}(z)\right) - \left(z \frac{d}{dz} \tilde{b}_i(z)\right) \tilde{b}_{-i}(z):,$$
$$\tilde{D}(z) = \sum_{i=1}^{l} \tilde{D}_i(z). \quad (I.3.16)$$

Definition (I.3.15) implies the following equality

$$[D(m), b_{\pm i}(n)] = -\sum_{k\in Z} (k - m/2)(b_i(k)\{b_{-i}(m-k), b_{\pm i}(n)\} \tag{I.3.17}$$
$$- \{b_i(k), b_{\pm i}(n)\}\, b_{-i}(m-k)) = (n + m/2)\, b_{\pm i}(m+n).$$

The construction of the affine Lie algebra $\hat{\mathfrak{o}}(2l)$ and the equality (I.3.17) imply immediately that the Virasoro operators define the representation of the derivation algebra $\mathfrak{D}$. As in Section 2 (I.2.21) we obtain that the are scalar multiples of the Segal's operators.

PROPOSITION I.3.18. (i) *$D(m) = -[1/2(h+1)]\, L(m)$, in which $h = 2l - 2$ is a Coxeter number of $\mathfrak{o}(2l)$.*

(ii) $[D(m), x(k)] = kx(m+k), \quad x \in \mathfrak{o}(2l).$

(iii) $[D(m), D(k)] = (k-m)\, D(m+k) + (l/12)(m^3 - m)\, \delta_{m,-k}.$

Proof. Statement (ii) follows from (I.3.17) and also from (i). We will prove now the statement (i)[2]), leaving the verification of (iii) to the reader. We will see in the next section that (iii) follows from (I.2.21 iii).

We verfy first the following equality

$$:(:\bar{b}_i(z)\, \bar{b}_{-i}(z)\, \bar{b}_{-i}(z): :\bar{b}_i(z)\, \bar{b}_{-i}(z):): = -2\, :\bar{b}_i(z) \left(z \frac{d}{dz}\right) \bar{b}_{-i}(z):. \tag{I.3.19}$$

Let us set $\bar{b}_i(z) = \bar{b}_i^+(z) + \bar{b}_i^-(z)$, in which

$$\bar{b}_i^-(z) = b_i(0)/2 + \sum_{n=1}^{\infty} b_i(-n)\, z^n, \qquad Z = \mathbb{Z},$$
$$= \sum_{n>0, n\in Z} b_i(-n)\, z^n, \qquad Z = \mathbb{Z} + \tfrac{1}{2}.$$

Then one has

$$\{\bar{b}_i^+(z), \bar{b}_{-i}^-(z_0)\} = \frac{\xi}{z - z_0}, \qquad \xi = \frac{1}{2}(z + z_0), \quad Z = \mathbb{Z}$$
$$\xi = (zz_0)^{1/2}, \quad Z = \mathbb{Z} + \tfrac{1}{2}$$

We can transform the left side of (I.3.19) as follows:

[2] The statement (i) probably was first proven by S. Mandelstam (see Footnote 12 in [15]).

$$\lim_{z\to z_0}\left[(\bar{b}_i(z)\,\bar{b}^{+}_{-i}(z)-\bar{b}_{-i}(z)\,\bar{b}_i(z))(\bar{b}_i(z_0)\,\bar{b}^{+}_{-i}(z_0)-\bar{b}_{-i}(z_0)\,\bar{b}_i(z_0))-\frac{zz_0}{(z-z_0)^2}\right]$$

$$=\lim_{z\to z_0}\{\bar{b}_i(z)\{\bar{b}^{+}_{-i}(z),\bar{b}_i(z_0)\}\,\bar{b}^{+}_{-i}(z_0)+\bar{b}^{-}_{-i}(z)\}\,\bar{b}_i(z),\bar{b}_{-i}(z_0)\}\,b_i(z_0)$$

$$+\bar{b}_i(z)\,\bar{b}^{-}_{-i}(z_0)\,\bar{b}^{+}_{-i}(z)\,\bar{b}_i(z_0)-\frac{zz_0}{(z-z_0)^2}\Bigg]$$

$$=\lim_{z\to z_0}\frac{\xi}{z-z_0}(\bar{b}_i(z)\,\bar{b}^{+}_{-i}(z_0)+\bar{b}^{-}_{-i}(z)\,\bar{b}_i(z_0)$$

$$-\bar{b}_{-i}(z_0)\,\bar{b}_i(z)-\bar{b}_i(z_0)\,\bar{b}^{+}_{-i}(z)+\frac{\xi^2}{(z-z_0)^2}-\frac{zz_0}{(z-z_0)^2}\Bigg]$$

$$=-2\,:\bar{b}_i(z)\left(z\frac{\vec{d}}{dz}\right)\bar{b}_{-i}(z):.$$

Similar transformations give another equality:

$$:(\bar{b}_i(z)\,\bar{b}_{\pm j}(z))(\bar{b}_{-i}(z)\,\bar{b}_{\mp j}(z):+:(\bar{b}_{-i}(z)\,\bar{b}_{\mp j}(z)\,\bar{b}_i(z)\,\bar{b}_{\pm j}(z)):$$
$$=-:(\bar{h}_i(z)\pm\bar{h}_j(z))^2:, \qquad \text{(I.3.20)}$$

where $i,j>0$, $i\neq j$. Now the simple identity $\sum_{i<j}(h_i+h_j)^2+\sum_{i<j}(h_i-h_j)^2=(2l-2)\sum_{i=1}^{l}h_i^2$ completes the proof of (i).

It follows from the proof of Proposition I.3.18 (i) that the operators $D_i(m)$ defined in (I.3.15) can be constructed from the operators of the algebra $\hat{\mathfrak{O}}(2l)$. This fact is essential in the proof of the next result.

THEOREM I.3.21. *Each of the spinor representations of* $\hat{\mathfrak{O}}(2l)$ *in the spaces* $V(\mathbb{Z}^l)$ *and* $V((\mathbb{Z}+1/2)^l)$ *is decomposed into two irreducible components according to its* $\mathbb{Z}_2$*-grading. One has* $V^0((\mathbb{Z}+1/2)^l)\cong V_0(D_l^{(1)})$ $V^1((\mathbb{Z}+1/2)^l)\cong V_1(D_l^{(1)})$, $V^0(\mathbb{Z}^l)\cong V(D_l^{(1)})$, $V^1({}^l)\cong V_{l-1}(D_l^{(1)})$ *and* v_0, $b_{-l}(0)\,v_0$, v_0, $b_{-l}(0)\,v_0$, *are, respectively, highest weight vectors.*

Proof. Let us consider the one-dimensional groups generated by the operators $D_j(0)$, $j=1,\ldots,l$. Their action on the elements of the Chevalley basis of $\hat{\mathfrak{O}}(2l)$ is the following:

$$e^{i\tau D_j(0)}\left(\sum_{n\in Z}b_{\pm j}(n)\,b_k(m-n)\right)e^{-i\tau D_j(0)}=\sum_{n\in Z}e^{\pm i\tau n}b_{\pm j}(n)\,b_k(m-n), \qquad \text{(I.3.22)}$$

where $k=\pm1,\ldots,\pm l$, $j=1,\ldots,l$; $k\neq\pm j$.

The Fourier transform by τ of the right side of (I.3.22), provides us with the elements $b_{\pm j}(n)\,b_k(m-n)$. The union of these elements for arbitrary $m\in\mathbb{Z}$, $n\in Z$, $k=\pm1,\ldots,\pm1$, certainly generates $C^0(Z^l)$ the even component of the Clifford algebra. This subalgebra obviously has two irreducible components in $V(Z^l)$ according to its $\mathbb{Z}_2$-grading. Now recalling the fact that the operators $D_i(0)$, $i = 1,\dots, l$, are generated by $\hat{\mathfrak{O}}(2l)$, and consequently the whole $C^0(Z^{2l})$, we obtain that $\hat{\mathfrak{O}}(2l)$ has at most two irreducible components in $V(Z^l)$. On the other hand, it obviously preserves $\mathbb{Z}_2$-grading; this proves the first part of the theorem.

To prove the second part of the theorem one has to verify (I.1.12) and compare the highest weight with (I.1.14). Note that

$$h_l(0)\, v_0 = \left(\sum_{k\in Z} :b_{-l}(k)\, b_l(-k): \right) v_0 = 0, \qquad Z = \mathbb{Z} + \tfrac{1}{2},$$
$$= \tfrac{1}{2} v_0, \qquad Z = \mathbb{Z}. \tag{I.3.23}$$

Therefore

$$hv_0 = \langle \hat{\omega}_0, h\rangle\, v_0,\, hb_{-l}(0)\, v_0 = \langle \hat{\omega}_1, h\rangle\, b_{-l}(0)\, v_0, \qquad Z = \mathbb{Z} + \tfrac{1}{2},$$
$$hv_0 = \langle \hat{\omega}_l, h\rangle\, v_0, \oplus\, hb_{-l}(0)\, v_0 = \langle \hat{\omega}_{l-1}, h\rangle\, b_{-l}(0)\, v_0; \; Z = \mathbb{Z},$$

where $h \in \hat{\mathfrak{h}} \oplus \mathbb{C}d$, $\hat{\omega}_1, \hat{\omega}_{l-1}, \hat{\omega}_l$ are given by (I.1.14). From the definition (I.3.8) one gets

$$x(m)\, v = 0, \qquad m > 0, \quad x \in \mathfrak{o}(2l),$$

where $v = v_0$, or $v = b_{-l}(0)\, v_0$, and also $x(0)\, v = 0$, $x \in \mathfrak{o}(2l)^\alpha$, $\alpha \in \Delta_+$. This completes the proof.

Propositions I.3.18 and I.3.11 imply that any element H of the scalar subalgebra $\mathfrak{o}(2l) \oplus \mathbb{C}\, d(0)$ acts as the operator

$$H = \int_{C_R} \left\{ a: \vec{e}(z) \left(z \frac{\vec{d}}{dz} \right) \vec{e}^{\,\prime}(z); + :\vec{e}(z)\, A \vec{e}^{\,\prime}(z): \right\} \frac{dz}{z},$$

where $\vec{e}(z) = \{\tilde{e}_1(z),\dots, \tilde{e}_l(z), \tilde{e}_{-1}(z),\dots, \tilde{e}_{-l}(z)\}$, $A \in \mathfrak{o}(2l)$. If the projection of H to $\mathfrak{o}(2l)$ is semisimple then we can find $g \in \mathfrak{o}(2l)$, $H' \in \mathfrak{h} \oplus \mathbb{C}\, d(0)$ so that $H = \operatorname{ad} g.\ H'$ and obtain a spectrum of H. The spectral decomposition of H also has a very simple form: Each operator $b_j(m)$, $m < 0$, has to be replaced by $(gb_j)(m)$, where gb_j denotes the natural representation of the group $O(2l)$. In the case $Z = \mathbb{Z}$, to complete the picture, we have to add that the product of $b_j(0)$ transformed by the 2^l-dimensional spinor representation.

3.2. In this section we will construct spinor representations of the affine Lie algebra $\hat{\mathfrak{O}}^{(2)}(2l + 2)$. In the preceding section we have already obtained the spinor representation of its subalgebra $\hat{\mathfrak{O}}(2l)$. The next step is the construction of the spinor representation of the intermidiate affine Lie algebra $\hat{\mathfrak{O}}(2l + 1)$. To this end we define the simple $C(Z^{2l+1})$-module, which we denote by $V(Z^l)$, as follows. Let $C(Z)$ be an infinite-dimensional Clifford algebra generated by elements $e(m)$, $m \in Z(e := e_{2l+1})$. We call $e(-m)$, $m > 0$, creation operators and $e(m)$, $m > 0$, annihilation operators, $m \in Z$. As above we define $V(Z_+)$ to be the free module generated by the creation operators. Then the action of $C(Z)$ is completely defined by the conditions

$$e(m)\, v_0 = 0, \qquad m > 0, \tag{I.3.24}$$

$$e(0)\, v_0 = i v_0. \tag{I.3.25}$$

Let $\bar{V}(Z^l) = V(Z^l) \otimes V(Z_+)$, and let the representation of $\hat{\mathfrak{O}}(2l+1)$ is constructed by the formula (I.3.8), where $x \in \mathfrak{o}(2l+1)$. The analogues of Propositions I.3.11, I.3.18 and Theorem I.3.21 and valid for $\hat{\mathfrak{O}}(2l+1)$. We obtain

THEOREM I.3.26. *The spinor representation of* $\hat{\mathfrak{O}}(2l+1)$ *in the space* $\bar{V}((\mathbb{Z}+1/2)^l)$ *is decomposed into two irreducible components according to its* $\mathbb{Z}_2$*-grading, and is irreducible in the space* $\bar{V}(\mathbb{Z}^l)$. *One has*

$$\bar{V}^0((\mathbb{Z}+1/2)^l) \cong V_0(B_l^{(1)}), \quad \bar{V}^1((\mathbb{Z}+1/2)^l) \cong V_1(B_l^{(1)}), \quad \bar{V}(\mathbb{Z}^l) \cong V_l(B_l^{(1)}),$$

and v_0, $b_{-l}(0)\, v_0$, v_0 *are, respectively, highest-weight vectors.*

Proof. Thanks to identity (I.3.19) the operators $D_j(0), j = 1, \dots, l$ preserve irreducible components of $\bar{V}(Z^l)$ and, by the same arguments as we used in the proof of Theorem I.3.21, the operators $b_{\pm j}(n)\, b_k(m-n)$, $b_{\pm j}(n)\, e(m-n)$ preserve it as well. These elements certainly generate $C^0(Z^{2l-1})$. For $Z = \mathbb{Z} + 1/2$, $C^0(Z^{2l+1})$ preserves Z_2-rading of $\bar{V}(Z^l)$, therefore the spinor representation of $\hat{\mathfrak{O}}(Z^{2l+1})$ preserves Z_2-grading of $\bar{V}(Z^l)$, therefore the spinor representation of $\hat{\mathfrak{O}}(2l+1)$ has two irreducible components. However, for $Z = \mathbb{Z}$, $C^0(Z^{2l+1})$ contains thje element $e(0)$ and mixes the two components of $\bar{V}(Z^l)$. In this case the spinor representation of $\hat{\mathfrak{O}}(2l+1)$ is irreducible.

The calculations of the highest weights of three irreducible representations are the same as for $\hat{\mathfrak{O}}(2l)$. We have $hv_0 = \langle \hat{\omega}_0, h \rangle\, v_0$, $hb_{-l}(0)\, v_0 = \langle \hat{\omega}_1, h \rangle\, b_{-l}(0)\, v_0$, $Z = \mathbb{Z} + 1/2$, and $hv_0 = \langle \hat{\omega}_l, h \rangle\, v_0$, $Z = \mathbb{Z}$, where $h \in \mathfrak{h} \oplus \mathbb{C}c$, $\hat{\omega}_1$, $\hat{\omega}_l$ are given by (I.1.18). Q.E.D.

The last step in the construction of the spinor representation of $\hat{\mathfrak{O}}^{(2)}(2l+2)$ is the construction of the action of elements $x(m)$, $x \in \mathfrak{e}$, $m \in \mathbb{Z} + \frac{1}{2}$. We enlarge again the Clifford algebra $C(Z^{2l+1})$ to the algebra $C(Z^{2l+1} \oplus Z') = C(Z^{2l+1}) \otimes C(Z')$, where Z' is the complement of Z to $\frac{1}{2}\mathbb{Z}$. We define the simple $C(Z^{2l+1} \oplus Z')$-module $\bar{\bar{V}}(Z^l)$ as the tensor product

$\bar{V}(Z') \otimes V(Z'_+)$. For $x = r \in \mathfrak{e}$, we define the action of $x(m) \in \hat{\mathfrak{O}}^{(2)}(2l+2)$, $m \in \mathbb{Z} + \frac{1}{2}$ in $\bar{V}(Z')$ by the formula

$$x(m) = \sum_{k \in \mathbb{Z}} r(k)\, e(m-k), \qquad m \in \mathbb{Z} + \tfrac{1}{2}. \tag{I.3.27}$$

PROPOSITION I.3.28. *One has a representation of* $\hat{\mathfrak{O}}(2l+2)$ *in* $\bar{V}(Z')$ *defined by* (I.3.8), (I.3.27), c acts as Id/2, *and also*

$$x^* = -\theta x, \qquad x \in \hat{\mathfrak{O}}^{(2)}(2l+2).$$

Proof. The proof is essentially the same as for (I.3.11). We will verify the commutation relations (I.1.1) for $x(m)$ defined by (I.3.27), $y(n) = \sum_{k \in \mathbb{Z}} s(k)\, e(n-k)$, $m, n \in \mathbb{Z} + \frac{1}{2}$. One has

$$[x(m), s(j)] = \sum_{k \in \mathbb{Z}} [r(k)\, e(m-k), s(j)] = \{r, s\}\, e(m+j), \qquad j \in \mathbb{Z},$$

$$[x(m), e(j)] = \sum_{k \in \mathbb{Z}} [r(k)\, e(m-k), e(j)] = -2r(m+j), \qquad j \in Z',$$

$$\begin{aligned}
[x(m), y(n)] &= \sum_{k \in \mathbb{Z}} [x(m), s(k)\, e(n-k)] \qquad\qquad (\text{I.3.29})\\
&= \sum_{k \in \mathbb{Z}} (-\{r, s\}\, e(m+k)\, e(n-k) + s(k)(-2r(m+n-k)))\\
&= \sum_{k \in \mathbb{Z}} :r(m+n-k)\, s(k) - s(k)\, r(m+n-k): + \delta_{m,-n} \cdot \text{Const}\\
&= [r, s](m+n) + \delta_{m,-n} \cdot \text{Const}.
\end{aligned}$$

We calculate the constant as in (I.3.14):

$$\begin{aligned}
(v_0, x(m)\, y(-m)\, v_0) &= (v_0, \sum_{0 \leqslant k \leqslant m} r(k)\, e(m-k) \sum_{0 \leqslant k \leqslant m} s(-k)\, e(-m+k) v_0)\\
&= (-2)\, m \{r, s\} = m\langle x, y \rangle, \qquad m > 0. \qquad (\text{I.3.30})
\end{aligned}$$

We now define the Virasoro operators in $V(Z')$

$$D(m) = \sum_{i=0}^{l} D_i(m), \qquad m \in \mathbb{Z}, \tag{I.3.31}$$

$$D_0(m) = \sum_{k \in \mathbb{Z}/2} (k - m/2) :e(k)\, e(m-k):, \qquad m \in \mathbb{Z}, \tag{I.3.32}$$

where $D_i(m)$, $i = 1,\ldots, l$, have been defined above. Then the analogue of Proposition I.3.18 is valid; in particular, we have

$$D(m) = -\frac{1}{2(h+1)} L(m), \tag{I.3.33}$$

where $h = 2l$ is a Coxeter number of $\mathfrak{o}(2l+1)$. Finally we state the analogue of (I.3.21), (I.3.26):

THEOREM I.3.34. *Each of the spinor representations of* $\hat{\mathfrak{O}}^{(2)}(2l+2)$ *in the space* $\bar{V}(\mathbb{Z}^l)$ *and* $\bar{V}((\mathbb{Z}+1/2)^l)$ *is irreducible. One has* $\bar{V}(\mathbb{Z}^l) \cong V_l(D_{l+1}^{(2)})$, $\bar{V}((\mathbb{Z}+1/2)^l) \cong V_0(D_{l+1}^{(2)})$ *and* v_0 *is the highest vector.*

Proof. The proof is the same as in I.3.21 except of the point that $C^0(\mathbb{Z}^{2l+1} \oplus \mathbb{Z}')$ contains the element $e(0)$, and therefore mixes $\mathbb{Z}_2$-components of $\bar{V}(\mathbb{Z}^l)$. It gives the irreducibility. The highest weights $\hat{\omega}_0$, $\hat{\omega}_l$ of $\hat{\mathfrak{O}}^{(2)}(2l+2)$ and of $\hat{\mathfrak{O}}(2l)$ coincide by (I.1.14), (I.1.17); this implies the equivalence.

4. *Isomorphism between the Two Constructions of Representations*

4.1. In the Sections 2 and 3 we have introduced two constructions of the fundamental representations of affine Lie algebras of type $D_l^{(1)}$, $D_{l+1}^{(2)}$ corresponding to endpoints of their Dynkin diagram. In this section we will establish isomorphism between these constructions.

We choose $\varepsilon: P \times P \to T$, where P is a weight lattice of $\mathfrak{o}(2l)$, as follows. We note first that $Q \subset \mathbb{Z}^l \subset P$, where $\mathbb{Z}^l$ is an integer lattice spanned by the orthonormal basis $\{h_i\}_{i=1}^l$ of $\mathfrak{h}$. We define

$$\begin{aligned} \varepsilon(h_i, h_j) &= 1, \qquad i<j, \\ &= -1, \qquad i \geqslant j, \end{aligned} \tag{I.4.1}$$

and extend it by the bilinear condition to $\varepsilon: \mathbb{Z}^l \times \mathbb{Z}^l \to \{\pm 1\}$. One can verify easily that ε satisfies the conditions (S.1) and (S.2) on $Q \times Q$. One can extend ε to $P \times P$ with values in T, in an arbitrary way.

Evidently $\mathbb{Z}^l$ has two orbits in P, which we denote by P_{12} and P_{34} ($P_{12} \cong \mathbb{Z}^l$). We consider the infinite-dimensional spaces $V_{12} = S(\mathfrak{H}^-) \otimes \mathbb{C}[P_{12}]$ and $V_{34} = S(\mathfrak{H}^-) \otimes \mathbb{C}[P_{34}]$, and define the vertex operators $X^\varepsilon(\pm h_i, z)$ by te formula (I.2.10) in these spaces. We can define the decomposition

$$X^\varepsilon(\pm h_i, z) = \sum_{n \in Z} X_n^\varepsilon(\pm h_i) z^{-n}, \tag{I.4.2}$$

where $Z = \mathbb{Z}$ for V_{12} (because $\langle P_{12}, \mathbb{Z}^l \rangle \subset \mathbb{Z}$) and $Z = \mathbb{Z} + \frac{1}{2}$ for V_{34}

(because $\langle P_{34}\, \mathbb{Z}^l \rangle \subset \mathbb{Z} + \frac{1}{2}$) as in Section 2. Proposition I.2.15 implies the following anticommutation relations

$$\begin{aligned} \{X_n^\varepsilon(h_i), X_m^\varepsilon(-h_j)\} &= -\delta_{ij}\,\delta_{n,-m}, \\ \{X_n^\varepsilon(h_i), X_m^\varepsilon(h_j)\} &= \{X_n^\varepsilon(-h_i), X_m^\varepsilon(-h_j)\} = 0. \end{aligned} \tag{I.4.3}$$

Let us set

$$v_0 = 1 \otimes 1 \text{ in } V_{12}, \qquad v_0 = 1 \otimes e^{\omega_l} \text{ in } V_{34}. \tag{I.4.4}$$

Then one can check easily that for $n > 0$, $1 \leqslant i \leqslant l$,

$$X_n^\varepsilon(\pm h_i)\, v_0 = 0, \qquad X_0^\varepsilon(h_l)\, v_0 = 0. \tag{I.4.5}$$

Since we have in V_{12} that $X_n^\varepsilon(\pm h_i)\, 1 \otimes 1 = \operatorname{Res}_{z=0}(\exp \sum_{m=1}^{\infty} (z^m/m)$ $(\pm h_i(-m))1 \otimes 1 \cdot z^{1/2} \cdot z^{n-1}) = 0$, for $n \in \mathbb{Z} + \frac{1}{2}$, $n > 0$, and similar for V_{34}. Therefore, if we set (for $n \in Z$)

$$b_i(n) = X_n^\varepsilon(h_i), \qquad b_{-i}(n) = X_n^\varepsilon(-h_i), \tag{I.4.6}$$

then the properties (I.4.3) and (I.4.5) of $X_n^\varepsilon(\pm h_i)$ imply that $b_i(n)$, $n \in \mathbb{Z}$, $i = \pm 1, \ldots, \pm l$, satisfy conditions (I.3.4) and (I.3.5) of the preceding section.

THEOREM I.4.7. *Under the identification* (I.4.4) *and* (I.4.6) *we obtain the isomorphism of* $\hat{\mathfrak{o}}(2l)$*-modules*

$$V((\mathbb{Z} + 1/2)^l) \cong V_{12}, \qquad V(\mathbb{Z}^l) \cong V_{34},$$

where the correspondence of the Chevalley basis elements of $\hat{\mathfrak{o}}(2l)$ *is given by the formulas*:

$$\sum_{k \in Z} b_i(k)\, b_{-j}(m-k) = X_m^\varepsilon(h_i - h_j), \qquad i < j, \tag{I.4.8}$$

$$\sum_{k \in Z} b_{\pm i}(k)\, b_{\pm j}(m-k) = X_m^\varepsilon(\pm(h_i + h_j)), \qquad i < j, \tag{I.4.9}$$

$$\sum_{k \in Z} :b_i(k)\, b_{-i}(m-k): = h_i(m). \tag{I.4.10}$$

Proof. Equation (I.4.8) follows immediately from the fact $X^\varepsilon(h_i, z)\, X^\varepsilon(-h_j, z) = \varepsilon(h_i, -h_j)\, X^\varepsilon(h_i - h_j, z)$ and definition (I.4.1) of ε; (I.4.9) is similar.

To prove (I.4.10) we note that

$$:\tilde{b}_i(z_0)\, \tilde{b}_{-i}(z_0): = \lim_{z \to z_0} (X^\varepsilon(h_i, z)\, X^\varepsilon(-h_i, z_0) - f(z, z_0)),$$

where

$$f(z, z_0) = (z^{-1}z_0)^{1/2} + (z^{-1}z_0)^{3/2} + \cdots$$
$$= (z^{-1}z_0)^{1/2}/(1 - z^{-1}z_0) = (zz_0)^{1/2}/(z - z_0),$$

when $n \in \mathbb{Z} + \frac{1}{2}$, and $f(z, z_0) = \frac{1}{2} + z^{-1}z_0 + (z^{-1}z_0)^2 + \cdots = 1/(1 - z^{-1}z_0) - \frac{1}{2} = \frac{1}{2}(z + z_0)/(z - z_0)$, $n \in \mathbb{Z}$. Using (I.2.15) (i) we obtain

$$\lim_{z \to z_0} \left(\frac{(zz_0)^{1/2}}{z - z_0} X(h_i, -h_i, z, z_0) - \frac{(zz_0)^{1/2}}{z - z_0} \right)$$
$$= \frac{d}{dz} ((zz_0)^{1/2} X(h_i, -h_i, z, z_0) - (zz_0)^{1/2} |_{z = z_0} = \tilde{h}_i(z_0),$$

when $n \in \mathbb{Z} + \frac{1}{2}$, and similar for $n \in \mathbb{Z}$.

Now the main statement of the theorem follows from the theorems (I.2.16) and (I.3.21), which claim that the spaces in consideration are pairwise decomposed in the same fundamental $\hat{\mathfrak{O}}(2l)$-modules.

PROPOSITION I.4.11. *The two definitions* (I.2.18) *and* (I.3.15) *of the Virasoro operators $D(m)$, $m \in \mathbb{Z}$, acting on $\hat{\mathfrak{O}}(2l)$-modules $V((\mathbb{Z} + \frac{1}{2})^l)$, $V(\mathbb{Z}^l)$ and V_{12}, V_{34}, respectively, coincide under the isomorphism given in* (I.4.7).

Proof. The proof follows from Theorem I.4.7 and the propositions (I.2.20) and (I.3.18).

Remark I.4.12. Note that the correspondence of the Chevalley bases (I.4.8)–(I.4.10) of $\hat{\mathfrak{O}}(2l)$ in the two realizations and the extension of this correspondence to $\hat{\mathfrak{H}}$ (I.4.11) does not imply (I.4.6). Let us introduce the parity operator

$$\begin{aligned} e(0)\, v \otimes e^{\lambda} &= iv \otimes e^{\lambda}, && \text{if } v \otimes e^{\lambda} \in V_0 \text{ (resp. } V_l), \\ &= -iv \otimes e^{\lambda}, && \text{if } v \otimes e^{\lambda} \in V_1 \text{ (resp. } V_{l-1}), \end{aligned} \tag{I.4.13}$$

where $v \in S(\mathfrak{H}^-)$, $\lambda \in P_{0,l}$ (resp. $P_{1,l-1}$); i.e., $e(0)$ is the multiplication by i when $\langle \lambda, \lambda \rangle$ (resp. $\langle \lambda - \omega_1, \lambda - \omega_1 \rangle$) is even and is the multiplication by $(-i)$ otherwise. One has $e(0)^2 = -1$ and $\{e(0), X_n(\pm h_i)\} = 0$. Thus the anticommutation relations (I.4.3) are valid also for the operators

$$X_n^{e\prime}(\pm h_i) = X_n(\pm h_i)\, e(0). \tag{I.4.14}$$

It follows immediately from the properties of $e(0)$ that all the quadratic expressions of the operators $X_n^{e\prime}(\pm h_i)$ are equal to the corresponding quadratic expressions of the operators $X_n^\varepsilon(\pm h_i)$. Therefore if we substitute (I.4.6) by

$$b_i(n)=X_n^{\varepsilon\prime}(h_i),\qquad b_{-i}(n)=X_n^{\varepsilon\prime}(-h_i). \tag{I.4.15}$$

Theorem I.4.7 and Proposition I.4.11 remain true.

Using Theorem I.4.7 and Proposition I.4.11 we can compare character formulas of the isomorphic representations and obtain some identities. First we recall the definition of the character of $\hat{\mathfrak{G}}$-module V (see [17]).

Let $V=\sum_{n\in N} V^n$ be a decomposition into the eigenspaces of the operator $2D(0)$ with eigenvalue n. We will consider only such modules V that $V^n=\{0\}$, for sufficiently large n, and that $\dim V^n<+\infty$. Every V^n is decomposed into the direct sum of the subspaces V^n_μ with respect to $\mathfrak{h}\subset\mathfrak{g}\subset\hat{\mathfrak{G}}$, μ is in the weight lattice P. By definition the character of V is a formal sum

$$\operatorname{ch} V=\sum_{n<N}\left(\sum_{\mu\in P}\dim V^{n,\mu}\cdot e^\mu\right)q^n, \tag{I.4.16}$$

where $\sum_{\mu\in P}\dim V^{n,\mu}\cdot e^\mu$ is an element of the group algebra $\mathbf{C}[P]$, for every $n<N$. Using the standard properties of characters $\operatorname{ch}(V_1\otimes V_2)=\operatorname{ch} V_1\cdot\operatorname{ch} V_2$, $\operatorname{ch}(V_1\oplus V_2)=\operatorname{ch} V_1+\operatorname{ch} V_2$ and the commutation relations $[h_j(0),b_{\pm i}(m)]=\pm\delta_{ij}b_{\pm i}(m)$, $i,j=1,\dots,l$, we can rewrite the equality for characters

$$\operatorname{ch} V((\mathbb{Z}+1/2)^l)=\operatorname{ch} V_{12},\qquad \operatorname{ch} V(\mathbb{Z}^l)=\operatorname{ch} V_{34} \tag{I.4.17}$$

in the form

$$\prod_{n=1}^{\infty}\prod_{i=1}^{l}(1+q^{2n-1}e^{h_i})(1+q^{2n-1}e^{-h_i})=\prod_{n=1}^{\infty}(1-q^{2n})^{-l}\Sigma q^{\langle\gamma,\gamma\rangle}e^\gamma$$

$$e^{\omega_l}\prod_{n=1}^{\infty}\prod_{i=1}^{l}(1+q^{2n-2}e^{h_i})(1+q^{2n}e^{-h_i})=\prod_{n=1}^{\infty}(1-q^{2n})^{-l}\Sigma q^{\langle\gamma,\gamma\rangle+2\langle\gamma,\omega_l\rangle}e^{\gamma+\omega_l}, \tag{I.4.18}$$

in which the sum is taken over all $\gamma=n_1h_1+\cdots+n_lh_l$, $h_k\in\mathbb{Z}$, $k=1,\dots,l$, $\omega_l=(h_1+\cdots+h_l)/2$.

It is clear that each expression in (I.4.18) is decomposed into l products. As a result we obtain famous Jacobi identity for elliptic θ-functions.

4.2. Now we proceed to the type $D_{l+1}^{(2)}$. Let us consider first the degenerate case $l=0$, which has special interest. In this case the corresponding infinite-dimensional Lie algebra is spanned by the elements $h_0(n)$, $n \in \mathbb{Z} + \frac{1}{2}$, with the commutation relations

$$[h_0(n), h_0(m)] = n\, \delta_{n,-m} c. \qquad (I.4.19)$$

We denote this algebra by $\mathfrak{A}$.

Let $V = S(\mathfrak{A}^-)$ be the space of the natural representation of this algebra, $c = \mathrm{Id}$. We also know that the irreducible representation of this algebra can be obtained in the space $V(\frac{1}{2}\mathbb{Z}_+)$ which is a simple module of the infinite-dimensional Clifford algebra $C(\frac{1}{2}\mathbb{Z})$. Therefore we can easily identify these spaces. The main question here is how to construct the generators of the Clifford algebra in the space $V = S(\mathfrak{A}^-)$. The answer is again given by the vertex operator but of a special form

$$X(h_0, \zeta) = \exp\left(2 \sum_{m=1}^{\infty} \frac{\zeta^{2m-1}}{2m-1} h_0\left(-\frac{2m-1}{2}\right)\right)$$
$$\times \exp\left(-2 \sum_{m=1}^{\infty} \frac{\zeta^{-2m+1}}{2m-1} h_0\left(\frac{2m-1}{2}\right)\right). \qquad (I.4.20)$$

PROPOSITION I.4.21.

(i) $\{X_{m/2}(h_0), X_{m/2}(h_0)\} = 2(-1)^n \delta_{n,-m}$, $\quad n, m \in \mathbb{Z}$,

(ii) $X_{n/2}(h_0)\, v_0 = 0$, $\quad h > 0, \quad n \in \mathbb{Z}$,
$X_0(h_0)\, v_0 = v_0$

(iii) $h_0(m - 1/2) = \frac{1}{2} \sum_{n \in \mathbb{Z}} X_{m-n-1/2}(h_0) X_n(h_0)$, $\quad m \in \mathbb{Z}$.

Proof. Part (ii) can be proven as (I.4.5). Part (i) follows from the standard calculations with vertex operators (cf. (I.2.15)).

$$X(h_0, \zeta) X(h_0, \zeta_0) = \frac{\zeta - \zeta_0}{\zeta + \zeta_0} X(h_0, h_0\, \zeta, \zeta_0), \qquad |\zeta| > |\zeta_0|.$$

$$X_{n/2}(h_0), X(h_0, \zeta_0) = \operatorname*{Res}_{\zeta = -\zeta_0} \left(\frac{\zeta - \zeta_0}{\zeta + \zeta_0} X(h_0, h_0, \zeta, \zeta_0)\, \zeta^{n-1}\right) = 2(-1)^n \zeta_0^n.$$

To prove (iii) we have to show that

$$\tilde{h}_0(\zeta) = \tfrac{1}{2} :(X(h_0, \zeta) - X(h_0, -\zeta))(X(h_0, \zeta) + X(h_0, -\zeta)):. \qquad (I.4.22)$$

The right side can be transformed in the following way:

$$\lim_{\zeta \to \zeta_0} \frac{1}{8} \left(\frac{\zeta - \zeta_0}{\zeta + \zeta_0} X(h_0, h_0, \zeta, \zeta_0) + \frac{\zeta + \zeta_0}{\zeta - \zeta_0} X(h_0, h_0, \zeta, -\zeta_0)\right.$$
$$\left. - \frac{-\zeta - \zeta_0}{-\zeta + \zeta_0} X(h_0, h_0, -\zeta, \zeta_0) - \frac{-\zeta + \zeta_0}{-\zeta - \zeta_0} X(h_0, h_0, -\zeta, -\zeta_0)\right)$$

$$= \frac{1}{8} \frac{\partial}{\partial \zeta} [(\zeta + \zeta_0)(X(h_0, h_0, \zeta, -\zeta_0) - X(h_0, h_0, -\zeta, \zeta_0)]_{\zeta = \zeta_0}$$

$$= \frac{\zeta_0}{2} \frac{\partial}{\partial \zeta} (X(h_0, h_0, \zeta, -\zeta_0))_{\zeta = \zeta_0} = \bar{h}_0(\zeta_0). \qquad \text{Q.E.D.}$$

Proposition I.4.21 implies a simple but striking correspondence between the infinite-dimensional Heisenberg algebra (I.4.19) and the algebra of anticommutation relations (11).

Proposition I.4.23. *The spaces of representations $S(\mathfrak{A}^-)$ and $V(\frac{1}{2}\mathbb{Z}_+)$ of algebra* (I.4.19) *and* (11) *are isomorphic under the identifications*

$$e(n) = X_n(h_0), \qquad n \in \mathbb{Z} + \tfrac{1}{2},$$
$$e(n) = iX_n(h_0), \qquad n \in \mathbb{Z},$$

which implies the reverse identification

$$h_0(m) = (i/2) \sum_{n \in \mathbb{Z}} e(k)\, e(m-k), \qquad m \in \mathbb{Z} + \tfrac{1}{2}.$$

Let us consider now afine Lie algebra $\mathfrak{O}^{(2)}(2l+2)$ for an arbitrary positive integer l. In this case we have $Q^0 = \mathbb{Z}^l \subset P^0$. The cocycle ε on $Q \times Q$ defined by (I.4.1) on $\mathbb{Z}^{l+1} \supset Q$ satisfies the condition (S.7) and its restriction ε^0 on Q^0 again is defined by (I.4.1); thus we will denote it by the same letter ε. The vertex operators $X^\varepsilon(\pm h_i, z)$ are defined as above or $i = 1, \ldots, l$. They satisfy the anticommutation relations (I.4.2) but they *commute* with the operators $X_n(h_0)$. In order to construct the generators of the Clifford algebra $C(Z^{2l} \oplus \mathbb{Z})$ we consider modified operators $X_m^{\varepsilon\prime}(\pm h_i)$, $m \in Z$, $i = 1, \ldots, l$, $X'_n(h_0) = X^\varepsilon_n(h_0)\, e(0)$, $n \in \frac{1}{2}Z$, then

$$\{X_m^{\varepsilon\prime}(\pm h_i), X'_n(h_0)\} = 0. \qquad (I.4.24)$$

Thus we can identify

$$e(n) = X'_n(h_0), \qquad n \in \mathbb{Z}, \qquad (I.4.25)$$
$$e(n) = iX'_n(h_0), \qquad n \in \mathbb{Z} + \tfrac{1}{2}, \qquad (I.4.26)$$
$$v_0 = 1 \otimes 1 \text{ in } V_0, \qquad v_0 = 1 \otimes e^{\omega_l} \text{ in } V_l. \qquad (I.4.27)$$

Now using Remark I.4.12 we obtain the analogue of Theorem I.4.7.

Theorem I.4.28. *Under the identifications* (I.4.4), (I.4.6), (I.4.25), (I.4.26) *we obtain the isomorphism of $\mathfrak{O}^{(2)}(2l+2)$-modules*

$$\bar{V}((\mathbb{Z} + \tfrac{1}{2})^l) \cong V_0, \qquad \bar{V}(\mathbb{Z}^l) \cong V_l,$$

where the correspondence of the Chevalley bases of the subalgebra $\mathfrak{O}(2l)$ is given by Theorem I.4.7, *and the correspondence between other elements is given by the formulas*

$$-\sum_{k\in Z} b_{\pm j}(k)\, e(m-k) = X^{\epsilon}_{m}(\pm h_j + h_{l+1}), \qquad m \in Z,$$
$$-i \sum_{k\in Z} b_{\pm}j(k)\, e(m-k) = X^{\epsilon}_{m}(\pm h_j + h_{l+1}), \qquad m \in Z', \tag{I.4.29}$$

$$\frac{i}{2}\sum_{k\in Z} e(k)\, e(m-k) = h_0(m), \qquad m \in Z + \frac{1}{2}. \tag{I.4.30}$$

Proof. Theorems I.2.25 and I.3.24 imply that the spaces under consideration are isomorphic $\hat{\mathfrak{O}}^{(2)}(2l+2)$-modules. Thus we can identify highest weight vectors by (I.4.4). The creation and annihilation operators in the spinor representation under the identifications (I.4.6), (I.4.25) satisfy the right anticommutation relations, which follow from (I.4.3), (I.4.2) (i), (I.4.24). Then the identity

$$X^{\epsilon}_{n}(\pm h_i + h_0) = \sum_{k\in Z} X^{\epsilon}_{n}(\pm h_i)\, X_{n-k}(h_0)$$

and the above identifications imply (I.4.29), and (I.4.30) follows from (I.4.2((iii). Q.E.D.

The analogue of Proposition I.4.10 is also valid for the isomorphic $\hat{\mathfrak{O}}^{(2)}(2l+2)$-modules and the equalities for characters gives the formulas (I.4.13) in which both sides are multiplied by $\prod_{n=1}^{\infty}(1+q^n)$.

In the conclusion of this section we consider one special isomorphism of the affine Lie algebras, namely, $\hat{\mathfrak{O}}^{(2)}(4) \cong \widehat{\mathfrak{sl}}(2)$. Moreover two pairs of vertex representations of these algebras discribed in Theorem I.2.16 and I.2.25 are isomorphic. In fact the homogeneous components of the operators $X^{\sigma}(h_1+h_0,\zeta)$, $X^{\sigma}(-h_1-h_0,\zeta)$, $h_1(\zeta^2)+h_0(\zeta)$ in $\hat{\mathfrak{O}}^{(2)}(4)$ satisfy the same commutation relations as the operators $X(\alpha,\zeta)$, $X(-\alpha,\zeta)$, $\frac{1}{2}\alpha(\zeta)$ in $\widehat{\mathfrak{sl}}(2)$, respectively. We write down these commutation relations, denoting the homogeneous components by $x_+(n)$, $x_-(n)$, $\frac{1}{2}h(n)$, $n \in Z$, respectively,

$$\begin{aligned}
[h(n), h(m)] &= 2n\,\delta_{n,-m},\\
[h(n), x_+(m)] &= 2x_+(n+m),\\
[h(n), x_-(m)] &= -2x_-(n+m),\\
[x_+(n), x_-(m)] &= h(n+m) + n\,\delta_{n,-m}.
\end{aligned} \tag{I.4.31}$$

In Section 3.2 we constructed the spinor representations for the affine Lie algbras of the type $D^{(2)}_{l+1}$. For $l = 1$, thanks to the above isomorphism, we obtain the spinor representation for the simplest affine Lie algebra $\widehat{\mathfrak{sl}}(2)$. We note that among with the vertex and spinor representations one can consider mixed constructions, which are vertex by some of the indexes and spinor by the other. Thus for $\widehat{\mathfrak{sl}}(2)$ we obtain another construction which is interesting for applications.

PROPOSITION I.4.32. *One has the following construction of the fundamental representations of* $\widehat{\mathfrak{sl}}(2)$

$$V_0 = S(\mathfrak{A}^-) \otimes V(\mathbb{Z} + \tfrac{1}{2}); \qquad V_1 = S(\mathfrak{A}^-) \otimes V(\mathbb{Z});$$

$$\tilde{x}_+(\zeta) = \tilde{b}_+(z)\, X(h_0, \zeta); \qquad \tilde{x}_-(z) = \tilde{b}_-(z)\, X(-h_0, \zeta);$$

$$\tilde{h}(\zeta) = \tilde{h}_0(\zeta) + :\tilde{b}_+(z)\, \tilde{b}_-(z):;$$

$$\tilde{d}(z) = -\frac{1}{2}:(\tilde{h}_0(\zeta))^2 - :\tilde{b}_+(z)\, z\frac{\vec{d}}{dz}\tilde{b}_-(z):, \qquad z = \zeta^2;$$

i.e., the homogeneous components of $\tilde{x}_+(\zeta)$, $\tilde{x}_-(\zeta)$, $\tilde{h}(\zeta)$ *satisfy the commutation relations* (I.4.31), *the homogeneous components of* $\tilde{d}(z)$ *represent the Virasoro algebra.*

Proof. The statement follows from the identifications $X^{\epsilon}(\pm h_1, \zeta^2) \leftrightarrow \tilde{b}_{\pm}(\zeta^2)$ applied to the vertex representations of $\mathfrak{O}^{(2)}(4)$.

If we compare the characters of $\widehat{\mathfrak{sl}}(2)$ in the last construction with the characters of the vertex representations we obtain the simple form of Jacobi identities

$$\prod_{n=1}^{\infty} (1 - q^{2n-1})^{-1}(1 + e^h q^{2n-1})(1 + e^{-h} q^{2n-1}) = \prod_{n=1}^{\infty} (1 - q^n)^{-1} \sum_{n \in \mathbb{Z}} e^{nh} q^{n^2},$$

$$e^{h/2} \prod_{n=1}^{\infty} (1 - q^{2n-1})^{-1}(1 + q^{2n-2} e^h)(1 + q^{2n} e^{-h}) \tag{I.4.33}$$

$$= \prod_{n=1}^{\infty} (1 - q^n)^{-1} \sum_{n \in \mathbb{Z}} e^{(n-1/2)h} q^{n^2 - n}.$$

PART II. BOSON–FERMION CORRESPONDENCE IN QUANTUM FIELD THEORY

1. *Current Algebras*

1.1. Current algebras play an important role in particle physics, especially in quantum field theory in one space and one time dimension (see e.g. [7]). They will be the basic object in the second part of our paper in just the same way as affine Lie algebras were in the first part. And what is more, the two kinds of algebras can be transformed into each other. With a view to the audience of physicists we will use in this part their notation when possible.

Let $\mathfrak{g}$ be an arbitrary finite dimensional Lie algebra with bilinear symmetric invariant form $\langle\ ,\ \rangle$. We choose a basis λ_k, normed by the condition $\langle \lambda_k, \lambda_m \rangle = 2\delta_{km}$, and we let

$$[\lambda_k, \lambda_m] = 2if^{kmn}\lambda_n, \tag{II.1.1}$$

where f^{kmn} denotes the structure constants of $\mathfrak{g}$, and the sign of summation over n is omitted. The current algebra with values in $\mathfrak{g}$ is defined by the following commutation relations

$$[J_k(u_1), J_m(u_2)] = 2if^{kmn}J_n(u_1)\,\delta(u_1 - u_2) + 2i(c/2\pi)\,\delta_{km}\,\delta'(u_1 - u_2), \tag{II.1.2}$$

where $u_1, u_2 \in \mathbb{R}$, $\delta(\cdot)$ and $\delta'(\cdot)$ denote the Dirac delta-function and its derivative on the real line, respectively, and c is a constant. Along with the current algebra (II.1.2) one introduces the second copy of the same current algebra depending on a second real variable v. One stipulates that the two algebras commute with each other (see (2.24) in [7]). Then the variables $t = (u + v)/2$, $x = (u - v)/2$ mean physical time and space. We will mainly restrict ourselves to one copy of the algebra (II.1.2). For an arbitrary element $x \in \mathfrak{g}$ we denote the corresponding current by

$$J_x(u) = x_k J_k(n), \qquad \text{where} \quad x = x_k \lambda_k. \tag{II.1.3}$$

Here again we assume the summation over k, and x_k denote the components of x with respect to the basis λ_k. It is technically necessary sometimes to consider the current algebra in a spatial box of length L with periodic boundary conditions (in this case $\delta(\cdot)$ in (II.1.2) is considered as the Dirac delta-function of the interval $[0, L]$ and similarly for $\delta'(\cdot)$). We obtain a decomposition

$$J_x(u) = (1/L) \sum_{n \in \mathbb{Z}} x(n)\, e^{-2\pi i n u/L}. \tag{II.1.4}$$

and we let $\mathfrak{g}$ be a simple Lie algebra of a type A_l, D_l, or E_l and $\langle\ ,\ \rangle$ be a canonical invariant form. Then the commutation relations (II.1.2) imply that the elements $x(n)$ satisfy the commutation relations (I.1.1) of the affine Lie algebra $\mathfrak{G}$. In the absence of the periodic conditions we denote

$$J_x(u) = \frac{1}{2\pi} \int_{-\infty}^{\infty} x\langle p \rangle\, e^{-ipu}\, dp \tag{II.1.5}$$

Then the elements $x\langle p \rangle$ satisfy the commutation relations

$$[x\langle p \rangle, y\langle q \rangle] = [x, y]\langle p + q \rangle + p\,\delta(p + q)\langle x, y \rangle\, c, \tag{II.1.6}$$

where $p, q \in \mathbb{R}$, $x, y \in \mathfrak{g}$. We call (II.1.6) the current algebra in the momentum representation. One can see that the algebra (II.1.6) is a continuous analogue of the afine Lie algebra (I.1.1); the Dirac delta-function here plays the role of the Kronecker symbol in the affine Lie algebra case. Thus one can expect that the analogy can be pushed further to the structural theory, representation theory and so on. However, on the way, one meets a number of divergences, which physicists skillfully bypass using a cutoff renormalization procedure. We will employ here another approach. We will consider $x\langle p\rangle$, $p \in \mathbb{R}$, $x \in \mathfrak{g}$, as elements in some completion of the affine Lie algebra $\hat{\mathfrak{G}}$, which will satisfy the commutation relations (II.1.6). Then we will apply all the results of Part I. On the other hand, the affine Lie algebra $\hat{\mathfrak{G}}$ will be realized with respect to the new continuous basis $x\langle p\rangle$, $p \in \mathbb{R}$, $x \in \mathfrak{g}$.

Our approach is based on the simple idea of the Cayley transform of the circle into the upper half-plane. For arbitrary fixed constant $M > 0$ one has:

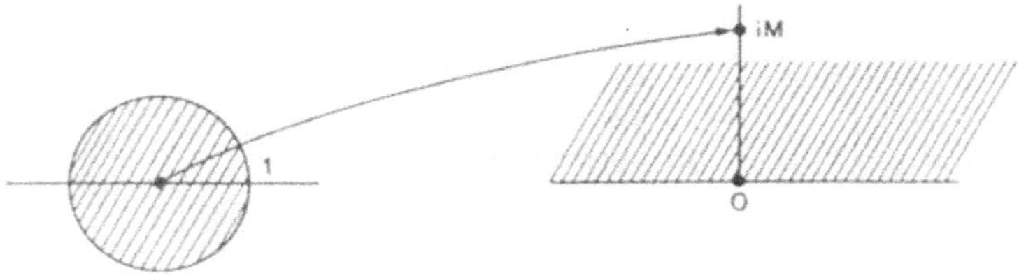

$$z = \frac{Mi - w}{Mi + w}; \qquad w = Mi\frac{1 - z}{1 + z}; \tag{II.1.7}$$

$$|z| < 1 \quad \text{iff} \quad Im w > 0, \tag{II.1.8}$$

$$|z| = 1 \quad \text{iff} \quad Im w = 0, \tag{II.1.9}$$

$$\frac{dz}{iz} = \frac{2M}{M^2 + w^2} dw.$$

Thanks to (II.1.7) we can transfer a current algebra defined on the circle to the real line and then extend this transform to the operators of representations. We will obtain a priori renormalized expressins for some field operators with the cutoff constant M, which have been invented by physicists from the other point of view.

We emphasize that *both our approaches to the quantum field theory in one space and time dimension are based on the representation theory of affine Lie algebras developed in Part I. In the first case we consider quantum fields in a spatial box with periodic boundary conditions; thus they depend on the fixed constant L, the length of the box. In the second case we consider renor-*

malized quantum fields which depend on the fixed renormalization constant M (cutoff). To obtain the final physical picture one can take limits $L \to +\infty$ and $M \to +\infty$.

We will see that the first approach allows us to apply the results of Part I directly. The second approach requires the study of one special transform which connects te elements $x\langle p\rangle$ and $x(n)$, $x \in \mathfrak{g}$, $p \in \mathbb{R}$, $n \in \mathbb{Z}$.

THEOREM II.1.11. *Let M be a fixed real positive number. Then the affine Lie algebra $\hat{\mathfrak{G}}$ is isomorphic to the Lie algebra consisting of the elements*

$$\int_0^{\infty} \left[\sum_{n=0}^{N} (x_n\langle p\rangle + x_{-n}\langle -p\rangle) p^n \right] e^{-Mp} \frac{dp}{p}, \qquad x_n \in \mathfrak{g}, \quad n \in \mathbb{Z},$$

with the Lie bracket defined by (II.1.1). *The isomorphism is defined by the following straight and inverse Laguerre transformations of bases:*

$$x\langle \pm p\rangle = \sum_{n=0}^{\infty} (-1)^n e^{-Mp} L_n^{-1}(2Mp)\, x(\pm n), \qquad p \geqslant 0, \qquad \text{(II.1.12)}$$

$$x(\pm n) + (-1)^{n-1} x(0) = (-1)^{n-1}\, 2M \int_0^{\infty} x\langle \pm p\rangle e^{-Mp} L_n^{-1}(2Mp)\, dp,$$

$$n > 0,$$

$$x(0) = \frac{1}{\pi i} \int_{-\infty}^{\infty} x\langle p\rangle e^{-M|p|} \frac{dp}{p - i0}, \qquad \text{(II.1.13)}$$

where L_n^{α} denotes Laguerre polynomials (see Appendix).

Proof. The inverse transform (II.1.13) follows from the orthogonality conditions for the Laguerre polynomials (A6). Commutation relations (II.1.6) follows from the properties of Laguerre polynomials, for $p \geqslant q > 0$, $z = [x, y]$, one has by (A14) and (A15)

$$\begin{aligned}
[x\langle p\rangle, y\langle -q\rangle] &= e^{-Mp-Mq} \sum_{n=0}^{\infty} \sum_{m=0}^{\infty} (-1)^{n+m} L_n^{-1}(2Mp) \\
&\quad \times L_m^{-1}(2Mq)[x(n), y(-m)] \\
&= e^{-Mp-Mq} \sum_{N=0}^{\infty} z(N) \left\{ \sum_{m=0}^{\infty} (-1)^N L_{m+N}^{-1}(2Mp)\, L_m^{-1}(2Mq) \right\} \\
&\quad + \sum_{N=-\infty}^{-1} z(N) \cdot 0 + \langle x, y\rangle^{e^{-Mp-Mq}} \sum_{n=1}^{\infty} n L_n^{-1}(2Mp)\, L_n^{-1}(2Mq) \\
&= z\langle p - q\rangle + p\langle x, y\rangle\, \delta(p + q).
\end{aligned}$$

Remark. Though the elements $x\langle p\rangle$, $p \in \mathbb{R}$, $x \in \mathfrak{g}$, by themselves do not belong to the affine Lie algebra $\hat{\mathfrak{G}}$ (they belong only to some extension of $\hat{\mathfrak{G}}$), we will formulate a number of results in terms of this continuous basis of $\hat{\mathfrak{G}}$.

The Laguerre transform (II.1.12), (II.1.13) are chosen in such a way that the following correspondence is valid

PROPOSITION II.1.14.

$$\frac{1}{2\pi}\tilde{x}(z) = \frac{M^2 + w^2}{2M} J_x(w), \qquad -M < Im w < M.$$

Proof. We use the Laplace transform (A5).

$$\begin{aligned} 2\pi J_x(w) &= \int_0^\infty x\langle p\rangle\, e^{-ipw}\, dp + \int_0^\infty x\langle -p\rangle\, e^{ipw}\, dp \\ &= \sum_{n=0}^{\infty} (-1)^n x(n) \left(\int_0^\infty L_n^{-1}(2Mp)\, e^{-p(M+iw)}\, dp \right. \\ &\quad \left. + \sum_{n=0}^{\infty} (-1)^n x(-n) \left(\int_0^\infty L_n^{-1}(2Mp)\, e^{-p(M-iw)}\, dp \right) \right. \\ &= \frac{2M}{M^2 + w^2} \sum_{n \in \mathbb{Z}} x(n) \left(\frac{M - iw}{M + iw} \right)^n \end{aligned}$$

We define now one special derivation $d\langle 0\rangle$ of Lie algebra $\hat{\mathfrak{G}}$ in the continuous basis

$$[d\langle 0\rangle, x\langle p\rangle] = p x\langle p\rangle. \tag{II.1.15}$$

The derivation $d\langle 0\rangle$ plays the role of the derivation $d(0)$ in a standard form of $\hat{\mathfrak{G}}$. Moreover $d\langle 0\rangle$ belongs to the principal subalgebra $\mathfrak{sl}(2, \mathbb{C})$ of the derivation algebra $\mathfrak{D}$.

PROPOSITION II.1.16.

(i) $$[d(0), x\langle p\rangle] = \frac{p}{2M}\left(-\frac{d^2}{dp^2} + M^2 \right) x\langle p\rangle,$$

$$[d(\pm 1), x\langle p\rangle] = \frac{p}{2M}\left(\frac{d}{dp} \pm M \right)^2 x\langle p\rangle.$$

(ii) $d\langle 0\rangle = (2\,d(0) + d(1) + d(-1))/2M$.

Proof. The identities follows from the differential equations for Laguerre polynomials (A.10), (A.12).

1.2. Let us consider now the twisted affine Lie algebra $\hat{\mathfrak{G}}_\sigma$. We can apply both our approaches, which have been described in the beginning of this section. First we consider the current algebra in a spatial box of length L. For $x \in \mathfrak{g}^0$ we again impose on $J_x(n)$ periodic boundary condition and we obtain the decomposition (II.1.4). However, for $x \in \mathfrak{g}^1$ we impose on $J_x(u)$ antiperiodic boundary conditions; i.e. $J_x(L) = -J_x(0)$. It implies a decomposition

$$J_x(u) = \frac{1}{L} \sum_{n \in \mathbb{Z}+1/2} x(n)\, e^{-2\pi i n u/L}. \qquad \text{(II.1.17)}$$

The twisted current algebra is defined by the same formula (II.1.2) as the usual current algebra; the distinction consists only in the type of boundary conditions. One can check that the elements $x(n)$, where $x \in \mathfrak{g}^0$, $n \in \mathbb{Z}$ or $x \in \mathfrak{g}^1$, $n \in \mathbb{Z} + \frac{1}{2}$, satisfy commutation relations (I.1.13) of the twisted affine Lie algebra $\hat{\mathfrak{G}}_\sigma$.

To apply the second approach we have to find a generalization of the Laguerre transform (II.1.12), (II.1.13) for the semi-integral values of the parameter n. One knows (see e.g. [4]) that the Laguerre polinomials are particular case of the Whittaker functions for integral values of one parameter. The Whittaker functions for the semi-integral values of this parameter and fixed another parameter have been considered in details in [3]. By means of these functions, which are called the Bateman k-functions (see the Appendix for definitions) we can define the following transforms for $x \in \mathfrak{g}^1$;

$$x\langle p\rangle = \sum_{n \in \mathbb{Z}+1/2} k_{2n}(Mp)\, x(n), \qquad \text{(II.1.18)}$$

$$x(n) = \int_{-\infty}^{\infty} k_{2n}(Mp) \frac{n}{p}\, x\langle p\rangle\, dp. \qquad \text{(II.1.19)}$$

The inverse transform follows from the orthogonality conditions. One should understand the integral in (II.1.19) in the sense of the principal value. As in (II.1.14) we obtain also the equality

$$\frac{1}{2}\tilde{x}(z) = \frac{M^2 + w^2}{2M} J_x(w), \qquad -M < Im w < M. \qquad \text{(II.1.20)}$$

It follows from the integral formula for k-functions (B4).

The analogue of Theorem II.1.11 is also valid, but the elements of $\hat{\mathfrak{G}}_\sigma$ will contain the integrals of the type(II.1.19), where $x\langle p\rangle$, $p \in \mathbb{R}$, $x \in \mathfrak{g}$, satisfying the commutation relations (II.1.6).

2. Boson Fields and Generalized Sine-Gordon Model

2.1. We proceed now to the construction of the representation of the current algebra in boson space. Boson space is the space of representation of the canonical Bose fields φ_j, which satisfy the standard commutation relations on the light cone

$$[\varphi_j(u_1), \varphi_k(u_2)] = -\frac{i}{4}\delta_{jk}\,\mathrm{sign}(u_1 - u_2). \tag{II.2.1}$$

The canonical space of representation of Bose fields φ_j is a symmetric Fock space generated by creation operators which are Fourier coefficients of the positive frequency part of φ_j. The exact definition requires a more careful description. Let us consider Bose fields φ_j defined in a spatial box of length L with periodic boundary conditions. Then we can set

$$\varphi_{L,j}(u) = \frac{1}{2i\sqrt{\pi}}\left(\sum_{n\neq 0} h_j(n)\frac{e^{-2\pi inu/L}}{n} + p_j\frac{2\pi u}{L} + q_j\right), \tag{II.2.2}$$

where $h_j(n)$, $n > 0$ are annihilation operators, while for $n < 0$ are creation operators satisfying (I.2.1), and q_j denotes charge, and p_j its conjugate momenta (see [1])

$$[q_j, p_j] = i. \tag{II.2.3}$$

Thus the space of representation $\tilde{\mathscr{H}}_b$ is the tensor product of Fock space $\mathscr{H}'_b$ generated by the creation operators and charge space $\tilde{\mathscr{H}}''_b \cong L_2(\mathbb{R}^l)$ in which p_j acts as a derivation $(1/i)(\partial/\partial x_j)$, q_j as a multiplication by x_j. We will always quantize charge, which defines the natural projection

$$P : \tilde{\mathscr{H}}''_b \to \mathscr{H}''_b, \tag{II.2.4}$$

where $\mathscr{H}''_b \simeq L_2(\mathbb{R}^l/\mathbb{Z}^l)$. Thus the operator q_j does not act in $\mathscr{H}''_b$, though e^{inq_j}, $n \in \mathbb{Z}$, are uniquely defined. We denote

$$\mathscr{H}_b = \mathscr{H}'_b \otimes \mathscr{H}''_b. \tag{II.2.5}$$

Now recalling the construction of Section I.2, one can easily recognize in $\mathscr{H}'_b$, $\mathscr{H}''_b$, $\mathscr{H}_b$ the completion of the spaces $S(\mathfrak{H}^-)$, $\mathbb{C}[\mathbb{Z}^l]$, $V_{\mathbb{Z}^l}$, respectively, in their Hermitian scalar product. It is not difficult to generalize our construction to an arbitrary lattice Γ (instead of $\mathbb{Z}^l$) containing the root lattice Q, if one imposes compatible integral conditions on the charges q_j, $j = 1,\ldots, l$. The representation of the current algebra (II.1.2) in a spatial box follows directly from the results of Part I. One has the following correspondences

$$\partial_u \varphi_{L,j}(u) = (\sqrt{\pi}/L)\, \bar{h}_j(e^{2\pi i u/L}), \tag{II.2.6}$$

$$:e^{2i\sqrt{\pi}\langle \alpha, \varphi_L(u)\rangle}: = X(\alpha, e^{2\pi i u/L}), \tag{II.2.7}$$

where $\langle \alpha, \varphi(u)\rangle = a_j \varphi_j(u)$, $\alpha = a_j h_j \in \Delta$ is decomposed with respect to an onthonormal basis in $\mathfrak{h}$, and in the realization (II.2.5) the middle exponent of the vertex operator $\exp(\alpha(0)(2\pi i u/L) + \alpha)$ turns into $\exp(a_j p_j(2\pi u/L) + a_j q_j)$; i.e., α assumes the meaning of charge and $\alpha(0)$ the meaning of momenta. The root system Δ is supposed to be of one of the types $A_l^{(1)}$, $D_l^{(1)}$, $E_l^{(1)}$, thus for $D_l^{(1)}$ we have

$$\langle \alpha, \varphi_L(u)\rangle = \pm\varphi_{L,j}(u) \pm \varphi_{L,j}(u), \qquad j \neq k.$$

We noted in [10] that for $|z| < 1$, $h_j(z)$, $X(\alpha, z)$ are bounded operators in the Hilbert space $\mathscr{H}_b$, which is not so for $|z| = 1$. Thus we will define the operators of the current algebra as a limit, when it exists, as

$$\lim_{\varepsilon \to +0} X(\alpha, e^{2\pi i(u + i\varepsilon)/L})\, v, \qquad v \in \mathscr{H}_b, \tag{II.2.8}$$

and similar for $\bar{h}_j(e^{2\pi i u/L})$. Then the formulas for the Wightman functions (mathematicians would say spherical functions) can be read off (I.2.15)

$$\frac{1}{L^{\langle \alpha\alpha\rangle}} \langle 0|\, :e^{2i\sqrt{\pi}\langle \alpha, \varphi_L(u_1)\rangle}:\ :e^{-2\sqrt{\pi}\langle \alpha, \varphi_L(u_2)\rangle}:\, |0\rangle$$

$$= \left(\frac{2\pi i}{L/\pi \sin[(\pi/L)(u_1 - u_2 - i\varepsilon)]}\right)^{\langle \alpha,\alpha\rangle}. \tag{II.2.9}$$

(In Part I we denoted $\langle 0| \cdot |0\rangle$ by $(v_0, \cdot\, v_0)$, v_0 is the vacuum vector). The symbol $\lim_{\varepsilon \to +0}$ in the right side is always omitted in physical notation. In the quantum field theory this form of he Wightman function is postulated from the beginning (see e.g. [1]). One can also write explicitely n-point Wightman functions. Now using the standard calculations with the generalized functions

$$\frac{1}{L/\pi \sin[\pi/L(u_1 - u_2 - i\varepsilon)]} - \frac{1}{L/\pi \sin[\pi/L(-u_1 + u_2 + i\varepsilon)]} = 2\pi i \delta(u_1 - u_2), \tag{II.2.10}$$

we obtain the commutation relations for the current algebra.
(I am indebted to I. Bars for this idea)

$$[\partial_{u_1} \varphi_{L,j}(u_1), \partial_{u_2} \varphi_{L,k}(u_2)] = \frac{i}{2} \delta_{jk}\, \delta'(u_1 - u_2), \tag{II.2.11}$$

$$\left[\frac{1}{\sqrt{\pi}}\partial_{u_1}\varphi_{L,j}(u_1), \frac{1}{L}:e^{2i\sqrt{\pi}\langle\alpha,\varphi_L(u_2)\rangle}:\varepsilon_\alpha\right]$$

$$= -\alpha_j\,\delta(u_1-u_2)\tfrac{1}{2}\,:e^{2i\sqrt{\pi}\langle\alpha,\varphi_L(u_2)\rangle}:\varepsilon_\alpha, \tag{II.2.12}$$

$$\left[\frac{1}{L}:e^{2i\sqrt{\pi}\langle\alpha,\varphi_L(u_1)\rangle}:\varepsilon_\alpha, \frac{1}{L}:e^{2i\sqrt{\pi}\langle\beta,\varphi_L(u_2)\rangle}:\varepsilon_\beta\right]$$

$$= \varepsilon(\alpha,\beta)\,\delta(u_1-u_2)\frac{1}{L}:e^{2i\sqrt{\pi}\langle\alpha+\beta,\varphi_L(u_1)\rangle}:\varepsilon_{\alpha+\beta}, \tag{II.2.13}$$

$$\text{if}\quad \langle\alpha,\beta\rangle=-1 \qquad \text{and} \qquad 0 \quad \text{if} \quad \langle\alpha,\beta\rangle>-1$$

$$\left[\frac{1}{L}:e^{2i\sqrt{\pi}\langle\alpha,\varphi_L(u_1)\rangle}:\varepsilon_\alpha, \frac{1}{L}:e^{-2i\sqrt{\pi}\langle\alpha,\varphi_L(u_2)\rangle}:\varepsilon_{-\alpha}\right]$$

$$= \frac{1}{\sqrt{\pi}}\delta(u_1-u_2)\langle\alpha,\partial_{u_1}\varphi_L(u_1)\rangle - \frac{i}{2\pi}\delta'(u_1-u_2), \tag{II.2.14}$$

where ε_α is an operator of multiplication by ± 1 depending on charge and $\varepsilon(\alpha,\beta)$ is an eigenvalue of ε_α on $|\beta\rangle$ (for details see (S.1), (S.2)). For example, in the case of the $\mathfrak{o}(2l)$ current algebra $\varepsilon_\alpha=\varepsilon_j\varepsilon_k$, $\alpha=\pm h_j\pm h_k$, $j\neq k$, and $\varepsilon_j=(-1)^{N_j}$ is the Klein transformation operator [15], N_j is the charge of the j-boson.

We can also make use of the second approach. Bose fields $\varphi_j(u)$ on the real line can be decomposed into the Fourier series,

$$\varphi_j(u)=\frac{1}{2i\sqrt{\pi}}\left(\int_{-\infty}^{\infty} h_j\langle p\rangle\, e^{-ipu}\frac{dp}{p}+q_j\right), \tag{II.2.15}$$

where the $h_j\langle p\rangle$ span a continuous Heisenberg algebra,

$$[h_j\langle p_1\rangle, h_k\langle p_2\rangle]=p_1\,\delta_{jk}\,\delta(p_1+p_2)\,c. \tag{II.2.16}$$

To avoid the divergences we have to normalize Bose fields fixing a cutoff constant $M>0$. The desired renormalization can be obtained immediately if we apply the Laguerre transform (II.1.12). In particular, (A5) implies the identity

$$\sum_{n\neq 0} h_j(n)\frac{z^{-n}}{n}+h_j(0)\ln z=\int_{-\infty}^{\infty} h_j\langle p\rangle(e^{ipw}-e^{-Mp})\frac{dp}{p}. \tag{II.2.17}$$

This allows us to define the analogue of the vertex operator (I.2.10) by the formula

$$X\langle\alpha, w\rangle = (2M)^{\langle\alpha,\alpha\rangle/2} \exp\left(\int_0^{\infty} \alpha\langle -p\rangle(e^{-ipw} - e^{-Mp}) \frac{dp}{p}\right) \exp\alpha$$

$$\times \exp\left(-\int_0^{\infty} \alpha\langle p\rangle(e^{-ipw} - e^{-Mp}) \frac{dp}{p}\right). \qquad \text{(II.2.18)}$$

We call $X\langle\alpha, w\rangle$ the Mandelstam operators. (Operators of this type for arbitrary α: $\langle\alpha,\alpha\rangle \in \mathbb{Z}_+$ have been introduced first by Mandelstam, when he considered bosonification of Thirring model [22]. He noticed that these operators satisfy commutation relations when $\langle\alpha,\alpha\rangle \in 2\mathbb{Z}_+$ and anticommutation relations when $\langle\alpha,\alpha\rangle \in 2\mathbb{Z}_+ + 1$.)

PROPOSITION II.2.19. *Let $z = (Mi - w)/(Mi + w)$, then*

$$X(\alpha, z) = \left(\frac{M^2 + w^2}{2M}\right)^{\langle\alpha,\alpha\rangle/2} X\langle\alpha, w\rangle.$$

Proof. One has the simple identity

$$\exp(\ln z\alpha(0) + \alpha) = \left(\frac{M^2 + w^2}{4M^2}\right)^{\langle\alpha,\alpha\rangle/2} \exp\left(\ln\frac{2M}{M - iw}\alpha(0)\right)$$

$$\times \exp\alpha \exp\left(-\ln\frac{2M}{M + iw}\alpha(0)\right).$$

Then the statement follows from the two halves of the identity (II.2.17); i.e.,

$$\ln\frac{2M}{M + iw}\alpha(0) + \sum_{n=1}^{\infty} \alpha(n)\frac{z^{-n}}{n} = \int_0^{\infty} \alpha\langle p\rangle(e^{-ipw} - e^{-Mp}) \frac{dp}{p}.$$

As in Section I.2 we introduce the following operators

$$X_p\langle\mu\rangle = \frac{1}{2\pi}\int_{-\infty}^{\infty} X\langle\mu, w\rangle e^{ipw}\, dw, \qquad \text{(II.2.20)}$$

$$X^{\varepsilon}\langle\mu, w\rangle = X\langle\mu, w\rangle\,\varepsilon_\mu, \qquad X^{\varepsilon}_p\langle\mu\rangle = X_p\langle\mu\rangle\,\varepsilon_\mu, \qquad \text{(II.2.21)}$$

$$X^{\varepsilon}\langle\lambda, \mu, w, w_0\rangle = :X\langle\lambda, w\rangle X\langle\mu, w_0\rangle:\, \varepsilon_{\lambda+\mu}. \qquad \text{(II.2.22)}$$

We also have the analogue of (I.2.15).

PROPOSITION II.2.23. *Let $\lambda, \mu \in \Gamma$ and $v \in \mathscr{H}_b$. One has*

(i) $X^{\varepsilon}\langle\lambda, w\rangle X^{\varepsilon}\langle\mu, w_0\rangle = \varepsilon(\lambda,\mu)(i(w - w_0))^{\langle\alpha,\mu\rangle} X^{\varepsilon}\langle\lambda, \mu, w, w_0\rangle$ *where* $\operatorname{Im} w > \operatorname{Im} w_0$

(ii) *if* $\varepsilon(\lambda,\mu)\,\varepsilon(\mu,\lambda)^{-1} = \pm(-1)^{\langle\lambda,\mu\rangle}$ *then*

$$[X^{\varepsilon}_p\langle\lambda\rangle, X^{\varepsilon}\langle\mu, w_0\rangle]_{\mp} = \varepsilon(\lambda,\mu)\, i \operatorname*{Res}_{w=w_0}\left((i(w - w_0))^{\langle\lambda,\mu\rangle} X^{\varepsilon}\langle\lambda, \mu, w, w_0\rangle e^{ipw}\right).$$

Proof. (i) follows from (I.2.16). Direct verification also gives the result:

$$I \cdot II := \exp\left(-\int_0^\infty \lambda\langle p\rangle(e^{-ip} - e^{-Mp})\frac{dp}{p}\right)$$

$$\times \exp\left(\int_0^\infty \mu\langle -p\rangle(e^{ipw_0} - e^{-Mp})\frac{dp}{p}\right.$$

$$= \exp\left(-\langle\lambda,\mu\rangle\int_0^\infty (e^{-ipw} - e^{-Mp})(e^{ipw_0} - e^{-Mp})\frac{dp}{p}\right) \cdot II \cdot I$$

$$= \left[\frac{2Mi(w - w_0)}{(M - iw_0)(M + iw)}\right]^{\langle\lambda,\mu\rangle} \cdot II \cdot I$$

$$I \cdot II := \exp\lambda \cdot \exp\left(\mu(0)\ln\frac{2M}{M - iw_0}\right) = \left(\frac{M - iw_0}{2M}\right)^{\langle\lambda,\mu\rangle} \cdot II \cdot I$$

$$I \cdot II := \exp\left(-\lambda(0)\ln\frac{2M}{M + iw}\right) \cdot \exp\mu = \left(\frac{M + iw}{2M}\right)^{\langle\lambda,\mu\rangle} \cdot II \cdot I.$$

Multiplying three factors in the right sides we obtain (i). The proof of (ii) is the same as in (I.2.15).

THEOREM II.2.24. *We define the action of the current algebra in momentum realization* (II.1.6) *in the Fock space* $\mathscr{H}_b$ *as follows: The element* $x^\epsilon_\alpha\langle p\rangle$, $\alpha \in \Delta$, $p \in \mathbb{R}$, *is represented by the operator* $X^\epsilon_p\langle\alpha\rangle$. *Then the representation of affine Lie algebra* $\mathfrak{G}$ *can be obtained from* (II.1.11). $\mathscr{H}_b$ *is decomposed into irreducible components according to the decomposition of the lattice* Γ *into the orbits of* Q.

Proof cf. (I.2.16). The commutation relations (II.1.6) follows from Proposition II.2.23 (ii).

Now we have got all the necessary technical tools and we can apply the second approach to the current algebra representation which comes out in quantum field theory as a result of a renormalization procedure. The current algebra can be constructed from the following operators

$$\partial_u\varphi_j(u) = \frac{1}{2\sqrt{\pi}}\int_{-\infty}^{\infty} h_j\langle p\rangle e^{-ipu}\,dp, \tag{II.2.25}$$

$$\frac{1}{(2M)^{\langle\alpha\alpha\rangle/2}} N_M e^{2i\sqrt{\pi}\langle\alpha,\varphi(u)\rangle} = X\langle\alpha, u\rangle, \tag{II.2.26}$$

where N_M symbolized this renormalization by introduction of cutoff constant M and ordering. The Cayley transform (II.1.7)–(II.1.10) transfers the unit circle to the upper half-plane, therefore the Mandelstam operator (II.2.18) is well defined when $Imw > 0$. On the real line we will define them as above in (II.2.8)

$$\lim_{\varepsilon\to+0} X\langle a, u + i\varepsilon\rangle\, v. \tag{II.2.27}$$

The Wightman functions can be found from (II.2.23)

$$\frac{1}{(2M)^{(a,a)}}\langle 0|\, N_M e^{2i\sqrt{\pi}(a,\varphi(u_1))} \cdot N_M e^{-2i\sqrt{\pi}(a,\varphi(u_2))}\,|0\rangle = \left(\frac{2\pi i}{u_1 - u_2 - i\varepsilon}\right)^{(a,a)} \tag{II.2.28}$$

and does not depend on M. The commutatin relations for the current algebra will be exactly the same as in the first approach (II.2.11)–(II.2.14), one should substitute there L by $2M$ and normal ordering: : by the renormalization N_M.

Proposition II.2.29. *Under the conditions of Theorem* II.2.24

$$\tfrac{1}{2}\, d\langle 0\rangle = -\int_{-\infty}^{\infty} N_M(\partial_u \varphi(u))^2\, du.$$

Proof. The identity (II.1.14) implies

$$\frac{1}{(2\pi)^2} :h_i^2(z): = \left(\frac{M^2 + w^2}{2M}\right)^2 N_M(J_{h_i}(w))^2.$$

Using the simple equality $2M/M^2 + w^2 = (1/2M)(2 + z + z^{-1})$ we obtain the identity

$$\int_{-\infty}^{\infty} N_M h_i^2\langle\zeta\rangle\, d\zeta = \frac{1}{2M}\int_C :h_i^2(z):\, (2 + z + z^{-1})\frac{dz}{iz}. \tag{II.2.30}$$

Now Proposition II.1.16 and (II.1.10) imply the statement. One can also prove the result by the direct computation of commutation relations (II.1.15).

Theorem II.2.24 implies that any element H of the scalar subalgebra $\mathfrak{g}$ can be written in terms of Bose fields

$$H = \int_{-\infty}^{\infty} \left\{\sum_{j=1}^{l} a_j \partial_u \varphi_j(u) + \sum_{\alpha\in\Delta} a_\alpha N_M e^{2i\sqrt{\pi}(\alpha,\varphi(u))}\right\} du, \tag{II.2.31}$$

where a_j, $a_\alpha \in \mathbb{C}$. We will call a fixed element H of the form (II.2.31) the interaction Hamiltonian, and we will say that H defines a light cone field model. The simplest example in which $\mathfrak{g} = \mathfrak{sl}(2)$, $a_\alpha = a_{-\alpha} = m$,

$$H = m\int_{-\infty}^{\infty} N_M \cos\sqrt{8\pi}\,\varphi(u)\,du, \tag{II.2.32}$$

is called the sine-Gordon model, $\varphi(u)$ satisfies (II.2.1). We will call the model of type (II.2.31) the *generalized light cone sine-Gordon model.*

We set

$$\varphi_j(u, v) = e^{ivH}\varphi_j(u)\,e^{-ivH}, \qquad j = 1,\dots, l. \tag{II.2.33}$$

Then the commutation relations (II.2.12) imply the differential equation

$$\partial_v\partial_v\varphi(u, v) = \sum_{\alpha\in\Delta} a_\alpha \alpha N_M e^{2i\sqrt{\pi}(\alpha,\varphi(u,v))}. \tag{II.2.34}$$

The simple example (II.2.32) gives the sine-Gordon equation:

$$\partial_v\partial_u\varphi(u, v) = mN_M \sin\sqrt{8\pi}\,\varphi(u, v). \tag{II.2.35}$$

Remark II.2.36. It is known that the classical analogue of (II.2.35) is a completely integrable Hamiltonian system [25]. However, it is not so in the general case of (II.2.34) (Shankar, private communication). And only in the case, when $a_\alpha = 0$ for $\alpha \in \Delta\backslash\Delta'$, and Δ' is the so-called admissible subsystem of roots, e.g., $\Pi \cup \{-\tilde{\alpha}\}$, the classical analogue of (II.2.34) is still completely integrable (see [23]). We hope that further understanding of the structures of affine Lie algebras will give an explanation of this phenomenon.

Finally we define one model which plays the most significant role in two-dimensional quentum field theory. Suppose we have a representation of the current algebra $J_k(u)$ satisfying (II.1.2) in the space $\mathscr{H}$. We consider a second copy of the same current algebra in the space isomorphic to $\mathscr{H}$, so that two algebras are commute. We define (up to a scalar factor the operator

$$\mathscr{L}(u, v) = J^{(1)k}(u)\,J_k^{(1)}(u) + J^{(2)k}(v)\,J_k^{(2)}(v) + gJ^{(1)k}(u)\,J_k^{(2)}(v) \tag{II.2.37}$$

in the space $\mathscr{H}^{(1)}\otimes\mathscr{H}^{(2)}$, where (1) (resp. (2)) denotes the first (resp. second) copy of current algebra and the space of representation. One can note that $\mathscr{L}(u, v)$ is the simplest G-invariant operator. It is called in physics, the *Lagrangian density*. We will say that (II.2.37) defines a *canonical model associated with the representation of the current algebra* in the space $\mathscr{H}$. The constant g is called the *interaction constant*, thus the theory is called free when $g = 0$. In the case of the representation (II.2.11)–(II.2.14) one can simplify the expression (II.2.37)

$$\mathscr{L}(u,v)=\frac{h+1}{\pi}\left((\partial_u\varphi^{(1)})^2+(\partial_v\varphi^{(1)})^2\right)+\frac{2g}{\pi}(\partial_u\varphi^{(1)})(\partial_v\varphi^{(2)})$$

$$+\frac{2g}{(2M)^2}N_M\left(\sum_{\alpha\in\Delta_+}2\cos 2\sqrt{\pi}\langle\alpha,\varphi^{(1)}-\varphi^{(2)}\rangle\right),\quad \text{(II.2.38)}$$

where h is a Coxeter number ($h=l$, $G=SL(l)$; $h=2l-2$, $G=SO(2l)$, $h=12, 18, 30$, G resp. of type E_6, E_7, E_8).

Remark II.2.39. In quantum field theory one usually sets

$$\begin{aligned}\varphi^{(1)}-\varphi^{(2)}&=\gamma\Phi,\\ \partial_t(\varphi^{(1)}-\varphi^{(2)})&=\partial_u\varphi^{(1)}-\partial_v\varphi^{(2)}=\gamma\partial_t\Phi,\\ \partial_x(\varphi^{(1)}-\varphi^{(2)})&=\partial_u\varphi^{(1)}+\partial_v\varphi^{(2)}=\gamma\partial_x\Phi,\end{aligned}\quad \text{(II.2.40)}$$

where γ is chosen in such a way that the coefficient before $(\partial_x\Phi)^2$ is $\frac{1}{2}$, the same as for a free field ($g=0$, $\gamma=1$). One has

$$(\partial_u\varphi^{(1)})^2+(\partial_v\varphi^{(2)})^2+\frac{2g}{h+1}(\partial_u\varphi^{(1)})(\partial_v\varphi^{(2)})$$

$$=\frac{\gamma^2}{2}\left(1-\frac{g}{h+1}\right)(\partial_t\Phi)^2+\frac{\gamma^2}{2}\left(1+\frac{g}{h+1}\right)(\partial_x\Phi)^2;$$

i.e., $\gamma=(1+g_v/2\pi)^{-1/2}$, where $g_v=2\pi g/(h+1)$ is a physical interaction constant (see below a Thirring model). For $G=SL(2)$ one obtains sine-Gordon model with the interaction part of Lagrangian

$$\mathscr{L}_I=\frac{\text{Const}}{(2M)^2}N_M\cos\beta\Phi,\qquad \beta=\sqrt{8\pi}\left(1+\frac{g_v}{2\pi}\right)^{-1/2}\quad \text{(cf. [1]).}\quad \text{(II.2.41)}$$

2.2. Now we turn to the twisted case. It will allow us to consider the generalized sine-Gordon model for arbitrary root systems (with one exception, G_2; to include this root system we have to consider another boundary conditions). We begin from the first approach. We consider two sets of Bose fields $\varphi_1,\dots,\varphi_l$ and $\phi_1,\dots,\phi_m$, which satisfy the commutation relations (II.2.1). The distinction between the two sets of fields is the boundary conditions, which are periodic for the first set and antiperiodic for the second. We let $m=\operatorname{rank}\mathfrak{g}-\operatorname{rank}\mathfrak{g}^0$. Bose fields $\varphi_{L,j}$ have decomposition (II.2.2) and we have

$$\phi_{L,j}(u) = \frac{1}{2i\sqrt{\pi}} \sum_{n \in \mathbb{Z}+1/2} h_j(n) \frac{e^{-2\pi i n u/L}}{n}, \qquad \text{(II.2.42)}$$

where $h_j(n)$ for $n > 0$ are annihilation operators and for $n < 0$ creation operators. The complete space of the representation is the Fock space $\mathscr{H}_b = \mathscr{H}_b^\varphi \otimes \mathscr{H}_b^\phi$, where $\mathscr{H}_b^\varphi$ is described above, and $\mathscr{H}_b^\phi$ is the space generated by creation operators (I.2.3) with $n \in \mathbb{Z} + \frac{1}{2}$, $n < 0$.

We again can get a representation o the current algebfa of ADE type if we set

$$J_x(u) = (1/L) :e^{2i\sqrt{\pi}(\langle \alpha^0, \varphi_L(u)\rangle + \langle \alpha^1, \phi_L(u)\rangle)}: \varepsilon_\alpha^0, \qquad \text{(II.2.43)}$$

where $\alpha = \alpha^0 + \alpha^1$, $\alpha^0 \in \mathfrak{g}^0$, $\alpha^1 \in \mathfrak{g}^1$, $x = x_\alpha^\varepsilon$ (see (S.3)) ε^0 described in Section I.2.2. This algebra will satisfy the commutation relations (II.2.11)–(II.2.14). Note that the only distinction is the boundary conditions for Bose fields.

However, the scalar subalgebra $\mathfrak{g}^0$ has type B_l, C_l or F_4. If $H_1 \in \mathfrak{g}^0$ then one has

$$H_1 = \int_0^L \left(\sum_{j=1}^{l} a_j \partial_u \varphi_{L,j}(u) + \sum_{\alpha^0 \in \Delta} a_\alpha :e^{i(\langle \alpha^0, \varphi_L(u)\rangle + \langle \alpha^1, \varphi_L(u)\rangle)}: \varepsilon_\alpha^0 \right) du, \qquad \text{(II.2.44)}$$

$$d\langle 0\rangle = -\tfrac{1}{2} \int_0^L \left(\sum_{j=1}^{l} (\partial_u \varphi_{L,j}(u))^2 + \sum_{j=1}^{m} (\partial_u \phi_{L,j}(u))^2 \right) du. \qquad \text{(II.2.45)}$$

We can also apply the second approach.Then the Bose fields $\varphi_j(u)$ on the real line can be obtained as above by the Laguerre transform; at that time for the Bose fields $\varphi_j(u)$ one should consider k-transformation (B.1). The last case is not so simple as the first, because k-transformation does not preserve the polarization. We deduce from (B.5) that

$$\int_{-\infty}^{\infty} f(p) x\langle p\rangle \frac{dp}{p}$$

is a creation (annihilation) operator, where $f(p)$ is a linear combination of $k_{2n}(Mp)$ with $n < 0$ ($n > 0$). The next step is to rewrite the vertex operator (I.2.23) in the continuous basis. It can be done with the help of (II.1.19), but the final formula is rather complicated and requires some calculations with k-functions, which we omit here.

3. Fermion Fields and Generalized Thirring Model

3.1. In this section we are going to discuss the spinor representation from the field theory point of view. It turns out that the spinor representation of affine Lie algebra $\hat{\mathfrak{o}}(2l)$ in the case $Z = \mathbb{Z} + \frac{1}{2}$ is known in physics as the dual-quark model [2]. The analogue of this spinor representation on the real line plays an especially important role in quantum field theory, where it is called free Fermi fields. The formal correspondence between dual-quark model and fermion fields has been noticed by Halpern (see Appendix B in [15]). However, this mathematical relation has no physical meaning.

Fermi fields $\psi_i(u)$, $\psi_i^+(u)$, $i = 1,\dots, l$, satisfy the standard anticommutation relations on the light cone

$$\begin{aligned} \{\psi_j(u_1), \psi_k^+(u_2)\} &= \delta_{jk}\, \delta(u_1 - u_2), \\ \{\psi_j(u_1), \psi_k(u_2)\} &= \{\psi_j^+(u_1), \psi_k^+(u_2)\} = 0. \end{aligned} \tag{II.3.1}$$

This is again only one-half of the picture. Physicists always introduce the second copy of fields (II.3.1) depending on another variable v; then the variables $t = (u + v)/2$, $x = (u - v)/2$ mean physical time and space. To apply the results of Section I.3 we can again use two approaches. Let us first consider ψ_j, ψ_j^+ Fermi fields defined in a spatial box of length L with antiperiodic boundary conditions. (Note that we deal with the case $Z = \mathbb{Z} + \frac{1}{2}$, in the case $Z = \mathbb{Z}$ we should consider periodic boundary conditions). Then

$$\begin{aligned} \psi_{L\,j}(u) &= \frac{1}{\sqrt{L}} \sum_{n \in \mathbb{Z}+1/2} b_{-j}(n)\, e^{-2\pi i n u/L}, \\ \psi_{L\,j}^+(u) &= \frac{1}{\sqrt{L}} \sum_{n \in \mathbb{Z}+1/2} b_j(n)\, e^{-2\pi i n u/L}, \end{aligned} \tag{II.3.2}$$

where $b_{\pm j}(n)$, $n > 0$, annihilation operators, $n < 0$, creation operators. The operators $b_{\pm j}(u)$ satisfy the relations (I.3.4). The canonical space of states of Fermi fields is an antisymmetric Fock space $\mathscr{H}_f$ generated by the creation operators. The action of the field operators $\psi_{L\,j}(u)$, $\psi_{L\,j}^+(u)$ is defined as in Section II.2 of (cf. II.2.8). The Wightman functions will be

$$\langle 0 |\, \psi_{L\,j}(u_1)\, \psi_{L\,j}^+(u_2)\, | 0 \rangle = -\frac{1}{2\pi i} \frac{1}{L/\pi \sin[(\pi/L)(u_1 - u_2 - i\varepsilon)]}. \tag{II.3.3}$$

The representation of $\mathfrak{gl}(l)$-current algebra in $\mathscr{H}_f$ is defined by (see e.g. [7])

$$J_k(u) = :\psi_L^+ \lambda_k \psi_L:, \tag{II.3.4}$$

where ψ_L^+ is the row $(\psi_{L,1}^+,..., \psi_{L,l}^+)$ and ψ_L the corresponding column, λ_k is a Hermitian $l \times l$ matrix. To extend this representation to the $\mathfrak{o}(2l)$-current algebra, one has to add the currents $:\psi_{L,j}^+\psi_{L,k}^+:$, $:\psi_{L,j}\psi_{L,k}:$, $j \neq k$.

We can also exploit the second approach and consider Fermi fields $\psi_j\langle u\rangle$, $\psi_j^+\langle u\rangle$ without periodic conditions. Then we have

$$\begin{aligned}\psi_j\langle u\rangle &= \frac{1}{\sqrt{2\pi}}\int_{-\infty}^{\infty} b_{-j}\langle p\rangle\, e^{-ipu}\, dp,\\ \psi_j^+\langle u\rangle &= \frac{1}{\sqrt{2\pi}}\int_{-\infty}^{\infty} b_j\langle p\rangle\, e^{-ipu}\, dp,\end{aligned} \tag{II.3.5}$$

where the elements $b_{\pm j}\langle p\rangle$ satisfy the anticommutation relations

$$\{b_j\langle p_1\rangle, b_k\langle p_2\rangle\} = -\delta_{j,-k}\delta(p_1+p_2). \tag{II.3.6}$$

We call the algebra $\mathfrak{B}$ spanned by the elements $b_{\pm j}\langle p\rangle$ the algebra of the anticommutation relations. It is certainly a continuous analogue of the algebra spanned by $b_{\pm j}(n)$, which generates the Clifford algebra $C(Z^{2l})$ (see (II.3.4)). In the case $Z = \mathbb{Z} + \frac{1}{2}$ one has:

THEOREM II.3.7. *Let M be a fixed real positive number. Then the algebra $\mathfrak{B}$ is isomorphic to the algebra consisting of the elements*

$$\int_0^{\infty}\left[\sum_{m=1}^{N} (r_m\langle p\rangle + r_{-m}\langle -p\rangle)\, p^m\right] e^{-Mp}\, dp, \qquad m \in \mathbb{Z},$$

where $r_m\langle p\rangle$ is a linear combination of $b_j\langle p\rangle$, $j = \pm 1,..., \pm l$, satisfying (II.3.6). *The isomorphism is defined by the following straight and inverse Laguerre transformations of bases.*

$$r\langle \pm p\rangle = (2M)^{1/2}\sum_{m=0}^{\infty} (-1)^m e^{-Mp} L_m^0(2Mp)\, r(\pm(m+1/2)), \qquad p \geqslant 0, \tag{II.3.8}$$

$$r\langle \pm(m+1/2)\rangle = (-1)^m (2M)^{1/2}\int_0^{\infty} r\langle \pm p\rangle\, e^{-Mp} L_m^0(2Mp)\, dp, \qquad m \geqslant 0, \tag{II.3.9}$$

where L_m^0 denotes Laguerre polynomils (see Appendix).

Proof. The inverse transformation (II.3.9) follows from the orthogonality conditions for the Laguerre polynomials (A.7). Anticommutation relations in $\mathfrak{B}$ follows as in Theorem II.1.11.

The Laguerre transformations (II.3.8), (II.3.9) have been introduced first by Halpern (Appendix B in [15]). They are chosen in such a way that the following corresponding is valid.

Proposition II.3.10.

$$\frac{1}{\sqrt{2\pi}}\mathfrak{b}_j(z)=\left(\frac{M^2+w^2}{2M}\right)^{1/2}\psi_j^+(w),\qquad \frac{1}{\sqrt{2\pi}}\mathfrak{b}_{-j}(z)=-\left(\frac{M^2+w^2}{2M}\right)^{1/2}\psi_j(w)$$

Proof. The proof is the same as in (II.1.14): now we use the Laplace transformation (A.5) with $\alpha=0$.

The standard representation of the algebra of the anticommutation relations (II.3.6) is the fermion Fock space $\mathscr{H}_f$, which is well defined if we let $r\langle p\rangle$, $p<0$, be the creation operators and $r\langle p\rangle$, $p>0$, be the annihilation operators. We denoted the Fock space $\mathscr{H}_f$ by the same symbol as in the first approach because one can see from (II.3.8), (II.3.9) that the polarization is preserved under the Laguerre transformation, thus $\mathscr{H}_f$ is in fact the same space. It is certainly isomorphic to the completion of the space $V((\mathbb{Z}+\frac{1}{2})^l)$ in its Hermitian scalar product.

Fermi fields on the real line can be defined again in $\mathscr{H}_f$ as $\lim_{\varepsilon\to+0}\psi_j(u+i\varepsilon)\,v$, which implies the standard form for the Wightman functions

$$\langle 0|\,\psi_j(u_1)\,\psi_j^+(u_2)\,|0\rangle=-\frac{1}{2\pi i}\frac{1}{u_1-u_2-i\varepsilon}. \tag{II.3.11}$$

Let us consider now all the possible currents formed from fermion and antifermion fields:

$$:\psi_j(u)\,\psi_k^+(u):,\ :\psi_j(u)\,\psi_k(u):\quad j\neq k,\qquad :\psi_j^+(u)\,\psi_k^+(u):\quad j\neq k; \tag{II.3.12}$$

then (I.3.18), (I.3.21) imply

Theorem II.3.13. (i) *The currents* (II.3.12) *give a representation of the* $\mathfrak{o}(2l)$*-current algebra in* $\mathscr{H}_f$. *This representation has two irreducible components: eigenspaces of the parity operator.*

(ii) *The representation is extended to the free energy operator* $d\langle 0\rangle$ *which is given by*

$$d\langle 0\rangle=\int_{-\infty}^{\infty}:\psi^+(u)\,i\overleftrightarrow{\partial}_u\,\psi(u):\,du.$$

It follows from Theorem II.3.13 that any element of the scalar subalgebra can be written as an integral of currents (II.3.12), e.g., $H\in\mathfrak{gl}(l)$, one has

$$H=\int_{-\infty}^{\infty}:\psi^+(u)\,A\,\psi(u):\,du,\qquad A\in\mathfrak{gl}(l). \tag{II.3.14}$$

The element H defines an exactly integrable model, which we call a light cone gluon model (see [15]). The simple example of this model for $l = 2$, $A = \begin{pmatrix} 0 & m \\ m & 0 \end{pmatrix}$ is the free massive light cone Thirring model,

$$H = m \int_{-\infty}^{\infty} :\psi_1^+ \psi_2 + \psi_2^+ \psi_1: du. \tag{II.3.15}$$

One can define the action of the one-parameter group generated by the element H by the formula

$$\psi_j(u, v) = e^{ivH} \psi_j(u)\, e^{-ivH}, \qquad j = 1, \ldots, l. \tag{II.3.16}$$

From the commutation relations of Fermi fields with currents we obtain the linear equation for the free fermion field

$$(\partial_v + A)\, \psi(u, v) = 0. \tag{II.3.17}$$

We consider now a canonical model associated with the representation of the current algebra in the fermion Fock space. For the subalgebra $\mathfrak{gl}(l) \subset \mathfrak{o}(2l)$ one has from (II.2.37)

$$\begin{aligned} \mathscr{L}(u, v) = {}& \psi^{(1)+}(u)\, i\overleftrightarrow{\partial}_u \psi^{(1)}(u): + :\psi^{(2)+}(v)\, i\overleftrightarrow{\partial}_v \psi^{(2)}(v): \\ & + g\pi/(l+1) :\psi^{(1)+}(u)\, \psi^{(1)}(u)\, \psi^{(2)+}(v)\, \psi^{(2)}(v):. \end{aligned} \tag{II.3.18}$$

This is a Lagrangian density of the $SU(l)$-Thirring model [7]. Note that we obtain scale invariance [7] when $g = 0$ or $g = 2$.

3.2. We could see from the above construction that the fermion field representation of the $\mathfrak{o}(2l)$-current algebra is essentially the spinor representation of affine Lie algebra $\hat{\mathfrak{o}}(2l)$. Both of our approaches lead to the same result. However we note that the choice $Z = \mathbb{Z} + \frac{1}{2}$ was especially important in the second approach, because in the Laguerre transformations (II.3.8) and (II.3.9) the polarization is strictly preserved. In the case $Z = \mathbb{Z}$ we have no transformation satisfying this property. It implies that in the generalization of the above construction to the twisted affine Lie algebra $\hat{\mathfrak{o}}^{(2)}(2l+2)$ we will meet the same difficulties as in Section 2.2, as the generators $e_i(m)$ in the spinor representation contain both integer and half-integer values of the parameter m.

Let us follow the procedure of Section I.3.2. Thus the first step will be to extend the representation of $\mathfrak{o}(2l)$-current algebra to $\mathfrak{o}(2l+1)$. To this end we wil admit another point of view on the Fermi fields introduced above.

Along with Dirac fermions $\psi_j(u)$, $j = 1,\ldots, l$ it is convenient to consider another basis $\check{\psi}_j(u)$, $j = 1,\ldots, 2l$, of Majorana–Fermi fields ([26, 24])

$$\begin{aligned} \check{\psi}_j(u) &= \psi_j(u) + \psi_j^+(u), \qquad & j = 1,\ldots, l, \\ \check{\psi}_{j+l}(u) &= i(\psi_j(u) + \psi_j^+(u)), \qquad & j = 1,\ldots, l. \end{aligned} \tag{II.3.19}$$

Certainly one has $\check{\psi}_i^+(u) = \check{\psi}_i(u)$, thus two Majorana fermions are equivalent to one Dirac fermion.

$$\check{\psi}(u) = (\check{\psi}_1(u),\ldots, \check{\psi}_{2l}(u)) \tag{II.3.20}$$

is by definition an $2l$-component Majorana–Fermi field. $\mathfrak{o}(2l)$-model has a simple form in terms of Majorana–Fermi field

$$H = \int_{-\infty}^{\infty} :\left(\frac{a}{2}\check{\psi}(u)\, i\vec{\partial}_u \check{\psi}'(u) + \check{\psi}(u)A\check{\psi}'(u)\right): du, \tag{II.3.21}$$

where $a \in \mathbb{C}$, $A \in \mathfrak{o}(2l)$, $H \in \mathfrak{o}(2l) \otimes \mathbb{C}\, d\langle 0\rangle$.

In the language of Majorana fermions there is no distinction between even and odd case of orthogonal algebra $\mathfrak{o}(n)$. Thus all the results of Section 3.1 can be immediately generalized to the case $n = 2l + 1$.

In Section I.3.2 we have completed the representation of $\hat{\mathfrak{O}}(2l+1)$ to those of $\hat{\mathfrak{O}}^{(2)}(2l+2)$. In field theory it requires the introduction of a new Majorana fermion. In a spatial box it satisfies the periodic boundary conditions and has a form

$$\check{\psi}_{2l+2}(u) = \frac{1}{\sqrt{2\pi}} \sum_{n \neq 0} e_{2l+2}(u), \tag{II.3.22}$$

where $e(0)$ is a parity operator. One can add the quadratic operators of the type $\check{\psi}_j\check{\psi}_{2l+2}$, $j = 1,\ldots, 2l+1$ and obtain representation of $\mathfrak{o}(2l+2)$-current algebra. However, the integral of these operators in the spatial box of length L is zero. Thus we still obtain $\mathfrak{o}(2l+1)$-models of the form (II.3.21). The second approach based on the Laguerre transformation meets here some technical difficulties as we mentioned in the beginning of this section.

4. Boson–Fermion Correspondence

4.1. Several years ago a remarkable correspondence has been noted by physicists between boson and fermion fields in one space dimension. First Coleman [6] noted that the Green's functions of the massive Thirring model and those governed by the sine-Gordon equation are related. Then Kogut and Susskind [19] wrote down the exact correspondence between sine-Gordon operators and bilinear functions of the Fermi operators. Soon afterward Mandelstram [22] expressed the Fermi field operator itself in terms of Bose field operators. Then Halpern [15] extended such correspondence to $\mathfrak{su}(l)$. Banks *et al.* [1] continued studying $SU(l)$-model working in a spatial box. Halpern as well as Banks et al considered an interaction picture, which is based on the free field correspondences: "If one knows free field correspondences, it is generally very easy to add many interactions in the interaction picture" (Halpern). Note that for a free field these correspondences split into two light cone halves and becomes quite simple.

In our approach these discoveries in quantum field theory gain especially striking meaning: *Boson–Fermion correspondence is nothing else but the canonical isomorphism between two realizations of the same representation of the affine Lie algebra* $\hat{\mathfrak{o}}(2l)$ *and in particular of its subalgebra* $\widehat{\mathfrak{gl}}(l)$. Then relations between Green's functions are just equalities of the spherical functions. Kogut–Susskind correspondence is in fact two forms of representations of affine Lie algebra. The Mandelstam operators are the intertwining operators between two representations. Two approaches to infrared regularization in two-dimensional field theory also found a natural explanation in the framework of affine Lie algebras. The regularization by working in a spatial box with periodic boundary conditions of Banks *et al.* is a result of the Fourier transform of affine Lie algebras. The regularization by introduction of a big auxiliary constant cutoff, which appers in all the calculations (see e.g. Halpern) comes out immediately from the Laguerre transformation.

We summarize here the statements about the boson–fermion correspondence, which follow from Theorem I.4.7 and Proposition I.4.11 if one applies the Laguerre transformations (II.1.13) and (II.3.9).

THEOREM II.4.1. (L) *Let L be a fixed length of spatial box*

(i) *Under the following identifications (due to Banks–Horn–Neuberger)*

$$\psi^{\pm}_{L,j}(u) = \frac{\pm 1}{\sqrt{L}} :e^{\pm 2i\sqrt{\pi}\Phi_{L,j}(u)}: \varepsilon_j \tag{II.4.2}$$

we obtain the isomorphism of two $\mathfrak{o}(2l)$*-current algebra representations in the Fock spaces*

$$\mathscr{H}_{\mathrm{b}} \cong \mathscr{H}_{\mathrm{f}}, \tag{II.4.3}$$

where the correspondence of the currents is given by the formulas

$$:\psi^{\pm}_{L,j}(u)\,\psi_{L,j}(u):=\frac{1}{\sqrt{\pi}}\partial_u\varphi_{L,j}(u) \tag{II.4.4}$$

$$:\psi^{\pm}_{L,j}(u)\,\psi^{\pm}_{L,k}(u):=\frac{1}{L}:e^{2i\sqrt{\pi}(\pm\varphi_{L,j}(u)\pm\varphi_{L,k}(u))}:\varepsilon_j\varepsilon_k,\qquad j\neq k. \tag{II.4.5}$$

(ii) *Lagrangian densities of free boson and fermion fields coincide; i.e.,*

$$:\psi^{+}_{L,j}(u)\,i\overleftrightarrow{\partial}_u\psi_{L,j}(u);=:(\partial_u\varphi_{L,j}(u))^2:. \tag{II.4.6}$$

(M) *Let M be a fixed cutoff constant.*

(i) *Under the following identifications (due to Mandelstam)*

$$\psi^{\pm}_j(u)=\frac{\pm 1}{\sqrt{2M}}N_M e^{\pm 2i\sqrt{\pi}\varphi_j(u)}\varepsilon_j \tag{II.4.2$'$}$$

we obtain the isomorphism of two $\mathfrak{o}(2l)$*-current algebra representations in the Fock spaces*

$$\mathscr{H}_{\mathrm{b}}\cong\mathscr{H}_{\mathrm{f}}, \tag{II.4.3$'$}$$

where the correspondence of the currents is given by the formulas

$$:\psi^{+}_j(u)\,\psi_j(u):=\frac{1}{\sqrt{\pi}}\partial_u\varphi_j(u), \tag{II.4.4$'$}$$

$$:\psi^{\pm}_j(u)\,\psi^{\pm}_k(u):=\frac{1}{2M}N_M e^{2i\sqrt{\pi}(\pm\varphi_j(u)\pm\varphi_k(u))}\varepsilon_j\varepsilon_k,\qquad j\neq k. \tag{II.4.5$'$}$$

(ii) *Lagrangian densities of free boson and fermion fields coincide; i.e.,*

$$:\psi^{\pm}_j(u)\,i\overleftrightarrow{\partial}_u\psi_j(u):=N_M(\partial_u\varphi_j(u))^2. \tag{II.4.6$'$}$$

(L–M) *Two approaches to the infrared regularization are equivalent in the following sense. Let*

$$e^{2\pi iu/L}=\frac{Mi-u'}{Mi+u'},\qquad u\in\left(-\frac{L}{2},\frac{L}{2}\right),\qquad u'\in\mathbb{R}, \tag{II.4.7}$$

defines a map of the spatial box $(-L/2,+L/2)$ *into the real line. Then*

Laguerre transformations (II.1.13), (II.3.9) *establish natural isomorphism between boson (resp. fermion) Fock spaces in* (II.4.3), (II.4.3)′. *One has*

$$\begin{aligned} \left(\frac{M^2+u'^2}{2M}\right)^{1/2} \cdot \psi_j^{\pm}(u') &= \left(\frac{L}{2\pi}\right)^{1/2} \psi_{L,j}^{\pm}(u), \\ \left(\frac{M^2+u'^2}{2M}\right) \cdot J_k(u') &= \frac{L}{2\pi} J_{L,k}(u). \end{aligned} \tag{II.4.8}$$

The formulation of the boson–fermion correspondence in a spatial box is almost identical and is just a reformulation of Theorem I.4.7 and Proposition I.4.11 in the field theoretical language.

We can deduce a number of corollaries about the equivalence of field models (II.2.38) and (II.3.18). Thus for $l=2$ Theorem II.4.1 implies the equivalence of the $SU(2)$-Thirring model and the theory of a massless scalar field and a sine-Gordon field, which has been noted first in [1]. Moreover, if one will write the complete correspondence for $l=2$ in terms of Majorana spinors, then one get the equivalence of the $\mathfrak{o}(4)$-model and of two decoupled sine-Gordon systems, which has been noted first in [26]. This fact is certainly a result of the classical isomorphism $\mathfrak{o}(4) \cong \mathfrak{sl}(2) \oplus \mathfrak{sl}(2)$. Another classical isomorphism $\mathfrak{o}(6) \cong \mathfrak{sl}(4)$ implies the equivalence of the $SU(4)$-Thirring model and the genralized sine-Gordon model, constructed from 3 (not 4!) Bose fields. In the next case $l=3$ the S_3 symmetry of the $\mathfrak{o}(8)$ Dynkin diagram leads to a triality symmetry in the dynamics and allows to transform ψ particles and two kinds of kinks into each other [24]. The generalization to an arbitrary l provides the equivalence of the $SU(l)$-Thirring model and generalized sine-Gordon model of type A_{l-1}, and $\mathfrak{o}(2l)$-model and generalized sine-Gordon model of type D_l.

4.2. The results of Section I.4.2 suggest to us that we can extend the boson–fermion correspondence to the twisted case. This approach leads to several additional observations in field theory models.

We begin with $O(n)$-models with odd n. This model as we could see in Section 3.2 contains n Majorana fermions. Boson–fermion correspondence discussed above allows us to substitute for each two Majorana fermions, which are equivalent to one Dirac fermion, by one boson. Thus we can obtain the theory consisting of $(n-1)/2$ bosons and one Majorana fermion, it is often called the supersymmetric theory. For example, the $O(3)$-model is equivalent to the supersymmetric form of the sine-Gordon model, which has been discovered in [26].

Let us consider now $O^{(2)}(2l+2)$-model in the spatial box approach. Then we have to add to $(2l+1)$ Majorana fermions one more fermion $\check{\psi}_{L,0}$ with *periodic* boundary conditions. The sum of $\check{\psi}_{L,0}$ with one of the Majorana fermions, e.g., $\check{\psi}_{L,1}$, can be expressed in terms of an odd boson ϕ_L (II.2.42)

$$\check{\psi}_{L,0}(u) + \check{\psi}_{L,1}(u) = (1/L)^{1/2} :\exp 2\sqrt{\pi}\, i\phi_L(u):. \tag{II.4.9}$$

This correspondence is a direct result of (I.4.23). Thus $O^{(2)}(l+2)$-models can be considered as the field theory with l Dirac fermions $\psi_1,\ldots,\psi_l$ and one odd boson ϕ or by boson–fermion correspondence as field theory with ordinary bosons $\varphi_1,\ldots,\varphi_l$ and one odd boson ϕ.

We consider now in details the case when $l=1$. Our set of particles is ψ, ψ^+, ϕ, with the antiperiodic boundary conditions in the spatial box of the length L. Thanks to the special isomorphism $\hat{\mathfrak{O}}^{(2)}(4) \cong \hat{\mathfrak{sl}}(2)$ we can construct the represenation of $\mathfrak{sl}(2)$-current algebra in the tensor product $\mathscr{H}_f \otimes \mathscr{H}_b$, where $\mathscr{H}_f$ (resp. $\mathscr{H}_b$) is the Fock space of states of Fermi particles ψ, ψ^+ (resp. of Bose particle ϕ). Proposition I.4.32 for V_0 can be reformulated now in the form of currents.

PROPOSITION II.4.10. *One has the following construction of the irreducible* $\mathfrak{sl}(2)$*-current algebra representation with the periodic boundary conditions in the spatial box of length* $2L$:

$$J_+(u) = \frac{1}{\sqrt{L}}\,\psi_L^+(u) :e^{2i\sqrt{\pi}\phi_L(u)}:, \qquad J_-(u) = \frac{1}{\sqrt{L}}\,\psi_L(u) :e^{-2i\sqrt{\pi}\phi_L(u)}:,$$
$$J_3(u) = \partial\phi_L(u) + :\psi_L^+(u)\,\psi_L(u):. \tag{II.4.11}$$

The currents are defined in the Fock space of states of the particles ψ, ψ^+, ϕ *and satisfy the anticommutation relations*

$$\begin{aligned} [J_3(u_1), J_3(u_2)] &= (i/\pi)\,\delta'(u_1-u_2),\\ [J_3(u_1), J_+(u_2)] &= 2J_+(u_2)\,\delta(u_1-u_2),\\ [J_3(u_1), J_-(u_2)] &= -2J_-(u_2)\,\delta(u_1-u_2),\\ [J_+(u_1), J_-(u_2)] &= J_3(u_1)\,\delta(u_1-u_2) + (i/2\pi)\,\delta'(u_1-u_2), \end{aligned} \tag{II.4.12}$$

where δ, δ' *are Dirac delta functions in the box of length* $2L$.

The free Hamiltonian density is given by

$$:(\partial_u\phi_L(u))^2: + :\psi_L^+(u)\, i\vec{\partial}_u\,\psi(u):. \tag{II.4.13}$$

Proof. The statement follows from Proposition I.4.32 and the commutation relations (I.4.31) if we set $J_\pm(u) = \tilde{x}_\pm(e^{2\pi i u/2L})$, $J_3(u) = \tilde{h}(e^{2\pi i u/2L})$.

Using Proposition II.4.10 we can write down the Lagrangian density (II.2.38) for the two-dimensional canonical model

$$\mathcal{L}(u, v) = \mathcal{L}_{\text{free}}(u, v) + \frac{g_v}{L}\,(\psi_L^{(1)+}\psi_L^{(2)} e^{2i\sqrt{\pi}(\phi^{(1)} - \phi^{(2)})}$$
$$+ \psi_L^{(1)}\psi_L^{(2)+} e^{-2i\sqrt{\pi}(\phi^{(1)} - \phi^{(2)})})$$
$$+ g_v : \left(\frac{1}{\sqrt{\pi}}\partial_u \phi_L^{(1)} - \psi_L^{(1)+}\psi_L^{(1)}\right)\left(\frac{1}{\sqrt{\pi}}\partial_v \phi_L^{(2)} - \psi_L^{(2)+}\psi_L^{(2)}\right):, \quad \text{(II.4.14)}$$

where $\mathcal{L}_{\text{free}}(u, v)$ is a Lagrangian density of free particles ψ, ψ^+, ϕ.

The two-dimensional model defined by (II.4.14) is quite similar to the quantum electrodynamics in $(1 + 1)$ dimensions (it is often called the Schwinger model, see details in [19]). On the other hand the special isomorphism $\hat{\mathfrak{O}}^{(2)}(4) \cong \hat{\mathfrak{sl}}(2)$ implies the equivalence of the model (II.4.14) and the ordinary sine-Gordon model. In fact one can introduce currents in the Fock sace of one bose particle $\varphi_{2L}(u)$ defined in the box of length $2L$ with periodic boundary conditions

$$J_\pm(u) = \frac{1}{2L} :e^{\pm\sqrt{8\pi}i\varphi_{2L}(u)}:, \qquad J_3(u) = \partial\varphi_{2L}(u). \qquad \text{(II.4.15)}$$

The commutation relations (II.4.12) are the particular case of (II.2.11)–(II.2.14) and we obtain our last

PROPOSITION II.4.15. *Under the following identifications*

$$\psi_L(u) = \frac{1}{\sqrt{L}} :\exp(i\sqrt{\pi}\,(\varphi(u) + \varphi(u \pm L))):,$$
$$\psi_L^+(u) = \frac{1}{\sqrt{L}} :\exp(-i\sqrt{\pi}\,(\varphi(u) + \varphi(u \pm L))):,$$
$$\phi_L(u) = (\varphi(u) - \varphi(u \pm L))/2,$$

one has the isomorphism of two irreducible $\mathfrak{sl}(2)$*-current algebra representations in the Fock spaces of the corresponding particles.*

The QED-like model (II.4.14) *is equivalent to the ordinary sine-Gordon model.*

We recall now that in the first work on boson–fermion correspondence in quantum field theory [6] Coleman found the equivalence between the sine-Gordon and the abelian Thirring model. Therefore the main models are closely related to each other and all together with the simplest representation of the simplest affine Lie algebra $\hat{\mathfrak{sl}}(2)$.

Finally, we would like to mention here a more superficial observation, which concerns two of the most important models in two-dimensional axiomatic quantum field theory, namely, $P(\varphi)_2$ and Yukawa$_2$ models [14]. The first one is considered in the boson Fock space of φ, the second in the Fock space $\mathscr{H}_f \otimes \mathscr{H}_b$, which is the same as in our QED-like model.

Summarizing Part II we can say that there is an evidence of a deep connection between quantum field theory in one space and time dimension and the representation theory of the affine Lie algebras. In this paper we have studied mainly one relation, namely, betwen boson–fermion correspondence and the isomorphism of two constructions of affine Lie algebra representations. We discussed certain papers in the immense literature of two-dimensional quantum field theory. We apologize to the authors whom we failed to mention here. We hope that the mathematicians who are working in the representation theory will appreciate (and maybe even use!) the rich ideas contained in the physical literature. On the other hand, we hope that the physicists who are interested in this subject will manage to look at the different field theory models from a unified point of view, and will use the powerful instrument of Lie algebras, in particular, affine Lie algebras, and their representations for their needs. We are certainly far from exhausting this connection. We can only suppose now that the mysterious question of which special quantum field models are completely integrable has its origin in the representation theory of affine Lie algebras. This would be analogous to the large class of completely integrable finite-dimensional systems explained by the orbital theory of affine Lie algebras (see e.g. [12]). Similarly, the inverse-scattering method has probably been so successful thanks to the presence of the global symmetry provided by the action of the corresponding infinite-dimensional groups.

Appendix: Laguerre Polynomials and Bateman Functions

We formulate here all the necessary facts concerning the properties of Laguerre polynomials and Bateman functions. The details can be found in [4] and [3].

Definition.

$$L_n^\alpha(x) = \frac{1}{n!} e^x x^{-\alpha} \frac{d^n}{dx^n} (e^{-x} x^{n+\alpha}), \qquad n \geqslant 0, \quad x \geqslant 0. \tag{A1}$$

We will consider only the polynomials with $\alpha = 1, 0, -1$. We note first the following relation

$$xL_{n-1}^1(x) + nL_n^{-1}(x) = 0. \tag{A2}$$

Definition A1 implies immediately the form of

Generating function

$$\sum_{n=0}^{\infty} L_n^\alpha(x)\, z^n = (1-z)^{-\alpha-1} \exp \frac{xz}{z-1}. \tag{A3}$$

The integration over the unit circle in the indeterminate z gives

Integral formula

$$(-1)^n e^{-x/2} L_n^\alpha(x) = \frac{1}{2\pi} \int_{-\infty}^{\infty} e^{ixu} \left(\frac{i/2+u}{i/2-u}\right)^{n+(\alpha+1)/2} \times \frac{du}{(1/4+u^2)^{(1-\alpha)/2}}, \qquad x > 0, \tag{A4}$$

for $\alpha = 1$ one should consider this integral in the sense of the principal value. Note that for $x < 0$ the right-hand side turns to zero. Tus the Fourier transform of (A4) becomes

Laplace transform

$$\int_0^{\infty} ((-1)^n e^{-x/2} L_n^\alpha(x))\, e^{-ixu}\, dx = \left(\frac{i/2+u}{i/2-u}\right)^{n+(\alpha+1)/2} \times \frac{1}{(1/4+u^2)^{(1-\alpha)/2}} + c, \tag{A5}$$

where $c = (-1)^n$ for $\alpha = 1$, $c = 0$ for $\alpha = 0$ or $\alpha = -1$, $n \neq 0$ and $c = 1/i(u + i/2)$ for $\alpha = -1$, $n = 0$.

Using the generating function for Laguerre polynomials one can deduce:

Orthogonality conditions

$$\int_0^{\infty} (e^{-x/2} L_n^{-1}(x))(e^{-x/2} L_n^{-1}(x)) \frac{dx}{x} = \frac{1}{n} \delta_{n,m}, \qquad n, m \geqslant 1, \tag{A6}$$

$$\int_0^{\infty} (e^{-x/2} L_n^0(x))(e^{-x/2} L_m^0(x))\, dx = \delta_{m,m}, \qquad n, m \geqslant 0. \tag{A7}$$

One knows that $\{e^{-x/2} L_n^{-1}(x)(n/x)^{1/2}\}_{n=1}^{\infty}$ and $\{e^{-x/2} L_n^0(x)\}_{n=0}^{\infty}$ are complete orthogonal systems on $[0, +\infty)$ (see e.g. [4]). It implies

$$\sum_{n=1}^{\infty} (e^{-x/2} L_n^{-1}(x))(e^{-y/2} L_n^{-1}(y)) \frac{n}{y} = \delta(x-y), \tag{A8}$$

$$\sum_{n=0}^{\infty} (e^{-x/2} L_n^0(x))(e^{-y/2} L_n^0(y)) = \delta(x-y). \tag{A9}$$

Using the generating function (A3) one can also deduce the following formulas:

Differential equations

$$\left(x\frac{d^2}{dx^2}+n-\frac{x}{4}\right)(e^{-x/2}L_n^{-1}(x))=0, \qquad \text{(A10)}$$

$$\left(x\frac{d^2}{dx^2}+\frac{d}{dx}+n+\frac{1}{2}-\frac{x}{4}\right)(e^{-x/2}L_n^0(x))=0. \qquad \text{(A11)}$$

Differential recursion relations:

$$x\frac{d}{dx}(e^{-x/2}L_n^{-1}(x))$$

$$=\left(n-\frac{x}{2}\right)(e^{-x/2}L_n^{-1}(x))-(n-1)(e^{-x/2}L_{n-1}^{-1}(x))$$

$$=\left(\frac{x}{2}-n\right)(e^{-x/2}L_n^{-1}(x)+(n+1)(e^{-x/2}L_{n+1}^{-1}(x)) \qquad \text{(A12)}$$

$$x\frac{d}{dx}(e^{-x/2}L_n^0(x))$$

$$=\left(n-\frac{x}{2}\right)(e^{-x/2}L_n^0(x))-n(e^{-x/2}L_{n-1}^0(x))$$

$$=\left(\frac{x}{2}-n-1\right)(e^{-x/2}L_n^0(x))+(n+1)\,e^{-x/2}L_{n+1}^0(x)). \qquad \text{(A13)}$$

Addition formulas

$$\sum_{n=0}^{\infty} L_n^{-1}(x)\,L_{n+N}^{-1}(y)=e^xL_N^{-1}(x-y), \qquad x>y, =0, \quad x\leqslant y, \qquad \text{(A14)}$$

$$\sum_{n=0}^{N} L_n^{-1}(x)\,L_{N-n}^{-1}(y)=L_N^{-1}(x+y).$$

The generalization of Laguerre polynomials for an arbitrary parameter n is given by Whittaker functions $W_{n,m}$ (see e.g. [4]), the case when $m=\frac{1}{2}$, has been considered in details by Bateman [3].

DEFINITION.

$$k_n(x)=\frac{2}{\pi}\int_0^{\pi/2}\cos(x\tan\theta-n\theta)\,d\theta. \qquad \text{(B1)}$$

We note that $k_{2n}(x/2) = \Gamma(n+1)^{-1} W_{n,1/2}(x)$, $x > 0$, in particular for an integer n, $k_{2n}(x/2) = (-1)^n e^{-x/2} L_n^{-1}(x)$. The properties of these functions have been considered before. We are interested now in the case of semiinteger n.

Definition (B1) implies the relations

$$k_{-n}(x) = k_n(-x), \tag{B2}$$

$$k_n(0) = \frac{2}{n\pi} \sin \frac{\pi n}{2}. \tag{B3}$$

One can also obtain the generalization of (A4).

Integral formula:

$$k_n(x/2) = \frac{1}{2\pi} \int_{-\infty}^{\infty} e^{ixu} \left(\frac{i/2 + u}{i/2 - u} \right)^n \frac{du}{u^2 + \frac{1}{4}}, \qquad x > 0. \tag{B4}$$

The generalization of (A6) is valid in the sense of principal value:

Orthogonality conditions

$$\int_{-\infty}^{\infty} k_{2n+1}(x/2)\, k_{2m+1}(x/2) \frac{dx}{x} = \frac{1}{n + \frac{1}{2}} \delta_{n,m}, \qquad n, m \in \mathbb{Z}. \tag{B5}$$

Note added in proof: Recently there appeared several papers which are closely connected with our results:

Segal [27] found another approach to the vertex representation. Kac and Peterson [28] announced construction of spinor representation for affine Lie algebras $\hat{D}(n)$. Date *et al.* [29] obtained solutions of Korteweg–de-Vries type equations using the construction of vertex representation given in [18]. They found independently the important components of the boson-fermion correspondence (cf. I.4.6) and (I.4.23)), which are crucial in their investigation. Drinfeld and Sokolov [30] announced a natural interpretation of the Hamiltonian structure of Korteweg–de-Vries and *classical* sine-Gordon type equations (see [23]) from the point of view of affine Lie algebras without using their representations.

Acknowledgments

I wish to thank I. Bars, H. Garland, Shankar, P. Trauber and G. Zuckerman for helpful discussions.

References

1. T. Banks, D. Horn, and H. Neuberger, Bosonization of the $SU(N)$ Thirring models, *Nucl. Phys. B* **108** (1976), 119–129.
2. K. Bardakci and M. B. Halpern, New dual quark models, *Phys. Rev. D* **3** (1971), 2493–2506.
3. H. Bateman, The k-function, a particular case of the confluent hypergeometric function, *Trans. Amer. Math. Soc.* **33** (1931), 817–831.
4. H. Bateman and A. Erdelyi, "Higher transcendental functions," Vol. 1, Mc Graw–Hill New York/Toronto/London, 1953.
5. N. Bourbaki, "Groupes et algebras de Lie," Chaps. 7, 8, Herman, Paris, 1975.
6. S. Coleman, Quantum sine-Gordon equation as the massive Thirring model, *Phys. Rev. D* **11** (1975), 2088–2097.
7. R. Dashen and Y. Frishman, Four-fermion interactions and scale invriance, *Phys. Rev. D* **11** (1975), 2781–2802.
8. A. Feingold and J. Lepowsky, The Weyl–Kac character formula and power series identities, *Adv. in Math.* **29** (1978), 271–309.
9. I. B. Frenkel, Spinor representation of affine Lie algebras, *Proc. Nat. Acad. Sci. USA* **77** (1980), 6303–6306.
10. I. B. Frenkel and V. G. Kac, Basic representations of affine Lie algebras and dual resonance models, *Invent. Math.* **62** (1980), 23–66.
11. I. B. Frenkel and V. G. Kac, Addenda to "Basic representations of affine Lie algebras and dual resonance model," to be published.
12. I. B. Frenkel, A. G. Reiman, and M. A. Semenov-Tjan-Šanskii, Graded Lie algebras and completely integrable dynamical systems, *Sov. Math. Dokl* **20** (1979), 811–814.
13. H. Garland, The arithmetic theory of loop algebras, *J. Algebra* **53** (1978), 480–551.
14. J. Glimm and A. Jaffe, *in* "Quantum Field Models in Statistical Mechanics and Quantum Field Theory, Les Houches, 1970" (C. DeWitt and R. Stora, Eds.), pp. 1–108, Gordon & Breach, New York, 1971.
15. M. B. Halpern, Quantum "solitons" which are $SU(N)$ fermions, *Phys. Rev. D* **12** (1975), 1684–1699.
16. V. G. Kac, Infinite-dimensional Lie algebras and Dedekind's η-function, *J. Funct. Anal. Appl.* **8** (1974), 68–70.
17. V. G. Kac, Infinite dimensional algebras, Dedekind's η-function, classical Möbius function and the very strange formula, *Adv. in Math.* **30** (1978), 85–136.
18. V. G. Kac, D. A. Kazhdan, J. Lepowsky, and R. L. Wilson, Realization of the basic representations of the Eudidean Lie lalgebras, *Aav. in Math.*
19. J. Kogut and L. Susskind, How quark confinement solves the $\eta - 3\pi$ problem, *Phys. Rev. D* **11** (1975), 3594–3610.
20. J. Lepowsky and R. L. Wilson, Construction of the affine Lie algebra $A_1^{(1)}$, *Commun. Math. Phys.* **62** (1978), 43–53.
21. S. Mandelstam, Dual-resonance models, *Phys. Rep.* Sect. C **13**, (1974), 259–353.
22. S. Mandelstam, Soliton operators for the quantized sine-Gordon equation, *Phys. Rev D* **11** (1974), 3026–3030.
23. A. V. Mikhailov, M. A. Olshanetsky, and A. M. Perelomov, Two dimensional generalized Toda lattice, preprint. *Commun. Math. Phys.* **79** (1981), 473–488.
24. R. Shankar, Some novel features of the Gross–Neveu model, *Phys. Lett. B* **92** (1980), 333–336.
25. L. A. Takhtadzhyan, Exact theory of propagation of ultrashort optical pulses in two-level media, *Sov. Phys.* JETP **39**, (1975), 228–233.
26. E. Witten, Some properties of the $(\bar\psi\psi)^2$ model in two dimensions, *Nucl. Phys. B* **142** (1978), 285–300.

27. G. Segal, Unitary representations of some infinite dimensional groups, *Commun. Math. Phys.* **80** (1981), 301–342.
28. V. G. Kac and D. H. Peterson, Spin and wedge representations of infinite-dimensional Lie algebras and groups, *Proc. Nat. Acad. Sci. USA* **78** (1981), 3308–3312.
29. E. Date, M. Jimbo, M. Kashiwara, and T. Miwa, Transformation groups for soliton equations, RIMS preprints, pp. 356–362 (1981).
30. V. G. Drinfeld and V. V. Sokolov, Korteweg–de-Vries type equations and simple Lie algebras, *Sov. Math. Dokl. ANSSSR* **258** (1981), 11–16. (Russian)

CHAPTER 4

THE VIRASORO ALGEBRA

Reprinted Papers

9. B.L. Feigin and D.B. Fuks, "Invariant Skew-Symmetric Differential Operators on the Line and Verma Modules over the Virasoro Algebra", Funct. Anal. Appl. **16** (1982) 114–127.

10. A. Rocha-Caridi, "Vacuum Vector Represenations of the Virasoro Algebra", in *Vertex Operators in Mathematics and Physics*, MSRI Publication #3 (Springer, Heidelberg, 1984) 451–473.

11. D. Friedan, Z. Qiu and S. Shenker, "Conformal Invariance, Unitarity, and Critical Exponents in Two Dimensions", Phys. Rev. Lett. **52** (1984) 1575–1578.

12. D. Friedan, Z. Qiu and S. Shenker, "Details of the Non-Unitarity Proof for Highest Weight Representations of the Virasoro Algebra", Commun. Math. Phys. **107** (1986) 535–542.

13. P. Goddard, A. Kent and D. Olive, "Unitary Representations of the Virasoro and Super-Virasoro Algebras", Commun. Math. Phys. **103** (1986) 105–119.

14. W. Boucher, D. Friedan and A. Kent, "Determinant Formulae and Unitarity for the N = 2 Superconformal Algebras in Two Dimensions or Exact Results on Superstring Compactification", Phys. Lett. 172B **(1986)** 316–322.

The Virasoro algebra occurs whenever we consider a two-dimensional theory with conformal symmetry. The conformal group in two dimensions is infinite-dimensional, unlike its counterparts in higher dimensions, and its Lie algebra consists of two commuting copies of the Virasoro algebra. More precisely, the Virasoro algebra is a central extension of the conformal algebra, but it is this algebra, with its extra c-number term, that is relevant in quantum-mechanical discussions of conformal symmetry. Virasoro [1] introduced the algebra into string theory in a discussion of conditions on physical states which might ensure the absence of ghosts (negative norm states), required to make Lorentz invariance manifest. But it was J.H. Weis who subsequently pointed out the existence of the crucial c-number term. (See note added in proof in [2].) It was eventually established [3,4] that the conditions do in fact ensure the absence of ghosts provided that the dimension of space-time does not exceed 26.

The representations of the Virasoro algebra that occur in string theory are typically representations in spaces which have scalar products that are not positive definite; there are negative norm states which it is the function of the Virasoro conditions to remove. Elsewhere, in two-dimensional field theory and in the theory of the behaviour of two-dimensional statistical systems at critical points, unitary representations in positive definite spaces are relevant. The representations which are interesting physically are the so-called "highest weight" ones which are those for which one of the Virasoro generators, L_0, has a spectrum which is bounded below. The irreducible unitary highest weight representations are labelled by the lowest eigenvalue, h, of L_0, and the value, c, of the central term. The question of whether such a unitary representation exists for given (c, h) amounts to deciding whether certain matrices are positive semi-definite. A remarkable formula for the determinant of these matrices was found by V.G. Kac [5] and a proof was provided by B.L. Feigin and D.B. Fuks [*Reprinted Paper* #9]. Following on this work, character formulas for these representations were calculated by A. Rocha-Caridi [*Reprinted Paper* #10], ignoring the question of unitarity.

It is not immediately evident from the formula of Kac that the set of values of (c, h) for which unitary highest weight representations exist has an interesting structure. By a detailed analysis, D. Friedan, Z. Qiu and S. Shenker [*Reprinted Papers* #11 *and* #12] were able to use Kac's formula to show that, apart from the continuum of representations that exists for $c \geq 1$ and $h \geq 0$, unitary representations could only exist for a discrete series of representations with $0 \leq c < 1$. The existence of these representations was demonstrated by the explicit construction of P. Goddard, A. Kent and D. Olive [*Reprinted Paper* #13] using relationships between affine Kac-Moody algebras and the Virasoro algebra.

The Virasoro algebra has supersymmetric extensions, the simplest of which, the super-Virasoro algebra, has $N = 1$ supersymmetry. It was discovered in the spinning string theory of Ramond, Neveu and Schwarz [6–8] and is responsible for the elimination of ghosts in that model, provided that the space-time dimension does not exceed 10. There are parallel results on unitary highest weight representations of the super-Virasoro algebra which are described in *Reprinted Papers* #11 *and* #13. Recently there have been discussions for the higher supersymmetric extensions, in particular *Reprinted Paper* #14 discusses the N = 2 case and [9] points out the equivalence of the twisted and untwisted N = 2 algebras and makes comments on the N = 3 and 4 algebras.

References

[1] M.A. Virasoro, "Subsidiary conditions and ghosts in dual resonance models", Phys. Rev. **D1** (1970) 2933–2936.

[2] S. Fubini and G. Veneziano, "Algebraic treatment of subsidiary conditions in dual resonance models", Ann. Phys. **63** (1971) 12–27.

[3] R.C. Brower, "Spectrum generating algebra and the no-ghost theorem for the dual model", Phys. Rev. **D6** (1972) 1655–1662.

[4] P. Goddard and C.B. Thorn, "Compatibility of the dual pomeron with unitarity and the absence of ghosts in the dual resonance models", Phys. Lett. **40B** (1972) 235–238.

[5] V.G. Kac, "Highest weight representations of infinite-dimensional Lie algebras and superalgebras" in *Proceedings of the International Congress of Mathematicians* (Helsinki, 1978) 299–304; "Contravariant form for infinite-dimensional Lie algebras and superalgebras", *Lecture Notes in Physics* Vol. **94** (Springer Verlag, New York, 1979) 441–445.

[6] P. Ramond, "Dual theory for free fermions", Phys. Rev. **D3** (1971) 2415–2418.

[7] A. Neveu and J.H. Schwarz, "Factorizable dual model of pions", Nucl. Phys.**B31** (1971) 86–112; "Quark model of dual pions", Phys. Rev. **D4** (1971) 1109–1111.

[8] A. Neveu, J.H. Schwarz and C.B. Thorn, "Reformulation of the dual pion model", Phys. Lett. **35B** (1971) 529–533.

[9] A. Schwimmer and N. Seiberg, "Comments on the N = 2, 3, 4 superconformal algebras in two dimensions', Phys. Lett. **B184** (1987) 191–196.

INVARIANT SKEW-SYMMETRIC DIFFERENTIAL OPERATORS ON THE LINE AND VERMA MODULES OVER THE VIRASORO ALGEBRA

B. L. Feigin and D. B. Fuks

UDC 517.43+519.46

The main result of this article is Theorem 1.1, which gives a complete classification of skew-symmetric differential operators, acting in tensor fields on the line and invariant with respect to diffeomorphisms of the line. The statement of this theorem was stated as a hypothesis in our note [1].

It turns out that the problem of enumeration of the invariant skew-symmetric differential operators of the line is closely connected with the problem of reducibility of the Verma modulus over a one-dimensional central extension of the Lie algebra of the polynomial vector fields on the circle – the so-called Virasoro algebra. This reducibility problem was solved in 1979 by Kac [2], and with the help of his results we could prove our main theorem. Moreover, the connection between the invariant differential operators and the Verma modules can be used also in the opposite direction: Simple arguments about invariant operators enable us to give a new proof of the Kac theorem. A combination of the methods of this article with the traditional methods of the theory of invariant differential operators leads also to a new series of combinatorial identities.

As an intermediate object between the modules of the tensor fields on the line and the Verma modules over the Virasoro algebra, we introduce some new modules over the Virasoro algebra: spaces of their own kind of skew-symmetric tensor fields of infinite rank. The construction of these modules, possibly of independent interest, overlaps the Kac–Peterson constructions [3].

This article is organized in the following manner. In Sec. 1 we recall necessary definitions from the theory of invariant differential operators and formulate the main theorem. The reader, acquainted with [1], will find here little new. Some progress on the way to the proof of the main theorem can be achieved by carrying out a direct investigation of the singular vectors in the modules of tensor fields; these results are set forth in Sec. 2. In Sec. 3 we formulate the Kac theorem and prove the main theorem with the help of this theorem and the results of Sec. 2. The last section, Sec. 4, contains a proof of the Kac theorem and also of the above-mentioned combinatorial identities.

1. Invariant Differential Operators on the Line

Let W_1 denote the algebra of the polynomial vector fields on the line, and for each $\lambda \in C$, let F_λ denote the W_1-module of the tensor fields of the form $\varphi(z)dz^{-\lambda}$, where $\varphi \in C[z]$. The action of the algebra W_1 in F_λ is described by the equation

$$\left(f(z)\frac{d}{dz}\right)(\varphi(z)\,dz^{-\lambda}) = [f(z)\varphi'(z) - \lambda f'(z)\varphi(z)]\,dz^{-\lambda}.$$

It is obvious that F_1 is the adjoint representation of the algebra W_1, F_0 is the module of polynomial functions, and F_{-1} is the module of the polynomial 1-forms.

A W_1-homomorphism of the form

$$F_{\lambda_1} \otimes \ldots \otimes F_{\lambda_n} \to F_{\lambda}. \tag{1}$$

Solid-State Physics Institute, Academy of Sciences of the USSR. Moscow State University. Translated from Funktsional'nyi Analiz i Ego Prilozheniya, Vol. 16, No. 2, pp. 47-63, April-June, 1982. Original article submitted September 15, 1981.

0016-2663/82/1602-0114$07.50

Reprinted from Funct. Anal. Appl. 16 (1982) 114–126, with permission.

is known as an n-ary invariant differential operator (on the line) (the number n is called the arity of the operator). Let us observe that this definition does not include the requirement that the operator is differential. As we will see in Sec. 2, operator (1) automatically turns out to be differential and, in addition, has order $\lambda_1 + \ldots + \lambda_n - \lambda_0$; in particular, it follows from the existence of a nontrivial invariant operator (1) that $\lambda_1 + \ldots + \lambda_n - \lambda_0$ is a nonnegative integer.

We restrict ourselves in this article to a consideration of skew-symmetric operators, i.e., W_1-homomorphisms $\Lambda^n F_\lambda \to F_{\lambda_0}$; by virtue of what we have said, $\lambda_0 = n\lambda - N$, where N is the order of the operator. It is obvious that each (even noninvariant) differential operator $\Lambda^n F_\lambda \to F_{\lambda_0}$ of order N acts by the formula

$$(\varphi_1 dz^{-\lambda}, \ldots, \varphi_n dz^{-\lambda}) \mapsto \sum_{\substack{0 \leqslant j_1 < \ldots < j_n \\ j_1 + \ldots + j_n = N}} a_{j_1 \ldots j_n} \begin{vmatrix} \varphi_1^{(j_1)} & \ldots & \varphi_1^{(j_n)} \\ \cdots & \cdots & \cdots \\ \varphi_n^{(j_1)} & \ldots & \varphi_n^{(j_n)} \end{vmatrix} dz^{-(n\lambda - N)}, \tag{2}$$

where $a_{j_1 \ldots j_n}$ are polynomials in z. Moreover, it is easily shown (and this will be done in Sec. 2) that if the operator is invariant, then all the coefficients $a_{j_1 \ldots j_n}$ are constant. It is obvious from Eq. (2) that if the operator is nontrivial, then $N \geq n(n-1)/2$. We call the difference $N - n(n-1)/2$ the true order of the operator.

Now we can formulate our main theorem.

THEOREM 1.1. For arbitrary $\lambda \in \mathbb{C}$, $n \geq 0$, and $k \geq 0$ there exists at most one, up to proportionality, invariant differential operator

$$\Lambda^n F_\lambda \to F_{n\lambda - \frac{n(n-1)}{2} - k} \tag{3}$$

(of true order k). This operator exists if and only if one of the following conditions is fulfilled:

(a) $k = 0$.

(b) $0 < k \leq n$ and λ satisfies the quadratic equation

$$\left[\left(\lambda + \frac{1}{2}\right)(k' + 1) - n\right]\left[\left(\lambda + \frac{1}{2}\right)(k'' + 1) - n\right] = \frac{(k'' - k')^2}{2}, \tag{4}$$

in which k' and k'' are arbitrary complementary divisors of k (i.e., $k' \in \mathbb{Z}_+$, $k'' \in \mathbb{Z}_+$, and $k'k'' = k$).

Theorem 1.1 shows, in particular, that if $k = 0$, then the invariant operator (3) exists and is unique up to proportionality for arbitrary λ. As we will see in Sec. 2, this statement becomes obvious in the language of "the singular vectors"; besides, it is quite obvious and immediate. The nontrivial operator (3) with $k = 0$ is defined by the formula

$$(\varphi_1 dz^{-\lambda}, \ldots, \varphi_n dz^{-\lambda}) \mapsto \begin{vmatrix} \varphi_1 & \varphi_1' & \ldots & \varphi_1^{(n-1)} \\ \varphi_2 & \varphi_2' & \ldots & \varphi_2^{(n-1)} \\ \cdots & \cdots & \cdots & \cdots \\ \varphi_n & \varphi_n' & \ldots & \varphi_n^{(n-1)} \end{vmatrix} dz^{-n\lambda - \frac{n(n-1)}{2}}; \tag{5}$$

it is denoted by $\Delta_{\lambda,n}$ and is called the operator of general position. It is also obvious from Theorem 1.1 that for $k > n$ nontrivial invariant operators (3) do not exist at all, and for $0 < k \leq n$ these operators exist only for specially chosen values of λ. These λ are found from Eq. (4). As computation shows, under the condition (b), Eq. (4) has real roots, which coincide if and only if $k'' = k'$. These roots are positive with one exception: If $k = n$ and $\min(k', k'') = 1$, then one of the roots is equal to zero. Further, let d be the number of the positive integral divisors of k, and $k_1 = 1, k_2, \ldots, k_{d-1}, k_d = k$ be the increasing sequence of these divisors. For indices i and j such that $i \leq j$ and $k_i k_j = k$, let λ_i and λ_j denote the roots of Eq. (4) arranged in nondecreasing order, in which $k' = k_i$ and $k'' = k_j$. Then $\lambda_1, \ldots, \lambda_d$ is also an increasing sequence. Of course, this sequence depends on n since Eq. (4) depends on n. The numbers $\lambda_1, \ldots, \lambda_d$ for certain k and n are given in Table 1.

Some of our operators have a rather simple description. For $n = 1$ we have the operator of general position

$$\Delta_{\lambda,1} : F_\lambda \to F_\lambda$$

(the identity operator) and one operator of first order $d : F_0 \to F_{-1}$, corresponding to the divisor 1 of the number 1 (the differentiation of functions). For $n = 2$ the operator of general position

$$\Delta_{\lambda,2} : \Lambda^2 F_\lambda \to F_{2\lambda-1}$$

TABLE 1

n	k	$\lambda_1, \dots, \lambda_d$	n	k	$\lambda_1, \dots, \lambda_d$
1	1	0	4	3	$(2\pm\sqrt{2})/2$
2	1	1/2	4	4	0, 5/6, 9/5
2	2	0, 2/3	. . .	. . .	. . . .
3	1	1	n	1	$\frac{n-1}{2}$
3	2	$(9\pm\sqrt{21})/12$			
3	3	0, 5/4	. . .	. . .	. . . .
4	1	3/2	n	n	$0, \dots, \frac{(n-1)(n+2)}{2(n+1)}$
4	2	$(7\pm\sqrt{7})/6$			
			. . .	. . .	. . . .

has order one. Moreover, there exist one operator of second order, i.e., of true order one,

$$\Lambda^2 F_{1/2} \to F_{-1},$$

corresponding to the divisor 1 of the number 1 (the composition $d \cdot \Delta_{1/2,2}$) and two operators of order three, i.e., of true order two (corresponding to the divisors 1 and 2 of the number 2)

$$\Lambda^2 F_0 \to F_{-3}, \quad \Lambda^2 F_{2/3} \to F_{-5/3}.$$

The first of these operators is the composition $\Delta_{-1,2} \cdot \Lambda^2 d$ and the second one was discovered in 1977 by P. Grozman, who classified all invariant binary differential operators in a space of arbitrary number of dimensions [4]. For each n the only n-ary operator of true order 1 is the composition of the operator of general position $\Delta_{(n-1)/2,n}$ that takes values in F_0 and the operator $d: F_0 \to F_{-1}$. In equally simple manner we can construct the operator of true order n corresponding to the divisor 1 of the number n for arbitrary n. This operator $\Lambda^n F_0 \to F_{-n(n+1)/2}$ is the composition $\Delta_{-1,n} \cdot \Lambda^n d$. *Our note [1] contains direct descriptions of some more* operators; nevertheless, many of them remain mysterious for us. An algorithm for the computation of their coefficients is given in Sec. 2.5, but we have not been able to write any visible general formula for them.

2. Invariant Operators and Singular Vectors

2.1. Auxiliary Algebras and Modules. Let $\mathcal{L}$ denote the Lie algebra over **C** with the additive basis $\{e_i \mid i \in \mathbf{Z}\}$ and the commutator $[e_i, e_j] = (j - i)e_{i+j}$. We can identify this algebra with the Lie algebra of the polynomial vector fields on the circle $|z| = 1$ ($e_i = z^i d/d \arg z$). For an integer $k \geq -1$ the subspace L_k^+ of the algebra $\mathcal{L}$ generated by e_i with $i \geq k$ and the subspace L_k^- of the algebra $\mathcal{L}$ generated by e_i with $i \leq -k$ constitute (isomorphic) subalgebras of $\mathcal{L}$; the symbol L_k^+ is usually abbreviated to L_k. The algebra L_{-1} is identified with W_1 by means of the formula $e_i = z^{i+1}d/dz$. The algebra $\mathcal{L}$ as well as the indicated subalgebras of it have a natural Z-gradation: $\deg e_i = i$.

Let $M = \oplus M_j$ be a graded $\mathcal{L}$-module. We associate three more modules with M: "The adjoint module" M', "the contragradient module" $\bar{M}$, and "the inverted module" M^0. Namely, we set $M' = \oplus (M')_j$, $\bar{M} = \oplus \bar{M}_j$, and $M^0 = \oplus M_j^0$, where

$$(M')_j = (M_{-j})' \,[= \mathrm{Hom}\,(M_{-j}, \mathbf{C})], \quad \bar{M}_j = (M_j)', \; M_j^0 = M_{-j},$$

and we define the operators

$$e_i: (M')_j \to (M')_{j+i}, \quad e_i: \bar{M}_j \to \bar{M}_{j+i}, \quad e_i: M_j^0 \to M_{j+i}^0$$

respectively, as follows:

$$-e_i': (M_{-j})' \to (M_{-j-i})', \quad e_{-i}': (M_j)' \to (M_{j+i})', \quad -e_{-i}: M_{-j} \to M_{-j-i}.$$

It is obvious that $\bar{M} = (M')^0$. It is also clear that a homogeneous $\mathcal{L}$-homomorphism $\varphi: M \to N$ induces homogeneous $\mathcal{L}$-homomorphisms $\varphi': N' \to M'$, $\bar{\varphi}: \bar{N} \to \bar{M}$, and $\varphi^0: M^0 \to N^0$.

We will sometimes take liberty to speak about the adjoint, the contragradient, and the inverted modules for modules not over the whole algebra $\mathcal{L}$, but over its subalgebras. For example, the module contragradient to an L_k^+-module is an L_k^--module.

Now we describe some concrete modules. For $\lambda, \mu \in \mathbf{C}$ we denote by $\mathcal{F}_{\lambda,\mu}$ the $\mathcal{L}$-module with the additive basis $\{f_j \mid j \in \mathbf{Z}\}$ and the action of the algebra

$$e_i f_j = [j + \mu - (i+1)\lambda] f_{i+j}.$$

The subspace $F^{+}_{\lambda,\mu}$ of the space $\mathcal{F}_{\lambda,\mu}$ generated by f_j with $j \geq 0$ inherits the structure of an L_0^{+} module from $\mathcal{F}_{\lambda,\mu}$, and the subspace $F^{-}_{\lambda,\mu}$ generated by f_j with $j \leq 0$ inherits the structure of an L_0^{-}-module. We will also consider the L_0-module $G_{\lambda,\mu} = \mathcal{F}_{\lambda,\mu}/F^{-}_{\lambda,\mu}$; here the coset of the element f_j of the space $\mathcal{F}_{\lambda,\mu}$ in $G_{\lambda,\mu}$ for $j \geq 0$ will also be denoted by f_j. The symbol $F^{+}_{\lambda,\mu}$ is abbreviated to $F_{\lambda,\mu}$, and the symbols $\mathcal{F}_{\lambda,0}, \ldots, G_{\lambda,0}$ are abbreviated to $\mathcal{F}_{\lambda}, \ldots, G_{\lambda}$. In this connection, the symbol F_λ is consistent with the same symbol of Sec. 1: $f_j = z^j dz^{-\lambda}$ (F_λ is an L^{+}_{-1}-module, because if $\mu = 0$, then $e_{-1}f_0 = 0$). Let us observe that the module $F_{\lambda,\mu}$ can also be given an analytical meaning: It is the space of "the tensor fields" of the form $p(z)z^{\mu}dz^{-\lambda}$, where p is a polynomial; in addition, $f_j = z^{\mu+j}dz^{-\lambda}$. Let us also observe that all the indicated modules are graded: $\deg f_j = j$.

The above-mentioned operations over graded modules do not lead out of the class of the modules : There exist obvious isomorphisms

$$\mathcal{F}'_{\lambda,\mu} = \mathcal{F}_{-1-\lambda,-\mu}, \quad \bar{\mathcal{F}}_{\lambda,\mu} = \mathcal{F}_{-1-\lambda,\mu-2\lambda-1}, \quad \mathcal{F}^{u}_{\lambda,\mu} = \mathcal{F}_{\lambda,-\mu+2\lambda+1}.$$

The second of these isomorphisms induces the isomorphism $F_{\lambda,\mu} = G_{-1-\lambda,\mu-2\lambda-1}$. Moreover, the formula $f_j \to f_{j+1}$ defines (shifting the gradation by 1) an isomorphism $\mathcal{F}_{\lambda,\mu} \to \mathcal{F}_{\lambda,\mu+1}$.

In conclusion, we describe one more, more complicated, isomorphism. The formula

$$f_{j_1} \wedge \cdots \wedge f_{j_n} \to f_0 \wedge f_{j_1+1} \wedge \cdots \wedge f_{j_n+1}$$

defines a homomorphism

$$\Lambda^n F_{\lambda,\mu} \to \Lambda^{n+1} F_{\lambda,\mu-1},$$

which is consistent with the "true gradation" in $\Lambda^n F_{\lambda,\mu}$ and $\Lambda^{n+1} F_{\lambda,\mu-1}$:

$$\deg(f_{j_1} \wedge \cdots \wedge f_{j_s}) = j_1 + \cdots + j_s - \frac{s(s-1)}{2}$$

(cf. with the true order of an operator in Sec. 1). This homomorphism is a monomorphism; in addition, it turns out to be an isomorphism of restricted by the elements, the (true) degrees of which do not exceed n. Thus, if we are not interested in the elements of excessively high degrees, we can identify $\Lambda^n F_{\lambda,\mu}$ with $\Lambda^{n+1} F_{\lambda,\mu-1}$ and, therefore, also $\Lambda^n G_{\lambda,\mu}$ with $\Lambda^{n+1} G_{\lambda,\mu-1}$. (This equitableness between n and μ, which is difficult to explain analytically, enables us in practice to forget that the "arity" n is a positive integer, and not an arbitrary complex number.)

2.2. Connection with Singular Vectors. Let M be a module over the algebra $\mathcal{T}$ or over a subalgebra of this algebra that contains L_1^{-}. Following A. N. Rudakov, we call a nonzero element m of the module M a singular vector of this module if $e_i m = 0$ for $i \leq -1$. For example, f_0 is a singular vector of the L_0^{-}-module $G_{\lambda,\mu}$ with arbitrary λ and μ. Let us observe that since the algebra L_1^{-} is generated by e_{-1} and e_{-2}, for the singularity of a nonzero vector m it is sufficient that $e_{-1}m = 0$ and $e_{-2}m = 0$.

The following construction establishes a one-to-one correspondence between the nontrivial W_1-homomorphisms of the module $\Lambda^n F_\lambda$ into a module of the form F_{λ_0} and the singular vectors of the $L_0^{|}$-module $\Lambda^n \times G_{-1-\lambda,-2\lambda-1}$. We start with the observation that a W_1-homomorphism $\varphi: \Lambda^n F_\lambda \to F_{\lambda_0}$ necessarily transforms $f_{j_1} \wedge \cdots \wedge f_{j_n}$ into an element of the module F_{λ_0}, a multiple of $f_{j_1+\ldots+j_n-n\lambda+\lambda_0}$: This follows from the permutability of φ with e_0. Thus, the homomorphism φ is homogeneous and, therefore, induces an L^{-}_{-1}-homomorphism $\bar{\varphi}: \bar{F}_{\lambda_0} \to \Lambda^n \bar{F}_\lambda$, i.e.,

$$\bar{\varphi}: G_{-1-\lambda_0,-2\lambda_0-1} \to \Lambda^n G_{-1-\lambda,-2\lambda-1}.$$

We set $s_\varphi = \bar{\varphi}(f_0)$. Since f_0 is a singular vector of the module $G_{-1-\lambda_0,-2\lambda_0-1}$, it follows that s_φ is a singular vector of the module $\Lambda^n G_{-1-\lambda,-2\lambda-1}$. Further, since $e_1^j f_0 = j! f_j$, it follows that $\bar{\varphi}(f_j) = e_1^j s_\varphi / j!$. The last formula shows that, in the first place, the homomorphism $\bar{\varphi}$ is completely determined by the vector s_φ and that, secondly, conversely each singular vector of the module $\Lambda^n G_{-1-\lambda,-2\lambda-1}$ that has the required degree (determined by the condition of permutability of $\bar{\varphi}$ with e_0) determines a certain homomorphism $\bar{\varphi}$ and, by means of it, the homomorphism φ. It is easy to write down an explicit formula for the last one: The homomorphism

$$\Lambda^n F_\lambda \to F_{\lambda_0},$$

determined by the singular vector

$$\Sigma b_{j_1 \ldots j_n} f_{j_1} \wedge \cdots \wedge f_{j_n} \in \Lambda^n G_{-1-\lambda,-2\lambda-1},$$

acts by the formula (2), in which

$$a_{j_1 \ldots j_n} = \frac{b_{j_1 \ldots j_n}}{j_1! \ldots j_n!}.$$

The three statements, promised in Sec. 2.1, follow from what we have said above. Firstly, each W_1-homomorphism $\Lambda^n F_\lambda \to F_{\lambda_0}$ is a differential operator of order $n\lambda - \lambda_0$ (the analogous statement for W_1-homomorphisms of the form $F_{\lambda_1} \otimes \cdots \otimes F_{\lambda_n} \to F_{\lambda_0}$ is proved in the same manner). Let us observe that in this connection the true order of the operator φ coincides with the true degree of the singular vector s_φ. Secondly, the coefficients $a_{j_1 \ldots j_n}$ of each invariant differential operator are also constant. The third statement is the part (a) of Theorem 1.1: Since $f_0 \wedge f_1 \wedge \cdots \wedge f_{n-1} \in \Lambda^n G_{-1-\lambda, -2\lambda-1}$ is a singular vector of true degree zero and is, at the same time, a unique, up to proportionality, vector of true degree zero, there exists a unique, up to proportionality, invariant operator $\Lambda^n F_\lambda \to F_{n\lambda - \frac{n(n-1)}{2}}$ of true order zero. It is also obvious that it is defined by Eq. (5).

2.3. Fundamental System. Thus, our problem consists in the determination of singular vectors of the module $\Lambda^n G_{-1-\lambda, -2\lambda-1}$. Let us set

$$s = \sum_{\substack{j_1 < \ldots < j_n \\ j_1 + \ldots + j_n = \frac{n(n-1)}{2} + k}} b_{j_1 \ldots j_n} f_{j_1} \wedge \cdots \wedge f_{j_n} \in \Lambda^n G_{-1-\lambda, -2\lambda-1}.$$

The vector s is singular if and only if $e_{-1}s = 0$ and $e_{-2}s = 0$, i.e.,

$$\sum_{i=1}^{n} (j_i - 2\lambda)\, b_{j_1 \ldots j_i+1 \ldots j_n} = 0 \quad \left(j_1 < \ldots < j_n,\ j_1 + \ldots + j_n = \frac{n(n-1)}{2} + k - 1\right), \tag{6_1}$$

$$\sum_{i=1}^{n} (j_i - 3\lambda)\, b_{j_1 \ldots j_i+2 \ldots j_n} = 0 \quad \left(j_1 < \ldots < j_n,\ j_1 + \ldots + j_n = \frac{n(n-1)}{2} + k - 2\right) \tag{6_2}$$

(this means that $b_{j_1 \ldots j_n}$ has sense not only when $j_1 < \ldots < j_n$, but is a skew-symmetric function of $j_1, \ldots, j_n$). We can adjoin the equations $e_{-l}s = 0$, i.e.,

$$\sum_{i=1}^{n} (j_i - (l+1)\lambda)\, b_{j_1 \ldots j_i+l \ldots j_n} = 0 \tag{6_l}$$

($l = 3, \ldots, k$) to Eqs. (6_1) and (6_2), but these equations can be linearly expressed in terms of (6_1) and (6_2).

Thus, to find the singular vectors of true degree k of the module $\Lambda^n G_{-1-\lambda, -2\lambda-1}$, it is necessary to solve the system of Eqs. (6_1) and (6_2). For example, for $k = 1$ this system reduces to the single equation

$$(n - 2\lambda - 1)\, b_{0 \ldots n-2, n} = 0;$$

for $k = 2, 3$ it has, respectively, the forms

$$\begin{cases} (n - 2\lambda)\, b_{0 \ldots n-2, n+1} + (n - 2\lambda - 2)\, b_{0 \ldots n-3, n-1, n} = 0, \\ (n - 3\lambda - 1)\, b_{0 \ldots n-2, n+1} - (n - 3\lambda - 2)\, b_{0 \ldots n-3, n-1, n} = 0; \end{cases}$$

$$\begin{cases} (n - 2\lambda + 1)\, b_{0 \ldots n-2, n+2} + (n - 2\lambda - 2)\, b_{0 \ldots n-3, n-1, n+1} = 0, \\ (n - 2\lambda)\, b_{0 \ldots n-3, n-1, n+1} + (n - 2\lambda - 3)\, b_{0 \ldots n-4, n-2, n-1, n} = 0, \\ (n - 3\lambda)\, b_{0 \ldots n-2, n+2} - (n - 3\lambda - 3)\, b_{0 \ldots n-4, n-2, n-1, n} = 0. \end{cases}$$

For arbitrary k the unknowns in this system are numbered by the partitions of the number k in a sum of natural numbers (the unknown $b_{0, \ldots, n-s-1, n-s+k_1, \ldots, n-1+k_s}$ corresponds to the partition $k = k_1 + \ldots + k_s$, $k_1 \le \ldots \le k_s$), and the equations of the system (6_l) are numbered by the partitions of the number $k - l$. Thus, Eqs. (6_1) and (6_2) form a system of $p(k-1) + p(k-2)$ equations with $p(k)$ unknowns (p is the number of partitions), whose coefficients depend (linearly) on λ and n. For $k \le 4$ the system is quadratic and for $k \ge 5$ the number of equations in this system is greater than the number of unknowns. Therefore, the condition of existence of a nontrivial solution of this system is a nontrivial restriction on λ and n.

We consider below this system, supposing that n (occurring in the coefficients and not in the indices) is an arbitrary complex number (this can be given sense by considering the more general problem of finding the singular vectors of the module $\Lambda^n G_{-1-\lambda, \mu-2\lambda-1}$). The condition of nontrivial solvability of the system selects on the plane λ, n an algebraic set, which either coincides with the whole plane (this possibility will be readily discarded in the next subsection) or is a union of algebraic curves and isolated points. We denote this set by S_k.

2.4. Asymptotic Solution of the Fundamental System. Proposition 2.1. The set S_k, after the removal of its isolated points, is the algebraic curve defined by an equation whose leading term is a divisor of the product

$$(n-(k_1+1)\lambda)\dots(n-(k_d+1)\lambda),$$

where $k_1, \dots, k_d$ are all the natural numbers that divide k.

Proof. Let Σ_k be the space of the skew-symmetric polynomials in n variables of degree $[n(n-1)/2]+k$; such polynomials can be uniquely represented in the form $\Delta P(N_1, N_2, \dots)$, where Δ is the product of all the differences of the variables; N_s, sum of the s-th powers of the variables, and P, polynomial. Further, let $\nu_s : \Sigma_{k-s} \to \Sigma_k$ be the multiplication in N_s.

Let us divide the coefficients of the system (6_l) by λ and take limit as n and λ approach to ∞ such that the ratio $\rho = n/\lambda$ remains constant. Then the matrix of the system (6_l) becomes in the limit the transposed matrix of the operator ν_l (in the basis composed of the alternating monomials), multiplied by $\rho-(l+1)$. Let us observe that, as a result of this limiting passage, systems (6_l) with $l \geq 3$ can no longer be linearly expressed in terms of (6_1) and (6_2).

Thus, the dimension of the solution space of the limiting system is equal to the codimension of the space

$$\sum_{l=1}^{k} \mathrm{Im}\,[\rho-(l+1)]\,\nu_l \tag{7}$$

in Σ_k. If ρ is not equal to 2, 3, ..., k + l, then the space (7) coincides with Σ_k; but if $\rho = l+1$ for some l, then (7) is the space of monomials of the form

$$\Delta\,(N_1P_1+\dots\widehat{N_lP_l}\dots+N_kP_k),$$

i.e., has codimension 1 in Σ_k, if l divides k (in this case, it does not contain the polynomial $\Delta N_l^{k/l}$), and coincides with Σ_k in the contrary case. Thus, the limiting system has a unique, up to proportionality, nontrivial solution if $\rho = l+1$, where l is a divisor of k, and does not have any solution in the contrary case. Consequently, S_k can have at most one single asymptote with the slope k_s+1 $(s = 1, \dots, d)$ and cannot have other asymptotes. The proposition is proved.

2.5. Transformation of the Fundamental System. Absence of Operators of True Order Greater Than n. An attempt of the explicit solution of the system (6) has turned out to be unsuccessful, but has led to some beautiful formulas and also to a proof of the absence of operators of true order greater than n (the last statement is specially valuable since the methods of Sec. 3 preserve only operators of true order at most n). We give certain results obtained in this way.

Let us identify the space of elements of true degree k of the module $\Lambda^n G_{-1-\lambda,-2\lambda-1}$ with the space Σ_k of skew-symmetric polynomials of degree $[n(n-1)/2]+k$ in the n variables $t_1, \dots, t_n$ by putting in correspondence with the product $f_{j_1} \wedge \dots \wedge f_{j_n}$ the alternating monomial $t_1^{j_1}\dots t_n^{j_n}$. Let us consider the polynomials

$$Q_{k_1k_2\dots k_n} = c_{k_1k_2\dots k_n}\Delta N_1^{k_1}N_2^{k_2}\dots N_n^{k_n} \qquad (k_1+2k_2+\dots+nk_n=k),$$

where Δ is the product of the differences of the variables, N_r is the sum of the r-th powers of the variables, and

$$c_{k_1k_2\dots k_n} = 2^{k_1}4^{k_2}\dots(2n)^{k_n}k_1!\,k_2!\dots k_n!.$$

(If the last few elements of the sequence $k_1, \dots, k_n$ are equal to zero, then we do not write then: $Q_{k_1\dots k_r} = Q_{k_1\dots k_r0\dots0}$.) For $k \leq n$ these polynomials form a basis in Σ_k. For $k > n$ these polynomials are linearly dependent and it is necessary to eliminate some of them to construct a basis of Σ_k. In this connection, we can select the polynomials to be deleted in various ways. We do not fix any definite way and restrict ourselves to the observation that for $k > n$ the polynomial Q_k can be expressed in terms of the remaining ones and can be deleted. This is a consequence of the following (unique) relation of degree n + 1:

$$n!\,N_{n+1}-\dots+(-1)^{n-1}N_1^{n+1}=0.$$

The following formula, which is proved by direct check, forms the basis of further computations:

$$(e_{-1}+\dots+e_{-1})\,Q_{k_1\dots k_r} = \sum_{s=1}^{r} sk_s\sum_{l=1}^{s-1} Q_{\dots k_l+1\dots k_s-1\dots} + 2\sum_{s=1}^{r} sk_s\alpha_s Q_{\dots k_s-1\dots} +$$

$$+4\sum_{s=1}^{r} s^2 k_s (k_s-1)\, Q_{\ldots k_s-2\ldots} + 2\sum_{1\le i<l\le r} s l k_s k_l\, Q_{\ldots k_s-1\ldots k_l-1\ldots} \tag{8}$$

where $\alpha_s = (s+1)(\lambda + 1/2)$ and the dots on the right-hand side of the equation denote the indices k_i occurring in their places. Ordering the elements of the basis $\{Q_{k_1\ldots k_r}\}$ in the spaces Σ_k and $\Sigma_{k-1} \oplus \ldots \oplus \Sigma_0$ lexicographically (in the decreasing order of the indices), we write down the matrix of the transformation $e_{-1} + \ldots + e_{-k}$. For $k \le n$ this matrix appears as follows:

	Q_k	$Q_{k-2,1}$	$Q_{k-3,1}$	$Q_{k-4,2}$	$Q_{k-4,0,0,1}$	$Q_{k-5,1,1}$	$Q_{k-6,0,0,0,1}$	
Q_{k-1}	$2k\alpha_1$	2						.
Q_{k-2}	$2k(k-1)$	$4\alpha_1$	3					.
$Q_{k-3,1}$		$2(k-2)\alpha_1$	3	4				.
Q_{k-3}		$4(k-2)$	$6\alpha_1$		4			.
$Q_{k-4,1}$		$(k-2)(k-3)$		$8\alpha_1$	4	3		.
$Q_{k-4,0,1}$			$2(k-3)\alpha_1$		4	2		.
Q_{k-4}			$12(k-3)$	16	$8\alpha_1$		5	.
. . .	. . .	. . .	. . .	. . .	. . .	. . .	. . .	.

(9)

[this matrix has $p(k)$ columns and $p(k-1) + p(k-2) + \ldots + p(0)$ rows]. If $k > n$, then it is necessary to delete certain columns and rows of this matrix. Let us observe that the deleted rows (more precisely, their parts that do not lie in the deleted columns) are linear combinations of the remaining rows.

System (6) is equivalent to the homogeneous system of linear equations with this matrix; in particular, for arbitrary λ and n in the dimensions of the solution spaces of these systems are the same. But the columns of the matrix (9), starting from the second column, are linearly independent. Hence, for $k \le n$ the dimension of this space does not exceed one – we have proved one of the assertions of Theorem 1.1. Further, if $k > n$, then the matrix of the new system is obtained from (9) by deleting parts of the columns, including that of the first one, and parts of the rows such that the deleted rows are linear combinations of the remaining rows. Therefore, for $k > n$ the matrix of the system has linearly independent columns, i.e., for $k > n$ the system does not have any nontrivial solution – we have proved another assertion of Theorem 1.1.

The remaining part of this section may be omitted without detriment to the understanding of the proofs of the main results. As before, we assume n to be an arbitrary complex number.

Let us examine the matrix (9) more closely. For $0 \le i \le k$ and $0 \le j \le k$, let B_{ij} denote the block of this matrix formed from the rows corresponding to $Q_{k-i,\ldots}$ and the columns corresponding to $Q_{k-j,\ldots}$. We see that (i) $B_{ij} = 0$ if $j > i+1$ or $j < i-2$; (ii) $B_{i,i+1}$ is independent of k, λ, and n; (iii) B_{ii} does not depend on k; (iv) $B_{i,i-1}$ is the product of the number $k-i+1$ by a matrix that does not depend on k; and (v) $B_{i,i-2}$ is the product of the number $(k-i+1)(k-i+2)$ by a matrix that is independent of k, λ, and n. The statements (i)-(v) enable us to assume that the matrix B_{ij} is defined for $i \ge 0$ and $j \ge 0$ and depends on the three parameters k, λ, and n. In its turn, this enables us to assume the matrix (9) to be the matrix of an infinite system of (finite) homogeneous linear equations with infinite number of unknowns. We denote the unknown, corresponding to the column $Q_{k-i,k_2\ldots k_r}$ (let us recall that $i = 2k_2 + \ldots + rk_r$), by $P_{k_2\ldots k_r}$.

From what we have said above, it is obvious that our infinite system has at most one, up to proportionality, solution. Its most remarkable property is that it has a nontrivial solution for arbitrary k, λ, and n. (We do not prove this fact; the reader, if he wishes, can deduce it from Theorem 1.1.) Here are some first components of the solution:

$$P = 1,$$
$$P_1 = -k\alpha_1,$$

$$P_{01} = \frac{4}{3}k\left(\alpha_1\alpha_2 - \frac{k-1}{2}\right),$$
$$P_2 = -\frac{1}{2}k\left(\alpha_1\alpha_2 - (k-1) - (k-3)\alpha_1^2\right),$$
$$P_{001} = -2k\alpha_2\left(\alpha_1\alpha_2 - (k-1)\right) + k(k-3)\alpha_1,$$
$$P_{11} = \frac{8}{3}k\alpha_2\left[\alpha_1\alpha_1 - \frac{3}{2}(k-1) - \frac{1}{2}(k-4)\alpha_1^2\right] + \frac{2}{3}k(k-3)(k-4)\alpha_1.$$

. . . .

The problem of determination of the n-ary invariant skew-symmetric operators of true order $k \le n$ thus reduces to the problem of determination of λ such that for these k, λ, and n we have

$$P_{i_2\ldots i_r} = 0, \quad \text{if} \quad 2i_2 + \ldots + ri_r > k. \tag{10}$$

The main assertion of Theorem 1.1 (the only assertion of it not proved by us) is that if we restrict ourselves to integers $n \geq k$, then (10) holds only when the condition (4), i.e., the equality $\alpha_{k'}\alpha_{k''} = (k'' - k')^2/2$, where k' and k'' are complementary divisors of k, is fulfilled. Indeed, this is valid without any restriction on n, but with a small modification: For each k the relation (10) is still fulfilled for a finite number of pairs λ, n. With regard to these "special values" of k, λ, and n, we will prove in Sec. 3 that n is either purely imaginary or real, but its modulus is less than k. We also see that λ and n, occurring in the special triples, satisfy two different equations (4) (with different parameters k' and k"). In all the examples known to us, n is purely imaginary. The first special triple appears for k = 5:

$$n = \pm \frac{8i}{\sqrt{3}}, \quad \lambda = -\frac{1}{2} \pm \frac{7i}{2\sqrt{3}}, \quad k = 5$$

[these n and λ satisfy Eq. (4) with $(k', k'') = (1, 3)$ and $(1, 4)$]. We could not give a complete list of the special triples; we have only proved that they are infinitely many. We could not also directly prove that (10) follows from the relation $\alpha_{k'}\alpha_{k''} = (k'' - k')^2/2$, $k'k'' = k$, except in the cases $k' = 1$ and 2 (the first result duplicates the main result of [1]). In order to formulate the result in these cases, we pass from the coordinates k, λ, n to new coordinates t, u, v that are connected with k, λ, n by the relations

$$k = uv, \quad \lambda = -t + \frac{1}{2t} - \frac{1}{2}, \quad n = -(u+1)t + \frac{v+1}{2t}.$$

In these coordinates $\alpha_s = (u - s)t - [(v - s)/2t]$ and the system, formed from Eq. (4) and the equation $k'k'' = k$, acquires the form $u = k'$, $v = k''$. (The meaning of these coordinates, to which we will return in Sec. 4, also remain unclear to us.) We set

$$P_{i_2 \ldots i_r}(\lambda, n, k) = P'_{i_2 \ldots i_r}(t, u, v)$$

and it is necessary for us to prove that for arbitrary natural numbers k' and k" we have

$$P'_{i_2 \ldots i_r}(t, k', k'') = 0 \qquad \text{for} \quad 2i_2 + \ldots + ri_r > k'k''. \tag{10'}$$

Computation shows that

$$P'_{i_2 \ldots i_r}(t, 1, v) = \frac{v(v-1)\ldots(v-i)}{2^{i_2}\ldots r^{i_r} i_2! \ldots i_r!}(-2t)^{i-i_2-\ldots-i_r},$$

where $i = 2i_2 + \ldots + ri_r$. Hence, (10') follows in the case $k' = 1$. With the help of more complicated computations, we can show that

$$P'_{i_2 \ldots i_r}(t, 2, v) = \frac{(-t)^{i-i_2-\ldots-i_r}}{2^{i_2}\ldots r^{i_r} i_2! \ldots i_r!} \mathcal{A}(T_2^{i_2} \ldots T_r^{i_r}),$$

where $T_j = (2v)^{j/2}\,\mathrm{Re}\,(\sqrt{2v} + it^{-1})^j$ and the operator $\mathcal{A}$ transforms the monomial $(2v)^p t^{-2q}$ into

$$\frac{2v(2v-1)\ldots(2v-(p+q-1))}{(2v-1)(2v-3)\ldots(2v-(2q-1))}(t^{-2}+1)(t^{-2}+3)\ldots(t^{-2}+(2q-1));$$

hence, (10') follows for $k' = 2$.

We could not extract much from the system (9). It remains to observe that the system (9) can be considered as a recurrence formula for the determination of the polynomials $P_{i_2 \ldots i_r}$ and, consequently, of the coefficients $a_{j_1 \ldots j_n}$ of our operators, when they exist.

3. Connection with the Verma Modules

3.1. Verma Modules. The algebra $\mathcal{L}$ has a unique, up to isomorphism, central extension $\hat{\mathcal{L}}$, which is described as follows in the traditional notation. A basis in the space $\hat{\mathcal{L}}$ is formed by the vectors e_i ($i \in \mathbb{Z}$) and the commutator z, described by the formulas

$$[z, e_i] = 0, \quad [e_i, e_j] = (j - i)e_{i+j} \quad \text{for} \quad i + j \neq 0,$$
$$[e_{-i}, e_i] = 2ie_0 + \frac{1}{12}(i^3 - i)z.$$

The two-dimensional cocycle, describing this extension, was discovered by Gel'fand and Fuks [5] in 1968. Later on this extension was rediscovered by physicists, and the algebra $\hat{\mathcal{L}}$ has been called the Virasoro algebra.

The Virasoro algebra has a natural gradation: $\deg e_i = 1$, $\deg z = 0$. We extend the notions of the adjoint, the contragradient, and the inverted modules (see Sec. 2.1) to graded modules over the Virasoro algebra in obvious manner.

The subalgebra of the Virasoro algebra generated by z and e_i with $i \geq 0$ ($i \leq 0$) is denoted by $\hat{L}_0^+$ ($\hat{L}_0^-$). Each of the algebras $\hat{L}_0^+$ and $\hat{L}_0^-$ is projected onto the two-dimensional subalgebra of $\hat{L}_0^+ \cap \hat{L}_0^-$ generated by z and e_0. Let h and c be two complex numbers. Let us consider the one-dimensional $(\hat{L}_0^+ \cap \hat{L}_0^-)$-module in which z and e_0 act by means of multiplication by c and h, respectively, turning it into an $\hat{L}_0^-$-module by means of the projection $\hat{L}_0^- \to (\hat{L}_0^+ \cap \hat{L}_0^-)$, and let us form the induced $\hat{\mathcal{L}}$-module. The last module is denoted by $V_{h,c}$ and is called the Verma module. As is easily seen, the Verma module is constructed in the following manner. As a $U(\hat{L}_0^+)$-module, this is a free module with one generator (U denotes the universal enveloping algebra). This generator v is called the vacuum vector. Thus, an additive basis in $V_{h,c}$ is formed by the vectors

$$v;\ e_1v;\ e_1^2v,\ e_2v;\ e_1^3v,\ e_2e_1v,\ e_3v;\ \ldots.$$

[We arrange them to the natural gradation in $V_{h,c}$; the homogeneous component of degree k has dimension p(k).] The action of the algebra $\hat{\mathcal{L}}$ in $V_{h,c}$ is carried out as follows: z acts by multiplication by c, and in order to compute $e_i(e_{j_1}\ldots e_{j_k}v)$, where $j_1 \geq \ldots \geq j_k$, we must, at first, using the commutation relations, transform this expression into a sum of terms of the form $e_{r_1}\ldots e_{r_l}v$, $e_{s_1}\ldots e_{s_m}zv$ with $r_1 \geq \ldots \geq r_l$ and $s_1 \geq \ldots \geq s_m$, and then use the equations $e_iv = 0$ for $i < 0$, $e_0v = hv$, and $zv = cv$. For example,

$$e_{-2}(e_2e_1v) = e_2e_{-2}e_1v + 4e_0e_1v + \frac{1}{2}ze_1v = e_2e_1e_{-2}v + 3e_2e_{-1}v + 4e_1e_0v + 4e_1v + \frac{1}{2}ce_1v = \left(4h + 4 + \frac{1}{2}c\right)e_1v.$$

The Verma modules are universal in the following sense. If W is an arbitrary $\hat{\mathcal{L}}$-module and w is a singular vector of it such that $e_0(w) = hw$ and $z(w) = cw$, then there exists a unique $\hat{\mathcal{L}}$-homomorphism $V_{h,c} \to W$ that transforms v into w. This homomorphism acts by the formula $e_{i_1}\ldots e_{i_r}v \to e_{i_1}\ldots e_{i_r}w$ ($i_1 \geq \ldots \geq i_r > 0$). If the Verma module $V_{h,c}$ is irreducible, then this homomorphism is a monomorphism.

3.2. Kac Theorem. One of the most important problems of the representation theory is to know what Verma modules are irreducible. This problem can be reformulated as the problem of determination of the singular vectors of the modules $V_{h,c}$: The vacuum vector is always singular; the condition of existence of other singular vectors is a necessary and sufficient condition for the reducibility of a module.

Let us consider the homogeneous component of degree k of the module $V_{h,c}$. This is a space of dimension p(k) with a basis

$$e_1^kv,\ e_2e_1^{k-2}v,\ \ldots,\ e_kv.$$

Let us apply the following p(k) linearly independent elements of the algebra $U(L_1^-)$ to the elements of the above basis:

$$e_{-1}^k,\ e_{-1}^{k-2}e_{-2},\ \ldots,\ e_{-k}.$$

We obtain $p(k)^2$ elements of the one-dimensional space Cv, i.e., a numerical square matrix of order p(k); let us denote this matrix by $M_{h,c}(k)$. (We can show that this matrix is symmetric; the quadratic form defined by this matrix is symmetric; the quadratic form defined by this matrix is called the Shapovalov form.) It is clear that the matrix $M_{h,c}(k)$ is nonsingular if and only if the module $V_{h,c}$ does not have singular vectors of degrees 1, 2, ..., k.

Kac has succeeded in computing the determinant of the matrix $M_{h,c}(k)$. Here is his result.

THEOREM 3.1 (see [2]).

$$\det{}^2 M_{h,c}(k) = C\prod_{i=1}^{k}\prod_{j|i}\Phi_{j,i/j}(h,c)^{p(k-i)},$$

$$\Phi_{\alpha,\beta}(h,c) = \left(h + \frac{1}{24}(\alpha^2-1)(c-13) + \frac{1}{2}(\alpha\beta-1)\right)\left(h + \frac{1}{24}(\beta^2-1)(c-13) + \frac{1}{2}(\alpha\beta-1)\right) + \frac{(\alpha^2-\beta^2)^2}{16}.$$

where C is a nonzero constant.

COROLLARY. The module $V_{h,c}$ has a singular vector of degree at most k if and only if $\Phi_{k',k''}(h,c) = 0$ for certain natural numbers k' and k'' with $k'k'' \leq k$.

As far as we know, the proof of Theorem 2.1 has not been published. We give its proof in Sec. 4.

3.3. Basic Construction. We fix n, λ, and μ and consider the space $H_{\lambda,\mu,n}$, a basis in which is formed by the infinite products

$$\cdots\wedge f_{m-2}\wedge f_{m-1}\wedge f_m\wedge f_{m_1}\wedge\cdots\wedge f_{m_r}, \tag{11}$$

where f_i are the basic elements of the module $\mathcal{F}_{\lambda,\mu}$; $m<m_1<\cdots<m_r$; and $m+r=n-1$. We graduate this space by setting the degree of the element (11) equal to

$$[m_1-(m+1)]+\ldots+[m_r-(m+r)];$$

It is obvious that the elements of degree k form a space of dimension p(k). The algebras L_1^+ and L_1^- act in $H_{\lambda,\mu,n}$ in a natural manner: To apply e_i with $i\neq 0$ to (11), it is sufficient to apply e_i to each factor and, if no repeated f_j are obtained, to rearrange the factors so that they go according to order (with proper change of sign), after which all the obtained results are added (the number of terms is obviously finite). It turns out that these actions of the algebras L_1^+ and L_1^- have a consistent extension to the action of the algebra $\hat{\mathfrak{L}}$. Let us define homomorphisms e_0: $H_{\lambda,\mu,n}\to H_{\lambda,\mu,n}$ and z: $H_{\lambda,\mu,n}\to H_{\lambda,\mu,n}$ by the equations

$$e_0(\alpha)=(h+\deg\alpha)\,\alpha,\qquad z(\alpha)=c\alpha,$$

where

$$h=\frac{1}{2}(\mu+n)(\mu+n-2\lambda-1),\qquad c=-2(6\lambda^2+6\lambda+1).$$

Proposition 3.1. The operators e_0 and z, together with the above-described actions of the algebras L_1^+ and L_1^-, form the structure of an $\hat{\mathfrak{L}}$-module in $H_{\lambda,\mu,n}$.

This proposition is proved by direct computation.

Let us enumerate the important properties of the modules $H_{\lambda,\mu,n}$. First of all, an increment of 1 in the indices of f_j leads to the isomorphism $H_{\lambda,\mu,n}\cong H_{\lambda,\mu-1,n+1}$. This isomorphism enables us to regard the parameter n to be unimportant. We set $H_{\lambda,\mu}=H_{\lambda,\mu,0}$. Further, the modules $H_{\lambda,\mu}$ and $H_{-1-\lambda,-\mu-2\lambda-1}$ are contragradient (because the corresponding modules $\mathcal{F}$ are contragradient), and the modules $H_{\lambda,\mu}$ and $H_{-1-\lambda,-\mu}$ are isomorphic: An isomorphism is established by the formula

$$\cdots f_{-r}\wedge f_{i_r-(r-1)}\wedge\cdots\wedge f_{i_2-1}\wedge f_{i_1}\leftrightarrow\cdots f_{-s}\wedge f_{j_s-(s-1)}\wedge\cdots\wedge f_{j_2-1}\wedge f_{j_1},$$

where $(i_1,\ldots,i_r)$ and $(j_1,\ldots,j_s)$ are two partitions, written in nonincreasing order, of the same number (i.e., $i_1+\ldots+i_r=j_1+\ldots+j_s$, $j_1=r$, $j_2=\max\{t\mid i_t\geq 2\},\ldots,j_s=\max\{t\mid i_t\geq s\}$).

Proposition 3.2. The following statements are equivalent:

(i) The Verma module $V_{h,c}$ has a singular vector of positive degree at most k.

(ii) At least one of the modules $H_{\lambda,\mu}$ and $H_{-1-\lambda,\mu-2\lambda-1}$ has a singular vector of positive degree at most k.

Proof. It is obvious that $\ldots f_{-3}\wedge f_{-2}\wedge f_{-1}\in H_{\lambda,\mu}$ is a singular vector, and e_0 and z multiply it by h and c, respectively. By virtue of what we have said in Sec. 3.2, there arises the $\hat{\mathfrak{L}}$-homomorphism

$$V_{h,c}\to H_{\lambda,\mu} \tag{12}$$

that is consistent with the gradations. Let us recall that the homogeneous components of the same degree of the modules $V_{h,c}$ and $H_{\lambda,\mu}$ have the same dimension.

Let us suppose that $V_{h,c}$ has a singular vector of positive degree at most k. If (12) is an isomorphism in degrees at most k, then the image of this vector is a singular vector of the module $H_{\lambda,\mu}$. But if (12) is not an isomorphism, then it is also not an epimorphism. Hence a certain element of degree at most k of the module $H_{\lambda,\mu}$ does not belong to $U(L_1^+)H_{\lambda,\mu}$. The last statement means precisely that the contragradient module $H_{-1-\lambda,\mu-2\lambda-1}$ has a singular vector of positive degree at most k.

Further, $H_{\lambda,\mu}$ has a singular vector of positive degree at most k, then either (12) is an isomorphism and $V_{h,c}$ also has a singular vector of positive degree at most k, or (12) is not an isomorphism, but then the module $V_{h,c}$ is reducible and, in addition, has a singular vector of positive degree at most k. Finally, the case where $H_{-1-\lambda,\mu-2\lambda-1}$ has a singular vector of positive degree at most k reduces to the preceding one, since h and c remain invariant under the substitution

$$(\lambda,\mu)\to(-1-\lambda,\mu-2\lambda-1). \tag{13}$$

The proposition is proved.

Let us now consider the homomorphism

$$\Lambda^n G_{\lambda,\mu} \to H_{\lambda,\mu,n}, \tag{14}$$

defined by the formula

$$f_{i_1} \wedge \cdots \wedge f_{i_n} \mapsto \cdots f_{-3} \wedge f_{-2} \wedge f_{-1} \wedge f_{i_1} \wedge \cdots \wedge f_{i_n}.$$

Proposition 3.3. The homomorphism (14) is a homogeneous L_0^--homomorphism of degree zero (the "true gradation" is taken in $\Lambda^n G_{\lambda,\mu}$). In addition, if only elements of degree at most n are considered in $\Lambda^n G_{\lambda,\mu}$ and $H_{\lambda,\mu}$, then (14) becomes an isomorphism.

This obvious proposition has the following convenient reformation: For each N, the part of the module $H_{\lambda,\mu}$, formed from the elements of degree at most N, is isomorphic to the analogous part of the module $\Lambda^N \times G_{\lambda,\mu-N}$.

3.4. Proof of Theorem 1.1. Let us set

$$B_{\alpha,\beta}(\lambda,\mu) = \left((\alpha+1)\left(\lambda+\tfrac{1}{2}\right)-\mu\right)\left((\beta+1)\left(\lambda+\tfrac{1}{2}\right)-\mu\right) - \frac{(\alpha-\beta)^2}{2}.$$

The following equations, which can be verified directly, lie at the basis of the proof of Theorem 1.1:

$$\Phi_{\alpha,\beta}\left(\tfrac{1}{2}\mu(\mu-2\lambda-1),\ -2(6\lambda^2+6\lambda+1)\right) = \tfrac{1}{4} B_{\alpha,\beta}(\lambda,\mu) B_{-\alpha,-\beta}(\lambda,\mu), \tag{15}$$

$$B_{-\alpha,-\beta}(\lambda,\mu) = B_{\alpha,\beta}(-1-\lambda,\ \mu-2\lambda-1). \tag{16}$$

Let S_k denote the set of the pairs of complex numbers λ, μ for which the module $\Lambda^n G_{-1-\lambda,\mu-n-2\lambda-1}$ has a singular vector of degree k. This symbol is consistent with the same symbol of Sec. 2.3, the only difference is that there we had denoted by n what we have denoted here by μ. Thus, we must show that if we restrict ourselves to integers μ greater than or equal to k, then S_k is given by the equation

$$\left(\prod_{k'|k} B_{k', k/k'}(\lambda,\mu)\right)^{1/k} = 0. \tag{17}$$

To begin with, we show that the whole set S_k (without restrictions on μ) consists of the curve (17) and, possibly, isolated points. By virtue of Proposition 3.3, we can replace the module $\Lambda^n G_{-1-\lambda,\mu-n-2\lambda-1}$ by the module $H_{-1-\lambda,\mu-n-2\lambda-1,n} = H_{-1-\lambda,\mu-2\lambda-1}$. Let $\bar{S}_k$ denote the set obtained from S_k by means of the transformation (13) and let T_k denote the set given by the equation $\det M_{h,c}(k) = 0$, where $h = \mu(\mu-2\lambda-1)/2$ and $c = -2(6\lambda^2+6\lambda+1)$. It follows from Proposition 3.2 that

$$T_k - T_{k-1} \subset S_k \cup \bar{S}_k \subset T_k.$$

Applying Theorem 3.1 and Eqs. (15) and (16), we see that the union $S_k \cup \bar{S}_k$ consists of the curves $B_{k',k''}(\lambda,\mu) = 0$ and $B_{k',k''}(-1-\lambda,\ \mu-2\lambda-1) = 0$ with $k'k'' = k$ and, possibly, more of such curves with $k'k'' < k$ or their isolated points. Finally, using Proposition 2.1, we conclude that S_k, after deletion of the isolated points, is the curve (17).

It remains to investigate the isolated points of the set S_k. It is easily seen that if (λ,μ) is such a point, then the corresponding h and c have the following properties: The module $V_{h,c}$ has a singular vector of degree less than k and the module generated by this singular vector has a singular vector of degree k (in the gradation of the module $V_{h,c}$). This means that there exist natural numbers s, t, u, and v such that $st + uv = k$ and

$$\begin{cases} \Phi_{s,t}(h,c) = 0, \\ \Phi_{u,v}(h+st,c) = 0. \end{cases} \tag{18}$$

The system (18) can be solved easily. If we consider its unknowns λ and μ, then there are eight possibilities for μ: two of them are described by the equation

$$\mu^2 = \frac{(2st-ut+sv \pm (s-u+t+v))^2}{2(u-s)(t+v)}, \tag{19}$$

the remaining ones are obtained by means of the permutations $s \leftrightarrow t$ and $u \leftrightarrow v$. If $u < s$, then the right-hand side of Eq. (19) is negative and μ is a purely imaginary number. But if $u > s$, then it is easily shown that $\mu^2 < (st+uv)^2$. Thus, if (λ,μ) is an isolated point of the set S_k, then μ cannot be an integer greater than k. Theorem 1.1 is proved.

In conclusion, we indicate one more desirable property of the polynomials Φ: If h and c satisfy the conditions (18), then $\Phi_{w,z}(h, c) = 0$, where (w, z) is one of the pairs (u, v + 2t), (u, v + 2s), (v, u + 2t), and (v, u + 2s). Thus, the isolated points of the set S_k necessarily belong to the intersection of the sets S_l or $\tilde{S}_l$ with the indices less than k.

4. Appendix

4.1. Explicit Construction of Singular Vectors. We indicate a method for the construction of an infinite number of singular vectors in the different modules $H_{\lambda,\mu}$.

Let $n \in \mathbb{Z}_+$ and $m \in \mathbb{Z}$. Let us define an element $\varphi_{m,n}$ of the space $\Lambda^n \mathcal{F}_{-\frac{(n-1)(n+2)}{2n}, \frac{m}{n}}$ by the equation

$$\varphi_{m,n} = \sum_{\substack{j_1+\ldots+j_n = -m-\frac{n(n+1)}{2}+1 \\ j_1<\ldots<j_n}} \Big[\prod_{1\leq s<t\leq n} (j_t - j_s)\Big] f_{j_1} \wedge \cdots \wedge f_{j_n}.$$

The following proposition is established by a direct check.

Proposition 4.1. $e_i \varphi_{m,n} = 0$ for each i.

Remark. The origin of the invariants $\varphi_{m,n}$ in each case when m = 0 has the following explanation in terms of the "operators of general position" $\Delta_{\lambda,n}: \Lambda^n F_\lambda \to F_{n\lambda - \frac{n(n-1)}{2}}$ (see Sec. 1). Equation (2), defining these operators, also defines an $\mathcal{L}$-homomorphism

$$\tilde{\Delta}_{\lambda,n}: \Lambda^n \mathcal{F}_\lambda \to \mathcal{F}_{n\lambda - \frac{n(n-1)}{2}}.$$

Let us consider the adjoint $\mathcal{L}$-homomorphism

$$\tilde{\Delta}_{\lambda,n}^{*}: \mathcal{F}_{-1-n\lambda+\frac{n(n-1)}{2}} \to \Lambda^n \mathcal{F}_{-1-\lambda}.$$

If $\lambda = \frac{(n+1)(n-2)}{2n}$, then $-1 - n\lambda + \frac{n(n-1)}{2} = 0$ and the module $\mathcal{F}_{-1-n\lambda+\frac{n(n-1)}{2}} = \mathcal{F}_0$ has an invariant f_0. The image $\tilde{\Delta}_{\lambda,n}^{*}(f_0) \in \Lambda^n \mathcal{F}_{-\frac{(n-1)(n+2)}{2n}}$ and is $\varphi_{0,n}$.

For even n and arbitrary k > 0 we set

$$\xi^k_{m,n} = (\cdots \wedge f_{-2} \wedge f_{-1}) \wedge \underbrace{\varphi_{m,n} \wedge \cdots \wedge \varphi_{m,n}}_{k} \in H_{-\frac{(n-1)(n+2)}{2n}, \frac{m}{n}, nk}.$$

Proposition 4.2. If $m \leq -\frac{n^2(k+1)}{2} + 1$, then $\xi^k_{m,n}$ is a singular vector of the module

$$H_{-\frac{(n-1)(n+2)}{2n}, \frac{m}{n}, nk} = H_{-\frac{(n-1)(n+2)}{2n}, \frac{m}{n}+nk}.$$

The degree of this vector is kl, where

$$l = -m - \frac{n^2(k+1)}{2} + 1. \tag{20}$$

This follows from Proposition 4.1 (the condition $m \leq -\frac{n^2(k+1)}{2} + 1$ ensures that the vector $\xi^k_{m,n}$ is nontrivial).

4.2. Proof of Theorem 3.1. As a direct check shows, for arbitrary m, n, and k the numbers $\lambda = -\frac{(n-1)(n+2)}{2n}$, and $\mu = \frac{m}{n} + nk$ satisfy the equation $B_{k,l}(-1-\lambda, \mu - 2\lambda - 1) = 0$, where l is found from Eq. (20). (Let us observe that this point of the curve $B_{k,l} = 0$ is described very simply in the coordinates t, u, v of Sec. 2.5: It has the coordinates t = 1/n, u = l, v = k.) Together with Propositions 4.2 and 3.2 and Eqs. (15) and (16), this shows that $\det M_{h,c}(kl)$ vanishes at infinitely many points (h, c) of the curve $\Phi_{k,l}(h, c) = 0$. This means that the determinant $\det M_{h,c}(kl)$, considered as a polynomials in h and c, is divisible by $\Phi_{k,l}(h, c)$ [or by the square root of $\Phi_{k,l}(h, c)$, if it can be extracted; this happens for k = l]. By induction, we can suppose that Theorem 3.1 has been proved for the homogeneous components of the Verma modules of degrees less than kl. Hence, we can suppose that if $k'l' < kl$, then the following two statements are valid for all, except a finite number of, the points (h, c) of the curve $\Phi_{k',l'}(h, c) = 0$: (i) the module $V_{h,c}$ has a singular vector of degree $k'l'$; (ii) the homogeneous component of degree kl of the submodule of $V_{h,c}$ generated by this vector has dimension $p(kl - k'l')$. Since each proper submodule of the Verma module annihilates the Shapovalov form, we conclude that $\det M_{h,c}(k,l)$ is divisible by $\Phi_{k'l'}(h,c)^{p(kl-k'l')}$ (by the square root of the last polynomial, if $k' = l'$). Thus, $\det^2 M_{h,c}(r)$ is divisible by $\prod_{j}\prod_{i}\Phi_{j,i}(h,c)^{p(r-ij)}$, and the obvious computation of degrees shows that the fraction is constant (the fact that it is nonzero follows from Propositions 3.2 and 2.1).

4.3. Combinatorial Identities. The construction of Sec. 4.1 gives the coordinates of the singular vectors of the modules $H_{\lambda,\mu}$ for infinitely many values of λ and μ. In order to write down an explicit formula for them, we set

$$\Psi_k(j_1,\dots,j_{2kn})=\sum_{\tau}\operatorname{sgn}(\tau)\prod_{1\le u<v\le 2n}\prod_{r=0}^{k-1}\left(j_{\tau(2rn+u)}-j_{\tau(2rn+v)}\right),$$

where the summation is carried over all $\tau\in\operatorname{Symm}(2kn)$ such that $\sum_{m=1}^{2n} j_{\tau(2rn+m)}$ does not depend on r. The coefficient of

$$\cdots f_{-2}\wedge f_{-1}\wedge f_{j_1}\wedge\cdots\wedge f_{j_{2nk}},$$

where $j_1+\dots+j_{2nk}=k\left(-m-\frac{n(n+1)}{2}+1\right)$, in the expression for $\xi^{k}_{m,2n}$ is equal to $\frac{1}{(2n!)^k}\Psi_k(j_1,\dots,j_{2nk})$. At the same time, the coordinates of the singular vectors can be found directly in certain cases. For example, the module $H_{\lambda,\mu}$ has a singular vector of degree 2 for $B_{2,1}(-1-\lambda,\mu-2\lambda-1)=0$, i.e., for $\mu(2\mu+2\lambda+1)=1$, and this singular vector can be found effortlessly:

$$(1-\mu)(\cdots\wedge f_{-3}\wedge f_{-2}\wedge f_1)+(1+\mu)(\cdots\wedge f_{-3}\wedge f_{-1}\wedge f_0).$$

Combining the last formula with the formula for $\xi^{2}_{-4n^2,2n}$, we get

$$\frac{\Psi_2(1,2,\dots,4n-1,4n+2)}{1-n}=\frac{\Psi_2(1,2,\dots,4n-2,4n,4n+1)}{1+n}. \tag{21}$$

In general, the module $H_{\lambda,\mu}$ with $B_{k,1}(-1-\lambda,\mu-2\lambda-1)=0$ has a singular vector of degree k, whose coordinates can be computed explicitly (an equivalent problem is solved in [1] and in Sec. 2.5). Comparing these coordinates with the coordinates of the vector $\xi^{k}_{-k(k+1)n^2,2n}\in H_{-(2n-1)(n+1)/2n,(k-1)n}$, we get a generalization of the identity (21): For $0\le j_1<\dots<j_k$, $j_1+\dots+j_k=k(k+1)/2$, the ratio

$$\frac{\Psi_k(1,2,\dots,2nk-k,2nk-k+1+j_1,\dots,2nk-k+1+j_k)}{\prod_{1\le s<s'\le k}(j_s-j_{s'})\prod_{u=1}^{k}(j_u+1)\prod_{t=1}^{k}\prod_{w=0}^{j_t}\frac{(k-1)(nk-nw+1)}{(w+1)}}$$

does not depend on $j_1,\dots,j_k$. Computing the coefficients of the invariant operators explicitly (e.g., by means of the formulas of Sec. 2.5) and comparing them with the coordinates of the vectors $\xi^{k}_{m,n}$, we can write out an infinite number of similar identities.

LITERATURE CITED

1. B. L. Feigin and D. B. Fuks, "Invariant differential operators on the line," Funkts. Anal. Prilozhen., 13, No. 4, 91-92 (1979).
2. V. G. Kac, "Contravariant form for infinite-dimensional Lie algebras and superalgebras," in: W. Beiglböck, A. Böhm, and E. Takasugi (eds.), Group Theoretical Methods in Physics, Lecture Notes in Physics, Vol. 94, Springer-Verlag, Berlin–New York (1979).
3. V. G. Kac and D. H. Peterson, "Spin and wedge representations of infinite-dimensional Lie algebras and groups," Proc. Nat. Acad. USA, 78, 3308-3312 (1981).
4. P. Ya. Grozman, "Classification of invariant bilinear operators on tensor fields," Funkts. Anal. Prilozhen., 14, No. 2, 58-59 (1980).
5. I. M. Gel'fand and D. B. Fuks, "Cohomologies of the Lie algebra of the vector fields on the circle," Funkts. Anal. Prilozhen., 2, No. 4, 92-93 (1968).

VACUUM VECTOR REPRESENTATIONS OF THE VIRASORO ALGEBRA

A. Rocha-Caridi[+]

1. INTRODUCTION

A lot of attention has been focused lately on certain infinite dimensional Lie algebras for their importance in some physical theories as well as the richness of their mathematical theories. One of these algebras is the Virasoro algebra. The Virasoro algebra is known to physicists in the theory of dual string models (cf. [25]). The first mathematical reference on the Virasoro algebra that is known to us is by Gelfand and Fuchs [9]. They proved that the second cohomology of the Lie algebra v of polynomial vector fields on the circle is one-dimensional. Using this one can show that the Virasoro algebra is the universal central extension $\hat{v}$ of v (see §4 below). The Virasoro algebra was later realized as an algebra of operators on the representation space of a Kac-Moody algebra (cf. [5, 3, 11, 17]), in a way reminiscent of its earlier introduction in dual models.

Among the first significant results on the representation theory of the Virasoro algebra was Kac's formula for the determinant of the contravariant form [14,15] (see also [1]). One of the main problems in representation theory is to determine the characters of the irreducible highest weight representations. Explicit character formulas for some of these representations have been determined according to the action of the center. In the case where the center acts by the scalar $c = 1$, the formulas were obtained by Kac [14,15], using the above mentioned realization of the Virasoro algebra as an algebra of operators. Wallach and the author [22,23] developed very general methods for the determination of irreducible characters, which were

[+] Partially supported by National Science Foundation Grant MCS-8201260.

Vertex Operators in Mathematics and Physics - Proceedings of a Conference November 10-17, 1983. Publications of the Mathematical Sciences Research Institute #3, Springer-Verlag, 1984.

applied to the cases $c = 0$, 25 and 26. In particular, the irreducible characters of the Lie algebra of vector fields on the circle were computed in [22,23].

Recently, Feigin and Fuchs [2][1] announced the results that completely describe the submodules of a Verma module. From their findings one can infer what the irreducible characters look like. There has also been an important development in Physics, where the representation theory of the Virasoro algebra plays a crucial role. In [6,7], Friedan, Qiu and Shenker introduced an infinite family of representations $L((h_i,c_i))$ with (h_i,c_i) in the domain $0 \leq c < 1$. The importance of these representations lies in the fact that the complementary set of the weights (h_i,c_i) in the domain $0 \leq c < 1$ corresponds to nonunitary representations [6,7]. The representations $L((h_i,c_i))$ that are unitary are intimately related to the unitary, conformally invariant, models in statistical mechanics [6,7]. The characters of these representations seem to be of interest in Physics. In this paper we calculate them explicitly.

In §2 we review the basics on the representation theory of the Virasoro algebra, including Kac's formula. In §3 we present the results of Goodman and Wallach [12] on the Segal operators and use them to indicate how to derive the irreducible characters in the case $c = 1$. In §4 we discuss the case of the Lie algebra of polynomial vector fields on the circle ($c = 0$). Although we only concern ourselves with the characters for certain special values of c, we feel that the methods presented in the exposition outlined above are of independent interest. This part of the paper is based on the lecture presented by the author at the Workshop. In §5 we state and discuss the implications of the results announced in [2]. In particular, we show how to derive the irreducible characters. In §6 we derive, using the results of [2], explicit formulas for the characters of the representations introduced in [6,7].

---

[1] We thank I. Frenkel for kindly translating the work [2] into English for us.

ACKNOWLEDGMENT

It is a pleasure to thank D. Friedan for many stimulating discussions.

2. REPRESENTATION THEORY

The *Virasoro algebra* is, by definition, the complex Lie algebra g with basis $\{d_k, d_0'\}_{k\in\mathbb{Z}}$, and bracket relations:

$$[d_k, d_\ell] = (\ell-k)d_{k+\ell} + \delta_{k,-\ell}\,\frac{k(k^2-1)}{12}\,d_0'. \tag{1}$$

$$[d_k, d_0'] = 0, \quad \text{for all } k \in \mathbb{Z}. \tag{2}$$

Let $h = \mathbb{C}d_0 \oplus \mathbb{C}d_0'$ and $\lambda \in h^*$ such that $\lambda(d_0) = h$, $\lambda(d_0') = c$. Given any h-module M we define its λ-weightspace:

$$M_\lambda = \{v \in M \mid d_0 v = hv,\ d_0' v = cv\}.$$

If $M = \bigoplus_{\lambda\in h^*} M_\lambda$ and $\dim M_\lambda < \infty$ for all $\lambda \in h^*$, we define the *formal character* of M, ch M by

$$\text{ch } M = \sum_{\lambda\in h^*} (\dim M_\lambda) z^\lambda$$

A *highest weight* (or vacuum vector) *representation* M of g, with highest weight λ, is one generated by a vector v_λ such that:

$$\begin{cases} d_m v_\lambda = 0, & m \geq 1 \\ d_0 v_\lambda = h v_\lambda & \text{and} \quad d_0' v_\lambda = c v_\lambda \end{cases} \tag{3}$$

v_λ is called a highest weight vector of M. Next we recall the definition of the universal highest weight representation with highest weight λ. Let $n = \bigoplus_{i\in\mathbb{N}} \mathbb{C}d_i$, $n^- = \bigoplus_{i\in\mathbb{N}} \mathbb{C}d_{-i}$, $b = h \oplus n$, and let $\mathbb{C}(\lambda)$ denote the one-dimensional b-module with trivial n-action and with h-action given by λ. Let $M(\lambda) = U(g) \otimes_{U(b)} \mathbb{C}(\lambda)$. (Here, we denote by $U(a)$ the universal enveloping algebra of a, where a is any subalgebra of g). $M(\lambda)$ is the *Verma module* associated with g, h, n and λ. We denote by $L(\lambda)$ its unique irreducible quotient. It is clear that $M(\lambda) = \bigoplus_{m\in\mathbf{Z}_+} M(\lambda)_{\lambda-m}$. Hence

(4) $$\mathrm{ch}\,(M(\lambda) = z^{\lambda}\sum_{i\in\mathbf{Z}_+} p(i)z^{-i} = z^{\lambda}\prod_{i=1}^{\infty}(1-z^{-i})^{-1}$$

where $p(i)$ is the number of partitions of i.

Let σ denote the linear antiautomorphism of $U(g)$ such that $\sigma(d_m) = d_{-m}$, $m \in \mathbf{Z}$, and $\sigma(d_0') = d_0'$. Let β be the projection of $U(g) = U(h)\oplus(n^-U(g) + U(g)n)$ onto $U(h)$. We write v_λ for $1\otimes 1$ in $M(\lambda)$ and set $(Xv_\lambda, Yv_\lambda)_\lambda = (\lambda\circ\beta)(\sigma(X)Y)$, for all X, $Y \in U(g)$. This defines a symmetric bilinear form on $M(\lambda)$ which is *contravariant* in the following sense:

(5) $$(Xv,w)_\lambda = (v,\sigma(X)w)_\lambda$$

for all $v,w \in M(\lambda)$, $X \in U(g)$. If $v \in M(\lambda)_\mu$, $w \in M(\lambda)_\nu$, with $\mu \neq \nu$, then $(v,w)_\lambda = 0$, by (5). It is easy to see that the radical of $(\ ,\ )_\lambda$, Rad $(\ ,\ )_\lambda$, is the unique maximal submodule of $M(\lambda)$, i.e., $L(\lambda) = M(\lambda)/\mathrm{Rad}\,(\ ,\ )_\lambda$.

We identify $\lambda \in h^*$ with $(h,c) \in \mathbf{C}^2$, where $\lambda(d_0) = h$, $\lambda(d_0') = c$ and let $(\ ,\ )_{h,c,m}$ denote the restriction of $(\ ,\ )_\lambda$ to $M(\lambda)_{\lambda-m}$. The following formula is due to Kac [14,15]:

(6) $$\det\,(\ ,\ )_{h,c,m} = \prod_{i=1}^{m}\left[\prod_{\substack{rs=i\\ r\leq s}} \psi_{r,s}(h,c)\right]^{p(m-i)}$$

where $\psi_{r,s}(h,c) = (h-\alpha^+_{r,s}(c))(h-\alpha^-_{r,s}(c))$, $r,s \in \mathbf{N}$, $r \neq s$, $\psi_{r,r}(h,c) = h + \frac{1}{24}(r^2-1)(1-c)$, and

$$\alpha^{\pm}_{r,s}(c) = -\frac{1}{48}[(13-c)(r^2+s^2)\pm\left\{c^2-26c+25\right\}^{1/2}(r^2-s^2) - 24rs - 2 + 2c].$$

Remarks: 1) Formula (6) is valid up to nonzero constant.

2) A proof of (6) is provided by Feigin and Fuchs [1].

Following [22], we let $T_\lambda\colon U(n^-) \longrightarrow M(\lambda)$ be the linear isomorphism such that $T_\lambda(X) = Xv_\lambda$. Set $V = U(n^-)$. If we let $\tau_\lambda(X)v = T_\lambda^{-1}(XT_\lambda(v))$ for $X \in g$, $v \in V$, then (τ_λ, V) is a representation of g. For each $t \in \mathbb{C}$, we let (π_t, V) be the representation $(\tau_{\lambda+t}, V)$ for some fixed $\lambda \in \mathbb{C}^2$. Let $B_t(v,w) = (T_{\lambda+t}(v), T_{\lambda+t}(w))_{\lambda+t}$ for all $v,w \in V$. Then $B_t(\pi_t(X)v,w) = B_t(v,\pi_t(\sigma(X))w)$, for all $X \in g$, $v,w \in V$. Let $V_\eta = \{X \in V \mid [H,X] = \eta X$, for all $H \in h\}$. Clearly $V = \bigoplus_{-\eta \in \mathbb{N}} V_\eta$. Let $\mathcal{O}(V)$ (resp. $\mathcal{O}(\mathbb{C})$) denote the space of germs $\mathbf{f}$ of holomorphic functions f at 0 with values in a finite sum of weightspaces V_η of V (resp. with values in $\mathbb{C}$). For each $k \in \mathbb{Z}_+$, set

$$\mathcal{O}(V)_{(k)} = \{\mathbf{f} \in \mathcal{O}(V) \mid B_t(f(t),w) \in t^k\mathcal{O}(\mathbb{C}), \text{ for all } w \in V\}.$$

Now, put $V_{(k)} = \{f(0) \mid \mathbf{f} \in \mathcal{O}(V)_{(k)}\}$ and $M(\lambda)_{(k)} = T_\lambda(V_{(k)})$.

Proposition 1 [22]. $M(\lambda) = M(\lambda)_{(0)} \supset M(\lambda)_{(1)} \supset \ldots$ is a g-module filtration satisfying:

(i) $M(\lambda)_{(1)} = \mathrm{Rad}\,(\ ,\ )_\lambda$.

(ii) $M(\lambda_{(k)}/M(\lambda)_{(k+1)}$ carries a nondegenerate (symmetric) contravariant form

(iii) $\sum_{k>0} \mathrm{ch}\, M(\lambda)_{(k)} = \sum_{m\in\mathbb{Z}_+} z^{\lambda-m}\, \mathrm{ord}_0(\det(\ ,\ )_{\lambda+t,m})$

Let $\varphi_i(h,c) = \prod_{\substack{rs=i \\ r\leqslant s}} \psi_{r,s}(h,c)$ and set $\mathbb{N}_+(\lambda) = \{i \in \mathbb{N} \mid \varphi_i(h,c) = 0\}$.

<u>Corollary</u> **2:** In the notation of Proposition 1,

$$\sum_{k>0} \mathrm{ch}\ M(\lambda)_{(k)} = \sum_{m\in \mathbb{N}_+(\lambda)} \mathrm{ch}\ M(\lambda - m).$$

We write $\mu \uparrow \lambda$ if there are $m_1,\dots,m_r \in \mathbb{N}$ such that $m_i \in \mathbb{N}_+(\lambda - m_1 - \dots - m_{i-1})$, $2 \leq i \leq r, m_1 \in \mathbb{N}_+(\lambda)$, and $\mu = \lambda - m_1 - \dots - m_r$.

Repeated applications of Corollary 2 give

<u>Theorem</u> 3 [14,15]. Let $\lambda, \mu \in \mathfrak{h}^*$. If $L(\mu)$ is a subquotient of $M(\lambda)$, then $\mu \uparrow \lambda$.

3. THE SEGAL OPERATORS[2] AND THE CASE $c = 1$

In this section we recall, following Goodman and Wallach [12], a realization of the Virasoro algebra. The difference in the statements below and the ones of [12] is explained by the use of a different invariant bilinear form.

Let $\mathfrak{a}$ be a simple Lie algebra over $\mathbb{C}$, and let (,) be a symmetric, invariant form on $\mathfrak{a}$, normalized so that the square length of a long root is equal to 2. Let $\mathfrak{t} \subset \mathfrak{a}$ be a Cartan subalgebra and let Δ (resp. $\Delta^\vee$) denote the root system (resp. dual root system) of $(\mathfrak{a}, \mathfrak{t})$. Let $\tilde{\alpha}$ be the highest root of Δ. We write $\tilde{\alpha} = \sum_{i=1}^{\ell} g_i \alpha_i$ where $\alpha_1,\dots,\alpha_\ell$ are the simple roots of Δ. Set $g = 1 + \sum_{i=1}^{\ell} g_i$. We denote by $B(\ ,\ )$ the Killing form of $\mathfrak{a}$. Clearly, $B(\ ,\ )$ is a nonzero scalar multiple of (,). The following result determines this scalar.[3]

---

[2] These operators first appeared in an unpublished manuscript by G. Segal (cf. [3]).

[3] We thank I. Frenkel for bringing this result to our attention.

Lemma 1 [18, Lemma 1.2]. $(X,Y) = \frac{1}{2g}B(X,Y)$, for every $X,Y \in \mathfrak{a}$.

Given $\gamma \in \mathfrak{t}^*$ let h_γ (resp. t_γ) be the unique element of $\mathfrak{t}$ such that $\gamma(H) = B(h_\gamma,H)$ (resp. $\gamma(H) = (t_\gamma,H)$) for all $H \in \mathfrak{t}$. If $\lambda, \mu \in \mathfrak{t}^*$, set $B(\lambda,\mu) = B(h_\lambda,h_\mu)$ and $(\lambda,\mu) = (t_\lambda,t_\mu)$.

Next is another version of Lemma 1:

Lemma 1': $(\lambda,\mu) = 2gB(\lambda,\mu)$, for all $\lambda,\mu \in \mathfrak{t}^*$.

Proof: We have $(t_\gamma,H) = \frac{1}{2g}B(t_\gamma,H)$ for all $H \in \mathfrak{t}$, $\gamma \in \mathfrak{t}^*$, by Lemma 1. Hence

$\frac{1}{2g}B(t_\gamma,H) = B(h_\gamma,H)$ for all $H \in \mathfrak{t}$, $\gamma \in \mathfrak{t}^*$. This implies that $t_\gamma = 2gh_\gamma$ for all $\gamma \in \mathfrak{t}^*$. Therefore, $(\lambda,\mu) = 4g^2(h_\lambda,h_\mu) = 2gB(h_\lambda,h_\mu) = 2gB(\lambda,\mu)$, for all $\lambda, \mu \in \mathfrak{t}^*$. q.e.d.

We let $\hat{\mathfrak{a}} = \mathfrak{a}\otimes\mathbb{C}[t,t^{-1}]\oplus\mathbb{C}Z$ and set $X(p) = X\otimes t^p$, for $X \in \mathfrak{a}$, $p \in \mathbb{Z}$. We define

$$[X(p),Y(q)] = [X,Y](p+q) + p\delta_{p,-q}(X,Y)Z$$

for all $X,Y \in \mathfrak{a}$, $p,q \in \mathbb{Z}$, and $[\hat{\mathfrak{a}},Z] = (0)$.

The above relations give $\hat{\mathfrak{a}}$ a Lie algebra structure, called the *affine Lie algebra* associated with $\mathfrak{a}$.

The element $d_0 \in \mathfrak{g}$ (see §2) acts on $\hat{\mathfrak{a}}$ as the degree derivation so that we can form

$$\hat{\mathfrak{a}}^e = \mathbb{C}d_0 \times \hat{\mathfrak{a}}.$$

Let $\mathfrak{u}$ (resp. $\mathfrak{u}^-$) denote the sum of the positive (resp. negative) rootspaces. Than $\mathfrak{a} = \mathfrak{u}^-\oplus\mathfrak{t}\oplus\mathfrak{u}$. We set $\hat{\mathfrak{t}} = \mathfrak{t}\oplus\mathbb{C}Z$, $\hat{\mathfrak{t}}^e = \hat{\mathfrak{t}}\oplus\mathbb{C}d_0$, $\hat{\mathfrak{u}}^- = \mathfrak{u}^-\oplus\mathfrak{a}\otimes t^{-1}\mathbb{C}[t^{-1}]$, and $\hat{\mathfrak{u}} = \mathfrak{u}\oplus\mathfrak{a}\otimes t\mathbb{C}[t]$. Then $\hat{\mathfrak{a}}^e = \hat{\mathfrak{u}}^-\oplus\hat{\mathfrak{t}}^e\oplus\hat{\mathfrak{u}}$.

A *highest weight* (or vacuum vector) *representation* V of $\hat{a}^e$ is one generated by a vector v_λ, $\lambda \in t^*$, such that $Hv_\lambda = \lambda(H)v_\lambda$ for all $H \in \hat{t}^e$, and $\hat{u}\cdot v_\lambda = 0$; λ is called the *highest weight* of V. It is clear that for any $v \in V$, there exists $n_0 \in \mathbb{N}$ (depending on v) such that $X(n)v = 0$ for all $n > n_0$ and $X \in a$.

The normal-ordered product of X(p) and Y(q), for $X,Y \in a$ and $p,q \in \mathbb{Z}$, is defined by

$$:X(p)Y(q): \begin{cases} X(p)Y(q), & \text{if } p < q \\ \frac{1}{2}(X(p)Y(q) + Y(q)X(p)), & \text{if } p = q \\ Y(q)X(p), & \text{if } p > q \end{cases}$$

Let $\{u_i\}$, $i = 1,\ldots,n$, be an orthonormal basis of a relative to $(\ ,\ )$. Then

$$T_p = \frac{1}{2}\sum_{k\in\mathbb{Z}}\sum_{i=1}^{n} :u_i(k)u_i(p-k):$$

defines an operator on any highest weight representation.

Using Lemmas 1 and 1' we can now state the results of [12] already in normalized form.

<u>Lemma</u> 2 [12]. Let V be a highest weight representation of $\hat{a}^e$ with highest weight λ. Then

(7) $$[X(p), T_q] = (\lambda(Z) + g)pX(p+q)$$

for all $X \in g$, $p,q \in \mathbb{Z}$.

Let V be a highest weight representation of $\hat{a}^e$, λ its highest weight. Assume that $\lambda(Z) \neq -g$. Set

$$D_p = -\frac{1}{\lambda(Z)+g}T_p$$

Lemma 3 [12]. $[D_p, D_q] = (q-p)D_{p+q} + \delta_{p+q,0}\,\frac{p(p^2-1)}{12}\,\tau$,

where $\tau = \frac{n\lambda(Z)}{\lambda(Z)+g}$.

Let g denote the Virasoro algebra (§2). It is easy to see (§4) that g acts on $\hat{\mathfrak{a}}$. Set $\mathfrak{m} = g \times \hat{\mathfrak{a}} \supset \hat{\mathfrak{a}}^e$.

Theorem 4 [12]. Let (π, V) be a highest weight representation of $\hat{\mathfrak{a}}^e$ with highest weight λ such that $\lambda(Z) \neq -g$. Then $\pi\big|_{\hat{\mathfrak{a}}}$ extends to a representation $\tilde{\pi}$ of $\mathfrak{m}$, with $\tilde{\pi}(d_p) = D_p$ and $\tilde{\pi}(d_0) = \tau$, when $\tau = \frac{n\lambda(Z)}{\lambda(Z)+g}$.

Remarks: 1) If $\mathfrak{a}$ is of type A, D or E, g is the Coxeter number h of $\mathfrak{a}$. Therefore, if in addition $\lambda(Z) = 1$, then $\tau = \frac{n}{1+h}$ = rank of $\mathfrak{g}$ (Compare [5,3]).

2) If V is standard (i.e. V is irreducible and λ is dominant integral) and $\lambda(d_0) \in \mathbb{R}$ then $(\tilde{\pi}, v)$ carries a positive definite, Hermitian contravariant form ([12, Corollary 2.4]).

3) In [17] a more general version of Theorem 4 is obtained, which includes the case of twisted affine algebras.

We now return to the study of the representations L((h,c)) or the Virasoro algebra g. We retain the notation of §2. Setting c = 1 in (6) we obtain:

$$\det(\ ,\ )_{h,1,m} = \prod_{i=1}^{m} \Big(\prod_{rs=i} \big(h + \tfrac{1}{4}(r-s)^2\big)\Big)^{p(m-i)} \tag{8}$$

It is obvious from (8) that M((h,1)) is irreducible if and only if $h \neq -\frac{m^2}{4}$ for all $m \in \mathbb{Z}_+$. We let $\mathfrak{a} = sl(2,\mathbb{C})$ with canonical basis

$$E = \begin{bmatrix} 0 & 1 \\ 0 & 0 \end{bmatrix},\quad H = \begin{bmatrix} 1 & 0 \\ 0 & -1 \end{bmatrix} \quad\text{and}\quad F = \begin{bmatrix} 0 & 0 \\ 1 & 0 \end{bmatrix}$$

Set $h_1 = H \otimes 1$ and $h_0 = H \otimes 1 + Z$ in $\hat{\mathfrak{a}}$. Then h_0, h_1 and d_0 generate $\hat{\mathfrak{t}}^e$ Let $\Lambda_i(h_j) = \delta_{ij}$, $0 \leq i,j \leq 1$. We denote by V_0 (resp. V_1) the irreducible highest weight representation of $\hat{\mathfrak{a}}$ with highest weight Λ_0 (resp. Λ_1). Applying Theorem 4 we obtain

$$(9) \qquad V_0 \oplus V_1 = \bigoplus_{m \in \mathbb{Z}_+} Q(m) \otimes T(m)$$

as a representation of $g \times sl(2,\mathbb{C})$, where $T(m)$ is the $(m+1)$-dimensional irreducible representation of $sl(2,\mathbb{C})$, and $Q(m) = \{v \in V_0 \oplus V_1 \mid Ev = 0,\ Hv = mv\}$. Using (9), Remark 2 and the representation theory of $sl(2,\mathbb{C})$ (see, e.g. [4, Theorem 3.1]) we obtain

<u>Theorem</u> 5. [14,15] For $m \in \mathbb{Z}_+$ one has

$$\operatorname{ch} L\left(\left(-\frac{m^2}{4},1\right)\right) = \left(\prod_{i=1}^{\infty} (1-z^{-i})\right)^{-1}\left(z^{-\frac{m^2}{4}} - z^{-\frac{1}{4}(m+2)^2}\right)$$

The above, together with (8) gives all the characters ch $L((h,1))$.

4. VECTOR FIELDS ON THE CIRCLE

Let v denote the Lie algebra of polynomial vector fields on the circle. It was proved by Gelfand and Fuchs [9] that $H^2(v,\mathbb{C}) \cong \mathbb{C}$. A nonzero cocycle on v is given by

$$(10) \qquad C\left(\frac{1}{i}e^{ki\theta}\frac{d}{d\theta}, \frac{1}{i}e^{\ell i\theta}\frac{d}{d\theta}\right) = \delta_{k,-\ell}\,\frac{k(k^2-1)}{12}$$

for all $k,\ell \in \mathbb{Z}$. The virasoro algebra g (§2) is clearly isomorphic to the central extension $\hat{v} = v \oplus \mathbb{C}$ of v, corresponding to the cocycle (10). The isomorphism sends d_k to $\left(\frac{1}{i}e^{ki\theta}\frac{d}{d\theta},1\right)$ and d_0' to $(0,1)$. We note that $\frac{1}{i}e^{ki\theta}\frac{d}{d\theta} = \frac{1}{k}\left[\frac{1}{i}\frac{d}{d\theta}, \frac{1}{i}e^{ki\theta}\frac{d}{d\theta}\right]$, if $k \neq 0$, and $\frac{1}{i}\frac{d}{d\theta} = \frac{1}{2}\left[\frac{1}{i}e^{-i\theta}\frac{d}{d\theta}, \frac{1}{i}e^{i\theta}\frac{d}{d\tau}\right]$, i.e., $v = [v,v]$. Therefore, by [8, lemma I.10] $\hat{v}$ is the universal covering of v.[4]

Let g, h, λ be as in §2. The highest weight representations of v correspond to the weights λ such that $\lambda(d_0') = 0$, i.e., $c = 0$.

---

[4] We thank H. Garland for pointing this out to us.

For simplicity we drop the variable c in the notation of §2. (6) becomes

$$(11) \qquad \det(\ ,\ )_{h,m} = \prod_{i=1}^{m} \Big(\prod_{rs=i} \big(h + \frac{1}{24}[(3r-2s)^2-1]\big)\Big)^{p(m-i)}$$

Corollary 1 [14,15]. $M(h) = L(h) \Leftrightarrow h \neq -\frac{1}{24}(m^2-1)$, for all $m \in \mathbb{N}$. By the corollary and formula (11) all the characters ch L(h) will be determined once we compute $\operatorname{ch} L\left(-\frac{1}{24}(m^2-1)\right)$ for all $m \in \mathbb{N}$.

The integers of the form $\frac{1}{24}(m^2-1)$, $m \in \mathbb{N}$, are *Euler's pentagonal numbers*, i.e., the numbers of the form $\frac{3k^2-k}{2}$, $k \in \mathbb{Z}$. The integral case, being the most difficult one, well illustrates the methods developed in [22,23]. For this reason, we discuss only this case in this paper, and refer the reader to [23] for the easier case of nonintegral highest weights. We now state the theorem that gives the integral irreducible characters. We set $s_k = -\frac{3k^2+k}{2}$ and $t_k = -\frac{3k^2-k}{2}$, $k \in \mathbb{Z}_+$.

Theorem 2 [22]. Let $\nu_k \in \{s_k, t_k\}$. There exists a resolution of g-modules:

$$(12) \qquad \cdots \xrightarrow{d_{k+2}} M(s_{k+1}) \oplus M(t_{k+1}) \xrightarrow{\eta_{k+1}} M(\nu_k) \xrightarrow{\epsilon_k} L(\nu_k) \longrightarrow 0$$

The following are immediate consequences of theorem 2:

Corollary 3 [22]:

$$H^p(n, L(\nu_k)) = \mathbb{C}(s_{k+p}) \oplus \mathbb{C}(t_{k+p})$$

as h-modules, $p \geq 1$.

Corollary 4 [22]:

$$\operatorname{ch} L(\nu_k) = \prod_{i=1}^{\infty} (1-z^{-i})^{-1} \big(z^{\nu_k} + (-1)^k \sum_{i>k} (-1)^i (z^{s_k} + z^{t_k})\big)$$

For k = 0, (12) is the resolution of the trivial module:

(13) $$\ldots M(s_i)\oplus M(t_i)\xrightarrow{d_i}\ldots\xrightarrow{d_2}M(-1)\oplus M(0)\xrightarrow{d_1}M(0)\xrightarrow{\epsilon_0}L(0)\longrightarrow 0$$

which was obtained in [21] (see also [19]) using Theorem 2.3, Goncharova's result on the cohomology of n^- with trivial coefficients [10,11], and standard arguments of homological algebra. The resolution (13) and the formula of Corollary 4 were conjectured by Kac [16].

We now indicate how to obtain (12) using (13) as a starting point. It was shown in [21] that $M(s_k) + M(t_k)$ is isomorphic to $M(s_{k-1}) \cap M(t_{k-1})$ for $k \geq 2$. Let $k \in \mathbb{N}$ be fixed and let $\nu_k \in \{s_k, t_k\}$. We define $\tilde{L}(\nu_k) = M(\nu_k)/(M(s_{k+1}) + M(t_{k+1}))$. (13) gives rise to a g-module resolution:

(12') $$\ldots \xrightarrow{d_{k+2}} M(s_{k+1})\oplus M(t_{k+1}) \xrightarrow{\eta_{k+1}} M(\nu_k) \xrightarrow{\epsilon_k'} \tilde{L}(\nu_k) \longrightarrow 0$$

In §2 we constructed a filtration of g-modules $M(\nu_k) = M(\nu_k)_{(0)} \supset M(\nu_k)_{(1)} \supset \ldots$. By Corollary 2.2 and (11) we have

(14) $$\sum_{i>0} \mathrm{ch}\, M(\nu_k)_{(i)} = \sum_{\ell\in\mathbb{N},\,\ell\ \mathrm{odd}} (\mathrm{ch}\, M(s_{k+\ell})+\mathrm{ch}\, M(t_{k+\ell}))$$

We set $N(\nu_k) = M(\nu_k)/M(t_{k+1})$. $N(\nu_k)_0$ also posseses a g-module filtation $N(\nu_k) = N(\nu_k)_{(0)} \supset N(\nu_k)_{(1)} \supset \ldots$ such that $N(\nu_k)_{(1)}$ is the largest proper submodule of $N(\nu_k)$, $N(\nu_k)_{(i)}/N(\nu_k)_{(i+1)}$ has a non-degenerate contravariant form, and

(15) $$\sum_{i>0} \mathrm{ch}\, N(\nu_k)_{(i)} = \sum_{i>0} \mathrm{ch}\, M(\nu_k)_{(i)} - \sum_{i>0} \mathrm{ch}\, M(t_{k+1})_{(i)} - \mathrm{ch}\, M(t_{k+1})$$

The construction of this filtation, although it follows the general lines desribed in §2, is of much deeper nature than that of $M(\lambda)$, and it involves the construction of a cross-section on the variety $\psi_{r,s}(h,c) = 0$, for appropriate $r, s \in \mathbb{N}$. Combining (14) and (15) we see that

(16) $$\sum_{i>0} \text{ch } N(\nu_k)_{(i)} = \text{ch } \tilde{L}(s_{k+1}).$$

Also, $\text{ch}((M(s_{k+1}) + M(t_{k+1}))/M(t_{k+1})) = \text{ch } \tilde{L}(s_{k+1})$. Since $(M(s_{k+1}) + M(t_{k+1}))/M(t_{k+1}) \subset N(\nu_k)_{(1)}$ we conclude that this inclusion is in fact an equality. Hence $\text{ch}(\tilde{L}(\nu_k)) = \text{ch } (N(\nu_k)/N(\nu_k)_{(1)})$, i.e., $\tilde{L}(\nu_k) = L(\nu_k)$. This concludes the proof of the existence of (12).

5. THE GENERAL CASE

A description of the subrepresentations of the Verma modules $M((h,c))$ over g, for arbitrary h and c, was recently announced in [2]. This description implies, in particular, the characters derived in §§3 and 4. In this section we state and discuss the results of [2]. We use the notation of §2.

Theorem 1 [2]. Let $\lambda \in h^*$ and let M be any submodule of $M(\lambda)$. There exists $\lambda_i \in h^*$, $v_i \in M_{\lambda_i}$, $i = 1,\ldots,n$ such that $n \cdot v_i = 0$, $i = 1,\ldots,n$ and $M = \sum_{i=1}^{n} V(g)v_i$.

Theorem 1 says that any submodule of $M(\lambda)$ is a sum of submodules that are Verma modules. This implies that the knowledge of all the embeddings among Verma modules enables one to describe all the submodules of a given Verma module.

We write $\mu \longleftarrow \lambda$ whenever there is an embedding $M(\mu) \longrightarrow M(\lambda)$, $\mu,\lambda \in h^*$. We consider the following types of diagrams of embeddings

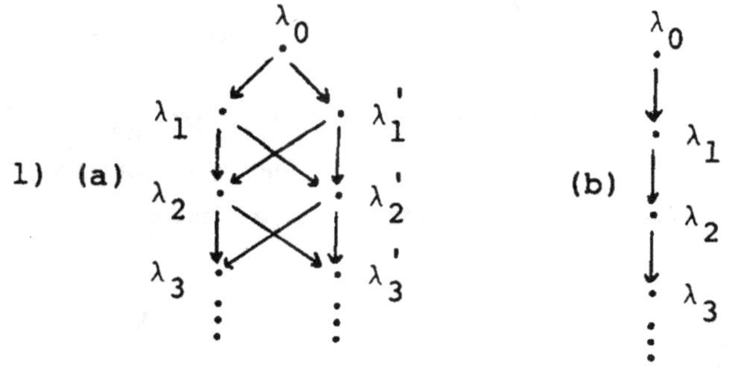

2) (a) 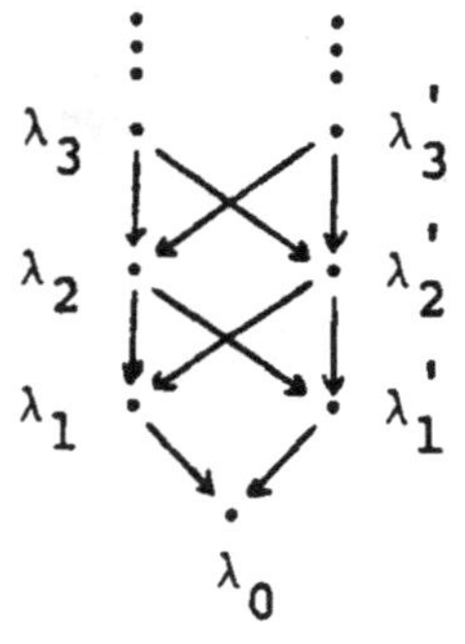 (b) 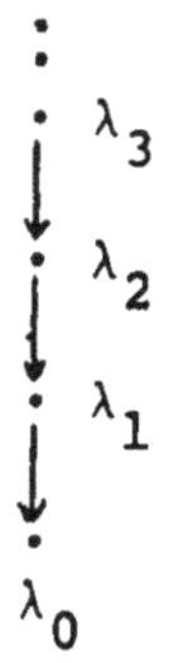

3) 

Theorem 3 of [2] states that any embedding among two Verma modules corresponds to an arrow in a diagram of one of the types above.

We now show how to use the above to write down the characters of the $L(\lambda)$. If $L(\lambda) = M(\lambda)$ this is given in §2 (4). So, assume that $L(\lambda) \neq M(\lambda)$. Let $\lambda \in \{\lambda_i, \lambda_i'\}$ in diagram 1) a). Then $M(\lambda_{i+1})$ and $M(\lambda_{i+1}')$ are both contained in $M(\lambda)$ and any other Verma module which is contained in $M(\lambda)$ is also contained in $M(\lambda_{i+1})$ and in $M(\lambda_{i+1}')$. By Theorem 1, $M(\lambda_{i+1}) + M(\lambda_{i+1}')$ is the largest proper submodule of $M(\lambda)$. The same reasoning shows that $M(\lambda_j) \cap M(\lambda_j') = M(\lambda_{j+1}) + M(\lambda_{j+1}')$, for all j It is now easy to see that $\mathrm{ch}\ L(\lambda) = \mathrm{ch}\ M(\lambda) + (-1)^i \sum_{j>i} (-1)^j(\mathrm{ch}\ M(\lambda_j) + \mathrm{ch}\ M(\lambda_1'))$.

Similarly, for $\lambda \in \{\lambda_i, \lambda_i'\}$ in diagram 2) a) we obtain $\mathrm{ch}\ L(\lambda) =$

$\mathrm{ch}\ M(\lambda) + (-1)^i \left[\sum_{0<j<i} (-1)^i(\mathrm{ch}\ M(\lambda_j) + \mathrm{ch}\ M(\lambda_j')) + \mathrm{ch}\ M(\lambda_0)\right]$. If $\lambda = \lambda_i$ in diagrams 1) b), 2) b) or 3) it is also clear that $\mathrm{ch}\ L(\lambda) = \mathrm{ch}\ M(\lambda) - \mathrm{ch}\ M(\lambda_{i+1})$. Summing up and using §2(4) we have:

$$\text{(17)} \qquad \mathrm{ch}\ L(\lambda) = \left[\prod_{h=1}^{\infty} (1-z^{-n})\right]^{-1} \left[z^{\lambda} + (-1)^i \sum_{j>i} (-1)^j (z^{\lambda_j} + z^{\lambda_j'})\right]$$

$$(18)\quad \mathrm{ch}\ L(\lambda) = \left[\prod_{h=1}^{\infty}(1-z^{-n})\right]^{-1}(z^{\lambda}+(-1)^{i}\left[\sum_{0<j<i}(-1)^{j}(z^{\lambda_j}+z^{\lambda'_j})+z^{\lambda_0}\right]$$

$$(19)\quad \mathrm{ch}\ L(\lambda) = \left[\prod_{n=1}^{\infty}(1-z^{-n})\right]^{-1}\left[z^{\lambda} - z^{\lambda_{i+1}}\right]$$

for $\lambda \in \{\lambda_i, \lambda'_i\}$ in diagram 1)(a), $\lambda \in \{\lambda_i, \lambda'_i\}$ in diagram 2)(b) and $\lambda = \lambda_i$ in any of the other diagrams, respectively.

In [2, Theorem 3] a procedure is given to find the vertices connected to a given vertex in diagrams 1) - 3). The full statement of Theorem 3 is somewhat long and it involves the separate treatment of various cases. For this reason we chose to state only their result in the situation that is the most interesting to us. This case (subcase $III_{\pm}$ of [2, Theorem 3]) corresponds to the diagrams 1)(a) and 2)(a).

Let h, c $\in \mathbb{C}$. It is easy to see that the set of points (r,s) of any of the four lines representing the equation $\psi_{r,s}(h,c) = 0$ in rs-plane yield the same set of products rs. Let $\ell_{h,c}$ be any one of these four lines and we denote by $S_{h,c}$ the set of points of $\ell_{h,c}$ with integral coordinates. Suppose that

(*) $\quad S_{h,c}$ is infinite and if $(r,s) \in S_{h,c}$ then $rs \neq 0$.

By (*) $ab \neq cd$ if $(a,b) \neq (c,d)$, $(a,b), (c,d) \in S_{h,c}$. We enumerate the elements of $S_{h,c}$: ... (r_{-1},s_{-1}), (r_0,s_0), (r_1,s_1), ... so that ... $<r_{-1}s_{-1} < 0 < r_0s_0 < r_1s_1 < \dots$. Now, let $\ell'_{h,c}$ be the line passing through $(-r_0,s_0)$ and parallel to $\ell_{h,c}$. Let $S'_{h,c}$ be the set of points of $\ell'_{h,c}$ with integral coordinates. The points of $S'_{h,c}$ can be ordered:

$$\dots (r'_{-1},s'_{-1}),\ (r'_0,s'_0) = (-r_0,s_0),\ (r'_1,s'_1),\ \dots ,$$

so that

$$\dots r'_{-1}s'_{-1} < r'_0s'_0 < 0 < r'_1s'_1 < \dots$$

By [2, Theorem 3] one has the following diagram of embeddings:

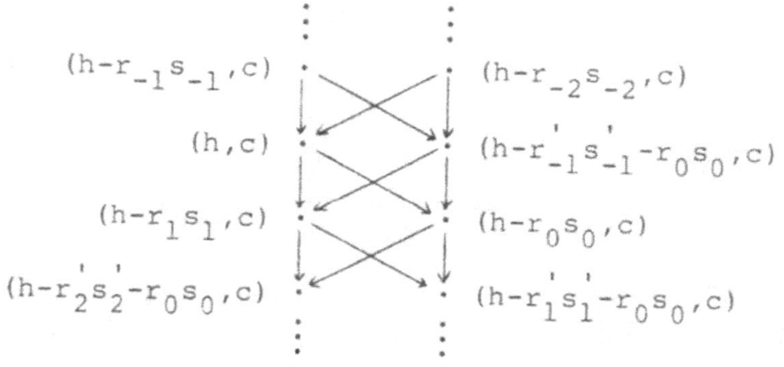

Remarks: 1) The assumption that $S_{h,c}$ is infinite implies that $\ell_{h,c}$ is real and so $c \in \mathbb{R}$ with either $c \leqslant 1$ or $c \geqslant 25$. The diagram above is of type 1)(a) if $c \leqslant 1$ and of type 2)(a) if $c \geqslant 25$.

2) Let $\ell''_{h,c}$ denote the line through $(-r_1, s_1)$ and parallel to $\ell_{h,c}$, and let $S''_{h,c}$ denote the set of points of $\ell''_{h,c}$ with integral coordinates. Then $\{h - r'_i s'_i - r_0 s_0\} = \{h - r''_i s''_i - r_1 s_1\}$, where $S''_{h,c} = \{(r''_i, s''_i)\}$.

3) $\{h - r'_i s'_i - r_0 s_0\} = \{h - r_0 s_0 - a_i b_i\}$ where $S_{h - a_0 s_0, c} = \{(a_i, b_i)\}$.

4) Using Remark 3) and the above diagram of embeddings we can see that for $\lambda = (h,c)$, $M(\mu) \subset M(\lambda)$ if and only if $\mu \uparrow \lambda$ (see §2).

6. NEW CHARACTER FORMULAS

If $\lambda = (h,c)$, $h,c \in \mathbb{R}$, one can construct a unique, up to scalar, *hermitian contravariant* form on $M(\lambda)$ as follows; let ω be the unique conjugate-linear antiautomorphism of $U(g)$ such that $\omega(d_i) = d_{-i}$ for all $i \in \mathbb{Z}$, and $\omega(d'_0) = d'_0$. Let β be as in §2, and set $\langle Xv_\lambda, Yv_\lambda \rangle_\lambda = (\lambda \circ \beta)(\omega(Y)X)$, for all $X, Y \in U(g)$. As in §2 we can show that distinct weightspaces are orthogonal relative to $\langle\ ,\ \rangle_\lambda$, and that $\langle\ ,\ \rangle_\lambda$ induces a nondegenerate hermitian contravariant form on $L(\lambda)$. $L(\lambda)$ is said to be *unitarizable* if the induced form is positive definite. It is easy to see that if $L(\lambda)$ is unitarizable then

$h \leqslant 0$ and $c \geqslant 0$. Furthermore, if $c = 0$ then $h = 0$. In [6,7], the authors investigate the unitarizable representations $L(\lambda)$, for $0 \leqslant c < 1$. They show that a necessary condition for the unitarizability of $L(\lambda)$ is given by the formulas:

$$(20)\quad \begin{cases} c = 1 - \dfrac{6}{m(m+1)}, & m \in \mathbb{N},\ m \geqslant 2 \\ h = h_{p,q} = -\dfrac{[(m+1)p-mq]^2-1}{4m(m+1)}, & p=1,\ldots,m-1,\ q=1,\ldots,p. \end{cases}$$

We set $x_{p,q}(z) = \mathrm{ch}\ L((h,c))$, where h,c are given in (20). We found that the characters $x_{p,q}(z)$ are given by the formulas:

$$(21)\quad x_{p,q}(z) = \left[\prod_{i=1}^{\infty}(1-z^{-i}))^{-1}\sum_{k\in\mathbf{Z}}(-z^{-a(k)} + z^{-b(k)}\right]$$

where

$$(22)\quad a(k) = \frac{[2m(m+1)k + (m+1)p + mq]^2-1}{4m(m+1)}$$

$$(23)\quad b(k) = \frac{[2m(m+1)k + (m+1)p - mq]^2-1}{4m(m+1)}$$

In order to convey the combinatorial flavor of (21) we give some examples.

If we set $m = 2$, (20) gives only the trivial module $L((0,0))$. (21) yields

$$(24)\quad x_{1,1}(z) = \mathrm{ch}\ L((0,0)) = \left[\prod_{i=1}^{\infty}(1-z^{-i})\right]^{-1}\sum_{k\in\mathbf{Z}}(-1)^k z^{-\frac{3k^2+k}{2}}$$

Since $\mathrm{ch}\ L((0,0) = 1$, (24) is simply the statement of Euler's pentagonal number theorem (see. e.g. [13]).

For $m = 3$, we obtain from (21):

$$(25)\quad x_{1,1}(z) = \mathrm{ch}\ L((0,\tfrac{1}{2})) =$$

$$\left[\prod_{i=1}^{\infty}(1-z^{-i})\right]^{-1}\sum_{\substack{r\equiv 0,3 (\mathrm{mod}\ 4)\\ s\equiv 0,1 (\mathrm{mod}\ 4)}}\left[(-1)^r z^{-\frac{3r^2-r}{4}}+(-1)^s z^{-\frac{3s^2+s}{4}}\right],\ r,s\in\mathbb{N}$$

(26) $$\chi_{2,1}(z) = \mathrm{ch}\ L((-\tfrac{1}{2},\tfrac{1}{2})) = \left(\prod_{i=1}^{\infty}(1-z^{-i})\right)^{-1}$$

$$\sum_{\substack{r\equiv 2,1\ (\mathrm{mod}\ 4)\\ s\equiv 2,3\ (\mathrm{mod}\ 4)}}\left[(-1)^{r+1} z^{-\frac{3r^2-r}{4}}+(-1)^{s+1} z^{-\frac{3s^2+s}{4}}\right],\ r,s\in\mathbb{N}$$

(27) $$\chi_{2,2}(z) = \mathrm{ch}\ L(-\frac{1}{16},\frac{1}{2})) =$$

$$\left[\prod_{i=1}^{\infty}(1-z^{-i})\right]^{-1}\sum_{k\in\mathbb{Z}}(-1)^k z^{-(3k^2+k)}$$

Using Jacobi's triple product identity (see e.g. [13]) we obtain:

(28) $$\mathrm{ch}\ L((0,\tfrac{1}{2})) \pm \mathrm{ch}\ L(-\tfrac{1}{2},\tfrac{1}{2})) = \prod_{n\in\mathbb{N}-\frac{1}{2}}(1\pm z^{-n})$$

(29) $$\mathrm{ch}\ L\ ((-\frac{1}{16},\frac{1}{2})) = z^{-\frac{1}{16}}\prod_{n\in\mathbb{N}}(1+z^{-n})$$

Remark: The formulas for ch $L((0,\frac{1}{2}))$ and ch $L((-\frac{1}{2},\frac{1}{2}))$ were known to physicists. The version (28) of them was communicated to the author by D. Friedan before we obtained (21). Inspired by the preceding formulas, I. Frenkel derived the simplified version (29) of (27) by different methods.

For m = 4, (21) gives the following set of formulas.

(30) $$\chi_{1,1}(z) = \mathrm{ch}\ L((0,\frac{7}{10})) = \left[\prod_{i=1}^{\infty}(1-z^{-i})\right]^{-1}$$

$$\sum_{\substack{r\equiv 0,1\ (\mathrm{mod}\ 4)\\ s\equiv 0,3\ (\mathrm{mod}\ 4)}}\left[(-1)^r z^{-\frac{5r^2-r}{4}} + (-1)^s z^{-\frac{5s^2+s}{4}}\right],\ r,s\in\mathbb{N}$$

(31) $$\chi_{3,1}(z) = \mathrm{ch}\, L((-\tfrac{3}{2},\tfrac{7}{10})) = \left(\prod_{i=1}^{\infty}(1-z^{-i})\right)^{-1}$$

$$\sum_{\substack{r\equiv 2,3 \pmod 4\\ s\equiv 2,1 \pmod 4}} \left[(-1)^{r+1} z^{-\frac{5r^2-r}{4}} + (-1)^{s+1} z^{-\frac{5s^2+s}{4}}\right],\quad r,\ s \in \mathbb{N}$$

(32) $$\chi_{3,3}(z) = \mathrm{ch}\, L((-\tfrac{1}{10},\tfrac{7}{10})) = \left(\prod_{i=1}^{\infty}(1-z^{-i})\right)^{-1}$$

$$z^{-\frac{1}{10}} \sum_{\substack{r\equiv 0,3 \pmod 4\\ s\equiv 0,1 \pmod 4}} \left[(-1)^{r} z^{-\frac{5r^2-3r}{4}} + (-1)^{s} z^{-\frac{5s^2+3s}{4}}\right],\quad r,\ s \in \mathbb{N}$$

(33) $$\chi_{3,2}(z) = \mathrm{ch}\, L((-\tfrac{3}{5},\tfrac{7}{10})) = \left(\prod_{i=1}^{\infty}(1-z^{-i})\right)^{-1}$$

$$z^{-\frac{1}{10}} \sum_{\substack{r\equiv 2,1 \pmod 4\\ s\equiv 2,3 \pmod 4}} \left[(-1)^{r+1} z^{-\frac{5r^2-3r}{4}} + (-1)^{s+1} z^{-\frac{5s^2+3s}{4}}\right],\quad r,\ s \in \mathbb{N}$$

(34) $$\chi_{2,1}(z) = \mathrm{ch}\, L((-\tfrac{7}{16},\tfrac{7}{10})) =$$

$$\left(\prod_{i=1}^{\infty}(1-z^{-i})\right)^{-1} z^{-\frac{7}{16}} \sum_{k\in\mathbb{Z}} (-1)^k z^{(5k^2+3k)}$$

(35) $$\chi_{2,2}(z) = \mathrm{ch}\, L((-\tfrac{3}{80},\tfrac{7}{10})) =$$

$$\left[\prod_{i=1}^{\infty}(1-z^{-i})\right]^{-1} z^{-\frac{3}{80}} \sum_{k\in\mathbb{Z}} (-1)^k z^{-(5k^2+3k)}$$

Using Jacobi's identity, we obtain

(36) $$\mathrm{ch}\, L((0,\tfrac{7}{10})) \pm \mathrm{ch}\, L((-\tfrac{3}{2},\tfrac{7}{10})) =$$

$$\prod_{n=0}^{\infty} \frac{(1\pm z^{-(5n+\frac{5}{2})})(1\pm z^{-(5n+\frac{3}{2})})(1\pm z^{-(5n+\frac{7}{2})})}{(1-z^{-(5n+2)})(1-z^{-(5n+3)})}$$

(37) $$\mathrm{ch}\ L\left(\left(-\frac{1}{10},\frac{7}{10}\right)\right) \pm \mathrm{ch}\ L\left(\left(-\frac{3}{5},\frac{7}{10}\right)\right) =$$

$$z^{-\frac{1}{10}} \prod_{n=0}^{\infty} \frac{(1 \pm z^{-(5n+\frac{5}{2})})(1 \pm z^{-(5n+\frac{1}{2})})(1 \pm z^{-(5n+\frac{9}{2})})}{(1 - z^{-(5n+1)})(1 - z^{-(5n+4)})}$$

(38) $$\mathrm{ch}\ L\left(\left(-\frac{7}{16},\frac{7}{10}\right)\right) =$$

$$z^{-\frac{7}{16}} \prod_{n=0}^{\infty} \frac{(1+z^{-(5n+1)})(1+z^{-(5n+4)})(1+z^{-(5n+5)})}{(1 - z^{-(5n+2)})(1 - z^{-(5n+3)})}$$

(39) $$\mathrm{ch}\ L\left(\left(-\frac{3}{80},\frac{7}{10}\right)\right) =$$

$$z^{-\frac{3}{80}} \prod_{n=0}^{\infty} \frac{(1+z^{-(5n+2)})(1+z^{-(5n+3)})(1+z^{-(5n+5)})}{(1 - z^{-(5n+1)})(1 - z^{-(5n+4)})}$$

We note the appearance of the product sides of the Rogers-Ramanujan identities (see e.g. [13]) as factors in (36) - (39).

We now prove (21). We use the notation and the results of §5. By (20) we may take $\ell_{h,c}$ to be:

(40) $$(m+1)r - ms = (m+1)p - mq$$

Let $r(k) = mk+p$, $s(k) = (m+1)k + q$. Then $S_{h,c} = \{(r(k), s(k)) \mid k \in \mathbf{Z}\}$. It is clear from (20) that $S_{h,c}$ satisfies condition (*) of §5. Also, $r(k)s(k) > 0$ for all $k \in \mathbf{Z}$. By remark 2) of §5 we may consider the line

(41) $$(m+1)r - ms = -(m+1)p - mq.$$

The points of (41) with integral coordinates are $(r'(k), s'(k))$, where $r'(k) = mk - p$, $s'(k) = (m+1)k + q$. It follows from §5 that one has the following diagram of embeddings:

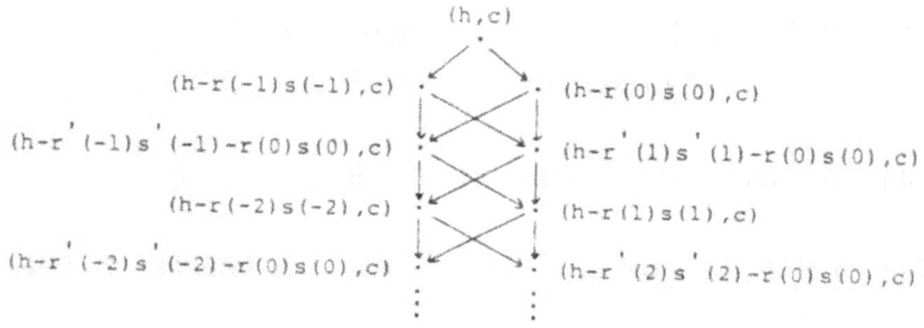

Formula (21) now follows from §5 (17).

REFERENCES

[1] B.L. Feigin and D.B. Fuchs, Functs. Anal. Prilozhen. 16, No. 2 (1982), 47-63.

[2] ————————, Functs. Anal. Prilozhen. 17, No. 3 (1983), 91-92.

[3] I. Frenkel, J. Funct. Anal. 44 (1981), 259-327.

[4] ————————, Representations of Kac-Moody Algebras and Dual Resonance Models, Proc. 1982 Summer Seminar on Applications of Group Theory in Physics and Mathematical Physics, Lectures in Applied Math., A.M.S. 21 (1984).

[5] I. Frenkel and V.G. Kac, Invent. Math. 62 (1980), 23-66.

[6] D. Friedan, Z. Qiu and S.H. Shenker, Conformal Invariance, Unitarity and Two dimensional critical exponents, these Proceedings.

[7] D. Friedan, Z. Qiu and S.H. Shenker, Phys. Rev. Lett. 52 (1984), 1575.

[8] H. Garland, Public. Math. No. 52, I.H.E.S.

[9] I.M. Gelfand and D.B. Fuchs, Functs. Anal. Prilozhen. 2 (1968), 92-93.

[10] L.V. Goncharova, Functs. Anal. Prilozhen. 7, No. 2 (1973), 6-14.

[11] ——————, Functs. Anal. Prilozhen. 7, No. 3 (1973), 33-44.

[12] R. Goodman and N.R. Wallach, Journal für die reine und angew. Math., 347 (1984), 69-133.

[13] G.H. Hardy and E.M. Wright, An Introduction to the theory of numbers, Clarendon Press, Oxford (1965).

[14]. V.G. Kac, Proceedings of the International Congress of mathematicians, Helsinki (1978).

[15] ——————, Lecture Notes in Physics 94 (1979), 441.

[16] ——————, Lecture Notes in Math. 933 (1982) 117-126.

[17] V.G. Kac and D.H. Peterson, Infinite-Dimensional Lie Algebras, Theta Functions and Modular Forms, preprint.

[18] E. Looijenga, Invent. Math. 38, (1976), 17-32.

[19] A. Rocha-Caridi, Lecture Notes in Math 933 (1982), 176-190.

[20] A. Rocha-Caridi and N.R. Wallach, Math. Z. 180 (1982), 151-177.

[21] ——————, Transactions of the A.M.S. 277, No. 1 (1983),

133-162.

[22] ——————, Invent. Math. 72 (1983). 57-75.

[23] ——————, Math. Z. 185 (1984), 1-21.

[24] G. Segal, Commun. Math. Phys. 80 (1981), 301-342.

[25] J. Schwarz, Phys. Rep. 8C (1973), 269-335.

VOLUME 52, NUMBER 18 PHYSICAL REVIEW LETTERS 30 APRIL 1984

Conformal Invariance, Unitarity, and Critical Exponents in Two Dimensions

Daniel Friedan, Zongan Qiu, and Stephen Shenker
Enrico Fermi and James Franck Institutes and Department of Physics, University of Chicago, Chicago, Illinois 60637
(Received 31 January 1984)

Conformal invariance and unitarity severely limit the possible values of critical exponents in two-dimensional systems.

PACS numbers: 05.70.Jk, 11.30.Na, 11.30.Pb, 64.60.-i

One of the most intriguing features of statistical mechanical systems is the existence of scale-invariant critical points. Polyakov[1] has shown that *local* scale invariance, i.e., conformal invariance, can be used to construct critical theories. In two dimensions the group of conformal transformations is exceptionally large (infinite dimensional) since any analytic function mapping the complex plane to itself is conformal. Belavin, Polyakov, and Zamolodchikov (BPZ)[2] have shown how the rich structure of the conformal group in two dimensions can be used to analyze conformally invariant field theories.

Many two-dimensional statistical mechanical systems can also be interpreted as $(1+1)$-dimensional quantum field theories. The distinguishing feature of the quantum theories is unitarity, equivalent to reflection positivity in the statistical systems. We point out here that unitarity, in the presence of the large conformal transformation group, puts a powerful constraint on the allowed physical systems.

Conformally invariant systems are described by correlation functions of a collection of conformal fields $\phi(z,\bar{z})$. The fields can be interpreted as operators and the correlation functions as vacuum expectation values. The infinitesimal conformal transformations $z \to z + \epsilon z^{n+1}$, $\bar{z} \to \bar{z} + \bar{\epsilon}\bar{z}^{n+1}$ are generated by operators L_n, $\bar{L}_n$:

$$[L_n,\phi] = z^{n+1}\partial_z\phi + h(n+1)z^n\phi,$$
$$[\bar{L}_n,\phi] = \bar{z}^{n+1}\partial_{\bar{z}}\phi + \bar{h}(n+1)\bar{z}^n\phi. \quad (1)$$

The correlation functions are invariant under the global conformal transformations generated by L_n, $\bar{L}_n$, for $n = -1, 0, 1$. In particular,

$$\langle \phi(re^{i\theta})\phi(0)\rangle = r^{-2(h+\bar{h})}e^{-2i\theta(h-\bar{h})}, \quad (2)$$

and so ϕ has scaling dimension $x = h + \bar{h}$ and spin $h - \bar{h}$.

We are using here a nonstandard operator interpretation in which radial ordering takes the place of time ordering. Reflection positivity in the radial and time directions are equivalent by global conformal invariance. In terms of coordinates $z = e^{\tau + i\theta}$ the radial quantization gives quantum field theory in imaginary "time" τ and periodic one-dimensional "space" θ.

Local scale invariance is equivalent to the existence of a conserved traceless stress energy tensor $T(z)dz^2 + \bar{T}(\bar{z})d\bar{z}^2$ which acts as the generator of conformal transformations:

$$T(z) = \sum_{n=-\infty}^{\infty} z^{-n-2}L_n,$$
$$\bar{T}(\bar{z}) = \sum_{n=-\infty}^{\infty} \bar{z}^{-n-2}\bar{L}_n. \quad (3)$$

It satisfies the self-adjointness condition $[T(1/\bar{z}) \times d(1/\bar{z})^2]^\dagger = T(z)dz^2$. Equivalently,

$$L_n^\dagger = L_{-n}. \quad (4)$$

The L_n obey the commutation relations

$$[L_m,L_n] = (m-n)L_{m+n} + c[(m^3-m)/12]\delta_{m,-n}. \quad (5)$$

The $\bar{L}_n$ commute with all the L's and satisfy the same Hermiticity and commutation relations (with the same c, under the assumption that the theory is $z \leftrightarrow \bar{z}$ invariant). The algebra (5) is the Virasoro algebra.[3] The central term with coefficient c describes the particular realization of conformal symmetry. It appears in a quantum theory because the composition law for transformations need only be satisfied by the operator representation up to a phase. The central term also measures the response of the ground-state energy to curving the underlying space—the trace anomaly.[4]

Our result is that unitarity restricts the allowed values of c, for $c < 1$, to

$$c = 1 - 6/m(m+1), \quad (6a)$$
$$m = 2, 3, 4, \ldots . \quad (6b)$$

VOLUME 52, NUMBER 18 PHYSICAL REVIEW LETTERS 30 APRIL 1984

For a model in which c takes one of these values, the scaling dimensions $x = h + \bar{h}$ are rational numbers with h and $\bar{h}$ limited to the values

$$h_{p,q}(c) = \frac{[(m+1)p - mq]^2 - 1}{4m(m+1)}, \tag{7a}$$

$$p = 1, 2, \ldots, m-1, \quad q = 1, 2, \ldots, p. \tag{7b}$$

This follows from a study of the representations of the Virasoro algebra.

The lowering operators for L_0 ($\bar{L}_0$) are the L_n ($\bar{L}_n$), $n > 0$. A state annihilated by all the lowering operators is called a highest-weight vector (HWV). The vacuum $|0\rangle$ is a HWV because it has the lowest eigenvalue of the "Hamiltonian" $L_0 + \bar{L}_0$. More generally, there is a one-to-one correspondence between the conformal fields and the HWV's, the field ϕ corresponding to $\phi(0)|0\rangle$, which is a HWV by Eq. (1). The invariance of the correlation functions under global conformal transformations implies that the vacuum is also annihilated by L_0, $\bar{L}_0$, L_{-1}, and $\bar{L}_{-1}$. It follows that $\phi(0)|0\rangle$ is an eigenstate of L_0 ($\bar{L}_0$) with eigenvalue h ($\bar{h}$). The space of states is a sum of irreducible representations of the product algebra (of L's and $\bar{L}$'s), each generated from one of the HWV's.

The L's and $\bar{L}$'s form identical commuting Virasoro algebras and so we can restrict our attention to representations of the L's. A representation of the Virasoro algebra is built from a HWV by applying the L_{-n}, $n \geq 1$. A state is in the nth level if its L_0 eigenvalue is $h + n$. The nth level is spanned by the vectors

$$L_{-k_1} \cdots L_{-k_m} \phi(0)|0\rangle, \tag{8}$$

for $k_1 \geq \ldots \geq k_m > 0$ and $\sum k_i = n$. There are $P(n)$ such states, where $P(n)$ is the number of ways of writing n as a sum of positive integers. The higher-level states correspond to operators of higher scaling dimension obtained by applying products of stress-energy tensors to ϕ.

Unitarity means that the inner product in the space of states is positive definite. The inner product of any two of the states (8) can be computed from Eqs. (4) and (5). A state $|\psi\rangle$ with $\langle\psi|\psi\rangle$ negative is called a "ghost." If a ghost is found on any level the representation cannot occur in a unitary theory.

Kac[5] has given a formula for the determinant of the matrix of inner products of the states (8):

$$\det M_{(n)}(c,h) = C \prod_{pq \leq n} [h - h_{p,q}(c)]^{P(n-pq)}, \tag{9}$$

where p,q range over the positive integers, $h_{p,q}(c)$ is given by Eqs. (6a) and (7a), and C is a positive constant. The matrix of inner products is manifestly positive definite for $1 < c$, $0 < h$.

We immediately eliminate all regions where the determinant is negative because they necessarily contain an odd number of ghosts. A straightforward examination of the determinant formula shows that any point c,h in the half-plane $c < 1$ which is not on an $h_{p,q}$ curve will have a negative determinant at some level.

In fact the unitary theories are far more limited. All points on vanishing curves have ghosts except possibly the "first intersections." A first intersection is an intersection of vanishing curves that on some level is the intersection closest to $c = 1$ on a given curve $h_{p,q}$. These are exactly the points listed in Eqs. (6a)–(7b). The proof[6] that there is a ghost in every open interval bounded by first intersections on a vanishing curve $h_{p,q}$ is based on the following observations: (1) When $h = h_{p,q}(c)$ there is a null HWV at level pq which generates a subrepresentation, and (2) whenever a first intersection appears on $h_{p,q}$ (at level n) the determinant $\det M_{(n-pq)}(c, h+pq)$ for the subrepresentation is nonzero, so that there can be no HWV inside the subrepresentation at level $n - pq$.

All unitary representations for $c < 1$ are contained in the list, Eqs. (6a)–(7b), but we have not proved that all representations on the list are in fact unitary. We have verified numerically that they are through level 12 and we have a heuristic argument that they remain ghost-free to all levels. The argument is based on the structure of subrepresentations on the $c = 1$ line[7] and the assumed existence of an analytic diagonalization of the matrix in the whole region of interest. Such a diagonalization might be provided by techniques[8] which give analytic deformations of correlation functions away from $c = 1$.

Models with Hermitian transfer matrices provide concrete examples of reflection-positive systems. Systems with continuously variable critical exponents like the Gaussian model have $c \geq 1$ where unitarity allows all $h \geq 0$. In the range $c < 1$ we find by matching scaling dimensions that the $m = 3$ representations describe the Ising model, $m = 4$ the tricritical Ising, $m = 5$ the three-state Potts, and $m = 6$ the tricritical three-state Potts.[9] Table I compares the known scaling dimensions for these models[10] with the allowed values of h. All of the known exponents are accounted for. We know of no models with $m \geq 7$. The fact that $h = 1$ is never unitary for $c < 1$ rules out marginal operators as well as continuous internal symmetries generated

VOLUME 52, NUMBER 18 PHYSICAL REVIEW LETTERS 30 APRIL 1984

TABLE I. Comparison with known scaling dimensions. The $h_{p,q}$ from Eqs. (6a)–(7b) are listed with p running horizontally from 1 to $m-1$ and q running vertically from 1 to p. The spins $h-\bar{h}$ are consistent with what is known. Some operators have alternative interpretations as derivatives.

model	x	h	$\bar{h}$
Ising	$\frac{1}{8}$	$\frac{1}{16}$	$\frac{1}{16}$
	1	$\frac{1}{2}$	$\frac{1}{2}$
	$\frac{1}{2}$	$\frac{1}{2}$	0

$h_{p,q}(c)$, $m=3$:

		$\frac{1}{16}$
	0	$\frac{1}{2}$

model	x	h	$\bar{h}$
tri-critical 3-state Potts	$\frac{2}{21}$	$\frac{1}{21}$	$\frac{1}{21}$
	$\frac{2}{7}$	$\frac{1}{7}$	$\frac{1}{7}$
	$\frac{20}{21}$	$\frac{10}{21}$	$\frac{10}{21}$
	$\frac{10}{7}$	$\frac{5}{7}$	$\frac{5}{7}$
	$\frac{17}{7}$	$\frac{12}{7}$	$\frac{5}{7}$
	$\frac{8}{3}$	$\frac{4}{3}$	$\frac{4}{3}$
	$\frac{23}{7}$	$\frac{22}{7}$	$\frac{1}{7}$
	5	5	0

$h_{p,q}(c)$, $m=6$:

				$\frac{1}{7}$
			$\frac{5}{56}$	$\frac{5}{7}$
		$\frac{1}{21}$	$\frac{33}{56}$	$\frac{12}{7}$
	$\frac{1}{56}$	$\frac{10}{21}$	$\frac{85}{56}$	$\frac{22}{7}$
0	$\frac{3}{8}$	$\frac{4}{3}$	$\frac{23}{8}$	5

model	x	h	$\bar{h}$
tri critical Ising	$\frac{3}{40}$	$\frac{3}{80}$	$\frac{3}{80}$
	$\frac{1}{5}$	$\frac{1}{10}$	$\frac{1}{10}$
	$\frac{7}{8}$	$\frac{7}{16}$	$\frac{7}{16}$
	$\frac{6}{5}$	$\frac{3}{5}$	$\frac{3}{5}$
	3	$\frac{3}{2}$	$\frac{3}{2}$

$h_{p,q}(c)$, $m=4$:

		$\frac{1}{10}$
	$\frac{3}{80}$	$\frac{3}{5}$
0	$\frac{7}{16}$	$\frac{3}{2}$

model	x	h	$\bar{h}$
3-state Potts	$\frac{2}{15}$	$\frac{1}{15}$	$\frac{1}{15}$
	$\frac{4}{5}$	$\frac{2}{5}$	$\frac{2}{5}$
	$\frac{4}{3}$	$\frac{2}{3}$	$\frac{2}{3}$
	$\frac{9}{5}$	$\frac{7}{5}$	$\frac{2}{5}$
	$\frac{14}{5}$	$\frac{7}{5}$	$\frac{7}{5}$
	3	3	0

$h_{p,q}(c)$, $m=5$:

			$\frac{1}{8}$
		$\frac{1}{15}$	$\frac{2}{3}$
	$\frac{1}{40}$	$\frac{21}{40}$	$\frac{13}{8}$
0	$\frac{2}{5}$	$\frac{7}{5}$	3

by local currents. The possibility of large discrete-symmetry groups seems worth exploring.

Although unitarity provides strong constraints on possible representations of conformal invariance, a sensible theory also requires closure of the operator-product expansion and crossing symmetry of correlation functions.[1,2] BPZ[2] have shown how to implement crossing symmetry when the space of states is made up of representations with null states, using the fact that such states give rise to linear differential equations on the correlation functions. It follows from our result that the differential equation technique applies to all unitary models with $c<1$. BPZ[11] have found finite sets of conformal fields that must close under the operator-product expansion for the special values of c corresponding to m rational, $m=r/(s-r)$, $r<s$, $3r\geq 2s$. The scaling dimensions are given by $h=h_{p,q}(c)$, $1\leq p<r$, $1\leq q<s$. Note that the unitary representations correspond to $s=r+1$. Dotsenko[12] has used the differential equation technique to find a closed operator algebra for the three-state Potts model and to construct some of its correlation functions. The representations which actually appear in the tricritical Potts model form a similar closed algebra by Dotsenko's argument.

Kac[5] has also written a determinant formula for the supersymmetric extension of the Virasoro algebra, the Ramond-Neveu-Schwarz algebra.[13] Only first intersections can be ghost-free. The allowed representations, for $2c/3<1$, are

$$2c/3=1-8/m(m+2),$$
$$h=\frac{[(p-q)m+2p]^2-4}{8m(m+2)}, \tag{10}$$

for $m=2,3,4,\ldots$ and p,q integers, both even or both odd, $0<p<m$, $0<q\leq p$. Note that $h=\frac{1}{2}$ does not occur, so that there are no internal supercurrents. The representations $c=\frac{7}{10}$, $h=0,\frac{1}{10}$ are composed of the Virasoro representations $c=\frac{7}{10}$, $h=0,\frac{3}{2},\frac{1}{10},\frac{3}{5}$. These are the representations occuring in the Z_2 invariant subalgebra of the tricritical Ising model. It seems that this subalgebra is supersymmetric.[14] Supersymmetric models at $m=4$, $c=1$ using $h=0,\frac{1}{6},1$ or $h=0,\frac{1}{16},1$ can be identified with special points in the Gaussian model.[14] The representations of the superalgebra might also be of interest in superstring theory.[15]

It should now be straightforward to sort out all possible unitary conformally invariant models with $c<1$ ($2c/3<1$ in the super case). The differential equation techniques of BPZ always apply because the only allowed representations have null states, and the problem is finite because only a finite number of representations are allowed for each possible m. Such a systematic construction of conformally invariant models would partially realize the bootstrap program initiated by Kadanoff[16] and Polyakov.[17] The extreme rigidity of the conformal bootstrap in two dimensions is a striking feature of the present result. It seems that unitarity is the crucial constraint for $c < 1$. For $c \geq 1$ crossing symmetry would have to limit the possible realizations of conformal invariance.

We thank L. P. Kadanoff and M. den Nijs for a number of helpful conversations, D. Arnett for making available the Chicago Astrophysics computer, and M. Crawford, P. Schinder, and G. Toomey for their generous help in using it. One of us (D.F.) is grateful to the U. S. National Academy of Science and to the Academy of Sciences of the U.S.S.R. for their support of the exchange which led to this work, to the L. D. Landau Institute for Theoretical Physics for its hospitality, to A. A. Migdal for interesting conversations, and to A. B. Zamolodchikov, A. M. Polyakov, and V.S. Dotsenko for many patient explanations of their ideas and much discussion. This work was supported in part by U. S. Department of Energy Contract No. DE-AC02-81ER-10957, National Science Foundation Grant No. NSF-DMR-82-16892, and the Alfred P. Sloan Foundation.

Note added.—D. A. Huse has informed us that G. E. Andrews, R. J. Baxter, and P. J. Forrester have solved two new infinite sets of two-dimensional models. The critical exponents in one of the sets suggest that the corresponding models provide examples for each value of c in Eqs. (6a)–(6b). The supersymmetry which we have noted at $c = 0.7$ has also been described by A. M. Zamolodchikov, Yad Fiz. (to be published).

The appearance of the Virasoro algebra in conformally invariant two-dimensional field theory was pointed out by F. Mansouri and Y. Nambu, Phys. Lett. **39B**, 375 (1972), and by S. Ferrara, A. F. Gatto, and R. Grillo, Nuovo Cimento **12A**, 959 (1972). We thank G. Parisi for the latter reference. Radial quantization and its relation to the Virasoro algebra was described by S. Fubini, A. J. Hansen, and R. Jackiw, Phys. Rev. D **7**, 1732 (1973). We thank R. Jackiw for bringing this reference to our attention. M. Lüscher and G. Mack (unpublished, 1976) considered the constraint of unitarity for representations of the Virasoro algebra. They showed that for $h = 0$ all nontrivial unitary representations must have $c \geq \frac{1}{2}$.

[1] A. M. Polyakov, Pis'ma Zh. Eksp. Teor. Fiz. **12**, 538 (1970) [JETP Lett. **12**, 381 (1970)], and Zh. Eksp. Teor. Fiz. **66**, 23 (1974) [Sov. Phys. JETP **39**, 10 (1974)].

[2] A. B. Zamolodchikov, A. M. Polyakov, and V. S. Dotsenko, private communication. After our work was completed we received Refs. 11 and 12.

[3] M. A. Virasoro, Phys. Rev. D **1**, 2933 (1970); I. M. Gel'fand and D. B. Fuks, Funkts. Anal. Prilozhen. **2**, 92 (1968).

[4] D. Friedan, in "Common Trends in Field Theory and Statistical Mechanics," edited by J. B. Zuber (to be published).

[5] V. G. Kac, in Proceedings of the International Congress of Mathematicians, Helsinki, 1978 (unpublished), and in *Group Theoretical Methods in Physics*, edited by W. Beiglbock, A. Bohm, Lecture Notes in Physics Vol. 94 (Springer-Verlag, New York, 1979), p. 441; B. L. Feigin and D. B. Fuchs, Funkts. Anal. Prilozhen. **16**, 47 (1982) [Funct. Anal. Appl. **16**, 114 (1982)].

[6] D. Friedan, Z. Qiu, and S. H. Shenker, Enrico Fermi Institute, University of Chicago, Report No. EFI 83-66, 1983 (unpublished), and in Proceedings of the Mathematical Sciences Research Institute Workshop in Vertex Operators, edited by J. Lepowsky (Springer-Verlag, New York, to be published). A detailed version is in preparation for submission elsewhere.

[7] G. Segal, Commun. Math. Phys. **80**, 301 (1981); I. B. Frenkel and V. G. Kac, Invent. Math. **62**, 23 (1980).

[8] B. L. Feigin and D. B. Fuchs, unpublished; A. B. Zamolodchikov, private communication; L. P. Kadanoff and B. Nienhuis, unpublished.

[9] L. P. Kadanoff has confirmed our assignments of c using four-point functions of energy operators in the q-state Potts model calculated by himself and B. Nienhuis. He finds $c = (3t-1)(3-4t)/t$ where $q^{1/2} = 2\cos(\pi/2t)$.

[10] M. den Nijs, unpublished; D. A. Huse, to be published. We thank Daniel S. Fisher for bringing the latter reference to our attention.

[11] A. A. Belavin, A. M. Polyakov, and A. B. Zamolodchikov, in Proceedings of STATPHYS 15, Edinburgh, 25–29 July 1983 (to be published).

[12] V. S. Dotsenko, unpublished, and to be published.

[13] P. Ramond, Phys. Rev. D **3**, 2415 (1971); A. Neveu and J. H. Schwarz, Nucl. Phys. B **31**, 86 (1971), and Phys. Rev. D **4**, 1109 (1971).

[14] D. Friedan, Z. Qiu, and S. H. Shenker, to be published.

[15] M. B. Green and J. H. Schwarz, Phys. Lett. **109B**, 444 (1982).

[16] L. P. Kadanoff, Phys. Rev. Lett. **23**, 1430 (1969).

[17] A. M. Polyakov, Zh. Eksp. Teor. Fiz. **57**, 271 (1969) [Sov. Phys. JETP **30**, 151 (1970)].

Commun. Math. Phys. 107, 535–542 (1986)

Communications in Mathematical Physics

Details of the Non-Unitarity Proof for Highest Weight Representations of the Virasoro Algebra

Daniel Friedan[1], Zongan Qiu[2], and Stephen Shenker[1]

[1] Enrico Fermi and James Franck Institutes, and Department of Physics, University of Chicago, Chicago, Il 60637, USA

[2] Institute for Advanced Studies, Princeton, NJ 08540, USA

Abstract. We give an exposition of the details of the proof that all highest weight representations of the Virasoro algebra for $c<1$ which are not in the discrete series are non-unitary.

The Virasoro algebra is the infinite dimensional Lie algebra with generators L_n, $n \in \mathbf{Z}$, satisfying the commutation relations

$$[L_m, L_n] = (m-n)L_{m+n} + \tfrac{1}{12}c(m^3-m)\delta_{m+n,0}\,. \tag{1}$$

The number c is called the central charge. The Verma module $V(c,h)$ is the representation of the Virasoro algebra generated by a vector $|h\rangle$ satisfying

$$L_0|h\rangle = h|h\rangle\,, \qquad L_n|h\rangle = 0\,, \qquad n>0\,, \tag{2}$$

and spanned by the linearly independent vectors $|h\rangle$ and

$$L_{-k_1}L_{-k_2}\dots L_{-k_n}|h\rangle\,, \qquad 1 \leqq k_1 \leqq k_2 \leqq \dots \leqq k_n\,. \tag{3}$$

We assume that both c and h are real. In this case, a hermitian inner product on $V(c,h)$ is defined by $\langle h|h\rangle = 1$, and $L_n^\dagger = L_{-n}$. Define, for p and q positive integers,

$$c(m) = 1 - \frac{6}{m(m+1)}\,, \qquad h_{p,q}(m) = \frac{((m+1)p-mq)^2-1}{4m(m+1)}\,. \tag{4}$$

The non-unitary theorem [1] is

Theorem 1. *For $c<1$ there are negative metric states in $V(c,h)$ if (c,h) does not belong to the discrete list*

$$c = c(m)\,, \qquad m = 2,3,4\dots\,, \qquad h = h_{p,q}(m)\,, \qquad p+q \leqq m\,. \tag{5}$$

This work was supported in part by DOE grant DE-FG02-84ER-45144, NSF grant PHY-8451285 and the Sloan Foundation

The proof of Theorem 1 was given in [1]. The present paper is an exposition of the details of that proof. We recommend the graphs in [1] as a visual aid.

There are analogous non-unitarity theorems for the $N=1$ supersymmetric extensions of the Virasoro algebra [1, 2]. The details of the proofs of the $N=1$ non-unitarity theorems are exactly parallel to the proof of the Virasoro theorem. Goddard et al. [3] proved that all representations in the discrete series allowed by the non-unitarity theorems for the Virasoro algebra and its $N=1$ extensions are in fact unitary. Boucher et al. [4] have given the non-unitarity theorems for the $N=2$ extensions. The $N=2$ proofs [5] are somewhat different from the $N<2$ proofs. Di Vecchia, Petersen, Yu, and Zheng have proved that the discrete series of representations allowed by the $N=2$ non-unitarity theorems are in fact unitary [6].

For N a nonnegative integer, define *level N* to be the eigenspace of the Verma module on which L_0 has eigenvalue $h+N$. Level 0 is spanned by $|h\rangle$, and level N, $N \geqq 1$, is spanned by the vectors listed in (3) which satisfy $\sum k_i = N$. Level N has dimension $P(N)$, the partition number of N. Clearly, the levels span $V(c,h)$ and are linearly independent. Since $L_0^\dagger = L_0$, levels N and N' are orthogonal if $N \neq N'$. Define the null subspace on level N to be the subspace of vectors in level N which are orthogonal to all of level N, and thus to all of $V(c,h)$.

The inner products of the states on level N listed in (3) form a $P(N) \times P(N)$ real symmetric matrix $M_N(c,h)$ whose entries are polynomials in c and h. An explicit formula for the determinant of this matrix was announced by Kac [7] and proved by Feigin and Fuchs [8]. Up to multiplication by a positive number independent of c and h,

$$\det M_N(c,h) = \prod_{\substack{p,q \geqq 1 \\ pq \leqq N}} (h - h_{p,q}(m))^{P(N-pq)}, \tag{6}$$

where $h_{p,q}(m)$ is given by Eq. (4). In Eq. (6) it does not matter which branch is chosen for m as a function of c. For $c<1$ we choose by convention the branch $0<m<\infty$. There is a nontrivial null subspace on level N if and only if $\det M_N(c,h)=0$.

Kac [9] showed that, for $c \geqq 1$, the metric on $V(c,h)$ is nonnegative if and only if $h \geqq 0$. Direct calculation gives the 1×1 matrix $M_1 = 2h$, so $h \geqq 0$ is necessary if the metric is to be nonnegative. It is straightforward to verify that, in the limit $h \to +\infty$, M_N goes to a diagonal matrix with positive entries. It is also straightforward to check that $\det M_N(c,h) \neq 0$ for $c>1$, $h>0$. Therefore $M_N(c,h)$ is nondegenerate and positive for $c>1$, $h>0$, and is non-negative for $c \geqq 1$, $h \geqq 0$. Since this is true for all levels N, the result follows.

The proof of Theorem 1 is entirely elementary. The strategy is to consider the matrices M_N, $N=1,2,\ldots$, one by one. For each N we find a subset G_N of the half-plane $c<1$ on which $M_N(c,h)$ has a negative eigenvalue. We then say that the subset G_N has been *eliminated*. Theorem 1 will follow from the fact that the discrete set (5) is the complement of $\bigcup_N G_N$ in the half-plane $c<1$.

Henceforth we write $h_{p,q}(c)$ in place of $h_{p,q}(m)$, with the understanding that, for $c<1$, we choose the branch of m with $0<m<\infty$. Write $C_{p,q}$ for the vanishing curve $h=h_{p,q}(c)$. Because $\det M_N(c,h)$ vanishes on the curve $C_{p,q}$ for $pq \leqq N$, we say that the vanishing curve $C_{p,q}$ *first appears* on level pq, and that the vanishing curves on level N are the $C_{p,q}$, $pq \leqq N$. The curve $C_{p,q}$ intersects the line $c=1$ at $h=h_{p,q}(1) = (p-q)^2/4$. Orient each vanishing curve so that $c=1$ is the initial point, and forward is the direction of decreasing c.

Proposition 1. *When the curve $C_{p,1}$ first appears on level $N=p$, it intersects no other vanishing curves in the half-plane $c<1$. When $C_{p,q}$, $q>1$, first appears on level $N=pq$, its first intersection, moving forward from $c=1$, is with $C_{q-1,p}$ at $m=p+q-1$.*

Proof. The proof is straightforward algebra. □

For $q=1$ define $C'_{p,1}$ to be all of $C_{p,q}$ in the half-plane $c<1$. For $q>1$ define $C'_{p,q}$ to be the part of $C_{p,q}$ for which $m>p+q-1$. That is, $C'_{p,q}$ is the open subset of $C_{p,q}$ between $c=1$ and the first intersection of $C_{p,q}$ on level $N=pq$. The first step in the proof of Theorem 1 is to eliminate all of the half-plane $c<1$ except the curves $C'_{p,q}$. For $N \geqq 1$ define

$$S_N = \bigcup_{q<p,\, pq \leqq N} \{(c,h): c<1,\ h_{q,p}(c) \leqq h \leqq h_{p,q}(c)\} \bigcup_{p^2 \leqq N} \{(c,h): c<1,\ h \leqq h_{p,p}(c)\}\,. \tag{7}$$

Proposition 2. $\lim_{N\to\infty} S_N$ *is the half-plane $c<1$.*

Define a *first intersection* F on $C'_{p,q}$ to be an intersection of $C'_{p,q}$ and $C_{p',q'}$, $p'q'>pq$, such that, on level $N'=p'q'$, (c,h) is the first intersection encountered on $C'_{p,q}$, starting from $c=1$.

Proposition 3. *The first intersections on $C'_{p,q}$ are the intersections $F_{p,q,k}$ of $C'_{p,q}$ and $C_{p',q'}=C_{q+k-1,p+k}$, $k\geqq 1$. $F_{p,q,k}$ is the point $h=h_{p,q}(m)$, $m=p+q+k-1$. Each of these first intersections is, at level $p'q'$, the intersection of exactly two vanishing curves.*

Proof. The proof is straightforward algebra. □

It immediately follows that

Proposition 4. *The discrete list* (5) *consists exactly of the first intersections, on all the vanishing curves $C'_{p,q}$.*

Define $R_{1,1}$ to be the open quadrant $c<1$, $h<0$. Define $R_{p,1}=R_{1,p}$, for $p>1$, to be the open region bounded by $C'_{p,1}$, $C'_{p-1,1}$, and $C'_{1,p}$. For $p,q>1$, define $R_{p,q}$ to be the open region bounded by $C'_{p,q}$, $C'_{p-1,q-1}$, and $C'_{q-1,p}$.

Proposition 5. *No vanishing curves on level $N=pq$ intersect $R_{p,q}$.*

Proof. A vanishing curve which did intersect $R_{p,q}$ would have to intersect its boundary. By Proposition 3, this does not happen. □

Proposition 6. $S_N - S_{N-1} = \bigcup_{pq=N} R_{p,q} \bigcup_{pq=N} C'_{p,q}$.

Proposition 7. *Except possibly for the curves $C'_{p,q}$, $pq \leqq N$, all of S_N is eliminated on levels $\leqq N$.*

Proof. The proof is by induction in N. The proposition is clearly true for $N=1$, because S_1 is the quadrant $c<1$, $h\leqq 0$, and $C'_{1,1}$ is the line $h=0$, $c<1$. Now suppose the proposition is true for $N-1$. We show that it is also true for N. By Proposition 6, we need to show that the $R_{p,q}$, $pq=N$, are eliminated on level N.

We say that two connected regions of the (c, h) plane are *contiguous on level N* if they can be connected by a path which does not intersect any vanishing curves on level N. If two regions are contiguous on level N, then the signature of M_N is the same in both regions, because the signature can only change when a vanishing curve is crossed. For each $C_{p,q}$ on level N, for $pq\leqq N$, choose a neighborhood U of $C_{p,q}$ small enough so that the only other vanishing curves on level N which intersect U also intersect $C_{p,q}$. $U-C_{p,q}$ has two connected components. Define the $c>1$ *side* of $C_{p,q}$ to be the connected component on the right of $C_{p,q}$, moving forward, if $p\geqq q$, and on the left, moving forward, if $p<q$. The other component is called the $c>1$ *side* of $C_{p,q}$. The motivation for this terminology is that the $c>1$ side of $C_{p,q}$, for c near 1, is contiguous on level $N=pq$ with the region $c>1$, $h>0$. This is easily verified by expanding $h_{p,q}(c)$ around $c=1$. It follows that $M_N(c,h)$ is a positive matrix on the $c>1$ side of $C_{p,q}$ for c near 1. $\det M_N$ vanishes to first order on $C_{p,q}$. Therefore $\det M_N(c,h)$ is negative on the $c<1$ side of $C_{p,q}$, for c near but not at 1. The sign of $\det M_N(c,h)$ can only change at a vanishing curve, so $\det M_N(c,h)$ is negative in the entire region of the $c<1$ half-plane which is contiguous on level pq to the $c<1$ side of $C_{p,q}$ for c near but not at 1. By Proposition 5, this region is $R_{p,q}$. So the region $R_{p,q}$ is eliminated. The induction step now follows from Proposition 6. □

Given Propositions 2 and 7, we are left with the task of eliminating the intervals on the curves $C'_{p,q}$ in between the points in the discrete list (5). Let $I_{p,q,k}$, $k\geqq 2$, be the open interval on $C'_{p,q}$ between $F_{p,q,k-1}$ and $F_{p,q,k}$. Let $I_{p,q,1}$ be the open subset of $C'_{p,q}$ beyond $F_{p,q,1}$. That is, $I_{p,q,1}$ is the open subset of $C'_{p,q}$ with $m<p+q$. Clearly,

Proposition 8.

$$C'_{p,q}=\bigcup_{k\geqq 0} I_{p,q,k} \bigcup_{k\geqq 1} F_{p,q,k}\,. \tag{8}$$

The goal is to eliminate the open intervals $I_{p,q,k}$, $k\geqq 1$. Recall that, when $C_{p',q'}=C_{q+k-1,p+k}$ first appears on level $N'=p'q'$, there is a negative metric state on its $c<1$ side, near $c=1$. We will show that this negative metric state continues to exist on the $c<1$ side of $C_{p',q'}$ moving away from $c=1$, and in particular exists on $C'_{p,q}$ on the $c<1$ side of $C_{p',q'}$. That part of $C'_{p,q}$ is a subset of $I_{p,q,k}$, and, by the definition of first intersections, there are no intersections on $I_{p,q,k}$ at level N'. It will then follow that there is a negative metric state on all of $I_{p,q,k}$, and we will be done.

Proposition 9. *On level $N'=p'q'$, the first k successive intersections on $C_{p',q'}$, are with $C'_{p+k-j,q+k-j}$, $1\leqq j\leqq k$. These are the first intersections $F_{p+k-j,q+k-j,j}$ on $C'_{p+k-j,q+k-j}$, occurring at $m=p+q+2k-j-1$.*

Proof. The proof is straightforward algebra. □

Proposition 10. *Suppose (c', h') is on some $C_{p,q}$, $pq=N$, but is not on an intersection of vanishing curves at level N. Then the null space on level N is one dimensional at (c', h').*

Proof. $\det M_N(c,h)$ vanishes to first order at $C_{p,q}$ near (c', h'). □

Proposition 11. *At $F_{p,q,k}$, the intersection of $C'_{p,q}$ and $C_{p',q'}=C_{q-1+k,p+k}$, $k\geqq 1$, occurring at $c=c(m)$, $h=h_{p,q}(c)$, $m=p+q+k-1$,*

$$\det M_{p'q'-pq}(c, h+pq)\neq 0. \tag{9}$$

Proof. If this determinant were zero, then $(c, h+pq)$ would be on a vanishing curve $C_{r,s}$ on level $rs=p'q'-pq$. Direct calculation of $p'q'-pq$ gives

$$rs=m(m+1)-(m+1)p-mq. \tag{10}$$

The condition that $(c, h+pq)$ lie on $C_{r,s}$ is

$$(m+1)p+mq=\pm((m+1)r-ms). \tag{11}$$

It follows from Eqs. (10, 11) that $r=m$ or $s=m+1$. But this gives a contradiction if we take Eq. (10) mod$m$ or mod$m+1$, since $1\leqq p<m$ and $1\leqq q<m+1$. □

Proposition 12. *For $j=1,2,\ldots,k$ there exists an open neighborhood $U_{p',q',j}$ of*

$$F_{p+k-j,q+k-j,j}=F_{q'-j,p'+1-j,j},$$

and a nowhere zero analytic function $v_j(c,h)$, defined on $U_{p',q',j}$ with values in level $N'=p'q'$ of $V(c,h)$, such that $v_j(c,h)$ is in the null space of level N' if and only if (c,h) is on $C_{p',q'}$.

Proof. Write $p''=p+k-j$, $q''=q+k-j$, $N''=p''q''<N'$. Let $U=U_{p',q',j}$ be a neighborhood of $F_{p+k-j,q+k-j,j}$ small enough that it intersects no vanishing curves but $C'_{p'',q''}$ and $C_{p',q'}$ on level N'. Choose coordinates (x,y) in U, analytic in (c,h) and real for c,h real, such that $C'_{p'',q''}$ is given by $x=0$ and $C_{p',q'}$ is given by $y=0$. This is possible because the intersection is transversal. At level N'', $x=0$ is the only vanishing curve in U. The one dimensional null spaces of level N'' form a line bundle over the vanishing curve $x=0$ near $y=0$. Let $v''_j(0,y)$ be a nowhere zero analytic section of this line bundle, and let $v''_j(x,y)$ be an analytic function on U with values in level N'', which extends this section. Define the subspace $V''(x,y)$ of level N' to be the span of the vectors

$$L_{-k_1}L_{-k_2}\ldots L_{-k_n}v''_j(x,y), \quad 1\leqq k_1\leqq k_2\leqq\ldots\leqq k_n, \quad \sum k_i=N'-N''. \tag{12}$$

The dimension of $V''(x,y)$ is $P(N'-N'')$. For $y\neq 0$, the order of vanishing of $\det M_{N'}(x,y)$ at $x=0$ is also $P(N'-N'')$. Therefore, for $y\neq 0$, $V''(0,y)$ is the null subspace of level N'. Let $V'(x,y)$ be a subspace of level N' complementary to $V''(x,y)$, so level N' is $V''\oplus V'$. The matrix of inner products $M_{N'}$ can now be written in block diagonal form:

$$M_{N'}(x,y)=\begin{pmatrix} xQ(x,y) & xR(x,y) \\ xR(x,y)^t & S(x,y)\end{pmatrix}, \tag{13}$$

where Q and S are symmetric matrices. Three blocks of $M_{N'}(x, y)$ are divisible by x, as in Eq. (13), because $V''(0, y)$ is in the null subspace of level N'.

The key point now is that $Q(0,0)$ is non-degenerate. To see this, first note that, for $n>0$, the vector $L_n v_j''(0, y)=0$, since $L_n v_j''(0, y)$ is in the null subspace of level $N'-n$, which is trivial. From this, and from the explicit basis (12) for $V''(x, y)$, we see that

$$Q(x, y)=M_{p'q'-p''q''}(c, h+p''q'')+O(x), \tag{14}$$

where (c, h) corresponds to $(0, y)$ under the change of coordinates. Since $(0,0)$ is the first intersection $F_{p'',q'',j}$, Proposition (11) gives $\det Q(0,0)\neq 0$.

Since $\det Q(0,0)\neq 0$, $Q(x, y)$ is non-degenerate on all of U, if necessary replacing U by a smaller neighborhood of $(0,0)$. Let W be the matrix

$$\begin{pmatrix} 1 & -Q^{-1}R \\ 0 & 1 \end{pmatrix}, \tag{15}$$

and make the change of basis

$$M_{N'}\to W^t M_{N'} W=\begin{pmatrix} xQ(x, y) & 0 \\ 0 & T(x, y) \end{pmatrix}. \tag{16}$$

Let $V'''(x, y)$ be the new complement to $V''(x, y)$, on which $T(x, y)$ is the inner product. The order of vanishing argument implies that $\det T(x, y)$ is nonzero for $y\neq 0$ and vanishes to first order at $y=0$. The one dimensional null space of $T(x, 0)$ is the null space of level N' for $x\neq 0$. At $x=y=0$, the one dimensional null space of $T(0, 0)$ spans, with $V''(x, y)$, the $P(N')-P(N'')+1$ dimensional null subspace of level N'. By the same argument which gave $v_j''(x, y)$, we can choose a nowhere zero analytic function $v_j(x, y)$ on U, with values in $V'''(x, y)$, such that $v_j(x, 0)$ is in the null space of $T(x, 0)$ and therefore in the null space of level N'. Since $T(x, y)$ is non-degenerate for $y\neq 0$, $v_j(x, y)$ is not in the null space of level N' if $y\neq 0$. □

Let $J_{p',q',j}$, $1<j\leq k$, be the open interval on $C_{p',q'}$ between

$$F_{p+k-j,q+k-j,j} \quad \text{and} \quad F_{p+k-j-1,q+k-j-1,j+1}\,.$$

Let $J_{p',q',1}$ be the open interval on $C_{p',q'}$ lying between $c=1$ and $F_{p+k-1,q+k-1,1}$. Let $W_{p',q',j}$, $1\leq j\leq k$, be a neighborhood in the plane which intersects no vanishing curves on level N' except $J_{p',q',j}$. For $j>1$, require

$$\begin{gathered} J_{p',q',j}\subset U_{p',q',j-1}\cup W_{p',q',j}\cup U_{p',q',j}, \\ W_{p',q',j}\cap U_{p',q',j}\neq\emptyset, \qquad W_{p',q',j}\cap U_{p',q',j-1}\neq\emptyset. \end{gathered} \tag{17}$$

For $j=1$ require only

$$W_{p',q',1}\cap U_{p',q',1}\neq\emptyset. \tag{18}$$

Proposition 13. *For each j, $1\leq j\leq k$, there is a nowhere zero analytic function $w_j(c, h)$ on $W_{p',q',j}$ with values in level N' such that $w_j(c, h)$ is in the null space of level N' if and only if (c, h) is on $J_{p',q',j}$. On the intersections of their neighborhoods of definition,*

$w_j = f_j v_j$, where f_j is a nonzero function, and $w_j = g_j v_{j-1}$, where g_j is a nonzero function.

Proof. Again, the null space of level N' is trivial on $W_{p',q',j}$ except on $J_{p',q',j}$, where it is one dimensional. □

Proposition 14. *The level N' metric is negative on the vectors $v_{p',q',j}(c,h)$ and on the vectors $w_{p',q',j}(c,h)$, on the $c<1$ side of $C_{p',q'}$.*

Proof. The matrix $M_{N'}$ is positive in $W_{p',q',1}$ on the $c>1$ side of $C_{p',q'}$, by the contiguity argument, since there are no intersections on $C_{p',q'}$ between $W_{p',q',1}$ and $c=1$. The inner product is thus positive on $w_{p',q',1}$ on the $c>1$ side of $C_{p',q'}$. The inner product vanishes to first order on $w_{p',q',1}$ on $C_{p',q'}$. Therefore the inner product is negative on $w_{p',q',1}$ on the $c<1$ side of $C_{p',q'}$. The proposition now follows by induction on the series $w_1, v_1, w_2, v_2, \ldots$, since neighboring vectors in the series differ by nonzero functions f_j or g_j, and since the $w_j(c,h)$ and $v_j(c,h)$ are in the level N' null space only for (c,h) on $C_{p',q'}$. □

Proposition 15. *$I_{p,q,k}$ is eliminated on level $N'=(q+k-1)(p+k)$.*

Proof. By the previous proposition, the metric is negative on $v_{p',q',k}(c,h)$, on the $c<1$ side of $C_{p',q'}$. But $I_{p,q,k}$ approaches arbitrarily close to $C_{p',q'}$ on the $c<1$ side within $U_{p',q',k}$. Therefore $M_N(c,h)$ has a negative eigenvalue at one end of $I_{p,q,k}$. But the signature of $M_N(c,h)$ cannot change along $I_{p,q,k}$, because there are no intersections at level N' on $I_{p,q,k}$. The proposition follows. □

Propositions 2, 7, 8, and 15 imply Theorem 1.

Acknowledgement. We thank Adrian Kent for a critical reading of the manuscript.

Note added in proof. A similar but not identical version of the details of the non-unitarity proof has been given by Langlands [10].

References

1. Friedan, D., Qiu, Z., Shenker, S.: In: Vertex operators in mathematics and physics. Lepowsky, J. et al. (eds.), Berlin, Heidelberg, New York: Springer 1984, p. 419; Conformal invariance, unitarity and critical exponents in two dimensions. Phys. Rev. Lett. **52**, 1575 (1984)
2. Friedan, D., Qiu, Z., Shenker, S.: Superconformal invariance in two dimensions and the tricritical Ising model. Phys. Lett. **151B**, 37 (1985)
3. Goddard, P., Olive, D.: Kac-Moody algebras, conformal symmetry and critical exponents. Nucl. Phys. B **257**, 83 (1985); Goddard, P., Kent, A., Olive, D.: Virasoro algebras and coset space models. Phys. Lett. **152B**, 88 (1985); Unitary representations of the Virasoro and super-Virasoro algebras. Commun. Math. Phys. **103**, 105 (1986)
4. Boucher, W., Friedan, D., Kent, A.: Determinant formulae and unitarity for the $N=2$ superconformal algebras in two dimensions or exact results in string compactification. Phys. Lett. **172B**, 316 (1986)
5. Boucher, W., Friedan, D., Kent, A., Shenker, S.: In preparation
6. Di Vecchia, P., Petersen, J.L., Yu, M., Zheng, H.B.: Explicit construction of unitary representations of the $N=2$ superconformal algebra. Phys. Lett. **174B**, 280 (1986)

7. Kac, V.: Lecture Notes in Physics, Vol. 94, p. 441. Berlin, Heidelberg, New York: Springer 1979; in Proc. Int. Congress of Math., Helsinki (1978)
8. Feigin, B.L., Fuchs, D.B.: Funct. Anal. Appl. **16**, 114 (1982)
9. Kac, V.: Lecture Notes in Mathematics, Vol. 933, p. 117. Berlin, Heidelberg, New York: Springer 1982
10. Langlands, R.P.: On unitary representations of the Virasoro algebra. In: Proceedings of the Montreal Workshop on infinite dimensional Lie algebras and their applications. Kass, S. (ed.) (to appear)

Communicated by A. Jaffe

Received May 23, 1986

Commun. Math. Phys. 103, 105–119 (1986)

Communications in Mathematical Physics

Unitary Representations of the Virasoro and Super-Virasoro Algebras

P. Goddard[1], A. Kent[1,*], and D. Olive[2]

[1] Department of Applied Mathematics and Theoretical Physics, University of Cambridge, Silver Street, Cambridge CB3 9EW, U.K.

[2] Blackett Laboratory, Imperial College, London SW7 2BZ, U.K.

Abstract. It is shown that a method previously given for constructing representations of the Virasoro algebra out of representations of affine Kac-Moody algebras yields the full discrete series of highest weight irreducible representations of the Virasoro algebra. The corresponding method for the super-Virasoro algebras (i.e. the Neveu-Schwarz and Ramond algebras) is described in detail and shown to yield the full discrete series of irreducible highest weight representations.

1. Introduction

In a recent letter [1] we described a method for constructing representations of the Virasoro algebra out of representations of affine Kac-Moody algebras. The Virasoro algebra occurs as the algebra of the conformal group in one dimension, or, in the form of two commuting copies, in two dimensions. Thus it is of importance in physical contexts where two-dimensional conformal invariance plays a crucial rôle, such as string theories or the behaviour at critical points of two-dimensional statistical systems [2, 3]. The Virasoro algebra is defined by the commutation relations

$$[L_m, L_n] = (m-n)L_{m+n} + \frac{c}{12}m(m^2-1)\delta_{m,-n}, \qquad m, n \in \mathbb{Z}, \tag{1.1}$$

where c is a central element, i.e. $[L_n, c] = 0$, so that c is assigned a numerical value in any irreducible representation. In this paper we shall be concerned with unitary representations of this algebra, that is representations satisfying the hermiticity conditions,

$$L_n^\dagger = L_{-n}, \tag{1.2}$$

and, more particularly, highest weight representations, that is ones in which all the

* Present address: Enrico Fermi Institute, University of Chicago, Chicago, IL 60637, USA

states can be generated from a highest-weight state $|h\rangle$ satisfying,

$$L_n|h\rangle = 0, \qquad n > 0, \tag{1.3}$$

$$L_0|h\rangle = h|h\rangle. \tag{1.4}$$

An irreducible highest-weight representation is specified by the pair of numbers (c, h). It is easy to show that [3], in this case, unitarity requires

$$c \geqq 0 \quad \text{and} \quad h \geqq 0. \tag{1.5}$$

Unitary representations exist for all values of (c, h) with $c \geqq 1$ and $h \geqq 0$, but Friedan, Qiu, and Shenker (FQS) [3] showed that the only values of (c, h) with $0 \leqq c < 1$ which might correspond to unitary representations are

$$c = 1 - \frac{6}{m(m+1)}, \qquad m = 2, 3, \ldots, \tag{1.6}$$

and

$$\begin{aligned} h = h_{p,q}(c) &\equiv \frac{[(m+1)p - mq]^2 - 1}{4m(m+1)}, \\ p = 1, 2, \ldots, m-1; &\qquad q = 1, \ldots, p. \end{aligned} \tag{1.7}$$

[We could extend the range of q up to m but this would only repeat one of the $\frac{1}{2}m(m-1)$ values above because the substitution $p \to m - p$ and $q \to m + 1 - q$ leaves $h_{p,q}(c)$ unchanged.]

Our construction [1], which generalises an earlier approach [4], is based on the affine Kac-Moody algebras, $\hat{g}$, $\hat{h}$, associated with a compact Lie group G and a subgroup H. For each unitary representation of $\hat{g}$, and induced representation of $\hat{h}$, we obtain a representation of the Virasoro algebra. All the values of c in the discrete series (1.6) can be obtained either with $G = \mathrm{Sp}(m-1)$, $H = \mathrm{Sp}(m-2) \times \mathrm{Sp}(1)$ or with $G = \mathrm{SU}(2) \times \mathrm{SU}(2)$ and H being the diagonal SU(2) subgroup (and using suitable representations of $\hat{g}$ in each case). These constructions should be thought of as complementary. We shall use the second point of view here, and show that it gives all the values of h given by (1.7). This demonstrates that the values of (c, h) listed by FQS do indeed correspond to unitary representations.

Friedan et al. [3, 5] also analysed the representations of the two supersymmetric extensions of the Virasoro algebra, the Ramond [6], and Neveu and Schwarz [7] algebras, defined by (1.1) together with

$$[L_m, G_r] = \left(\frac{m}{2} - r\right) G_{m+r}, \tag{1.8a}$$

$$\{G_r, G_s\} = 2L_{r+s} + \frac{c}{3}\left(r^2 - \frac{1}{4}\right)\delta_{r,-s}, \tag{1.8b}$$

where $m \in \mathbb{Z}$ and *either* $r, s \in \mathbb{Z}$ [Ramond case] *or* $r, s \in \mathbb{Z} + \frac{1}{2}$ [Neveu-Schwarz case]. They found that the only possible unitary highest weight representations, i.e.

representations generated from a state $|h\rangle$ satisfying (1.3), (1.4) and

$$G_r|h\rangle = 0, \quad r>0, \tag{1.9}$$

are characterised by (c,h) where either $c \geqq \frac{3}{2}$, $h \geqq 0$ or

$$c = \frac{3}{2}\left[1 - \frac{8}{m(m+2)}\right], \quad m = 2, 3, \ldots \tag{1.10}$$

and

$$h = h_{p,q}(c) \equiv \frac{[(m+2)p - mq]^2 - 4}{8m(m+2)} + \frac{\varepsilon}{8}, \tag{1.11}$$

where $p = 1, 2, \ldots, m-1$ and $q = 1, 2, \ldots, m+1$. Here $p-q$ even or odd corresponds to the Neveu-Schwarz and Ramond cases respectively, with $\varepsilon = 0$ or $\frac{1}{2}$ correspondingly in Eq. (1.11).

A construction giving all the values of c in the discrete series (1.10) was sketched in [1]. It is described in more detail in Sect. 3. In Sect. 4 we demonstrate that it too produces all the corresponding values of h in Eq. (1.11), completing the classification of unitary highest weight representations of the Neveu-Schwarz and Ramond algebras.

2. Unitary Representation of the Virasoro Algebra

The methods of constructing representations of Virasoro algebras given in [1, 4] start from the Virasoro algebras one can construct from affine Kac-Moody algebras. If g is a simple Lie algebra,

$$[T^a, T^b] = if^{abc}T^c, \tag{2.1}$$

written in a basis in which the structure constants f^{abc} are totally antisymmetric, the associated affine Kac-Moody algebra $\hat{g}$ takes the form

$$[T^a_m, T^b_n] = if^{abc}T^c_{m+n} + km\delta^{ab}\delta_{m,-n}, \tag{2.2}$$

$m, n \in \mathbb{Z}$, where k is a central element (and so is assigned a numerical value in any irreducible representation). We shall be concerned with unitary highest weight representations of $\hat{g}$ also, that is representations in a positive definite Hilbert space satisfying

$$T^{a\dagger}_n = T^a_{-n}, \tag{2.3}$$

and with the representation space generated from vacuum states Ψ satisfying

$$T^a_n\Psi = 0, \quad n>0. \tag{2.4}$$

The central element k is quantised in multiples of $\frac{1}{2}\psi^2$, where ψ is a long root of g. The integer $2k/\psi^2$ is called the level.

From $\hat{g}$ can be constructed [8] a Virasoro algebra L^g_n defined by

$$L^g_n = \frac{1}{2\beta}\sum_{a,m} {}^\circ_\circ T^a_{m+n}T^a_{-m} {}^\circ_\circ, \tag{2.5}$$

where the normal ordering operation is defined by

$$\overset{\circ}{\circ} T_n^a T_{-n}^a \overset{\circ}{\circ} = T_{-n}^a T_n^a \quad \text{if} \quad n \geqq 0, \tag{2.6}$$

and

$$\beta = k + \tfrac{1}{2} c_\psi^g. \tag{2.7}$$

Here c_ψ^g denotes the quadratic Casimir operator for the adjoint representation of g,

$$f^{abc} f^{abd} = c_\psi^g \delta^{cd}. \tag{2.8}$$

The L_n^g satisfy (1.1) with c taking the value

$$c^g = k \dim g / \beta = \frac{2k \dim g}{c_\psi^g + 2k}. \tag{2.9}$$

Clearly if we start with a unitary representation of $\hat{g}$, this provides a unitary representation of the Virasoro algebra.

If g is semisimple there is a central element k_i in $\hat{g}$ for each simple factor g_i, $1 \leqq i \leqq N$, of g. Then

$$L_n^g = \sum_{i=1}^{N} L_n^{g_i} \tag{2.10}$$

is a Virasoro algebra with central element

$$c^g = \sum_{i=1}^{N} c^{g_i} = \sum_{i=1}^{N} \frac{2k_i \dim g_i}{c_\psi^{g_i} + 2k_i}. \tag{2.11}$$

The main idea in [1] was to consider not only an algebra g but also a subalgebra $h \subset g$. We label the basis for g so that the first $\dim h$ generators form a basis for h. In this situation we have two Virasoro algebras L_m^g, L_n^h and we can consider their difference

$$K_n = L_n^g - L_n^h, \tag{2.12}$$

which satisfies the Virasoro algebra with central element

$$c = c^g - c^h, \tag{2.13}$$

and commutes with $\hat{h}$,

$$[K_m, T_n^a] = 0, \qquad 1 \leqq a \leqq \dim h, \tag{2.14}$$

$m, n \in \mathbb{Z}$. It is not difficult to see that a unitary highest weight representation of $\hat{g}$ provides a unitary highest weight representation of K_n and so, necessarily,

$$c^g \geqq c^h. \tag{2.15}$$

Let us now consider what are the possibilities for highest weight irreducible unitary representations of $\hat{g}$. (For reviews see [9–11].) In such a representation the vacuum states Ψ satisfying (2.4) form a finite-dimensional irreducible representation of $g \cong \{T_0^a\}$ whose highest weight λ must satisfy

$$|\alpha \cdot \lambda| \leqq k \tag{2.16}$$

for all roots α of g, which is assumed simple for the moment. This taken, together with the condition $2k/\psi^2 \in \mathbb{Z}$, is a necessary and sufficient condition for the existence of an irreducible unitary representation of $\hat{g}$. This representation can be labelled by $(2k/\psi^2, \lambda)$. If g is not simple, the irreducible representations of $\hat{g}$ can be constructed from those of $\hat{g}_i$, where the g_i are the factors of g.

In [1] it was remarked that the values of c in the discrete series (1.6) could be obtained by taking $g = \mathrm{su}(2) \oplus \mathrm{su}(2)$ and h to be the diagonal su(2) subalgebra. The su(2) (affine untwisted) Kac-Moody algebra can be written in the form

$$[T_m^a, T_n^b] = i\varepsilon_{abc} T_{m+n}^c + \frac{N}{2} m \delta^{ab} \delta_{m,-n}, \tag{2.17}$$

and then $\psi^2 = 1$. It follows that $N \in \mathbb{Z}$ for a unitary highest weight representation and is the level of that representation. Such representations are labelled by (N, l) where l is the highest helicity (i.e. the largest eigenvalue of $T = T_0^3$) for a vacuum state Ψ. We shall call this the level N, spin l representation; condition (2.16) implies that $0 \leq 2l \leq N$. A level N representation of $\hat{\mathrm{su}}(2)$ gives rise to a Virasoro algebra with central element

$$c = \frac{3N}{N+2}. \tag{2.18}$$

If we construct a representation of $\hat{g}$ for $g = \mathrm{su}(2) \oplus \mathrm{su}(2)$ by taking a level N representation of the first factor and a level 1 representation of the second factor, we obtain a level $N+1$ representation of the diagonal $h = \mathrm{su}(2)$. Thus K_n given by Eq. (2.12) has a value of c, given by (2.13),

$$c = 1 - \frac{6}{(N+2)(N+3)}. \tag{2.19}$$

Taking $N = 0, 1, \ldots$ gives the sequence of Eq. (1.6).

Our aim now is to show that from such representations of K_n we can obtain all the values of h given by Eq. (1.7). To do this we decompose the level $(N, 1)$ representation of $\hat{g}$ with respect to $\hat{h} \times V$, where V denotes the Virasoro algebra $\{K_n\}$. We shall show that the representation obtained by taking the (N, l) representation of the first su(2) factor and the $(1, \varepsilon)$, $\varepsilon = 0$ or $\frac{1}{2}$, representation of the second su(2) factor decomposes into the direct sum of representations $(N', l') \times (c, h)$ of $\hat{h} \times V$,

$$\bigoplus_q (N+1, \tfrac{1}{2}[q-1]) \times (c, h_{p,q}(c)), \tag{2.20}$$

where c is given by Eq. (2.19); $p = 2l + 1$, so that $1 \leq p \leq N+1$, and the sum is taken over q such that $p - q$ is even or odd, depending on whether $\varepsilon = 0$ or $\frac{1}{2}$, and $1 \leq q \leq N+2$. This gives all the values required. Note that this implies that the representation $(N, l) \times (1, \varepsilon)$ of $\hat{g}$ is finitely and simply reducible in terms of $\hat{h} \times V$.

To establish (2.20) we introduce characters for the algebras we are studying, though it turns out that we shall only need to make a very limited appeal to the theory of characters. Essentially all we need are the Kac-Moody formula [9] and a formula for the characters of the Virasoro algebra [12], but we shall explain these as we develop the argument.

We define the character $\chi_{N,l}$ for the level N spin l representation of the su(2) Kac-Moody algebra by

$$\chi_{N,l}(z,\theta)=\mathrm{tr}(z^{L_0}e^{i\theta T_0^3}). \tag{2.21}$$

In the level 1 case, where $\varepsilon=0,\frac{1}{2}$, there is a rather explicit formula for $\chi_{1,\varepsilon}$ which can be deduced from the vertex operator construction of these representations [13, 14]. An account of this and its relation to string theory is given in [15]. Because these representations can be constructed irreducibly in a Fock space defined by the annihilation and creation operators α_m acting on momentum eigenstates $|\gamma\rangle$, $\gamma/\sqrt{2}\in\mathbb{Z}+\varepsilon$, where

$$[\alpha_m,\alpha_n]=m\delta_{m,-n}, \qquad m,n\in\mathbb{Z}, \tag{2.22a}$$

$$\alpha_m^\dagger=\alpha_{-m}, \qquad \alpha_m|\gamma\rangle=0, \qquad m>0, \tag{2.22b}$$

$$p|\gamma\rangle=\gamma|\gamma\rangle, \qquad \alpha_0\equiv p, \tag{2.22c}$$

with

$$L_0=\tfrac{1}{2}p^2+\sum_{n>0}\alpha_{-n}\alpha_n, \qquad T=p/\sqrt{2}, \tag{2.22d}$$

we have

$$\chi_{1,\varepsilon}(z,\theta)=\sum_{m\in\mathbb{Z}+\varepsilon} z^{m^2}e^{im\theta}\prod_{n=1}^{\infty}(1-z^n)^{-1}. \tag{2.23}$$

To obtain the characters for other representations of the su(2) Kac-Moody algebra we need to resort to the Kac-Weyl formula, of which some explanation is given in Appendix A. In the su(2) case this gives the expression

$$\chi_{N,l}(z,\theta)=\Delta_{N,l}(z,\theta)\prod_{n=1}^{\infty}(1-z^n)^{-1}(1-z^ne^{i\theta})^{-1}(1-z^{n-1}e^{-i\theta})^{-1}, \tag{2.24}$$

where the numerator

$$\Delta_{N,l}(z,\theta)=z^{l(l+1)/\lambda}\sum_{n\in\mathbb{Z}} z^{\lambda n^2+(2l+1)n}\{e^{i(l+\lambda n)\theta}-e^{-i(l+1+\lambda n)\theta}\} \tag{2.25}$$

with

$$\lambda=N+2. \tag{2.26}$$

We can regard Eq. (2.25) as defining a function for all values of N and l. With this extension, it is straightforward to establish directly the symmetry properties

$$\Delta_{N,l+\lambda}(z,\theta)=\Delta_{N,\lambda}(z,\theta), \tag{2.27}$$

$$\Delta_{N,-l-1}(z,\theta)=-\Delta_{N,l}(z,\theta), \tag{2.28}$$

which follow from the symmetry of characters under the action of the Weyl group of $\hat{g}$.

The other character formulae we need are those for representations of the Virasoro algebra. The character of the representation (c,h) is defined by the trace

$$\chi^V_{c,h}(z)=\mathrm{tr}(z^{L_0}). \tag{2.29}$$

For the discrete series of Eq. (1.6), (1.7) these have been given by Rocha-Caridi [12],

$$\chi^V_{c,h}(z)=\Delta^m_{p,q}(z)\prod_{n=1}^{\infty}(1-z^n)^{-1}, \tag{2.30}$$

where

$$\Delta^m_{p,q}(z)=\sum_{n\in\mathbb{Z}}\{z^{\alpha^m_{p,q}(n)}-z^{\beta^m_{p,q}(n)}\}, \tag{2.31}$$

with

$$\alpha^m_{p,q}(n)=\frac{[2m(n+1)n-qm+p(m+1)]^2-1}{4m(m+1)}, \tag{2.32a}$$

$$\beta^m_{p,q}(n)=\frac{[2m(m+1)n+qm+p(m+1)]^2-1}{4m(m+1)}. \tag{2.32b}$$

These expressions follow from the general results of Feigin and Fuchs [16] on the structure of Virasoro algebra representations. For a proof and discussion of these results see [17].

Let us consider what (2.20) is equivalent to in terms of characters. Because the T for $\hat{h}$ is the sum of the T's for the two su(2) factors in $\hat{g}$, and because

$$L^1_0+L^2_0=L^h_0+K_0, \tag{2.33}$$

where L^i_n, $i=1,2$, denotes the Virasoro algebra for the two su(2) factors of $\hat{g}$, the decomposition (2.20) implies

$$\chi_{N,\ell}(z,\theta)\chi_{1,\varepsilon}(z,\theta)=\sum_q\chi_{N+1,\frac{1}{2}(q-1)}(z,\theta)\chi^V_{c,h}(z) \tag{2.34}$$

with c, h, p, q specified as in Eq. (2.20). On the other hand, if Eq. (2.34) holds we can use it to decompose the given representation of $\hat{g}$ into irreducible representations of $\hat{h}\times V$, because from it we can successively isolate highest weight states for the algebra $\hat{h}\times V$, that is, states Ψ^{jm}_h satisfying

$$T^i_n\Psi^{jm}_h=0, \quad K_n\Psi^{jm}_h=0, \quad n>0, \tag{2.35a}$$

$$K_0\Psi^{jm}_h=h\Psi^{jm}_h, \tag{2.35b}$$

$$T^3_0\Psi^{jm}_h=m\Psi^{jm}_h, \quad (T^i_0)^2\Psi^{jm}_k=j(j+1)\Psi^{jm}_h. \tag{2.35c}$$

To do this we look for the lowest powers of z on the right-hand side of (2.34). These will be a sum of terms of the form

$$z^{h+\frac{1}{4}L(L+1)}\frac{\sin(L+\frac{1}{2})\theta}{\sin\frac{1}{2}\theta}, \tag{2.36}$$

where $L=\frac{1}{2}(q-1)$. Such a term indicates the presence of highest weight states Ψ^{jm}_h with $j=L$ and a representation $(N+1,L)\times(c,h)$ of $\hat{h}\times V$. Removing the states of this representation corresponds to subtracting the term

$$\chi_{N+1,\frac{1}{2}(q-1)}(z,\theta)\chi^V_{c,h}(z) \tag{2.37}$$

from (2.34). Proceeding inductively in this way one establishes (2.20).

It remains to prove (2.34) using the expressions (2.23)–(2.25) and (2.30)–(2.32). Because the denominators match, it is equivalent to

$$\left\{\sum_{m\in\mathbb{Z}+\varepsilon} z^{m^2}e^{im\theta}\right\}\Delta_{N,l}(z,\theta)=\sum_q \Delta_{N+1,\frac{1}{2}(q-1)}(z,\theta)\Delta_{p,q}^{N+2}(z)\,. \tag{2.38}$$

Our strategy in proving this is to rewrite the left-hand side of (2.38) as a sum over functions $\Delta_{N+1,j}(z,\theta)$, multiplied by functions of z only, and then use the symmetry properties (2.27), (2.28) to bring j into the appropriate range that $\frac{1}{2}(q-1)$ goes over. Using (2.25), this left-hand side equals

$$z^{l(l+1)/\lambda}\sum_{m\in\mathbb{Z}+\varepsilon}\sum_{n\in\mathbb{Z}} z^{\lambda n^2+(2l+1)n+m^2}\{e^{i(l+\lambda n+m)\theta}-e^{-i(l+1+\lambda n+m)\theta}\}\,. \tag{2.39}$$

Now put $m'=l+m-n$ so that $m'\in\mathbb{Z}+\varepsilon'$, where $\varepsilon'=0$ or $\frac{1}{2}$ as $l+\varepsilon\in\mathbb{Z}$ or $\mathbb{Z}+\frac{1}{2}$. Then

$$l+m+\lambda n=m'+(\lambda+1)n \tag{2.40a}$$

and

$$\lambda n^2+(2l+1)n+m^2=(\lambda+1)n^2+(2m'+1)n+(m'-l)^2\,. \tag{2.40b}$$

Thus

$$\left\{\sum_{m\in\mathbb{Z}+\varepsilon} z^{m^2}e^{im\theta}\right\}\Delta_{N,l}(z,\theta)=\sum_{m'\in\mathbb{Z}+\varepsilon'} z^{A}\Delta_{N+1,m'}(z,\theta)\,, \tag{2.41}$$

where

$$A=l(l+1)/\lambda-m'(m'+1)/(\lambda+1)+(m'-l)^2\,. \tag{2.42}$$

Now put

$$m'=-(\lambda+1)M+l'\,, \tag{2.43}$$

where $M\in\mathbb{Z}$ and $0\leqq l'\leqq\lambda$. Also put

$$p=2l+1\,,\qquad q=2l'+1 \tag{2.44}$$

so that $p,q\in\mathbb{Z}$, $1\leqq q\leqq 2\lambda+1$, $1\leqq p\leqq\lambda-1$ and $p-q$ is even or odd as $\varepsilon=0$ or $\frac{1}{2}$. Then

$$A=\alpha^{\lambda}_{p,q}(M)\,, \tag{2.45}$$

as defined by Eq. (2.32a). Using the periodicity property (2.27), the character becomes

$$\sum_{1\leqq q\leqq 2\lambda+1}\Delta_{N+1,l'}(z,\theta)\left\{\sum_M z^{\alpha^{\lambda}_{p,q}(M)}\right\}. \tag{2.46}$$

If we use the reflection property (2.28), substituting $(\lambda+1)$ for λ here, together with the relation

$$\alpha^{\lambda}_{p,2\lambda+2-q}(M)=\beta^{\lambda}_{p,q}(M-1)\,, \tag{2.47}$$

we obtain the right-hand side of Eq. (2.30), thus completing our proof.

3. Construction of the Super-Virasoro Algebras

In this section we describe in more detail the construction, sketched in a special case in [1], of operators satisfying the super-Virasoro algebra of Eqs. (1.1) and (1.8). The framework within which we work is again that of the Kac-Moody algebras $\hat{g}, \hat{h}$ associated with the Lie algebras $g \supset h$. We shall take g to have the form $g = h_T \oplus h_v$, where h_T and h_v are isomorphic, and h is their diagonal subalgebra, i.e. the form used in the last section with $h = \mathrm{su}(2)$ to obtain all the discrete series representations of the Virasoro algebra.

The representations of $\hat{h}_v$ that we shall use will be defined in fermionic Fock spaces [18] built up from fermionic fields in the adjoint representation of h; for $\hat{h}_T$ we can use any highest weight unitary representation. In this situation, the Virasoro algebra

$$K_n = L_n^g - L_n^h \tag{3.1}$$

can be extended by fermionic generators G_r, satisfying

$$[K_n, G_r] = \left(\frac{n}{2} - r\right) G_{n+r} \tag{3.2}$$

and

$$\{G_r, G_s\} = 2K_{r+s} + \frac{c}{3}(r^2 - \tfrac{1}{4})\delta_{r,-s}, \tag{3.3}$$

where *either* $r, s \in \mathbb{Z}$ *or* $r, s \in \mathbb{Z} + \frac{1}{2}$, consistently, thus giving the Ramond and Neveu-Schwarz algebras respectively.

In the context where $g = h_T \oplus h_v$, with both $\hat{h}_T$ and $\hat{h}_v$ being represented by fermion fields (in the latter case in the adjoint representations) there are a number of super-Virasoro algebras that can be constructed. Of these, the supersymmetric extension of K_n is the least expected. We shall discuss these constructions and their significance elsewhere [19]. Here our objective is to show that for $h = \mathrm{su}(2)$ the K_n, G_r algebra produces the full discrete series of representations of the Ramond and Neveu-Schwarz algebras.

To construct the representation of h_v that we need we introduce $\dim h$ fermion fields

$$H(z) = \sum_r b_r^a z^{-r}, \qquad 1 \leqq a \leqq \dim h, \tag{3.4}$$

where the sum is over *either* $r \in \mathbb{Z}$ (Ramond [R] case) *or* $r \in \mathbb{Z} + \frac{1}{2}$ (Neveu-Schwarz [NS] case), and the fermi oscillators satisfy

$$\{b_r^a, b_s^b\} = \delta^{ab}\delta_{r,-s}, \tag{3.5a}$$

$$b_r^a|0\rangle = 0, \qquad r > 0, \qquad b_r^{a\dagger} = b_{-r}^a. \tag{3.5b}$$

In the NS case the vacuum $|0\rangle$ is non-degenerate, whilst in the R case the vacua form a representation of the Dirac algebra b_0^a, $1 \leqq a \leqq \dim h$. We define fermion normal ordering by

$$:b_r^a b_s^b: = b_r^a b_s^b, \qquad r<0, \tag{3.6a}$$

$$= -b_s^b b_r^a, \qquad r>0, \tag{3.6b}$$

$$= \tfrac{1}{2}[b_r^a, b_s^b], \qquad r=0. \tag{3.6c}$$

The representation V_m^a of $\hat{h}_v$ is defined by

$$V^a(z) \equiv \sum V_n^a z^{-n} = -\frac{i}{2} f^{abc} : H^b(z) H^c(z):. \tag{3.7}$$

From this it follows that [4]

$$[V_m^a, V_m^b] = i f^{abc} V_{m+n}^c + v m \delta^{ab} \delta_{m,-n}, \tag{3.8}$$

where $v = \frac{1}{2} c_\psi^h$. Thus we have a (reducible) representation of $\hat{h}_v \cong \hat{h}$ of level c_ψ^h/ψ^2.

For h_T we take a highest weight irreducible representation labelled by $(2k/\psi^2, \lambda)$,

$$T^a(z) \equiv \sum T_n^a z^{-n}. \tag{3.9}$$

Then the Virasoro algebra K_n is defined by Eq. (3.1) where

$$L^g(z) = L^{h_T}(z) + L^{h_v}(z) = \frac{1}{2(k+v)} \,{}^\circ_\circ T^a(z) T^a(z) {}^\circ_\circ + \frac{1}{4v} \,{}^\circ_\circ V^a(z) V^a(z) {}^\circ_\circ, \tag{3.10}$$

and

$$L^h(z) = \frac{1}{2(k+2v)} \,{}^\circ_\circ [T^a(z) + V^a(z)][T^a(z) + V^a(z)] {}^\circ_\circ. \tag{3.11}$$

It has central element

$$c = C^{h_T} + C^{h_v} - C^h = \frac{k(k+3v)\dim h}{2(k+v)(k+2v)}. \tag{3.12}$$

From Eq. (2.14) it follows that

$$[K_m, T_n^a + V_n^a] = 0, \qquad 1 \leqq a \leqq \dim h, \tag{3.13}$$

and it is natural to seek super-Virasoro operators G_r which also commute with $\hat{h}$. Candidate building-blocks for $G(z)$ are $T^a(z) H^a(z)$ and

$$\frac{1}{3} : V^a(z) H^a(z): = -\frac{i}{6} : f^{abc} H^a(z) H^b(z) H^c(z):. \tag{3.14}$$

Straightforward calculation shows that

$$G(z) \equiv \sum_r G_r z^{-n} = [v(k+v)(k+2v)]^{-1/2} \left[k T^a(z) H^a(z) - \frac{v}{3} : V^a(z) H^a(z): \right] \tag{3.15}$$

provides a supersymmetric extension of K_n satisfying (3.2), (3.3) and commuting with $\hat{h}$.

The construction we have given in this section can be applied to any Lie algebra h, using in its adjoint representation and any representation of $\hat{h} \cong \hat{h}_T$. One can equally regard this as an irreducible representation of the super-affine Kac-Moody algebra based on h [20]. However, we have found no underlying reason for the existence of this particular super-Virasoro algebra. For further discussion see [19].)

4. Unitary Representations of the Super-Virasoro Algebras

To find the whole of the discrete series of unitary representations we use the construction of the last section in the particular case in which $h \cong \text{su}(2)$. For the su(2) Kac-Moody algebra $\hat{h}_T$ we use the level N, spin l irreducible representation, whilst for $\hat{h}_v$ we must use the adjoint fermion representation described in Sect. 3. This is a level 2 representation which in the NS case decomposes into spin 0 and spin 1 irreducible components, with highest weight states $|0\rangle$ and $b^i_{-1/2}|0\rangle$, whilst in the R case the vacuum can be taken to be 2-fold degenerate, and then the representation will be an irreducible level 2, spin $\frac{1}{2}$ representation. From Sect. 3, it follows that, acting on the resulting representation space for $\hat{g}$, we have an su(2) Kac-Moody algebra $\hat{h}$ and the super-Virasoro algebras $s = \{K_n, G_r\}$. Thus we can parallel the discussion of Sect. 2 by decomposing the resulting representations of $\hat{g}$ with respect to $\hat{h} \times s$. The central element in s obtained in this way is [1]

$$c = \frac{3}{2}\left(1 - \frac{8}{(N+2)(N+4)}\right), \tag{4.1}$$

providing the whole of the discrete series of values of c. We need to show that we also obtain all the corresponding values of $h_{p,q}$ listed in Eq. (1.11).

The characters of the adjoint fermion representations, as defined by (2.20), can be written down directly using their Fock space construction,

$$\chi_2^{\text{NS}}(z,\theta) = \prod_{n=1}^{\infty} (1 + z^{n-1/2})(1 + z^{n-1/2}e^{i\theta})(1 + z^{n-1/2}e^{-i\theta}), \tag{4.2}$$

and

$$\chi_2^{R}(z,\theta) = z^{3/16}(e^{i\theta/2} + e^{-i\theta/2}) \prod_{n=1}^{\infty} (1 + z^n)(1 + z^n e^{i\theta})(1 + z^n e^{-i\theta}). \tag{4.3}$$

Here we have used the fact that for this representation [4], $L^g(z)$ is the same as

$$L(z) = \frac{i}{2} : \frac{dH^a}{dz} H^a(z) : + \frac{3\varepsilon}{8}. \tag{4.4}$$

Using the Jacobi triple product formula

$$\prod_{n=1}^{\infty} (1 - x^n)(1 + x^{n-1}w)(1 + x^n/w) = \sum_{n \in \mathbf{Z}} x^{1/2n(n-1)} w^n, \tag{4.5}$$

we can rewrite these characters in the forms

$$\chi_2^{\text{NS}}(z,\theta) = \sum_{m \in \mathbf{Z}} z^{1/2m^2} e^{im\theta} \prod_{n=1}^{\infty} \left(\frac{1 + z^{n-1/2}}{1 - z^n}\right) \tag{4.6}$$

and

$$\chi_2^R(z,\theta)=z^{1/16}\sum_{m\in\mathbb{Z}+\frac{1}{2}} z^{1/2m^2}e^{im\theta}\prod_{n=1}^{\infty}\left(\frac{1+z^n}{1-z^n}\right). \tag{4.7}$$

If we denote these representations of su(2) by (2, NS) and (2, R), as we have just remarked,

$$(2,\mathrm{NS})=(2,0)\oplus(2,1), \tag{4.8a}$$

$$(2,R)=(2,\tfrac{1}{2}). \tag{4.8b}$$

In order to repeat the argument of Sect. 2, we also need to know the characters of the discrete series of super-Virasoro algebra representations. For the representations $(c, h_{p,q})$ of Eqs. (1.10), (1.11) these are given by [17]

$$\chi_{c,h}^{\mathrm{NS}}(z)=\Gamma_{p,q}^m(z)\prod_{n=1}^{\infty}\left(\frac{1+z^{n-1/2}}{1-z^n}\right) \tag{4.9}$$

and

$$\chi_{c,h}^{R}(z)=\Gamma_{p,q}^m(z)z^{1/16}\prod_{n=1}^{\infty}\left(\frac{1+z^n}{1-z^n}\right), \tag{4.10}$$

where $1\leqq p\leqq m-1$, $1\leqq q\leqq m+1$ and $p-q$ is even in the NS case and odd in the R case. The functions $\Gamma_{p,q}^m(z)$ are defined by an equation similar to (2.31),

$$\Gamma_{p,q}^m(z)=\sum_{m\in\mathbb{Z}}\{Z^{\gamma_{p,q}^m(n)}-Z^{\delta_{p,q}^m(n)}\}, \tag{4.11}$$

where

$$\gamma_{p,q}^m(n)=\frac{[2m(m+1)n-p(m+2)+mq]^2-4}{8m(m+2)} \tag{4.12a}$$

and

$$\delta_{p,q}^m(n)=\frac{[2m(m+1)n+p(m+2)+mq]^2-4}{8m(m+2)}. \tag{4.12b}$$

We can now establish the identity

$$\left\{\sum_{m\in\mathbb{Z}+\varepsilon} z^{m^2}e^{im\theta}\right\}\Delta_{N,\frac{1}{2}(p-1)}(z,\theta)=\sum_q \Delta_{N+2,\frac{1}{2}(q-1)}(z,\theta)\Gamma_{p,q}^{N+2}(z), \tag{4.13}$$

where the sum over q is over values lying in the range $1\leqq q\leqq N+3$ and with $p-q\in 2\mathbb{Z}+2\varepsilon$, using exactly the same sort of manipulation that were used in Sect. 2 to prove Eq. (2.38). Thus it follows that

$$\chi_{N,\frac{1}{2}(p-1)}(z,\theta)\chi_2^F(z,\theta)=\sum_q \chi_{N+2,\frac{1}{2}(q-1)}(z,\theta)\chi_{c,h}^F(z), \tag{4.14}$$

where F stands either for NS or R, $h=h_{p,q}(c)$, c is as in Eq. (4.1) and the range of q is as in Eq. (4.13) with $\varepsilon=0$ in the NS case and $\varepsilon=\frac{1}{2}$ in the R case. By the same arguments used in Sect. 2, Eq. (4.14) implies the decompositions

$$(N,\tfrac{1}{2}[p-1])\times(F,2)=\bigoplus_q(N+2,\tfrac{1}{2}[q-1])\times(c,h_{p,q}(c))_F, \tag{4.15}$$

where $(c,h)_F$ denotes the appropriate representation of the NS or R super-Virasoro algebras, as $F=\mathrm{NS}$ or R, and the rest of the notation is as in Eq. (4.14). This provides the whole of the discrete series of unitary representations of the super-Virasoro algebras and establishes the unitarity.

5. Comments

In this paper we have established the existence of the full discrete series of representations of the Virasoro and super-Virasoro algebras starting from irreducible representations of the (affine untwisted) su(2) Kac-Moody algebra. In [1] we based our construction on the representations of this algebra which could be constructed using NS or R fermion fields transforming under spin $\frac{1}{2}$ or spin 1 representations of su(2). We did not show that all the values of h given by Eqs. (1.7) and (1.11) could be found in this context (thus leaving the existence of all the discrete series representations still in doubt) but in fact they can [17].

Another approach to the construction of Virasoro algebra, but not super-Virasoro algebra construction, described in [1], was to use the inclusion

$$\mathrm{Sp}(n-1)\times\mathrm{Sp}(1)\subset\mathrm{Sp}(n), \tag{5.1}$$

and this can also be used to obtain all the representations for the Virasoro algebra. This approach has interesting applications and it will be described in [17]. Altschuler [21] has verified that within this latter construction the Ramond vacuum yields the eigenvalues $h_{p,p}$ and $h_{p,p-1}$ $(1\leqq p\leqq m-1)$.

Appendix. The Weyl-Kac Character Formula

The aim of this appendix is to write down and motivate the Weyl-Kac formula for the character of a highest weight unitary irreducible representation of the affine Kac-Moody algebra, indicating that it is the natural generalisation of the Weyl formula for the character of a unitary irreducible representation of the finite dimensional Lie algebra g. We will show how, when g is chosen to be su(2), it yields the character formula (2.24) used in the text.

Let λ be the highest weight of a unitary irreducible representation of g (so that it is integral and dominant). Then the character of this representation is given by

$$\chi^{\lambda}(\theta)=\mathrm{tr}(e^{i\theta\cdot T})=\sum_{\sigma\in W(g)}\varepsilon(\sigma)e^{i\sigma(\lambda+\varrho)\cdot\theta}e^{-i\varrho\cdot\theta}\prod_{\alpha>0}(1-e^{-i\alpha\cdot\theta})^{-1}, \tag{A.1}$$

where $\varepsilon(\sigma)=\det\sigma$, $W(g)$ denotes the Weyl group of g, and ϱ denotes half the sum of the positive roots of g. As $W(\mathrm{su}(2))$ has just two elements, ±1, (A.1) reduces to the familiar formula

$$\chi^{L}(\theta)=\frac{\sin(L+\frac{1}{2})\theta}{\sin\frac{1}{2}\theta} \tag{A.2}$$

occurring in Eq. (2.36).

Both sides of (A.1) can be assigned a meaning for the Kac-Moody algebra $\hat{g}$, since $\hat{g}$ still has finite rank, a system of positive roots and a Weyl group $W(\hat{g})$ generated by reflections in real roots, which is in fact isomorphic to the semidirect product of $W(g)$ with the co-root lattice of g, Λ_r^{v}. For further explanation of these results and the notation we are about to adopt we refer to a review by one of us [11].

The Cartan subalgebra of $\hat{g}$, extended by derivation $d = -L_0$, is taken to have basis (T^i, $1 \leqq i \leqq \operatorname{rank} g, k, d$). In an irreducible unitary highest weight representation, k takes a fixed value, $\frac{1}{2}x\psi^2$, where x is the level, while L_0 exceeds its smallest value, $-\delta$, by positive integers. Thus it is meaningful to adopt the following definition of the character.

$$\chi_{x,\iota}(z, \theta) \equiv \operatorname{tr}(z^{L_0} e^{i\theta \cdot T}) \tag{A.3}$$

with convergence expected for $|z| < 1$. Formally, we can rewrite this as follows in order to compare with (A.1),

$$\chi_{x,\iota}(z, \theta) = \operatorname{tr}\{e^{i(\theta \cdot T + i(\ln z)d + o \cdot k)}\}. \tag{A.4}$$

Now the Kac-Moody analogue in (A.4) of $\lambda + \varrho$ in (A.1) is

$$y = (\lambda + \varrho, \tfrac{1}{2}\psi^2(x + \tilde{h}), \delta), \tag{A.5}$$

where $\tilde{h}$, the "level" of ϱ, is the dual Coxeter of g, and δ is yet to be determined by comparing the leading z behaviour of the character. The action of a Weyl reflection S_a in a root of $\hat{g}$ is, according to Sect. 8 of [11],

$$S_a(y) = (\sigma_a(\lambda + \varrho + \iota), \tfrac{1}{2}\psi^2(x + \tilde{h}), \delta + [(\lambda + \varrho)^2 - (\lambda + \varrho + \iota)^2]/(x + \tilde{h})\psi^2),$$

where

$$\iota = n\alpha(x + \tilde{h})\psi^2/\alpha^2, \tag{A.5}$$

and so is proportional to an element of the co-root lattice of g.

Thus we expect the Kac-Moody analogue of (A.1) to read

$$\chi_{x,\lambda}(z, \theta) = \Delta_{x,\lambda}(z, \theta)/\Delta_0(z, \theta), \tag{A.6}$$

where

$$\Delta_{x,\lambda}(z, \theta) = \sum_{\iota \in \Lambda_r^\vee} z^{-\delta + [(\lambda + \varrho + \iota)^2 - (\lambda + \varrho)^2]/(x + \tilde{h})\psi^2} \sum_{\sigma \in W(g)} \varepsilon(\sigma) e^{i\sigma(\lambda + \varrho + \iota) \cdot \theta} \tag{A.7}$$

and

$$\Delta_0(z, \theta) = e^{i\varrho \cdot \theta} \prod_{n=1}^{\infty} \left\{(1 - z^n)^r \prod_{\alpha > 0} (1 - z^{n-1} e^{-i\alpha \cdot \theta})(1 - z^n e^{i\alpha\theta})\right\} \tag{A.8}$$

using the structure of the root system and Weyl group of $\hat{g}$. It follows that the leading term in z occurring in $\chi_{x,\lambda}(z, \theta)$ is $z^{-\delta}\chi_\lambda(\theta)$. Since the states of lowest L_0 value form an irreducible representation of g with highest weight λ, this is as it should be. When $L_0 = L_0^g$, given by the Sugawara formula (2.5), we know that its value on these states is $\lambda(\lambda + 2\varrho)/(x + \tilde{h})\psi^2$ and so this must be the value of $-\delta$. Using this, the expression for $\Delta_{x,\lambda}(z, \theta)$ given by (A.7) simplifies to

$$\Delta_{\lambda,x}(z, \theta) = \sum_{\iota \in \Lambda_r^\vee} z^{[(\lambda + \varrho + \iota)^2 - \varrho^2]/(x + \tilde{h})\psi^2} \sum_{\sigma \in W(g)} \varepsilon(\sigma) e^{i\sigma(\lambda + \varrho + \iota) \cdot \theta}. \tag{A.9}$$

Formulae (A.6), (A.8), and (A.9) constitute the Weyl-Kac formula.

For $g = \mathrm{su}(2)$, the dual Coxeter number $\tilde{h} = 2$ and it is customary to take $\psi = 1$ so that $\varrho = \frac{1}{2}$. Then replacing x by N and λ by l, we obtain Weyl-Kac formula in the form of Eq. (2.36).

Acknowledgements. We are grateful to E. Corrigan, D. Fairlie, Z. Qiu, S. Shenker, and especially D. Friedan for interesting and useful conversations and communications. A.K. is grateful to the U.K. Science and Engineering Research Council for a Studentship.

References

1. Goddard, P., Kent, A., Olive, D.: Virasoro algebras and coset space models. Phys. Lett. **152** B, 88 (1985)
2. Polyakov, A.M.: Conformal symmetry of critical fluctuations. JETP Lett. **12**, 381 (1970); Belavin, A.A., Polyakov, A.M., Zamolodchikov, A.B.: Infinite conformal symmetry in two-dimensional quantum field theory. Nucl. Phys. B **241**, 333 (1980)
3. Friedan, D., Qiu, Z., Shenker, S.: In: Vertex operators in mathematics and physics. Lepowsky, J. et al. (eds.). MSRI publications No. 3, Berlin, Heidelberg, New York: Springer 1984, p. 419; Conformal invariance unitarity and critical exponents in two dimensions. Phys. Rev. Lett. **52**, 1575 (1984)
4. Goddard, P., Olive, D.: Kac-Moody algebras, conformal symmetry and critical exponents. Nucl. Phys. B **257** [FS 14], 226 (1985)
5. Friedan, D., Qiu, Z., Shenker, S.: Superconformal invariance in two dimensions and the tricritical Ising model. Phys. Lett. **151** B, 37 (1985)
6. Ramond, P.: Dual theory for free fermions. Phys. Rev. D **3**, 2415 (1971)
7. Neveu, A., Schwarz, J.H.: Factorizable dual model of pions. Nucl. Phys. B **31**, 86 (1971); Quark model of dual pions. Phys. Rev. D **4**, 1109 (1971)
8. Sugawara, H.: A field theory of currents. Phys. Rev. **170**, 1659 (1968)
   Sommerfield, C.: Currents as dynamical variables. Phys. Rev. **176**, 2019 (1968)
   Coleman, S., Gross, D., Jackiw, R.: Fermion avatars of the Sugawara model. Phys. Rev. **180**, 1359 (1969);
   Bardakci, K., Halpern, M.: New dual quark models. Phys. Rev. D **3**, 2493 (1971)
   Dashen, R., Frishman, Y.: Four-fermion interactions and scale invariance. Phys. Rev. D **11**, 2781 (1975)
9. Kac, V.G.: Infinite dimensional Lie algebras. Boston: Birkhäuser 1983
10. Goddard, P.: Kac-Moody algebras: representations and applications. DAMTP preprint 85/7
11. Olive, D.: Kac-Moody algebras: an introduction for physicists. Imperial College preprint TP/84–85/14
12. Rocha-Caridi, A.: In: Vertex operators in mathematics and physics. Lepowsky, J. et al. (eds.). MSRI publications No. 3, p. 451. Berlin, Heidelberg, New York: Springer 1984
13. Frenkel, I.B., Kac, V.G.: Basic representations of Lie algebras and dual resonance models. Invent. Math. **62**, 23 (1980)
14. Segal, G.: Unitary representations of some infinite dimensional groups. Commun. Math. Phys. **80**, 301 (1981)
15. Goddard, P., Olive, D.: In: Vertex operators in mathematics and physics. Lepowsky, J. et al. (eds.). MSRI publications No. 3, p. 51. Berlin, Heidelberg, New York: Springer 1984
16. Feigin, B.L., Fuchs, D.B.: Funct. Anal. Appl. **17**, 241 (1983)
17. Kent, A.: Papers in preparation
18. Bardakci, K., Halpern, M.B.: New dual quark models. Phys. Rev. D3, 2493 (1971)
19. Goddard, P., Kent, A., Olive, D.: In preparation
20. Kac, V., Todorov, I.: Superconformal current algebras and their unitary representations. Commun. Math. Phys. **102**, 337–347 (1985)
21. Altschuler, D.: University of Geneva preprint (1985) UGVA-DPT 1985/06–466

Communicated by A. Jaffe

Received September 16, 1985

Volume 172, number 3,4 PHYSICS LETTERS B 22 May 1986

DETERMINANT FORMULAE AND UNITARITY FOR THE $N = 2$ SUPERCONFORMAL ALGEBRAS IN TWO DIMENSIONS OR EXACT RESULTS ON STRING COMPACTIFICATION

Wayne BOUCHER [1], Daniel FRIEDAN [2,3] and Adrian KENT [4,3]

Enrico Fermi and James Franck Institutes, and Department of Physics, University of Chicago, Chicago, IL 60637, USA

Received 3 March 1986

Determinant formulae are presented for the periodic, antiperiodic and twisted $N = 2$ superconformal algebras in two dimensions, and a classification is derived of the unitary highest weight representations. Physical realisations of several of these representations are discussed. In particular, it is noted that the unitarity constraints apply to string compactification, giving results which are nonperturbative in the compactification radius.

The infinite-dimensional conformal algebra of the two-dimensional cylinder, the Virasoro algebra, is the gauge algebra of the world surface of the covariantly quantized bosonic string [1]. The Virasoro algebra also acts in the operator representation of two-dimensional critical phenomena [2], where its unitary representation theory gives constraints on the possible values of critical indices [3]. In the investigation of unitarity, the basic technical tool is the determinant formula conjectured by Kac [4] and proved by Feigin and Fuchs [5].

The gauge algebras of the supersymmetric string, the Ramond [6] and Neveu–Schwarz [7] algebras, are the $N = 1$ superconformal algebras on the cylinder. The $N = 1$ algebras are realized in supersymmetric critical phenomena, and unitarity again restricts the possible values of critical indices, permitting the identification of physical systems with supersymmetric critical behavior [3,8]. The Neveu–Schwarz determinant formula [4] and the Ramond formula [8,9] were proved by Meurman and Rocha-Caridi [10] and by Thorn [9].

In this paper we give the determinant formulae and unitarity constraints for the $N = 2$ extensions of the $N = 1$ superconformal algebras. The proofs will be given elsewhere [11]. The $N = 2$ algebras first appeared as gauge algebras of the U(1) fermionic string [12]. Preliminary calculations towards determinant formulae were performed by Di Vecchia, Petersen and Zheng [13], Qiu and Shenker [14], and Thorn [15]. While this paper was being completed, we received preprints by Di Vecchia, Petersen and Yu [16] and Nam [17], which describe partial results overlapping with some of our work.

$N = 2$ superconformal invariance has applications to supersymmetric critical phenomena in two dimensions [18]. It also arises in the classical compactifications of the supersymmetric string. A priori, string compactifications are given by $N = 1$ superconformal nonlinear models [19,20]. But all known nontrivial examples actually have $N = 2$ superconformal symmetry [21,22]. The $N = 2$ supersymmetry of the nonlinear model is associated with spacetime supersymmetry in the string ground state after compactification [20,23]. This spacetime supersymmetry should persist to all orders in the string coupling [24].

To date, all results on the nonlinear model, including $N = 2$ superconformal symmetry, are known at best to all orders in the inverse radius of the compact dimensions. The present work is, inter alia, the first step in a program to study classical compactifications

[1] Supported in part by NSF grant PHY-84-16691.
[2] Supported in part by the Alfred P. Sloan Foundation.
[3] Supported in part by DOE grant DE-FG02-84ER-45144.
[4] Enrico Fermi Postdoctoral Fellow.

0370-2693/86/$ 03.50 © Elsevier Science Publishers B.V. (North-Holland Physics Publishing Division)

Volume 172, number 3,4 PHYSICS LETTERS B 22 May 1986

nonperturbatively in the compactification scale, assuming world sheet $N = 2$ superconformal symmetry. Surprisingly, the unitarity constraints already have consequences for string compactification. These first results are not particularly dramatic, but they suggest that the algebraic approach to compactification is worth pursuing.

The anti-commutation relations of the $N = 2$ algebras are

$$[L_m, L_n] = (m-n)L_{m+n} + \tfrac{1}{4}\tilde{c}(m^3 - m)\delta_{m,-n},$$
$$[L_m, G^i_n] = (\tfrac{1}{2}m - n)G^i_{m+n},$$
$$[L_m, T_n] = -nT_{m+n},$$
$$[T_m, T_n] = \tilde{c}m\delta_{m,-n},$$
$$[T_m, G^i_n] = i\epsilon^{ij}G^j_{m+n},$$
$$[G^i_m, G^j_n]_+ = 2\delta^{ij}L_{m+n} + i\epsilon^{ij}(m-n)T_{m+n} + \tilde{c}(m^2 - \tfrac{1}{4})\delta^{ij}\delta_{m,-n}. \quad (1)$$

The Virasoro generators L_m are the Fourier coefficients of the traceless stress–energy tensor of a conformally invariant quantum field theory on the cylinder. Equivalently (by $z = e^w$) they are the Laurent coefficients of the traceless stress–energy tensor $T(z) = \Sigma L_m z^{-m-2}$ on the plane. The T_m are the coefficients of a U(1) current algebra $J(z) = \Sigma T_m z^{-m-1}$. The G^i_m are the coefficients of the fermionic partner fields $G^i(z) = \Sigma G^i_m z^{-m-3/2}$ which complete the $N = 2$ super stress–energy multiplet. All of the generators satisfy hermiticity conditions of the form $A^\dagger_m = A_{-m}$, which follow from reality of the super stress–energy tensor. The central charge $\tilde{c}$ is related to the usual Virasoro central charge c and the usual $N = 1$ charge $\hat{c}$ by $\tilde{c} = c/3 = \hat{c}/2$. The normalization is fixed so that $\tilde{c} = 1$ for the free $N = 2$ superfield, consisting of two scalars each with $c = 1$ and two Majorana fermions each with $c = \frac{1}{2}$.

The three $N = 2$ algebras are given by three modings of the generators, corresponding to three ways of choosing boundary conditions on the cylinder. The L_m are always integrally moded, because the bosonic stress–energy tensor is always periodic on the cylinder. The P (for periodic) algebra has integer modes for T_m and G^i_m. The A (for anti-periodic) algebra has integer modes for T_m, but half-integer ($m \in \mathbf{Z} + \frac{1}{2}$) modes for G^i_m. The T (for twisted [‡1]) algebra has integer modes for G^1_m, half-integer modes for T_m and G^2_m.

[‡1] Twisted scalar fields were first described in ref. [25].

We consider highest weight representations of the $N = 2$ algebras. These representations are generated from a vector of lowest L_0 eigenvalue h, called the *highest weight vector* (hwv). The hwv is necessarily annihilated by all the lowering operators L_m, G^i_m, T_m $(m > 0)$. The remaining generators consist of the raising operators L_{-m}, T_{-m}, G^i_{-m} $(m > 0)$ together with the zero modes. The hwv must be an eigenstate of a maximal commuting set of zero modes.

The basic technique for studying a highest weight representation is to construct it as a quotient of the Verma module, which is the largest possible representation generated from the hwv. A natural basis for the Verma module is a maximal independent set of states which are given by ordered monomials of the generators acting on the hwv. The Verma modules of all three algebras can be decomposed into eigenspaces of L_0, called *levels*. Level n is the eigenspace with L_0 eigenvalue $h + n$.

For the A algebra the zero modes are L_0 and T_0, so an hwv $|h, q\rangle$ is characterized by its energy (L_0 eigenvalue) h and charge (T_0 eigenvalue) q. Each level n of the Verma module can be further decomposed into T_0 eigenspaces with eigenvalue $q + m$, when m is called the *relative charge*. The counting of states is summarized by the partition function $P_A(n, m)$ defined by

$$\sum_{n,m} P_A(n,m)x^n y^m = \prod_{k=1}^{\infty} \frac{(1 + x^{k-1/2}y)(1 + x^{k-1/2}y^{-1})}{(1-x^k)^2}. \quad (2)$$

In the P algebra the zero modes are L_0, T_0, G^i_0. There are two kinds of hwv, $|h, q \mp \frac{1}{2}\rangle_\pm$, with energy $L_0 = h$ and charge $T_0 = q \mp \frac{1}{2}$. Each satisfies an additional highest weight condition with respect to the charge, $(G^1_0 \mp iG^2_0)|h, q \mp \frac{1}{2}\rangle_\pm = 0$. These two representations $P^\pm$ are isomorphic under charge conjugation ($T_m \to -T_m$, $G^2_m \to -G^2_m$). Again, each level n is decomposed by relative charge m, with partition function

$$\sum_{n,m} P_P(n,m)x^n y^m = (y^{1/2} + y^{-1/2}) \times \prod_{k=1}^{\infty} \frac{(1 + x^k y)(1 + x^k y^{-1})}{(1-x^k)^2}. \quad (3)$$

Volume 172, number 3,4 PHYSICS LETTERS B 22 May 1986

We will also need partition functions in the presence of single charged fermions:

$$\sum_{n,m} \tilde{P}_{\mathrm{X}}(n,m;k)x^n y^m = (1 + x^{|k|}y^{\mathrm{sgn}(k)})^{-1} \sum_{n,m} P_{\mathrm{X}}(n,m)x^n y^m \,, \tag{4}$$

for X = A, P. We define $\mathrm{sgn}(k) = 1$ for $k > 0$, $\mathrm{sgn}(k) = -1$ for $k < 0$ and $\mathrm{sgn}(0) = \pm 1$ for the representation $\mathrm{P}^{\pm}$.

In the T algebra the zero modes are L_0 and G_0^1. An hwv $|h\rangle$ is characterized by its energy h. Each level n splits into two equal subspaces of fermion parity $(-)^F = \pm 1$, where $(-)^F$ is the operator which commutes with L_m and T_m, anticommutes with G_m^i and is 1 on $|h\rangle$. The partition function for each fermion parity is

$$\sum_n P_{\mathrm{T}}(n)x^n = \prod_{k=1}^{\infty} \frac{(1+x^k)(1+x^{k-1/2})}{(1-x^k)(1-x^{k-1/2})} \,. \tag{5}$$

As usual, the inner product on the Verma module is defined by the hermiticity conditions, the commutation relations and the highest weight conditions. Subspaces of different level and relative charge or fermion parity are orthogonal. A powerful tool in the representation theory of these algebras is the determinant formula. This is a polynomial expression in $\tilde{c}, h, q$ for the determinant of the matrix of inner products of a basis of ordered monomials for a given eigenspace (up to a basis-dependent positive constant). The vanishing surfaces, or curves, of the determinant formula describe the Verma modules which contain null vectors. The presence of such null vectors can be used to solve for correlation functions in conformal quantum field theories [2,26,27,8,28]. The vanishing surfaces also mark changes in the signature of the metric, so they give crucial information about unitarity [29,3].

For the P and A algebras let $M^{\mathrm{P}}_{n,m}$ and $M^{\mathrm{A}}_{n,m}$ be the inner product matrices for level n, relative charge m. The determinant formula for the A algebra is

$$\det M^{\mathrm{A}}_{n,m}(\tilde{c},h,q) = \prod_{\substack{1 \leq rs \leq 2n \\ s \text{ even}}} (f^{\mathrm{A}}_{r,s})^{P_{\mathrm{A}}(n-rs/2,m)} \times \prod_{k \in \mathbf{Z}+1/2} (g^{\mathrm{A}}_k)^{\tilde{P}_{\mathrm{A}}(n-|k|,m-\mathrm{sgn}(k);k)} \,, \tag{6}$$

where

$$f^{\mathrm{A}}_{r,s}(\tilde{c},h,q) = 2(\tilde{c}-1)h - q^2 - \tfrac{1}{4}(\tilde{c}-1)^2 + \tfrac{1}{4}[(\tilde{c}-1)r+s]^2 \quad (s \text{ even}) \,,$$

$$g^{\mathrm{A}}_k(\tilde{c},h,q) = 2h - 2kq + (\tilde{c}-1)(k^2 - \tfrac{1}{4}) \quad (k \in \mathbf{Z} + \tfrac{1}{2}) \,. \tag{7}$$

For the P algebra,

$$\det M^{\mathrm{P}}_{n,m}(\tilde{c},h,q) = \prod_{\substack{1 \leq rs \leq 2n \\ s \text{ even}}} (f^{\mathrm{P}}_{r,s})^{P_{\mathrm{P}}(n-rs/2,m)} \times \prod_{k \in \mathbf{Z}} (g^{\mathrm{P}}_k)^{\tilde{P}_{\mathrm{P}}(n-|k|,m-\mathrm{sgn}(k),k)} \,, \tag{8}$$

where

$$f^{\mathrm{P}}_{r,s}(\tilde{c},h,q) = 2(\tilde{c}-1)(h - \tfrac{1}{8}\tilde{c}) - q^2 + \tfrac{1}{4}[(\tilde{c}-1)r+s]^2 \quad (s \text{ even}) \,,$$

$$g^{\mathrm{P}}_k(\tilde{c},h,q) = 2h - 2kq + (\tilde{c}-1)(k^2 - \tfrac{1}{4}) - \tfrac{1}{4} \quad (k \in \mathbf{Z}) \,. \tag{9}$$

For the T algebra let $M^{\mathrm{T}}_{\pm,n}$ be the matrix of inner products for level n and fermion parity $(-)^F = \pm 1$. On level 0 the determinant formulae are $\det M^{\mathrm{T}}_{+,0} = 1$, $\det M^{\mathrm{T}}_{-,0} = h - \tilde{c}/8$. For $n > 0$,

$$\det M^{\mathrm{T}}_{\pm,n}(\tilde{c},h) = (h - \tfrac{1}{8}\tilde{c})^{P_{\mathrm{T}}(n)/2} \times \prod_{\substack{1 \leq rs \leq 2n \\ s \text{ odd}}} (f^{\mathrm{T}}_{r,s})^{P_{\mathrm{T}}(n-rs/2)} \,, \tag{10}$$

where

$$f^{\mathrm{T}}_{r,s}(\tilde{c},h) = 2(\tilde{c}-1)(h - \tfrac{1}{8}\tilde{c}) + \tfrac{1}{4}[(\tilde{c}-1)r+s]^2 \quad (s \text{ odd}). \tag{11}$$

A vanishing of the determinant formula signals a new hwv generating a submodule inside the Verma module. For the P algebra, along the quadratic vanishing surface $f^{\mathrm{P}}_{r,s} = 0$ there is an hwv on level $rs/2$ with relative charge $-\frac{1}{2}\mathrm{sgn}(0)$. Along the vanishing plane $g^{\mathrm{P}}_k = 0$ there is an hwv at level $|k|$ and relative charge $-\frac{1}{2}\mathrm{sgn}(0) + \mathrm{sgn}(k)$. For $k = 0$ this reflects the unbroken supersymmetry of states with $h = \tilde{c}/8$, and the possibility of a non-zero Witten index [30,8].

Volume 172, number 3,4 PHYSICS LETTERS B 22 May 1986

For the A algebra, along the quadratic surfaces $f^{\mathrm{A}}_{r,s} = 0$ there is an hwv of level $rs/2$ and relative charge 0. Along the vanishing plane $g^{\mathrm{A}}_k = 0$ there is an hwv of level $|k|$ and relative charge sgn(k). For the T algebra, along the quadratic vanishing curve $f^{\mathrm{T}}_{r,s} = 0$, there are two hwvs at level $rs/2$ with fermion parity ± 1. Along the vanishing line $h = \tilde{c}/8$, the state at level 0 and fermion parity -1 becomes an hwv, again allowing a nonzero index.

The quadratic vanishing surfaces for the P and A algebras are roughly analogous to the vanishing curves of the $N = 0$ or $N = 1$ superconformal formulae. The vanishing planes $g^{\mathrm{P,A}}_k = 0$ are something new. For the A algebra, there are precisely two at each half-integral level; for the P algebra there is one at level 0 and two at each of the higher levels. The hwv on a vanishing plane does not generate a full Verma submodule of states, because there exist *raising* operators which annihilate the hwv. (A simple example is given by the vanishing plane $g^{\mathrm{A}}_{1/2} = 0$, where the hwv $G_{-1/2}|h, q\rangle$ is annihilated by $G_{-1/2}$.) This explains the need for the modified partition functions $\widetilde{P}_{\mathrm{P}}, \widetilde{P}_{\mathrm{A}}$ (eq. (4)); their precise form is justified in the proof of the determinant formulae [11].

We stress that there are no further vanishings. In particular, the formulae imply that the A algebra has hwvs only for relative charge $m = -1, 0, 1$, and the P algebra only for $m = -\frac{1}{2}, \frac{1}{2}, -3\,\mathrm{sgn}(0)/2$.

Determinant formulae give a great deal of information about the representation theory of a Lie algebra. For the $N = 0$ and $N = 1$ superconformal algebras, determinant formulae were used to prove the *non*-unitarity of a large class of representations; the remaining representations were conjectured to be unitary [3,8]. These conjectures were based on explicit low-level calculations, and on a small number of examples from statistical mechanics. The unitarity conjectures were subsequently verified by manifestly unitary constructions of all the allowed representations [31].

An interesting picture emerges when the same strategy is applied to the $N = 2$ algebras. In the region $\tilde{c} < 1$, we find a discrete series of possibly unitary representations, while the non-unitarity of the remaining representations follows from the determinant formulae. (This much is familiar from the $N = 0, 1$ results.) However, for the P and A algebras, it is no longer the case that all $\tilde{c} \geq 1$, $h \geq 0$ representations are unitary, since the vanishing surfaces impinge on this region. Unitarity is obvious for h above the upper boundary of the vanishing surfaces, which is composed of plane segments. But the determinant formulae indicate that some representations between this boundary and the plane $h = 0$ survive the non-unitarity proof. These lie on segments of vanishing planes.

The precise results are as follows. For the A algebra, the only possible unitary representations fall into three classes:

A_3: $\tilde{c} \geq 1$, $(\tilde{c}, h, q)$ such that $g^{\mathrm{A}}_n \geq 0$, for all $n \in \mathbf{Z} + \frac{1}{2}$.

A_2: $\tilde{c} \geq 1$, $(\tilde{c}, h, q)$ such that $g^{\mathrm{A}}_n = 0$, $g^{\mathrm{A}}_{n+\mathrm{sgn}(n)} < 0$, $f^{\mathrm{A}}_{1,2} \geq 0$, for some $n \in \mathbf{Z} + \frac{1}{2}$.

A_0: $\tilde{c} < 1$, $\tilde{c} = 1 - 2/\tilde{m}$, $h = (jk - \frac{1}{4})/\tilde{m}$, $q = (j - k)/\tilde{m}$, for integer $\tilde{m} \geq 2$, and $j, k \in \mathbf{Z} + \frac{1}{2}$, $0 < j, k, j + k \leq \tilde{m} - 1$.

For the P algebra, the only possible unitaries (with $\mathrm{sgn}(0) = \pm 1$) are

$\mathrm{P}^{\pm}_3$: $\tilde{c} \geq 1$, $(\tilde{c}, h, q)$ such that $g^{\mathrm{P}}_n \geq 0$ for all $n \in \mathbf{Z}$.

$\mathrm{P}^{\pm}_2$: $\tilde{c} \geq 1$, $(\tilde{c}, h, q)$ such that $g^{\mathrm{P}}_n = 0$, $g^{\mathrm{P}}_{n+\mathrm{sgn}(n)} < 0$, $f^{\mathrm{P}}_{1,2} \geq 0$, for some $n \in \mathbf{Z}$.

$\mathrm{P}^{\pm}_0$: $\tilde{c} < 1$, $\tilde{c} = 1 - 2/\tilde{m}$, $h = \tilde{c}/8 + jk/\tilde{m}$, $q = \mathrm{sgn}(0)(j - k)/\tilde{m}$, for integer $\tilde{m} \geq 2$, and $j, k \in \mathbf{Z}$, $0 \leq j - 1, k, j + k \leq \tilde{m} - 1$.

For the T algebra the only possible unitaries are

T_2: $\tilde{c} \geq 1$, $h \geq \tilde{c}/8$.

T_0: $\tilde{c} < 1$, $\tilde{c} = 1 - 2/\tilde{m}$, $h = \tilde{c}/8 + (\tilde{m} - 2r)^2/16\tilde{m}$ for integers $\tilde{m}, r$ such that $2 \leq \tilde{m}$ and $1 \leq r \leq \tilde{m}/2$.

(The subscripts indicate the dimension of the moduli spaces.)

Observe that P^{+}_3 and P^{-}_3 are identical, while $\mathrm{P}^{+}_{2,0}$ and $\mathrm{P}^{-}_{2,0}$ differ only at the supersymmetric value $h = \tilde{c}/8$, with the allowed charges asymmetric around $q = 0$. Note also that, for the T_0 discrete series, only even $\tilde{m}$ values allow $h = \tilde{c}/8$. Such information constrains the spontaneous breaking of supersymmetry in finite volume and in the presence of supersymmetric mass perturbations [8,18].

Volume 172, number 3,4 PHYSICS LETTERS B 22 May 1986

The representations A_3, $P_3^\pm$, T_2 are obviously unitary; we conjecture that all of the remaining possibly unitary representations are indeed unitary. This conjecture is supported by examination of the inner product matrices at low levels. Further support for the conjecture is provided by manifestly unitary constructions of some of the discrete series of representations, which we now describe. The $\tilde{c} = \frac{1}{3}$ representations are realised by a single periodic scalar field, at special values of the period, demonstrating $N = 2$ supersymmetry in the gaussian model [18]. A similar construction, applied to the free $N = 1$ scalar superfield, produces $\tilde{c} = \frac{1}{2}$ representations. Tensor products of these yield $\tilde{c} = \frac{2}{3}$ and $\tilde{c} = \frac{5}{6}$.

Another construction method, which has proved useful in the context of the $N = 0, 1$ superconformal algebras [31,32], uses "quark model" fermionic oscillators. For any semisimple Lie algebra g, let $H^{i,a}(z)$ $(1 \leqslant i \leqslant 2,\ 1 \leqslant a \leqslant \dim(g))$ be $2 \cdot \dim(g)$ Fermi fields. Define

$$V^{i,a}(z) = \tfrac{1}{2}\, \mathrm{i} f_{abc} :H^{i,b}(z) H^{i,c}(z): \quad \text{(no sum on } i\text{)}\,,$$

$$W^i(z) = \tfrac{1}{3} V^{i,a}(z) H^{i,a}(z) \quad \text{(no sum on } i\text{)}\,,$$

$$S^1(z) = V^{2,a}(z) H^{1,a}(z)\,, \quad S^2(z) = V^{1,a}(z) H^{2,a}(z)\,,$$

$$G^i(z) = (1/\sqrt{6\upsilon})\, [W^i(z) - S^i(z)]\,,$$

$$J(z) = \tfrac{1}{3} H^{1,a}(z) H^{2,a}(z)\,,$$

$$T(z) = (1/4\upsilon) :V^{i,a}(z) V^{i,a}(z): - (1/6\upsilon) :(V^{1,a} + V^{2,a})(z)(V^{1,a} + V^{2,a})(z):\,. \qquad (12)$$

The operators $T(z)$, $G^i(z)$, $J(z)$ define a representation of the P, A or T algebras (depending on the choice of moding). Here f_{abc} are the structure constants of g, while the constant $\upsilon = c_\psi/2$ is defined by $f_{abc} f_{bcd} = -2\upsilon \delta_{ad}$. The (manifestly unitary) representation on the fermionic Fock space has $\tilde{c} = \dim(g)/9$, giving constructions for $\tilde{c} = \frac{1}{3}, \frac{2}{3}, \frac{8}{9}$. In addition, a similar construction using two sets of Fermi fields transforming under the fundamental representation of G_2 and replacing f_{abc} by the invariant anti-symmetric three-tensor, yields $\tilde{c} = \frac{7}{9}$.

An explicitly unitary construction of the entire discrete series would be of great value. One intriguing avenue of investigation is suggested by an apparent relationship between the $N = 2$ discrete series and the unitary representations of the affine algebra su(2) with

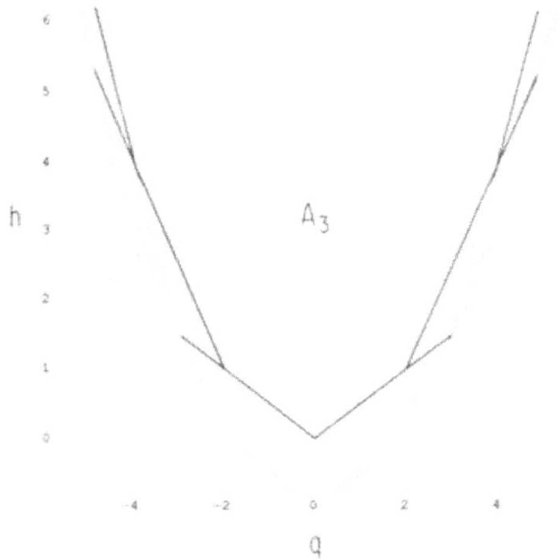

Fig. 1. Unitary representations of the A algebra at $\tilde{c} = 3.1$. The unitary representations in A_3 form the convex region bounded by solid lines. The unitaries in A_2 lie on the solid lines outside A_3. The parabola is $f^{\mathrm{A}}_{1,2}(3, h, q) = 0$. The lines are segments of $g^{\mathrm{A}}_k(3, h, q) = 0$, $|k| \leqslant 7/2$.

generators T_n^i. The discrete series of $N = 2$ central charges $c = 3\tilde{c} = 3 - 6/\tilde{m}$, $\tilde{m} = 2, 3, \ldots$, are precisely the central charges c of the Virasoro algebra $L_n^{\mathrm{su}(2)}$ associated with the unitary representations of $\widehat{\mathrm{su}(2)}$. The h, q weights of the $N = 2$ representations are simple functions of the possible $\widehat{\mathrm{su}(2)}$ highest weights ($L_0^{\mathrm{su}(2)}$ and T_0^3 eigenvalues).

The unitary representations A_2 and $P_2^\pm$, which lie on vanishing planes in the region $\tilde{c} \geqslant 1$, are a novel feature in the representation theory of infinite-dimensional Lie algebras. They imply, for fixed $\tilde{c}$ and q, a discrete spectrum of h-values lying below the continuum of unitary representations. The unitary representations for $\tilde{c} = 3$ are pictured in figs. 1 and 2.

Representations with $\tilde{c} \geqslant 1$ arise in the compactification of supersymmetric string. $N = 2$ supersymmetric nonlinear models based on Calabi–Yau spaces are superconformally invariant, at least to all orders in perturbation theory [21]. They give representations of the A and P algebras with $\tilde{c} = d/2$, where d is the real dimension of the Calabi–Yau space. The value $\tilde{c} = 3$ corresponds to compactification from ten to four dimensions. The A representations determine the boson-

Volume 172, number 3,4 PHYSICS LETTERS B 22 May 1986

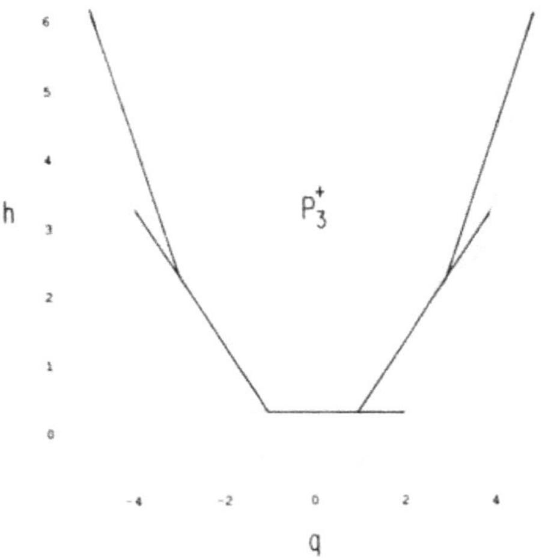

Fig. 2. Unitary representations of the P algebra at $\tilde{c} = 3$, sgn(0) = 1. The unitary representations in P_3^+ form the convex region bounded by solid lines. The unitaries in P_2^+ lie on the solid lines outside P_3^+. The parabola is $f_{1,2}^{P}(3, h, q) = 0$. The lines are segments of $g_k^{P}(3, h, q) = 0$, $|k| < 3$. The equivalent diagram for sgn(0) = -1 is given by the reflection $q \to -q$.

ic spectrum of the compactified string; the P representations determine the fermionic spectrum. Certain special operators in the nonlinear models correspond to unitary representations in the classes A_2, $P_2^{\pm}$. In particular, the (anti-)holomorphic ϵ-tensors correspond to the A_2 representations with $q = \pm\tilde{c}$, $h = \tilde{c}/2$, at intersections of $g_{\pm 1/2}^{A}$ with $f_{1,2}^{A}$. The covariantly constant spinors corresponds to the $P_0^{\pm}$ representations with $q = \mathrm{sgn}(0)\,\tilde{c}/4$, $h = \tilde{c}/8$, which are at intersections of g_0^{P} with $f_{1,2}^{P}$. Any compactification (Calabi–Yau or not [22]) with unbroken spacetime supersymmetry must have a superconformal field with these quantum numbers. A key problem in the study of string compactification is to show non-perturbatively that these operators are present and that they satisfy the operator product relations needed for the construction of the gravitino vertex and the world sheet spacetime supersymmetry current [33].

The $N = 2$ unitary representation theory has two immediate consequences for compactification. First, if operators with the quantum numbers of the covariantly constant spinors and the (anti-)holomorphic ϵ-tensors occur in a compactification, then they must keep those quantum numbers in any compactification nearby, because of the gap in allowed h values. This assumes that q is not renormalised (even nonperturbatively). Second, the fact that the interesting (h, q) values lie at the intersection of vanishing surfaces can be used, by the methods of refs. [2,26,27,8,28], to prove the operator-product identities which imply spacetime supersymmetry [34] [‡2].

These algebraic results are interesting because they are nonperturbative in the compactification radius. They suggest that further algebraic investigation of string compactification could be fruitful.

We are grateful to J. Cohn, D. Kastor and Z. Qiu for their insights and most especially S.H. Shenker for many helpful discussions. A.K. thanks P. Goddard and D. Olive for useful conversations about the $N = 2$ algebras. Exploratory computer calculations were carried out using Macsyma.

‡2 Note that, at special values of $\tilde{c}$, namely $\tilde{c} = 1 + 2/n$, $n = 1, 2, \ldots$, there is a discrete series of algebraically special representations, identified by a triple intersection of vanishing surfaces. It is not clear whether this condition is significant. However, for $n = 1$ and $n = 2$ it does apply to six- and four-dimensional Calabi–Yau compactifications.

References

[1] M. Virasoro, Phys. Rev. D 1 (1970) 2933.
[2] A.A. Belavin, A.M. Polyakov and A.B. Zamolodchikov, J. Stat. Phys. 34 (1984) 763; Nucl. Phys. B 241 (1984) 333.
[3] D. Friedan, Z. Qiu and S. Shenker, in: Vertex operators in mathematics and physics, eds. J. Lepowsky et al. (Springer, Berlin, 1984); Phys. Rev. Lett. 52 (1984) 1575.
[4] V. Kac, Lecture Notes in Physics, Vol. 94 (Springer, Berlin, 1979) p. 441; in: Proc. Intern. Congress of Mathematics (Helsinki, 1978).
[5] B.L. Feigin L. Feigin and D.B. Fuchs, Funct. Anal. Appl. 16 (1982) 114.
[6] P. Ramond, Phys. Rev. D 3 (1971) 2415.
[7] A. Neveu and J. Schwarz, Nucl. Phys. B 31 (1971) 86.
[8] D. Friedan, Z. Qiu and S. Shenker, Phys. Lett. B 151 (1985) 37.
[9] C. Thorn, unpublished; Nucl. Phys. B 248 (1984) 551.

Volume 172, number 3,4 PHYSICS LETTERS B 22 May 1986

[10] A. Meurman and A. Rocha-Caridi, MSRI preprint (1985).
[11] W. Boucher et al., in preparation.
[12] M. Ademollo et al., Phys. Lett. B 62 (1976) 105; Nucl. Phys. B 111 (1976) 77.
[13] P. Di Vecchia, in: Proc. Santa Barbara Workshop on Unified string theories, eds. M.B. Green and D. Gross (World Scientific, Singapore, 1986); P. Di Vecchia, J. Petersen and H. Zheng, Phys. Lett. B 162 (1985) 327.
[14] S.H. Shenker and Z. Qiu, private communication.
[15] C. Torn, unpublished.
[16] J. Petersen, Niels Bohr Institute preprint NBI-HE-85-31 (1985); P. Di Vecchia, J. Petersen and M. Yu, Niels Bohr Institute preprint NBI-HE-86-04 (1986).
[17] S. Nam, Yale preprint (1986).
[18] D. Friedan and S. Shenker, EFI preprint 86-17 (1986).
[19] D. Friedan and S. Shenker, unpublished talk Aspen Summer Institute (1984); D. Friedan, Z. Qiu and S. Shenker, Proc. APS Division of Particles and Fields Conf. (Santa Fe, 1984), eds. T. Goldman and M. Nieto (World Scientific, Singapore).
[20] P. Candelas, G. Horowitz, A. Strominger and E. Witten, Nucl. Phys. B 258 (1985) 46.
[21] L. Alvarez-Gaumé and D. Freedman, Phys. Rev. D 22 (1980) 846; L. Alvarez-Gaumé, S. Coleman and P. Ginsparg, Harvard preprint HUTP 85/A037 (1985); C. Hull, MIT preprints (1985).
[22] E. Witten, Princeton preprint (1985); A. Strominger, ITP preprint 86-16 (1986).
[23] C. Hull and E. Witten, Phys. Lett. B 160 (1985) 398.
[24] E. Martinec, Princeton preprint (1986).
[25] E. Corrigan and D. Fairlie, Nucl. Phys. B 91 (1975) 527.
[26] V.L. Dotsenko and V. Fateev, Nucl. Phys. B 240 (1984) 312.
[27] M. Bershadsky, V. Knizhnik and M. Teitelman, Phys. B 151 (1985) 31.
[28] Z. Qiu, University of Chicago Ph. D. dissertation (1985), Nucl. Phys. B., to be published.
[29] V. Kac, Lecture Notes in Mathematics, Vol. 933 (Springer, Berlin, 1982) p. 117.
[30] E. Witten, Nucl. Phys. B 202 (1982) 253.
[31] P. Goddard and D. Olive, Nucl. Phys. B 257 (1985) 83; P. Goddard, A. Kent and D. Olive, Phys. Lett. B 152 (1985) 88; Commun. Math. Phys. 103 (1986) 105.
[32] A. Kent, EFI preprint 85-62 (1985).
[33] D. Friedan, E. Martinec and S. Shenker, Phys. Lett. B 160 (1985) 55; EFI preprint 85-89; and Princeton prepritn (1985), Nucl. Phys. B., to be published.
[34] W. Boucher et al., in preparation.

CHAPTER 5

CONFORMAL SYMMETRY

Reprinted Papers

15. A.A. Belavin, A.M. Polyakov and A.B. Zamolodchikov, "Infinite Conformal Symmetry in Two-Dimensional Quantum Field Theory", Nucl. Phys. **B241** (1984) 333–380.

16. J.L. Cardy, "Operator Content of Two-Dimensional Conformally Invariant Theories", Nucl. Phys. **B270 [FS16]** (1986) 186–204.

The history of the conformal symmetry of field equations dates back to soon after the discovery of special relativity [1–3] and has been studied ever since [4]. In two dimensions the conformal group is infinite-dimensional and has an algebra which consists of two commuting copies of the Virasoro algebra. The idea that the Virasoro algebra might play an important role in two-dimensional conformal field theory is not a recent one [5,6]. A.M. Polyakov [7] put forward the point of view that the requirements of conformal invariance might so severely constrain a theory that a "conformal bootstrap" could be performed to determine the Green functions of the theory. This approach was developed much later in *Reprinted Paper* #15 to show that, under suitable circumstances, the correlation functions satisfy linear differential equations.

A key idea which must be added along side conformal invariance in order to determine the whole structure of a theory is that of modular invariance: in certain applications of two- dimensional conformal field theories there are physical reasons why the partition function of the theory should be invariant under the modular group (the group of fractional linear transformations with integral coefficients). The potential significance of modular invariance as a constraint in string theory was realised by W. Nahm [8] over ten years ago. J. Cardy [*Reprinted Paper* #16] showed that modular invariance provides severe constraints on two-dimensional field theories in the context of their application to critical behaviour in statistical systems. Similar ideas were developed in the context of string theories, specifically the motion of strings on group manifolds, by D. Gepner and E. Witten [9]. Important progress towards solving the constraints resulting from modular invariance has been made by A. Cappelli, C. Itzykson and J.B. Zuber; see e.g. [10].

References

[1] E. Cunningham, "The principle of relativity in electrodynamics and an extension thereof", Proc. London Math. Soc. **8** (1909) 77–98.

[2] H. Bateman, "The transformation of the electrodynamical equations", Proc. London Math. Soc. **8** (1910) 223–264.

[3] P.A.M. Dirac, "Wave equations in conformal space", Ann. Math. **37** (1936) 429–442.

[4] I.T. Todorov, M.C. Mintchev and V.B. Petkova, *Conformal invariance in quantum field theory* (Scuola Normale Superiore, Pisa, 1978).

[5] S. Ferrara, A.F. Grillo and R. Gatto, "Conformal algebra in two space-time dimensions and the Thirring model", Nuovo Cim. **12A** (1972) 959–968.

[6] S. Fubini, A.J. Hanson and R. Jackiw, "New approach to field theory", Phys. Rev. **D7** (1973) 1732–1760.

[7] A.M. Polyakov, "A non-Hamiltonian approach to conformal field theory", Sov. Phys. JETP **39** (1974) 10–18.

[8] W. Nahm, "The mass spectra of dual strings", Nucl. Phys. **B114** (1976) 174–188.

[9] D. Gepner and E. Witten, "String theory on group manifolds", Nucl. Phys. **B278** (1987) 493–549.

[10] A. Cappelli, C. Itzykson and J.B. Zuber, "Modular invariant partition functions in two dimensions", Nucl. Phys. **B280** **[FS18]** (1987) 445–465.

Nuclear Physics B241 (1984) 333–380

INFINITE CONFORMAL SYMMETRY IN TWO-DIMENSIONAL QUANTUM FIELD THEORY

A.A. BELAVIN, A.M. POLYAKOV and A.B. ZAMOLODCHIKOV

L.D. Landau Institute for Theoretical Physics, Academy of Sciences, Kosygina 2, 117334 Moscow, USSR

Received 22 November 1983

We present an investigation of the massless, two-dimensional, interacting field theories. Their basic property is their invariance under an infinite-dimensional group of conformal (analytic) transformations. It is shown that the local fields forming the operator algebra can be classified according to the irreducible representations of Virasoro algebra, and that the correlation functions are built up of the "conformal blocks" which are completely determined by the conformal invariance. Exactly solvable conformal theories associated with the degenerate representations are analyzed. In these theories the anomalous dimensions are known exactly and the correlation functions satisfy the systems of linear differential equations.

1. Introduction

Conformal symmetry was introduced into quantum field theory about twelve years ago due to the scaling ideas in the second-order phase transition theory (see [1] and references therein). According to the scaling hypothesis, the interaction of the fields of the order parameters in the critical point is invariant with respect to the scale transformations

$$\xi^a \to \lambda \xi^a, \tag{1.1}$$

where ξ^a are the coordinates, $a = 1, 2, \ldots, D$. In the quantum field theory the scale symmetry (1.1) takes place provided the stress-energy tensor is traceless

$$T_a^a(\xi) = 0. \tag{1.2}$$

Under the condition (1.2) the theory possesses not only the scale symmetry but is also invariant with respect to the coordinate transformations

$$\xi^a \to \eta^a(\xi) \tag{1.3}$$

having the property that the metric tensor transforms as

$$g_{ab} \to \frac{\partial \xi^{a'}}{\partial \eta^a} \frac{\partial \xi^{b'}}{\partial \eta^b} g_{a'b'} = \rho(\xi) g_{ab}, \tag{1.4}$$

where $\rho(\xi)$ is a certain function. Coordinate transformations of this type constitute the *conformal group*. These transformations can be easily described, properties of the conformal group being different for the cases $D>2$ and $D=2$. If $D>2$, the conformal group is finite-dimensional and consists of translations, rotations, dilatations and special conformal transformations (see [2, 3]). Kinematic manifestation of this symmetry and its dynamical realization in the quantum field theory has been investigated in many papers (see for example, [2–4]). In particular, it has been shown that the local fields $A_j(\xi)$, involved in the conformal theory, should possess anomalous scale dimensions d_j, i.e. they transform as follows under the transformation (1)

$$A_j \to \lambda^{-d_j} A_j, \tag{1.5}$$

where the parameters d_j are non-negative. Computation of the spectrum $\{d_j\}$ of the anomalous dimensions is the most important problem of the theory since these quantities determine the critical exponents.

To solve this problem, in [4] the bootstrap approach based on the operator algebra hypothesis has been proposed. Let us describe it in some detail since it is most suitable for our purposes. The operator algebra is a strong version of the Wilson operator product expansion [5]; namely, if the existence of an infinite set of local fields $A_j(\xi)$ is assumed, then the set of operators $\{A_j(0)\}$ is assumed to be complete in the sense specified below. The set $\{A_j\}$ contains the identity operator I as well as all coordinate derivatives of each field involved. The completeness of the set $\{A_j(0)\}$ means that any state can be generated by the linear action of these operators. This condition is equivalent to the operator algebra

$$A_i(\xi)A_j(0) = \sum_k C_{ij}^k(\xi)A_k(0), \tag{1.6}$$

where the structure constants $C_{ij}^k(\xi)$ are the c-number functions which should be single-valued so that local properties be taken into account. The relation (1.6) is understood as an exact expansion of the correlation functions

$$\langle A_i(\xi)A_j(0)A_{l_1}(\xi_1)\dots A_{l_N}(\xi_N)\rangle = \sum_k C_{ij}^k(\xi)\langle A_k(0)A_{l_1}(\xi_1)\dots A_{l_N}(\xi_N)\rangle,$$

which is convergent in some finite domain of ξ, the domain being certainly dependent on the location of $\xi_1,\dots,\xi_N$. The most rigid requirement, considered as the main dynamical principle of this approach, is associativity of the operator algebra (1.6). This requirement leads to an infinite system of equations for the structure constants $C_{ij}^k(\xi)$. Since the conformal symmetry fixes the form of the functions $C_{ij}^k(\xi)$ up to some numerical parameters (which are the anomalous dimensions and numerical factors), this system of equations has to determine these parameters. However in the multidimensional theory ($D > 2$) this system proves to be too complicated to be solved exactly, the main difficulty being the classification of the fields A_j entering the algebra.

The situation is somewhat better in two dimensions. The main reason is that the conformal group is infinite-dimensional in this case; it consists of the conformal analytical transformations. To describe this group, it is convenient to introduce the complex coordinates

$$z = \xi^1 + i\xi^2, \qquad \bar{z} = \xi^1 - i\xi^2, \tag{1.7}$$

the metric having the form

$$ds^2 = dz\, d\bar{z}. \tag{1.8}$$

The conformal group of the two-dimensional space which will be denoted by $\mathcal{G}$, consists of all substitutions of the form

$$z \to \zeta(z), \qquad \bar{z} \to \bar{\zeta}(\bar{z}), \tag{1.9}$$

where ζ and $\bar{\zeta}$ are arbitrary analytical functions.

For our purposes it will be convenient to consider the space coordinates ξ^1, ξ^2 as complex variables, i.e. to deal with the complex space $\mathbf{C}^2$. Therefore in general we shall treat the coordinates (1.7) not as complex conjugated but as two independent complex variables; the same is supposed for the functions (1.9). This space $\mathbf{C}^2$ has the complex metric (1.8). The euclidean plane and Minkowski space-time can be obtained as appropriate real sections of this complex space.

In the complex case it is clear from (1.9) that the conformal group $\mathcal{G}$ is a direct product

$$\mathcal{G} = \Gamma \otimes \bar{\Gamma}, \tag{1.10}$$

where Γ ($\bar{\Gamma}$) is a group of the analytical substitutions of the variable z ($\bar{z}$). In what follows we shall often concentrate on properties of the group Γ, keeping in mind that the same properties hold for $\bar{\Gamma}$.

Infinitesimal transformations of the group Γ are

$$z \to z + \varepsilon(z), \tag{1.11}$$

where $\varepsilon(z)$ is an infinitesimal analytical function. It can be represented as an infinite Lourant series

$$\varepsilon(z) = \sum_{n=-\infty}^{\infty} \varepsilon_n z^{n+1}. \tag{1.12}$$

Therefore the Lie algebra of the group Γ coincides with the algebra of differential operators

$$l_n = z^{n+1}\frac{\mathrm{d}}{\mathrm{d}z}, \qquad n = 0, \pm 1, \pm 2, \ldots, \tag{1.13}$$

the commutation relations having the form

$$[l_n, l_m] = (n-m)l_{n+m}. \tag{1.14}$$

The generators $\bar{l}_n$ of the group $\bar{\Gamma}$ satisfy the same commutation relations, the operators l_n and $\bar{l}_m$ being commutative. We shall denote the algebra (1.14) as $\mathfrak{L}_0$.

The generators l_{-1}, l_0, l_{+1} form the subalgebra $\mathrm{sl}(2, \mathbf{C}) \subset \mathfrak{L}_0$. The corresponding subgroup $\mathrm{SL}(2, \mathbf{C}) \subset \Gamma$ consists of the projective transformations

$$z \to \zeta = \frac{az+b}{cz+d}, \qquad ad - bc = 1. \tag{1.15}$$

Note that the projective transformations are uniquely invertible mappings of the whole z-plane on itself and these are the only conformal transformations with this property.

This is the first paper of the series we intend to devote to the general properties of the two-dimensional quantum field theory, invariant with respect to the conformal group $\mathcal{G}$*. In this paper we give the general classification of the fields $A_j(\xi)$ entering the operator algebra (1.6) according to the representations of the conformal group and investigate special "exactly solvable" cases of the conformal quantum field theory associated with degenerate representations. In more detail we shall show the following.

(i) The components of the stress-energy tensor $T_{ab}(\xi)$ (satisfying (1.2)) represent the generators of the conformal group $\mathcal{G}$ in the quantum field theory. The algebra of these generators is the central extension of the algebra $\mathfrak{L}_0$ (1.14) and coincides with the *Virasoro algebra* $\mathfrak{L}_c$. The value of the central charge c is the parameter of the theory.

(ii) Among the fields $A_j(\xi)$ forming the operator algebra, there are some *primary fields* $\phi_n(\xi)$ which transform in the simplest way

$$\phi_n(z, \bar{z}) \to \left(\frac{\mathrm{d}\zeta}{\mathrm{d}z}\right)^{\Delta_n}\left(\frac{\mathrm{d}\bar{\zeta}}{\mathrm{d}\bar{z}}\right)^{\bar{\Delta}_n}\phi_n(\zeta, \bar{\zeta}) \tag{1.16}$$

* Although the projective group (1.15) and the complete conformal group $\mathcal{G}$ are both consequences of (1.2) and therefore appear in the quantum field theory together, we found it instructive to consider first the general consequences of the projective symmetry. The corresponding formulae, which are certainly no other than the particular case $D = 2$ of the results of refs. [2–4], are presented in appendix A.

under the substitutions (1.9). Here Δ_n and $\bar{\Delta}_n$ are real non-negative parameters. In fact, the combinations $d_n = \Delta_n + \bar{\Delta}_n$ and $s_n = \Delta_n - \bar{\Delta}_n$ are the anomalous scale dimension and the spin of the field ϕ_n, respectively*. We shall often refer to the quantities Δ_n and $\bar{\Delta}_n$ as to the *dimensions* of the field. The simplest example of the primary field is the identity operator I. A nontrivial theory involves more than one primary field and the index n is introduced to distinguish between them.

(iii) A complete set of the fields $A_j(\xi)$ consists of *conformal families* $[\phi_n]$, each corresponding to a certain primary field ϕ_n. The primary field ϕ_n belongs to the *conformal family* $[\phi_n]$ and, in some sense, serves as the ancestor of the family. Each conformal family also contains infinitely many other secondary fields (descendants). Dimensions of these secondary fields form integer spaced series

$$\Delta_n^{(k)} = \Delta_n + k\,, \qquad \bar{\Delta}_n^{(\bar{k})} = \bar{\Delta}_n + \bar{k}\,, \tag{1.17}$$

where $k, \bar{k} = 0, 1, 2, \ldots$. Variations of any secondary field $A \in [\phi_n]$ under the infinitesimal conformal transformations (1.11) are expressed linearly in terms of representations of the same conformal family $[\phi_n]$. So, each conformal family corresponds to some representation of the conformal group $\mathcal{G}$. In accordance with (1.10), this representation is a direct product $[\phi_n] = V_n \otimes \bar{V}_n$, where V_n and $\bar{V}_n$ are representations of the Virasoro algebra $\mathfrak{L}_c$**; in general, these representations are irreducible.

(iv) Correlation functions of any secondary fields can be expressed in terms of the correlators of the corresponding primary fields by means of special linear differential operators. Therefore, all information about the conformal quantum field theory is accumulated in the correlators of the primary field ϕ_n.

(v) The structure constants $C_{ij}^k(\xi)$ of the operator algebra (1.6) can, in principle, be computed in terms of the coefficients C_{nm}^l of the primary field ϕ_l in the operator product expansion of $\phi_n \phi_m$. Therefore, the bootstrap equations (i.e. the associativity condition for the operator algebra) can be reduced to equations imposing constraints upon these coefficients and the dimensions Δ_n of the primary field.

(vi) At a given value of the charge c there are infinitely many special values of the dimension Δ such that the representation $[\phi_\Delta]$ proves to be degenerate. The most important property of the corresponding "degenerate" primary field ϕ_Δ is that the correlation functions involving this field, satisfy special linear differential equations, the simplest example of which is the hypergeometry equation.

(vii) If the parameter c satisfies the equation:

$$\frac{\sqrt{25-c} - \sqrt{1-c}}{\sqrt{25-c} + \sqrt{1-c}} = \frac{p}{q}\,, \tag{1.18}$$

* The spin s_n of a local field can take an integer or half-integer value only.

** The representation V_n is known as the Verma modulus over the Virasoro algebra (see, for example, [6]). This representation is evidently characterized by the parameter Δ_n only.

where p and q are positive integers, the "minimal" conformal quantum field theory can be constructed so that it be exactly solvable in the following sense. (i) A finite number of conformal families $[\phi_n]$ is involved in the operator algebra, each of them being degenerate, (ii) all anomalous dimensions Δ_n are known exactly, (iii) all correlation functions of the theory can be computed as solutions of special systems of linear partial differential equations. There are infinitely many conformal quantum field theories of this type, each associated with a certain solution of (1.18), the simplest nontrivial example ($c = \frac{1}{2}$) describing the critical theory of the two-dimensional Ising model. We suppose that other "minimal" conformal theories describe second-order phase transitions in some two-dimensional spin systems with discrete symmetry groups.

Apart from second-order phase transitions in two dimensions, there is another application of the conformal quantum field theory. This is the dual theory. From the mathematical point of view dual models are no other than special kinds of the two-dimensional conformal quantum field theory. This is natural in view of their association with the string theory. Quantum fields describe the degrees of freedom associated with the string, the conformal symmetry being a manifestation of the reparametrization invariance of the world surface swept out by the string. In fact, the dual amplitudes are expressed in terms of correlation functions of some local fields (vertex operators). In standard models (like the Veneziano model) vertex operators are related in a simple way to free massless fields. We suppose that if considerably interacting fields are incorporated into the theory, it can produce new types of dual models with more suitable physical properties.

2. Stress-energy tensor in the conformal quantum field theory

Consider an arbitrary correlation function of the form

$$\langle X\rangle = \langle A_{j_1}(\xi_1)\dots A_{j_N}(\xi_N)\rangle, \tag{2.1}$$

where $A_{j_k}(\xi)$ are local fields, and perform an infinitesimal coordinate transformation

$$\xi^a \to \xi^a + \varepsilon^a(\xi). \tag{2.2}$$

As is well known in quantum field theory, the following relation is valid:

$$\sum_{k=1}^{N} \langle A_{j_1}(\xi_1)\dots A_{j_{k-1}}(\xi_{k-1})\delta_\varepsilon A_{j_k}(\xi_k)A_{j_{k+1}}(\xi_{k+1})\dots A_{j_N}(\xi_N)\rangle$$

$$+\int d^2\xi\, \partial^a\varepsilon^b(\xi)\langle T_{ab}(\xi)X\rangle = 0, \tag{2.3}$$

where the field $T_{ab}(\xi)$ is the stress-energy tensor and $\delta_\varepsilon A_j$ denotes variations of the fields A_j under the transformation (2.2). Due to their local properties, these variations are linear combinations of a finite number of derivatives of the function $\varepsilon(\xi)$ taken at the point $\xi = \xi_k$, the coefficients being certain local fields. It follows from (2.3) that

$$\partial_a \langle T^{ab}(\xi) X \rangle = 0 \tag{2.4}$$

everywhere but at the points $\xi_1, \xi_2, \ldots, \xi_N$. In the conformal quantum field theory the trace of the stress-energy tensor vanishes, $T_a^a = 0$. Therefore in two dimensions this tensor has only two independent components which can be chosen in the form

$$T(\xi) = T_{11} - T_{22} + 2iT_{12},$$

$$\bar{T}(\xi) = T_{11} - T_{22} - 2iT_{12}. \tag{2.5}$$

Combining relations (1.2) and (2.4), it is easy to find that these components satisfy the Cauchy-Riemann equations

$$\partial_{\bar{z}} \langle T(\xi) X \rangle = 0,$$

$$\partial_z \langle \bar{T}(\xi) X \rangle = 0, \tag{2.6}$$

where z and $\bar{z}$ are defined by (1.7). So, each of the fields T and $\bar{T}$ is an analytic function of the single variable (z and $\bar{z}$, respectively) and we shall write

$$T = T(z), \qquad \bar{T} = \bar{T}(\bar{z}). \tag{2.7}$$

Take now the correlation function*

$$\langle T(z) X \rangle. \tag{2.8}$$

It is the analytic function of z that is single-valued (due to its local properties) and regular everywhere but at the points $z = z_k$, $z_k = \xi_k^1 + i\xi_k^2$, where it has poles, the orders and residues of these poles being determined by the conformal properties of the fields $A_{j_k}(\xi)$. Actually, for the conformal coordinate transformations (1.11) the relation (2.3) can be reduced to the form

$$\langle \delta_\varepsilon X \rangle = \oint_C d\zeta\, \varepsilon(\zeta) \langle T(\zeta) X \rangle, \tag{2.9}$$

* Here and below we generally consider correlation functions in the complex space $\mathbf{C}^2$, see the introduction.

where $\delta_\varepsilon X$ is a variation of the product $X = A_{j_1}(\xi_1)\ldots A_{j_N}(\xi_N)$ under the transformation (1.11) and the contour C encloses all singular points z_k, $k=1,\ldots,N$. Equivalently, the following relation is valid

$$\delta_\varepsilon A_j(z,\bar z) = \oint_{C_z} d\zeta\, \varepsilon(\zeta) T(\zeta) A_j(z,\bar z), \tag{2.10}$$

where the contour C_z surrounds the point z. The same formula (with the substitution $T \to \bar T$) holds for the variation $\delta_{\bar\varepsilon} A_j$ of the field A_j under the infinitesimal transformation

$$\bar z \to \bar z + \bar\varepsilon(z) \tag{2.11}$$

of the group $\bar\Gamma$. Therefore the fields $T(z)$ and $\bar T(\bar z)$ represent the generators of the conformal group $\Gamma \otimes \bar\Gamma$ in the quantum field theory.

The conformal transformation laws for general fields A_j will be considered in the next section. Now we are interested in the conformal properties of the fields $T(z)$ and $\bar T(\bar z)$ themselves which are obviously related to the algebra of the conformal group generators. The variations $\delta_\varepsilon T$ and $\delta_\varepsilon \bar T$ should be expressed linearly in terms of the same fields T and $\bar T$ and their derivatives and may also include the c-number Schwinger terms. Taking into account tensorial properties of the field $T(z)$ and the locality condition, write down the following most general expression for the variation $\delta_\varepsilon T$:

$$\delta_\varepsilon T(z) = \varepsilon(z) T' + 2\varepsilon'(z) T(z) + \tfrac{1}{12} c\, \varepsilon'''(z), \tag{2.12}$$

where the prime denotes the z-derivative*. For the variation $\delta_{\bar\varepsilon} T$ it is possible to get

$$\delta_{\bar\varepsilon} T(z) = 0. \tag{2.13}$$

* Formula (2.12) corresponds to the following transformation of $T(z)$ under the finite conformal substitution (1.9):

$$T(z) \to \left(\frac{d\zeta}{dz}\right)^2 T(\zeta) + \tfrac{1}{12} c \{\zeta, z\},$$

where $\{\zeta, z\}$ is the Schwartz derivative [12]

$$\{\zeta, z\} = \left(\frac{d^3\zeta}{dz^3} \Big/ \frac{d\zeta}{dz}\right) - \frac{3}{2}\left(\frac{d^2\zeta}{dz^2} \Big/ \frac{d\zeta}{dz}\right)^2.$$

Note, that the Schwartz derivative satisfies the following composition law:

$$\{w, z\} = \left(\frac{d\zeta}{dz}\right)^2 \{w, \zeta\} + \{\zeta, z\}.$$

The numerical constant c in the relation (2.12) is not determined by the general principles; it should be treated as the parameter of the theory. The variation $\delta_\varepsilon \bar{T}$ satisfies the same relation (2.1), the respective constant $\bar{c}$ being equal to c. The constant c can take real positive values. These statements result from the reality condition for the stress-energy tensor in euclidean space and Minkowski space-time.

If none of the points z_k, $k = 1, 2, \ldots, N$ in (2.1) is equal to infinity, the correlation function $\langle T(z)X\rangle$ should be regular at $z = \infty$. This means that, as can be easily verified by means of the transformation law (2.12), that the function $\langle T(z)X\rangle$ decreases as

$$T(z) \sim \frac{1}{z^4} \quad \text{at} \quad z \to \infty . \tag{2.14}$$

In the quantum field theory the correlation functions (2.1) are represented as vacuum expectation values of the time-ordered products of the local field operators $A_j(\xi)$. In our case it is convenient to introduce the coordinates σ and τ according to the formulae:

$$z = \exp(\tau + i\sigma), \qquad \bar{z} = \exp(\tau - i\sigma). \tag{2.15}$$

Choosing σ and τ as real, σ being an angular variable, $0 < \sigma \leqslant \pi$, one gets the euclidean real section. Correlation functions in this euclidean space can be represented as

$$\langle X\rangle = \langle 0|T\left[A_{j_1}(\sigma_1, \tau_1) \ldots A_{j_N}(\sigma_N, \tau_N)\right]|0\rangle, \tag{2.16}$$

where the chronological ordering should be performed with respect to the "euclidean time" τ. In the operator formalism the variations $\delta_\varepsilon A_j$ can be expressed in terms of equal time commutators

$$\delta_\varepsilon A_j(\sigma, \tau) = \left[T_\varepsilon, A_j(\sigma, \tau)\right], \tag{2.17}$$

where the generators T_ε are defined by the formula

$$T_\varepsilon = \oint_{\log|z| = \tau} \varepsilon(z) T(z)\,\mathrm{d}z. \tag{2.18}$$

Note that due to eqs. (2.7) these operators are in fact τ-independent.

The relation (2.12) becomes

$$\left[T_\varepsilon, T(z)\right] = \varepsilon(z)T'(z) + 2\varepsilon'(z)T(z) + \tfrac{1}{12}c\varepsilon'''(z). \tag{2.19}$$

It is useful to introduce the operators L_n, $\bar{L}_n$, $n = 0, \pm 1, \pm 2, \ldots$ as coefficients of the Lourant expansions

$$T(z)=\sum_{n=-\infty}^{\infty}\frac{L_n}{z^{n+2}},\qquad \bar{T}(\bar{z})=\sum_{n=-\infty}^{\infty}\frac{\bar{L}_n}{\bar{z}^{n+2}}. \tag{2.20}$$

It follows from (2.19) that the operators L_n satisfy the commutation relations:

$$[L_n, L_m]=(n-m)L_{n+m}+\tfrac{1}{12}c(n^3-n)\delta_{n+m,0}. \tag{2.21}$$

Clearly, the same relations are satisfied by the $\bar{L}_n$'s, the operators L_n and $\bar{L}_m$ being commutative. The algebra (2.21) of the conformal generators L_n is the central extension of the algebra (1.14)*. This is well known in the dual theory and the algebra (2.21) is called the Virasoro algebra [11]; we shall denote it as $\mathcal{L}_c$.

Like the algebra $\mathcal{L}_0$, the Virasoro algebra $\mathcal{L}_c$ contains a subalgebra sl(2, **C**), generated by the operators L_{-1}, L_0, L_{+1} (note that the c-number term in (2.21) vanishes for $n=0,\pm1$). In particular, the operators L_{-1} and $\bar{L}_{-1}$ generate translations whereas L_0 and $\bar{L}_0$ generate infinitesimal dilatations of the coordinates z and $\bar{z}$. In the coordinate system σ, τ defined by (2.15) the operator

$$H=L_0+\bar{L}_0, \tag{2.22}$$

is a generator of "time" shifts. It plays the role of the hamiltonian. Note, that the "infinite past" $\tau\to-\infty$ and the "infinite future" $\tau\to\infty$ correspond to the points $z=0$ and $z=\infty$, respectively.

The vacuum $|0\rangle$ in (2.16) is the ground state of the hamiltonian (2.22). The vacuum must satisfy the equations

$$L_n|0\rangle=0,\qquad \text{if}\qquad n\geqslant-1, \tag{2.23}$$

since otherwise the stress-energy tensor would have been singular at $z=0$. Note that the operators L_n with $n\geqslant-1$ generate the conformal transformations which are regular at $z=0$. Therefore eqs. (2.23) are manifestations of the conformal invariance of the vacuum. The transformations generated by the operators L_n with $n\leqslant-2$ are singular at $z=0$; these operators distort the vacuum

$$L_n|0\rangle=\text{new states}\qquad \text{if}\qquad n\leqslant-2. \tag{2.24}$$

The field $T(z)$ should also be regular at $z=\infty$. Similarly to (2.23), it implies that

$$\langle 0|L_n=0\qquad \text{if}\qquad n\leqslant 1. \tag{2.25}$$

Since in the Minkowski space-time (which can be obtained if imaginary values of τ

* This central extension has been discovered by Gelfand and Fuks [10].

are dealt with), the field $T(z)$ must be real, the operators L_n satisfy the conjugation relation

$$L_n^+ = L_{-n}. \tag{2.26}$$

Note that the generators L_{-1}, L_0, L_1 annihilate both the "in" and "out" vacuua

$$\langle 0|L_s = L_s|0\rangle = 0, \qquad s = 0, \pm 1. \tag{2.27}$$

These equations are manifestation of the regularity of projective transformations mentioned in the introduction. Eqs. (2.27) are self-consistent because the c-number term in (2.21) vanishes for $n = 0, \pm 1$.

Eqs. (2.23), (2.25) and the commutation relations (2.21) enable one to compute any correlation function of the form*

$$\langle T(\zeta_1)\dots T(\zeta_N)\bar{T}(\eta_1)\dots\bar{T}(\eta_M)\rangle = \langle T(\zeta_1)\dots T(\zeta_N)\rangle\langle\bar{T}(\eta_1)\dots\bar{T}(\eta_M)\rangle. \tag{2.28}$$

In particular, a two-point function is given by the formula

$$\langle T(\zeta_1)T(\zeta_2)\rangle = c(\zeta_1 - \zeta_2)^{-4}, \tag{2.29}$$

which shows that $c > 0$.

3. Ward identities and conformal families

Consider the variation $\delta_\varepsilon A_j(\xi)$ of a certain local field A_j under the infinitesimal conformal transformation (1.11). Due to its local properties, this variation is a linear combination of the function $\varepsilon(z)$ and a finite number of its derivatives taken at the point $z = \xi^1 + i\xi^2$:

$$\delta_\varepsilon A_j(z) = \sum_{k=0}^{\nu_j} B_j^{(k-1)}(z)\frac{\mathrm{d}^k}{\mathrm{d}z^k}\varepsilon(z), \tag{3.1}$$

where $B_j^{(k-1)}$ are local fields belonging to the set $\{A_j\}$ and ν_j is a certain integer. In

* It can be shown that these correlators coincide with those of the fields

$$T^{(0)} = \varphi_z\varphi_z + 2\alpha_0\varphi_{zz}$$

where φ is a free massless boson field and the parameter α_0 is defined by the formula

$$c = 1 + 24\alpha_0^2.$$

(3.1) we have omitted the argument $\bar{z}$ which is not important here. The study of infinitesimal translations and dilatations of the variable shows that the first and second coefficients in (3.1) are

$$B_j^{(-1)}(z)=\frac{\partial}{\partial z}A_j(z), \qquad B_j^0(z)=\Delta_j A_j(z), \tag{3.2}$$

where Δ_j is the dimension of the field A_j. It is evident that the dimensions of the fields $B_j^{(k-1)}$ in (3.1) are equal to

$$\Delta_{j,(k-1)}=\Delta_j+1-k, \qquad k=0,1,\dots,\nu_j. \tag{3.3}$$

Let us take again the correlation function (2.8). As has already been mentioned in the previous section, this correlator is a single-valued analytic function of z, possessing the poles at $z=z_k$, $k=1,2,\dots N$. In virtue of (2.10) and (3.1) it is possible to write down the relation

$$\langle T(z)A_{j_1}(z_1)\dots A_{j_N}(z_N)\rangle=\sum_{l=1}^{N}\sum_{k=0}^{\nu_l}k!(z-z_l)^{-k-1}\langle A_{j_1}(z_1)\dots$$
$$\dots A_{j_{l-1}}(z_{l-1})B_{j_l}^{(k-1)}(z_l)A_{j_{l+1}}(z_{l+1})\dots A_{j_N}(z_N)\rangle. \tag{3.4}$$

This formula is a general form of the conformal Ward identities.

In a physically suitable theory the dimensions Δ_j of all the fields A_j should satisfy the inequality

$$\Delta_j\geqslant 0, \tag{3.5}$$

since otherwise the theory will possess correlations increasing with distance. In what follows we shall suppose that the only field with zero dimensions $\Delta=\bar{\Delta}=0$ is the identity operator I. Comparing (3.3) with condition (3.5) we see that the sum in (3.1) contains a finite number of terms $\nu_j\leqslant\Delta_j+1$. Another important conclusion following from (3.3) is that the spectrum of dimensions $\{\Delta_j\}$ in any two-dimensional conformal quantum field theory consists of the infinite integer spaced series

$$\Delta_n^{(k)}=\Delta_n+k, \qquad k=0,1,2,\dots \tag{3.6}$$

Here Δ_n denotes the minimal dimension of each series, whereas the index n labels the series. The same is obviously valid for the dimensions $\bar{\Delta}_j$, i.e. the spectrum $\{\bar{\Delta}_j\}$ also consists of the series

$$\bar{\Delta}_n^{(k)}=\bar{\Delta}_n+k, \qquad k=0,1,2,\dots \tag{3.7}$$

Let ϕ_n be the field with the dimensions Δ_n and $\bar{\Delta}_n$. The variation (3.1) of this field has the simplest possible form

$$\delta_\varepsilon \phi_n(z) = \varepsilon(z) \frac{\partial}{\partial z} \phi_n(z) + \Delta_n \varepsilon'(z) \phi_n(z), \tag{3.8}$$

since the corresponding fields $B^{(k-1)}$ with $k > 0$ would have dimensions smaller than Δ_n. A similar formula holds for the variation $\delta_{\bar{\varepsilon}} \phi_n$. The finite form of this conformal transformation law is given by (1.16). We shall call the operators ϕ_n having the transformation laws (1.16) the *primary fields*. Note that formula (3.8) is equivalent to the commutation relation:

$$[L_m, \phi_n(z)] = z^{m+1} \frac{\partial}{\partial z} \phi_n(z) + \Delta_n (m+1) z^m \phi_n(z), \tag{3.9}$$

which are satisfied by the vertex operators of the dual theory [8, 9].

If all the fields $A_j(\xi)$ entering the correlation function (2.8) are primary, the general relation (3.4) is reduced to the form

$$\langle T(z) \phi_1(z_1) \dots \phi_N(z_N) \rangle = \sum_{i=1}^{N} \left\{ \frac{\Delta_i}{(z - z_i)^2} + \frac{1}{z - z_i} \frac{\partial}{\partial z_i} \right\} \langle \phi_1(z_1) \dots \phi_N(z_N) \rangle, \tag{3.10}$$

where $\Delta_1, \Delta_2, \dots, \Delta_N$ are dimensions of the primary fields $\phi_1, \phi_2, \dots, \phi_N$, respectively. Note that this Ward identity explicitly relates the correlation functions $\langle T(z) \phi_1 \dots \phi_N \rangle$ to the correlators $\langle \phi_1 \dots \phi_N \rangle$. It is also noteworthy that the projective conformal Ward identities (A.6) can be directly derived from (3.10) if one takes into account the asymptotic condition (2.14).

The primary fields themselves cannot form the closed operator algebra. In fact, there are infinitely many other fields associated with each of the primary fields ϕ_n. We shall refer to these fields as to the secondary fields with respect to the primary fields ϕ_n. The dimensions of the secondary fields form the integer spaced series, mentioned above. These fields together with the primary field ϕ_n constitute a *conformal family* $[\phi_n]$. It is essential that under the transformations every member of each conformal family transforms in terms of the representatives of the same conformal family. So, each conformal family forms some irreducible representation of the conformal algebra. The complete set of the fields $\{A_j\}$ consists of some number (which can be infinite) of the conformal families

$$\{A_j\} = \bigoplus_n [\phi_n]. \tag{3.11}$$

To understand the nature of these secondary fields, consider the product

$T(\zeta)\phi_n(z,\bar{z})$. This product can be expanded according to (1.6), the coefficients C_{ij}^k being single-valued analytic functions of $(\zeta - z)$ in virtue of relation (2.7) and the local properties of the fields $T(\zeta)$ and $\phi_n(z,\bar{z})$. Therefore this product can be represented as

$$T(\zeta)\phi_n(z) = \sum_{k=0}^{\infty} (\zeta - z)^{-2+k} \phi_n^{(-k)}(z), \tag{3.12}$$

where we have again omitted the dependences of the fields on the variable $\bar{z}$. The dimensions of the fields $\phi_n^{(-k)}$ are given by (3.7). The singular terms in (3.12) are completely determined by the transformation law (3.8) (remember (2.10)). Thus the first two coefficients in (3.12) are

$$\phi_n^{(-1)}(z) = \frac{\partial}{\partial z}\phi_n(z), \qquad \phi_n^{(0)}(z) = \Delta_n \phi_n(z). \tag{3.13}$$

The coefficients $\phi_n^{(-k)}$, $k = 2, 3, \ldots$, of the regular terms in (3.12) are new local fields. To make sure of the existence of these fields, it is possible to expand the Ward identity (3.10) in power series, say, in $z - z_1$. These new fields are representatives of the conformal family $[\phi_n]$, $\phi_n^{(-k)} \in [\phi_n]$. The conformal properties of these secondary fields $\phi_n^{(-k)}$ are more complicated than those of the primary field ϕ_n. The infinitesimal conformal transformation and comparison of both sides of (3.12) yield

$$\begin{aligned}\delta_\varepsilon \phi_n^{(-k)}(z) &= \varepsilon(z)\frac{\partial}{\partial z}\phi_n^{(-k)}(z) + (\Delta_n + k)\varepsilon'(z)\phi_n^{(-k)}(z) \\ &+ \sum_{l=1}^{k} \frac{k+l}{(l+1)!}\left[\frac{\mathrm{d}^{l+1}}{\mathrm{d}z^{l+1}}\varepsilon(z)\right]\phi_n^{(l-k)}(z) \\ &+ \tfrac{1}{12}c\frac{1}{(k-2)!}\left[\frac{\mathrm{d}^{k+1}}{\mathrm{d}z^{k+1}}\varepsilon(z)\right]\phi_n(z).\end{aligned} \tag{3.14}$$

The fields $\phi_n^{(-k)}$ are not the only ones belonging to the conformal family $[\phi_n]$. Consider, for instance, the operator product expansion

$$\begin{aligned}T(\zeta)\phi_n^{(-k_2)}(z) &= \tfrac{1}{12}c(\zeta - z)^{-k_2-2}(k_2^3 - k_2)\phi_n(z) \\ &+ \sum_{l=1}^{k_2} (\zeta - z)^{-l-2}(l + k_2)\phi_n^{(l-k_2)}(z) \\ &+ \sum_{k_1=0}^{\infty} (\zeta - z)^{-2+k_1}\phi_n^{(-k_1,-k_2)}(z).\end{aligned} \tag{3.15}$$

The operators accompanying the singular terms in (3.15) are unambiguously determined by formula (3.14). In particular

$$\phi_n^{(-1,-k)}(z)=\frac{\partial}{\partial z}\phi_n^{(-k)}(z),\qquad \phi_n^{(0,-k)}(z)=(\Delta_n+k)\phi_n^{(-k)}(z). \tag{3.16}$$

The new local fields $\phi_n^{(-k_1,-k_2)}$ with $k_1>1$ also belong to the conformal family $[\phi_n]$. The variations $\delta_\epsilon\phi^{(-k_1,-k_2)}$ are expressed in terms of the fields $\phi_n^{(-l_1,-l_2)}$, $\phi_n^{(-l)}$ and ϕ_n.

Considering the operator products $T(\zeta)\phi_n^{(-k_1,-k_2)}(z),\dots$ etc., one can discover an infinite set of the secondary fields

$$\phi_n^{(-k_1,-k_2,\dots,-k_N)}(z), \tag{3.17}$$

where $k_i\geqslant 1$ and $N=1,2,\dots$. The fields (3.17) can be defined by the explicit formula

$$\phi_n^{(-k_1,\dots,-k_N)}(z)=L_{-k_1}(z)\dots L_{-k_N}(z)\phi_n(z), \tag{3.18}$$

where the operators $L_{-k}(z)$ are given by the contour integrals

$$L_{-k}(z)=\oint\frac{\mathrm{d}\zeta\, T(\zeta)}{(\zeta-z)^{k+1}}. \tag{3.19}$$

The integration contours associated with each of the operators $L_{-k_i}(z)$ in (3.18) enclose the point z as well as the points $\zeta_{i+1},\zeta_{i+2},\dots,\zeta_N$, which are the integration variables, corresponding to the operators L to the right of L_{-k_i}*. The dimensions of the fields (3.17) are

$$\Delta_n^{(k_1,\dots,k_N)}=\Delta_n+k_1+\cdots+k_N. \tag{3.20}$$

An infinite set of the fields (3.17) constitutes the conformal family $[\phi_n]$. These fields are not linearly independent (see below). In fact, in general the fields (3.17) with $k_1\leqslant k_2\leqslant\cdots\leqslant k_N$ form the basis**. Note that

$$\phi_n^{(-1,-k_1,-k_2,\dots,-k_N)}=\frac{\partial}{\partial z}\phi_n^{(-k_1,-k_2,\dots,-k_N)}. \tag{3.21}$$

Therefore the conformal family $[\phi_n]$ naturally includes all the derivatives of each field involved. It can be derived from (3.18) that the variations $\delta_\epsilon\phi_n^{\{k\}}$, $\{k\}=(-k_1,\dots,-k_N)$ are expressed in terms of the fields, belonging to the same conformal family $[\phi_n]$, and therefore each conformal family corresponds to some representation of the conformal algebra.

* One can easily verify that the operators (3.19), where $=0,\pm1,\pm2,\dots$, satisfy the Virasoro algebra (2.21). Obviously, the operators L_n introduced in sect. 2 are no other than $L_n(0)$.

** This statement does not hold for some special values of Δ_n, see sect. 5.

To describe the structure of the representation it is convenient to turn again to the operator formalism. Let us introduce the vectors (primary states)

$$|n\rangle = \phi_n(0)|0\rangle . \tag{3.22}$$

Using the properties (2.23) of the vacuum and the commutation relations (3.9) one can get

$$L_m|n\rangle = 0 \quad \text{if} \quad m > 0,$$
$$L_0|n\rangle = \Delta_n|n\rangle . \tag{3.23}$$

It follows from (3.18) that

$$\phi_n^{(-k_1,\dots,-k_N)}(0)|0\rangle = L_{-k_1}\dots L_{-k_N}|n\rangle . \tag{3.24}$$

So, the conformal family $[\phi_n]$ is isomorphic to the space of states, generated from the primary state $|n\rangle$ by the negative components L_m, $m<0$*. In the representation theory this space is known as the Verma modulus V_n (see, for example, [6]). Due to the relations (2.21), there are linear dependences between the vectors (3.24). As has been mentioned above, in all cases, excluding certain special values of Δ_n (see sect. 5), the states (3.24) with $k_1 \leqslant k_2 \leqslant \cdots \leqslant k_N$ form the basis in V_n. Note that the vectors (3.24) are the eigenstates of the operator L_0, the eigenvalues being given by (3.20).

So far we have dealt only with the subgroup Γ of the conformal group $\mathcal{G}$. Actually, more precise definitions are required. Since the complete conformal group is the direct product (1.10), the representations $[\phi_n]$ are, in fact, the direct products of the representations of Γ and $\bar{\Gamma}$.

$$[\phi_n] = V_n \otimes \bar{V}_n . \tag{3.25}$$

This means that it contains not only the vectors (3.24) but also all the states of the form

$$\phi_n^{\{k\}\{\bar{k}\}}(0)|0\rangle = L_{-k_1}\dots L_{-k_N}\bar{L}_{-\bar{k}_1}\dots \bar{L}_{-\bar{k}_M}|n\rangle , \tag{3.26}$$

where

$$\{k\} = (-k_1, -k_2, \dots, -k_N); \qquad \{\bar{k}\} = (-\bar{k}_1, -\bar{k}_2, \dots, -\bar{k}_M).$$

k_i and $\bar{k}_j$ are independent positive integers. Remember that the operators L and $\bar{L}$

* This statement is not precise because we neglected the $\bar{z}$ dependence of the fields; the correct definition is given below.

are commutative. According to (1.16), the primary state $|n\rangle$ satisfies, besides (3.23), the equations

$$\bar{L}_m|n\rangle = 0, \qquad \text{if} \qquad m>0,$$
$$\bar{L}_0|n\rangle = \bar{\Delta}_n|n\rangle. \tag{3.27}$$

Therefore each conformal family $[\phi_n]$ is characterized by two parameters Δ_n and $\bar{\Delta}_n$.

Because of the conformal invariance, the two-point functions $\langle\phi_n(\xi_1)\phi_m(\xi_l)\rangle$ vanish unless the fields ϕ_n and ϕ_m have the same dimensions (see appendix A). Moreover, the system of the primary fields can always be chosen to be orthonormal

$$\langle\phi_n(z_1,\bar{z}_1)\phi_m(z_2,\bar{z}_2)\rangle = \delta_{nm}(z_1-z_2)^{-2\Delta_n}(\bar{z}_1-\bar{z}_2)^{-2\bar{\Delta}_n}. \tag{3.28}$$

Let us define the "out" primary states by the formula

$$\langle n| = \lim_{z,\bar{z}\to\infty}\langle 0|\phi_n(z,\bar{z})z^{2L_0}\bar{z}^{2\bar{L}_0}. \tag{3.29}$$

These vectors satisfy the equations

$$\langle n|L_m = 0, \qquad \text{if} \qquad m<0,$$
$$\langle n|L_0 = \Delta_n < n|, \tag{3.30}$$

and the same equation with the substitution $L\to\bar{L}$, $\Delta_n\to\bar{\Delta}_n$. Like in (3.26), we have

$$\lim_{z,\bar{z}\to\infty}\langle 0|\phi_n^{\{k\}\{\bar{k}\}}(z,\bar{z})z^{2L_0}\bar{z}^{2\bar{L}_0} = \langle n|L_{k_N}L_{k_{N-1}}\cdots L_{k_1}\bar{L}_{\bar{k}_M}\cdots\bar{L}_{\bar{k}_1}. \tag{3.31}$$

The orthonormality condition (3.28) can be rewritten as

$$\langle n|m\rangle = \delta_{nm}. \tag{3.32}$$

The conformal Ward identities make it possible to express explicitly any correlation function as

$$\langle T(\zeta_1)\dots T(\zeta_M)\phi_1(z_1)\dots\phi_N(z_N)\rangle, \tag{3.33}$$

in terms of the correlator

$$\langle\phi_1(z_1)\dots\phi_N(z_N)\rangle. \tag{3.34}$$

Here $\phi_1,\dots,\phi_N$ are certain primary fields. This can be done by successively applying the relation

$$\langle T(\zeta)T(\zeta_1)\dots T(\zeta_M)\phi_1(z_1)\dots\phi_N(z_N)\rangle$$
$$=\left\{\sum_{i=1}^{N}\left[\frac{\Delta_i}{(\zeta-z_i)^2}+\frac{1}{\zeta-z_i}\frac{\partial}{\partial z_i}\right]+\sum_{j=1}^{M}\left[\frac{2}{(\zeta-\zeta_j)^2}+\frac{1}{\zeta-\zeta_j}\frac{\partial}{\partial \zeta_j}\right]\right\}$$
$$\times\langle T(\zeta_1)\dots T(\zeta_M)\phi_1(z_1)\dots\phi_N(z_N)\rangle$$
$$+\sum_{j=1}^{M}\frac{c}{(\zeta-\zeta_j)^4}\langle T(\zeta_1)\dots T(\zeta_{j-1})T(\zeta_{j+1})\dots T(\zeta_M)\phi_1(z_1)\dots\phi_N(z_N)\rangle. \tag{3.35}$$

The first term in (3.35) is of the same origin as (3.10), whereas the second term is due to the c-number term in the transformation law (2.12)*.

Using the correlation functions (3.33) one can also compute any correlators of the form

$$\langle\phi_1^{\{k_1\}}(z_1)\dots\phi_N^{\{k_N\}}(z_N)\rangle, \tag{3.36}$$

where $\phi_i^{\{k_i\}}$ are some secondaries of the field ϕ_i, since these secondary fields are no other than the coefficients in the operator product expansions like (3.12), (3.15), etc. Actually in this way the correlators (3.36) are expressed in terms of the correlation functions (3.34) by means of linear differential operators. The general expression is rather cumbersome and we present the simplest example only**

$$\langle\phi_n^{(-k_1,-k_2,\dots-k_M)}(z)\phi_1(z_1)\dots\phi_N(z_N)\rangle$$
$$=\hat{\mathfrak{L}}_{-k_M}(z,z_i)\hat{\mathfrak{L}}_{-k_{M-1}}(z,z_i)\dots\hat{\mathfrak{L}}_{-k_1}(z,z_k)\langle\phi_n(z)\phi_1(z_1)\dots\phi_N(z_N)\rangle, \tag{3.37}$$

where the differential operators $\hat{\mathfrak{L}}_{-k}$ are given by the formula

$$\hat{\mathfrak{L}}_{-k}(z,z_i)=\sum_{i=1}^{N}\left[\frac{(1-k)\Delta_i}{(z-z_k)^k}-\frac{1}{(z-z_i)^{k-1}}\frac{\partial}{\partial z_i}\right]. \tag{3.38}$$

* Obviously, the fields $T(z)$ and $\bar{T}(\bar{z})$ are not primary fields: they belong to the conformal family $[I]$ of the identity operator.

** To obtain (4.5) in the simplest way one can substitute the explicit formula (3.18) and deform the integration contours so as to enclose them around the singularities $z_1, z_2, \dots, z_N$.

Thus the conformal Ward identities enable one to express any correlation functions in terms of the correlators of the primary fields (3.34). Hence, all the information about the conformal quantum field theory is contained in these correlators.

4. Conformal properties of the operator algebra

In the quantum field theory the correlation functions (2.1) should obey the operator algebra (1.6). The conformal symmetry imposes hard restrictions on the coefficients $C_{ij}^{k}(\xi)$. Consider the product of two primary fields $\phi_n(\xi)\phi_m(0)$. The operator product expansion can be represented as

$$\phi_n(z,\bar{z})\phi_m(0,0) = \sum_p \sum_{\{k\}} \sum_{\{\bar{k}\}} C_{nm}^{p;\{k\},\{\bar{k}\}}$$

$$\times z^{\Delta_p - \Delta_n - \Delta_m + \Sigma_i k_i} \bar{z}^{\bar{\Delta}_p - \bar{\Delta}_n - \bar{\Delta}_m + \Sigma \bar{k}_i} \phi_p^{\{k\}\{\bar{k}\}}(0,0), \qquad (4.1)$$

where $\phi_p^{\{k\}\{\bar{k}\}}$ are the secondary fields, belonging to the conformal family $[\phi_p]$. Both sides of (4.1) should exhibit the same conformal properties. The transformation law of the left-hand side is determined by (3.8); the conformal properties of each term in the right-hand side can be derived, in principle, from (3.18). The requirement of the conformal invariance of (4.1) leads to the relations for the numerical constants $C_{nm}^{p\{k\}\{\bar{k}\}}$ with different $\{k\}$'s but with the same index (see appendix B). In principle, these relations can be solved recurrently, the solution being represented as

$$C_{nm}^{p\{k\}\{\bar{k}\}} = C_{nm}^{p}\beta_{nm}^{p,\{k\}}\bar{\beta}_{nm}^{p\{\bar{k}\}}, \qquad (4.2)$$

where C_{nm}^{p} are the constants of the primary fields ϕ_p themselves and the factors β ($\bar{\beta}$) are expressed unambiguously in terms of the dimensions $\Delta_n, \Delta_m, \Delta_p$ ($\bar{\Delta}_n, \bar{\Delta}_m, \bar{\Delta}_p$) only; the condition $\beta_{nm}^{p\{0\}} = \bar{\beta}_{nm}^{p\{0\}} = 1$ is implied. The factorized (in terms of β) form of (4.2) is a consequence of (3.25). The expansion (4.1) can be rewritten as

$$\phi_n(z,\bar{z})\phi_m(0,0) = \sum_p C_{nm}^{p} z^{\Delta_p - \Delta_n - \Delta_m} \bar{z}^{\bar{\Delta}_p - \bar{\Delta}_n - \bar{\Delta}_m} \Psi_p(z,\bar{z}|0,0), \qquad (4.3)$$

where

$$\Psi_p(z,\bar{z}|0,0) = \sum_{\{k\}\{\bar{k}\}} \beta_{nm}^{p\{k\}}\bar{\beta}_{nm}^{p\{\bar{k}\}} z^{\Sigma k_i} \bar{z}^{\Sigma \bar{k}_i} \phi_p^{\{k\}\{\bar{k}\}}(0,0) \qquad (4.4)$$

is the contribution of the conformal family $[\phi_p]$. Let us stress that the conformal properties of the "bilocal" operators (4.4) coincide with those of the product $\phi_n(z,\bar{z})\phi_m(0,0)$, all the coefficients in the power series (4.4) being unambiguously determined by this requirement. Unfortunately, equations, determining these coefficients are too complicated to be solved exactly. The first few coefficients β are presented in appendix B for the particular case $\Delta_n = \Delta_m$.

The constants C^p_{nm} in (4.3) and the values of the dimensions $\Delta_n, \bar{\Delta}_n$ are not determined by the conformal symmetry itself. These numerical parameters are the most important dynamical characteristics of the conformal quantum field theory. Note that under the orthonormality condition (3.28) the coefficients $C^l_{nm} = C_{nml}$ are symmetric functions of the indices n, m, l and coincide with the numerical factors in the three-point functions:

$$\langle n|\phi_m(z,\bar{z})|l\rangle = C_{nml} z^{\Delta_n - \Delta_m - \Delta_l} \bar{z}^{\bar{\Delta}_n - \bar{\Delta}_m - \bar{\Delta}_l}, \tag{4.5}$$

where for simplicity we put two points equal to 0 and ∞. To determine the parameters C^l_{nm} and Δ_n it is necessary to apply some dynamical principle. In the bootstrap approach described in the introduction, the associativity of the operator algebra (1.6) is taken as the main dynamical principle. As is shown in appendix C, the associativity condition is equivalent to the crossing symmetry of the four-point correlation functions

$$\langle A_{j_1}(\xi_1) A_{j_2}(\xi_2) A_{j_3}(\xi_3) A_{j_4}(\xi_4)\rangle . \tag{4.6}$$

Thanks to the relations discussed at the end of the previous section, it is sufficient to consider the four-point functions of the primary fields

$$\langle \phi_k(\xi_1)\phi_l(\xi_2)\phi_n(\xi_3)\phi_m(\xi_4)\rangle . \tag{4.7}$$

Due to the projective invariance (see appendix A), the four-point functions essentially depend only on two anharmonic quotients

$$x = \frac{(z_1 - z_2)(z_3 - z_4)}{(z_1 - z_3)(z_2 - z_4)}, \qquad \bar{x} = \frac{(\bar{z}_1 - \bar{z}_2)(\bar{z}_3 - \bar{z}_4)}{(\bar{z}_1 - \bar{z}_3)(\bar{z}_2 - \bar{z}_4)} . \tag{4.8}$$

Therefore it is convenient to set $z_1 = \bar{z}_1 = \infty$, $z_2 = \bar{z}_2 = 1$, $z_3 = x$, $\bar{z}_3 = \bar{x}$, $z_4 = \bar{z}_4 = 0$ and to define the functions

$$G^{lk}_{nm}(x,\bar{x}) = \langle k|\phi_l(1,1)\phi_n(x,\bar{x})|m\rangle . \tag{4.9}$$

In terms of these functions the crossing symmetry condition is

$$G^{lk}_{nm}(x,\bar{x}) = G^{mk}_{nl}(1-x, 1-\bar{x}) = x^{-2\Delta_n}\bar{x}^{-2\bar{\Delta}_n} G^{lm}_{nk}\left(\frac{1}{x}, \frac{1}{\bar{x}}\right). \tag{4.10}$$

Substituting the expansion (4.3) for the product $\phi_n(x,\bar{x})\phi_m(0,0)$ one can rewrite (4.9) as

$$G^{lk}_{nm}(x,\bar{x}) = \sum_p C^p_{nm} C_{klp} A^{lk}_{nm}(p|x,\bar{x}), \tag{4.11}$$

where each of the "partial waves"

$$A^{lk}_{nm}(p|x,\bar{x}) = (C^{p}_{kl})^{-1} x^{\Delta_p-\Delta_n-\Delta_m} \bar{x}^{\bar{\Delta}_p-\bar{\Delta}_m-\bar{\Delta}_n} \times \langle k|\phi_l(1,1)\Psi_p(x,\bar{x}|0,0)|0\rangle \tag{4.12}$$

represents the "s-channel" contribution of the conformal family $[\phi_p]$ to the four-point function (4.9). It is convenient to introduce the diagrams associated with these amplitudes

$A^{lk}_{nm}(p|x,\bar{x}) =$ 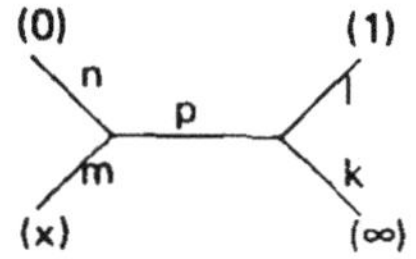

(4.13)

Then the "partial wave" decomposition (4.11) can be represented as

$G^{lk}_{nm}(x,\bar{x}) =$ 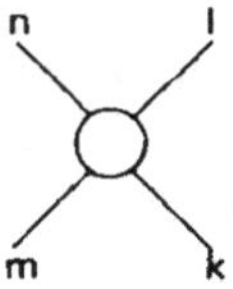

$= \sum_p C^p_{nm} C_{lkp}$ 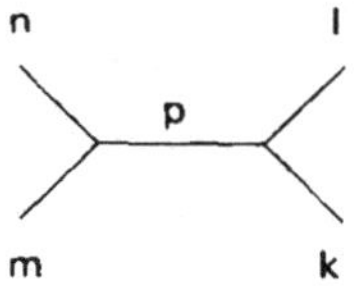

(4.14)

It is clear from (4.4) that the amplitudes (4.12) have the following factorized form

$$A^{lk}_{nm}(p|x,\bar{x}) = \mathcal{F}^{lk}_{nm}(p|x)\bar{\mathcal{F}}^{lk}_{nm}(p|\bar{x}), \tag{4.15}$$

where, for instance, the function $\mathcal{F}$ is given by the power series

$$\mathcal{F}^{lk}_{nm}(p|x) = x^{\Delta_p-\Delta_n-\Delta_m} \sum_{\{k\}} \beta^{p\{k\}}_{nm} x^{\Sigma k_i} \frac{\langle k|\phi_l(1,1)L_{-k_1}\dots L_{-k_N}|p\rangle}{\langle k|\phi_l(1,1)|p\rangle}. \tag{4.16}$$

The matrix elements in the right-hand side of (4.16) can be computed exactly with the use of the commutation relations (3.9) and eqs. (3.30). Therefore, the functions (4.16) are completely determined by the conformal symmetry. These functions depend on six parameters: five dimensions Δ_n, Δ_m, Δ_k, Δ_l, Δ_p and the central charge c. We shall call (4.16) the *conformal blocks*, because any correlation function (4.7) is built up of these functions $\mathcal{F}$.

The crossing symmetry conditions for the four-point functions (4.9) can be represented as the following diagrammic equations:

$\sum_p C^p_{nm} C_{lkp}$ 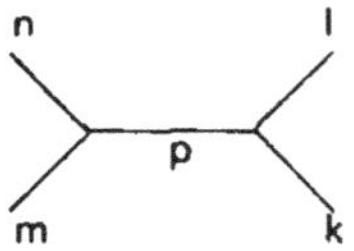

$= \sum_q C^q_{nl} C_{mkq}$ 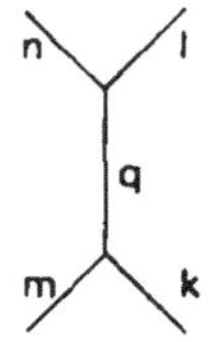

(4.17)

The analytic form of these equations is

$$\sum_p C^p_{nm} C_{lkp} \mathcal{F}^{lk}_{nm}(p|x) \bar{\mathcal{F}}^{lk}_{nm}(p|\bar{x}) = \sum_q C^q_{nl} C_{mkq} \mathcal{F}^{mk}_{nl}(q|1-x) \bar{\mathcal{F}}^{mk}_{nl}(q|1-\bar{x}) . \tag{4.18}$$

If the conformal blocks $\mathcal{F}$ are known, (4.18) yields a system of equations, determining the constants C^l_{nm} and the dimensions $\Delta_n, \bar{\Delta}_n$. Therefore, the computation of the conformal blocks (4.16) for general values of Δ_n's is the problem of principle importance for the conformal quantum field theory. The first few terms of the power expansion for these functions are given in appendix B, where the case $\Delta_n = \Delta_m = \Delta_k = \Delta_l = \Delta$ is considered for the sake of simplicity. Although the conformal blocks are not yet known for the general case, there are the special values of the dimensions Δ (associated with the degenerate representation of the Virasoro algebra, see sect. 5) such that the corresponding conformal blocks can be computed exactly, being the solutions of certain linear differential equations. The simplest example is the hypergeometric function. In these special cases the bootstrap eq. (4.18) can be solved completely.

5. Degenerate conformal families

The representation V_Δ of the Virasoro algebra is irreducible unless the dimension Δ takes some special values [6,7]. For these values the vector space V_Δ proves to contain a special vector (*the null vector*) $|\chi\rangle \in V_\Delta$ satisfying the equations

$$L_n|\chi\rangle = 0, \quad \text{if} \quad n > 0,$$
$$L_0|\chi\rangle = (\Delta + K)|\chi\rangle, \tag{5.1}$$

characteristic of the primary fields. Here K is some positive integer. For example, one can easily verify that the vector

$$|\chi\rangle = \left[L_{-2} + \frac{3}{2(2\Delta+1)} L^2_{-1} \right] |\Delta\rangle, \tag{5.2}$$

(where $|\Delta\rangle$ denotes the primary state of the dimension Δ) satisfies (5.1) with $K=2$, provided Δ takes any of the two values

$$\Delta = \tfrac{1}{16}\left[5 - c \pm \sqrt{(c-1)(c-25)}\right]. \tag{5.3}$$

In general, the jessenull vector $|\chi\rangle$ can be considered as the primary state of its own Verma modulus $V_{\Delta+K}$. Therefore the representation V_Δ proves to be reducible. One obtains the irreducible representation $V_\Delta^{(ir)}$ if the null vector $|\chi\rangle$ (together with all the states belonging to $V_{\Delta+K}$) is formally put equal to zero:

$$|\chi\rangle = 0. \tag{5.4}$$

Note that eq. (5.4) does not lead to contradictions since due to (5.1) the null vector is orthogonal to any state of V_Δ and, in particular, has the zero norm

$$\langle\psi|\chi\rangle = 0, \qquad |\Psi\rangle \in V_\Delta,$$

$$\langle\chi|\chi\rangle = 0. \tag{5.5}$$

In the conformal quantum field theory the meaning of this phenomenon is the following. If the dimension Δ of some primary field ϕ_Δ happens to take one of the special values mentioned above, then the conformal family $[\phi_\Delta]$, formally computed according to (3.18) proves to contain the special secondary field $\chi_{\Delta+K} \in [\phi_\Delta]$, which possesses the conformal properties of a primary field, i.e. satisfies the commutation relations of the type (3.9). This field corresponds to the null vector $|\chi\rangle \in V_\Delta$ and we call it the *null field*. For example, if Δ is given by (5.3) the operator

$$\chi_{\Delta+2} = \phi_\Delta^{(-2)} + \frac{3}{2(2\Delta+1)}\frac{\partial^2}{\partial z^2}\phi_\Delta \tag{5.6}$$

is the null field.

Formally, the extra primary field $\chi_{\Delta+K}$ originates from the conformal family $[\chi_{\Delta+K}]$ which is imbedded into $[\phi_\Delta]$. Note, however, that any correlation functions of the form

$$\langle\chi_{\Delta+K}(z)\phi_1(z_1)\ldots\phi_N(z_N)\rangle$$

vanishes. So, the null field $\chi_{\Delta+K}$ can be self-consistently regarded as zero:

$$\chi_{\Delta+K} = 0. \tag{5.7}$$

This condition obviously kills all the secondary fields of the null field

$$[\chi_{\Delta+K}] = 0. \tag{5.8}$$

If eq. (5.7) is applied, one gets the true irreducible conformal family $[\phi_\Delta]$ of the original primary field ϕ_Δ. In this case the conformal family contains "less" fields than usual and we call it a *degenerate conformal family*. We shall also call degenerate the corresponding primary field ϕ_Δ.

All the special values of Δ, corresponding to the reducible representations V_Δ, have been listed by Kac [7] (see also [6]). These values, which can be labelled by two positive integers n and m, are given by the formula

$$\Delta_{(n,m)} = \Delta_0 + \left(\tfrac{1}{2}\alpha_+ n + \tfrac{1}{2}\alpha_- m\right)^2, \tag{5.9}$$

where

$$\Delta_0 = \tfrac{1}{24}(c-1), \tag{5.10}$$

$$\alpha_\pm = \frac{\sqrt{1-c} \pm \sqrt{25-c}}{\sqrt{24}}. \tag{5.11}$$

If $\Delta = \Delta_{(n,m)}$, then the corresponding null vector has the dimension

$$\Delta_{(n,m)} + nm. \tag{5.12}$$

Let us denote the degenerate primary field $\phi_{\Delta_{(n,m)}}$ having the dimension $\Delta_{(n,m)}$ as $\psi_{(n,m)}$*. Note that

$$\Delta_{(1,1)} = 0. \tag{5.13}$$

It can be shown that the field $\psi_{(1,1)}$ is z-independent, i.e.**

$$\frac{\partial}{\partial z}\psi_{(1,1)} = 0. \tag{5.14}$$

The dimensions $\Delta_{(1,2)}$ and $\Delta_{(2,1)}$ are just the two values given by (5.3).

Consider the correlation functions of the form

$$\langle \psi_{(n,m)}(z)\phi_1(\xi_1)\dots\phi_N(\xi_N)\rangle. \tag{5.15}$$

* This notation is not complete because it says nothing about the second dimension $\bar{\Delta}$ of the primary field. This fact, which should be always kept in mind, does not violate the conclusions we make below.

** If both dimensions Δ and $\bar{\Delta}$ of the field ψ are zero this field does not depend on the coordinates at all and coincides with the identity operator I.

An important property of these correlation functions is that they satisfy the linear partial differential equations, the maximal order of derivatives being nm*. To make this evident let us recall that the correlation functions of any secondary fields

$$\langle \psi_{(n,m)}^{(-k_1,\dots,-k_L)}(z)\phi_1(\xi_1)\dots\phi_N(\xi_N)\rangle \tag{5.16}$$

can be expressed in terms of the correlation function (5.15) by means of the linear differential operators (see (3.37)). The null field $\chi_{\Delta+nm}$ is a certain linear combination of the secondary fields $\Psi_{(n,m)}^{(-k_1,\dots,-k_L)}$. Therefore, the differential equation for (5.15) follows directly from eq. (5.7). For example, taking into account (5.6) and (3.37), for the degenerate field $\psi_{(1,2)}(z)$ one gets

$$\left\{\frac{3}{2(2\delta+1)}\frac{\partial^2}{\partial z^2}-\sum_{i=1}^{N}\frac{\Delta_i}{(z-z_i)^2}-\sum_{i=1}^{N}\frac{1}{z-z_i}\frac{\partial}{\partial z_i}\right\}$$
$$\times\langle\Psi_{(1,2)}(z)\phi_1(z_1)\dots\phi_N(z_N)\rangle=0, \tag{5.17}$$

where $\delta=\Delta_{(1,2)}$ and $\Delta_1,\dots,\Delta_N$ are the dimensions of the primary fields $\phi_1,\dots,\phi_N$, respectively. The correlation function, involving the field $\psi_{(2,1)}$, satisfies the same differential equation, the only difference being $\delta=\Delta_{(2,1)}$**. The differential equation, satisfied by the degenerate fields $\psi_{(1,3)}$ and $\psi_{(3,1)}$, is presented in appendix D as another example.

In the case of the four-point functions

$$\Psi_{(n,m)}(z|z_1,z_2,z_3)=\langle\psi_{(n,m)}(z)\phi_1(z_1)\phi_2(z_2)\phi_3(z_3)\rangle, \tag{5.18}$$

the partial differential equations can be reduced to ordinary ones. Actually in this

* The simplest example of these equations is (5.14).

** The following interpretation of eq. (5.17) is worth noting. Let $\psi(z)$ stand for one of the fields $\psi_{(1,2)}$ or $\psi_{(2,1)}$, δ being the corresponding dimension $\Delta_{(1,2)}$ or $\Delta_{(2,1)}$. Then the field $\psi(z)$ satisfies the operator equation

$$\frac{\partial^2}{\partial z^2}\psi(z)=\gamma\,{:}T(z)\psi(z){:}\,, \tag{$\bullet$}$$

where $\gamma=\frac{2}{3}(2\delta+1)$, whereas the singular operator product is regularized by means of the subtractions

$${:}T(z)\psi(z){:}=\lim_{\zeta\to z}\left\{T(\zeta)\psi(z)-\frac{\delta}{(\zeta-z)^2}\psi(z)-\frac{1}{\zeta-z}\frac{\partial}{\partial z}\psi(z)\right\}.$$

The classical limit of eq. ($\bullet$) (which corresponds to the choice $\psi=\psi_{(1,2)}$ and $c\to\infty$) is an essential part of classical theory of the Liouville equation (see, for example, [13]). We suppose that eq. ($\bullet$) plays the analogous role in the quantum theory of this equation, which is apparently associated with the string theory [14]. We intend to discuss this point in another paper.

case the relations (A.) can be solved for the derivatives $\partial/\partial z_i$, $i=1,2,3$. For example substituting these derivatives into (5.17) one gets the Riemann ordinary differential equation

$$\left\{\frac{3}{2(2\delta+1)}\frac{\mathrm{d}^2}{\mathrm{d}z^2}+\sum_{i=1}^{3}\left[\frac{1}{z-z_i}\frac{\mathrm{d}}{\mathrm{d}z}-\frac{\Delta_i}{(z-z_i)^2}\right]\right.$$

$$\left.+\sum_{j<i}\frac{\delta+\Delta_{ij}}{(z-z_i)(z-z_j)}\right\}\Psi(z|z_1,z_2,z_3)=0, \tag{5.19}$$

where $\Delta_{12}=\Delta_1+\Delta_2-\Delta_3$, etc., $\delta=\Delta_{(1,2)}$, $\Psi=\Psi_{(1,2)}$ or $\delta=\Delta_{(2,1)}$, $\Psi=\Psi_{(2,1)}$. So, for the cases $(n,m)=(1,2)$ or $(2,1)$ the four-point function (5.18) can be expressed in terms of the hypergeometric function.

Consider the operator algebra containing the degenerate fields. Some important information about this operator algebra can be obtained from the differential equations discussed above. For example, consider the product $\psi(z)\phi_\Delta(z_1)$ where ϕ_Δ is some primary field of the dimension Δ whereas $\psi(z)$ temporarily stands for one of the degenerate fields $\psi_{(1,2)}(z)$ or $\psi_{(2,1)}(z)$. Let us substitute the expansion

$$\psi(z)\phi_\Delta(z_1)=\mathrm{const}(z-z_1)^{\kappa}\left[\phi_{\Delta'}(z_1)+\beta^{(-1)}(z-z_1)\phi_{\Delta'}^{(-1)}(z_1)+\cdots\right], \tag{5.20}$$

into the differential eq. (5.17). In (5.20) $\phi_{\Delta'}$ denotes some primary field of the dimension Δ', $\kappa=\Delta'-\Delta-\delta$ where δ is the dimension of the field ψ, i.e. one of the values given by (5.3). Considering the most singular term at $z\to z_1$, one immediately obtains the characteristic equation, determining the exponent

$$\frac{3\kappa(\kappa-1)}{2(2\delta+1)}-\Delta+\kappa=0. \tag{5.21}$$

To describe the solutions of this equation it is convenient to introduce the following parametrization of the dimensions

$$\delta(\alpha)=\Delta_0+\tfrac{1}{4}\alpha^2, \tag{5.22}$$

where Δ_0 is defined by (5.10). If $\Delta=\Delta(\alpha)$, the two solutions of (5.21) are given by the formulae

$$\Delta'_{(1)}=\Delta_0+\tfrac{1}{4}(\alpha+\alpha_{\pm})^2,$$

$$\Delta'_{(2)}=\Delta_0+\tfrac{1}{4}(\alpha-\alpha_{\pm})^2, \tag{5.23}$$

where $\alpha_\pm$ are given by (5.11) and α_+ (α_-) is chosen if $\psi = \psi_{(1,2)}$ ($\psi = \psi_{(2,1)}$). Let $\phi_\alpha(z)$ be the primary field with the dimension (5.22). The result of the above calculation can be represented by the following symbolic formulae

$$\psi_{(1,2)}\phi_{(\alpha)} = \left[\phi_{(\alpha-\alpha_+)}\right] + \left[\phi_{(\alpha+\alpha_+)}\right],$$

$$\psi_{(2,1)}\phi_{(\alpha)} = \left[\phi_{(\alpha-\alpha_-)}\right] + \left[\phi_{(\alpha+\alpha_-)}\right]. \tag{5.24}$$

Here the square brackets denote the contributions of the corresponding conformal families to the operator product expansion of $\psi(z)\phi_{(\alpha)}(z_1)$. In (5.24) overall factors, standing in front of these contributions are omitted. These factors cannot certainly be determined by simple calculations like the one performed above*. As we shall see in the next section, some of these coefficients could vanish.

It can be shown that the "fusion rule" (5.24) is generalized to the cases of arbitrary degenerate fields $\psi_{(n,m)}$ as follows:

$$\psi_{(n,m)}\phi_\alpha = \sum_{l=1-m}^{1+m} \sum_{k=1-n}^{1+n} \left[\phi_{(\alpha + l\alpha_- + k\alpha_+)}\right], \tag{5.25}$$

where the variable k runs through the even (odd) values provided the index n is odd (even); the same is valid for the variable l and the index m. So in the general case the sum in (5.25) contains nm terms in agreement with the fact that the degenerate field $\psi_{(n,m)}$ satisfies the nm-order differential equation.

We see that the differential equations satisfied by the degenerate fields impose hard constraints on the operator algebra. Certainly, in the general case these differential equations do not provide enough information to determine the correlation functions (5.15) completely. Even in the cases of the four-point functions (5.18) one has to take into account the $\bar{z}$-dependence of the fields and local properties. In the next section we shall study the "minimal models" of the conformal quantum field theory in which all primary fields involved are degenerate.

6. Minimal theories

Consider the "fusion rule" (5.24). The substitution $\phi_{(\alpha)} = \psi_{(1,2)}$ yields

$$\psi_{(1,2)}\psi_{(1,2)} = \left[\psi_{(1,1)}\right] + \left[\psi_{(1,3)}\right]. \tag{6.1}$$

Here (5.9) is taken into account. Similarly, one gets for $m > 1$

$$\psi_{(1,2)}\psi_{(1,m)} = \left[\psi_{(1,m-1)}\right] + \left[\psi_{(1,m+1)}\right]. \tag{6.2}$$

* To determine these factors in the quantum field theory one should take into account the associativity condition for the operator algebra and local properties of the fields.

So, if the degenerate field $\psi_{(1,2)}$ is involved in the operator algebra, in the general case this algebra includes also all the degenerate fields $\psi_{(1,m)}$. Moreover, assuming that the operator algebra also includes the degenerate field $\psi_{(2,1)}$ and using (5.24), one can obtain all the degenerate fields $\psi_{(n,m)}$. In the "fusion rule" (5.24) the fields $\psi_{(1,2)}$ and $\psi_{(2,1)}$ act as the "shift operators"

$$\psi_{(1,2)}\psi_{(n,m)} = \left[\psi_{(n,m-1)}\right] + \left[\psi_{(n,m+1)}\right], \tag{6.3a}$$

$$\psi_{(2,1)}\psi_{(n,m)} = \left[\psi_{(n-1,m)}\right] + \left[\psi_{(n+1,m)}\right]. \tag{6.3b}$$

The following remark is necessary. Using the rules (8.3) formally, one would get as a result all the fields of dimension $\Delta_{(n,m)}$ given by (5.9) where the integers n, m take the zero and negative values as well as positive values. In fact, the fields of dimension $\Delta_{(n,m)}$ with the zero and negative n, m drop out from the algebra, i.e. the operator algebra developed by "fusing" the fields $\psi_{(1,2)}$ and $\psi_{(2,1)}$. $\psi_{(2,1)}$ proves to contain the degenerate fields $\psi_{(n,m)}$ $(n, m > 0)$ only. To understand the nature of this phenomenon, consider, for instance, the product $\psi_{(1,2)}\psi_{(2,1)}$. Analyzing the differential equation for the degenerate field $\psi_{(1,2)}$, one gets, according to (6.3a),

$$\psi_{(1,2)}\psi_{(2,1)} = C_1\left[\phi_{(2,0)}\right] + C_2\left[\Psi_{(2,2)}\right], \tag{6.4}$$

where $\phi_{(2,0)}$ denotes the primary field of the dimension $\Delta_{(2,0)} = \Delta_0 + (\alpha_+)^2$. In (6.4) we have explicitly written out the numerical coefficients C_1 and C_2 of the corresponding primary fields in the operator product expansion. In the above symbolic formulae like (6.1)–(6.3) such coefficients are omitted. On the other hand, the field $\psi_{(2,1)}$, also being degenerate, satisfies the differential eq. (5.17) which leads to the expansion

$$\psi_{(1,2)}\psi_{(2,1)} = C_1'\left[\phi_{(0,2)}\right] + C_2'\left[\Psi_{(2,2)}\right], \tag{6.5}$$

where the field $\phi_{(0,2)}$ has the dimension $\Delta_{(0,2)} = \Delta_0 + (\alpha_-)^2$ and C_1', C_2' are some numerical coefficients. The comparison of this formula with (6.4) yields that $C_1 = C_1' = 0$ and $C_2 = C_2'$. Hence, the expansion of the product $\psi_{(1,2)}\psi_{(2,1)}$ contains the contribution of only one conformal family

$$\psi_{(1,2)}\psi_{(2,1)} = \left[\psi_{(2,2)}\right]. \tag{6.6}$$

We shall call the phenomenon described above the *truncation* of the operator algebra*. It can be shown that for the degenerate fields $\psi_{(n,m)}$ this is the general

* It is interesting to understand the connection of the truncation phenomenon with the monodromy properties of the differential equations satisfied by the correlation functions. This problem can be most easily investigated for the four-point differential equations. If all the fields involved are degenerate, the space of solutions of the differential equations proves to contain the subspace invariant under the monodromy transformations. The solutions, belonging to this subspace, correspond to the degenerate fields $\psi_{(k,l)}$ $(k, l > 0)$ in (6.7) and these very solutions contribute to the correlation function.

situation: the degenerate conformal families $[\psi_{(n,m)}]$ with $n, m > 0$ actually appear only in the "fusion rules" like (6.3). The general "fusion rules" for the degenerate fields have the form*

$$\psi_{(n_1,m_1)}\psi_{(n_2,m_2)} = \sum_{k=|n_1-n_2|+1}^{n_1+n_2-1} \sum_{l=|m_1-m_2|+1}^{m_1+m_2-1} [\psi_{(k,l)}], \tag{6.7}$$

where the variable k (l) runs over the even integers, provided $n_1 + n_2$ ($m_1 + m_2$) is odd and vice versa.

So, the degenerate fields (more precisely, the degenerate conformal families) form the closed operator algebra. This observation gives rise to the idea of conformal quantum field theory in which all the primary fields are degenerate. To examine this possibility let us concentrate once again on the Kac formula (5.9). It is clear that there are three distinct domains of the parameter c. If $c \geqslant 25$ the second term in (5.9) is negative and the dimensions $\Delta_{(n,m)}$ become negative for sufficiently large n and m. If $25 > c > 1$, the dimensions $\Delta_{(n,m)}$ are, in general, complex. Neither possibility is acceptable in the quantum field theory**. Therefore in what follows we shall consider the domain

$$0 < c \leqslant 1 . \tag{6.8}$$

To understand the properties of the spectrum (5.9) clearly, let us consider the "diagram of dimensions" shown in fig. 1. The vertical and horizontal axes in this figure correspond to the values of the parameters n and m in (5.9). The "physical" (i.e. the positive integer) values of these parameters are shown by dots. The dotted line has the slope:

$$\operatorname{tg}\theta = -\frac{\alpha_+}{\alpha_-} = \frac{\sqrt{25-c}-\sqrt{1-c}}{\sqrt{25-c}+\sqrt{1-c}} . \tag{6.9}$$

The value (5.22) of the dimension is associated with each point of the plane in fig. 1, the parameter α being proportional to the distance between the point and the dotted line.

* The "fusion rule" (6.7) can be obtained from the following formula

$$\psi_{(n,m)} \sim (\psi_{(1,2)})^{m-1}(\psi_{(2,1)})^{n-1},$$

for the degenerate field $\psi_{(n,m)}$. Although this formula scarcely has a precise mathematical meaning, one can use it to derive (6.7) assuming the associativity and taking into account the truncation phenomenon.

** To avoid misunderstanding let us stress that these statements by no means exclude the possibility of quantum field theory existing at $c > 1$, but rather prevent from including the degenerate fields in the operator algebra.

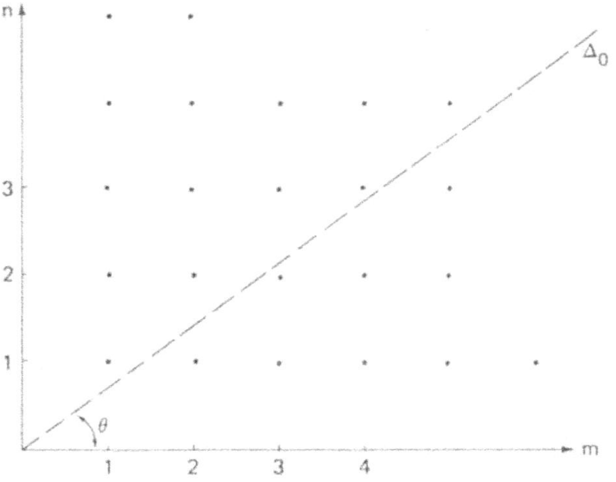

Fig. 1. "Diagram of dimensions". The dimension $\Delta = \Delta_0 + \frac{1}{4}\alpha^2$ is associated with each point of the plane, α being proportional to the distance between the point and the dotted line. The dots with coordinates (n, m) corresponds to the dimensions $\Delta_{(n,m)}$ described by Kac formula (5.9).

If the slope (6.9) takes an arbitrary irrational value, the dotted line in fig. 1 passes arbitrarily close to some of the dots. Since at $c < 1$, Δ_0 is negative, we meet again with the problem of negative dimensions. Let us consider, however, the cases of the rational slope

$$\operatorname{tg}\theta = -\alpha_-/\alpha_+ = p/q, \tag{6.10}$$

where p and q are positive integers. The characteristic feature of the corresponding values of c is that each degenerate representation $V_{\Delta_{(n,m)}}$ contains not only one but infinitely many null vectors of different dimension. This is evident from (5.9) and (6.10). In these cases the irreducible conformal families $[\psi_{(n,m)}]$ obtained by nullification of all the null fields, contain considerably fewer fields than the usual families and we call the conformal quantum field theories, corresponding to (6.10) and involving these degenerate fields $\psi_{(n,m)}$, minimal theories. It is important that in the minimal theories the correlation functions satisfy infinitely many differential equations, obtained by nullification of all the corresponding null fields*. This fact enables one to prove that the operator algebra of degenerate fields in the minimal theories possesses not only "truncation from below", described in the beginning of the section, but also the "truncation from above". Namely, if one starts with the fields $\psi_{(n,m)}$ with $0 < n < p$, $0 < m < q$, the degenerate fields with $n \geqslant p$ or $m \geqslant q$ drop out from the "fusion rules" (6.7) (like the fields $\phi_{(2,0)}$ and $\phi_{(0,2)}$ in (6.4),(6.5)). In other words, the conformal families $[\psi_{(n,m)}]$ with $0 < n < p$, $0 < m < q$ form the

* In fact, these differential equations are not all independent: they follow from two "basic" equations.

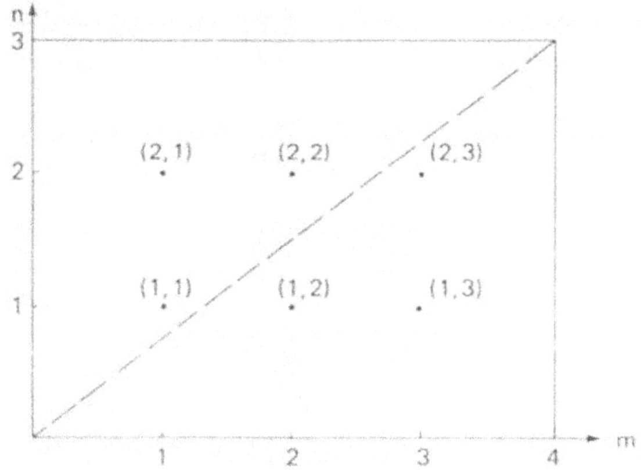

Fig. 2. Diagram of dimensions corresponding to the case tg $\theta = \frac{3}{4}$ $(c = \frac{1}{2})$. The degenerate conformal families associated with the dots inside the rectangle form the closed operator algebra.

closed algebra which can be treated as the operator algebra of the quantum field theory. Note that (under the condition (6.10)) $n = p$, $m = q$ are the coordinates of the nearest dot in fig. 1 which the dotted line passes through. The degenerate fields with the dimensions associated with the dots inside the rectangle $0 < n < p, 0 < m < q$, shown in figs. 2 and 3, form the closed operator algebra. Due to the diagonal symmetry of this rectangle there are $\frac{1}{2}(p-1)(q-1)$ different dimensions.

Consider in more detail the simplest nontrivial example of the minimal theory corresponding to the case

$$p/q = \tfrac{3}{4}, \tag{6.11}$$

which occurs if

$$c = \tfrac{1}{2}. \tag{6.12}$$

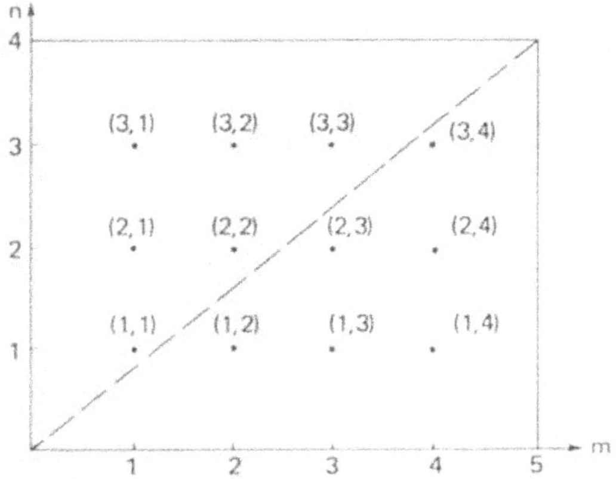

Fig. 3. Diagram of dimensions for the case tg $\theta = \frac{4}{5}$ $(c = \frac{7}{10})$.

The "diagram of dimensions" for this case is shown in fig. 2. Let us demonstrate the "truncation from above", using this example. The dimensions corresponding to the dots in fig. 2 are

$$\Delta_{(1,1)} = \Delta_{(2,3)} = 0\,,$$
$$\Delta_{(2,1)} = \Delta_{(1,3)} = \tfrac{1}{2}\,,$$
$$\Delta_{(1,2)} = \Delta_{2,2)} = \tfrac{1}{16}\,. \tag{6.13}$$

Respectively, there are three degenerate fields* which we shall denote by

$$I = \psi_{(1,1)} = \psi_{(2,3)}\,,$$
$$\varepsilon = \Psi_{(2,1)} = \psi_{(1,3)}\,,$$
$$\sigma = \psi_{(1,2)} = \psi_{(2,2)}\,. \tag{6.14}$$

Consider, for instance, the product $\varepsilon \cdot \varepsilon$. The field ε, being equal to $\psi_{(2,1)}$, satisfies the second order differential eq. (5.17). Therefore, according to (6.36), one gets

$$\varepsilon \cdot \varepsilon = \psi_{(2,1)}\psi_{(2,1)} = c_1[I] + c_2[\Psi_{(3,1)}]\,, \tag{6.15}$$

where the field $\psi_{(3,1)}$ has the dimension $\Delta_{(3,1)} = \frac{5}{3}$. On the other hand, since $\varepsilon = \psi_{(1,3)}$, this field satisfies the third order differential equation (D.8) and hence

$$\varepsilon \cdot \varepsilon = \Psi_{(1,3)}\psi_{(1,3)} = c_1'[I] + c_2'[\psi_{(1,3)}] + c_3'[\psi_{F(1,5)}]\,, \tag{6.16}$$

where the field $\psi_{(1,5)}$ has the dimension $\Delta_{(1,5)} = \frac{5}{2}$. Comparing (6.16) and (6.15), one concludes that in fact

$$\varepsilon \cdot \varepsilon = [I]\,. \tag{6.17}$$

By similar considerations the following "fusion rules" for the fields (6.14) can be obtained:

$$I \cdot \varepsilon = [\varepsilon]\,, \qquad \varepsilon \cdot \varepsilon = [I]\,,$$
$$I \cdot \sigma = [\sigma]\,, \qquad \varepsilon \cdot \sigma = [\sigma]\,,$$
$$I \cdot I = [I]\,, \qquad \sigma \cdot \sigma = [I] + [\varepsilon]\,. \tag{6.18}$$

* Certainly, the analysis of the dimensions (6.13) does not prove that the operator algebra contains only three primary fields. To elucidate the structure of the fields constituting the operator algebra one should take into account the $\bar{z}$-dependence and the local properties of the fields. For the model under consideration this is done in appendix E.

It is shown in appendix E that this minimal theory describes the critical point of the two-dimensional Ising model, the primary fields σ, ε and I being identified with the local spin, energy density and identity operators, respectively.

In fig. 3 the "diagram of dimensions" for the minimal theory characterized by the values

$$p/q=\tfrac{4}{5}, \qquad c=\tfrac{7}{10}, \tag{6.19}$$

is presented as another example. The corresponding numerical values of the dimensions are

$$\begin{aligned}
&\Delta_{(1,1)}=\Delta_{(3,4)}=0,\\
&\Delta_{(1,2)}=\Delta_{(3,3)}=\tfrac{1}{10},\\
&\Delta_{(1,3)}=\Delta_{(3,2)}=\tfrac{3}{5},\\
&\Delta_{(1,4)}=\Delta_{(3,1)}=\tfrac{3}{2},\\
&\Delta_{(2,2)}=\Delta_{(2,3)}=\tfrac{3}{80},\\
&\Delta_{(2,4)}=\Delta_{(2,1)}=\tfrac{7}{16}.
\end{aligned} \tag{6.20}$$

Note that due to the inequalities (6.8) the integers p and q in (6.10) are restricted as follows

$$\tfrac{2}{3}<p/q<1. \tag{6.21}$$

Nevertheless, there are infinitely many rational numbers, satisfying (6.21) and each of them corresponds to some minimal model of the conformal quantum field theory. We suppose that the minimal theories describe second order phase transitions in two-dimensional systems with discrete symmetry groups*. In any case each of the minimal models seems to deserve a most detailed investigation. Note that the anomalous dimensions associated with each of the minimal model are known exactly (they are given by the Kac formula (5.9)), whereas the correlation functions can be computed in the following way. At first one has to derive the corresponding conformal blocks as solutions of the respective differential equations with the

* V. Dotzenko has noticed that the spectrum of dimensions associated with the minimal model

$$p/q=\tfrac{5}{6}, \qquad c=\tfrac{4}{5}$$

contains some dimensions characteristic of the three-state Potts model.

appropriate initial conditions. Then, substituting these conformal blocks into the bootstrap eq. (4.18) and taking into account the local properties of the fields, one should calculate the structure constants C^l_{nm} of the operator algebra, which provide enough information to construct the correlation functions. For the minimal theory (6.11) this computation is presented in appendix E. In the general case it has not yet been performed.

We are obliged to B. Feigin for numerous consultations about the representations of the Virasoro algebra and to A.A. Migdal for useful discussions. The two of us (AB and AZ) are very grateful to D. Makagonenko and A.A. Anselm for the kind hospitality in the Scientific Center in Komarovo during January 1983 where this work was completed.

Appendix A

Let L_{-1}, L_0, L_{+1} and $\bar{L}_{-1}, \bar{L}_0, \bar{L}_{+1}$ be generators of the infinitesimal projective transformations

$$z \to z + \varepsilon_{-1} + \varepsilon_0 z + \varepsilon_1 z^2,$$

$$\bar{z} \to \bar{z} + \bar{\varepsilon}_{-1} + \bar{\varepsilon}_0 \bar{z} + \bar{\varepsilon}_1 \bar{z}^2, \tag{A.1}$$

where ε and $\bar{\varepsilon}$ are infinitesimal parameters. The operators L_s, $s = 0, \pm 1$ satisfy the commutation relations

$$[L_0, L_{\pm 1}] = \pm L_{\pm 1},$$

$$[L_1, L_{-1}] = 2L_0. \tag{A.2}$$

The same relations are satisfied by the $\bar{L}$'s, the L's and $\bar{L}$'s being commutative. The operators $P^0 = L_{-1} + \bar{L}_{-1}$ and $P^1 = -i(L_{-1} - \bar{L}_{-1})$ are components of the total momentum, whereas $M = i(L_0 - \bar{L}_0)$ and $D = L_0 + \bar{L}_0$ are generators of the rotations (Lorentz boosts in the Minkowski space-time) and dilatations, respectively. The operators L_1 and $\bar{L}_1$ correspond to the special conformal transformations. The vacuum of the conformal quantum field theory satisfies the relations:

$$\langle 0 | L_s = L_s | 0 \rangle = 0, \qquad s = 0, \pm 1, \tag{A.3}$$

which are equivalent to the asymptotic condition (2.14).

We shall call the local field $O_l(z, \bar{z})$ *quasiprimary*, provided it satisfies the commutation relations:

$$[L_s, O_l(z, \bar{z})] = \left[z^{s+1} \frac{\partial}{\partial z} + (s+1)\Delta_l z^s \right] O_l(z, \bar{z}),$$

$$[\bar{L}_s, O_l(z, \bar{z})] = \left[\bar{z}^{s+1} \frac{\partial}{\partial \bar{z}} + (s+1)\bar{\Delta}_l \bar{z}^s \right] O_l(z, \bar{z}), \tag{A.4}$$

where $s = 0, \pm 1$. The constants Δ_l and $\bar{\Delta}_l$ are dimensions of the field O_l. These relations mean that the fields $O_l(z, \bar{z})$ transform according to formula (1.16) under the projective transformations (1.15). This distinguishes them from the primary fields ϕ_n which transform according to (1.16) with respect to all conformal transformations (1.9)*. In the conformal quantum field theory the complete set of local fields A_j, forming the algebra (1.6), can be constituted by an infinite number of quasiprimary fields and their coordinate derivatives of all orders

$$\{A_j\} = \left\{ O_l, \frac{\partial}{\partial z} O_l, \frac{\partial}{\partial \bar{z}} O_l, \frac{\partial^2}{\partial z^2} O_l, \ldots \right\}. \tag{A.5}$$

Consider an N-point correlation function of the quasiprimary fields. It follows from (A.3) and (A.4) that this correlation function satisfies the equations:

$$\hat{\Lambda}_s \langle O_{l_1}(z_1, \bar{z}_1) \ldots O_{l_N}(z_N, \bar{z}_N) \rangle = 0, \tag{A.6}$$

where $s = 0, \pm 1$ and $\hat{\Lambda}_s$ are the differential operators

$$\hat{\Lambda}_{-1} = \sum_{i=1}^{N} \frac{\partial}{\partial z_i},$$

$$\hat{\Lambda}_0 = \sum_{i=1}^{N} \left(z_i \frac{\partial}{\partial z_i} + \Delta_i \right),$$

$$\hat{\Lambda}_1 = \sum_{i=1}^{N} \left(z_i^2 \frac{\partial}{\partial z_i} + 2 z_i \Delta_i \right), \tag{A.7}$$

where $\Delta_1, \Delta_2, \ldots, \Delta_N$ are dimensions of the fields $O_{l_1}, \ldots O_{l_N}$, respectively. Eqs. (A.6) are the projective Ward identities. Note that these Ward identities follow directly from the general relation (2.9). For the infinitesimal projective transformations the function $\varepsilon(z)$ is regular in the finite part of the z-plane and due to the asymptotic condition (2.14) the contour integral in (2.9) vanishes. Let us stress that for the general conformal transformations the analytic function $\varepsilon(z)$ has singularities. Therefore the corresponding Ward identities cannot be reduced to the closed equations for the correlation functions like (A.6). The general solution of eqs. (A.6) (and the analogous equations obtained by the substitution $z_i \to \bar{z}_i$, $\Delta_i \to \bar{\Delta}_i$) is

$$\langle O_{l_1}(z_1, \bar{z}_1) \ldots O_{l_N}(z_N, \bar{z}_N) \rangle = \prod_{i<j} (z_i - z_j)^{\gamma_{ij}} (\bar{z}_i - \bar{z}_j)^{\bar{\gamma}_{ij}} Y(x_{ij}^{kl}, \bar{x}_{ij}^{kl}), \tag{A.8}$$

* Obviously, any primary field is quasiprimary whereas there are infinitely many quasiprimary fields which are secondaries.

where γ_{ij} and $\bar{\gamma}_{ij}$ are arbitrary solutions of the equations

$$\sum_{j\neq i}\gamma_{ij}=2\Delta_i,\qquad \sum_{j\neq i}\bar{\gamma}_{ij}=2\Delta_i, \tag{A.9}$$

whereas Y is an arbitrary function of $2(N-3)$ anharmonic quotients

$$x_{ij}^{kl}=\frac{(z_i-z_j)(z_k-z_l)}{(z_i-z_l)(z_k-z_j)},\qquad \bar{x}_{ij}^{kl}=\frac{(\bar{z}_i-\bar{z}_j)(\bar{z}_k-\bar{z}_l)}{(\bar{z}_i-\bar{z}_l)(\bar{z}_k-\bar{z}_j)}. \tag{A.10}$$

In the particular cases $N=2$ and $N=3$ the correlation functions are determined by formulae (A.8)–(A.10) completely up to the numerical factor. Namely,

$$\langle O_{l_1}(z_1,\bar{z}_1)O_{l_2}(z_2,\bar{z}_2)\rangle=\begin{cases}0 \quad \text{if} \quad \Delta_{l_1}\neq\Delta_{l_2} \quad \text{or} \quad \bar{\Delta}_{l_1}\neq\bar{\Delta}_{l_2}\\ (z_1-z_2)^{-2\Delta_{l_1}}(\bar{z}_1-\bar{z}_2)^{-2\bar{\Delta}_{l_1}}D_{l_1} \quad \text{if}\\ \Delta_{l_1}=\Delta_{l_2} \quad \text{and} \quad \bar{\Delta}_{l_1}=\bar{\Delta}_{l_2},\end{cases} \tag{A.11}$$

for $N=2$ and

$$\langle O_{l_1}(z_1,\bar{z}_1)O_{l_2}(z_2,\bar{z}_2)O_{l_3}(z_3,\bar{z}_3)\rangle=Y_{l_1l_2l_3}\prod_{i<j}(z_i-z_j)^{-\Delta_{ij}}(\bar{z}_i-\bar{z}_j)^{-\bar{\Delta}_{ij}}, \tag{A.12}$$

for $N=3$ where D_l and $Y_{l_1l_2l_3}$ are constants and

$$\Delta_{12}=\Delta_1+\Delta_2-\Delta_3 \quad \text{etc.},$$

$$\bar{\Delta}_{12}=\bar{\Delta}_1+\bar{\Delta}_2-\bar{\Delta}_3 \quad \text{etc.} \tag{A.13}$$

Note that the functions (A.11) and (A.12) are single-valued in the euclidean space (obtained by the substitution $\bar{z}_i=z_i^*$), provided the spins $S_l=\Delta_l-\bar{\Delta}_l$ of all the fields involved take integer or half-integer values.

In the conformal quantum field theory the expansion (1.6) can be represented in the form

$$O_{l_1}(z,\bar{z})O_{l_2}(0,0)=\sum_{l_3}\sum_{k,\bar{k}=0}^{\infty}Y_{l_1l_2}^{l_3,k,\bar{k}}z^{\Delta_3+k-\Delta_1-\Delta_2}\bar{z}^{\bar{\Delta}_3+\bar{k}-\bar{\Delta}_1-\bar{\Delta}_2}$$

$$\times\left[\frac{\partial^{k+\bar{k}}}{\partial\zeta^k\partial\bar{\zeta}^{\bar{k}}}O_{l_3}(\zeta,\bar{\zeta})\right]_{\zeta,\bar{\zeta}=0}, \tag{A.14}$$

where $Y_{l_1l_2}^{l_3,k,\bar{k}}$ are constants, k and $\bar{k}$ being integers. The transformation properties of the both sides of this equation with respect to the projective transformations (A.1) must coincide. Commuting both sides of (A.14) with the projective generators L_s, $s=0, \pm 1$ and using (A.4), one gets equations relating the coefficients $Y_{l_1 l_2}^{l', k, \bar{k}}$ with different values of k. Solving these equations, one can rewrite (A.14) as

$$O_l(z,\bar{z})O_l(0.0)=\sum_{l'} G_{ll}^{l'} z^{\Delta'-2\Delta}\bar{z}^{\bar{\Delta}'-2\bar{\Delta}}$$

$$\times F\left(\Delta', 2\Delta', z\frac{\partial}{\partial\zeta}\right)F\left(\bar{\Delta}', 2\bar{\Delta}', \bar{z}\frac{\partial}{\partial\bar{\zeta}}\right)O_{l'}(\zeta,\bar{\zeta})|_{\zeta,\bar{\zeta}=0} \quad (A.15)$$

where the case $l_1 = l_2$ is considered for the sake of simplicity: $\Delta_{l_1}=\Delta_{l_2}=\Delta$, $\Delta_{l'}=\Delta'$. In (A.15) $G_{ll}^{l'}$ are the constants, coinciding with $Y_{ll}^{l', 0, 0}$ in (A.14) and $F(a, c, x)$ denotes the degenerate hypergeometric function.

Obviously, each conformal family $[\phi_n] = V_n \times \bar{V}_n$ (see sect. 3) contains infinitely many quasiprimary fields. These fields correspond to the states satisfying the equations

$$L_1|l\rangle = \bar{L}_1|l\rangle = 0,$$

$$L_0|l\rangle = \Delta_l|l\rangle, \qquad \bar{L}_0|l\rangle = \bar{\Delta}_l|l\rangle. \quad (A.16)$$

It can be shown that the basis in $[\phi_n]$ can be constituted by the states

$$(L_{-1})^n(\bar{L}_{-1})^{\bar{n}}|l\rangle, \quad (A.17)$$

where $n, \bar{n} = 0, 1, 2, \ldots$ and $|l\rangle$ are the quasiprimary states, belonging to $[\phi_n]$. This statement is equivalent to (A.5) because the operators L_{-1} and $\bar{L}_{-1}$ are associated with the derivatives $\partial/\partial z$ and $\partial/\partial\bar{z}$.

Appendix B

Here we shall demonstrate that the coefficients $\beta_{nm}^{l\{k\}}$ in (4.2) are determined completely by the requirement of the conformal symmetry of the expansion (4.1), considering the particular case $\Delta_n = \Delta_m = \Delta$ for the sake of simplicity. Applying both sides of (4.1) to the vacuum state, one gets the equation

$$\phi_\Delta(z,\bar{z})|\Delta\rangle = \sum_l C_{\Delta\Delta}^{\Delta_l} z^{\Delta_l - 2\Delta}\bar{z}^{\bar{\Delta}_l - 2\bar{\Delta}}\varphi_\Delta(z)\bar{\varphi}_{\bar{\Delta}}(\bar{z})|\Delta_l\rangle, \quad (B.1)$$

where $|\Delta\rangle$ is the primary state of the dimensions $\Delta, \bar{\Delta}$ and the operator $\varphi_\Delta(z)$ is given by the series

$$\varphi_\Delta(z) = \sum_{\{k\}} z^{\Sigma k_i}\beta_{\Delta\Delta}^{\Delta_l, \{k\}} L_{-k_1}\ldots L_{-k_N}. \quad (B.2)$$

The same formula with the substitution $z \to \bar{z}$, $\beta \to \bar{\beta}$, $L \to \bar{L}$ holds for $\bar{\varphi}_{\Delta}(\bar{z})$. Let us consider the state

$$|z, \Delta'\rangle = \varphi_{\Delta}(z)|\Delta'\rangle. \tag{B.3}$$

It can be represented as the power series

$$|z, \Delta'\rangle = \sum_{N=0}^{\infty} z^N |N, \Delta'\rangle, \tag{B.4}$$

where the vectors $|N, \Delta'\rangle$ satisfy the equations:

$$L_0|N, \Delta'\rangle = (\Delta_l + N)|N, \Delta'\rangle. \tag{B.5}$$

To compute these vectors let us apply the operators L_n to both sides of (B.1). This leads to the equations

$$\left[z^{n+1}\frac{\mathrm{d}}{\mathrm{d}z} + \Delta(n+1)z^n\right]|z, \Delta'\rangle = L_n|z, \Delta'\rangle. \tag{B.6}$$

Substituting the power series (B.4) one gets

$$L_n|N+n, \Delta'\rangle = [N + (n-1)\Delta + \Delta']|N, \Delta'\rangle. \tag{B.7}$$

Actually, one can consider eqs. (B.7) with $n = 1, 2$ only because in virtue of (2.21) the remaining equations follow from these two. Solving these equations one can compute the power series (B.4) order by order. In the first three orders the result is

$$|z, \Delta'\rangle = \left[1 + \tfrac{1}{2}zL_{-1} + \tfrac{1}{4}z^2\frac{\Delta'+1}{2\Delta'+1}L_{-1}^2 + z^2\frac{\Delta'(\Delta'-1) + 2\Delta(2\Delta'+1)}{c(2\Delta'+1) + 2\Delta'(8\Delta'-5)}\right.$$
$$\left.\times\left(L_{-2} + \frac{3}{2(2\Delta'+1)}L_{-1}^2\right) + \cdots\right]|\Delta'\rangle. \tag{B.8}$$

This formula gives the first three coefficients β in (B.2).

Obviously the conformal block $\mathcal{F}(\Delta, \Delta', x) \equiv \mathcal{F}_{\Delta\Delta}^{\Delta\Delta}(\Delta'|x)$ is given by the scalar product

$$\mathcal{F}(\Delta, \Delta', x) = x^{\Delta'-2\Delta}\langle 1, \Delta'|x, \Delta'\rangle. \tag{B.9}$$

The first few terms of the power expansion of this function can be directly obtained from (B.8)

$$\mathcal{F}(\Delta, \Delta', x) = x^{\Delta'-2\Delta}\left\{1 + \tfrac{1}{2}\Delta' x + \frac{\Delta'(\Delta'+1)^2}{4(2\Delta'+1)}x^2\right.$$
$$\left. + \frac{[\Delta'(1-\Delta') - 2\Delta(2\Delta'+1)]^2}{2(2\Delta'+1)[c(2\Delta'+1) + 2\Delta'(8\Delta'-5)]}x^2 + \cdots\right\}. \tag{B.10}$$

Appendix C

Consider the associative algebra determined by the relations

$$A_I A_J = \sum_K C_{IJ}^K A_K . \tag{C.1}$$

Eq. (1.6) is just (C.1) where each of the indices, say I, combines the space coordinate ξ and the index i, labelling the fields. Let the algebra (C.1) be supplied with the symmetric bilinear form

$$D_{IJ} = \langle A_I A_J \rangle , \tag{C.2}$$

which is no other than a set of all two-point correlation functions. Let us introduce the form

$$C_{IJK} = \sum_{K'} D_{KK'} C_{IJ}^{K'} , \tag{C.3}$$

and assume that this is a symmetric function of the indices I, J, K. Evidently, (C.3) coincides with the three-point correlation function

$$C_{IJK} = \langle A_I A_J A_K \rangle , \tag{C.4}$$

and it can be conveniently represented by the "vertex" diagram

C_{IJK} = 

(C.5)

Also introduce the diagram

D^{IJ} = 

(C.6)

for the "inverse propagator" D^{IJ} defined by the equation

$$\sum_K D^{IK} D_{KJ} = \delta_J^I . \tag{C.7}$$

The associativity condition of the algebra (C.1)

$$\sum_K C_{IJ}^K C_{KL}^M = \sum_K C_{IK}^M C_{JL}^K \tag{C.8}$$

can be represented by the diagrammatic equation

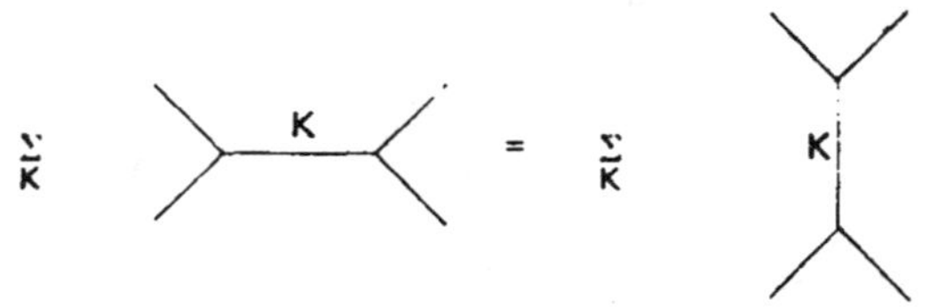

(C.9)

which coincides with the "crossing symmetry" condition by the four-point functions

$$\langle A_I A_J A_L A_M \rangle . \tag{C.10}$$

Appendix D

In this appendix we shall derive the differential equation satisfied by the correlation function

$$\langle \psi(z)\phi_1(z_1)\ldots\phi_N(z_N)\rangle , \tag{D.1}$$

where $\psi(z)$ denotes any of the degenerate fields $\psi_{(1,3)}(z)$ and $\psi_{(3,1)}(z)$, whereas $\phi_i(z)$ are arbitrary primary fields with the dimensions Δ_i, $i = 1, 2, \ldots, N$. First of all, note that the state

$$|\chi_3\rangle = \left[(\Delta + 2) L_{-3} - 2 L_{-1} L_{-2} + \frac{1}{(\Delta + 1)} L_{-1}^3 \right] |\Delta\rangle , \tag{D.2}$$

(where $|\Delta\rangle$ is the primary state with the dimension Δ) is the null vector (with the dimension $\Delta + 3$), provided Δ takes any of the values $\Delta_{(1,3)}$ or $\Delta_{(3,1)}$, i.e.

$$\Delta = \tfrac{1}{6}\left[7 - c \pm \sqrt{(1-c)(25-c)} \right] . \tag{D.3}$$

The equivalent statement is that the operator

$$\chi_{\Delta+3}(z) = (\Delta + 2)\psi^{(-3)}(z) - 2\frac{\partial}{\partial z}\psi^{(-2)}(z) + \frac{1}{\Delta + 1}\frac{\partial^3}{\partial z^3}\psi(z) \tag{D.4}$$

is the null field of the dimension $\Delta + 3$. In (D.4) are the secondaries of the degenerate field $\psi(z)$ ($= \psi_{(1,3)}$ or $\psi_{(3,1)}$) and Δ is given by (D.3). The differential equation for the correlation function (D.1) follows from the condition

$$\chi_{\Delta+3} = 0 . \tag{D.5}$$

It follows that

$$\langle\psi^{(-2)}(z)\phi_1(z_1)\dots\phi_N(z_N)\rangle$$

$$=\left\{\sum_{i=1}^{N}\frac{\Delta_i}{(z-z_i)^2}+\sum_{i=1}^{N}\frac{1}{z-z_i}\frac{\partial}{\partial z_i}\right\}\langle\psi(z)\phi_1(z_1)\dots\phi_N(z_N)\rangle, \quad \text{(D.6)}$$

$$\langle\psi^{(-3)}(z)\phi_1(z_1)\dots\phi_N(z_N)\rangle$$

$$=-\left\{\sum_{i=1}^{N}\frac{2\Delta_i}{(z-z_i)^3}+\sum_{i=1}^{N}\frac{1}{(z-z_i)^2}\frac{\partial}{\partial z_i}\right\}\langle\psi(z)\phi_1(z_1)\dots\phi_N(z_N)\rangle. \quad \text{(D.7)}$$

Substituting (D.4) into (D.5) and taking into account (D.6) and (D.7), one gets the third order differential equation

$$\left\{\frac{1}{\Delta+1}\frac{\partial^3}{\partial z^3}-\sum_{i=1}^{N}\frac{2\Delta\Delta_i}{(z-z_i)^3}-\sum_{i=1}^{N}\frac{\Delta}{(z-z_i)^2}\frac{\partial}{\partial z_i}\right.$$

$$\left.-\sum_{i=1}^{N}\frac{2\Delta_i}{(z-z_i)^2}\frac{\partial}{\partial z}-\sum_{i=1}^{N}\frac{2}{z-z_i}\frac{\partial^2}{\partial z\,\partial z_i}\right\}\langle\psi(z)\phi_1(z_1)\dots\phi_N(z_N)\rangle=0. \quad \text{(D.8)}$$

In the particular case $N=3$, the derivatives can be excluded by means of the projective Ward identities (A.7). Simple calculations lead to the following ordinary differential equation

$$\left\{\frac{1}{\Delta+1}\frac{\mathrm{d}^3}{\mathrm{d}z^3}+\sum_{i=1}^{3}\frac{1}{z-z_i}\frac{\mathrm{d}^2}{\mathrm{d}z^2}+\sum_{i=1}^{3}\frac{\Delta-2\Delta_i}{(z-z_i)^2}\frac{\mathrm{d}}{\mathrm{d}z}\right.$$

$$-\sum_{i=1}^{3}\frac{2\Delta\Delta_i}{(z-z_i)^3}+\sum_{i<j}^{3}\frac{2\Delta+2+\Delta_{ij}}{(z-z_i)(z-z_j)}$$

$$\left.+\sum_{i<j}^{3}\frac{\Delta+\Delta_{ij}}{(z-z_i)(z-z_j)}\left(\frac{1}{(z-z_i)}+\frac{1}{(z-z_j)}\right)\right\}\langle\psi(z)\phi_1(z_1)\phi_2(z_2)\phi_3(z_3)\rangle=0, \quad \text{(D.9)}$$

where

$$\Delta_{12}=\Delta_1+\Delta_2-\Delta_3 \quad \text{etc.}$$

Appendix E

As is well known (see, for instance, [15] and references therein), the two-dimensional Ising model is equivalent to the theory of free Majorana fermions. In the continuous limit this theory is described by the lagrangian density

$$\mathcal{L} = \tfrac{1}{2}\psi \frac{\partial}{\partial \bar{z}} \psi + \tfrac{1}{2}\bar{\psi} \frac{\partial}{\partial z} \bar{\psi} + m\bar{\psi}\psi , \tag{E.1}$$

where m is the mass parameter, proportional to $T - T_c$, and $(\psi, \bar{\psi})$ is the two-component Majorana field*. In what follows we shall consider the critical point only, where this field is massless:

$$m = 0 . \tag{E.2}$$

According to (E.1), in this case the fields $\psi, \bar{\psi}$ satisfy the equation of motion

$$\frac{\partial}{\partial \bar{z}} \psi = 0 , \qquad \frac{\partial}{\partial z} \bar{\psi} = 0 , \tag{E.3}$$

and therefore these fields are analytic functions of the variables z and $\bar{z}$, respectively. We shall write

$$\psi = \psi(z) , \qquad \bar{\psi} = \bar{\psi}(\bar{z}) . \tag{E.4}$$

The stress-energy tensor corresponding to this theory can be computed directly. In the case (E.2) it is traceless and the components (2.5) are given by the formulae

$$T(z) = -\tfrac{1}{2} :\psi(z) \frac{\partial}{\partial z} \psi(z): $$

$$\bar{T}(\bar{z}) = -\tfrac{1}{2} :\bar{\psi}(\bar{z}) \frac{\partial}{\partial \bar{z}} \bar{\psi}(\bar{z}): \tag{E.5}$$

It can be easily verified that the fields (E.5) satisfy the Virasoro algebra (2.21), the central charge c being

$$c = \tfrac{1}{2} . \tag{E.6}$$

The fundamental fields ψ and $\bar{\psi}$ satisfy the relations (1.16), i.e. these fields are primary. The dimensions of the field $\psi(z)$ $(\bar{\psi}(\bar{z}))$ are $\Delta = \frac{1}{2}$, $\bar{\Delta} = 0$ $(\Delta = 0, \bar{\Delta} = \frac{1}{2})$. It can be shown that four conformal families $[I], [\psi], [\bar{\psi}], [:\bar{\psi}\psi:]$ contitute a complete set of fields $\{A_j\}$, forming the operator algebra (1.6).

* The field $\bar{\psi}$ is an independent component but in general it is not the complex conjugated value of the field ψ.

Let us take, for instance, the field $\psi(z)$. This primary field proves to coincide with the degenerate field $\psi_{(2,1)}(z)$ (see (6.13)). Actually, the operator product expansion for $T(\zeta)\psi(z)$ (which is easily computed if (E.5) is employed) is given (up to the first three terms) by the formula

$$T(\zeta)\psi(z)=\frac{1}{2}\frac{1}{(\zeta-z)^2}\psi(z)+\frac{1}{\zeta-z}\frac{\partial}{\partial z}\psi(z)+\frac{3}{4}\frac{\partial^2}{\partial z^2}\psi(z)+O(\zeta-z), \quad \text{(E.7)}$$

which shows that the secondary field (5.2) vanishes. Therefore, the correlation functions, involving the degenerate field $\psi(z)$, satisfy the differential equation

$$\left\{\frac{3}{4}\frac{\partial^2}{\partial z^2}-\sum_{i=1}^{N}\frac{\Delta_i}{(z-z_i)^2}-\sum_{i=1}^{N}\frac{1}{z-z_i}\frac{\partial}{\partial z_i}\right\}\langle\psi(z)\phi_1(z_1)\dots\phi_N(z_N)\rangle=0, \quad \text{(E.8)}$$

where $\phi_i(z)$ are arbitrary primary fields (which are local themselves but not necessarily local with respect to $\psi(z)$). In particular, the correlation functions

$$\langle\psi(z)\psi(z_1)\dots\psi(z_N)\rangle, \quad \text{(E.9)}$$

(which can be computed if the Wick rules are used) satisfy (E.6).

On the other hand, the critical Ising model can be described in terms of either the order-parameter field $\sigma(z,\bar{z})$ or the disorder-parameter field $\mu(z,\bar{z})$*. Obviously, the fields σ and μ are primary. These fields have zero spins, i.e. $\Delta_\sigma=\bar{\Delta}_\sigma$, $\Delta_\mu=\bar{\Delta}_\mu$ and in virtue of the Krammers-Wanier symmetry, have the same scale dimensions

$$\Delta_\sigma=\Delta_\mu=\Delta. \quad \text{(E.10)}$$

The fields $\sigma(z,\bar{z})$ and $\mu(z,\bar{z})$ are neither local with respect to the fields $\psi(z)$ and $\bar{\psi}(\bar{z})$ nor mutually local. In fact, the correlation function

$$\langle\psi(z)\sigma(\xi_1)\dots\sigma(\xi_{2N-1})\mu(\xi_{2N})\dots\mu(\xi_{2M})\rangle \quad \text{(E.11)}$$

is a double-valued analytic function of z which acquires the phase factor (-1) after the analytical commutation around any of the singular points $z_k=\xi_k^1+i\xi_k^2$, $k=1,\dots,2M$. It follows from the definition that the products $\psi(\zeta)\sigma(z,\bar{z})$ and $\psi(\zeta)\mu(z,\bar{z})$ can be expanded as

$$\psi(\zeta)\sigma(z,\bar{z})=(\zeta-z)^{-1/2}\{\mu(z,\bar{z})+O(\zeta-z)\},$$

$$\psi(\zeta)\mu(z,\bar{z})=(\zeta-z)^{-1/2}\{\sigma(z,\bar{z})+O(\zeta-z)\}. \quad \text{(E.12)}$$

* The fields σ and μ are the scaling limit of the lattice spin $\sigma_{n,m}$ and the dual spin $\mu_{n-1/2,m+1/2}$, respectively. See ref. [15] for the detailed definition.

Substituting these expansions into the differential eq. (E.8), one gets the characteristic equation, determining the parameter Δ:

$$\Delta = \tfrac{1}{16} \tag{E.13}$$

in agreement with the known value of the scale dimension of the spin field $d_\sigma = 2\Delta = \frac{1}{8}$ [15]. So, the differential eq. (E.8) together with the qualitative properties (E.12) of the operator algebra enables one to compute exactly the dimension of the field $\sigma(z, \bar{z})$.

Now we are to compute the correlation functions of the order and disorder fields

$$\langle \sigma(\xi_1) \dots \sigma(\xi_{2N}) \mu(\xi_{2N+1}) \dots \mu(\xi_{2M}) \rangle . \tag{E.14}$$

Note that the double-valued function (E.11) can be represented by

$$\langle \psi(z) \sigma(\xi_1) \dots \mu(\xi_{2M}) \rangle = \prod_{i=1}^{2M} (z - z_i)^{-1/2} P(z | z_i, \bar{z}_i), \tag{E.15}$$

where $P(z | z_i, \bar{z}_i)$ is a polynomial in z:

$$P(z | z_i, \bar{z}_i) = \sum_{k=0}^{2M-1} (z - z_{2N})^K G_k(z_i, \bar{z}_i). \tag{E.16}$$

The order $2M - 1$ of this polynomial is determined by the asymptotic condition

$$\psi(z) \sim z^{-1}, \qquad z \to \infty . \tag{E.17}$$

The coefficients G_k are some functions of $z_1, \dots, z_{2M}, \bar{z}_1, \dots, \bar{z}_{2M}$. In virtue of (E.12), the coefficient $G_0(z_i, \bar{z}_i)$ coincides with the correlation function (E.14). Substituting (E.15) into the differential eq. (E.8), one gets the differential equations for the coefficients $G_k(z_i, \bar{z}_i)$ which enables one to compute the correlation function (E.14).

In fact, the differential equations for the correlation functions (E.14) can be obtained in a simpler way. Note that comparing (E.13) with (6.13), the field $\sigma(z, \bar{z})$ is the degenerate field $\psi_{(1,2)}$ with respect to the both variables z and $\bar{z}$. The same is valid for the field $\mu(z, \bar{z})$. Therefore, the correlation functions (E.14) satisfy the differential equations

$$\left\{ \frac{4}{3} \frac{\partial^2}{\partial z_i^2} - \sum_{j \neq i}^{2M} \frac{\frac{1}{16}}{(z_i - z_j)^2} - \sum_{j \neq i}^{2M} \frac{1}{z_i - z_j} \frac{\partial}{\partial z_j} \right\}$$
$$\times \langle \sigma(z_1, \bar{z}_1) \dots \sigma(z_{2N}, \bar{z}_{2N}) \mu(z_{2N+1}, \bar{z}_{2N+1}) \dots \mu(z_{2M}, \bar{z}_{2M}) \rangle = 0, \tag{E.18}$$

(where $i = 1, 2, \dots, 2M$) and the differential equations obtained from (E.18) by the substitution $z_i \to \bar{z}_i$.

Let us consider, for example, the four-point correlation function

$$G(\xi_1, \xi_2, \xi_3, \xi_4) = \langle \sigma(\xi_1)\sigma(\xi_2)\sigma(\xi_3)\sigma(\xi_4)\rangle$$
$$= \left[(z_1 - z_3)(z_2 - z_4)(\bar{z}_1 - \bar{z}_3)(\bar{z}_2 - \bar{z}_4)\right]^{-1/8} Y(x, \bar{x}), \tag{E.19}$$

where $Y(x, \bar{x})$ is some function of the anharmonic quotients

$$x = \frac{(z_1 - z_2)(z_3 - z_4)}{(z_1 - z_3)(z_2 - z_4)}, \qquad \bar{x} = \frac{(\bar{z}_1 - \bar{z}_2)(\bar{z}_3 - \bar{z}_4)}{(\bar{z}_1 - \bar{z}_3)(\bar{z}_2 - \bar{z}_4)}, \tag{E.20}$$

(we took into account (A.8)). In this case the differential eq. (E.18) is reduced to the following form:

$$\left\{\frac{4}{3}\frac{d^2}{dx^2} - \frac{1}{16}\left[\frac{1}{x^2} + \frac{1}{(x-1)^2}\right] + \frac{1}{8}\frac{1}{x(x-1)} + \left[\frac{1}{x} + \frac{1}{x-1}\right]\frac{d}{dx}\right\} Y(x, \bar{x}) = 0. \tag{E.21}$$

The same equation with respect to $\bar{x}$ is also valid. Substituting

$$Y(x, \bar{x}) = \left[x\bar{x}(1-x)(1-\bar{x})\right]^{-1/8} u(x, \bar{x}), \tag{E.22}$$

one gets the following equation for

$$\left\{x(1-x)\frac{\partial^2}{\partial x^2} + \left(\tfrac{1}{2} - x\right)\frac{\partial}{\partial x} + \tfrac{1}{16}\right\} u(x, \bar{x}) = 0. \tag{E.23}$$

The change of variables

$$x = \sin^2\theta, \qquad \bar{x} = \sin^2\bar{\theta}, \tag{E.24}$$

reduces (E.23) to

$$\left(\frac{\partial^2}{\partial\theta^2} + \tfrac{1}{4}\right) u(\theta, \bar{\theta}) = 0. \tag{E.25}$$

The equation obtained from (E.25) by the substitution $\theta \to \bar{\theta}$ is also valid. Therefore, the general solution of these differential equations has the form

$$u(\theta, \bar{\theta}) = u_{11}\cos\tfrac{1}{2}\theta\cos\tfrac{1}{2}\bar{\theta} + u_{12}\cos\tfrac{1}{2}\theta\sin\tfrac{1}{2}\bar{\theta}$$
$$+ u_{21}\sin\tfrac{1}{2}\theta\cos\tfrac{1}{2}\bar{\theta} + u_{22}\sin\tfrac{1}{2}\theta\sin\tfrac{1}{2}\bar{\theta}, \tag{E.26}$$

where $u_{\alpha\beta}$ $(\alpha, \beta = 1, 2)$ are arbitrary constants.

Note that two independent solutions of (E.21) coincide with the conformal blocks (see (B.9))

$$\mathcal{F}(\tfrac{1}{16},0,x)=[x(1-x)]^{-1/8}\cos\tfrac{1}{2}\theta,$$

$$\mathcal{F}(\tfrac{1}{16},\tfrac{1}{2},x)=[x(1-x)]^{-1/8}\sin\tfrac{1}{2}\theta, \tag{E.27}$$

and therefore the formula (E.26) can be considered as the decomposition (4.11), the coefficients $u_{\alpha\beta}$ being the structure constants.

Since the field $\sigma(z,\bar{z})$ is local, the correlation function (E.20) should be single-valued in the euclidean domain

$$\bar{x}=x^{*}, \tag{E.28}$$

where the asterisk denotes complex conjugation. As it is clear from (E.24), the analytical continuation of the variables x and $\bar{x}$ around the singular point $x=\bar{x}=0$ corresponds to the substitution

$$\theta\to-\theta,\qquad \bar{\theta}\to-\bar{\theta}. \tag{E.29}$$

The function (E.26) is unchanged under this transformation provided

$$u_{12}=u_{21}=0. \tag{E.30}$$

The same investigation of the singular point $x=\bar{x}=1$ (or, equivalently, imposing the crossing-symmetry condition) leads to the relation

$$u_{11}=u_{22}. \tag{E.31}$$

The overall factor in (E.26) depends on the σ-field normalization. We shall normalize this field so that

$$\langle\sigma(z,\bar{z})\sigma(0,0)\rangle=[z\bar{z}]^{-1/8}. \tag{E.32}$$

Then

$$u(\theta,\bar{\theta})=\cos\tfrac{1}{2}(\theta-\bar{\theta}). \tag{E.33}$$

The four-point function given by the formulae (E.20), (E.22) and (E.33) is in agreement with the previous result (see ref. [16]) obtained by a different method.

Note that in virtue of (E.27) the four-point function (E.20) can be represented as

$$G=\mathcal{F}(\tfrac{1}{16},0,x)\bar{\mathcal{F}}(\tfrac{1}{16},0,\bar{x})+\mathcal{F}(\tfrac{1}{16},\tfrac{1}{2},x)\bar{\mathcal{F}}(\tfrac{1}{16},\tfrac{1}{2},\bar{x}). \tag{E.34}$$

It is evident from this formula that only two conformal families contribute to the operator product expansion of $\sigma(\xi)\sigma(0)$. The corresponding primary fields have the dimensions $\Delta = \bar{\Delta} = 0$ and $\Delta = \bar{\Delta} = \frac{1}{2}$. The first of them is obviously identified with the identity operator I whereas the second is known as the energy density field

$$\varepsilon(z, \bar{z}) = \bar{\psi}(\bar{z})\psi(z). \tag{E.35}$$

The four-point correlation function

$$H(\xi_1, \xi_2, \xi_3, \xi_4) = \langle \sigma(\xi_1)\mu(\xi_2)\sigma(\xi_3)\mu(\xi_4)\rangle \tag{E.36}$$

can be represented in the form

$$H = \left[(z_1 - z_3)(z_2 - z_4)(\bar{z}_1 - \bar{z}_3)(\bar{z}_2 - \bar{z}_4)\right]^{-1/8}\tilde{Y}(x, \bar{x}), \tag{E.37}$$

where the function $\tilde{Y}$ satisfies the same differential equation (E.21). The investigation similar to the one performed above leads to the result

$$\tilde{Y}(x, \bar{x}) = \left[x\bar{x}(1-x)(1-\bar{x})\right]^{-1/8}\sin\tfrac{1}{2}(\theta + \bar{\theta}). \tag{E.38}$$

Therefore the function (E.36) is

$$H = \mathcal{F}(\tfrac{1}{16}, 0, x)\bar{\mathcal{F}}(\tfrac{1}{16}, \tfrac{1}{2}, \bar{x}) + \mathcal{F}(\tfrac{1}{16}, \tfrac{1}{2}, x)\bar{\mathcal{F}}(\tfrac{1}{16}, 0, \bar{x}). \tag{E.39}$$

This formula corresponds to the following operator product expansion

$$\sigma(z, \bar{z})\mu(0,0) = z^{3/8}\bar{z}^{-1/8}\{\Psi(z) + O(z, \bar{z})\} + z^{-1/8}\bar{z}^{3/8}\{\bar{\Psi}(\bar{z}) + O(z, \bar{z})\}, \tag{E.40}$$

which is in accordance with the idea of the field ψ as the regularized product $:\sigma\mu:$.

To avoid misunderstanding, let us stress that there are three different sets of fields

$$\{A_j\} = \{[I], [\Psi], [\bar{\Psi}], [\varepsilon]\},$$

$$\{B_j\} = \{[I], [\sigma], [\varepsilon]\},$$

$$\{C_j\} = \{[I], [\mu], [\varepsilon]\}. \tag{E.41}$$

Each of these sets forms the closed operator algebra and it is appropriate to describe the critical Ising field theory. All the fields entering the same set are mutually local whereas the fields entering different sets are in general nonlocal with respect to each other.

References

[1] A.Z. Patashinskii and V.L. Pokrovskii, Fluctuation theory of phase transitions (Pergamon, Oxford, 1979)
[2] A.M. Polyakov, ZhETF Lett. 12 (1970) 538
[3] A.A. Migdal, Phys. Lett. 44B (1972) 112
[4] A.M. Polyakov, ZhETF, 66 (1974) 23
[5] K.G. Wilson, Phys. Rev. 179 (1969) 1499
[6] B.L. Feigin and D.B. Fuks, Funktz. Analiz 16 (1982) 47
[7] V.G. Kac, Lecture notes in phys. 94 (1979) 441
[8] S. Mandelstam, Phys. Reports 12C (1975) 1441
[9] J.H. Schwarz, Phys. Reports 8C (1973) 269
[10] I.M. Gelfand and D.B. Fuks, Funktz. Analiz 2 (1968) 92
[11] M. Virasoro, Phys. Rev. D1 (1969) 2933
[12] H. Bateman and A. Erdelyi, Higher transcendental functions (McGraw-Hill, 1953)
[13] A. Poincaré, Selected works, vol. 3 (Nauka, Moscow, 1974)
[14] A.M. Polyakov, Phys. Lett. 103B (1981) 207
[15] B. McKoy and T.T. Wu, The two-dimensional Ising model (Harvard Univ. Press, 1973)
[16] A. Luther and I. Peschel, Phys. Rev. B12 (1975) 3908

Nuclear Physics B270 [FS16] (1986) 186–204
North-Holland, Amsterdam

OPERATOR CONTENT OF TWO-DIMENSIONAL CONFORMALLY INVARIANT THEORIES

John L. CARDY

Department of Physics, University of California, Santa Barbara CA 93106, USA

Received 22 November 1985
(Revised 3 January 1986)

It is shown how conformal invariance relates many numerically accessible properties of the transfer matrix of a critical system in a finite-width infinitely long strip to bulk universal quantities. Conversely, general properties of the transfer matrix imply constraints on the allowed operator content of the theory. We show that unitary theories with a finite number of primary operators must have a conformal anomaly number $c < 1$, and therefore must fall into the classification of Friedan, Qiu and Shenker. For such theories, we derive sum rules which constrain the numbers of operators with given scaling dimensions.

1. Introduction

The fact that a statistical system with short range interactions at a critical point should be conformally invariant has many interesting consequences, particularly in two dimensions [1]. A simple example is the mapping of the plane into a finite-width strip, from which the correlation functions [2] and other quantities accessible to numerical calculation may be determined. They are related to properties of the transfer matrix along the strip, which we shall denote by $e^{-a\hat{H}}$, where a is the lattice spacing. In the continuum limit, $\hat{H}$ may be thought of as the hamiltonian operator of a quantum field theory in (1 + 1) dimensions.

Two particularly useful results of this mapping, which have already been discussed elsewhere [3, 2], relate to the eigenvalues E_n of $\hat{H}$: for a strip whose width $l \to \infty$, with periodic boundary conditions,

$$E_0 \sim fl - \frac{\pi c}{6l}, \tag{1.1}$$

$$E_n - E_0 \sim \frac{2\pi x_n}{l}. \tag{1.2}$$

Eq. (1.1) relates the finite size correction to the lowest eigenvalue E_0 (the ground state energy) to the value of the conformal anomaly number c, which plays a central

0550-3213/86/$03.50 © Elsevier Science Publishers B.V.
(North-Holland Physics Publishing Division)

role in the analysis of conformal invariance, and which may be used to label different universality classes [4,5]. Eq. (1.2) relates the energy gaps of the excited states (which are the inverse correlation lengths in the strip) to the scaling dimensions x_n of the scaling operators of the theory, and gives a very accurate way of measuring them [6,7].

In the first part of this paper, we show that conformal invariance predicts a great deal more about the structure of the transfer matrix. In particular, matrix elements of operators between eigenstates of $\hat{H}$ are related to the universal coefficients of the operator product expansion. We are also able to determine the form of the corrections to the results in eqs. (1.1),(1.2) and to demonstrate the existence of universal ratios of their amplitudes.

In the second part, we exploit the fact that the transfer matrix for an infinitely long strip of width l also yields the partition function for a rectangle with periodic boundary conditions (a torus) of dimensions $l \times l'$:

$$Z(l, l') = \mathrm{Tr}\, e^{-l'\hat{H}}. \tag{1.3}$$

A similar result holds for a parallelogram (see eq. (3.8)) if l'/l is generalized to a complex number. The condition that $Z(l, l') = Z(l', l)$ then implies nontrivial constraints on the eigenvalues of $\hat{H}$ and their degeneracy, and hence on the allowed number of independent operators with a given scaling dimension in the theory. In the general theory of conformal invariance, it is shown that to each scaling operator ϕ with scaling dimension x corresponds an infinite number of other operators $L_n\phi$, with dimensions $x-n$, which are generated in the short distance expansion of ϕ with the stress tensor T. Since the scaling dimensions cannot be negative in a unitary theory (for example, one in which $\hat{H}$ is hermitian,) there exjst so-called *primary* operators for which $L_n\phi = 0$ for all $n > 0$. Belavin, Polyakov and Zamolodchikov [4] showed that if we parametrize c by

$$c = 1 - \frac{6}{m(m+1)} \tag{1.4}$$

and m is rational, the set of operators whose scaling dimensions $(h, \bar{h})$ (where $x = h + \bar{h}$) are given by the Kac formula [8] $h = h_{p,q}, \bar{h} = h_{\bar{p},\bar{q}}$, where

$$h_{p,q} = \frac{(p(m+1) - qm)^2 - 1}{4m(m+1)} \tag{1.5}$$

form a *finite* set of primary operators, in the sense that no more are generated in the operator product expansion. Friedan, Qiu and Shenker [5] showed that in a unitary theory with $c < 1$, m must be an integer > 2, and that the only possible scaling dimensions of the primary operators are given by the Kac formula with $1 \leqslant q \leqslant p \leqslant m-1$. Goddard, Kent and Olive [8a] showed that these conditions are also sufficient for unitarity. Our first result complements the results of these papers. We show that in a unitary theory with a finite number of primary operators, c must necessarily be less than one, and the theory must therefore fall into the classification of Friedan, Qiu and Shenker.

For such theories, it turns out that the eigenvalue structure of the transfer matrix may be completely determined. The essential results are contained in the character formulas for the appropriate representations of the Virasoro algebra, which were derived by Rocha-Caridi [9]. We show that the condition that $Z(l, l')$ be symmetric under the interchange $l \leftrightarrow l'$ may be satisfied as long as the quantities $\mathcal{N}(p, q; \bar{p}, \bar{q})$, defined as the number of operators with $h = h_{p,q}$ and $\bar{h} = h_{\bar{p},\bar{q}}$, satisfy certain sum rules.

This is of interest because although the Friedan, Qiu and Shenker [5] classification dictates the allowed values of the scaling dimensions of operators, it does not determine which operators may actually appear in a given theory. Indeed, there may be more than one possible set of operators. For example, both the universality classes of the 3-state Potts model [5, 10] and that of a "generic" tetracritical point [11] have been identified with $m = 5$. However, not all scaling dimensions allowed by the Kac formula appear in the Potts model, and some of the others should be doubled. In the tetracritical model, on the other hand, it seems as though all values appear. We show that both these examples satisfy the sum rules, and are able to exhibit the complete set of primary operators in both cases. In fact, these are the *only* unitary models with $m = 5$. The general solution of the sum rules, for arbitrary m, has however eluded us.

2. Structure of the transfer matrix

We begin by recalling the form of the general two-point function in a strip of width l with periodic boundary conditions [2]. In the infinite plane, the two-point function of an operator ϕ with scaling dimensions $(h, \bar{h})$ is

$$\langle \phi(z, \bar{z}) \phi(z', \bar{z}') \rangle = (z - z')^{-2h} (\bar{z} - \bar{z}')^{-2\bar{h}} . \tag{2.1}$$

If ϕ is a primary operator, under the conformal mapping $w = f(z)$ the correlation function transforms according to

$$\langle \phi(z, \bar{z}) \phi(z', \bar{z}') \rangle = (f'(z))^{h} \left(\overline{f'(z)}\right)^{\bar{h}} (f'(z'))^{h} \left(\overline{f'(z')}\right)^{\bar{h}} \langle \phi(w, \bar{w}) \phi(w', \bar{w}') \rangle . \tag{2.2}$$

Choosing $f(z) = (l/2\pi)\ln z$ and using (2.1), we obtain the correlation function in the strip:

$$\langle \phi(w,\bar{w})\phi(w',\bar{w}')\rangle = \frac{(\pi/l)^{2x}}{(\sinh \pi(w-w')/l)^{2h}(\sinh \pi(\bar{w}-\bar{w}')/l)^{2\bar{h}}}. \quad (2.3)$$

Putting $w = u + iv$, $w' = u' + iv'$, this has the expansion, for $u > u'$,

$$\left(\frac{2\pi}{l}\right)^{2x} \sum_{N,\bar{N}=0}^{\infty} a_N a_{\bar{N}} \exp[-2\pi(x+N+\bar{N})(u-u')/l]$$
$$\times \exp[2\pi i(s+N-\bar{N})(v-v')/l], \quad (2.4)$$

where $x = h + \bar{h}$ is the scaling dimension of ϕ, $s = h - \bar{h}$ is its spin, and the coefficients a_N are given by

$$a_N = \frac{\Gamma(x+N)}{\Gamma(x)N!}. \quad (2.5)$$

On the other hand, the correlation function in the strip may be evaluated using transfer matrix techniques. In that case the scaling "operators" $\phi(u,v)$ become true operators $\hat{\phi}(v)$ acting on the same Hilbert space as does the transfer matrix. The correlation function may be written

$$\langle \phi(u,v)\phi(u',v')\rangle = \sum_n \langle 0|\hat{\phi}(v)|n,k\rangle e^{-(E_n-E_0)(u-u')}\langle n,k|\hat{\phi}(v')|0\rangle, \quad (2.6)$$

where $|n,k\rangle$ is a complete set of eigenstates of $\hat{H}$ of energy E_n and momentum k (quantized in units of $(2\pi/l)$, so that the matrix elements depend on v and v' as $e^{ik(v-v')}$. Comparing with (2.4), we see that to each primary operator of dimension x and spin s there correspond an infinite number of eigenstates of $\hat{H}$, labelled by $(N,\bar{N})$, with energy $E_0 + 2\pi(x+N+\bar{N})/l$ and momentum $2\pi(s+N+\bar{N})/l$. The lowest such state must be non-degenerate, and we denote it by $|\phi\rangle$. From (2.4) we see that

$$\langle 0|\hat{\phi}(v)|\phi\rangle = (2\pi/l) \ . \quad (2.7)$$

Associated with each primary operator ϕ is an infinite number of other scaling operators in the conformal block [4,5] of ϕ. The operators $L_{-n}\phi$ are defined by the short-distance expansion of ϕ with the (zz) component of the stress tensor T:

$$T(z)\phi(z_1,\bar{z}_1)+\sum_{n=0}^{\infty}(z-z_1)^{-2-n}L_{-n}\phi(z_1,\bar{z}_1). \tag{2.8}$$

In the same way, the operators $\bar{L}_{-n}\phi$ are defined via the short-distance expansion with $\bar{T}$. Further operators may be generated by repeated short-distance expansions with T and $\bar{T}$. The most general operator ψ at level $(N,\bar{N})$ has the form

$$L_{-k_1}\dots L_{-k_m}\bar{L}_{-k'_1}\dots\bar{L}_{-k'_{m'}}\phi, \tag{2.9}$$

where $k_1 \leqslant \cdots k_m$, $k'_1 \leqslant \cdots \leqslant k'_{m'}$, and $\Sigma k_j = N$, $\Sigma k'_j = \bar{N}$. This operator has scaling dimensions $(h+N, \bar{h}+\bar{N})$, and therefore corresponds to an eigenstate $|\psi\rangle$ of $\hat{H}$ of energy $x+N+\bar{N}$ and momentum $s+N-\bar{N}$, both measured in units of $(2\pi/l)$. Since the two-point function between ψ and the primary operator ϕ is non-vanishing, this eigenstate must be identified with one of those appearing in eq. (2.6). However, not all the operators at a given level are independent. Indeed, for those operators whose scaling dimensions are given by the Kac formula, there is considerable degeneracy. The number of independent operators at level $(N,\bar{N})$ will be equal to the degeneracy of the appropriate eigenstate of $\hat{H}$.

The correlation function $\langle\phi\psi\rangle$ is given in the infinite plane in terms of differential operators [4] acting on $\langle\phi\phi\rangle$. This can be conformally transformed to the strip, yielding all the matrix elements of the form $\langle 0|\phi|\psi\rangle$. Only matrix elements involving the lowest states $|\phi\rangle$ in a given block have a very simple form, however. Further matrix elements of this type may be obtained by transforming the 3-point function to the strip. In the infinite plane the general 3-point function of primary operators has the form [12]

$$\langle\phi_i(z_1,\bar{z}_1)\phi_j(z_2,\bar{z}_2)\phi_k(z_3,\bar{z}_3)\rangle = c_{ijk}z_{12}^{-h_i-h_j+h_k}z_{23}^{-h_j-h_k+h_i}z_{31}^{-h_k-h_i+h_j}$$
$$\times\bar{z}_{12}^{-\bar{h}_i-\bar{h}_j+\bar{h}_k}\bar{z}_{23}^{-\bar{h}_j-\bar{h}_k+\bar{h}_i}\bar{z}_{31}^{-\bar{h}_k-\bar{h}_i+\bar{h}_j}, \tag{2.10}$$

where c_{ijk} is the operator product expansion coefficient of ϕ_k in the short-distance expansion of ϕ_i and ϕ_j. In the strip one finds

$$\langle\phi_i(u_1,v_1)\phi_j(u_2,v_2)\phi_k(u_3,v_3)\rangle = \left(\frac{2\pi}{l}\right)^{x_i+x_j+x_k} c_{ijk}e^{-2\pi x_i(u_1-u_2)/l}e^{-2\pi x_k(u_2-u_3)/l}$$
$$\times e^{2\pi i s_i(v_1-v_2)/l}e^{2\pi i s_k(v_2-v_3)/l} \tag{2.11}$$

for $u_1 \gg u_2 \gg u_3$. The same correlation function evaluated in the transfer matrix formalism is

$$\langle 0|\hat{\phi}_i(v_1)|\phi_i\rangle e^{-2\pi x_i(u_1-u_2)/l}\langle\phi_i|\hat{\phi}_j(v_2)|\phi_k\rangle e^{2\pi x_k(u_2-u_3)/l}\langle\phi_k|\hat{\phi}_k(v_3)|0\rangle . \tag{2.12}$$

Comparing these expressions, and using (2.7), one finds

$$\langle\phi_i|\hat{\phi}_j(v)|\phi_k\rangle = \left(\frac{2\pi}{l}\right)^{x_j} c_{ijk} e^{2\pi i(s_i - s_k)v/l} . \tag{2.13}$$

Thus the universal operator product expansion coefficients c_{ijk} are measurable in terms of matrix elements of operators between low-lying states. In practice, the operators $\hat{\phi}_j$ will not be normalized, in which case c_{ijk} may be obtained from

$$c_{ijk} = \frac{\langle\phi_i|\hat{\phi}_j(v)|\phi_k\rangle}{\langle\phi_j|\hat{\phi}_j(v)|0\rangle} e^{2\pi i s_k v/l} \tag{2.14}$$

2.1. CORRECTIONS TO FINITE-SIZE SCALING

One of the applications of the above result concerns the corrections to the result (1.2) for finite values of l. These occur because at a critical point the hamiltonian will differ from the fixed-point hamiltonian by terms involving irrelevant operators. If we assume that this departure is small, we can write the infinitesimal transfer matrix as

$$\hat{H} = \hat{H}^* + \sum_j a_j \int \mathrm{d}v\, \hat{\phi}_j(v) , \tag{2.15}$$

where the a_j are unknown parameters. To first order in the perturbation,

$$E_n - E_0 = \frac{2\pi x_n}{l} + \sum_j a_j \int \mathrm{d}v\, \langle\phi_n|\hat{\phi}_j(v)|\phi_n\rangle . \tag{2.16}$$

Using (2.13) this may be written

$$E_n - E_0 = \frac{2\pi x_n}{l}\left(1 + \sum_j a_j' c_{nnj} (2\pi/l)^{x_j - 2} + \cdots\right). \tag{2.17}$$

This shows the typical form of correction to scaling terms, since we may identify $2 - x_j$ with the renormalization group eigenvalue of ϕ_j. For ϕ_j to be irrelevant, $x_j > 2$, and the correction term becomes negligible as as $l \to \infty$, as expected. Eq. (2.17) shows that ratios of correction to scaling amplitudes are universal, and related to ratios of operator product expansions coefficients.

The conformal block of the identity operator $\mathbf{1}$ is present in all theories. It contains the operators $L_{-2}\mathbf{1} \propto T$ and $\bar{L}_{-2}\mathbf{1} \propto \bar{T}$. These operators are not allowed to appear in $\hat{H}$ since they are not scalars, but $L_{-2}\bar{L}_{-2}\mathbf{1}$, which has $x = 4$, is allowed. We thus expect corrections to finite size scaling of order l^{-2} to be present in any theory. (These are often referred to as "analytic" corrections, but note that terms $O(l^{-1})$ are not allowed.) On a lattice, non-scalar operators are also allowed. For example, on a square lattice operators of spin ± 4 will appear. The most relevant operators with $s = 4$ are $L_{-4}\mathbf{1}$ and $L^2_{-2}\mathbf{1}$, which both have $x = 2$ also, and therefore lead to $O(l^{-2})$ corrections. For a self-dual Ising model, these will be no other "non-analytic" corrections, since the only other blocks are those of the energy density ε, which is odd under duality, and the magnetization σ, which is odd under spin reversal. In other models, there will of course be non-analytic corrections, although they may be hard to disentangle [13].

3. Operator content of unitary theories

In the last section, we defined a primary operator as one annihilated by the lowering operators L_n and $\bar{L}_n$ for all $n > 0$. In a unitary theory, in which scaling dimensions must be positive, it is possible, given a list of operators in the theory, to construct all the primary ones by repeatedly applying the lowering operators. We now show that if the number of primary operators so constructed is *finite*, then the conformal anomaly number c must be less than one.

We consider the partition function for a theory defined on an $l \times l'$ rectangle, with toroidal boundary conditions, in the limit that $l, l' \to \infty$ with $l'/l \equiv \delta$ fixed. From (1.3) and (1.1) this has the form

$$Z(l, l') = e^{-fA + \pi c\delta/6} \sum_n e^{-(E_n - E_0)l'}, \tag{3.1}$$

where A is the area. In the limit under consideration, only those energy gaps which scale like l^{-1} contribute. By eq. (1.2), they are given by the dimensions of all the independent scaling operators in the theory. The sum over n may be broken into a sum over conformal blocks, and a sum over the operators in each block. At level $(N, \bar{N})$ in a block, the general operator has the form (2.9). There are $P(N)P(\bar{N})$ such operators, where $P(N)$ is the number of partitions of N into positive integers, not necessarily distinct. In general , however, some of these operators may not be independent. In a unitary theory, each term in (3.1) is positive. Therefore an upper bound on the contribution to the sum in (3.1) from one conformal block, whose primary operator has scaling dimension x, is

$$e^{-2\pi x\delta} \sum_{N, \bar{N}} P(N)P(\bar{N}) e^{-2\pi(N + \bar{N})\delta}. \tag{3.2}$$

This is just the square of the generating function for $P(N)$. It is equal to $f(\delta)^{-2}$, where

$$f(\delta)=\prod_{n=1}^{\infty}(1-e^{-2\pi n\delta}). \tag{3.3}$$

We therefore have the upper bound on the partition function

$$Z(l,l')\leqslant e^{-fll'+\pi c\delta/6}f(\delta)^{-2}\sum_{\substack{\text{primary}\\ \text{operators}}} e^{-2\pi x_n\delta}. \tag{3.4}$$

Now consider the limit $\delta\to 0$. In appendix B we show that $f(\delta)$ satisfies the inversion relation

$$f(\delta)=\delta^{-1/2}e^{\pi(\delta-\delta^{-1})/12}f(\delta^{-1}). \tag{3.5}$$

As $\delta\to 0$, the fact that $Z(\delta)=Z(\delta^{-1})$ implies from (3.1) that $Z(\delta^{-1})\sim e^{-fA+\pi c/6\delta}$, and hence, comparing with (3.4) that

$$e^{-fA+\pi c/6\delta}\leqslant \delta\mathcal{N}e^{-fA+\pi/6\delta}, \tag{3.6}$$

where $\mathcal{N}$ is the number of primary operators. If $\mathcal{N}$ is finite, we see that c must be strictly less than one. In that case, every primary operator must be degenerate at some level, and hence its scaling dimensions are given by the Kac formula.

This result must be interpreted carefully. It is clearly possible to consider theories consisting of several decoupled models, each of which has $c<1$. Since c is additive, the resulting theory may well have $c>1$, yet may appear at first glance to have finite $\mathcal{N}$. This is not so, however. As an example consider the Ashkin-Teller model. This consists of two Ising models, each with $c=\frac{1}{2}$, with a four-spin coupling between them. Consider the decoupling point, where this vanishes. Within each Ising model, the magnetization operators $\sigma^{(1)}$ and $\sigma^{(2)}$ are primary, being annihilated by $L_n^{(1)}$ and $L_n^{(2)}$, for $n>0$, respectively. In the composite model, a primary operator is one annihilated by $L_n\equiv L_n^{(1)}+L_n^{(2)}$. It is easy to show that there is an infinity of such operators, for example

$$\sigma^{(1)}\left(L_{-1}^{(1)}-L_{-1}^{(2)}\right)^k\sigma^{(2)}\qquad (k=1,2,\dots). \tag{3.7}$$

While this example may seem somewhat pedantic, it is important to realize that away from the decoupling point such primary scaling operators become non-trivial, although the number of them remains infinite, consistent with the fact that c remains at one.

Another case occurs if there is some additional symmetry which may relate the infinite number of primary operators back to a finite set. This happens, for example,

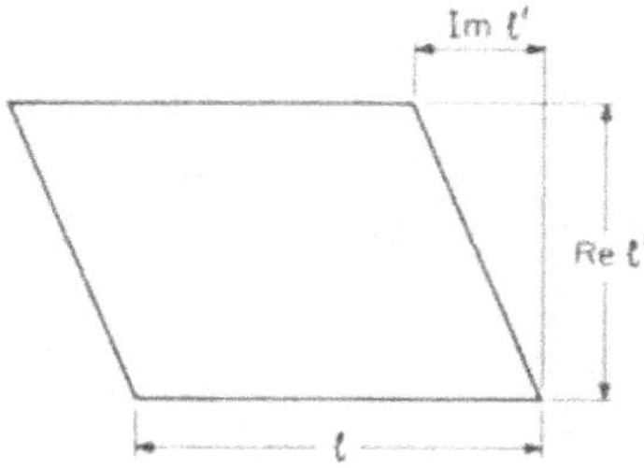

Fig. 1. Definition of l and l' for an arbitrary parallelogram. The partition function is invariant under the interchange $l \leftrightarrow l'$.

in supersymmetric theories [5, 14, 15], where in addition to the operators L_n we have their fermionic partners G_n. Associated with each primary operator is an infinite set of other primary operators formed by acting with the G_n. If we redefine the concept of primary to refer to operators annihilated by both the L_n and the G_n, then the above analysis is modified. Theories with a finite number of such primary operators must have $c < \frac{3}{2}$. This is consistent with the result of refs. [5, 14, 15]. (The relative factor of $\frac{1}{2}$ for the fermions can be traced to the fact that $G_n^2 = 0$.)

3.1. INVERSION SUM RULES

We now restrict ourselves to the case of unitary theories with a finite number of primary operators, so that they fall into the classification of Friedan, Qiu and Shenker [5]. It is first necessary to generalize (3.1) to the case of an arbitrarily shaped parallelogram. The shape may be specified by a single *complex* number $\delta = l'/l$, as is illustrated in fig. 1. Toroidal boundary conditions are once again assumed. The partition function in such a geometry is given in terms of the infinitesimal transfer matrix $\hat{H}$ and the momentum operator $\hat{k}$ of an infinitely long strip by

$$Z(l, l') = \mathrm{Tr}(e^{-\hat{H}})^{\mathrm{Re}\, l'}(e^{-i\hat{k}})^{\mathrm{Im}\, l'}, \tag{3.8}$$

where we have used the fact that $\hat{H}$ and $\hat{k}$ commute, and that $e^{ia\hat{k}}$ translates through a distance a. Inserting a complete set of eigenstates of $\hat{H}$ and $\hat{k}$,

$$Z(l, l') = e^{-fA + \pi c\,\mathrm{Re}\,\delta/6} \sum_n \exp[-E_n \mathrm{Re}\, l' - ik_n \mathrm{Im}\, l']. \tag{3.9}$$

Thus an operator with scaling dimensions $(h, \bar{h})$ will contribute to the sum in (3.9) a term $e^{-2\pi(h\delta + \bar{h}\delta^*)}$. Now consider the contribution of one conformal block, whose primary operator has dimensions $(h, \bar{h})$, to the sum. This will be of the form $\chi(\delta)\bar{\chi}(\delta^*)$, where

$$\chi(\delta) = e^{-2\pi h\delta} \sum_N d(N) e^{-2\pi\delta}, \tag{3.10}$$

together with a similar expression for $\bar{\chi}(\delta^*)$. Here $d(N)$ is the degeneracy at level N of the operator in question.

Eq. (3.10) has precisely the form of the character formula for the appropriate representation of the Virasoro algebra. These formulas have been derived by Rocha-Caridi [9]. The result is that, for an operator corresponding to $h = h_{p,q}$ in the Kac formula, $\delta(\delta) = \delta_{p,q}(\delta)$, where

$$\chi_{p,q}(\delta) = f(\delta)^{-1} g_{p,q}(\delta), \tag{3.11}$$

$f(\delta)$ is as given in eq. (3.5), and

$$g_{p,q}(\delta) = \sum_{k=-\infty}^{\infty} \left(\exp\left(-\frac{2\pi\delta}{4m(m+1)} \times \left[(2m(m+1)k + (m+1)p - mq)^2 - 1 \right] \right) - \{q \to -q\} \right). \tag{3.12}$$

If we denote the number of primary operators with $h = h_{p,q}$ and $\bar{h} = h_{\bar{p},\bar{q}}$ by $\mathfrak{N}(p,q;\bar{p},\bar{q})$, then (3.9) becomes

$$Z(l,l') = e^{-fA + \pi c \operatorname{Re}\delta/6} \sum_{p,q;\bar{p},\bar{q}} \mathfrak{N}(p,q;\bar{p},\bar{q}) \chi_{p,q}(\delta) \chi_{\bar{p},\bar{q}}(\delta^*). \tag{3.13}$$

The sum over (p,q) is over the values

$$1 \leqslant q \leqslant p \leqslant m-1. \tag{3.14}$$

Note that the expansion in (3.12) is very rapidly convergent for $\operatorname{Re}\delta > 1$. However, it is actually convergent for all $\operatorname{Re}\delta > 0$. One of the remarkable features of (3.12) is that it allows the symmetry of $Z(l,l')$ under $l \to l'$ to be respected, provided the $\mathfrak{N}(p,q;\bar{p},\bar{q})$ satisfy certain constraints. These we now describe.

The first step is to apply the Poisson sum formula to (3.11). This has the effect of bringing δ into the denominator of the exponent. After a little algebra, one obtains

$$g_{p,q}(\delta) = \left(\frac{2}{m(m+1)\delta} \right)^{1/2} \exp\left(\frac{\pi(\delta - \delta^{-1})}{2m(m+1)} \right) \times \sum_{r=-\infty}^{\infty} \exp\left(\frac{-\pi(r^2-1)}{2\delta m(m+1)} \right) \sin\frac{r\pi p}{m} \sin\frac{r\pi q}{m+1}. \tag{3.15}$$

Substituting in (3.13), and using (3.5), (1.4),

$$Z(\delta) = e^{-fA + \pi c \operatorname{Re} \delta^{-1}/6} \frac{2|f(\delta^{-1})|^{-2}}{m(m+1)} \sum_{p,q;\bar{p},\bar{q}} \mathfrak{N}(p,q;\bar{p},\bar{q})$$

$$\times \left[\sum_{r=-\infty}^{\infty} \exp\left(-\frac{\pi(r^2-1)}{2\delta m(m+1)} \right) \sin\frac{r\pi p}{m} \sin\frac{r\pi q}{m+1} \right]$$

$$\times \left[\sum_{\bar{r}=-\infty}^{\infty} \exp\left(-\frac{\pi(\bar{r}^2-1)}{2\delta^* m(m+1)} \right) \sin\frac{\bar{r}\pi \bar{p}}{m} \sin\frac{\bar{r}\pi \bar{q}}{m+1} \right]. \tag{3.16}$$

This is to be equated to

$$Z(\delta^{-1}) = e^{-fA + \pi c \operatorname{Re} \delta^{-1}/6} |f(\delta^{-1})|^{-2} \sum_{p,q;\bar{p},\bar{q}} \mathfrak{N}(p,q;\bar{p},\bar{q})$$

$$\times \Bigg[\sum_{k=-\infty}^{\infty} \Bigg(\exp\Bigg(-\frac{\pi}{2\delta m(m+1)}$$

$$\times \left[(2m(m+1)k + (m+1)p - mq)^2 - 1 \right] \Bigg) - \{ q \to -q \} \Bigg) \Bigg]$$

$$\times \Bigg[\sum_{\bar{k}=-\infty}^{\infty} \Bigg(\exp\Bigg(-\frac{\pi}{2\delta^* m(m+1)}$$

$$\times \left[(2m(m+1)\bar{k} + (m+1)\bar{p} - m\bar{q})^2 - 1 \right] \Bigg) - \{ \bar{q} \to -\bar{q} \} \Bigg) \Bigg]. \tag{3.17}$$

We now proceed to equate coefficients of powers of $e^{1/\delta}$ and e^{1/δ^*}. The fact that the exponents of the leading terms agree is a check on the validity of the result in (1.1). The first sum rule comes from these terms, which correspond to $r, \bar{r} = \pm 1$ in (3.16), and $k = \bar{k} = 0$, $p = q = \bar{p} = \bar{q} = 1$ in (3.17). The result is

$$\sum_{p,q;\bar{p},\bar{q}} \mathfrak{N}(p,q;\bar{p},\bar{q}) \sin\frac{\pi p}{m} \sin\frac{\pi q}{m+1} \sin\frac{\pi \bar{q}}{m+1} = \frac{m(m+1)}{8}, \tag{3.18}$$

where we have set $\mathfrak{N}(1,1:1,1) = 1$, corresponding to the fact that the identity operator should appear just once.

It would appear that an infinite number of other constraints follow from equating the non-leading powers. It is part of the magic of (3.12) that this is not so. First note that values of r and $\bar{r}$ in (3.16) such that

$$r, \bar{r} \equiv \begin{cases} 0 & (\bmod m), \quad \text{or} \\ 0 & (\bmod m+1), \end{cases} \tag{3.19}$$

do not contribute to the sums in (3.16). Thus we may run these sums from zero to infinity, including a factor 4 on the right-hand side. Now we use the following

Lemma. The square of any integer r^2 such that r is not divisible by m or $m+1$ can be written uniquely as

$$r^2 = (2m(m+1)k' + (m+1)p' - mq')^2, \tag{3.20}$$

where the integers (k', p', q') satisfy $1 \leqslant |q'| \leqslant p' \leqslant m-1$.

We shall refrain from giving a general proof. The case $m = 3$ is illustrated in fig. 2. The lemma implies that the sums over r and $\bar{r}$ can be converted into sums over k', p', q' and $\bar{k}', \bar{p}', \bar{q}'$ of the form

$$\sum_{k'=-\infty}^{\infty} \left(\exp\left(- \frac{\pi}{2\delta m(m+1)} \left[(2m(m+1)k' + (m+1)p' - mq')^2 - 1 \right] \right) \right.$$
$$\left. \times (-1)^{(p+1)(p'+q')} \sin\frac{\pi pp'}{m} \sin\frac{\pi qq'}{m+1} + \{q' \to -q'\} \right), \tag{3.21}$$

with a similar expression for the sum over $\bar{r}$. This now has almost the form of (3.17). The sums over k are seen to be redundant, and we end up with a finite number of constraints which we call the *inversion sum rules*:

$$\sum_{p,q;\bar{p},\bar{q}} \mathcal{N}(p,q:\bar{p},\bar{q})(-1)^{(p+q)(p'+q')+(\bar{p}+\bar{q})(\bar{p}'+\bar{q}')}$$
$$\times \sin\frac{\pi pp'}{m} \sin\frac{\pi qq'}{m+1} \sin\frac{\pi \bar{p}\bar{p}'}{m} \sin\frac{\pi \bar{q}\bar{q}'}{m+1}$$
$$= \tfrac{1}{8} m(m+1) \mathcal{N}(p',q';\bar{p}',\bar{q}'). \tag{3.22}$$

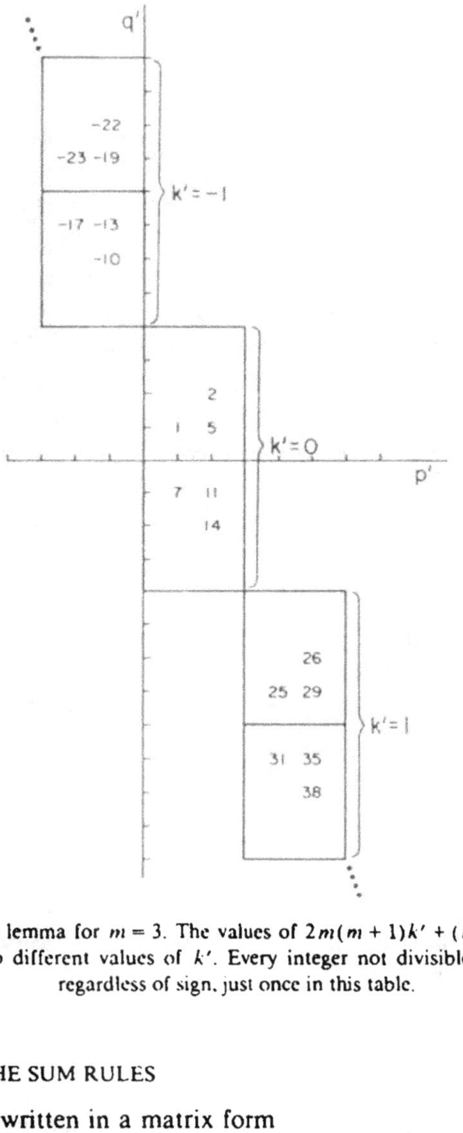

Fig. 2. Illustration of the lemma for $m = 3$. The values of $2m(m+1)k' + (m+1)p' - mq'$ are shown. The blocks correspond to different values of k'. Every integer not divisible by m or $m+1$ appears, regardless of sign, just once in this table.

3.2. SOLUTIONS TO THE SUM RULES

Eq. (3.22) may be written in a matrix form

$$M\mathcal{N} = \mathcal{N}, \tag{3.23}$$

where M is a direct product of two $\frac{1}{2}m(m-1) \times \frac{1}{2}m(m-1)$ matrices. We are interested in eigenvectors of M with eigenvalue one. In general, this eigenspace is multidimensional, so there is a whole manifold of solutions. However, we require only those in which the $\mathcal{N}(p, q; \bar{p}, \bar{q})$ are non-negative integers. This Diophantine nature makes the enumeration of the solutions difficult.

One solution which may always be found is

$$\mathcal{N}(p,q;\bar{p},\bar{q}) = \delta_{p,\bar{p}}\delta_{q,\bar{q}}. \tag{3.24}$$

This corresponds to the case of all possible scalar operators being present in the theory, and no others. Presumably such theories correspond to the generic multicritical points in the sequence of models of Andrews, Baxter and Forrester [16], analysed further by Huse [11]. He showed that each model could be associated with a value of m, and he identified exponents corresponding to all scalar operators with dimension $x < 1$. Presumably all the other scalar operators will then be generated in the operator product expansion. The first sum rule (3.18) shows that no other operators may then appear in the theory, because each term in (3.18) is non-negative.

To proceed further, we have examined low values of m. The analysis is simplified by the fact that, because of the periodic boundary conditions, only operators with integer spin may occur. That is, $\mathcal{N}(p,q;\bar{p},\bar{q}) = 0$ unless $h_{p,q} - h_{\bar{p},\bar{q}}$ is an integer. Also, for the partition function to be real,

$$\mathcal{N}(p,q;\bar{p},\bar{q}) = \mathcal{N}(\bar{p},\bar{q};p,q). \tag{3.25}$$

This considerably restricts the dimension of the space to be searched for solutions.

m = 3. In this case the allowed values of h and $\bar{h}$ are $0, \frac{1}{16}, \frac{1}{2}$, and there are no non-scalar integer spin operators. Truncated to the space of scalar operators, the matrix is

$$M = \frac{1}{4}\begin{pmatrix} 1 & 1 & 2 \\ 1 & 1 & 2 \\ 2 & 2 & 0 \end{pmatrix}, \tag{3.26}$$

from which it follows trivially that the only solution is of the form (3.24). The three primary scalar operators are the unit operator, the energy density and the magnetization operators of the Ising model.

m = 4. Once again there are no non-scalar integer spin operators. The truncated matrix is

$$M = \frac{1}{5}\begin{pmatrix} t & 2t & t & 2t' & t' & t' \\ 2t & 0 & 2t & 0 & 2t' & 2t' \\ t & 2t & t & 2t' & t' & t' \\ 2t' & 0 & 2t' & 0 & 2t & 2t \\ t' & 2t' & t' & 2t & t & t \\ t' & 2t' & t' & 2t & t & t \end{pmatrix}, \tag{3.27}$$

where $t = \sin^2(\frac{1}{5}\pi) = \frac{1}{8}(5-\sqrt{5})$ and $t' = \sin^2(\frac{2}{5}\pi) = \frac{1}{8}(5+\sqrt{5})$. Because $\sqrt{5}$ is irrational, this leads to a system of 12 equations. It is straightforward to show that the only solution is once again of the form (3.24). This case has been identified with the universality class of the tricritical Ising model [5].

$m = 5$. This case is more interesting because there exists the possibility of non-scalar operators, corresponding to $(p, q; \bar{p}, \bar{q}) = (2, 1; 3, 1), (3, 1; 2, 1), (1, 1; 4, 1), (4, 1; 1, 1)$; that is $(h, \bar{h}) = (\frac{2}{5}, \frac{7}{5}), (\frac{7}{5}, \frac{2}{5}), (0, 3), (3, 0)$ respectively. The resulting truncated 14×14 matrix has as its elements rational multiples of t or t'. This leads to a system of 28 equations, which we shall refrain from giving in detail. They simplify to

$$\begin{aligned}
\mathfrak{N}(1,1;1,1) = \mathfrak{N}(2,1;2,1) = \mathfrak{N}(3,1;3,1) = \mathfrak{N}(4,1;4,1) &\equiv a\,,\\
\mathfrak{N}(2,2;2,2) = \mathfrak{N}(3,2;3,2) = \mathfrak{N}(4,2;4,2) = \mathfrak{N}(4,4;4,4) &\equiv f\,,\\
\mathfrak{N}(3,3;3,3) = \mathfrak{N}(4,3;4,3) &\equiv i\,,\\
\mathfrak{N}(3,1;2,1) = \mathfrak{N}(4,1;1,1) &\equiv k\,,
\end{aligned} \tag{3.28}$$

where

$$\begin{aligned}
a &= k + f\,,\\
5a &= 3f + 2i + k\,,\\
i &= a + k\,,\\
5k &= a - 3f + 2i\,.
\end{aligned}$$

Since a is normalized to 1, the crucial equation is (3.28). It has two solutions, leading to the two possibilities

$$a = f = i = 1\,, \qquad k = 0\,, \tag{3.29}$$

or

$$a = k = 1\,, \qquad i = 2\,, \qquad f = 0\,. \tag{3.30}$$

The first possibility (3.29) corresponds to the solution (3.24). This model has been identified with a generic tetracritical point [11]. (In field theory this corresponds to a scalar field with a ϕ^8 interaction.) The second solution (3.30) is the universality class of the 3-state Potts model. This is so, because some of the known scaling dimensions have been already identified [5, 10], and it was observed that operators corresponding to q even are absent. This means $f = 0$. Also, the magnetization operators were found to correspond to $q = 3$. Since in the 3-state Potts model the order parameter has two components, we would expect to find $i = 2$. The complete list of scaling dimensions of primary operators is then

$$\begin{aligned}
&(0,0), (\tfrac{2}{5}, \tfrac{2}{5}), (\tfrac{7}{5}, \tfrac{7}{5}), (3,3)\,, && \text{energy;}\\
&(\tfrac{1}{15}, \tfrac{1}{15}) \times 2, (\tfrac{2}{3}, \tfrac{2}{3}) \times 2\,, && \text{magnetization;}\\
&(\tfrac{2}{5}, \tfrac{7}{5}), (\tfrac{7}{5}, \tfrac{2}{5}), (0,3), (3,0)\,, && \text{chiral.}
\end{aligned} \tag{3.31}$$

All previously found operators [5] appear in this list.

4. Summary and further remarks

In the first part of this paper, we have completed the program begun in ref. [2]. We have shown how all important universal properties of conformally invariant two-dimensional theories, including critical exponents and operator product expansion coefficients, may be related to numerically accessible properties of the transfer matrix of a finite width strip. The value of these results will lie in the investigation of new models, rather than in reproducing already known results. The multicritical points in the models obtained by Andrews, Baxter and Forrester [16] whose exponents do not [11] appear to fit the Kac formula are of the first type.

Second, we showed that unitary models with a finite number of primary operators (in the narrow sense defined by Belavin, Polyakov and Zamolodchikov [4]) have $c < 1$. This result partially fills a gap in the line of reasoning which picks out those models in the Friedan, Qiu and Shenker [5] classification as being special. For these models, we showed how the character formulas of Rocha-Caridi [9] give the partition function in an arbitrarily shaped parallelogram, once the number of operators with given scaling dimensions are known. Exploiting the symmetry of the parallelogram, we then derived sum rules which must be satisfied by these numbers. It is remarkable how the scaling dimensions allowed in the models in the Friedan, Qiu and Shenker [5] classification enable this symmetry to be realized. An arbitrary list of scaling dimensions would not have this property. This is another argument pointing to the special role of degenerate theories. We note that the symmetry of the parallelogram, which corresponds to the invariance of $Z(\delta)$ under the modular group, has recently been exploited to limit the possible gauge groups in heterotic string theories [22].

Finally, we obtained all solutions of the sum rules for $m = 3, 4, 5$, and showed that only the models which have been previously identified (Ising, tricritical Ising, 3-state Potts, and generic tetracritical point,) are in fact allowed. We gave for the first time a complete list of primary operators for these models. Solution of the sum rules for larger values of m will require greater effort or sophistication. However, it would appear that the number of solutions should grow with m. This points to the existence of as yet unexplored models, even with $c > 1$. However, it is important to realize that existence of a solution to the sum rules does not imply existence of a corresponding model, since the sum rules are only a necessary condition for the model to be consistent.

The sum rules form a more severe constraint on a theory than closure of the operator product expansion and crossing symmetry, which in some cases does determine the operator product expansion coefficients [17]. For example, in the case $m = 3$, the operator product expansion closes with the operators $\mathbf{1}$ and ε, the energy density. However, the sum rules show that the magnetization σ must be included to get a consistent theory. Once the solution of the sum rules is obtained, the expression (3.16) gives the shape dependence of the free energy at criticality in an arbitrary parallelogram. It would be interesting to generalize this to other quantities such as the susceptibility.

This work was supported by the National Science Foundation under grant no. PHY83-13324.

Appendix A

Several interesting properties of a model in a finite width, infinitely long strip can be obtained approximately if the infinitesimal transfer matrix $\hat{H}$ is truncated to a finite set of low-lying states. The simplest case is to consider just two states, which will be a reasonable approximation for some quantities if the gap to the first excited state is small, followed by a larger gap to the second excited state. Such is the case for the Ising model, which has gaps of $\pi/4l, 2\pi/l$ to the lowest excited states [2]. In general, if the magnetization operator has scaling dimension x, in the truncated basis

$$\hat{H} = \frac{2\pi x}{l}\begin{pmatrix} 0 & 0 \\ 0 & 1 \end{pmatrix}, \tag{A.1}$$

where $\hat{H}$ is subtracted so that $E_0 = 0$. An external magnetic field h corresponds to adding a term

$$\hat{H}_1 = h\left(\frac{2\pi}{l}\right)^x \begin{pmatrix} 0 & 1 \\ 1 & 0 \end{pmatrix}. \tag{A.2}$$

It is trivial to diagonalize the sum and obtain the h-dependent part of the free energy per unit area:

$$f \simeq \frac{\pi x}{l} - \left[\left(\frac{\pi x}{l}\right)^2 + \left(\frac{2\pi}{l}\right)^{2x} h^2\right]^{1/2} \tag{A.3}$$

From this follow the susceptibilities $\chi^{(n)} \equiv \partial^n f/\partial h^n|_{h=0}$. In particular

$$\chi^{(2)} \simeq (2\pi/l)^{2x}(l/\pi x), \tag{A.4}$$

and the dimensionless coupling constant [18]

$$g \equiv \chi^{(4)}/l^2(\chi^{(2)})^2 \simeq -3/\pi x. \tag{A.5}$$

Using eq. (2.3) it is possible to show that the corrections to (A.4) are in fact $O(x^2)$.

Both $\chi^{(2)}$ and g have been measured [19,20] for the Ising model, showing good agreement with the above crude estimates, if we take $x = \frac{1}{8}$.

Appendix B

We derive the inversion relation (3.5) for $f(\delta)$. This can be written, using the Euler pentagonal number theorem [21] as

$$f(\delta) = \sum_{n=-\infty}^{\infty} e^{-\pi\delta(3n^2+n)-i\pi n}. \tag{B.1}$$

Applying the Poisson sum formula, one obtains

$$f(\delta) = (3\delta)^{-1/2} e^{(\pi/12)(\delta-\delta^{-1})} \sum_{r=-\infty}^{\infty} e^{-r(r+1)\pi/3\delta} \cos\tfrac{1}{6}(2r+1)\pi. \tag{B.2}$$

Since

$$\cos\tfrac{1}{6}(2r+1)\pi = \begin{cases} \tfrac{1}{2}\sqrt{3} \quad (-1)^n, & r=3n \\ \tfrac{1}{2}\sqrt{3} \quad (-1)^n, & r=3n-1 \\ 0, & r=3n+1, \end{cases} \tag{B.3}$$

the sum over r may be rewritten as a sum over n. After a little manipulation, this has the same form as the sum in (B.1), with δ replaced by δ^{-1}.

References

[1] J.L. Cardy, in Phase transitions and critical phenomena, vol. 11, ed. C. Domb and J.L. Leibowitz (Academic Press, London) to be published
[2] J.L. Cardy, J. Phys. A16 (1984) L385
[3] H.W. Blöte, J.L. Cardy and M.P. Nightingale, Phys. Rev. Lett. 56 (1986) 742
[4] A.A. Belavin, M.M. Polyakov, and A.B. Zamolodchikov, J. Stat. Phys. 34 (1984), 763; Nucl. Phys. B241 (1984) 333
[5] D. Friedan, Z. Qui and S. Shenker, Phys. Rev. Lett. 52 (1984) 1575; *in* Vertex operators in mathematics and physics, Proc. Conf., November 10–17, 1983, ed. J. Lepowsky, S. Mandelstam and I.M. Singer (Springer, New York 1984) p. 419
[6] B. Derrida and L. de Seze, J. Physique 43 (1982), 475
[7] M.P. Nightingale, and H.W. Blöte, J. Phys. A16 (1983), L657
[8] V.G. Kac, *in* Group theoretical methods in physics, ed. W. Beiglbock and A. Bohm, Lecture Notes in Phys. 94 (1979) 441
[8a] P. Goddard, A. Kent, and D. Olive, Phys. Lett. 152B (1985) 88; DAMTP preprint 85-21
[9] A. Rocha-Caridi, *in* Vertex operators in mathematics and physics, Proc. Conf., November 10–17, 1983, ed. J. Lepowsky, S. Mandelstam and I.M. Singer (Springer, New York, 1984) p. 451
[10] Vl.S. Dotsenko, Nucl Phys. B235 (1984) 54
[11] D.A. Huse, Phys. Rev. B30 (1984) 3908
[12] A.M. Polyakov, Zh. Eksp. Teor. Fiz. 57 (1970) 271 [Sov. Phys. JETP 30 (1970) 151]

[13] V. Privman and M.E. Fisher, J. Phys. A16 (1983) L295
[14] D. Friedan, Z. Qiu, and S. Shenker, Phys. Lett. 151B (1985) 37
[15] M.A. Bershadsky, V.G. Kniznik and M.G. Teitelman, Phys. Lett. 151B (1985) 31
[16] G.E. Andrews, R.J. Baxter and P.J. Forrester, J. Stat. Phys. 35 (1984) 193
[17] Vl.S. Dotsenko and V.A. Fateev, Phys. Lett. 154B (1985) 291
[18] K. Binder, Z. Phys. B43 (1981) 119
[19] R. Hentschke, P. Kleban and G. Akinci, to appear
[20] T.W. Burkhardt and B. Derrida, Phys. Rev. B32 (1985) 7273
[21] M. Abramowitz and I.A. Stegun, eds., Handbook of mathematical functions (Dover, New York 1965) p. 825
[22] D. Gross, J. Harvey, E. Martinec and R. Rohm, Phys. Rev. Lett. 54 (1985) 502

CHAPTER 6

THE WESS-ZUMINO MODEL

Reprinted Papers

17. E. Witten, "Non-Abelian Bosonization in Two Dimensions", Commun. Math. Phys. **92** (1984) 455–472.

18. V.G. Knizhnik and A.B. Zamolodchikov, "Current Algebra and Wess-Zumino Model in Two Dimensions", Nucl. Phys. **B247** (1984) 83–103.

19. P. Goddard, W. Nahm and D. Olive, "Symmetric Spaces, Sugawara's Energy Momentum Tensor in Two Dimensions and Free Fermions", Phys. Lett. **160B** (1985) 111–116.

The equivalence between bosonic and fermionic field theories, mentioned in Chapter 3, was developed in the context of theories with abelian symmetry. E. Witten made a major advance in our understanding of fermion-boson equivalence and the role of affine Kac-Moody algebras in quantum field theory. He showed [*Reprinted Paper* #17] that isomorphic algebraic structures arise in the two-dimensional non-linear field theory of a bosonic field taking values on a group manifold G (the so-called principal chiral model) and the theory of free fermions transforming under G, provided that the Lagrangian of principal chiral model has added to it a suitable multiple of a certain topological term, the Wess-Zumino term. This term was originally introduced by J. Wess and B. Zumino in the context of four-dimensional field theory [1]. The corresponding model is called the Wess-Zumino model. Witten established the equivalence of these models for $G = SO(n)$. His results were extended by V.G. Knizhnik and A.B. Zamolodichikov [*Reprinted Paper* #18], using the techniques of Belavin, Polyakov and Zamolodichikov [*Reprinted Paper* #15], discussing also the case of $U(n)$.

In general, the equivalence of the Wess-Zumino model to a free fermion theory requires not only the isomorphism of the algebraic structures, that is the semi-direct product of the Kac-Moody and Virasoro algebras, but also that the Sugawara form for the energy-momentum tensor, calculated in the fermion theory, agree with the free fermion form. (Whilst this is a necessary condition, it is not in general sufficient [2].) A characterisation of this condition in terms of symmetric spaces was given by P. Goddard, W. Nahm and D. Olive [*Reprinted Paper* #19]. This result has consequences for the reduction of representations of affine algebras in terms of subalgebras.

References

[1] J. Wess and B. Zumino, "Consequences of anomalous Ward identities", Phys. Lett. **37B** (1971) 95–97.

[2] R.M. Ashworth, "The inequivalence of fermionic and bosonic theories sharing the same energy–momentum tensor", Nucl. Phys. **B280** [**FS18**] (1987) 321–339.

Commun. Math. Phys. 92, 455–472 (1984)

Communications in Mathematical Physics

Non-Abelian Bosonization in Two Dimensions

Edward Witten*

Joseph Henry Laboratories, Princeton University, Princeton, NJ 08544, USA

Abstract. A non-abelian generaljzation of the usual formulas for bosonization of fermions in $1+1$ dimensions is presented. Any fermi theory in $1+1$ dimensions is equivalent to a local bose theory which manifestly possesses all the symmetries of the fermi theory.

One of the most startling aspects of mathematical physics in $1+1$ dimensions is the existence of a (non-local) transformation from local fermi fields to local bose fields. Thus, consider the theory of a massless Dirac fermion:

$$\mathcal{L}_D = \bar{\psi} i \not{\partial} \psi . \tag{1}$$

This theory is equivalent [1] to the theory of a free massless scalar field:

$$\mathcal{L}_S = \tfrac{1}{2} \partial_\mu \phi \partial^\mu \phi . \tag{2}$$

The fermi field ψ has a relatively complicated and non-local expression [2] in terms of ϕ. However, fermion bilinears such as $\bar{\psi}\gamma_\mu\psi$ or $\bar{\psi}\psi$ take a simple form in the bose language. For example, the current $J_\mu = \bar{\psi}\gamma_\mu\psi$ becomes in terms of ϕ

$$J_\mu = \frac{1}{\sqrt{\pi}} \varepsilon_{\mu\nu} \partial^\nu \phi . \tag{3}$$

Similarly the chiral densities $\mathcal{O}_\pm = \bar{\psi}(1 \pm \gamma_5)\psi$ become

$$\mathcal{O}_\pm = M \exp \pm i \sqrt{4\pi} \phi , \tag{4}$$

where the value of the mass M depends on the precise normal ordering prescription that is used to define the exponential in (4).

By means of formulas like (3) and (4), the equivalence between the free Dirac theory and the free scalar theory can be extended to interacting theories. A perturbation of the free Dirac Lagrangian can be translated, via (3) and (4), into an equivalent perturbation of the free scalar theory. This procedure is remarkably

* Supported in part by NSF Grant PHY-80-19754

useful for elucidating the properties of $1+1$ dimensional theories. Many phenomena that are difficult to understand in the fermi language have simple, semiclassical explanations in the bose language. A major limitation of the usual bosonization procedure, however, is that in the case of fermi theories with non-abelian symmetries, these symmetries are not preserved by the bosonization. For instance, a theory with N free Dirac fields has a $U(N)\times U(N)$ chiral symmetry [actually $O(2N)\times O(2N)$, as we will see later]. Upon bosonization, this becomes a theory with N free scalar fields. The *diagonal* fermi currents can be bosonized conveniently, as in Eq. (3), but the *off-diagonal* currents are complicated and non-local in the bose theory. [Although the free scalar theory with N fields has an $O(N)$ symmetry, this $O(N)$ does not correspond to any subgroup of the fermion symmetry group.] For this reason, it is rather difficult [3] to bosonize non-abelian theories by the usual procedure. It is also sometimes difficult to understand via bosonization the realization of non-abelian global symmetries.

In this paper, an alternative bosonization procedure will be described which generalizes the usual one and can be used to bosonize *any* theory in a local way, while manifestly preserving *all* of the original symmetries. Unfortunately, the resulting bose theories are somewhat complicated.

First, we rewrite Eq. (3) for the currents in a way susceptible of generalization.

We define an element U of the $U(1)$ or $O(2)$ group by $U=\exp i\sqrt{4\pi}\phi$. Then (3) can be written

$$J_\mu = -\frac{i}{2\pi}\varepsilon_{\mu\nu}U^{-1}\partial^\nu U = -\frac{i}{2\pi}\varepsilon_{\mu\nu}(\partial^\nu U)\cdot U^{-1}. \tag{5}$$

We have emphasized in (5) that the ordering of factors does not matter, because the group $U(1)$ is abelian. In generalizing (5) we will have to be careful about factor ordering.

It is convenient to rewrite (5) in light cone coordinates. Let $x^\pm=(x^0\pm x^1)/\sqrt{2}$. In these coordinates the Lorentz invariant inner product is $A_\mu B^\mu = A^+B^- + A^-B^+ = A_+B_- + A_-B_+$; the components of a vector obey $A_+=A^-$, $A_-=A^+$. If we normalize the Levi-Civita symbol so that $\varepsilon_{01}=+1=-\varepsilon_{+-}$, then (3) and (5) become

$$\begin{aligned} J_+ &= -\frac{1}{\sqrt{\pi}}\partial_+\phi = \frac{i}{2\pi}U^{-1}\partial_+ U, \\ J_- &= +\frac{1}{\sqrt{\pi}}\partial_-\phi = -\frac{i}{2\pi}(\partial_- U)U^{-1}. \end{aligned} \tag{6}$$

Of course, the ordering of factors in (6) is still arbitrary.

For the massless Dirac particle, the vector and axial vector currents $\bar\psi\gamma^\mu\psi$ and $\bar\psi\gamma^\mu\gamma_5\psi$ are both conserved.[1] But in $1+1$ dimensions $\bar\psi\gamma^\mu\gamma_5\psi=\varepsilon^{\mu\nu}\bar\psi\gamma_\nu\psi$. So the current conservation equations are $0=\partial_\mu J^\mu=\varepsilon^{\mu\nu}\partial_\mu J_\nu$. In light cone coordinates this means $0=\partial_- J_+ = \partial_+ J_-$. The bosonization formula (6) is compatible with that strong condition because the free massless ϕ field obeys $0=\nabla^2\phi=2\partial_+\partial_-\phi$.

1 As usual, we define $\{\gamma_\mu,\gamma_\nu\}=2\eta_{\mu\nu}$, $\gamma_5=\gamma^0\gamma^1$ (so $\gamma_5^2=+1$), and $\bar\psi=\psi^*\gamma^0$. A convenient basis is $\gamma^0 = \begin{pmatrix}0&1\\1&0\end{pmatrix}$, $\gamma^1=\begin{pmatrix}0&-1\\1&0\end{pmatrix}$, $\gamma_5=\begin{pmatrix}1&0\\0&-1\end{pmatrix}$. We define light cone components $\psi_\pm$ of ψ by requiring $\gamma_5\psi_- = \psi_-$, $\gamma_5\psi_+=-\psi_+$. (The sign convention may seem odd but is useful.) Thus $\psi=\begin{pmatrix}\psi_+\\\psi_-\end{pmatrix}$. ψ_+ and ψ_- are left movers and right movers, respectively, as one may see from Eq. (8) later

We wish to generalize this to fermion theories with non-abelian symmetries. As we wish to be general, we will consider a theory with N Majorana fermions ψ^i, $i=1\ldots N$. [If one prefers, one can choose N even and consider this to be a theory of $N/2$ Dirac fields. If so, in much of the subsequent discussion one can consider the chiral group $U(N/2)\times U(N/2)$ instead of $O(N)\times O(N)$.] The conventional Lagrangian for free Majorana fields is

$$\mathcal{L}=\int d^2x \tfrac{1}{2}\bar{\psi}_k i\partial\!\!\!/ \psi^k . \tag{7}$$

The conserved vector currents are $V^a_\mu=\bar{\psi}\gamma_\mu T^a\psi$, T^a being any generator of $O(N)$. The axial currents are $A^a_\mu=\varepsilon_{\mu\nu}V^{\nu a}=\bar{\psi}\gamma_\mu\gamma_5 T^a\psi$. These currents generate chiral $O(N)\times O(N)$.

Since $\bar{\psi}\gamma^\mu\partial_\mu\psi=\psi^T(\partial_0+\gamma^0\gamma^1\partial_1)\psi$, the free Lagrangian, in terms of the light cone components of ψ, is

$$\mathcal{L}=\tfrac{1}{2}i\int d^2x\left[\psi^k_-\left(\frac{\partial}{\partial t}+\frac{\partial}{\partial x}\right)\psi^k_-+\psi^k_+\left(\frac{\partial}{\partial t}-\frac{\partial}{\partial x}\right)\psi^k_+\right]. \tag{8}$$

Instead of vector and axial vector currents, it is more useful to work with chiral components. We define $J^{ij}_+(x,t)=-i\psi^i_+\psi^j_+(x,t)$ and $J^{ij}_-(x,t)=-i\psi^i_-\psi^j_-(x,t)$. Note that $J^{ij}_\pm$ are hermitian and that by fermi statistics they obey $J^{ij}_+=-J^{ji}_+$, $J^{ij}_-=-J^{ji}_-$. J^{ij}_+ and J^{ij}_- generate chiral $O(N)_R$ and $O(N)_L$, respectively. [By $O(N)_R$ and $O(N)_L$ we mean $O(N)$ transformations for right-moving and left-moving fermions.] The conservation laws for $J_\pm$ are very simple

$$\partial_- J^{ij}_+=\partial_+ J^{ij}_-=0 . \tag{9}$$

Thus, J_+ is a function only of x^+, and J_- is a function only of x^-.

We wish to find an ansatz writing J_+ and J_- in terms of suitable bose fields. In the usual bosonization procedure, one considers a current $\bar{\psi}\gamma_\mu\psi$ that generates an abelian or U(1) symmetry; it is written [Eq. (5)] in terms of a field that takes values in the U(1) group. Now we are dealing with currents J^{ij}_- and J^{ij}_+ that generate $O(N)_L\times O(N)_R$, and it is natural to try to express these currents in terms of a suitable field g that takes values in the $O(N)$ group. $O(N)_L\times O(N)_R$ will act on g by $g\to AgB^{-1}$, $A, B\in O(N)$.

What is a suitable expression for the currents in terms of g? One is tempted to try $J_+\sim g^{-1}\partial_+ g$, $J_-\sim g^{-1}\partial_- g$. However, this is incompatible with (9) because in a *non-abelian* group the equations $0=\partial_-(g^{-1}\partial_+ g)$ and $0=\partial_+(g^{-1}\partial_- g)$ are inconsistent. Instead, we generalize the factor ordering of Eq. (6) and write

$$J_+=\frac{i}{2\pi}g^{-1}\partial_+ g , \qquad J_-=-\frac{i}{2\pi}(\partial_- g)g^{-1} . \tag{10}$$

[The ij indices are suppressed, it being understood that J_+ and J_- are elements of the $O(N)$ Lie algebra.] Notice that the equations $0=\partial_-(g^{-1}\partial_+ g)$ and $0=\partial_+((\partial_- g)g^{-1})$ are compatible and in fact equivalent.

What Lagrangian will govern g? The obvious guess is

$$\mathcal{L}=\frac{1}{4\lambda^2}\operatorname{Tr}\partial_\mu g\partial_\mu g^{-1} . \tag{11}$$

This is the unique renormalizable and manifestly chirally invariant Lagrangian for g. However, for many reasons, (11) is wrong.

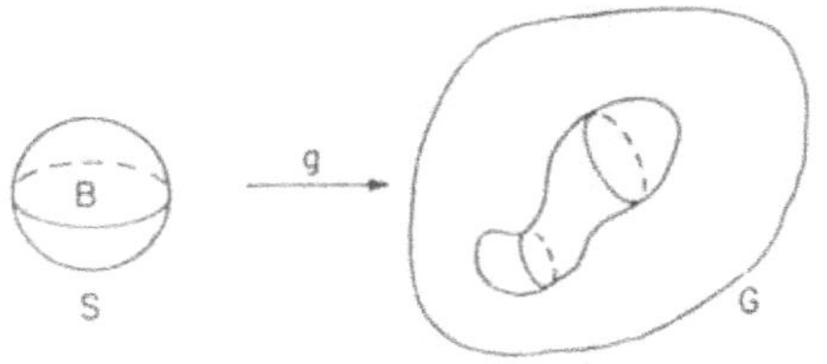

Fig. 1. A mapping g from a two sphere S (representing space-time) into a group manifold G. Since $\pi_2(G)=0$, any mapping of the *surface* S into G can be extended to a mapping into G of the solid sphere B (S and its interior)

First of all, (11) describes an asymptotically free theory with interactions that become strong in the infrared. It is certainly not equivalent to the conformally invariant free massless fermi field theory. Second, (11) leads to the equation of motion $0=\partial_\mu(g^{-1}\partial_\mu g)$ rather than the desired $0=\partial_-(g^{-1}\partial_+ g)=\partial_+((\partial_- g)g^{-1})$. Third, by analogy with similar considerations in QCD current algebra, [8] it may be shown that (11) has *more* discrete symmetries than the free, massless fermi theory.

Although (11) is the only renormalizable interaction for the non-linear sigma model that is *manifestly* chirally invariant, there is another one that is chirally invariant but not manifestly so. This is the two-dimensional analogue of the Wess-Zumino term [4], which has figured in various recent discussions of two dimensional models [5–7].

The two dimensional Wess-Zumino term can be constructed by analogy [8] with a similar treatment in four dimensions. Working in Euclidean space, we imagine space time to be a large two sphere S^2. Since π_2 (O(N))=0, a mapping g from S into the O(N) manifold can be extended to a mapping $\tilde{g}$ of a solid ball B whose boundary is S into O(N) (Fig. 1). If y_1, y_2, and y_3 are coordinates for B, the Wess-Zumino functional is

$$\Gamma=\frac{1}{24\pi}\int_B d^3y\varepsilon^{ijk}\,\mathrm{Tr}\,\tilde{g}^{-1}\frac{\partial\tilde{g}}{\partial y^i}\tilde{g}^{-1}\frac{\partial\tilde{g}}{\partial y^j}\tilde{g}^{-1}\frac{\partial\tilde{g}}{\partial y^k}. \tag{12}$$

As in four dimensions, the Wess-Zumino functional has a very essential property [8, 9]: it is well-defined only modulo a constant. Equation (12) has been normalized so that if g is a matrix in the fundamental representation of O(N), (12) is well-defined modulo $\Gamma\to\Gamma+2\pi$. The ambiguity in Γ arises because of the existence of topologically inequivalent ways to extend g into a mapping from B into O(N); the topologically distinct possibilities are classified by π_3 (O(N))$\simeq Z$.

In what sense is Γ an ordinary Lagrangian – an integral over space-time? This question is answered in the appendix, where it is shown that (locally in field space) Γ can be written as the integral over space-time of an ordinary but not manifestly chirally invariant Lagrangian which under a chiral transformation changes by a total divergence.

2 In Minkowski space, we instead consider *space* to be compact. We then consider finite time transition amplitudes between specified initial and final states of the g field. This "ties down" the fields at the boundary of space-time and leads to a similar quantization argument for Γ

Making use of Γ, we can consider a more general action for the field g:

$$I=\frac{1}{4\lambda^2}\int d^2x\,\mathrm{Tr}\,\partial_\mu g\partial^\mu g^{-1}+n\Gamma\,. \tag{13}$$

Here n must be an integer [19], since Γ is well-defined only modulo 2π. The theory (13) is renormalizable, since the new coupling constant is a dimensionless integer. Perhaps it should be stressed that (13) is not invariant under naive parity $x\to -x$, but is invariant under $x\to -x$, $g\to g^{-1}$.

We wish to ask whether for some values of λ and n this theory might be equivalent to the free massless fermi theory.

The first step is to calculate the equations of motion from (13). As has been discussed previously [6, 8], the variation of Γ is a simple, local functional. We find from (13) that the change of I under $g\to g+\delta g$ is

$$\delta I=\frac{1}{2\lambda^2}\int d^2\mathrm{Tr}\,g^{-1}\delta g\partial_\mu(g^{-1}\partial_\mu g)-\frac{n}{8\pi}\int d^2x\,\mathrm{Tr}\,g^{-1}\delta g\varepsilon^{\mu\nu}\partial_\mu(g^{-1}\partial_\nu g)\,. \tag{14}$$

The variational equations are therefore

$$\begin{aligned}0&=\frac{1}{2\lambda^2}\partial_\mu(g^{-1}\partial_\mu g)-\frac{n}{8\pi}\varepsilon^{\mu\nu}\partial_\mu(g^{-1}\partial_\nu g)\\ &=\left(\frac{1}{2\lambda^2}+\frac{n}{8\pi}\right)\partial_-(g^{-1}\partial_+g)+\left(\frac{1}{2\lambda^2}-\frac{n}{8\pi}\right)\partial_+(g^{-1}\partial_-g)\,.\end{aligned} \tag{15}$$

We see therefore that if $\lambda^2=\dfrac{4\pi}{n}$ the equation is as desired, $0=\partial_-(g^{-1}\partial_+g)$. Of course, λ^2 must be positive for stability, so this is only possible for $n>0$. For $n<0$ the parity conjugate equation $0=\partial_+(g^{-1}\partial_-g)$ arises at $\lambda^2=-\dfrac{4\pi}{n}$.

At $\lambda^2=\dfrac{4\pi}{n}$ the equations of motion of the theory can easily be solved in closed form. The general solution of $0=\partial_-(g^{-1}\partial_+g)$ is

$$g(x^+,x^-)=A(x^-)B(x^+)\,, \tag{16}$$

where $A(x^-)$ and $B(x^+)$ are arbitrary $O(N)$ valued functions of one coordinate. [At $\lambda^2=-4\pi/n$ the factorization is instead $g(x^+,x^-)=B(x^+)A(x^-)$.] Equation (16) means that left-moving and right-moving waves pass through each other without any interference. This property is strongly reminiscent of the fermion free field theory, in which the left- and right-moving waves are the γ_5 eigenstates. Combining this analogy with the fact that at $\lambda^2=4\pi/n$ the equation of motion for g reproduces the behavior of the fermion currents, we are led to conjecture that at $\lambda^2=4\pi/n$ and some value of n the non-linear sigma model is equivalent to the fermion free field theory.

What are the renormalization group properties of the theory with action (13)? Being an integer, n must not be subject to renormalization. This can be established in the background field method; in that method the counter-terms are local and *manifestly chinally invariant* functionals of the background field, so there is no

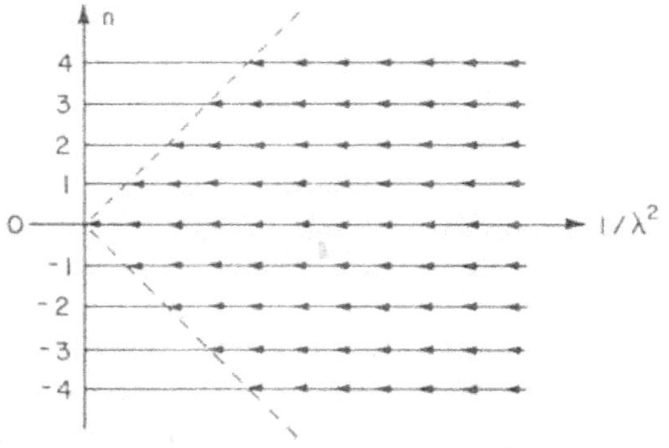

Fig. 2. Renormalization group flows. Plotted is the behavior of $1/\lambda^2$ in coming from high-energy to low-energy. Weak coupling is at the right, strong coupling is at the left. Assuming there are no non-trivial fixed points apart from the one found in the text, the $n=0$ theory flows to strong coupling at long distances, while for $n \neq 0$ the theory flows to $\lambda^2 = |4\pi/n|$. The behaviour of the non-asymptotically free theory with bare coupling bigger than $|4\pi/n|$ is not considered

counterterm proportional to Γ^3. We will illustrate this shortly at the one loop level.

The theory therefore requires only renormalization of λ. However, the renormalization of λ depends on both λ and n. For any n, the theory is asymptotically free, just as at $n=0$. This is so because for λ so small that $\frac{1}{\lambda^2} \gg n$, (13) is dominated by the first term, and the renormalization group calculation coincides with the standard calculation at $n=0$. However, as λ becomes large, the effects of the Wess-Zumino term can become important. We will argue that the beta function always has a zero at $\lambda^2 = \pm\frac{4\pi}{n}$. We will first illustrate this point with a one loop calculation; then we will establish the point by showing that the theory at $\lambda^2 = \pm 4\pi/n$ is equivalent to a known exactly soluble, conformally invariant theory. The existence of a non-trivial zero of the beta function at $\lambda^2 = \frac{4\pi}{n}$ means that the physical content of the weakly coupled theory with $n \neq 0$ is dramatically different from what it is for $n=0$. Instead of flowing in the infrared to strong coupling, the coupling constant flows (Fig. 2) to $\sqrt{4\pi/n}$ (or perhaps to another zero of the beta function closer to the origin).

Let us now calculate the one loop beta function of the theory. We will use the background field and expand around an arbitrary solution g_0 of the classical field equations. We write $g = g_0 \exp i\lambda T^a\pi^a$, where the T^a (normalized so $\mathrm{Tr}\, T^a T^b = 2\delta^{ab}$) are the generators of O(N) and π^a are the small fluctuation fields. The action becomes

$$I = \int d^2x \left[\frac{1}{4\lambda^2}\mathrm{Tr}\,\partial_\mu g_0 \partial_\mu g_0^{-1} + \frac{1}{2}\sum_a (\partial_\mu \pi^a)^2 + \frac{\eta^{\mu\nu}}{4}\mathrm{Tr}\, g_0^{-1}\partial_\mu g_0 [T\cdot\pi, \partial_\nu T\cdot\pi] \right.$$
$$\left. -\frac{1}{4}\frac{\lambda^2 n}{4\pi}\varepsilon^{\mu\nu}\mathrm{Tr}\, g_0^{-1}\partial_\mu g_0 [T\cdot\pi, \partial_\nu T\cdot\pi]\right] \qquad (17)$$

3 Similar reasoning has been given in discussion of the θ angle in four dimensions by Novikov, Shifman, Vainshtain, and Zakharov (private communication)

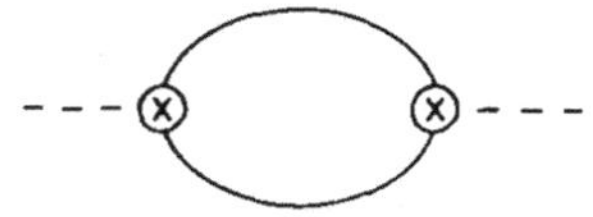

Fig. 3. The one loop renormalization calculation in the nonlinear sigma model. The dotted line is the background field; the solid line represents quantum fluctuations. The divergence arises only if the two vertices contain both $\eta^{\mu\nu}$ or both $\varepsilon^{\mu\nu}$

up to terms cubic or higher order in π. By power counting, a one loop divergence will be quadratic in $g_0^{-1}\partial_\mu g_0$. The only possible quadratic term is $\mathrm{Tr}(g_0^{-1}\partial_\mu g_0)^2 = -\mathrm{Tr}\,\partial_\mu g_0 \partial_\mu g_0^{-1}$, since $\varepsilon^{\mu\nu}\mathrm{Tr}(g_0^{-1}\partial_\mu g_0)(g_0^{-1}\partial_\nu g_0)=0$. This shows at the one loop level that only renormalization of λ is necessary – as was asserted earlier.

Since the sought for counterterm $\mathrm{Tr}\,\partial_\mu g_0 \partial_\mu g_0^{-1}$ is even under naive parity $x\to -x$, $g\to g$, the divergent one loop diagrams (Fig. 3) have two vertices both proportional to $\eta^{\mu\nu}$ or both proportional to $\varepsilon^{\mu\nu}$. The $\eta^{\mu\nu}$ vertex is the usual one that give asymptotic freedom [10]. The $\varepsilon^{\mu\nu}$ vertex is known from an old calculation in a different model [11] to give a positive contribution to the beta function. Actually even without evaluating the diagrams it is easy to see that they cancel if $\lambda^2 = \pm\dfrac{4\pi}{n}$. Apart from a factor of $\left(\dfrac{\lambda^2 n}{4\pi}\right)^2$, they differ in that one diagram has a factor of $\eta^{\alpha}_{\mu}\eta_{\alpha\nu}=\eta_{\mu\nu}$ while the other has $\varepsilon^{\alpha}_{\mu}\varepsilon_{\nu\alpha}=-\eta_{\mu\nu}$. Actual evaluation of the diagrams of Fig. 3 is not difficult. The divergent term in the effective action is

$$\int \frac{i(N-2)}{16\pi}\,\mathrm{Tr}\,\partial_\mu g_0 \partial_\mu g_0^{-1} \ln\left(\frac{\Lambda^2}{\mu^2}\right) d^2x\,, \tag{18}$$

where Λ is a momentum space cut-off and μ is a renormalization mass. From this we read off the one loop beta function

$$\beta(\lambda, n) = -\frac{\lambda^2(N-2)}{4\pi}\left[1-\left(\frac{\lambda^2 n}{4\pi}\right)^2\right] \tag{19}$$

which, as claimed, vanishes for $\lambda^2 = \left|\dfrac{4\pi}{n}\right|$.

If n is very large, say $n=10^{10}$, this perturbative calculation reliably shows the existence of a zero of the beta function, since the computed zero is at a very small coupling for which higher order terms are negligible. Of course, this reasoning does not show that the zero of the beta function is *precisely* at $|4\pi/n|$; and for n of order one the lowest order calculation does not reliably show even the *existence* of a zero. To show that the beta function vanishes for $\lambda^2=|4\pi/n|$ and that the theory at the zero is exactly soluble requires more information.

Let us return to the fermion currents, $J_+^{ij} = -i\psi_+^i\psi_+^j$, $J_-^{ij} = -i\psi_-^i\psi_-^j$, and to our hypothesis that these currents can be equated with suitable expressions constructed from g. What commutation relations do the fermion currents obey? The canonical anticommutation relations for the fermi fields are $\{\psi_+^i(x), \psi_+^j(y)\} = \{\psi_-^i(x), \psi_-^j(y)\} = \delta^{ij}\delta(x-y)$, $\{\psi_-^i(x), \psi_-^j(y)\} = 0$. Using these equations one can readily work out the canonical commutation rules for $J_\pm$. These canonical

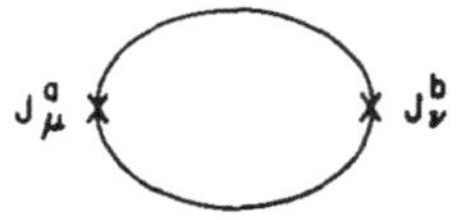

Fig. 4. The one loop diagram that yields the Schwinger anomaly in 1+1 dimensions

relations, however, are not valid quantum mechanically. The proper quantum mechanical formulas contain a c-number anomaly term, the Schwinger term. It arises [12] from diagram (4)[4]. The quantum mechanical commutation relations can be compactly written

$$\begin{aligned}
[\operatorname{Tr} AJ_-(x), \operatorname{Tr} BJ_-(y)] &= 2i\delta(x-y)\operatorname{Tr}[A,B]J_-(x) + \frac{i}{\pi}\delta'(x-y)\operatorname{Tr} AB\,,\\
[\operatorname{Tr} AJ_+(x), \operatorname{Tr} BJ_+(y)] &= 2i\delta(x-y)\operatorname{Tr}[A,B]J_+(x) - \frac{i}{\pi}\delta'(x-y)\operatorname{Tr} AB\,,\\
[J^{ij}_-, J^{kl}_+] &= 0\,,
\end{aligned} \tag{20}$$

where A and B are arbitrary antisymmetric matrices [generators of O(N)]. The terms proportional to $\delta'(x-y)$ originate from the anomaly.

Consider the following generalization of the first line of Eq. (20):

$$[\operatorname{Tr} AJ_-(x), \operatorname{Tr} BJ_-(y)] = 2i\delta(x-y)\operatorname{Tr}[A,B]J_-(x) + k\frac{i}{\pi}\delta'(x-y)\operatorname{Tr} AB\,. \tag{21}$$

Here we allow the coefficient of the anomaly to be rescaled by an arbitrary constant k. This algebra is known in the mathematical literature as the Kac-Moody algebra with a central extension, the central extension being $k \neq 0$ [13]. In the mathematical literature it is shown that this algebra has well-behaved unitary representations if and only if k is an integer. Actually, in quantum field theory one can easily find a system in which the anomaly has an arbitrary integer strength k. Consider a theory with k "flavors" and N "colors" of fermions ψ^{ia}, $a=1\ldots k$, $i=1\ldots N$, and define $J^{ij}_- = -i\sum_{a=1}^{k}\psi^{ia}_-\psi^{jk}_-$. The anomaly is then k times as large, coming from a sum over the flavor index in Fig. 4. [This gives an arbitrary positive integer k in (21); if a negative integer is desired, one may consider J_+ instead.] The fact that the Kac-Moody representation theory is well behaved only for integral k is another aspect of the a priori quantization of anomalies, a phenomenon that can also be seen from instanton physics [14] or from the multivaluedness of the Wess-Zumino term [8].

Our one flavor theory obeys (20) with $k=\pm 1$. The following very important facts are known about the Kac-Moody algebra. The unitary irreducible representation for $k=\pm 1$ is *essentially unique*. For $k>1$, there are a finite number of irreducible representations, obtained by taking tensor products of the $k=1$ representation with different symmetry or antisymmetry conditions. To prove the equivalence of a boson theory to the one flavor fermion theory it is sufficient to show that the boson theory gives currents that obey a Kac-Moody algebra with $k=\pm 1$.

---

4 The evaluation of the anomaly is standard. One derivation of this formula is described in detail by Coleman et al. [12, Eqs. (3.4), (4.19), and (4.28)]. They use, however, a notation based on Dirac fermions

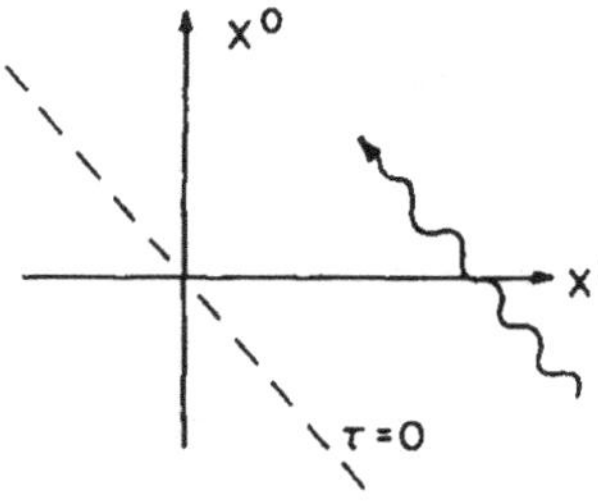

Fig. 5. This diagram is meant to illustrate the limitations of a canonical formalism based on light cone coordinates in 1+1 dimensions. The dotted line is an "initial value surface", $\tau=0$ $(\tau=(x^0+x^1)/\sqrt{2})$. The wave line is a massless particle traveling to the left at the speed of light. Its world path never meets $\tau=0$, so its existence cannot be predicted from initial data at $\tau=0$

We are thus led to try to calculate the canonical commutation relations of the currents $g^{-1}\partial_+ g$ and $(\partial_- g)g^{-1}$. Actually, we will calculate the purely classical Poisson bracket (PB). This calculation may appear formidable because of the complexity of the Wess-Zumino term but in fact at the critical coupling $\lambda^2=\left|\frac{4\pi}{n}\right|$ it can be carried out more or less simply.

First of all, since $g^{-1}\partial_+ g$ is only a function of x^+ and $(\partial_- g)g^{-1}$ is only a function of x^-, their Poisson bracket $[(g^{-1}\partial_+ g)_{ij}, ((\partial_- g)g^{-1})_{kl}]_{PB}$ vanishes. It is enough to calculate the Poisson bracket of $(\partial_- g)g^{-1}$ with itself; the Poisson bracket of $g^{-1}\partial_+ g$ can be deduced from the symmetry under $x\leftrightarrow -x$, $g\leftrightarrow g^{-1}$.

We will carry out the canonical analysis in a "light-cone frame." This means that we will regard $\sigma=x^-=\frac{x^0-x^1}{\sqrt{2}}$ as "space" while regarding $\tau=x^+=\frac{x^0+x^1}{\sqrt{2}}$ as "time." Actually, in 1+1 dimensions the light cone framework has a drawback. A left-moving massless degree of freedom may be unpredictable on the basis of initial data at $\tau=0$ (Fig. 5). For this reason, the light cone treatment fails to give the Poisson bracket of operators like $g^{-1}\partial_+ g$ that contain τ derivatives. (Their Poisson brackets can be obtained from an opposite light cone treatment in which τ is regarded as space and σ as time.) But the light cone framework yields straightforwardly the Poisson brackets of operators like $(\partial_- g)\cdot g^{-1}$ that do not contain τ derivatives.

With $\lambda^2=4\pi/n$, the action, in light cone coordinates, is

$$I=\frac{n}{16\pi}\int d\sigma d\tau\,\mathrm{Tr}\,\partial_\tau g\partial_\sigma g^{-1}+n\Gamma\,. \qquad (22)$$

Γ is rather complicated but it has one simple property: it is first order in time derivatives. Therefore, the whole action (22) has this property.

To introduce a canonical formalism it is necessary to formulate a theory with an action that is first order in time derivatives. Usually this requires introducing momenta that are independent of the coordinates, passing for instance from $\frac{1}{2}\dot{q}^2-V$ to $p\dot{q}-\frac{1}{2}p^2-V$. The case at hand is an exception. Equation (22) is *already* in Hamiltonian form; that is, it is already of first order in time derivatives.

In one other way, (22) differs from usual experience. In the light cone non-linear sigma model it is not convenient to split the dynamical variables into coordinates and momenta. Let us discuss, therefore, how Poisson brackets may in general be computed without making an explicit choice of p's and q's.

Consider a theory with dynamical variables ϕ^i and an arbitrary action that is first order in time derivatives:

$$I = \int dt A_i(\phi)\frac{d\phi^i}{dt}. \tag{23}$$

(The action may also contain terms independent of time derivatives; such terms are ignored in computing Poisson brackets.) We calculate the change in I under an arbitrary infinitessimal variation $\phi^i \to \phi^i + \delta\phi^i$:

$$\begin{aligned}\delta I &= \int dt\left(\frac{\partial A_i}{\partial\phi^j}\delta\phi^j\frac{d\phi^i}{dt} + A_i\frac{d}{dt}\delta\phi^i\right)\\ &= \int dt\left(\frac{\partial A_i}{\partial\phi^j}\delta\phi^j\frac{d\phi^i}{dt} - \frac{dA_i}{dt}\delta\phi^i\right)\\ &= \int dt\left(\frac{\partial}{\partial\phi^i}A_j - \frac{\partial}{\partial\phi^j}A_i\right)\delta\phi^i\frac{d\phi^j}{dt}.\end{aligned} \tag{24}$$

Define a matrix $F_{ij} = \partial_i A_j - \partial_j A_i$ as the coefficient of $\delta\phi^i \frac{d\phi^j}{dt}$. Notice that F_{ij} is always antisymmetric. Let F^{jk} be the universe matrix of F_{ij} (so $F^{jk}F_{ki} = \delta^j_i$).[5] Then the Poisson bracket of any two functions on phase space X and Y is defined by

$$[X, Y]_{PB} = \sum_{i,j} F^{ij}\frac{\partial X}{\partial\phi^i}\frac{\partial Y}{\partial\phi^j}. \tag{25}$$

In the simple case in which the ϕ^i *are* decomposed into coordinates and momenta q^i and p^i, and in which the part of the action containing time derivatives is $\int dt \sum_i p^i \frac{dq^i}{dt}$, (25) agrees with the usual definition of Poisson brackets.

In this calculation it is unnecessary to choose an explicit set of coordinates ϕ^i for the classical phase space. (Such a choice would be very awkward in the non-linear sigma model becuase of the nonlinearity of the phase space.) It is enough to have a basis of tangent vectors to the phase space (analogous to the tetrad in general relativity). The matrices F_{ij} and F^{jk} may be constructed relative to any such basis. In the non-linear sigma model a very convenient basis of tangents to the phase space are the matrices $g^{-1}\delta g(\sigma)$. The matrix F must act both on the Lie algebra index of $g^{-1}\delta g(\sigma)$ and on σ.

In this basis, it is very easy to calculate the matrix F in the non-linear sigma model. From (14), with $\lambda^2 = 4\pi/n$, the variation of the action is

$$\delta I = \frac{n}{4\pi}\int d\sigma d\tau \operatorname{Tr} g^{-1}\delta g \frac{d}{d\sigma} g^{-1}\frac{dg}{dt}. \tag{26}$$

5 If this inverse does not exist, one must introduce "constraints.") This does not occur in the case at hand

From (26) we see that F is $1\otimes\frac{n}{4\pi}\frac{d}{d\sigma}$, where "1" acts on the Lie algebra index and $\frac{n}{4\pi}\frac{d}{d\sigma}$ acts on the spatial coordinate. The inverse matrix is, of course $1\otimes\frac{4\pi}{n}\left(\frac{d}{d\sigma}\right)^{-1}$.

We now wish to apply definition (25) of the Poisson bracket with $X=\operatorname{Tr}A\frac{dg}{d\sigma}g^{-1}(\sigma)$, $Y=\operatorname{Tr}B\frac{dg}{d\sigma}g^{-1}(\sigma')$. Note that (25) can be understood as follows. First calculate $\delta X\delta Y=\frac{\partial X}{\partial\phi^i}\frac{\partial Y}{\partial\phi^j}\delta\phi^i\delta\phi^j$; then replace $\delta\phi^i\delta\phi^j$ by F^{ij}. So we calculate δX:

$$\begin{aligned}\delta X&=\operatorname{Tr}A\left(\frac{d}{d\sigma}\delta g\right)g^{-1}(\sigma)+\operatorname{Tr}A\frac{dg}{d\sigma}\delta g^{-1}\\&=\operatorname{Tr}A\left(\frac{d}{d\sigma}\delta g\right)g^{-1}-\operatorname{Tr}A\frac{dg}{d\sigma}g^{-1}\delta g g^{-1}\\&=\operatorname{Tr}g^{-1}(\sigma)Ag(\sigma)\frac{d}{d\sigma}(g^{-1}\delta g(\sigma)).\end{aligned}\tag{27}$$

δY is evaluated similarly, so

$$\delta X\delta Y=\operatorname{Tr}g^{-1}(\sigma)Ag(\sigma)\frac{d}{d\sigma}(g^{-1}\delta g(\sigma))\cdot\operatorname{Tr}g^{-1}(\sigma')Bg(\sigma')\frac{d}{d\sigma'}(g^{-1}\delta g(\sigma')).\tag{28}$$

After evaluating $\delta X\delta Y=\frac{\partial X}{\partial\phi^i}\frac{\partial Y}{\partial\phi^j}\delta\phi^i\delta\phi^j$, the next step is to replace $\delta\phi^i\delta\phi^j$ with F^{ij}. In our problem the role of $\delta\phi^i$ and $\delta\phi^j$ is played by $(g^{-1}\delta g(\sigma))^a$ and $(g^{-1}\delta g(\sigma'))^b$ (here we explicitly exhibit – temporarily – the Lie algebra indices a and b carried by these matrices). In view of our previous determination of F_{ij} and F^{ij}, we are to replace $(g^{-1}\delta g(\sigma))^a(g^{-1}\delta g(\sigma))^b$ by $\delta^{ab}\frac{4\pi}{n}\theta(\sigma,\sigma')$ where $\theta(\sigma,\sigma')$ is an inverse of $\frac{d}{d\sigma}$. Hence $\frac{d}{d\sigma}(g^{-1}\delta g(\sigma))^a\cdot\frac{d}{d\sigma'}(g^{-1}\delta g(\sigma'))^b$ is replaced by $\delta^{ab}\frac{4\pi}{n}\frac{d}{d\sigma}\cdot\frac{d}{d\sigma'}\theta(\sigma-\sigma')=-\delta^{ab}\frac{4\pi}{n}\delta'(\sigma-\sigma')$. For the Poisson bracket of X and Y we get therefore

$$\begin{aligned}[X,Y]_{\rm PB}&=-\frac{4\pi}{n}\delta'(\sigma-\sigma')\operatorname{Tr}g^{-1}(\sigma)Ag(\sigma)g^{-1}(\sigma')Bg(\sigma')\\&=-\frac{4\pi}{n}\delta(\sigma-\sigma')\operatorname{Tr}[A,B]\frac{dg}{d\sigma}g^{-1}-\frac{4\pi}{n}\delta'(\sigma-\sigma')\operatorname{Tr}AB.\end{aligned}\tag{29}$$

Bearing in mind the definition of X and Y and the relation between Poisson brackets and quantum mechanical commutation relations, this corresponds to the commutation relations

$$\left[\operatorname{Tr} A \frac{dg}{d\sigma} g^{-1}(\sigma), \operatorname{Tr} B \frac{dg}{d\sigma} g^{-1}(\sigma')\right]$$
$$= \frac{4\pi}{n} i\delta(\sigma-\sigma')\operatorname{Tr}[A,B]\frac{dg}{d\sigma}g^{-1} + \frac{4\pi}{n} i\delta'(\sigma-\sigma')\operatorname{Tr} AB. \tag{30}$$

Now, let us compare this to the Kac-Moody algebra (21). We see that they coincide if $k=n$ and if J_- is identified with $\frac{n}{2\pi}\frac{dg}{d\sigma}g^{-1}$.

Now, that conclusions can we draw? In the nonlinear sigma model and in Eq. (30), n is an integer because of the multivaluedness of the Wess-Zumino coupling. In (21) k is an integer because only then does the Kac-Moody algebra have well-behaved unitary representations. We see that single valuedness of e^{iI}, required in quantum mechanics for mathematical consistency, leads to a Kac-Moody algebra with properly normalized central charge.

Second, the theory at $\lambda^2 = 4\pi/n$ really does have a vanishing β function, because it is known [15] that the irreducible representation of the Kac-Moody algebra is conformally invariant (can be extended to the semi-direct product of the Kac-Moody algebra with the conformal algebra).

Third, and most important, it follows from Eq. (30) that the non-linear sigma model with $n=1$ and $\lambda^2=4\pi$ is equivalent to the free field theory of N massless Majorana fermions. For, with the identifications

$$\begin{aligned} J_-^{ij} &= i\psi_-^i\psi_-^j = \frac{1}{2\pi}\left(\frac{dg}{d\sigma}g^{-1}\right)^{ij}, \\ J_+^{ij} &= i\psi_+^i\psi_+^j = \frac{1}{2\pi}\left(g^{-1}\frac{dg}{d\tau}\right)^{ij}, \end{aligned} \tag{31}$$

the currents of these theories obey the same algebra [Eqs. (20) and (30)], and, as has been mentioned, this algebra has an essentially unique irreducible representation. (The Hilbert spaces of the quantum theories in question furnish irreducible representations of the current algebras because no operators commute with all the currents. This has been proved [13] in the fermi case and also [16] in the bose case.[6]) The Kac-Moody representation is not quite unique, but the non-uniqueness just refers to superselection rules and boundary conditions in the quantum field theory.

Moreover, the Hamiltonian H and the momentum operator P of the free fermi theory coincide with those of the non-linear sigma model at the special values of

6 Our discussion of the non-linear sigma model is closely related to the discussion of the Kac-Moody representations in [16]. The phase space of our theory in the light cone frame is a complex manifold, the loop space Z of $O(N)$. (Actually, this is only half the phase space of the theory, since it omits left-moving waves. The full phase space is $Z \times Z$.) The operator F_{ij} that we have constructed represents the first Chern class of a holomorphic line bundle E over Z. The Hilbert space of the theory is the space of holomorphic sections of E. This construction generalizes some classical theorems about representations of finite dimensional Lie groups to the Kac-Moody system. The novelty of our discussion of the non-linear sigma model is to show that the construction of Kac-Moody representations just mentioned can be realized by canonical quantization of a quantum field theory

the couplings under discussion. This can be seen in various ways. Because the Hilbert spaces form irreducible representations of the current algebras, H and P are uniquely determined in each case by their commutation relations with the currents. These commutation relations are suitable; H and P generate translations in both the bose and fermi theories. A more explicit argument is the following. In the fermi theory one can show $H+P=\text{const}\lim_{\varepsilon\to 0}\int dx J^{ij}_{+}(x+\varepsilon)J^{ij}_{+}(x)$, and a similar formula for $H-P$ in terms of J_{-}. (This is not a canonical equation. One must study the short distance behavior of the product of currents, and subtract an infinite c-number.) On the other hand, in the bose theory one can calculate canonically an equivalent formula

$$H+P=\text{const}\int dx\,\text{Tr}\left(g^{-1}\frac{dg}{d\tau}\right)^2$$

(and similarly for $H-P$). In view of (31) these relations show that H and P of the bose theory equal those of the fermi theory.

By introducing several fermion flavors, it is possible to make a bose-fermi translation also for $n\neq 1$. Consider a theory with N "colors" and n "flavors" of Majorana fermions ψ^{ia}, $i=1\ldots N$ and $a=1\ldots n$. We can define currents $\tilde{J}^{ij}_{\pm}=-i\sum_a\psi^{ia}_{\pm}\psi^{ja}_{\pm}$, $\tilde{J}^{ab}_{\pm}=-i\sum_i\psi^{ia}_{\pm}\psi^{ib}_{\pm}$. These currents generate $O(N)_L\times O(N)_R\times O(n)_L\times O(n)_R$. The $O(N)_L\times O(N)_R$ current commutators have anomalies of strength n. The $O(n)_L\times O(n)_R$ current commutators have anomalies of strength N. The free field theory Hilbert space is an irreducible representation [17] of the combined current algebra.

An equivalent bose theory is a theory with two fields g and h; g takes values in $O(N)$ and h in $O(n)$. For g we take a Wess-Zumino coupling n and $\lambda^2=4\pi/n$; for h we take a Wess-Zumino coupling N and $\lambda^2=4\pi/N$. g and h are decoupled, corresponding to the fact that at the fermion level amplitudes with a product of $O(N)_L\times O(N)_R\times O(n)_L\times O(n)_R$ currents factorize [18] as a product of $O(N)_L\times O(N)_R$ amplitudes and $O(n)_L\times O(n)_R$ amplitudes. The Hamiltonian and momentum operators of ψ^{ia} can be identified with the sum of those constructed from g and h.

As we have discussed, Eq. (31) generalizes to the non-abelian case the equation $\bar{\psi}\gamma_\mu\psi=\frac{1}{\sqrt{\pi}}\varepsilon_{\mu\nu}\partial^\nu\phi$ of conventional bosonization. In conventional bosonization, there also are formulas $\bar{\psi}\psi=M\cos\sqrt{4\pi}\,\phi$, $\bar{\psi}i\gamma_5\psi=M\sin\sqrt{4\pi}\,\phi$ (M is a renormalization mass). What are the analogues of those formulas here?

Let $Q^i_k=-i\psi^i_{-}\psi_{+k}$. We would like to translate Q^i_k into the bose language. The commutators of Q^ik with the fermion currents are as follows:

$$\begin{aligned}
[J^-_{ij}(x),Q^k_l(y)] &= -i\delta(x-y)(\delta_{jk}Q^i_l(y)-\delta_{ik}Q^j_l(y)],\\
[J^+_{ij}(x),Q^k_l(y)] &= -i\delta(x-y)(\delta_{jl}Q^k_i(y)-\delta_{il}Q^k_j(y)).
\end{aligned}\tag{32}$$

We must therefore find in the bose theory operators obeying the algebra (32). One eed not look far. The matrix elements $g^i_j(x)$ of the matrix g are the required operators. Canonically,

$$\begin{aligned}\left[\frac{1}{2\pi}\left(\frac{dg}{d\sigma}g^{-1}(x)\right)^{ij}, g_l^k(y)\right] &= -i\delta(x-y)(\delta^{jk}g_l^i(y)-\delta^{ik}g_l^j(y)),\\ \left[\frac{1}{2\pi}\left(g^{-1}\frac{dg}{d\tau}(x)\right)^{ij}, g_i^k(y)\right] &= -i\delta(x-y)(\delta^{jl}g_i^k(y)-\delta^{il}g_j^k(y)).\end{aligned} \tag{33}$$

[The evaluation of (33) is simple for the following reason. From (30) we know that $\int dx \frac{dg}{d\sigma}g^{-1}$ and $\int dx g^{-1}\frac{dg}{d\tau}$ generate chiral O(N) × O(N). Therefore, (33) holds up to total derivatives that vanish after integrating over x. By virtue of locality and dimensional analysis – g is dimensionless – there are no such possible terms. For the same reason, there can be no anomaly in (33) quantum mechanically.]

At least heuristically, it appears that $Q_k^i(x)$ and $g_k^i(x)$ are uniquely characterized (up to normalization) by the relations (32) and (33). For instance, in the free fermi theory one cannot find another operator that transforms like $Q_k^i(x)$, so we are led to conclude

$$Q_j^i(x) = -i\psi_-^i\psi_{j+}(x) = Mg_j^i(x), \tag{34}$$

where (as in the conventional bosonization) M is a mass that depends on the renormalization procedure for the bosonic operator. It should be noted that – while $-i\psi_-^i\psi_+^j$ has canonical dimension one – g_j^i is dimensionless classically. So Eq. (34) is possible only if g_j^i has anomalous dimension one in the fixed point theory with $n=1$.

Equation (34) is a generalization of the conventional bosonization formulas, in which $\bar\psi\psi$ and $\bar\psi i\gamma_5\psi$ are identified with matrix elements of the O(2) matrix $\begin{pmatrix}\cos\sqrt{4\pi}\,\phi & \sin\sqrt{4\pi}\,\phi\\ -\sin\sqrt{4\pi}\,\phi & \cos\sqrt{4\pi}\,\phi\end{pmatrix}$. Of course, in the free bose field theory, it is easy to see that $\cos\sqrt{4\pi}\,\phi$ and $\sin\sqrt{4\pi}\,\phi$ do have anomalous dimension one.

Equation (34) can be tested in the following way. We have identified $i\psi_i^-\psi_j^-$ with $\frac{1}{2\pi}\left(\frac{dg}{dx^-}g^{-1}\right)_j^i = \frac{1}{2\pi}\sum_k \frac{d}{dx^-}g_k^i(g^{-1})_j^k$. Since g is orthogonal, $(g^{-1})_j^k = g_k^j$. So $i\psi_i^-\psi_j^- = \frac{1}{2\pi}\sum_k \frac{dg_k^i}{dx^-}g_k^j$. If we identify g_j^i with $\frac{-\mathrm{i}}{\mathrm{M}}\psi_-^i\psi_{j+}$, we are led to require

$$i\psi_i^-\psi_j^-(x) = -\frac{1}{2\pi M^2}\sum_k\left(\frac{d}{dx^-}(\psi_i^-\psi_k^+(x))\right)(\psi_j^-\psi_k^+(x)). \tag{35}$$

At first sight Eq. (35) looks preposterous. It relates an operator quadratic in spinors to one quartic in spinors. However, to understand the right-hand side of (35), we must study the small Δ behavior of

$$\frac{1}{\Delta^2}\int_{\substack{|y^0-x^0|<\Delta/2\\ |y^1-x^1|<\Delta/2}} d^2y\,\frac{\partial}{\partial y^-}\,T(\psi_i^-\psi_k^+(y)\psi_j^-\psi_k^+(x)). \tag{36}$$

Since $\sum_{k=1}^{N} \frac{\partial}{\partial y^-} T(\psi_k^+(y)\psi_k^+(x)) = N\delta^2(x-y)$, (36) has a piece $-\frac{N}{\Delta^2}\psi_i^-\psi_j^-(x)$. This is the most singular part of (36), as $\Delta \to 0$. Equation (35) must be understood to mean that the operator on the left-hand side equals the most singular part of the operator on the right-hand side. Note that while Δ is cut-off dependent, the mass M appearing in $-i\psi_i^-\psi_j^+(x) = Mg_j^i(x)$ is also cutoff dependent [since $-i\psi_i^-\psi_j^+$ has no anomalous dimension, but as already noted g_j^i must have anomalous dimension one in the equivalent bose theory, if the relation $-i\psi_i^-\psi_j^+(x) = Mg_j^i(x)$ holds]. Evidently, in view of (35) and (36), the product $M\Delta$ is cut-off independent. Equation (35) holds in the limit as $M \to \infty$ and $\Delta \to 0$ with $M\Delta$ fixed. While these manipulations are somewhat bizarre, the same bizarre manipulations are needed in conventional bosonization to show the consistency of the relations

$$\bar{\psi}\gamma^\mu\psi = \frac{1}{\sqrt{\pi}}\varepsilon^{\mu\nu}\partial_\nu\phi, \qquad \bar{\psi}\psi = M\cdot\cos\sqrt{4\pi}\phi, \qquad \bar{\psi}i\gamma_5\psi = M\sin\sqrt{4\pi}\phi.$$

This completes the dictionary of bosonization of fermi-bilinears. Given the dictionary, it should be clear that the bosonization can be carried out also for arbitrary massive or interacting fermi theories. For instance, a fermion bare mass $m\bar{\psi}\psi = mi\sum_{k=1}^{N}\psi_-^k\psi_+^k$ can be included by adding to the Lagrangian a term proportional to $\sum_{i=1}^{N} g_i^i = \mathrm{Tr}g$. A $(\bar{\psi}\psi)^2$ coupling becomes $(\mathrm{Tr}g)^2$. One can likewise study gauge theories in this way. For instance, in the fermion language one may choose to gauge an arbitrary anomaly free subgroup H of chiral $O(N)\times O(N)$. The corresponding theory can be studied in the bose language by gauging the same subgroup H of the symmetry group of the nonlinear sigma model. One will be limited to anomaly free groups H – as one should be – because only for anomaly free groups does the Wess-Zumino term have a gauge invariant generalization [8].

Various applications of the present work can be imagined, but will not be explored here. The non-abelian bosonization may help in understanding $1+1$ dimensional field theories. It may be helpful in understanding the Callan-Rubakov effect, which is described by an effective $1+1$ dimensional s-wave field theory. And the conformally invariant theory with $\lambda^2 = 4\pi/n$ may provide a starting point for constructing generalizations of the usual string theories.

Appendix: Explicit Form of the Wess-Zumino Functional

In this appendix we will work out an explicit formula for the Wess-Zumino functional in the simplest case of an SU(2) non-linear sigma model in two space-time dimensions. We will use an index free notation, so antisymmetric tensors ω_{ijk} are denoted simply as ω, and the curl of ω, $(\partial_i\omega_{jkl} \pm \text{cyclic permutations})$ is denoted $d\omega$. Differentials are considered to anticommute, so $dxdy = -dydx$ and $(dx)^2 = 0$, if x and y are functions.

First let us write the Wess-Zumino functional in an abstract form. On the group manifold of any simple, non-abelian group G, there is a $G\times G$ invariant third rank tensor field ω. ω obeys $d\omega=0$, and locally but not globally $\omega=d\lambda$ for some second rank tensor field λ. ω may be normalized so that its integral over any three sphere in G is an integral multiple of 2π.

Let B be a three-dimensional ball whose boundary, the two sphere S, is identified with space time. Given a mapping g from S into G, which has been extended to a mapping (also denoted g) from B into G, the Wess-Zumino functional is defined as

$$\Gamma=\int_B g^*\cdot\omega=\int_B g^*\cdot d\lambda=\int_{\partial B} g^*\cdot\lambda=\int_S g^*\cdot\lambda. \tag{37}$$

Here g^* is the "pull-back" of differential forms. In the third step of (37) $\partial B=S$ is the boundary of B; Stokes' theorem has been used. The last formula in (37) exhibits Γ as the integral of an ordinary two-dimensional Lagrangian. In concrete terms, the meaning of this formula as follows. Let ϕ^i be a set of coordinates for the group manifold G. Let λ_{ij} be the components of the anti-symmetric tensor λ. Then the mapping $g:S\to G$ can be described by means of functions ϕ^i, and

$$\Gamma=\int d^2x\varepsilon^{\mu\nu}\lambda_{ij}(\phi^k(x))\partial_\mu\phi^i\partial_\nu\phi^j. \tag{38}$$

Equation (38) has "Dirac string" type singularities, because the defining equation of λ, $\omega=d\lambda$, can be solved only locally on the group manifold.

Equation (38) is $G\times G$ invariant, although not manifestly so. Under a $G\times G$ transformation λ transforms as $\lambda_{ij}\to\lambda_{ij}+\dfrac{\partial\beta_j}{\partial\phi^i}-\dfrac{\partial\beta_i}{\partial\phi^j}$ for some $\beta_i(\phi^k)$. So

$$\delta\Gamma=\int d^2x\varepsilon^{\mu\nu}(\partial_i\beta_j-\partial_j\beta_i)\partial_\mu\phi^i\partial_\nu\phi^j=2\int d^2x\frac{\partial}{\partial x^\mu}(\varepsilon^{\mu\nu}\beta_j\partial_\nu\phi^j), \tag{39}$$

and Γ changes by a total divergence whose integral vanishes.

Now let us construct explicit formulas for the simplest non-abelian group SU(2). The SU(2) manifold is a three sphere; it can be described by polar angles ψ, θ, ϕ with line element $ds^2=d\psi^2+\sin^2\psi(d\theta^2+\sin^2\theta d\phi^2)$. The only SU(2)×SU(2) invariant third rank antisymmetric tensor is the Levi-Civita tensor or volume form, so

$$\omega=\frac{1}{\pi}\sin^2\psi\sin\theta\, d\psi d\theta d\phi. \tag{40}$$

Note the normalization of (40). Since the volume of the SU(2) manifold is $2\pi^2$, (40) is chosen so that the integral of ω over the whole manifold is 2π.

The equation $\omega=d\lambda$ can be solved in many ways, for instance

$$\lambda=\frac{1}{\pi}\phi\sin^2\psi\sin\theta\, d\psi d\theta. \tag{41}$$

[Recall $(d\psi)^2=(d\theta)^2=0$.] So given a mapping of space-time into SU(2), the properly normalized Wess-Zumino interaction is

$$\Gamma=\frac{1}{\pi}\int d^2x\phi(x)\sin^2\psi(x)\sin\theta(x)\varepsilon^{\mu\nu}\partial_\mu\psi(x)\partial_\nu\theta(x). \tag{42}$$

[In this parametrization, the Dirac-string type singularities occur at $\theta(x)=0$ or π. For at $\theta=0$ or π, everything should be independent of ϕ. This is not true in Eq. (42). In the case of the group SU(2), it is possible to choose another parametrization that is singular only at a single point on the group manifold.]

Formulas similar to (42) can be constructed for other groups and also for the Wess-Zumino interaction in four dimensions. These formulas are not very enlightening, however.

Acknowledgements. I wish to thank A. Feingold and I. Frenkel for discussions about Kac-Moody algebras, and D. J. Gross for suggesting a test of Eq. (34). I also wish to acknowledge discussions of bosonization and related matters with S. Coleman and D. Olive.

References

1. Coleman, S.: Quantum sine-Gordon equation as the massive Thirring model. Phys. Rev. D **11**, 2088 (1975)
2. Mandelstam, S.: Soliton operators for the quantized sine-Gordon equation. Phys. Rev. D **11**. 3026 (1975)
3. Baluni, V.: The Bose form of two-dimensional quantum chromodynamics. Phys. Lett. **90** B, 407 (1980)
   Steinhardt, P.J.: Baryons and baryonium in QCD_2. Nucl. Phys. B **176**, 100 (1980)
   Amati, D., Rabinovici, E.: On chiral realizations of confining theories. Phys. Lett. **101** B, 407 (1981)
4. Wess, J., Zumino, B.: Consequences of anomalous word identities. Phys. Lett. **37** B, 95 (1971)
5. D'Adda, A., Davis, A.C., DiVecchia, P.: Effective actions in non-abelian theories. Phys. Lett. **121** B, 335 (1983)
6. Polyakov, A.M., Wiegmann, P.B.: Landau Institute preprint (1983)
7. Alvarez, O.: Berkeley preprint (1983)
8. Witten, E.: Global aspects of current algebra. Nucl. Phys. B (to appear)
9. Novikov, S.P.: Landau Institute preprint (1982)
10. Polyakov, A.M.: Interaction of Goldstone particles in two dimensions. Applications to ferromagnets and massive Yang-Mills fields. Phys. Lett. **59** B, 79 (1975)
    Belavin, A.A., Polyakov, A.M.: Metastable states of two-dimensional isotropic ferromagnets. JETP Lett. **22**, 245 (1975)
11. Nappi, C.R.: Some properties of an analog of the chiral model. Phys. Rev. D **21**, 418 (1980)
12. Goto, T., Imamura, I.: Note on the non-perturbation-approach to quantum field theory. Prog. Theor. Phys. **14**, 396 (1955)
    Schwinger, J.: Field-theory commutators. Phys. Rev. Lett. 3, 296 (1959)
    Jackiw, R.: In: Lectures on current algebra and its applications, Treiman S.B., et al. (eds.): Princeton, NJ: Princeton University Press 1972
    Coleman, S., Gross, D., Jackiw, R.: Fermion avatars of the Sugawara model. Phys. Rev. **180**, 1359 (1969)
13. Kac, V.G.: J. Funct. Anal. Appl. **8**, 68 (1974)
    Lepowsky, J., Wilson, R.L.: Construction of the affine Lie algebra. $A_1(1)$. Commun. Math. Phys. **62**, 43 (1978)
    Frenkel, I.B.: Spinor representations of affine Lie algebras. Proc. Natl. Acad. Sci. USA **77**, 6303 (1980); J. Funct. Anal. **44**, 259 (1981)
    Feingold, A.J., Frenkel, I.B.: IAS preprint (1983)
14. Belavin, A.M., Polyakov, A.M., Schwar, A.S., Tyupkin, Yu.S.: Pseudoparticle solutions of the Yang-Mills equations. Phys. Lett. **59** B, 85 (1975)
    't Hooft, G.: Symmetry breaking through Bell-Jackiw anomalies. Phys. Rev. Lett. **37**, 8 (1976); Computation of the quantum effects due to a four-dimensional pseudoparticle. Phys. Rev. D **14**, 3432 (1976)
    Callan, C.G., Jr., Dashen, R., Gross, D.J.: The structure of the gauge theory vacuum. Phys. Lett. **63** B, 334 (1976)
    Jackiw, R., Rebbi, C.: Vacuum periodicity in a Yang-Mills quantum theory. Phys. Rev. Lett. **37**, 172 (1976)

15. Segal, G.: Unitary representations of some infinite-dimensional groups. Commun. Math. Phys. **80**, 301 (1981)
Frenkel, I., Kac, V.G.: Basic representations. Invent Math. **62**, 23 (1980)
16. Kac, V.G., Peterson, D.H.: Spin and wedge representations of infinite-dimensional Lie algebras and groups. Proc. Natl. Acad. Sci. USA **78**, 3308 (1981)
17. Frenkel, I.: Private communication
18. Frishman, Y.: Quark trapping in a model field theory. Mexico City 1973. Berlin, Heidelberg, New York: Springer 1975
19. Deser, S., Jackiw, R., Templeton, S.: Three-dimensional massive gauge theories. Phys. Rev. Lett. **48**, 975 (1982); Topologically massive gauge theories. Ann. Phys. (NY) **140**, 372 (1982)

Communicated by A. Jaffe

Received September 29, 1983

Nuclear Physics B247 (1984) 83–103

CURRENT ALGEBRA AND WESS-ZUMINO MODEL IN TWO DIMENSIONS

V.G. KNIZHNIK and A.B. ZAMOLODCHIKOV

Landau Institute of Theoretical Physics, Moscow, USSR

Received 6 June 1984

We investigate quantum field theory in two dimensions invariant with respect to conformal (Virasoro) and non-abelian current (Kac-Moody) algebras. The Wess-Zumino model is related to the special case of the representations of these algebras, the conformal generators being quadratically expressed in terms of currents. The anomalous dimensions of the Wess-Zumino fields are found exactly, and the multipoint correlation functions are shown to satisfy linear differential equations. In particular, Witten's non-abelean bosonisation rules are proven.

1. Introduction

In recent papers [1–3] some novel important properties of the two-dimensional σ-model with Wess-Zumino action

$$S_{\lambda,k}(g)=\frac{1}{4\lambda^2}\int \mathrm{tr}\left(\partial_\mu g^{-1}\partial_\mu g\right)\mathrm{d}^2\xi+k\Gamma(g) \tag{1.1}$$

have been discovered. The matrix field $g(\xi)$ in (1.1) is taken to be an element of some semisimple group G, $\xi^\mu=(\xi^1,\xi^2)$ are the coordinates of two-dimensional space, λ^2 and k are dimensionless coupling constants, k being necessarily integer [1, 2]. The Wess-Zumino term $\Gamma(g)$ is defined by the integral

$$\Gamma(g)=\frac{1}{24\pi}\int \mathrm{d}^3X e^{\alpha\beta\gamma}\mathrm{tr}\left(g^{-1}\partial_\alpha g g^{-1}\partial_\beta g g^{-1}\partial_\gamma g\right) \tag{1.2}$$

over the three-dimensional ball with coordinates X^α; the boundary being identified with two-dimensional space [2]. The boundary values $g(\xi)$ determine (1.2) modulo 2π [1].

If $k=0$, the action (1.1) reduces to the usual σ-model which is well known to be asymptotically free and effectively massive. This model has been exactly solved in [4, 5]. Under the choice $k=1,2,\ldots$ the character of the theory changes drastically as has been shown by Witten [2], Polyakov and Wiegmann [3]. The renormalization group possesses the infrared-stable fixed point

$$\lambda^2 = \frac{4\pi}{k}, \tag{1.3}$$

and therefore the effective theory is massless and its large-distance behavior is governed by the action

$$S_k(g) = kW(g), \tag{1.4}$$

where

$$W(g) = \left\{ \frac{1}{16\pi} \int \mathrm{tr}\left(\partial_\mu g^{-1} \partial_\mu g\right) \mathrm{d}^2\xi + \Gamma(g) \right\}. \tag{1.5}$$

In further discussions this theory (1.4) will be referred to as the Wess-Zumino model*.

The most important property of the action (1.4) is its invariance with respect to infinite-dimensional current (Kac-Moody) algebra [2, 3, 8]. The action (1.4) remains unchanged under the transformations

$$g(\xi) \to \Omega(z) g(\xi) \bar{\Omega}^{-1}(\bar{z}), \tag{1.6}$$

where $\Omega(z)$ and $\bar{\Omega}(\bar{z})$ are arbitrary G-valued matrices analytically depending on the complex coordinates

$$z = \xi^1 + i\xi^2,$$

$$\bar{z} = \xi^1 - i\xi^2, \tag{1.7}$$

respectively. (Here we imply the euclidean version of the theory; in Minkowski space-time the variables (1.7) are the light-cone coordinates.) One can easily ensure symmetry (1.6) using the following remarkable relation [3, 8]:

$$W(gh^{-1}) = W(g) + W(h) + \frac{\mathrm{tr}}{16\pi} \int \left(g^{-1}\partial_{\bar{z}} g h^{-1} \partial_z h\right) \mathrm{d}^2\xi, \tag{1.8}$$

satisfied by the functional (1.5). Note that the group (1.6) generalizes the usual G × G symmetry of the chiral field, and it can also be represented as the direct product of the "left" and "right" gauge groups; we shall denote it by $G(z) \times G(\bar{z})$.

* The fixed-point theory (1.4) deserves special interest because of its analogies with the quantum Liouville theory related to the Polyakov string [7].

The symmetry (1.6) gives rise to an infinite number of conserved currents which can be derived from the equations

$$\partial_{\bar{z}} J = 0, \qquad \partial_z \bar{J} = 0, \tag{1.9}$$

where the basic currents J and $\bar{J}$:

$$J = J^a t^a = -\tfrac{1}{2} k \partial_z g g^{-1},$$

$$\bar{J} = \bar{J}^a t^a = -\tfrac{1}{2} k g^{-1} \partial_{\bar{z}} g, \tag{1.10}$$

correspond to the generators of the groups $G(z)$ and $G(\bar{z})$ respectively [2, 3]. Here t^a are the antihermitian matrices representing (for the field $g(\xi)$) the Lie algebra

$$[t^a, t^b] = f^{abc} t^c \tag{1.11}$$

of the group G; f^{abc} are the structure constants. Due to (1.9) we can write

$$J^a = J^a(z), \qquad \bar{J}^a = \bar{J}^a(\bar{z}). \tag{1.12}$$

The variations of the fields (1.10) under the infinitesimal transformations (1.6) with

$$\Omega(z) = I + \omega(z) = I + \omega^a(z) t^a, \tag{1.13a}$$

$$\bar{\Omega}(\bar{z}) = I + \bar{\omega}(\bar{z}) = I + \bar{\omega}^a(\bar{z}) t^a, \tag{1.13b}$$

are described by the formulae

$$\delta_\omega J(z) = [\omega(z), J(z)] + \tfrac{1}{2} k \omega'(z),$$

$$\delta_{\bar{\omega}} \bar{J}(\bar{z}) = [\bar{\omega}(\bar{z}), \bar{J}(\bar{z})] + \tfrac{1}{2} k \bar{\omega}'(\bar{z}), \tag{1.14}$$

which shows that the generators $J(\bar{J})$ of the group $G(z)$ $(G(\bar{z}))$ represent the Kac-Moody algebra [2] with the central charge k*. Since $\delta_\omega \bar{J} = \delta_{\bar{\omega}} J = 0$, the generators J and $\bar{J}$ are commutative.

Under the choice $G = O(N)$ the group $G(z) \times G(\bar{z})$ describes the symmetry of the free massless N-component Majorana fermion theory with the action

$$S_f(\psi, \bar{\psi}) = \tfrac{1}{2} \int \sum_{a=1}^{N} [\psi_a \partial_{\bar{z}} \psi_a + \bar{\psi}_a \partial_z \bar{\psi}_a] \, d^2\xi, \tag{1.15}$$

* Obviously, there are no divergent renormalizations of the integer-valued "coupling constant" k in the theory (1.1). However, finite renormalization of the type $k_0 \to k = k_0 + \Delta k$, where k_0 stands in front of the "bare" action (1.4) and Δk is some integer, cannot be excluded a priori. The one-loop computation shows that $\Delta k = 0$.

where ψ and $\bar{\psi}$ are the "left" and "right" components of the Fermi fields. Actually, the theories (1.4) and (1.15) are related. Witten [2] has shown that the Wess-Zumino model with $G = O(N)$ and $k = 1$ is equivalent to the free fermion theory (1.15), the fields (1.10) being equal to the corresponding currents of (1.15):

$$J_{\alpha\alpha'}(z) = :\psi_\alpha(z)\psi_{\alpha'}(z):, \qquad \bar{J}_{\beta\beta'}(\bar{z}) = :\bar{\psi}_\beta(\bar{z})\bar{\psi}_{\beta'}(\bar{z}):. \tag{1.16}$$

This result follows directly from the observation that the current of the models (1.14) and (1.15) satisfies the same algebra. Moreover, Witten suggested the local expression for the field $g(\xi) = g_{\alpha\beta}(\xi)$ in terms of the Fermi fields $\psi, \bar{\psi}$:

$$Mg_{\alpha\beta}(z,\bar{z}) = :\psi_\alpha(z)\bar{\psi}_\beta(\bar{z}):, \tag{1.17}$$

where M is the mass parameter dependent on the regularization scheme. The same formulae (with slight specifications, see [8] and sect. 4) relate the model (1.4) with $G = U(N)$, $k = 1$ to the theory of N-component charged Fermi fields. Formulae (1.16) and (1.17) are Witten's non-abelian bosonization rules.

Since the conformal anomaly vanishes at the fixed point (1.3), the Wess-Zumino theory (1.4) is invariant also with respect to the infinite-dimensional group of coordinate transformations:

$$z \to \zeta(z), \qquad \bar{z} \to \bar{\zeta}(\bar{z}), \tag{1.18}$$

with arbitrary analytic functions ζ and $\bar{\zeta}$; these transformations constitute the conformal group of two-dimensional space. In conformal quantum theory the local fields, like $g(z,\bar{z})$ in (1.4), can acquire anomalous dimensions, i.e. they are transformed as

$$g(z,\bar{z}) \to \left(\frac{d\zeta}{dz}\right)^{\Delta}\left(\frac{d\bar{\zeta}}{d\bar{z}}\right)^{\bar{\Delta}} g(\zeta,\bar{\zeta}) \tag{1.19}$$

(with real positive Δ and $\bar{\Delta}$) under the substitutions (1.18). Obviously, for the spinless field $g(z,\bar{z})$ of (1.4) Δ and $\bar{\Delta}$ must be equal.

In this paper we investigate the Wess-Zumino model with an arbitrary integer k. Using the infinite-dimensional symmetry (1.6), we compute exactly the anomalous dimensions and develop the method for computing the multipoint correlation functions (Green's functions)

$$\langle g(z_1,\bar{z}_1)\dots g(z_N,\bar{z}_N)\rangle; \tag{1.20}$$

some of them will also be constructed explicitly. We apply the technique similar to that proposed in [9] for the conformal field theories in two dimensions. In particular, we will find the field $g(z,\bar{z})$ (as well as some other "composite" fields of the theory) to be associated with the degenerate representation of symmetry algebra (semidirect product of current algebra and Virasoro algebra) of the model (1.4); therefore the correlation functions (1.20) satisfy special linear differential equations. Together with the general requirements of crossing symmetry, these equations determine the functions (1.20) completely. In the particular case $G = U(N)$ or $O(N)$ and $k = 1$, the relations (1.17) follow from our result. At $k > 1$ the correlation functions (1.20) turn out to be more complicated and the theory (1.4) can hardly be connected with free fields in any local way. In fact, the field $g(\xi)$ and other local fields possess nontrivial anomalous dimensions in the theory with $k > 1$.

Polyakov and Wiegmann [3] have managed to solve the model (1.1) (with arbitrary λ^2) exactly by means of the Bethe ansatz technique. Our approach is based completely on the symmetry (1.6) and conformal symmetry and therefore is restricted to the fixed-point theory (1.4). However, our approach provides much more detailed information about the theory (1.4); in particular, the computation of correlation functions like (1.20) remain beyond the powers of the Bethe ansatz method. It is also worth noting that the correlation functions of the model (1.4) studied in this paper describe exact infrared asymptotics of the general model (1.1).

2. General properties of conformal quantum field theory invariant with respect to current algebra

The stress-energy tensor $T_{\mu\nu}(\xi)$ of a conformal quantum field theory satisfies, besides the usual equation $\partial_\mu T^{\mu\nu}(\xi) = 0$, the zero trace condition $T^\mu_\mu(\xi) = 0$. In two dimensions these two equations can be reduced to

$$\partial_{\bar z} T = 0, \qquad \partial_z \bar T = 0, \tag{2.1}$$

where

$$T = T_{11} - T_{22} + 2iT_{12},$$
$$\bar T = T_{11} - T_{22} - 2iT_{12}. \tag{2.2}$$

In view of (2.1) we shall write

$$T = T(z), \qquad \bar T = \bar T(\bar z). \tag{2.3}$$

The fields (2.2) represent the generators of the infinitesimal conformal transformations

$$z \to z + \varepsilon(z), \tag{2.4a}$$

$$\bar z \to \bar z + \bar\varepsilon(\bar z), \tag{2.4b}$$

in the field theory, the field T being associated with the infinitesimal substitutions

(2.4a) of the variable z and $\bar{T}$ plays the same role for $\bar{z}$ [9]. If, besides the conformal symmetry, the field theory is invariant with respect to the transformations (1.6), there are also the local fields

$$J(z)=J^a(z)t^a, \qquad \bar{J}(z)=\bar{J}^a(\bar{z})t^a, \tag{2.5}$$

satisfying eqs. (1.9) and representing the generators of G(z) and G($\bar{z}$) respectively. These statements have the following precise meaning. Consider correlation functions of the form

$$\langle T(z)A_{j_1}(z_1,\bar{z}_1)\dots A_{j_N}(z_N,\bar{z}_N)\rangle, \tag{2.6a}$$

$$\langle J^a(z)A_{j_1}(z_1,\bar{z}_1)\dots A_{j_N}(z_N,\bar{z}_N)\rangle, \tag{2.6b}$$

where $A_{j_k}(z_k,\bar{z}_k)$ are the arbitrary local fields. These correlators are single-valued analytic functions of z, possessing poles at $z=z_1,z_2,\dots,z_N$. The order and residue of each of these poles, say z_k, are determined by the transformation properties of the corresponding field $A_{j_k}(z_k,\bar{z}_k)$ with respect to the conformal (2.4a) and gauge (1.13a) transformations*. In fact, the following relations are valid:

$$\delta_\varepsilon A_j(z,\bar{z})=\oint_{C_z} T(\zeta)\varepsilon(\zeta)A_j(z,\bar{z})\,d\zeta,$$

$$\delta_\omega A_j(z,\bar{z})=\oint_{C_z} J^a(\zeta)\omega^a(\zeta)A_j(z,\bar{z})\,d\zeta, \tag{2.7}$$

where $\delta_\varepsilon A_j$ and $\delta_\omega A_j$ are the variations of the field A_j under the infinitesimal transformations (2.4a) and (1.13a); the integration contour surrounds the point $\zeta=z$. Formulae (2.7) are understood as the relations between correlation functions. The same equations, with the substitutions $T\to\bar{T}$, $J\to\bar{J}$, are valid for the variations $\delta_{\bar{\varepsilon}}$ and $\delta_{\bar{\omega}}$.

Generally, the variations of the fields $T(z)$ and $J^a(z)$ themselves are given by the formulae

$$\delta_\varepsilon T(z)=\varepsilon(z)T'(z)+2\varepsilon'(z)T(z)+\tfrac{1}{12}c\varepsilon'''(z), \tag{2.8a}$$

$$\delta_\varepsilon J^a(z)=\varepsilon(z)J^{a\prime}(z)+\varepsilon'(z)J^a(z), \tag{2.8b}$$

$$\delta_\omega J^a(z)=f^{abc}\omega^b(z)J^c(z)+\tfrac{1}{2}k\omega^{a\prime}(z), \tag{2.8c}$$

* Here and below we constantly consider the correlators functions of the complex coordinates $\xi\in\mathbf{C}^2$. In the space $\mathbf{C}^2$ the coordinates z and $\bar{z}$ (1.7) are independent complex variables and the conformal group (like the group (1.6)) can be considered as the direct product $\Gamma(z)\times\Gamma(\bar{z})$ of two (identical) groups Γ of analytic substitutions of one variable.

where the prime denotes the derivative. The variations $\delta_{\bar\varepsilon}$ and $\delta_{\bar\omega}$ of the fields $\bar T$ and $\bar J$ are given by the same equations, whereas the variations $\delta_{\bar\varepsilon}$ and $\delta_{\bar\omega}$ of T and J vanish. The values of the real parameters c and k in (2.8) are not completely fixed by general principles. There are, however, the strong limitations of the values of these parameters: c must be positive and k must be an integer. It can be shown that otherwise the positivity condition of quantum field theory cannot be satisfied*. The equations (2.8) determine the algebra of the generators of the symmetry group in field theory.

According to (2.7), eqs. (2.8) can be rewritten in the form of operator product expansions:

$$T(z)T(z') = \frac{c}{2(z-z')^4} + \frac{2}{(z-z')^2}T(z') + \frac{1}{z-z'}T'(z') + \cdots, \tag{2.9a}$$

$$T(z)J^a(z') = \frac{1}{(z-z')^2}J^a(z') + \frac{1}{z-z'}J^{a\prime}(z') + \cdots, \tag{2.9b}$$

$$J^a(z)J^b(z') = \frac{k\delta^{ab}}{(z-z')^2} + \frac{f^{abc}}{z-z'}J^c(z') + \cdots, \tag{2.9c}$$

where the terms regular at $z \to z'$ are omitted from r.h.s.'s. The definition of the fields $T(z)$ and $J^a(z)$ (as well as $\bar T$ and $\bar J$) should be supplemented with the requirement of regularity at $z = \infty$, which is equivalent to the asymptotic conditions

$$T(z) \sim z^{-4}, \qquad J^a(z) \sim z^{-2} \qquad \text{as } z \to \infty. \tag{2.10}$$

Any local field $A_j(z, \bar z)$ of the theory is an "isotopic" tensor corresponding to some finite-dimensional representation of the "left" and "right" (global) groups G. Besides that, it is characterized by the anomalous dimensions $(\Delta_j, \bar\Delta_j)$ describing its transformation

$$A_j \to \lambda^{\Delta_j}\bar\lambda^{\bar\Delta_j}A_j \tag{2.11}$$

under the (complex) dilatations $z \to \lambda z$, $\bar z \to \bar\lambda \bar z$. In fact, the difference $s_j = \Delta_j - \bar\Delta_j$ is the spin of the field A_j (the spin s_j of the local fields can take integer or half-integer values only) whereas the sum $d_j = \Delta_j + \bar\Delta_j$ coincides with the conventional anomalous dimension. Applying arguments similar to those presented in [9] for the conformal theory, one can prove that there are the fields (like in [9] we shall call them "primary fields") which transform according to (1.19) and (1.6) under

* At $0 < c < 1$ the positivity condition selects also an infinite discrete set of allowed values of c [11]. This limitation is not significant here because in the Wess-Zumino model $c > 1$, see sect. 3.

arbitrary conformal and gauge transformations. Let us note that the matrices $\Omega(z)$ and $\bar{\Omega}(\bar{z})$ may correspond, in the general case, to different representations of G. Introducing the notation $\phi_l(z,\bar{z})$ for the primary fields and $(\Delta_l, \bar{\Delta}_l)$ for the corresponding dimensions, one can write down the singular terms of the operator product expansions:

$$T(\zeta)\phi_l(z,\bar{z}) = \frac{\Delta_l}{(\zeta - z)^2}\phi_l(z,\bar{z}) + \frac{1}{\zeta - z}\frac{\partial}{\partial z}\phi_l(z,\bar{z}) + \cdots, \tag{2.12a}$$

$$J^a(\zeta)\phi_l(z,\bar{z}) = \frac{t_l^a}{\zeta - z}\phi_l(z,\bar{z}) + \cdots, \tag{2.12b}$$

which are determined by the transformation properties of the field ϕ_l with respect to the infinitesimal transformations (2.4a) and (1.3a). Here t_l^a is the "left" representation of the generators of G for the field ϕ_l. Similar formulae (with the substitutions $\Delta_l \to \bar{\Delta}_l$ and $t_l^a \to \bar{t}_l^a$, where $\bar{t}_l^a$ corresponds to the "right" representation) are valid for the expansions of the products $\bar{T}\phi_l$ and $\bar{J}\phi_l$. Eqs. (2.12) allow one to determine explicitly the z-dependence of the correlation functions (2.6) provided all the fields A_j involved are primary ones:

$$\langle T(z)\phi_1(z_1,\bar{z}_1)\dots\phi_N(z_N,\bar{z}_N)\rangle = \sum_{j=1}^{N}\left\{\frac{\Delta_j}{(z-z_j)^2} + \frac{1}{z-z_j}\frac{\partial}{\partial z_j}\right\} \times \langle\phi_1(z_1,\bar{z}_1)\dots\phi_N(z_N,\bar{z}_N)\rangle, \tag{2.13a}$$

$$\langle J^a(z)\phi_1(z_1,\bar{z}_1)\dots\phi_N(z_N,\bar{z}_N)\rangle = \sum_{j=1}^{N}\frac{t_j^a}{z-z_j}\langle\phi_1(z_1,\bar{z}_1)\dots\phi_N(z_N,\bar{z}_N)\rangle, \tag{2.13b}$$

where the matrices t_j^a are applied to the "left" isotopic indices of the field $\phi_j(z,\bar{z})$. The relations (2.13) are the Ward identities corresponding to the conformal and gauge symmetries. Combining (2.13) with asymptotic conditions (2.10) one can easily derive the well-known Ward identities:

$$\sum_{j=1}^{N}\left\{z_j^{n+1}\frac{\partial}{\partial z_j} + (n+1)\Delta_j z_j^n\right\}\langle\phi_1(z_1,\bar{z}_1)\dots\phi_N(z_N,\bar{z}_N)\rangle = 0, \tag{2.14a}$$

where $n = -1, 0, +1$ and

$$\sum_{j=1}^{N} t_j^a\langle\phi_1(z_1,\bar{z}_1)\dots\phi_N(z_N,\bar{z}_N)\rangle = 0, \tag{2.14b}$$

which are manifestations of the invariance with respect to the regular subgroups

$SL_2 \subset \Gamma(z)$ of projective conformal transformations and $G \subset G(z)$ of global gauge transformations.

The variations δ_ε and δ_ω can be understood as some operators, defined by the r.h.s.'s of eqs. (2.7), acting on the fields A_j. It is convenient to expand the functions $\varepsilon(\zeta)$ and $\omega(\zeta)$ as a power series in $(\zeta - z)$ and to introduce the corresponding operators:

$$L_n A_j(z,\bar{z}) = \oint_{C_z} T(\zeta)(\zeta - z)^{n+1} A_j(z,\bar{z})\, d\zeta,$$

$$J_n^a A_j(z,\bar{z}) = \oint_{C_z} J^a(\zeta)(\zeta - z)^n t_j^a A_j(z,\bar{z})\, d\zeta. \tag{2.15}$$

The primary fields ϕ_l satisfy the equations

$$L_n\phi_l = J_n^a\phi_l = 0 \qquad \text{for } n>0,$$

$$L_0\phi_l = \Delta_l\phi_l, \qquad J_0^a\phi_l = t_l^a\phi_l, \tag{2.16}$$

which are the direct consequence of (2.12). Eqs. (2.15) define the operators L_n and J_n^a with negative values of n as well as with positive ones. In general, the local fields $L_{-n}A_j$ and $J_{-n}^a A_j$ with $n>0$ do not vanish. In particular, the operators L_{-1} and $\bar{L}_{-1}$ reduce to simple differential ones:

$$L_{-1}\phi_l(z,\bar{z}) = \frac{\partial}{\partial z}\phi_l(z,\bar{z}), \qquad \bar{L}_{-1}\phi_l(z,\bar{z}) = \frac{\partial}{\partial \bar{z}}\phi_l(z,\bar{z}). \tag{2.17}$$

Evidently, the regular terms omitted in the operator product expansions (2.12) can be expressed in terms of the fields $L_{-n}\phi_l$, $J_{-n}^a\phi_l$ with $n = 2,3,\ldots$.

Due to the singular terms in (2.9), the operators L_n and J_n^a are not commutative but satisfy the relations

$$[L_n, L_m] = (n-m)L_{n+m} + \tfrac{1}{12}c(n^3 - n)\delta_{n+m}, \tag{2.18a}$$

$$[L_n, J_m^a] = -mJ_{n+m}^a, \tag{2.18b}$$

$$[J_n^a, J_m^b] = f^{abc}J_{n+m}^c + \tfrac{1}{2}kn\delta^{ab}\delta_{n+m,0}. \tag{2.18c}$$

The commutation relations (2.18a) and (2.18c) are known as Virasoro algebra and Kac-Moody algebra respectively. The complete algebra (2.18), which is the semidirect product of these algebras, will be denoted here as $\mathcal{A}$. Obviously, the operators $\bar{L}_n$ and $\bar{J}_n^a$ defined by the same formulae as (2.15) except with the fields $\bar{T}$ and $\bar{J}$ constitute the same algebra which will be denoted by $\bar{\mathcal{A}}$.

The complete system of local fields $\{A_j\}$ involved in the theory includes, besides the primary fields ϕ_l, all the fields of the form

$$L_{-n_1}\dots L_{-n_N}\bar{L}_{-\bar{n}_1}\dots\bar{L}_{-\bar{n}_{N'}}J^{a_1}_{-m_1}\dots J^{a_M}_{-m_M}\bar{J}^{b_1}_{-\bar{m}_1}\dots\bar{J}^{b_{M'}}_{-\bar{m}_{M'}}\phi_l, \tag{2.19}$$

with arbitrary positives $n, \bar{n}, m, \bar{m}$. Like in [9], we shall denote by $[\phi_l]_{\mathscr{Q}}$ the totality of the fields (2.19) associated with some primary field ϕ_l. This infinite set of fields corresponds, obviously, to the highest weight representation (Verma modulus) of the algebra $\mathscr{Q}$ (more precisely, $[\phi_l]_{\mathscr{Q}}$ is the direct product of the highest weight representations of $\mathscr{Q}$ and $\bar{\mathscr{Q}}$), the primary field ϕ_l being associated with the highest weight vector. The dimensions of the fields (2.19) are

$$\Delta_l^{\{n,m\}} = \Delta_l + \sum_{i=1}^{N} n_i + \sum_{i=1}^{M} m_i, \qquad \bar{\Delta}_l^{\{\bar{n},\bar{m}\}} = \bar{\Delta}_l + \sum_{i=1}^{N'} \bar{n}_i + \sum_{i=1}^{M'} \bar{m}_i, \tag{2.20}$$

and their isotropic properties are self-evident from (2.19). The complete set of fields

$$\{A_j\} = \bigoplus_l [\phi_l]_{\mathscr{Q}} \tag{2.21}$$

form the closed operator algebra [9].

It is worth noting that the fields $T(z), J(z), \bar{T}(\bar{z}), \bar{J}(\bar{z})$ are not primary ones; they belong to $[I]_{\mathscr{Q}}$, where I is the identity operator. Namely,

$$T(z) = L_{-2}I, \qquad J^a(z) = J^a_{-1}I,$$

$$\bar{T}(\bar{z}) = \bar{L}_{-2}I, \qquad \bar{J}^a(\bar{z}) = \bar{J}^a_{-1}I. \tag{2.22}$$

3. Wess-Zumino model

The relations presented in the previous section concern any quantum field theory with the symmetry algebra $\mathscr{Q}\times\bar{\mathscr{Q}}$. The expression (1.10) of currents in terms of the g-field can be considered as the peculiarity of the Wess-Zumino model. In the quantum theory these expressions have the following meaning. We assume that the complete set of fields $\{A_j\}$ contains the spinless primary field $g(z,\bar{z})$ (which corresponds to the representation t^a of the left and right global groups G and have the dimensions $\Delta_g = \bar{\Delta}_g = \Delta$) satisfying the equations

$$\kappa\frac{\partial}{\partial z}g(z,\bar{z}) = :J^a(z)t^a g(z,\bar{z}):, \tag{3.1a}$$

$$\kappa\frac{\partial}{\partial \bar{z}}g(z,\bar{z}) = :\bar{J}^a(\bar{z})g(z,\bar{z})t^a:, \tag{3.1b}$$

where κ is the numerical factor (which will be computed later) and the local products of fields on the r.h.s. of (3.1) are regularized in a particular way. Eqs. (3.1) can be understood as the special property of the operator product expansions of $J^a(\zeta)t^a g(z,\bar z)$ and $\bar J^a(\bar z) g(z,\bar z) t^a$. Consider, for example, the first of these products which has the general form

$$J^a(\zeta)t^a g(z,\bar z) = \frac{c_g}{\zeta - z} g(z,\bar z) + \sum_{n=1}^{\infty} (\zeta - z)^{n-1} t^a J^a_{-n} g(z,\bar z), \tag{3.2}$$

where the constant c_g is defined as

$$t^a t^a = c_g I. \tag{3.3}$$

The operator coefficient accompanying the zero power of $(\zeta - z)$ in (3.2) has to coincide (up to a numerical factor) with the derivative $\partial_z g$, i.e.

$$J^a(\zeta)t^a g(z,\bar z) = \frac{c_g}{\zeta - z} g(z,\bar z) + \kappa \frac{\partial}{\partial z} g(z,\bar z) + O(\zeta - z). \tag{3.4}$$

So the product in (3.1a) is defined as

$$:J^a(z)t^a g(z,\bar z): \stackrel{\text{def}}{=} \lim_{\zeta\to z}\left[J^a(\zeta) - \frac{t^a}{\zeta - z}\right] t^a g(z,\bar z). \tag{3.5}$$

A similar definition applies to the r.h.s of (3.16).

The peculiarity of the operator product expansion (3.4) allows one to determine immediately the anomalous dimension Δ of the field g. Comparing (3.2) and (3.4) one gets

$$\chi \equiv (J^a_{-1} t^a - \kappa L_{-1}) g = 0, \tag{3.6}$$

where (2.17) is taken into account. From the mathematical point of view, this relation means that the representation $[g]_{\mathfrak{g}}$ of the algebra (2.18) is degenerate, the field χ (defined by (3.6)) being associated with the "null vector". The field χ should satisfy the equations

$$L_0\chi = (\Delta + 1)\chi, \qquad J^a_0\chi = t^a\chi, \tag{3.7a}$$

$$L_n\chi = J^a_n\chi = 0 \qquad \text{for } n > 0, \tag{3.7b}$$

since otherwise eq. (3.6) does not make sense. Note that eqs. (3.7a) are satisfied identically whereas (3.7b) has to be solved for $n = 1$ only because the other eqs. (3.7b) can be derived from these by means of (2.18). Using (2.16) and (2.18) one can ascertain that eqs. (3.7b) are satisfied provided

$$c_g + 2\Delta\kappa = 0,$$

$$c_V + k + 2\kappa = 0, \tag{3.8}$$

where c_V is defined as

$$f^{acd} f^{bcd} = c_V \delta^{ab}. \tag{3.9}$$

Eqs. (3.8) provide the values of the anomalous dimension of the field g:

$$\Delta = \frac{c_g}{c_V + k}, \tag{3.10}$$

and of the parameter κ in (3.1):

$$\kappa = -\tfrac{1}{2}(c_V + k). \tag{3.11}$$

There is another way of deriving eq. (3.6). In the classical Wess-Zumino theory the stress-energy tensor is expressed quadratically in terms of the currents (1.10). Let us assume that a similar relation holds in the quantum theory, i.e.

$$2\kappa T(z) = :J^a(z) J^a(z):,$$

$$2\kappa \bar{T}(\bar{z}) = :\bar{J}^a(\bar{z}) \bar{J}^a(\bar{z}):, \tag{3.12}$$

where the numerical parameter κ coincides, as we shall see below, with (3.11). This assumption can be considered as another definition of the Wess-Zumino model, equivalent to (3.1). Like (3.1), the relations (3.12) have to be understood in terms of the operator product expansions. For instance, the expansion of the product $J^a(z)J^a(z')$ has the following nonvanishing terms at $z \to z'$:

$$J^a(z) J^a(z') = \frac{kD}{(z-z')^2} + 2\kappa T(z') + O(z-z'), \tag{3.13}$$

where $D = \delta^{aa}$ is the dimension of the group G. Evaluating the multipoint correlation functions

$$\langle J^{a_1}(z_1) \ldots J^{a_N}(z_N) \rangle, \tag{3.14}$$

and performing the expansion (3.13) one can find that the definition (3.13) of the stress-energy tensor is consistent with (2.9) only if

$$c = \frac{kD}{c_V + k}, \tag{3.15}$$

and the parameter κ is given by (3.11). Formula (3.13) is equivalent to the following relation between the generators (2.15):

$$2\kappa L_n = \sum_{m=-\infty}^{\infty} :J^a_m J^a_{n-m}:, \tag{3.16}$$

where the symbol : : denotes the conventional normal ordering: the operators J_n with negative n are always placed to the left of the operators with $n > 0$*. Applying the eq. (3.16) with $n = -1$ to the field g one gets exactly (3.6). Note that in this way of reasoning no particular properties of the field g are specified. Therefore, the equation

$$(J^a_{-1} t^a_l - \kappa L_{-1})\phi_l = 0 \tag{3.17}$$

is satisfied with any primary field ϕ_l as well. Hence any primary field ϕ_l in the Wess-Zumino theory is degenerate and its dimensions are given by

$$\Delta_l = \frac{c_l}{c_V + k}, \qquad \bar{\Delta}_l = \frac{\bar{c}_l}{c_V + k}, \tag{3.18}$$

where $c_l = t^a_l t^a_l$, $\bar{c}_l = \bar{t}^a_l \bar{t}^a_l$. In particular, any primary field which is scalar with respect to the left and right gauge group both have vanishing dimensions and therefore it is proportional to the identity operator I. It is also worth noting that all the fields (2.19) belonging to any representation $[\phi_l]_{\mathcal{Q}}$ are expressed (by means of (3.16)) in terms of the fields

$$J^{a_1}_{-m_1} \dots J^{a_N}_{-m_N} \bar{J}^{b_1}_{-\bar{m}_1} \dots \bar{J}^{b_M}_{-\bar{m}_M} \phi_l. \tag{3.19}$$

Therefore in the Wess-Zumino theory the representations $[\phi_l]_{\mathcal{Q}}$ of the algebra $\mathcal{Q} \times \bar{\mathcal{Q}}$ are in fact the highest weight representations of the current algebra (2.18c) only; they can be denoted by $[\phi_l]_J$.

As in general conformal theory [9], the correlation functions (1.20) of the degenerate field g satisfy some linear differential equations. Consider the relation

$$t^a_i \langle J^a(z) g(z_1, \bar{z}_1) \dots g(z_N, \bar{z}_N) \rangle = \left\{ \frac{c_g}{z - z_i} + \sum_{j \neq i}^{N} \frac{t^a_i t^a_j}{z - z_j} \right\} \langle g(z_1, \bar{z}_1) \dots g(z_N, \bar{z}_N) \rangle, \tag{3.20}$$

which is a direct consequence of the Ward identity (2.13b); here the matrices t^a_i act

* The fact that the Virasoro algebra belongs to the enveloping algebra of a Kac-Moody algebra is well known by mathematicians [12]. Earlier, it had been discovered in the study of two-dimensional field theory models [10].

on the left indices of the field $g(z_i, \bar{z}_i)$. Substituting the operator product expansion (3.4) on the l.h.s. of (3.20) and taking the limit $z \to z_i$ one gets the equation

$$\left\{\kappa \frac{\partial}{\partial z_i} - \sum_{j \neq i}^{N} \frac{t_i^a t_j^a}{z_i - z_j}\right\}\langle g(z_1, \bar{z}_1) \dots g(z_N, \bar{z}_N)\rangle = 0. \qquad (3.21)$$

Since i can take any of N values, $i = 1, 2, \dots, N$, we have in fact the system of linear differential equations. The same equations with substitutions $z \to \bar{z}$, $t^a \to \bar{t}^a$ ($\bar{t}^a$ acting on g from the right) are also valid. Clearly, the equations (3.21) with obvious modifications are satisfied by any correlation function

$$\langle \phi_{l_1}(z_1, \bar{z}_1) \dots \phi_{l_N}(z_N, \bar{z}_N)\rangle \qquad (3.22)$$

of primary fields ϕ_l in the Wess-Zumino theory. The correlation functions (1.20) can be computed as the solutions of these differential equations with appropriate analytical characteristics; an example of this computation is given in the next section.

Let us consider the operator expansion of the product $g(z, \bar{z})g(0,0)$. One could expect, as in [9], the following form of this expansion:

$$g(z, \bar{z})g(0,0) = \sum_l C_{gg}^{(l)} z^{\Delta_l - 2\Delta} \bar{z}^{\bar{\Delta}_l - 2\Delta}[\phi_l(0,0) + \cdots], \qquad (3.23)$$

where $C_{gg}^{(l)}$ are numerical "structure constants" and ϕ_l denotes some primary fields of the theory having the dimensions $(\Delta_l, \bar{\Delta}_l)$. Here we omitted, in each term of the sum (3.23), the infinite power series in z and $\bar{z}$ with the fields belonging to $[\phi_l]_J$ as the coefficients. What kinds of primary fields could appear in (3.23)? The product $g(z, \bar{z})g(0,0)$ transforms as the tensor product of two irreducible representations of (for instance) the left global group G. It can be decomposed in the set of some irreducible representations:

$$g(z, \bar{z})g(0,0) = \sum_l P_l\{g(z, \bar{z})g(0,0)\}, \qquad (3.24)$$

where P_l are the projectors (acting on the left indices of the product $g \otimes g$) each selecting the subspace of l's representation. Let us substitute the expansion (3.23) for some pair of fields, say $g(z_1, \bar{z}_1)g(z_2, \bar{z}_2)$, in (3.21) and take into account the most singular (at $z_1 \to z_2$) contribution of each term of (3.23). Assuming $P_l\phi_l = \phi_l$ and using the identity

$$\{t^a g(z, \bar{z})\}\{t^a g(0,0)\} = \sum_l \tfrac{1}{2}(c_l - 2c_g)P_l\{g(z, \bar{z})g(0,0)\}, \qquad (3.25)$$

where $c_l = t_l^a t_l^a$, one gets the characteristic equation

$$-\kappa(\Delta_l - 2\Delta) = \tfrac{1}{2}(c_l - 2c_g) \tag{3.26}$$

for (3.21) and recovers (3.18). So the primary fields ϕ_l corresponding to the representations l entering the decomposition (3.24) could only appear in the operator product expansion (3.23). Surely this result was obvious before, and we presented the above computation as the self-consistency check.

Note that some of the coefficients $C^{(l)}$ in the expansions like (3.23) could vanish. In fact, there are special selection rules which forbid most of the a priori conceivable primary fields to appear in the operator algebra, generated by the fields g*. Apparently, in the general case a finite number of primary fields ϕ_l form the closed operator algebra of the Wess-Zumino model (a similar phenomenon takes place in the "minimal" conformal theories introduced in [9]). In the next section we shall meet an example ($k = 1$) of this situation.

It is worth saying some words about the composite fields of the model (1.4). Let us consider the primary field $\phi_1^{ab}(z, \bar{z})$, which transforms as adjoint representations of left and right groups G. According to (3.18) its dimensions are

$$\Delta_1 = \bar{\Delta}_1 = \frac{c_V}{c_V + k}. \tag{3.27}$$

This field is naturally identified with the composite field

$$\phi_1^{ab} = \mathrm{tr}(g^{-1} t^a g t^b). \tag{3.28}$$

The fields

$$K^a = J_{-1}^b \phi_1^{ba}, \qquad \bar{K}^a = \bar{J}_{-1}^b \phi_1^{ab}, \tag{3.29}$$

have the dimensions $(\Delta_1 + 1, \Delta_1)$ and $(\Delta_1, \Delta_1 + 1)$ respectively. They transform according to the formulae

$$\delta_{\bar{\omega}} K^a(z, \bar{z}) = f^{abc} \bar{\omega}^b(\bar{z}) K^c(z, \bar{z}),$$

$$\delta_{\omega} K^a(z, \bar{z}) = \tfrac{1}{2} k \omega^b(z) \phi_1^{ba}(z, \bar{z}). \tag{3.30}$$

Therefore they apparently coincide with the "wrong currents" of the model (1.4)

$$K^a \sim \mathrm{tr}(t^a g^{-1} \partial_z g), \qquad \bar{K}^a \sim \mathrm{tr}(t^a \partial_{\bar{z}} g g^{-1}), \tag{3.31}$$

* These selection rules have the following origin. In a general integer case of k, the highest weight representation $[g]_l$ contains null vectors. Like (3.6), these null-vectors give rise to some extra matrix equations for the correlation functions (1.20) which, contrary to (3.21), contain no derivatives. The selection rules come from the consistency condition of these matrix equations and (3.21).

Note that the "currents" (3.31) possess anomalous dimensions and therefore in conformal theory they cannot be conserved. Finally, the field

$$S(z,\bar{z})=J^{a}_{-1}\bar{J}^{b}_{-1}\phi^{ab}_{1}, \tag{3.32}$$

which has the dimensions (Δ_1+1,Δ_1+1), corresponds to the lagrangian density

$$S\sim \mathrm{tr}\left(\partial_\mu g^{-1}\partial_\mu g\right). \tag{3.33}$$

This enables one to predict the slope of the β-function of the model (1.1) at the fixed point (1.3):

$$\left.\frac{\mathrm{d}\beta(\lambda^2,k)}{\mathrm{d}\lambda^2}\right|_{\lambda^2=4\pi/k}=\frac{2c_V}{c_V+k}. \tag{3.34}$$

At $k\to\infty$ this equation is in agreement with the one-loop result [2].

4. Correlation functions

The projective Ward identities (2.14a) determine the two- and three-point correlation functions up to a numerical factor. Here we shall compute the four-point functions

$$G(z_i,\bar{z}_i)=\langle g(z_1,\bar{z}_1)g^{-1}(z_2,\bar{z}_2)g^{-1}(z_3,\bar{z}_3)g(z_4,\bar{z}_4)\rangle \tag{4.1}$$

for the Wess-Zumino model, combining the differential equations (3.21) with the general requirement of crossing symmetry. In fact, the crossing symmetry of four-point functions ensures the associativity of the complete operator algebra* and therefore the self-consistency of the field theory in general.

Firstly, let us note that the correlation function (4.1) depends essentially on two variables; due to the Ward identities (2.14a) it can be represented in the form

$$G(z_i,\bar{z}_i)=\left[(z_1-z_4)(z_2-z_3)(\bar{z}_1-\bar{z}_4)(\bar{z}_2-\bar{z}_3)\right]^{-2\Delta}G(x,\bar{x}), \tag{4.2}$$

where x and $\bar{x}$ are the anharmonic quotients

$$x=\frac{(z_1-z_2)(z_3-z_4)}{(z_1-z_4)(z_3-z_2)},\qquad \bar{x}=\frac{(\bar{z}_1-\bar{z}_4)(\bar{z}_3-\bar{z}_4)}{(\bar{z}_1-\bar{z}_4)(\bar{z}_3-\bar{z}_2)}, \tag{4.3}$$

and Δ is the dimension (3.10) of the field g. Further computation depends on the choice of the group G. Here we elaborate the case of unitary groups.

* See ref. [9] for the details. The bootstrap approach based on the operator algebra was originally proposed by Polyakov [14].

Let $G = SU(N)$, $g(z, \bar{z})$ being the $N \times N$ unitary matrix, $\det g = 1$, which transforms as the fundamental representation of $SU(N) \times SU(N)$. Let us denote by $\alpha_i(\beta_i)$ the tensor indices of the fields $g(z_i, \bar{z}_i)$ corresponding to the left (right) $SU(N)$ group; this way, in (4.1) we imply $g(z_i, \bar{z}_i) = g_{\alpha_i}^{\beta_i}(z_i, \bar{z}_i)$ and $g^{-1}(z_i, \bar{z}_i) = g_{\beta_i}^{-1\alpha_i}(z_i, \bar{z}_i)$. The correlation function (4.2) enjoys the $SU(N) \times SU(N)$ invariant decomposition

$$G(x, \bar{x}) = \sum_{A, B=1,2} (I_A)(\bar{I}_B) G_{AB}(x, \bar{x}), \tag{4.4}$$

with the scalar coefficients $G_{AB}(x, \bar{x})$. The matrices I and $\bar{I}$ are defined as

$$\begin{aligned} I_1 &= \delta_{\alpha_1}^{\alpha_2} \delta_{\alpha_3}^{\alpha_4}, & \bar{I}_1 &= \delta_{\beta_2}^{\beta_1} \delta_{\beta_3}^{\beta_4}, \\ I_2 &= \delta_{\alpha_1}^{\alpha_4} \delta_{\alpha_3}^{\alpha_2}, & \bar{I}_2 &= \delta_{\beta_2}^{\beta_4} \delta_{\beta_3}^{\beta_1}. \end{aligned} \tag{4.5}$$

The correlation function (4.1) satisfies the differential equations (3.21) and the same equations with respect to $\bar{z}$. By means of direct computation these equations can be converted to the form*

$$\begin{aligned} \frac{\partial G}{\partial x} &= \left[\frac{1}{x} P + \frac{1}{x-1} Q \right] G, \\ \frac{\partial G}{\partial \bar{x}} &= G \left[\frac{1}{\bar{x}} P^t + \frac{1}{\bar{x}-1} Q^t \right], \end{aligned} \tag{4.6}$$

where G denotes 2×2 matrix G_{AB}, the matrices P and Q are given by

$$P = \frac{1}{2N\kappa} \begin{pmatrix} N^2 - 1 & N \\ 0 & -1 \end{pmatrix}, \qquad Q = \frac{1}{2N\kappa} \begin{pmatrix} -1 & 0 \\ N & N^2 - 1 \end{pmatrix}, \tag{4.7}$$

and the mark t means the matrix transposition. The parameter κ, the same as in (3.11), in the case under consideration is equal to

$$\kappa = -\tfrac{1}{2}(N + k). \tag{4.8}$$

The general solution of eqs. (4.6) can be given in terms of hypergeometric functions; it is conveniently represented in the form

$$G_{AB}(x, \bar{x}) = \sum_{p, q=0,1} U_{pq} \mathcal{F}_A^{(p)}(x) \mathcal{F}_B^{(q)}(\bar{x}), \tag{4.9}$$

* Note that precisely eqs. (4.6) appeared earlier in the paper by Dashen and Frishman [10], in their study of the conformally invariant solution of the $SU(N)$ Thirring model.

with arbitrary constants U_{pq} and the functions $\mathcal{F}$ given by

$$\mathcal{F}_1^{(0)}(x)=x^{-2\Delta}(1-x)^{\Delta_1-2\Delta}F\left(-\frac{1}{2\kappa},\frac{1}{2\kappa},1+\frac{N}{2\kappa},x\right),$$

$$\mathcal{F}_2^{(0)}(x)=-(2\kappa+N)^{-1}x^{1-2\Delta}(1-x)^{\Delta_1-2\Delta}F\left(1-\frac{1}{2\kappa},1+\frac{1}{2\kappa},2+\frac{N}{2\kappa},x\right), \tag{4.10a}$$

$$\mathcal{F}_1^{(1)}(x)=x^{\Delta_1-2\Delta}(1-x)^{\Delta_1-2\Delta}F\left(-\frac{N-1}{2\kappa},-\frac{N+1}{2\kappa},1-\frac{N}{2\kappa},x\right),$$

$$\mathcal{F}_2^{(1)}(x)=-Nx^{\Delta_1-2\Delta}(1-x)^{\Delta_1-2\Delta}F\left(-\frac{N-1}{2\kappa},-\frac{N+1}{2\kappa},-\frac{N}{2\kappa},x\right). \tag{4.10b}$$

Here Δ is the dimension (3.10) of the field g:

$$\Delta=\frac{N^2-1}{2N(N+k)}, \tag{4.11}$$

whereas

$$\Delta_1=\frac{N}{N+k} \tag{4.12}$$

is the dimension of the composite field (3.28). Note that the functions $\mathcal{F}_A^{(p)}(x)$ play here a role similar to the "conformal blocks" in conformal theory [9]; the functions $\mathcal{F}^{(0)}$ and $\mathcal{F}^{(1)}$ in (4.9) describe the "s-channel" contribution of all the fields belonging to the representations $[I]_J$ and $[\phi_1^{ab}]_J$, respectively. It is not out of place to name them the current blocks.

Now we have to take into account the local properties of the field g and impose the crossing symmetry (4.2); this requirement proves to determine the constants U_{pq} in (4.9). According to the local properties, the correlation functions must be single-valued while considered in the euclidian domain $\bar{x}=x^*$, where the star denotes the complex conjugation. Concentrating attention in the vicinity of the point $x=\bar{x}=0$ one immediately recognizes that (4.9) is compatible with this requirement only if

$$U_{10}=U_{01}=0. \tag{4.13}$$

The crossing symmetry of the four-point function (4.1) requires

$$G_{AB}(x,\bar{x})=\sum_{A',B'=1,2}E_{AA'}G_{A'B'}(1-x,1-\bar{x})E_{B'B}, \tag{4.14}$$

where $E_{12}=E_{21}=1$, $E_{11}=E_{22}=0$. Let us substitute (4.9) into (4.14) and use the relation

$$\mathcal{F}_A^{(p)}(x) = \sum_{q,A'} C_q^p E_{AA'} \mathcal{F}_{A'}^{(q)}(1-x), \tag{4.15}$$

which can be verified directly. The elements of the "crossing matrix" C_q^p in (4.15) are given by

$$C_0^0 = -C_1^1 = N\frac{\Gamma(-N/2\kappa)\Gamma(N/2\kappa)}{\Gamma(-1/2\kappa)\Gamma(1/2\kappa)},$$

$$C_0^1 = -N\frac{\Gamma^2(-N/2\kappa)}{\Gamma(-(N-1)/2\kappa)\Gamma(-(N+1)/2\kappa)}, \qquad C_0^1 C_1^0 + C_0^0 C_1^1 = 1. \tag{4.16}$$

Eqs. (4.14) are satisfied provided $U_{11} = hU_{00}$:

$$h = \frac{1}{N^2}\frac{\Gamma((N-1)/(N+k))\Gamma((N+1)/(N+k))}{\Gamma((k+1)/(N+k))\Gamma((k-1)/(N+k))}\frac{\Gamma^2(k/(N+k))}{\Gamma^2(N/(N+k))}. \tag{4.17}$$

Finally we obtain

$$G_{AB}(x,\bar{x}) = M^{-8\Delta}\{\mathcal{F}_A^{(0)}(x)\mathcal{F}_B^{(0)}(\bar{x}) + h\mathcal{F}_A^{(1)}(x)\mathcal{F}_B^{(1)}(\bar{x})\}, \tag{4.18}$$

where the overall factor is the matter of the g-field normalization; the normalization in (4.18) corresponds to the two-point function

$$\langle g_{\alpha_1}^{\beta_1}(z,\bar{z}) g_{\beta_2}^{-1\alpha_1}(0,0)\rangle = M^{-4\Delta}\delta_{\alpha_1}^{\alpha_2}\delta_{\beta_1}^{\beta_2}(z\bar{z})^{-2\Delta}. \tag{4.19}$$

In the general case the function (4.18) possesses the power-like singularities at $x=\infty$, $\bar{x}=\infty$, which correspond to the contributions of the composite fields

$$\phi_A = g_{\{\alpha_1}^{\{\beta_1} g_{\alpha_2\}}^{\beta_2\}}(z,\bar{z}), \tag{4.20a}$$

$$\phi_S = g_{[\alpha_1}^{[\beta_1} g_{\alpha_2]}^{\beta_2]}(z,\bar{z}), \tag{4.20b}$$

in the operator product expansion of $g(z_1,\bar{z}_1)g(z_2,\bar{z}_2)$ in (1.1). Here the braces (square brackets) denote the antisymmetrization (symmetrization). The dimensions of these fields are

$$\Delta_A = \frac{(N-2)(N+1)}{N(N+k)}, \qquad \Delta_S = \frac{(N+2)(N-1)}{N(N+k)}, \tag{4.21}$$

in agreement with (3.18).

Note that at $k=1$ the second term in (4.18) vanishes and the function (4.4) becomes*

$$G(x,\bar{x})=[x\bar{x}(1-x)(1-\bar{x})]^{1/N}$$
$$\times\left[(I_1)\frac{1}{x}+(I_2)\frac{1}{1-x}\right]\left[(\bar{I}_1)\frac{1}{\bar{x}}+(\bar{I}_2)\frac{1}{1-\bar{x}}\right]. \qquad (4.22)$$

Obviously, this result has the following meaning. Let us consider the field

$$\tilde{g}(z,\bar{z})=e^{i\sqrt{4\pi/N}\gamma\varphi(z,\bar{z})}g(z,\bar{z}), \qquad (4.23)$$

where g is the above SU(N) Wess-Zumino field and φ is the free massless boson field so that (4.23) corresponds to the U(N) = SU(N) × U(1) group. In fact, the parameter γ can be chosen arbitrarily and the dimension of the field $\tilde{g}$ is

$$\tilde{\Delta}=\Delta+\Delta(\gamma), \qquad (4.24)$$

where

$$\Delta(\gamma)=\gamma^2/2N. \qquad (4.25)$$

The four-point function $\tilde{G}(z_i,\bar{z}_i)$ of the fields (4.23) is given by

$$\tilde{G}(x,\bar{x})=[x\bar{x}(1-x)(1-\bar{x})]^{-2\Delta(\gamma)}G(x,\bar{x})M^{-4\Delta(\gamma)}, \qquad (4.26)$$

and under the choice

$$\gamma=1 \qquad (4.27)$$

coincides with that of the bilinears of the free massless charged Fermi fields:

$$M\tilde{g}_{\alpha}^{\beta}(z,\bar{z})=:\psi_{\alpha}(z)\bar{\psi}^{+\beta}(\bar{z}):,$$
$$M\tilde{g}_{\beta}^{-1\alpha}(z,\bar{z})=:\bar{\psi}_{\beta}(\bar{z})\psi^{+\alpha}(z):, \qquad (4.28)$$

governed by the action

$$S_f(\psi,\bar{\psi})=\int\left[\psi^{+\alpha}\partial_{\bar{z}}\psi_{\alpha}+\bar{\psi}^{+\beta}\partial_{z}\bar{\psi}_{\beta}\right]d^2\xi. \qquad (4.29)$$

* In this case the operators (3.28) and (4.20b) decouple from the operator algebra generated by the field g; this is the simplest example of the selection rules mentioned in the previous section.

Clearly, this result holds for any multipoint correlation functions (1.20). In fact, there is no need in computing the multipoint functions to prove this. It is sufficient to verify that the stress-energy tensor

$$T_f(z) = \sum_{\alpha=1}^{N} :\psi^{+\alpha}\partial_z\psi_\alpha: \tag{4.30}$$

is related to the fermion currents $J^a = :\psi^+ t^a\psi:$ and $J = :\psi^+\psi:$ as [10]

$$-T_f(z) = \frac{1}{N+1}:J^a(z)J^a(z): + \frac{1}{2N}:J(z)J(z):, \tag{4.31}$$

and to note that the singlet current J is expressed in terms of the free massless boson field

$$J(z) = i\sqrt{N}\,\partial_z\varphi(z,\bar{z}). \tag{4.32}$$

Note also that (4.28) remains valid at arbitrary values of γ in (4.23) provided the ψ's are understood as the fermions of the N-component Thirring model with the isoscalar current coupling

$$S_f^{(\gamma)}(\psi,\bar{\psi}) = S_f(\psi,\bar{\psi}) + \frac{\gamma^2-1}{2\gamma^2}\int J(z)\bar{J}(\bar{z})\,d^2\xi. \tag{4.33}$$

Clearly, the relation (1.17) corresponding to the model (1.4) with $G = O(N)$ and $k = 1$ can be proved in the same way.

We thank A.A. Belavin, A.M. Polyakov and P.B. Wiegmann for many interesting comments. V.G.K. is obliged to A.Yu. Morozov for helpful discussions.

References

[1] S.P. Novikov, Usp. Mat. Nauk 37 (1982) 3
[2] E. Witten, Comm. Math. Phys. 92 (1984) 455
[3] A.M. Polyakov and P.B. Wiegmann, Phys. Lett. B131 (1983) 121
[4] A.M. Polyakov and P.B. Wiegmann, Phys. Lett. B (1984), to be published
[5] P.B. Wiegmann, Pisma ZhETF 39 (1984) 180;
E.I. Ogievetsky, N.Yu. Reshetikhind and P.B. Wiegmann, Nucl. Phys. B, to be published
[6] L.D. Faddeev and N.Yu. Reshetikhind, in Problems of nonlinear and nonlocal quantum field theory (Dubna, 1984)
[7] A.M. Polyakov, Phys. Lett. 103B (1981) 207
[8] P. Di Vecchia and P. Rossi, preprint TH-3808 CERN (1984)
[9] A.A. Belavin, A.M. Polyakov and A.B. Zamolodchikov, Nucl. Phys. B241 (1984) 333
[10] R. Dashen and Y. Frishman, Phys. Rev. D11 (1975) 2781
[11] D. Friedan, Z. Qui and S. Shenker, Chicago preprint EFI-83-66 (1983)
[12] V.G. Kac, Infinite dimensional Lie algebras, Progress in mathematics, vol. 44 (Birkhäuser, 1984)
[13] A.M. Polyakov, ZhETF 66 (1974) 23

Volume 160B, number 1,2,3 PHYSICS LETTERS 3 October 1985

SYMMETRIC SPACES, SUGAWARA'S ENERGY MOMENTUM TENSOR IN TWO DIMENSIONS AND FREE FERMIONS

P. GODDARD
DAMTP, Silver Street, Cambridge CB3 9EW, UK

W. NAHM
Physikalisches Institut der Universität, D53 Bonn 1, West Germany

and

D. OLIVE
Blackett Laboratory, Imperial College, Prince Consort Road, London SW7 2BZ, UK

Received 14 June 1985

It is shown that Sugawara's energy momentum tensor, bilinear in fermionic currents associated with a group G, equals the energy momentum tensor for free fermions if there exists a symmetric space G′/G with the symmetric space generators transforming under G as the fermions do. This result provides a list of chiral field theories with Wess–Zumino term that are equivalent to free fermion theories and specifies which representations of Kac–Moody algebras bilinear in fermions are finitely reducible.

The theory of Kac–Moody algebras [1,2] and their associated Virasoro algebras provides a precise mathematical formulation of certain conformally invariant quantum field theories in two spacetime dimensions [3–5] provided one pays the price (hopefully temporary) of considering fields periodic in space. The theory of free, massless fermions provides an example as does a chiral model in which the field is confined to the manifold of a Lie group G with algebra g and in which a Wess–Zumino term is included with the kinetic energy term appropriately normalised with respect to it [6]. The latter model has two variable parameters, the choice of the Lie group, G, and the value, x, of the coefficient of the Wess–Zumino term which has to be an integer for topological reasons. We shall call x the "level" and take it to be greater than or equal to unity. Because of a coincidence of the algebraic structure Witten has argued that the level 1 SO(N) model is equivalent to a theory with N real free massless fermions [6]. Similarly a level 1 U(N) chiral model could be equivalent to a theory with $2N$ real free fermions [3]. Both theories, fermionic and chiral, possess energy momentum tensors $\Theta^{\mu\nu}$ of the Sugawara form [7], by which we mean traceless and bilinear in conserved currents $j^{\mu}{}_{a}$ (whose dual currents $\epsilon^{\nu}{}_{\mu}$ $j^{\mu}{}_{a}$ are also conserved). The algebraic coincidence in the above examples is that the fermionic $\Theta^{\mu\nu}$ equals that for the constituent free fermions.

In this note, we show that a necessary and sufficient condition for this algebraic coincidence is that there exists a (compact) group $G' \supset G$ such that G'/G is a symmetric space with the fermions transforming under G just as the tangent space to G'/G does. Thus, using Cartan's classification of symmetric spaces [8], we can enumerate all the possible cases. The two examples above correspond to the sphere $S^N \cong SO(N+1)/SO(N)$ and the complex projective space $CP(N) \cong SU(N+1)/U(N)$, respectively. We can predict that many more chiral models are integrable in the sense of

0370-2693/85/$ 03.30 © Elsevier Science Publishers B.V.
(North-Holland Physics Publishing Division)

being equivalent to free fermion theories, for example the SO(N) and U(N) models with levels ($N-2$) and ($N+2$) as well as 1, corresponding to the symmetric spaces SO(N)×SO(N)/SO(N), SU(N)/SO(N), SO($2N$)/U(N) and Sp(N)/U(N) respectively. Our result also establishes a precise condition for the finite reducibility of the representations of Kac–Moody algebras bilinear in fermions.

The algebraic structure of $\Theta^{\mu\nu}$ and j^{μ}_{a}, in the two models coincides, and the two light cone components of j^{μ}_{a}, namely j^{+}_{a} and j^{-}_{a} commute with one another. Thus the structure consists of two mutually commuting copies of a semidirect product of a Kac–Moody algebra $\hat{g}$ (based on g), constructed from j^{+}_{a} with a Virasoro algebra constructed from Θ^{++}.

The generators T^{i}_{m} of $\hat{g}$ satisfy:

$$[T^{i}_{m}, T^{j}_{n}] = i f^{ijk} T^{k}_{m+n} + km\delta^{ij}\delta_{m+n,0}, \tag{1}$$

where f^{ijk} are the totally antisymmetric structure constants of the compact algebra g whose generators are T^{i}_{0}. k is the central term (the c-number coefficient of the Schwinger term). Assuming for the time being that g is simple, and therefore has a unique highest root ψ, we know that in a highest weight representation of $\hat{g}$ (i.e. one that can be built up from a vacuum by the action of step operators) that $2k/\psi^2$ has to be a positive integer which we shall also call x. Mathematicians call x the level of the representation [1,2]. In the chiral model Witten showed that this x indeed equals the coefficient of the Wess–Zumino term, at least for the classical. Poisson bracket analogue of (1) [6]. The energy momentum tensor components correspond to the Virasoro generators which can therefore be constructed from bilinears in the T^{i}_{m} by Sugawara's construction [7]:

$$\mathscr{L}(z) = \sum_{m\in Z} z^{-m}\mathscr{L}_m = \frac{1}{2\beta}\sum_{i=1}^{\dim g} {}^{\times}_{\times} T^{i}(z)T^{i}(z) {}^{\times}_{\times}, \tag{2}$$

where

$$\sum_{m\in Z} z^{-m} T^{i}_{m} = T^{i}(z). \tag{3}$$

The crosses denote normal-ordering of the T^{i}_{m} whereby quantities with positive suffixes are moved to the right of quantities with negative suffixes to ensure that $\mathscr{L}_m$ has finite matrix elements in the highest weight representations under consideration. The renormalisation factor β is given by [9,3–5]

$$2\beta = 2k + c_\psi = \psi^2(x + c_\psi/\psi^2) = \psi^2(x + \tilde{h}), \tag{4}$$

where c_ψ is the quadratic Casimir in the adjoint representation of g,

$$\delta_{ij} c_\psi = \sum_{m,n=1}^{\dim g} f_{imn} f_{jmn}, \tag{5}$$

and $c\psi/\psi^2$ is an integer $\tilde{h}$ called the dual Coxeter number of g. The choice of β given by eq. (4) ensures the conformal invariance of the theory by virtue of the relations

$$[\mathscr{L}_m, T^{i}_{n}] = -n T^{i}_{m+n} \tag{6}$$

and the fact that $\mathscr{L}_m$ satisfies the Virasoro algebra

$$[\mathscr{L}_m, \mathscr{L}_n] = (m-n)\mathscr{L}_{m+n} + \tfrac{1}{12}cm(m^2-1)\delta_{m+n,0} \tag{7}$$

with the central term c in eq. (7) given in terms of k in eq. (1) as the rational number

$$c = 2k \dim g/(2k + c_\psi) = x \dim g/(x + \tilde{h}). \tag{8}$$

Eqs. (1), (2), (6) and (7) encapsulate the algebraic structure consisting of a semidirect product of the Kac–Moody algebra with a Virasoro algebra. Witten has shown that this structure applies to the chiral model described for any G and non-zero integral x, at least classically, and conjectured that it can be extended to quantum mechanics [6].

We now investigate whether the same algebraic structure can be realised quantum mechanically in terms of free massless fermions. Following [4] we construct

$$T^{i}(z) = \tfrac{1}{2} i H^{\alpha}(z) M^{i}_{\alpha\beta} H^{\beta}(z), \tag{9}$$

where M^{i} are real matrices satisfying

$$[M^{i}, M^{j}] = f^{ijk} M^{k} \tag{10}$$

and generate a real, irreducible representation of g. Then the T^{i}_{m} given by (9) and (3) satisfy the Kac–Moody algebra (1) with the level x given by

$$x = 2k/\psi^2 = \kappa/\psi^2 ; \ \delta^{ij}\kappa = -\mathrm{Tr}(M^{i}M^{j}). \tag{11}$$

remembering g is, for the time being, supposed simple. Given that the representation is real, x is indeed an integer, called the Dynkin index. We take the Virasoro algebra to be given by the Sugawara construction (2) in order that we can assert that the algebraic structure of the fermion theory coincides with that of the chiral theory. However there is another Virasoro algebra that we can construct with dim M free real fermion fields $H^a(z)$, namely

$$L(z)=\sum_{m\in\mathbf{Z}} z^{-m}L_m = z\,{\circ\atop\circ} \sum_{a=1}^{\dim M} \frac{\mathrm{d}H^a}{\mathrm{d}z} H^a {\circ\atop\circ} + \varepsilon \dim M, \qquad (12)$$

where open dots indicate normal ordering with respect to the fermion operators and ε equals 0 or 1/16 according as the fermion fields are of Neveu–Schwarz or Ramond type respectively. Expression (12) corresponds to the energy momentum tensor for dim M free, massless, real fermions and hence yields c-number

$$c=\tfrac{1}{2}\dim M. \qquad (13)$$

In order to assert that the fermion theory is indeed a free, massless theory we must take (12) to give its energy momentum tensor. Hence we want the two alternative constructions of the Virasoro algebra for the fermion theory, namely (12) and Sugawara's construction (2) with (9), to coincide, and our task is to enumerate for what choices of g and M this coincidence occurs.

In general L_m, (12), and $\mathscr{L}_m$, (2), are unequal since their c-numbers are, and their difference, K_m, constitutes a third Virasoro algebra commuting with the Kac–Moody generators $T^i{}_m$ and hence by (2) with $\mathscr{L}_m$ [4]:

$$L_m=\mathscr{L}_m+K_m, \qquad (14)$$

where

$$[K_m, T^i_n]=0, \quad [K_m, \mathscr{L}_n]=0. \qquad (15)$$

The c-number for the K_m Virasoro algebra equals the difference of those for L_m and $\mathscr{L}_m$ i.e. by eqs. (8) and (13)

$$c_k=\tfrac{1}{2}\dim M - x\dim g/(x+\tilde{h}) \geqslant 0 \qquad (16)$$

and has to be non-negative as the fermionic Fock space must decompose into highest weight representations of L_m. Further c_k vanishes if and only if K_m vanishes. Thus, irrespective of whether Neveu–Schwarz or Ramond fermions are considered the vanishing of c_k is the precise numerical condition that the Sugawara and free energy momentum tensors coincide and has been used to obtain many new examples of the phenomenon [4]. It was also shown by Goddard and Olive [4] that an alternative way of writing this condition is

$$\sum_{i=1}^{\dim g} \left(M^i_{\alpha\beta}M^i_{\gamma\delta} + M^i_{\beta\gamma}M^i_{\alpha\delta} + M^i_{\gamma\alpha}M^i_{\beta\delta} \right) = 0. \qquad (17)$$

The observation that this resembles a Jacobi or Bianchi identity is the key to recognizing the classification of its solutions in terms of symmetric spaces. In fact (17) is the Bianchi identity for the Riemann tensor of the symmetric space we are about to construct.

Our main result is that the equality of the Sugawara (2) and free fermion (12) Virasoro generators occurs if and only if there exists a group G′ containing G such that G′/G is a symmetric space and the fermions transform with respect to G just as the symmetric space generators do. This result holds even if G is not simple, although eqs. (2), (10) and (13), amongst others, have to be modified accordingly, as we see later.

Suppose we are given the symmetric space G′/G and hence can decompose the algebra g′

$$g' = g \oplus p, \qquad (18)$$

with g and p denoting the even and odd parts of g′. We already have natural orthonormal bases in g and p separately so that ($T^i = T^i{}_0$)

$$[T^i, T^j] = \mathrm{i} f^{ijk} T^k, \quad [T^i, p^\alpha] = \mathrm{i}\, p^\beta M^i_{\beta\alpha}. \qquad (19)$$

All Jacobi identities are automatically satisfied because the p^α transform according to (19) just as the fermion fields $H^a(z)$ do. Because we have a symmetric space the commutator of p^α and p^β must involve T^i only with no p's occurring:

$$[p^\alpha, p^\beta] = \mathrm{i}\, X_k^{\alpha\beta} T_k .$$

We are free to scale the p^α so that

$$\mathrm{Tr}(T^iT^j)=y\delta^{ij}, \quad \mathrm{Tr}(p^\alpha p^\beta)=y\delta^{\alpha\beta}. \qquad (20)$$

Then we can evaluate the structure constants $X_k^{\alpha\beta}$

Volume 160B, number 1,2,3 PHYSICS LETTERS 3 October 1985

as

$$\mathrm{i}\,yX_k^{\alpha\beta} = \mathrm{Tr}\left(T^k\left[p^\alpha, p^\beta\right]\right) = \mathrm{Tr}\left(\left[T^k, p^\alpha\right] p^\beta\right) = \mathrm{i}\,yM^k_{\beta\alpha}.$$

Hence

$$\left[p^\alpha, p^\beta\right] = -\mathrm{i}M^k_{\alpha\beta}T^k. \qquad (21)$$

The Jacobi identity for three p's immediately yields condition (17). Conversely, given (17), we can consistently add eq. (21) to (19) thereby constructing a symmetric space as described.

Notice that when g and g′ are both simple and we have an orthonormal basis (20) for g′ we can evaluate y in two different ways using the adjoint representation of g′.

$$y\delta^{ij} = \mathrm{Tr}(T^iT^j) = f^{ikm}f^{jkm} + M^i_{\alpha\beta}M^j_{\alpha\beta} = \delta^{ij}(c_\psi + \kappa),$$

$$y\delta^{\alpha\beta} = \mathrm{Tr}(p^\alpha p^\beta) = -2M^i_{\alpha\gamma}M^i_{\gamma\beta} = \delta_{\alpha\beta}\frac{2\kappa \dim \mathrm{g}}{\dim M},$$

The agreement of these two calculations yields the vanishing of c_k in (16) which was the alternative condition for the vanishing of K_n.

Let us illustrate our theorem with some of the examples previously found [4] of equality between Sugawara and free energy momentum tensors. For fermions in the adjoint representation of any simple algebra g the level is the dual Coxeter number $\tilde{h}$ of g and the relevant symmetric space G × G/G corresponding to the type II of Cartan's classification [8]. The remaining compact symmetric spaces are called type I and have G′ simple. They are listed in table 1. If G = SO(N), for example, and the fermions belong to the N (defining), $(N\otimes N)_a$ (adjoint) or $(N\otimes N)_s$ (symmetric tensor) representations the corresponding levels are 1, $N-2$ and $N+2$ respectively, and the symmetric spaces SO($N+1$)/SO(N), SO(N) × SO(N)/SO(N) and SU(N)/SO(N). The table yields one other possibility of G = SO(N) previously overlooked, namely fermions in the spinor (128) of SO(16) corresponding to the symmetric space E_8/D_8.

It will be seen from the table that g need not be simple nor even semisimple. Indeed we mentioned G = U(N) at the beginning. To complete our analysis we must explain how our previous

Table 1
Type I symmetric spaces.

Name	g′	g	dim M = dim p	Level
AI	SU(N), $N \geqslant 4$	SO(N)	$\frac{1}{2}(N+2)(N-1)$	$N+2$
	SU(3)	SO(3)	5	10
AII	SU($2N$)	SP(N)	$(2N+1)(N-1)$	$N-1$
AIII	SU($M+N$)	SU(M) × SU(N) × U(1)	$(M, N, 2)$	$(N, M, -)$
BDI	SO($M+N$)	SO(M) × SO(N)	(M, N)	(N, M)
CI	SP(N)	U(N)	$(\frac{1}{2}N(N+1), 2)$	$(N+2, -)$
CII	SP($M+N$)	SP(M) × SP(N)	$(2M, 2N)$	(N, M)
DIII	SO($2N$)	U(N)	$(\frac{1}{2}N(N-1), 2)$	$(N-2, -)$
EI	E_6	C_4	42	7
EII	E_6	$A_5 + A_1$	(20, 2)	(6, 10)
EIII	E_6	$D_5 + U_1$	(16, 2)	(4, −)
EIV	E_6	F_4	26	3
EV	E_7	A_7	70	10
EVI	E_7	$D_6 + A_1$	(32, 2)	(8, 16)
EVII	E_7	$E_6 + U_1$	(27, 2)	(6, −)
EVIII	E_8	D_8	128	16
EIX	E_8	$E_7 + A_1$	(56, 2)	(12, 28)
FI	F_4	$C_3 + A_1$	(14, 2)	(5, 7)
FII	F_4	B_4	16	2
G	G_2	$A_1 + A_1$	(4, 2)	(10, 2)

Volume 160B, number 1,2,3 PHYSICS LETTERS 3 October 1985

formulae are modified if the compact Lie algebra g is not simple but composed out of factors $g_1 \oplus g_2 \oplus \cdots \oplus g_N$. Then we can construct mutually commuting Virasoro generators of the Sugawara type (2) for each g_j. The condition that their sum equals the free Virasoro algebra (12) can be alternatively expressed as

$$\sum_{i=1}^{\dim g} \frac{1}{\kappa + c_\psi}\left(M^i_{\alpha\beta}M^i_{\gamma\delta} + M^i_{\beta\gamma}M^i_{\alpha\delta} + M^i_{\gamma\alpha}M^i_{\beta\delta}\right) = 0, \tag{22}$$

or

$$\sum_{j=1}^{N} x_j \dim g_j/(x_j + \tilde{h}_j) = \tfrac{1}{2}\dim M. \tag{23}$$

Notice that in eq. (22) the index i is summed over all dim g values with the understanding that the κ_i and c_{ψ_i} take the value appropriate to the factor g_j of g to which the label i refers.

Instead of the normalisation (20) we can only now expect to have

$$\mathrm{Tr}(T^iT^j) = y^i\delta^{ij}, \quad \mathrm{Tr}(p^\alpha p^\beta) = y\delta^{\alpha\beta}, \tag{24}$$

where y^i takes the same value for all labels i referring to a single factor of g. As a consequence, if we can form a symmetric space G′/G eq. (21) is replaced by

$$[p^\alpha, p^\beta] = -\mathrm{i}(y/y^k)M^k_{\alpha\beta}T_k, \tag{25}$$

so that the Jacobi identity (17) for three p's now reads instead as

$$\sum_{i=1}^{\dim g} (1/y^i)\left(M^i_{\alpha\beta}M^i_{\gamma\delta} + M^i_{\beta\gamma}M^i_{\alpha\delta} + M^i_{\gamma\alpha}M^i_{\beta\delta}\right) = 0. \tag{26}$$

Using the adjoint representation g′ to evaluate the trace in (24) we find $y^i = c_{\psi_i} + \kappa_i$. Thus condition (22) is indeed guaranteed and the Sugawara and free energy momentum tensors again coincide. Conversely if eq. (26) is satisfied we can consistently construct the commutation relations (25) and hence the symmetric space.

We now turn to discussing the table listing the type I symmetric spaces [8]. The first three columns specify respectively the mathematical label of the symmetric space, g′ and g (choosing whichever is most convenient between the Lie group or algebra nomenclature). In column four we have calculated the dimensionality of the representation of g carried by the fermions or equivalently the p^α. Thus in BDI we have an MN component real fermion field transforming as an M of SO(M) and an N of SO(N). In column five we have calculated the levels. Thus for BDI the SO(M) Kac–Moody algebra has level N since it is constructed with N copies of the M multiplet (which alone has level 1). The corresponding SO(M)×SO(N) chiral model with the same algebraic structure has an SO(M) Wess–Zumino term with coefficient N plus an SO(N) Wess–Zumino term with coefficient M. $M = 1$ is a degenerate case discussed by Witten [6]. In EVII, for example, the fermions form an E_6 complex 27 or a real 54 which forms 27 U(1) doublets. There is no concept of U(1) level and indeed as c_ψ vanishes for U(1), x cancels out of equations like (23). Correspondingly the U(1) Wess–Zumino term vanishes.

In making up this table we noted simple Dynkin diagram rules for identifying symmetric spaces and the fermion representation. We now list the dual Coxeter numbers $\tilde{h}$ so that the reader has sufficient data to check eq. (23) A_n: $n+1$, C_n: $n+1$, E_6: 12, E_7: 18, E_8: 30, F_4: 9, G_2: 4 and SO(N): $N-2$.

Our result answers the important mathematical question as to when the "quark model" representation (9) of the Kac–Moody algebra (1) is finitely reducible (i.e. decomposes into a finite number of irreducible highest weight representations). Provided g is semisimple it is easy to see [4] that the representation (9) is finitely reducible if and only if K_n, (14), vanishes and hence, by our result, if and only if our criterion for the construction of a symmetric space G′/G from g and M is satisfied. As the Virasoro algebra K_n has only infinite dimensional representations (when K_n is non-zero), the action of the K_n on any highest weight state of g automatically yields an infinite number of degenerate ones. By general theory [1,2] (when g is semisimple) there are only a finite number of possible highest weights Λ corresponding to the level (9) and by (2) these yield $\mathscr{L}_0$ eigenvalues $\Lambda(\Lambda + 2\delta)/2\beta$ (if g is simple and with a trivial

Volume 160B, number 1,2,3 PHYSICS LETTERS 3 October 1985

modification if g is semisimple). But if K_n vanishes $\mathscr{L}_0$ equals L_0 (12) whose eigenspectrum is known to be either $Z_+/2$ (N–S) or $Z_+ + \dim M/16$ (R), with finite multiplicity. The finite reducibility then follows if $K_n = 0$ and g is semisimple.

Thus we have generalised a result of Kac and Peterson [10] who verified finite reducibility of what they called "spin representations", when g and g′ have equal ranks, g is semisimple and Ramond fields are used. Their method used characters rather than Virasoro algebras but when applicable gave the valuable extra information that the irreducible representations occurred with unit (or zero) multiplicity.

We should like to argue, following Witten [6] that those chiral models whose algebraic structure coincides with that of a free fermion theory is indeed quantum equivalent to it. Then table 1 would give a list of those chiral models (labelled by group G and level) whose currents equalled ones bilinear in free fermions and whose energy momentum tensor equalled the free fermion one. Unfortunately this argument is incomplete (even in the case considered by Witten) because the space of states inevitably has more than one irreducible component with respect to $\hat{g}$, though it obviously helps that the number is finite as we have just seen.

We close with two comments: (1) as yet we have found no physical interpretation or definite role for the symmetric space of our construction; (2) we have concentrated on Kac–Moody currents bilinear in free fermions yet we know one case, E_8 level 1, in which a transcendental construction exists [11] involving 16 real free fermions. The level 1 E_8 Kac–Moody algebra has a unique highest weight representation [1,2] so that we can confidently assert that the level 1 E_8 chiral model is quantum equivalent to a theory of 16 real free fermions.

We should like to thank B. Julia and C. Thorn for discussions, and V. Kac and D. Peterson for explaining their paper to us. We are grateful to the Max Planck Institute for mathematics in Bonn for hospitality.

References

[1] V.G. Kac, Infinite dimensional Lie algebras (Birkhäuser, Basel, 1983).
[2] D.I. Olive, Kac–Moody algebras: an introduction for physicists, Imperial College preprint TP/84-85/14; P. Goddard, Kac–Moody algebras, representations and applications, DAMTP 85/7.
[3] V.G. Knizhnik and A.B. Zamolodchikov, Nucl. Phys. B247 (1984) 83
[4] P. Goddard and D. Olive, Nucl. Phys. B257 [FS14] (1985) 226.
[5] I.T. Todorov, Phys. Lett. 153B (1985) 77.
[6] E. Witten, Commun. Math. Phys. 92 (1984) 455.
[7] H. Sugawara, Phys. Rev. 170 (1968) 1659.
[8] S. Helgason, Differential geometry, Lie groups and symmetric spaces (Academic Press, New York, 1978) p 518.
[9] S. Coleman, D. Gross and R. Jackiw, Phys. Rev. 180 (1969) 1359; K. Bardakci and M. Halpern, Phys. Rev. D3 (1971) 249; R. Dashen and Y. Frishman, Phys. Rev. D11 (1975) 278.
[10] V.G. Kac and D.H. Peterson, Proc. Natl. Acad. Sci. USA, 78 (1981) 3308.
[11] P. Goddard, D. Olive and A. Schwimmer, Phys. Lett. 157B (1985) 393.

CHAPTER 7

THE MONSTER GROUP

Reprinted Papers

20. I.B. Frenkel, J. Lepowsky and A. Meurman, "An Introduction to the Monster", in *Unified String Theories*, ed. M. Green and D. Gross (World Scientific, Singapore, 1986) 533–546.

21. I.B. Frenkel, J. Lepowsky and A. Meurman, "A Moonshine Module for the Monster", in *Vertex Operators in Mathematics and Physics*, MSRI Publication #3 (Springer, Heidelberg, 1984) 231–274.

As soon as it was realised [1,2] that the quantum mechanics of a relativistic string is only consistent in 26-dimensional space-time, with the degrees of freedom corresponding to vibrations in the 24 dimensions transverse to the worldsheet of the string, it was natural to wonder if there was any relation between these results on the consistency of a physical theory and the remarkable mathematical structures that are known to exist in 24 dimensions, such as the Leech latttice [3]. The work of I.B. Frenkel, J. Lepowsky and A. Meurman, outlined in *Reprinted Papers #20 and #21*, using vertex operators to construct the so-called "moonshine module" for the Monster group of Fischer and Griess [4] (the largest of the sporadic simple finite groups), demonstrated that these physical and mathematical ideas are indeed deeply interrelated. No doubt much more remains to be understood. In particular, it remains to be seen whether the Monster group has any direct role to play in any physical application of string theory.

References

[1] C. Lovelace, "Pomeron form factors and dual Regge cuts", Phys. Lett. **34B** (1971) 500–506.

[2] P. Goddard, J. Goldstone, C. Rebbi and C.B. Thorn, "Quantum dynamics of a massless relativistic string", Nucl. Phys. **B56** (1973) 109–135.

[3] J. Leech, "Notes on sphere packings", Canadian J. Math. **19** (1967) 252–267.

[4] R.L. Griess, "The Friendly Giant", Invent. Math. **69** (1982) 1–102.

AN INTRODUCTION TO THE MONSTER

I. B. Frenkel[1]

Department of Mathematics, Yale University
New Haven, CT 06520, USA

J. Lepowsky[2]

Department of Mathematics, Rutgers University
New Brunswick, NJ 08903, USA

A. Meurman[3]

Department of Mathematics, University of Stockholm
Stockholm, SWEDEN

ABSTRACT

In a brief exposition intended for string theorists, we discuss the Monster and its natural infinite-dimensional representation.

1. INTRODUCTION

The Fischer-Griess Monster, often denoted F_1, is a finite simple group with about 8×10^{53} elements. Predicted to exist in 1973 by B. Fischer and R. Griess, it was constructed by Griess [18]. We have found a natural infinite-dimensional Fock space representation of F_1, incorporating vertex operators and suggesting fundamental connections with string theory (refs. [9], [10], [11]). One could argue that the Monster is the most perfect structure in mathematics, and it appears that the miracles of string theory are very closely related to the miracles which allow the Monster to exist. Attempting to understand these miracles should lead to deeper interactions between mathematics and physics. Here we give a brief sketch of F_1 for string theorists, including a definition of F_1 based on vertex operators. This exposition complements an earlier one [10], which emphasized the infinite-

[1]Partially supported by a Sloan Foundation Fellowship.

[1,2]Partially supported by NSF Grant MCS83-01664.

[3]Partially supported by a grant from the Swedish Natural Sciences Research Council.

dimensional representation of F_1. See refs. [8], [10], and [19] for further discussion.

2. FINITE SIMPLE GROUPS

A group G is <u>simple</u> if it has no subgroups H other than {1} and G such that $gHg^{-1} = H$ for all $g \in G$. The smallest nonabelian finite simple group is the group, say I, of rotations of the regular icosahedron. Compact Lie groups without center, such as the group E_8, are also simple, but are infinite. The nonabelian finite simple groups are the following:

(1) the alternating group A_n on n letters (the group of even permutations) for $n \geq 5$; the group A_5 is isomorphic to I

(2) the groups of "Lie type" over finite fields, for example, SL(n) with entries in the 2-element field, for $n \geq 3$

(3) the sporadic groups, which have no apparent unifying pattern or general construction; F_1 is the largest of these, involving 20 or 21 sporadic groups as quotients of subgroups; string theorists will be amused that the number of sporadic groups is 26.

The precise description of all these groups is a long task. The classification theorem for finite simple groups - the assertion that this list is complete - is a landmark achievement of twentieth century mathematics. Its recently completed proof covers 10,000 to 15,000 journal pages and represents the work of over 100 mathematicians. See refs. [14], [15] for surveys. No one person has checked all the details of the proof, but outsiders to the classification project become more confident that the theorem is true every year that elapses without the discovery of a new finite simple group.

Both the search for sporadic groups and the classification effort made finite group theory a relatively isolated branch of mathematics for a long time. But suddenly the situation changed in 1978-79.

3. MONSTROUS MOONSHINE

Even before F_1 was proved to exist, it was strongly suspected to have a 196883-dimensional irreducible module (= representation), which would be the smallest possible nontrivial module.

Consider the modular function $j(e^{2\pi iz})$, which maps the quotient of the upper half plane by the standard action of $SL(2,\mathbb{Z})$, with the point at infinity adjoined, one-to-one onto the Riemann sphere. When this function is expanded in powers of $q = e^{2\pi iz}$, the coefficients are positive integers:

$$j(q) = q^{-1} + 744 + 196884q + 21493760q^2 + \cdots .$$

(cf. ref. [35]). The constant term 744 can be changed arbitrarily without destroying any of the fundamental properties of $j(q)$.

It was J. McKay who noticed the near coincidence $196884 = 196883 + 1$. Was there indeed a relationship between two traditionally such distant parts of mathematics? Interpreting the "1" as the dimension of the trivial F_1-module, J. Thompson extended the coincidence by checking that the first several coefficients of $j(q)$ (except for the constant term) are simple positive integral linear combinations of the conjectured dimensions of irreducible F_1-modules. These observations would be explained, he pointed out [37], if there were a natural infinite-dimensional $\mathbb{Z}$-graded F_1-module

$$V = \bigoplus_{n=-1}^{\infty} V_{-n}$$

such that the dimension of the F_1-module V_{-n} is the nth coefficient of $J(q) = j(q)-744$; in particular, $V_0 = 0$. (We have negated the subscripts because mathematicians conventionally consider modules with grading bounded above rather than below.) Defining the <u>character</u> of V to be

$$\text{ch } V = \sum_{n\geq -1} (\dim V_{-n})q^n,$$

we require

$$\text{ch } V = J(q). \tag{3.1}$$

By analyzing the expected action of nontrivial Monster elements on V, J. Conway and S. Norton dramatically amplified the conjectured relationship between finite group theory and modular function theory, calling it "Monstrous Moonshine" [5].

The numerology started people thinking that V might be some kind of analogue of the basic module of an affine Kac-Moody algebra (cf. refs. [22], [26]). This turned out to be true, but only in a very subtle sense.

It is interesting to note that modular invariance plays a crucial role in string theory. In particular, in the heterotic string [20], [21], this invariance limits the gauge group to either $E_8 \times E_8$ or SO(32), in agreement with the discovery of Green-Schwarz [17].

4. THE CONSTRUCTION OF F_1

In 1980, Griess [18] constructed F_1 as a group of automorphisms of a 196883-dimensional commutative nonassociative algebra, which we designate B_0. Very strange-looking, the Griess algebra B_0 was not destined to be stumbled upon by the usual axiomatic approach of non-associative algebraists, since it satisfies no low-degree identities besides the commutativity identity $xy-yx = 0$. For instance, B_0 is not a Jordan algebra. Nevertheless, B_0 looked to us a little like the Lie algebra E_8, in some as yet unknown presentation. This impression became the analogy explained below.

J. Tits has shown that F_1 is the full automorphism group of B_0 (see refs. [39], [40]). We shall define B_0 below, thereby giving the reader a precise definition of the Monster. Our definition is based on vertex operators, bringing another subject into the network of ideas.

5. AFFINE LIE ALGEBRAS

Let $\underline{g}$ be any Lie algebra and let $\langle\cdot,\cdot\rangle$ be a symmetric bilinear form on $\underline{g}$, invariant in the sense that

$$\langle [x,y],z\rangle = \langle x,[y,z]\rangle, \quad x,y,z \in \underline{g}. \tag{5.1}$$

(For instance, $\underline{g}$ might be a finite-dimensional semisimple Lie algebra and $\langle\cdot,\cdot\rangle$ a multiple of the Killing form. Or $\underline{g}$ might be an abelian Lie algebra and $\langle\cdot,\cdot\rangle$ an arbitrary symmetric form.) The corresponding (untwisted) affine Lie algebra is the infinite-dimensional Lie algebra

$$\hat{\underline{g}} = \underline{g}\otimes\mathbb{C}[t,t^{-1}]\oplus\mathbb{C}c$$

with brackets given by:

$$[c,\hat{\underline{g}}] = 0, \tag{5.2}$$

$$[x(m),y(n)] = [x,y](m+n) + \langle x,y\rangle m\delta_{m+n,0}c \tag{5.3}$$

for $x,y \in \underline{g}$ and $m,n \in \mathbb{Z}$. Here $\mathbb{C}[t,t^{-1}]$ designates the algebra of

Laurent polynomials in t, and $x(m) = x\otimes t^m$. If we write t as $e^{i\theta}$, then $\underline{g}\otimes\mathbb{C}[t,t^{-1}]$ becomes the space of maps $f: \mathbb{R} \to \underline{g}$ with periodicity $f(x+2\pi) = f(x)$ and finite Fourier expansion.

Suppose that ν is an automorphism of $\underline{g}$ such that $\langle \nu x, \nu y\rangle = \langle x,y\rangle$ for $x,y \in \underline{g}$ and $\nu^p = 1$ for some integer $p > 0$. For $n \in \mathbb{Z}$, let $\underline{g}_{(n)}$ be the $e^{2\pi in/p}$-eigenspace of ν in $\underline{g}$. The corresponding ν-twisted affine algebra is the Lie algebra

$$\hat{\underline{g}}[\nu] = \bigoplus_{n\in\mathbb{Z}} \underline{g}_{(n)}(n/p)\oplus\mathbb{C}c.$$

Here the space $\underline{g}_{(n)}(n/p)$ is spanned by the elements $x(n/p) = x\otimes t^{n/p}$ for $x \in \underline{g}_{(n)}$, and the brackets are again given by (5.2) and (5.3), this time for $m,n \in \mathbb{Z}/p$ and $x \in \underline{g}_{(pm)}$, $y \in \underline{g}_{(pn)}$. Then with $t = e^{i\theta}$, $\oplus\underline{g}_{(n)}(n/p)$ becomes the space of Fourier polynomial maps $f: \mathbb{R} \to \underline{g}$ with "twisted periodicity" $f(x+2\pi) = \nu f(x)$. For instance if $\underline{g}$ is abelian, $p = 2$ and $\nu = -1$, then f satisfies the antiperiodic boundary conditions $f(x+2\pi) = -f(x)$, and f involves only odd powers of $e^{i\theta/2} = t^{1/2}$.

For $\underline{g}$ semisimple, $\hat{\underline{g}}$ and $\hat{\underline{g}}[\nu]$ are examples of what mathematicians have termed Kac-Moody algebras, which are defined in terms of certain generators from a generalized Cartan matrix, and whose detailed study was begun by V. G. Kac, I. L. Kantor and R. L. Moody. The expression for the coefficient of c (the cocycle) in the bracket formula (5.3) for an untwisted affine Kac-Moody algebra seems not to have been recognized by mathematicians until around 1977. If ν is an inner automorphism of $\underline{g}$, then $\hat{\underline{g}}$ and $\hat{\underline{g}}[\nu]$ are isomorphic Lie algebras. See ref. [27] for further discussion.

6. VERTEX OPERATOR CONSTRUCTIONS

In 1977, one wanted to construct affine Kac-Moody algebras by means of concrete operators of some kind. The first result [30] was a Fock space realization of $\underline{s\ell}(2,\mathbb{C})^\wedge$, in a twisted form, based on an apparently new kind of differential operator which H. Garland recognized as similar to a vertex operator in string theory. This construction was generalized to a natural family of twisted affine algebras in ref. [24]. Subsequently, an analogous construction of the corresponding untwisted affine algebras $\hat{\underline{g}}$ was found (refs. [7], [34]), using the vertex operator of string theory. This is often called the homogeneous construction. See also refs. [42], [43].

We now describe still another vertex operator construction [8], corresponding to a different twisting. Let L be a lattice (the integral span of a basis) in a finite-dimensional Euclidean space with complexification $\underline{h}$. Denote the inner product by $\langle\cdot,\cdot\rangle$ and assume that L is an even lattice, i.e., that $\langle\alpha,\alpha\rangle$ is even for all $\alpha \in L$. View $\underline{h}$ as an abelian Lie algebra and let ν be the automorphism of $\underline{h}$ which multiplies each element by -1. Then the corresponding twisted affine algebra

$$\hat{\underline{h}}[-1] = \bigoplus_{n\in\mathbb{Z}+1/2} \underline{h}(n)\oplus\mathbb{C}c$$

can be realized as an algebra of half-integrally moded bosonic string oscillators acting by the canonical realization of the Heisenberg commutation relations on a Fock space S, with c acting as the identity operator. Here S is the space of polynomials on a basis of the space $\bigoplus_{n<0}\underline{h}(n)$, which we view as the space of creation operators. The space S has a natural nonpositive $\frac{1}{2}\mathbb{Z}$-grading which we denote as follows:

$$S = \bigoplus S_n \quad (n \in \tfrac{1}{2}\mathbb{Z},\ n \le 0).$$

(If we want the character of S to have modular transformation properties, we can shift the degrees n by adding a suitable uniform constant.)

A lattice is unimodular or self-dual if it contains one point per unit volume. For $n \in \mathbb{Z}$, set

$$L_n = \{\alpha \in L \mid \langle\alpha,\alpha\rangle = n\}.$$

We shall be especially interested in three lattices: $L = \mathbb{Z}\alpha_0$ where $\langle\alpha_0,\alpha_0\rangle = 2$, the root lattice of $s\ell(2,\mathbb{C})$ (which is not unimodular); $L = \Gamma$, the root lattice of E_8 - the unique (up to isometry) even unimodular lattice in 8 dimensions; and $L = \Lambda$, the Leech lattice - the unique even unimodular lattice in 24 dimensions such that Λ_2 is empty (see refs. [3], [25]; cf. refs. [29], [38]).

Set $\bar{L} = L/2L$, the lattice L with points differing by twice a lattice element identified. Then the set $\bar{L}$ has 2^ℓ elements, where $\ell = \dim \underline{h}$, and $\bar{L}$ can be viewed as a vector space over the 2-element field $\mathbb{Z}/2\mathbb{Z}$. Denote by $\bar{\alpha}$ the image of $\alpha \in L$ in $\bar{L}$. It is easy to construct a (not necessarily symmetric) bilinear map $\varepsilon_0: \bar{L}\times\bar{L} \to \mathbf{Z}/2\mathbf{Z}$ such that $\varepsilon_0(\bar{\alpha},\bar{\alpha}) = \langle\alpha,\alpha\rangle/2 \bmod 2$ for all $\alpha \in L$. Set $\varepsilon(\bar{\alpha},\bar{\beta}) = (-1)^{\varepsilon_0(\bar{\alpha},\bar{\beta})}$ for $\alpha,\beta \in L$ and define a multiplication on the set $F = \{\pm e_{\bar{\alpha}} | \bar{\alpha} \in \bar{L}\}$, the $e_{\bar{\alpha}}$ being a new set of symbols indexed by $\bar{L}$, by

$$e_{\bar{\alpha}}e_{\bar{\beta}} = \varepsilon(\bar{\alpha},\bar{\beta})e_{\bar{\alpha}+\bar{\beta}} \quad \text{for } \alpha,\beta \in L.$$

Then F is a finite group which is a "finite Heisenberg group" if $L = \Gamma$ or Λ, and

$$e_{\bar{\alpha}}e_{\bar{\beta}} = (-1)^{\langle\alpha,\beta\rangle}e_{\bar{\beta}}e_{\bar{\alpha}} \quad \text{for } \alpha,\beta \in L.$$

Let T be an irreducible representation of F such that $-e_0 = -1$ in F acts as -1 on T. Then for $L = \mathbf{Z}\alpha_0$, Γ and Λ, we have $\dim T = 1$, 2^4 and 2^{12}, respectively.

Set $W = S\otimes T$ and give W the $\frac{1}{2}\mathbf{Z}$-grading

$$W = \oplus W_n \quad (n \in \tfrac{1}{2}\mathbf{Z},\ n \leq 0)$$

defined by: $W_n = S_n\otimes T$. The space W has a unique (up to constant multiple) symmetric bilinear form determined by the condition $h(n)^* = h(-n)$ $(h \in \underline{h},\ n \in \mathbf{Z}+\frac{1}{2})$ on S and by F-invariance on T. The restriction of this form to each W_n is nonsingular.

For $\alpha \in L$ define a "vertex operator" $X(\alpha,\zeta)$, ζ being a formal parameter, as follows:

$$X(\alpha,\zeta) = \exp(\textstyle\sum\alpha(-n)\zeta^{-n}/n)\exp(-\sum\alpha(n)\zeta^{n}/n)\otimes e_{\bar{\alpha}}, \tag{6.1}$$

where both sums range over $n \in \mathbf{Z}+\frac{1}{2}$, $n > 0$. Then the expansion coefficients $x_\alpha(n)$ defined by

$$X(\alpha,\zeta) = \textstyle\sum_{n\in(1/2)\mathbf{Z}} x_\alpha(n)\zeta^n$$

are operators on W. (It turns out that the operators (6.1), without the tensor factor $e_{\bar{\alpha}}$, had been written down in ref. [6] in connection with electromagnetic currents.)

The first main theorem about these operators [8] states: Suppose that L is the root lattice of a semisimple Lie algebra $\underline{g}$ with all root lengths equal and normalized to be $\sqrt{2}$, so that L is the lattice generated by L_2. Then under brackets, the operators $x_\alpha(n)$ for $\alpha \in L_2$ and $n \in \frac{1}{2}\mathbf{Z}$ generate a copy of the twisted affine Lie algebra $\hat{\underline{g}}[\nu]$, where ν is an automorphism of order 2 of $\underline{g}$ extending -1 on $\underline{h}$, which is identified with a Cartan subalgebra of $\underline{g}$. Moreover, $\hat{\underline{g}}[\nu]$ is spanned by the $x_\alpha(n)$ and the infinite-dimensional "Heisenberg algebra" $\hat{\underline{h}}[-1]$. In particular, the canonical realization of the

Heisenberg commutation relations on S can be extended naturally using (6.1) to a realization of $\hat{\underline{g}}[\nu]$ on $W = S\otimes T$.

This result generalizes the original construction [30] - the case $L = \mathbb{Z}\alpha_0$ - in a direction different from that of ref. [24].

Let $L = \Gamma$. Then the degree zero operators $x_\alpha(0)$, $\alpha \in \Gamma_2$, span a Lie algebra $\underline{k}$ isomorphic to $\underline{so}(16)$, and the invariant bilinear form on $\underline{k}$ is determined from the vertex operator brackets (see the constant term in (5.2)). (Note that the E_8-symmetry of the untwisted construction (refs. [7], [34]) of $\hat{E}_8$ is broken down to $\underline{so}(16)$. In addition to symmetry breaking, we also have a "dimensional reduction", corresponding to the fact that the number of zero modes in the harmonic oscillators of refs. [7], [34] has been reduced down to 0.) Now the degree $-\frac{1}{2}$ operators $x_\alpha(-\frac{1}{2})$ and $\underline{h}(-\frac{1}{2})$ span a $120 + 8 = 128$-dimensional space which transforms under $\underline{so}(16)$ as one of the half-spin modules. On the other hand, the subspace $W_{-1/2} = S_{-1/2}\otimes T \simeq \underline{h}\otimes T$ of W is $8\cdot 16 = 128$-dimensional, and is a copy of the other half-spin module of $\underline{so}(16)$. In particular, the space $\underline{k}\oplus\underline{h}\otimes T$ carries a natural E_8-Lie algebra structure, in which the subalgebra $\underline{k}$ and its bracket action on its orthogonal complement $\underline{h}\otimes T$ are defined entirely by means of vertex operators. The brackets of pairs of elements of $\underline{h}\otimes T$ into $\underline{k}$ are then canonically determined from the symmetric bilinear form and the invariance condition (5.1). (The symmetric form on $\underline{k}\oplus\underline{h}\otimes T$ is a multiple of the Killing form of E_8.) To construct E_8, it does not matter which half-spin module (i.e., which chirality of spinors) one should adjoin to $\underline{so}(16)$, although to pass between the two realizations of E_8 one would need to use a complicated outer automorphism of $\underline{so}(16)$.

By contrast, the analogue of $\underline{k}$ for the Leech lattice (see below) admits a distinguished "chirality". Only the second, nonstandard, version of E_8, namely, $\underline{k}\oplus\underline{h}\otimes T$, leads by analogy to the Griess algebra, which is in a sense an even more unique structure than E_8.

7. THE CROSS BRACKET OPERATION

While the vertex operators (6.1) are defined for $L = \Lambda$ (the Leech lattice), there is no affine Lie algebra $\hat{\underline{g}}[\nu]$ because Λ_2 is empty. But it turns out in this case that the components $x_\alpha(n)$ of the vertex operators $X(\alpha,\zeta)$ for $\alpha \in \Lambda_4$ do generate something interesting under a new commutative nonassociative operation [8]: For $\alpha,\beta \in \Lambda_4$ and $m,n \in \frac{1}{2}\mathbf{Z}$, define

$$[x_\alpha(m)\times x_\beta(n)] = \frac{1}{2}([x_\alpha(m+1),x_\beta(n-1)]+[x_\beta(n+1),x_\alpha(m-1)]). \qquad (7.1)$$

We call this the <u>cross-bracket</u> of $x_\alpha(m)$ and $x_\beta(n)$, because it is made up of two brackets which "cross"; strictly speaking it depends on $X(\alpha,\zeta)$ and $X(\beta,\zeta)$, not just on $x_\alpha(m)$ and $x_\beta(n)$. Of course, the definition (7.1) makes sense for any two sequences of operators.

The Leech lattice analogue [8] the above theorem on brackets of vertex operators says that under cross-brackets, the operators $x_\alpha(n)$ for $\alpha \in \Lambda_4$ and $n \in \frac{1}{2}\mathbf{Z}$ generate a cross-bracket algebra spanned by the identity operator and the components of $X(\alpha,\zeta)$ $(\alpha \in \Lambda_4)$, $\zeta\frac{d}{d\zeta}\alpha(\zeta)$ $(\alpha \in \underline{h})$ and $:\alpha(\zeta)\beta(\zeta):$ $(\alpha,\beta \in \underline{h})$, where $\alpha(\zeta) = \sum_{n\in\mathbf{Z}+1/2}\alpha(n)\zeta^n$ and $:\ :$ denotes normal ordering. In particular, the Λ_4 vertex operators generate a canonical copy of the Virasoro algebra among the operators quadratic in the half-integrally moded creation and annihilation operators. The degree zero operators in the cross-bracket algebra span a commutative nonassociative algebra $\underline{k}$, the quadratic operators $:\alpha(\zeta)\beta(\zeta):$ giving a Jordan subalgebra. The new product operation $\times$ on $\underline{k}$ and a symmetric bilinear form $\langle\cdot,\cdot\rangle$ on $\underline{k}$ are determined by the formula, analogous to (5.3),

$$[x(m)\times y(n)] = (x\times y)(m+n)+\langle x,y\rangle m^2\delta_{m+n,0} \qquad (7.2)$$

for $x,y \in \underline{k}$ and $m,n \in \mathbf{Z}$. (Note the m^2 in (7.2).) In place of (5.1), we have the invariance condition

$$\langle x\times y,z\rangle = \langle x,y\times z\rangle \qquad (7.3)$$

for $x,y,z \in \underline{k}$.

The operators $\underline{k}$ preserve the subspace $W_{-1/2} = S_{-1/2}\otimes T \simeq \underline{h}\otimes T$ of W. This provides the space

$$\underline{B} = \underline{k}\oplus\underline{h}\otimes T$$

with a canonical commutative nonassociative algebra structure, with an identity element equal to the suitably normalized degree zero Virasoro generator. (Just as in the E_8 case, the products of pairs of elements of $\underline{h}\otimes T$ into $\underline{k}$ are canonically determined by the invariance condition (7.3) on $\underline{B}$.) The dimension of $\underline{B}$ is 196884.

The algebra $\underline{B}$ is precisely the Griess algebra with a natural identity element adjoined: $\underline{B} = B_0 \oplus \mathbb{C}\cdot 1$ (see ref. [8]). In particular, the Monster may be defined as the automorphism group of $\underline{B}$. Note that the most distinguished sporadic group of all - the Monster - may be defined canonically using only the vertex operators (6.1) and the cross-bracket operation.

8. THE INFINITE-DIMENSIONAL REPRESENTATION OF F_1

The algebra $\underline{B}$ (or B_0) has many relatively easy automorphisms, preserving $\underline{k}$ and $\underline{h}\otimes T$ and forming the centralizer C of an involution (= automorphism of order 2) in F_1. The group C is related to the Conway group $\cdot 0$ of automorphisms of the Leech lattice [2]. The existence of the Monster hinges on the very subtle fact that $\underline{B}$ admits additional automorphisms [18].

In order to construct such automorphisms and hence F_1 in a conceptual way, we consider the involution θ of the space $W = S\otimes T$ (for the Leech lattice Λ) defined as follows: If p is a monomial of degree r in the generators $h(n)$ ($h \in \underline{h}$, $n \in \mathbb{Z}+\frac{1}{2}$, $n < 0$) of the polynomial algebra S, then θ multiplies $p\otimes t$ ($t \in T$) by $(-1)^r$. Let W^- be the -1-eigenspace of θ in W. Now consider the vertex operator construction of untwisted affine Lie algebras $\hat{\underline{g}}$ of refs. [7], [34], a construction incorporated in the heterotic string [20], [21] in the cases $\underline{g} = E_8\times E_8$ and $\underline{g} = \underline{so}(32)$. This construction is based on integrally-moded bosonic string oscillators in the same sense that W is based on half-integrally moded oscillators. We modify this construction by starting with the Leech lattice in place of a root lattice, and we define an involution analogous to θ on the resulting space, say U, analogous to W. Let U^+ be the +1-eigenspace of θ in U, and form the $\mathbb{Z}$-graded space

$$V^\natural = U^+\oplus W^- = \bigoplus_{n=-\infty}^{1} V_n^\natural$$

($\natural$ = "natural"). There is a natural action of the group C on $V^\natural$. See refs. [9], [10] for the details. Cf. ref. [23], which describes an apparently similar space based on formulas in ref. [5]. This space seems, however, not to provide a natural F_1-module.

We construct a uniformly defined extra automorphism σ on each $V_n^{\natural}$ (refs. [9], [10]). The source of σ is a "triality" for $\underline{\underline{s\ell}}(2,\mathbb{C})^{\wedge}$ which intertwines the twisted and untwisted vertex operator constructions. In the E_8 "practice case" of our construction of $V^{\natural}$, the restriction of σ to a "bosonic" subspace V_1 of U^+ and a "fermionic" subspace V_2 of W^- gives a "supersymmetry" implementing and explaining the numerology pointed out in ref. [13] which led to the combination of the Neveu-Schwarz and Ramond models into superstrings. (See ref. [10].) The same phenomenon in the Leech lattice case underlies our construction of σ (see ref. [10]) and in particular gives, by analogy, a corresponding new kind of 24-dimensional "supersymmetry". See also ref. [1].

We obtain the action by cross bracket of a "commutative affinization"

$$\underline{\hat{B}} = \underline{B}\otimes\mathbb{C}[t,t^{-1}]\oplus\mathbb{C}c$$

of $\underline{B}$ with multiplication defined by (7.2), and a proof of the fact that σ defines an algebra automorphism of $\underline{B}$ (see refs. [9], [10]). In particular $V^{\natural}$ is an F_1-module. With respect to the canonical Virasoro algebra mentioned earlier, the "conformal fields" comprising $\underline{\hat{B}}$ have conformal weight 2. The module $V^{\natural}$ has character $J(q)$, as in (3.1). The finite-dimensional algebra $\underline{B}$ is better understood from the cross bracket operation than from its original finite-dimensional definition. See ref. [10] for a detailed exposition of $V^{\natural}$, including discussions of more analogies with E_8 and more links with string theory.

We remark that many of the properties of $V^{\natural}$ predicted in ref. [5] remain unproved, including a deeper connection with the theory of Riemann surfaces. Other recent (finite-dimensional) treatments of the construction of an extra automorphism of the algebra B_0 are given in refs. [4], [39].

Just as twisted affine Lie algebras can be constructed by means of twisted vertex operators, we anticipate that twistings of $\underline{\hat{B}}$ can be constructed using the general methods of ref. [28], leading to a subtle kind of dimensional reduction and symmetry breaking.

It would be very important to place all the elements of F_1 on an equal footing with one another. In the present formulation, which is analogous to the noncovariant light-cone quantization in string theory, the elements of C preserve a twisted and an untwisted sector of $V^\natural$, and the remaining elements, like σ, mix the two sectors. Such a unification would probably take place in some kind of "covariant" 26-dimensional formulation of the Monster and its natural module. It is interesting to note that the original 26-dimensional string model [41], which was later overshadowed by 10-dimensional models [13], [16], [31], [32], [33], has recently been revived (cf. refs. [12], [36]) in the heterotic string [20], [21].

The existence of the Monster, properly interpreted, might conceivably lead to a unique physically correct four-dimensional string theory.

REFERENCES

1. Chapline, G., Unification of gravity and elementary particle interactions in 26 dimensions?, Phys. Lett. 158B, 393-396 (1985).

2. Conway, J.H., A group of order 8,315,553,613,086,720,000, Bull. London Math. Soc. 1, 79-88 (1969).

3. Conway, J.H., A characterization of Leech's lattice, Inventiones Math. 7, 137-142 (1969).

4. Conway, J.H., A simple construction for the Fischer-Griess Monster group, Invent. Math. 79, 513-540 (1985).

5. Conway, J.H. and Norton, S.P., Monstrous moonshine, Bull. London Math. Soc. 11, 308-339 (1979).

6. Corrigan, E. and Fairlie, D.B., Nuc. Phys. B91, 527 (1975).

7. Frenkel, I.B. and Kac, V.G., Basic representations of affine Lie algebras and dual resonance models, Inventiones Math. 62, 23-66 (1980).

8. Frenkel, I.B., Lepowsky, J. and Meurman, A., An E_8-approach to F_1, in: Finite Groups - Coming of Age, Proc. 1982 Montreal Conference, ed. by J. McKay, Contemporary Math. 45, 99-120 (1985).

9. Frenkel, I.B., Lepowsky, J. and Meurman, A., A natural representation of the Fischer-Griess Monster with the modular function J as character, Proc. Nat. Acad. Sci. U.S.A. 81, 3256-3260 (1984).

10. Frenkel, I.B., Lepowsky, J. and Meurman, A., A Moonshine Module for the Monster, in: Vertex Operators in Mathematics and Physics-Proceedings of a Conference November 10-17, 1983, ed. by J. Lepowsky, S. Mandelstam and I.M. Singer, Publications of the Mathematical Sciences Research Institute #3, Springer-Verlag, New York, 231-273 (1985).

11. Frenkel, I.B., Lepowsky, J. and Meurman, A., Vertex operators and the Monster, to appear.

12. Freund, P.G.O., Phys. Lett. 151B, 387 (1985).

13. Gliozzi, F., Olive, D. and Scherk, J., Supersymmetry, supergravity theories and the dual spinor model, Nuclear Physics B122, 253-290 (1977).

14. Gorenstein, D., Finite simple groups, Plenum Press, New York, (1982).

15. Gorenstein, D., Classifying the finite simple groups, Anaheim Colloquium Lectures, January, 1985, Amer. Math. Soc., to appear.

16. Green, M.B. and Schwarz, J.H., Phys. Lett. 109B, 444 (1982).

17. Green, M.B. and Schwarz, J.H., Phys. Lett. 149B, 117 (1984).

18. Griess, R.L., Jr., The Friendly Giant, Invent. Math. 69, 1-102 (1982).

19. Griess, R.L., Jr., A brief introduction to the finite simple groups, in: Vertex Operators in Mathematics and Physics - Proceedings of a Conference November 10-17, 1983, ed. by J. Lepowsky, S. Mandelstam and I.M. Singer, Publications of the Mathematical Sciences Research Institute #3, Springer-Verlag, New York, 217-229 (1985).

20. Gross, D.J., Harvey, J.A., Martinec, E. and Rohm, R., Heterotic string theory (I). The free heterotic string, Nucl. Phys. B256, 253-284 (1985).

21. Gross, D.J., Harvey, J.A., Martinec, E. and Rohm, R., Heterotic string theory (II). The interacting heterotic string, to appear.

22. Kac, V.G., An elucidation of "Infinite-dimensional...and the very strange formula" $E_8^{(1)}$ and the cube root of the modular invariant j, Advances in Math. 35, 264-273 (1980).

23. Kac, V.G., A remark on the Conway-Norton conjecture about the "Monster" simple group, Proc. Natl. Acad. Sci. USA 77, 5048-5049 (1980).

24. Kac, V.G., Kazhdan, D.A., Lepowsky, J. and Wilson, R.L., Realization of the basic representations of the Euclidean Lie algebras, Advances in Math. 42, 83-112 (1981).

25. Leech, J., Notes on sphere packings, Canadian J. Math. 19, 252-267 (1967).

26. Lepowsky, J., Euclidean Lie algebras and the modular function j, Amer. Math. Soc. Proc. Symp. Pure Math., 37, 567-570 (1980).

27. Lepowsky, J., Introduction, in: Vertex Operators in Mathematics and Physics - Proceedings of a Conference November 10-17, 1983, ed. by J. Lepowsky, S. Mandelstam and I.M. Singer, Publications of the Mathematical Sciences Research Institute #3, Springer-Verlag, New York, 1-13 (1985).

28. Lepowsky, J., Calculus of twisted vertex operators, Proc. Nat. Acad. Sci. USA (1985), to appear.

29. Lepowsky, J. and Meurman, A., An E_8-approach to the Leech lattice and the Conway group, J. Algebra 77, 484-504 (1982).

30. Lepowsky, J. and Wilson, R.L., Construction of the affine Lie algebra $A_1^{(1)}$, Comm. Math. Phys. 62, 43-53 (1978).

31. Neveu, A. and Schwarz, J.H., Nucl. Phys. B31, 86 (1971).

32. Neveu, A. and Schwarz, J.H., Phys. Rev. D4, 1109 (1971).

33. Ramond, P., Phys. Rev. D3, 2415 (1971).

34. Segal, G., Unitary representations of some infinite-dimensional groups, Comm. Math. Phys. 80, 301-342 (1981).

35. Serre, J.-P., A course in arithmetic, Springer-Verlag, New York (1973).

36. Thierry-Mieg, J., Phys. Lett. 156B, 199 (1985).

37. Thompson, J., Some numerology between the Fischer-Griess Monster and the elliptic modular function, Bull. London Math. Soc. 11, 352-353 (1979).

38. Tits, J., Four presentations of Leech's lattice, in "Finite Simple Groups II", Proceedings of a London Math. Soc. Research Symposium, Durham, 1978, Academic Press, London/New York, 303-307 (1980).

39. Tits, J., Le monstre, Séminaire Bourbaki, 36e année, 1983/84, no. 620 (1983), Astérisque, to appear.

40. Tits, J., On R. Griess' "Friendly Giant", Invent. Math. 78, 491-499 (1984).

41. Veneziano, G., Nuovo Cim. 57A, 190 (1968).

42. Halpern, M.B., Quantum solitons which are SU(N) fermions, Phys. Rev. D12, 1684-1699 (1975).

43. Halpern, M.B., Prehistory of internal symmetry on the string, preprint (1985).

A MOONSHINE MODULE FOR THE MONSTER

Igor B. Frenkel,[†] James Lepowsky[†]
and Arne Meurman[†]

0. INTRODUCTION

The theory of finite simple groups was for a long time a rather isolated and unusual branch of mathematics. It achieved its goal in 1981 when a proof of the classification theorem was completed. This unique proof, comprising several thousand pages of published articles and preprints, leads to the following list of finite simple groups (see [12] for the details): the groups of Lie type, the alternating groups and the 26 sporadic groups. While each of the first two classes has a uniform description, the groups in the third class still have quite different constructions. The largest sporadic group, called the Monster and denoted F_1, was predicted independently by B. Fischer and R. Griess in 1973. It contains 20 or 21 of the sporadic groups and has order $>8\cdot 10^{53}$. This group gave rise to many mysteries even before its actual appearance, promising deep connections with different areas of mathematics.

Many amazing discoveries about the Fischer-Griess Monster were collected in the highly unusual paper "Monstrous Moonshine" by J. Conway and S. Norton [4]. Most of the discoveries were based on the existence of an irreducible F_1-module of dimension 196883. B. Fischer, D. Livingstone and M. Thorne computed the entire character table using this assumption, thus providing the numbers for the "Monstrous game". This started with J. McKay's observation that the modular function j, in its expansion

[†] Partially supported by the National Science Foundation through the Mathematical Sciences Research Institute.

Vertex Operators in Mathematics and Physics - Proceedings of a Conference November 10-17, 1983. Publications of the Mathematical Sciences Research Institute #3, Springer-Verlag, 1984.

(0.1) $j(q) = q^{-1} + 744 + 196884q + 21493760q^2 + \ldots = \sum_{n \geqslant -1} a_n q^n,$

has a coefficient 196884, which exceeds by only 1 the dimension of the minimal conjectured nontrivial F_1-module. Then Thompson observed [24] that the first few coefficients a_n (for $n \neq 0$!) are also simple linear combinations of the dimensions d_n of irreducible representations of the Monster, e.g., $a_{-1} = d_1$, $a_1 = d_1 + d_2$, $a_2 = d_1 + d_2 + d_3$, $a_3 = 2d_1 + 2d_2 + d_3 + d_4$, etc. The coefficient 744 is inessential for the modular property, so one can consider instead the normalized modular function

(0.2) $J(q) = j(q) - 744.$

Thompson also proposed replacing the coefficients a_n in the q-series for J by the representations V_n of F_1 that they suggested and considering the series

(0.3) $T_m = q^{-1} + 0 + \mathrm{tr}\, m|_{V_1} q + \mathrm{tr}\, m|_{V_2} q^2 + \ldots$

for arbitrary elements m of the Monster (not just $m = 1$). A great deal of evidence concerning these "Thompson series" was then collected and they appeared to be the normalized generators of function fields of genus 0. Conway and Norton wrote down a list of such series -- one for each conjugacy class in F_1 -- and conjectured that there exist representations V_n of F_1 compatible with this list via (0.3). A. O. L. Atkin, P. Fong and S. Smith (see [23]), using a computer, produced overwhelming evidence for this conjecture, supporting the extreme likelihood that there should exist a natural graded module for F_1 with the J-function as character.

About a year after these observations, Griess constructed the 196883-dimensional representation of F_1 ([14], [15]). However, instead of resolving the mysteries of "Monstrous Moonshine" he added a new one: He gave a construction of the Monster as an automorphism group of a peculiar commutative nonassociative algebra, say B_0, of dimension 196883.

Although very far from a Lie algebra, Griess's algebra appeared to us to be tantalizingly close to a Lie algebra, especially the Lie algebra E_8, in some as yet unknown presentation. Also, $j(q)^{1/3}$ had already appeared as the character of the basic representation of the affine Lie algebra $\hat{E}_8$ ([16], [18]). Our work started as an attempt to understand the appearance of J(q) and B_0 using vertex operators and the representation theory of affine Lie algebras, and it now includes the construction of a natural module for the Monster with character J (see [7], [8]).

These notes serve as an introduction to the main ideas and steps of our construction, which is purely conceptual. We do not include detailed proofs, which will appear in a separate publication. Griess's construction was very important during our work but we can now obtain major results of his as corollaries of ours. In spite of the fact that our module is infinite-dimensional it is the most natural and simplest representation of the Monster.

We hope that it will now be possible to develop a unified theory of the finite simple groups, based on lattices and vertex operators. In such a theory, H. Garland's arithmetic theory of loop algebras and groups ([9], [10]), as simplified and generalized in [20], [26], would probably be important.

1. LIE ALGEBRAS OF TYPE A, D, E

We first recall several facts about the simple Lie algebras $\mathfrak{g}$ of type A, D, E (that is, type A_ℓ, D_ℓ, E_6, E_7 or E_8). Such an algebra $\mathfrak{g}$ has a root space decomposition with respect to a Cartan subalgebra $\mathfrak{h}$:

$$(1.1) \qquad \mathfrak{g} = \mathfrak{h} \oplus \bigoplus_{\alpha\in\Delta} \mathbb{C}x_\alpha .$$

Let $\langle\cdot,\cdot\rangle$ be a nonsingular symmetric invariant bilinear form on $\mathfrak{g}$. We identify $\mathfrak{g}$ with its dual $\mathfrak{g}^*$ and $\mathfrak{h}$ with $\mathfrak{h}^*$ by means of this form and we normalize the form so that $\langle\alpha,\alpha\rangle = 2$ for $\alpha\in\Delta$. Let $Q = \mathbb{Z}\Delta$ be the root lattice of $\mathfrak{g}$. There exists a bilinear map

(1.2) $\epsilon: Q \times Q \longrightarrow \langle \pm 1 \rangle$

(the 2-element group $\langle \pm 1 \rangle$ being written multiplicatively) with the property

(1.3) $\epsilon(\alpha,\alpha) = (-1)^{\langle \alpha,\alpha \rangle /2}$ for $\alpha \in Q$.

The basis in (1.1) can be chosen to satisfy the bracket relations

$$(1.4) \quad [x_\alpha, x_\beta] = \begin{cases} 0 & \text{if } \langle \alpha,\beta \rangle \geq 0 \\ \epsilon(\alpha,\beta) x_{\alpha+\beta} & \text{if } \langle \alpha,\beta \rangle = -1 \\ -\alpha & \text{if } \langle \alpha,\beta \rangle = -2. \end{cases}$$

(See [6], [22].)

For many questions it is convenient to consider another basis of $\mathfrak{g}$: We set

(1.5) $k_\alpha = x_\alpha + x_{-\alpha}, \; p_\alpha = x_\alpha - x_{-\alpha}.$

Then we have

(1.6) $\mathfrak{g} = \mathfrak{k} \oplus \mathfrak{p}$,

where

$$(1.7) \quad \mathfrak{k} = \bigoplus_{\alpha \in \Delta_+} \mathbb{C}(x_\alpha + x_{-\alpha}), \; \mathfrak{p} = \mathfrak{h} \oplus \bigoplus_{\alpha \in \Delta_+} \mathbb{C}(x_\alpha - x_{-\alpha}),$$

Δ_+ being a system of positive roots. This decomposition is fundamental in the theory of compact forms of semisimple Lie algebras.

2. AFFINE LIE ALGEBRAS AND VERTEX REPRESENTATIONS

We have two canonical affine Lie algebras associated with the simple Lie algebra $\mathfrak{g}$ and its decomposition (1.6):

(2.1) $\hat{\mathfrak{g}}_{\mathbf{Z}} = \mathfrak{k} \otimes \mathbb{C}[t,t^{-1}] \oplus \mathfrak{p} \otimes \mathbb{C}[t,t^{-1}] \oplus \mathbb{C}c,$

(2.2) $\hat{\mathfrak{g}}_{\mathbf{Z}+1/2} = \mathfrak{k} \otimes \mathbb{C}[t,t^{-1}] \oplus \mathfrak{p} \otimes t^{1/2}\mathbb{C}[t,t^{-1}] \oplus \mathbb{C}c$

(t being a formal variable), with the Lie bracket

(2.3) $[x\otimes t^m, y\otimes t^n] = [x,y] \otimes t^{m+n} + \langle x,y\rangle m\delta_{m+n,0}c,$

where x,y belong to $\mathfrak{k}$ or $\mathfrak{p}$ and m,n belong to $\mathbf{Z}$ or $\mathbf{Z} + \frac{1}{2}$ appropriately. We shall sometimes denote $x \otimes t^m$ by x(m). We remark that the algebras $\hat{\mathfrak{g}}_{\mathbf{Z}}$ and $\hat{\mathfrak{g}}_{\mathbf{Z}+1/2}$ are isomorphic if and only if the involution defining (1.6) is inner. (We shall not need this fact.) These algebras are $\frac{1}{2}$ $\mathbf{Z}$-graded by the conditions

$$\deg x\otimes t^m = m$$
$$\deg c = 0.$$

Now we recall the construction of the vertex representations of the Lie algebras $\hat{\mathfrak{g}}_{\mathbf{Z}}$ and $\hat{\mathfrak{g}}_{\mathbf{Z}+1/2}$ (see [6], [7], [22]). Let $\hat{Q}$ be the central extension of Q by the group $\langle \pm 1\rangle$ given by the bilinear cocycle ϵ. Note that ϵ is trivial on $2Q \times Q$ and $Q \times 2Q$. Define

$$\bar{Q} = Q/2Q,$$

and let $\bar{Q}^{\wedge}$ be the group $\hat{Q}/2Q$, or equivalently, the extension of $\bar{Q}$ by $\langle \pm 1\rangle$ with cocycle given by ϵ. Let $\mathbb{C}[Q]$ be the group algebra of Q with the natural basis $\{e^{\alpha} \mid \alpha \in Q\}$, and let T be a faithful irreducible representation of $\bar{Q}^{\wedge}$. Define

(2.4) $e_{\alpha}: \mathbb{C}[Q] \longrightarrow \mathbb{C}[Q],\ \alpha \in Q$

(2.5) $e_{\alpha} \cdot e^{\beta} = \epsilon(\alpha,\beta)e^{\alpha+\beta}$

for $\beta \in Q$. Then

(2.6) $e_\alpha e_\beta = \epsilon(\alpha,\beta)e_{\alpha+\beta}$, $\alpha,\beta\in Q$.

One also has operators

(2.7) $e'_\alpha: T \longrightarrow T$, $\alpha\in Q$

so that

(2.8)
$$e'_\alpha = e'_{-\alpha},\ \alpha\in Q,$$
$$e'_\alpha e'_\beta = \epsilon(\alpha,\beta)e'_{\alpha+\beta},\ \alpha,\beta\in Q.$$

The vertex representations of $\hat{g}_{\mathbf{Z}}$ and $\hat{g}_{\mathbf{Z}+1/2}$ are constructed in the spaces

(2.9) $V_{\mathbf{Z}} = S(\hat{h}^-_{\mathbf{Z}}) \otimes \mathbb{C}[Q]$, $\hat{h}^-_{\mathbf{Z}} = h \otimes t^{-1}\mathbb{C}[t^{-1}]$

(2.10) $V_{\mathbf{Z}+1/2} = S(\hat{h}^-_{\mathbf{Z}+1/2}) \otimes T$, $\hat{h}^-_{\mathbf{Z}+1/2} = h \otimes t^{-1/2}\mathbb{C}[t^{-1}]$

by means of the vertex operators

(2.11)
$$X_{\mathbf{Z}}(\alpha,z) = z^{\langle\alpha,\alpha\rangle/2} \exp\Big(\sum_{\substack{n\in\mathbf{Z}\\ n>0}} \frac{\alpha(-n)z^n}{n}\Big) z^{\alpha(0)} \cdot$$
$$\cdot \exp\Big(-\sum_{\substack{n\in\mathbf{Z}\\ n>0}} \frac{\alpha(n)z^{-n}}{n}\Big)e_\alpha$$

(2.12)
$$X_{\mathbf{Z}+1/2}(\alpha,z) = 2^{-\langle\alpha,\alpha\rangle} \exp\Big(\sum_{\substack{n\in\mathbf{Z}+1/2\\ n>0}} \frac{\alpha(-n)z^n}{n}\Big) \cdot$$
$$\cdot \exp\Big(-\sum_{\substack{n\in\mathbf{Z}+1/2\\ n>0}} \frac{\alpha(n)z^{-n}}{n}\Big) e'_\alpha$$

for $\alpha\in Q$. Here the Heisenberg subalgebras

$$\hat{h}_{\mathbf{Z}} = h \otimes (\bigoplus_{\substack{n\in\mathbf{Z}\\ n\neq 0}} \mathbf{C}t^n) \oplus \mathbf{C}c,$$

$$\hat{h}_{\mathbf{Z}+1/2} = h \otimes t^{1/2}\mathbf{C}[t,t^{-1}] \oplus \mathbf{C}c$$

of $\hat{g}_{\mathbf{Z}}$ and $\hat{g}_{\mathbf{Z}+1/2}$, respectively, act irreducibly on the symmetric algebras $S(\hat{h}^-_{\mathbf{Z}})$ and $S(\hat{h}^-_{\mathbf{Z}+1/2})$, respectively, in the standard way, with c acting as the identity operator. The Fourier components of these vertex operators together with the Heisenberg algebras provide representations $\pi_{\mathbf{Z}}$ and $\pi_{\mathbf{Z}+1/2}$ of the Lie algebras $\hat{g}_{\mathbf{Z}}$ and $\hat{g}_{\mathbf{Z}+1/2}$, respectively. In particular, one has for $\alpha\in\Delta$

(2.13) $$\sum_{n\in\mathbf{Z}} \pi_Z(k_\alpha(n))z^{-n} = X_Z(\alpha,z) + X_Z(-\alpha,z)$$

(2.14) $$\sum_{n\in Z} \pi_Z(p_\alpha(n))z^{-n} = X_Z(\alpha,z) - X_Z(-\alpha,z)$$

where Z denotes $\mathbf{Z}$ or $\mathbf{Z}+1/2$. Both vertex representations are irreducible.

The spaces (2.9), (2.10) are $\frac{1}{2}\mathbf{Z}$-graded by setting

$$\deg\,(S(\hat{h}^-_{\mathbf{Z}})_n \otimes e^\alpha) = n - \frac{1}{2}\langle\alpha,\alpha\rangle$$

$$\deg\,(S(\hat{h}^-_{\mathbf{Z}+1/2})_n \otimes T) = n$$

for $n\in\frac{1}{2}\mathbf{Z}$, $\alpha\in Q$; here the subscripts n denote the obvious homogeneous subspaces. For $n\in\frac{1}{2}\mathbf{Z}$, $h\in h$ and $\alpha\in Q$, the operators $h(n)$, $k_\alpha(n)$ and $p_\alpha(n)$ all have degree n on both spaces (2.9), (2.10).

We can generalize the "homogeneous" construction (2.9), (2.11) by replacing Q in (2.9) by any coset of Q in the weight lattice (the dual lattice of Q, in our A, D, E situation). We can even replace Q in (2.9) by a more general coset Q' of Q in h. However, in this case, the expansion of the vertex operator (2.11) is not given by integral powers of z. In particular, when $\langle Q',Q\rangle = \frac{1}{2}\mathbf{Z}$, the expansion is given by half-integral powers. By choosing an appropriate ϵ, one obtains various constructions of the Lie algebras

(2.1), (2.2), among others. The "twisted" construction (2.10), (2.12) can be generalized analogously.

Note that the subalgebra $\hat{\mathfrak{k}} = \mathfrak{k} \otimes \mathbb{C}[t,t^{-1}] \oplus \mathbb{C}c$ is the same in both $\hat{\mathfrak{g}}_{\mathbf{Z}}$ and $\hat{\mathfrak{g}}_{\mathbf{Z}+1/2}$. We now consider the decomposition of the vertex representations with respect to $\hat{\mathfrak{k}}$. It is easy to see that $V_{\mathbf{Z}}$ and $V_{\mathbf{Z}+1/2}$ are no longer irreducible. Let

(2.15) $\theta_Z\colon V_Z \to V_Z$

be the involution such that

(2.16) $\theta_Z(h(n))\theta_Z^{-1} = -h(n),\ n\in Z,\ h\in\mathfrak{h}$

(2.17) $\theta_{\mathbf{Z}}(e^{\alpha}) = e^{-\alpha},\ \alpha\in Q$

(2.18) $\theta_{\mathbf{Z}+1/2}(t) = t,\ t\in T.$

We write

(2.19) $V_Z = V_Z^+ \oplus V_Z^-$,

where $V_Z^{\pm}$ are the eigenspaces of θ_Z corresponding to the eigenvalues ± 1. These $\hat{\mathfrak{k}}$-subrepresentations are decomposable in general. However, in the most important special case, when $\mathfrak{g}$ is of type E_8, they are irreducible.

3. CROSS PRODUCT ALGEBRAS AND THEIR REPRESENTATIONS

The root lattice Q of a simple Lie algebra of type A, D, E is *integral* in the sense that

$$<\alpha,\beta>\in\mathbf{Z},\quad \alpha,\beta\in Q$$

and *even* in the sense that

$$<\alpha,\alpha>\in 2\mathbf{Z},\quad \alpha\in Q.$$

A lattice L spanning $\mathfrak{h}$ is *unimodular* if it coincides with its $\mathbb{Z}$-dual

$$\{\alpha\in\mathfrak{h} \mid \langle\alpha,L\rangle\subset\mathbb{Z}\}.$$

Even unimodular lattices can exist only in dimensions divisible by 8. In dimension 8, there is only one such lattice, up to isometry -- the E_8-lattice. We shall denote this lattice by Γ.

We would like to note that the above vertex operator constructions remain valid for an arbitrary even (necessarily integral) lattice L spanning $\mathfrak{h}$. In particular, the spaces V_Z and the operators $X_Z(\alpha,z)$ are well defined. For $n\in\mathbb{Z}$, we set

$$L_n = \{\alpha\in L \mid \langle\alpha,\alpha\rangle=n\}.$$

If rank $\mathbb{Z}L_2$ = rank L, then V_Z decomposes into finitely many irreducible components with respect to $\hat{\mathfrak{g}}_Z$. It can happen however that L_2 is empty. The smallest example of an even unimodular lattice of this kind is the rank 24 Leech lattice Λ, which is in fact characterized by these properties (see [3], [17]). The Leech lattice gives the densest sphere packing in 24 dimensions. In the absence of an affine Lie algebra, we would still like to get some kind of vertex operator algebra from Λ.

The shortest nonzero vectors of Λ are those in Λ_4, but contrary to the root lattice case, the corresponding vertex operators generate an infinite family of vertex operators. Nevertheless, there is a way to stop the generation of new vertex operators. The resulting algebra will not be a Lie algebra but a commutative nonassociative algebra. We proceed now to its description.

Let $\mathfrak{g}$ be a commutative nonassociative algebra with a symmetric bilinear form $\langle\cdot,\cdot\rangle$, associative in the sense that

(3.1) $\langle x\times y,z\rangle = \langle x,y\times z\rangle$, $x,y,z\in\mathfrak{g}$,

where $\times$ denotes the product in $\mathfrak{g}$. We define

(3.2) $\hat{\mathfrak{g}} = \hat{\mathfrak{g}}_{\mathbf{Z}} = \mathfrak{g} \otimes \mathbb{C}[t,t^{-1}] \oplus \mathbb{C}c,$

an "affinization" of $\mathfrak{g}$, with the new commutative nonassociative product $\times$ determined by

$$(3.3)\quad \begin{aligned} &(x\otimes t^m) \times (y\otimes t^n) = (x\times y) \otimes t^{m+n} + \langle x,y\rangle m^2 \delta_{m+n,0}c \\ &c \times c = (x\otimes t^m) \times c = 0 \end{aligned}$$

where $x,y\in\mathfrak{g}$, $m,n\in\mathbf{Z}$. We shall also denote $x\otimes t^m$ by $x(m)$. We call a linear map π: $\hat{\mathfrak{g}}_{\mathbf{Z}} \longrightarrow \mathrm{End}\, V$ a *representation of* $\hat{\mathfrak{g}}_{\mathbf{Z}}$ (and V a $\hat{\mathfrak{g}}_{\mathbf{Z}}$-*module*) if

$$(3.4)\quad \pi(x(m)\times y(n)) = \\ = \frac{1}{2}\left([\pi(x(m+1)),\pi(y(n-1))] + [\pi(y(n+1)),\pi(x(m-1))]\right)$$

for $x,y\in\mathfrak{g}$, $m,n\in\mathbf{Z}$. We call the right-hand side a "cross bracket" because it is made up of two brackets that "cross". If V is of the form $\oplus_{n\in\mathbf{Z}} V_n$ and if $\pi(x(n))$ is homogeneous of degree n for all $x\in\mathfrak{g}$, $n\in\mathbf{Z}$, we say that V is a *graded* module.

To get a complete analogy with the Lie algebra case we consider $\mathfrak{g}$ with an involution preserving the product $\times$. Let

(3.5) $\mathfrak{g} = \mathfrak{k} \oplus \mathfrak{p}$

be the decomposition with respect to the involution. We define another commutative algebra

(3.6) $\hat{\mathfrak{g}}_{\mathbf{Z}+1/2} = \mathfrak{k}\otimes\mathbb{C}[t,t^{-1}] \oplus \mathfrak{p}\otimes t^{1/2}\mathbb{C}[t,t^{-1}] \oplus \mathbb{C}c$

with the product $\times$ again given by (3.3), for $x,y\in\mathfrak{k}$ or $\mathfrak{p}$, $m,n\in\mathbf{Z}$ or $\mathbf{Z}+\frac{1}{2}$ appropriately. The notion of representation for $\hat{\mathfrak{g}}_{\mathbf{Z}+1/2}$ is defined as for $\hat{\mathfrak{g}}_{\mathbf{Z}}$.

It turns out that for an even lattice L, the vertex operators $X_Z(\alpha,z)$, $\alpha\in L_4$, under the cross-bracket operation (3.4), generate commutative algebras of the types (3.2), (3.6), with a *finite-dimensional* underlying algebra g. In addition to the vertex operators the algebra $\hat{g}_Z$ contains operators which are the Fourier coefficients of operators of the form

(3.7) $:g(z)h(z):,\ g,h\in h,$ where $g(z)=\Sigma g(n)z^{-n}$

(3.8) $z\frac{d}{dz}h(z),\ h\in h$

(3.9) $:X_Z(\alpha,z)h(z):,\ \alpha\in L_2,\ h\in h.$

Here the double dots denote the normal ordering operation, determined by the properties

$$:g(m)h(n): = \begin{cases} g(m)h(n) & \text{if } m\leq n \\ h(n)g(m) & \text{if } m\geq n, \end{cases}$$

$$:h(0)e_\alpha: = \frac{1}{2}(h(0)e_\alpha+e_\alpha h(0))$$

for $g,h\in h$, $m,n\in\frac{1}{2}\mathbf{Z}$ and $\alpha\in Q$.

Now take $L = \Lambda$. Defining operators $k_\alpha(n)$, $p_\alpha(n)$ using (2.13), (2.14) for the vertex operators with $\alpha\in\Lambda_4$, one finds a finite-dimensional commutative nonassociative algebra of the form (3.5) with

(3.10) $$\mathfrak{f} = S^2(h) \oplus \sum_{\alpha\in\Lambda_4} \mathbb{C}k_\alpha$$

(3.11) $$p = h \oplus \sum_{\alpha\in\Lambda_4} \mathbb{C}p_\alpha.$$

Here h corresponds to (3.8) and $S^2(h)$ to (3.7).

We can already see part of Griess's algebra at this point, represented by cross-brackets on both $V_{\mathbf{Z}}$ and $V_{\mathbf{Z}+1/2}$: The commutative nonassociative algebra $\mathfrak{f}$ is isomorphic to a subalgebra of Griess's algebra.

4. WEYL GROUPS AND THE INVOLUTION CENTRALIZER

When L is an even unimodular lattice of rank 24 (a "Niemeier lattice"), the corresponding commutative nonassociative algebra analogous to (3.5), (3.10), (3.11), has dimension 196884. Moreover, if we translate the $\mathbf{Z}$-gradation of $V_{\mathbf{Z}}$ so that deg $(1\otimes 1) = 1$, then

$$\text{(4.1)} \quad \text{ch } V_{\mathbf{Z}} = q^{-1} + \dim \mathfrak{g} + 196884q + \ldots = J(q) + \dim \mathfrak{g},$$

where $\mathfrak{g}$ is the rank 24 semisimple or abelian Lie algebra corresponding to the root system L_2 and the character of the graded module

$$\text{(4.2)} \quad V = \bigoplus_{n \geqslant -1} V_{-n}$$

is defined by

$$\text{(4.3)} \quad \text{ch } V = \sum_{n \geqslant -1} (\dim V_{-n})\, q^n.$$

(The indices n are the negatives of those used in the introduction.) Note that the character of the homogeneous vertex representation, up to an added constant, is equal to the modular function which is the central issue of the "Monstrous game". Formula (4.1) follows from the fact that

$$\text{(4.4)} \quad \text{ch } V_{\mathbf{Z}} = \frac{\theta_L(q)}{\eta(q)^{24}},$$

which in turn follows from the construction of $V_{\mathbf{Z}}$. Here

$$\text{(4.5)} \quad \theta_L(q) = \sum_{\alpha \in L} q^{\langle \alpha, \alpha \rangle /2}$$

is the θ-function of L, and

$$\text{(4.6)} \quad \eta(q) = q^{1/24} \prod_{n>0} (1-q^n)$$

is Dedekind's η-function.

Unfortunately, though, none of the 24 algebras corresponding to different Niemeier lattices is Griess's algebra, the algebra that we are looking for. To see that the Monster does not belong to Aut $\mathfrak{g}$ we need to study the subgroups of Aut $\mathfrak{g}$.

Let us look first at the case when $\mathfrak{g}$ is a simple Lie algebra of type A, D, E. Set $W = \mathrm{Aut}\, Q$. Then W can be extended to an automorphism group $\tilde{W}$ of $\mathfrak{g}$ so that $\tilde{W}|_{\mathfrak{h}} = W$. The group $\tilde{W}$ is isomorphic to $2^{\ell}W$, 2^{ℓ} denoting the abelian group $\langle\pm 1\rangle^{\ell} \simeq \bar{Q}$ and $2^{\ell}W$ denoting an extension of W by 2^{ℓ}. For instance, for E_8, W is the Weyl group and $\tilde{W}$ is the normalizer of the 2^8-element group of elements with square 1 in a maximal torus in the corresponding Lie group. When $\mathfrak{g}$ is the above commutative algebra, W and $\tilde{W}$ can be defined analogously. For example, for the Leech lattice we obtain

$$(4.7) \qquad \tilde{W} = 2^{24}(\cdot 0)$$

where $\cdot 0$ is the Conway group, the automorphism group of the Leech lattice (see [2]).

At this point we need to inquire into finite group theory. The latter tells us that there is an important involution centralizer in the Monster which in fact characterizes the Monster among the finite simple groups. This group has the form

$$(4.8) \qquad C = \langle\pm 1\rangle 2^{24}(\cdot 0/\langle\pm 1\rangle) = 2_{+}^{1+24}(\cdot 1).$$

(The subgroup $2_{+}^{1+24} = \langle\pm 1\rangle 2^{24}$ of C is a finite Heisenberg group, called an "extraspecial group" in group theory, and the quotient group $\cdot 1 = \cdot 0/\langle\pm 1\rangle$ of C is a simple group of Conway.) This fact suggests that the Leech lattice is the most appropriate one among the 24 Niemeier lattices for our purposes. However, the discrepancy between (4.7) and (4.8) is very serious, and we need to modify our construction. We note first that though $\tilde{W} \neq C$, their actions on $\mathfrak{f}$ are identical. Moreover, $\mathfrak{f}$ constructed from the Leech lattice (see (3.10)) is isomorphic to the subalgebra of Griess's algebra on which the center of C acts trivially. Thus the problem is to extend the construction of

$\mathfrak{f}$ to the correct $\mathfrak{g}$.

We recall here that the root lattice Γ of E_8 and the Leech lattice Λ are unique in many respects. In particular, Γ is the smallest even unimodular lattice, and Λ is the smallest even unimodular lattice without vectors of square length 2. We have learned to expect many analogous phenomena from these two lattices.

5. THE MOONSHINE SPACE

In the search for the correct commutative algebra we return again to the Lie algebra case. We ask the following natural question: Is there another construction of $\hat{E}_8$ besides the straightforward constructions we considered in Section 2? Keeping in mind that $\hat{\mathfrak{f}}$ is the correct subalgebra in the commutative case, we concentrate our attention on the corresponding subalgebra of $\hat{E}_8'$. It is a Lie algebra of type $\hat{D}_8$. Each of the two vertex representations of $\hat{E}_8$ decomposes into two irreducible components with respect to $\hat{D}_8$ as in (2.19). One can easily identify these as follows:

(5.1) $V_{\mathbf{Z}}^{+} \simeq V_{\text{basic}}$

(5.2) $V_{\mathbf{Z}}^{-} \simeq V_{\text{half-spin}}$

(5.3) $V_{\mathbf{Z}+1/2}^{+} \simeq V_{\text{natural}}$

(5.4) $V_{\mathbf{Z}+1/2}^{-} \simeq V_{\text{half-spin}}$

Thus we obtain all 4 level 1 standard representations of $\hat{D}_8$. The name of the representation indicates the action of the scalar subalgebra D_8 of $\hat{D}_8$ on the top graded subspace. In the case of (5.1)-(5.4), this subspace has corresponding dimensions 1, 128, 16, 128. The finite-dimensional Lie algebra D_8 has one (up to conjugacy) outer involution, and this can be lifted to $\hat{D}_8$. The induced automorphism of the spin representation of $\hat{D}_8$ interchanges the two half-spin subrepresentations. Thus the two half-spin representations of $\hat{D}_8$ are distinguishable only after we choose a basis of this Lie algebra.

Hence we can construct another Lie algebra of type $\hat{E}_8$ if we interchange the two half-spin representations. This leads us to the definition of the following modified vertex representations of $\hat{\mathfrak{f}}$:

$$(5.5)\quad V_{\mathbb{Z}}^{\natural} = V_{\mathbb{Z}}^{+} \oplus V_{\mathbb{Z}+1/2}^{-}$$

$$(5.6)\quad V_{\mathbb{Z}+1/2}^{\natural} = V_{\mathbb{Z}+1/2}^{+} \oplus V_{\mathbb{Z}}^{-}.$$

Of course, we now lose the explicit construction of the second part of the Lie algebra $\hat{E}_8$. However, what is important for us is that it exists. An immediate corollary of the construction is that the dimensions of the corresponding graded components of $V_{\mathbb{Z}}$ and $V_{\mathbb{Z}}^{\natural}$ are equal. This can be expressed as the equality of the two characters, after we suitably translate the $\frac{1}{2}\mathbb{Z}$-gradations of $V_{\mathbb{Z}}$ and $V_{\mathbb{Z}+1/2}$:

$$(5.7)\quad \text{ch } V_{\mathbb{Z}} = \text{ch } V_{\mathbb{Z}}^{\natural}.$$

Now let us define $V_{\mathbb{Z}}^{\natural}$ in the case of the Leech lattice and compare its character with the character of $V_{\mathbb{Z}}$. Setting deg $(1\otimes 1) = 1$ and deg $(1\otimes T) = -\frac{1}{2}$, we obtain

$$(5.8)\quad \text{ch } V_{\mathbb{Z}}^{\natural} = q^{-1} + 0 + 196884q + \dots = J(q).$$

The constant term in (4.1) has disappeared! This discrepancy in the constant term with ch $V_{\mathbb{Z}}$ shows that the Leech lattice case has new features in comparison with the E_8 case; in particular, the algebra $\mathfrak{g}^{\natural}$ acting in $V_{\mathbb{Z}}^{\natural}$ (if such exists) need not be isomorphic to the original $\mathfrak{g}$ acting on $V_{\mathbb{Z}}$.

The space $V_{\mathbb{Z}}^{\natural}$ will be our natural module for the Monster. (This justifies our use of the symbol $\natural$ (natural).)

We can also check that the group C defined by (4.8) acts naturally on $V_{\mathbb{Z}}^{\natural}$. In fact, one can identify $\tilde{W}$ with a group of automorphisms of the group $\hat{\Lambda}$. This action preserves 2Λ and induces an action on the extraspecial group $\hat{\Lambda}/2\Lambda \simeq 2_{+}^{1+24}$. The extraspecial group acts on T (we denote this action by π), and yields a projective representation of $\tilde{\tilde{W}}$. We denote by $\hat{C}$ the linearization of this action. It has the following presentation:

$$\hat{C} = \{(a,a_T) \mid a\in\tilde{\tilde{W}},\ a_T\in GL(T),\ a=\mathrm{int}(a_T) \text{ on } \pi(\hat{\Lambda}/2\Lambda)\},$$

where $\mathrm{int}(a_T)$ denotes conjugation by a_T. The group $\hat{C}$ is a central extension of $\tilde{\tilde{W}}$ by the subgroup $(Id,\pm Id_T)$. The group $\hat{C}$ acts naturally on $V_{\mathbf{Z}} \oplus V_{\mathbf{Z}+1/2}$ as follows: Let $g = (a,a_T) \in \hat{C}$, and require that g fixes $1\otimes 1 \in V_{\mathbf{Z}}$, $g\big|_{1\otimes T} = 1\otimes a_T$ and

$$g\alpha(n)g^{-1} = (\bar{a}\alpha)(n)$$
$$ge_{\alpha}g^{-1} = a(e_{\alpha}).$$

for $\alpha\in\Lambda$, $n\in\frac{1}{2}\mathbf{Z}$; here $\bar{a}$ is the automorphism of Λ induced by a. Thus we obtain a natural action of $\hat{C}$ on $V_{\mathbf{Z}}^{\natural} \oplus V_{\mathbf{Z}+1/2}^{\natural}$. But again $\hat{C}$ does not act faithfully on the first summand, as the subgroup $\pm(Id,Id_T)$ acts trivially there. We denote the factor group by C, which has exactly the structure given by (4.8). Thus the two quite different groups $\tilde{W}$ and C can be viewed as factor groups of one group $\hat{C}$ by two different two-element central subgroups.

6. TRIALITY FOR D_4 AND THE CONSTRUCTION OF $\hat{E}_8$

We proceed now to the construction of $g^{\natural}$ starting from the E_8 case. As one might expect, the construction of explicit operators from $V_{\mathbf{Z}}^{+}$ to $V_{\mathbf{Z}+1/2}^{-}$ and vice versa would be very complicated. However, if we restrict $\hat{D}_8$ to $(D_4 \times D_4)^{\wedge}$ the principle of triality enters the picture. The decomposition of $V_{\mathbf{Z}}^{\natural}$ into irreducible components becomes the following:

(6.1) $\quad V_{\mathbf{Z}}^{\natural} = V_0 \oplus V_1 \oplus V_2 \oplus V_3$

where

(6.2) $\quad V_{\mathbf{Z}}^{+} = V_0 \oplus V_1.$

(6.3) $V^-_{\mathbf{Z}+1/2} = V_2 \oplus V_3$

and V_0 contains the highest weight vector of $V^+_{\mathbf{Z}}$.

Let U_0, U_1, U_2, U_3 be the four fundamental representations of $\hat{D}_4$ of level 1, with U_0 the basic representation. One can choose bases of the two $\hat{D}_4$'s and the indexing of the U_i's in such a way that

(6.4) $V_i = U^1_i \otimes U^2_i$, $i = 0,1,2,3$,

where the upper index refers to one of the two algebras $\hat{D}_4$. The representations U_1, U_2, U_3 have the same character. This fact is a corollary of the S_3 symmetry of the Dynkin diagram D_4

(6.5)

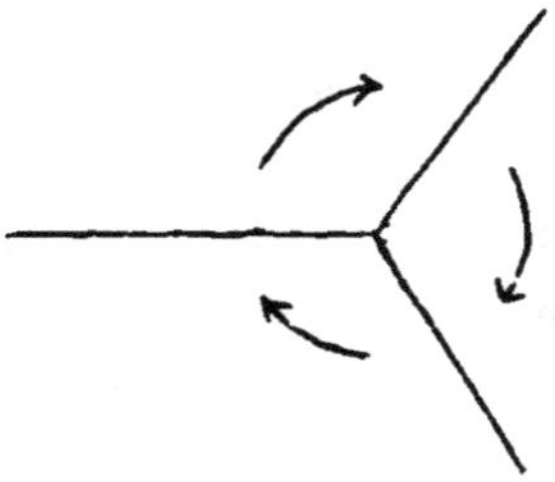

and is equivalent to the "aequatio identica satis abstrusa" of Jacobi

(6.6) $$\frac{1}{2}\left(\prod_{n=1}^{\infty}(1 + q^{n-1/2})^8 - \prod_{n=1}^{\infty}(1 - q^{n-1/2})^8\right) =$$

$$= 8q^{1/2}\prod_{n=1}^{\infty}(1 + q^n)^8,$$

which has already arisen in supersymmetric string models [11][1]. The group S_3 permutes U_1, U_2, U_3, and correspondingly, V_1, V_2, V_3. Let

(6.7) $\mathfrak{k} = \mathfrak{k}_0 \oplus \mathfrak{p}_1$,

where $\mathfrak{k}$ has type D_8, $\mathfrak{k}_0$ has type $D_4 \times D_4$, and $\mathfrak{p}_1$ is the orthogonal complement of $\mathfrak{k}_0$ in $\mathfrak{k}$. Let

---

[1]We thank Bernard Julia for bringing this reference to our attention.

$$\bar{\mathfrak{t}} = \mathfrak{t} \otimes \mathbb{C}[t,t^{-1}], \text{ etc.,}$$

be their affinizations. We define the following operators on $V_{\mathbb{Z}} \oplus V_{\mathbb{Z}+1/2}$:

$$(6.8) \quad \begin{aligned} p_2 &= \sigma_2\, p_1\, \sigma_2^{-1},\ p_3 = \sigma_3\, p_1\, \sigma_3^{-1}, \\ \bar{p}_2 &= \sigma_2\, \bar{p}_1\, \sigma_2^{-1},\ \bar{p}_3 = \sigma_3\, \bar{p}_1\, \sigma_3^{-1}, \end{aligned}$$

where σ_2, σ_3 are two involutions in S_3. Then the spaces of operators

$$(6.9) \quad g^{\natural} = \mathfrak{t}_0 \oplus p_1 \oplus p_2 \oplus p_3,$$

$$(6.10) \quad \hat{g}^{\natural} = \bar{\mathfrak{t}}_0 \oplus \bar{p}_1 \oplus \bar{p}_2 \oplus \bar{p}_3 \oplus \mathbb{C}c$$

form Lie algebras of type E_8 and $\hat{E}_8$, respectively.

Now we want to repeat this construction in the Leech lattice case. The first question we need to answer is: What is the analogue of $\mathfrak{t}_0$? To this end we shall look at $\mathfrak{t}_0$ for the E_8 case from a new point of view.

7. LATTICES FROM CODES. DEFINITION OF $\mathfrak{t}_0$

As we have mentioned, the E_8 and Leech lattices are unique in many respects. If we want to look at the more detailed structure of these lattices, we find that they can be built from no less unique binary codes, namely, the Hamming and Golay codes. The Hamming code $\mathcal{C}_\Gamma$ can be characterized as a 4-dimensional subspace of $\mathbb{F}_2^8$ whose smallest nonzero weight (number of nonzero coordinates) is 4. Similarly, the Golay code $\mathcal{C}_\Lambda$ is characterized by the property that it is a 12-dimensional subspace of $\mathbb{F}_2^{24}$ whose smallest zero weight is 8. (These properties characterize the codes up to a permutation of the different factors $\mathbb{F}_2$.) It is easy to see that there are exactly 14 code words of weight 4 in the Hamming code $\mathcal{C}_\Gamma$, and in addition this code contains the trivial code word and the code word of weight 8. Thus the weight distribution is

(7.1) $0^1\ 4^{14}\ 8^1.$

The Golay code $\mathcal{C}_\Lambda$ has the weight distribution

(7.2) $0^1\ 8^{759}\ 12^{2576}\ 16^{759}\ 24^1.$

Now let us consider the "square" lattices $\sum_{i=1}^{8} \mathbb{Z}(\frac{1}{4}\alpha_i)$, $\sum_{i=1}^{24} \mathbb{Z}(\frac{1}{4}\alpha_i)$ with the symmetric form $<\cdot,\cdot>$ determined by the property

(7.3) $<\alpha_i,\alpha_j> = 2\delta_{ij},$

and their sublattices Γ, Λ spanned by the vectors

(7.4) $\frac{1}{2}\sum_{i=1}^{8} \pm \epsilon_i\alpha_i, \quad \epsilon \in \mathcal{C}_\Gamma, \quad \text{weight } \epsilon = 4$

(7.5) $\frac{1}{2}\sum_{i=1}^{8} (\epsilon_i - \frac{1}{2})\ \alpha_i \pm \alpha_k, \quad \epsilon \in \mathcal{C}_\Gamma, \quad k=1,\ldots,8$

and respectively by the vectors

(7.6) $\frac{1}{2}\sum_{i=1}^{24} \pm \epsilon_i\alpha_i, \quad \epsilon \in \mathcal{C}_\Lambda, \quad \text{weight } \epsilon = 8$

(7.7) $\frac{1}{2}\sum_{i=1}^{24} (\epsilon_i - \frac{1}{2})\alpha_i \pm \alpha_k, \quad \epsilon \in \mathcal{C}_\Lambda, \quad k=1,\ldots,24,$

where the number of plus signs in (7.4), (7.6) is even, and the sign in (7.5), (7.7) is chosen in such a way that one gets $\pm\frac{3}{4}\alpha_k$. One can check easily that the lattices Γ and Λ have the characteristic properties of the E_8 and Leech lattices, respectively.

The vectors (7.4), (7.5) have square length 2 and constitute the root system of type E_8. Moreover, the division of the root system into two sets corresponds exactly to the decomposition (6.7), i.e., $\mathfrak{k}_0$ is constructed from the root subsystem (7.4), and $\mathfrak{p}_1$ from (7.5).

Similarly, the vectors (7.6), (7.7) have square length 4. Thus one can define p_1 from (7.7) for the Leech lattice algebra in exactly the same way as for E_8. However, Λ contains additional elements of square length 4:

(7.8) $\pm\alpha_i \pm \alpha_j$, $i,j = 1,\ldots,24$, $i \neq j$.

Thus (7.1) and (7.2) imply in particular that

$$|\Gamma_2| = 240,$$

$$|\Lambda_4| = 196560.$$

Moreover, in addition to the subspace constructed from the vectors (7.6) and (7.8), $\mathfrak{f}_0$ should contain $S^2(\mathfrak{h})$ (see (3.10)). With $\mathfrak{f}_0$ defined this way, we obtain a decomposition of $V_{\mathbf{Z}}^{\natural}$ into four irreducible components (6.1)-(6.3) exactly as in the E_8 case. One can show that V_1, V_2, V_3 have the same character. This suggests the existence of a group S_3 which permutes the V_i and induces a group of automorphisms of $\mathfrak{f}_0$. Precise formulas for the action of a group S_3 on $\mathfrak{f}_0$ were in fact obtained by Griess [15]. However, those calculations do not help us prove the closure of the algebra $\mathfrak{g}^{\natural}$. Thus we need to find the underlying mechanism of triality.

8. TRIALITY FOR $\hat{A}_1$

The principle of triality is very important in the construction of the exceptional Lie and Jordan algebras. It also turns out to be crucial for our construction of the commutative algebra $\mathfrak{g}^{\natural}$. The simplest manifestation of this principle is the existence of a symmetrical basis $\{y_1, y_2, y_3\}$ of $sl_2 = A_1$ satisfying

(8.1) $[y_1, y_2] = 2y_3$, $[y_2, y_3] = 2y_1$, $[y_3, y_1] = 2y_2$.

The group S_3 acts naturally on sl_2 by permuting the three indices of the basis elements. In particular, S_3 contains three involutions

(8.2) σ_i: $y_i \longmapsto -y_i$ $\quad$ σ_i: $y_{i+1} \longmapsto y_{i+2}$.

where the indices are thought of as periodic mod 3.

We can associate to sl_2 with the fixed basis four affine Lie algebras as in Section 2:

$$\hat{\mathfrak{a}}_0 = \sum_{m\in\mathbb{Z}} \mathbb{C}y_1(m) \oplus \sum_{m\in\mathbb{Z}} \mathbb{C}y_2(m) \oplus \sum_{m\in\mathbb{Z}} \mathbb{C}y_3(m) \oplus \mathbb{C}c$$

$$(8.3)\quad \hat{\mathfrak{a}}_i = \sum_{m\in\mathbb{Z}} \mathbb{C}y_i(m) \oplus \sum_{m\in\mathbb{Z}+1/2} \mathbb{C}y_{i+1}(m) \oplus$$

$$\oplus \sum_{m\in\mathbb{Z}+1/2} \mathbb{C}y_{i+2}(m) \oplus \mathbb{C}c,\ i=1,2,3.$$

Clearly the action of the group S_3 extends naturally to $\hat{\mathfrak{a}}_0$, while $\hat{\mathfrak{a}}_i$ is preserved only by the involution σ_i.

Each of these algebras is of type $\hat{A}_1$ and has two fundamental representations. Both of these admit the vertex operator constructions described in Section 2.

We recall that the vertex operator construction is based on restriction to the Heisenberg subalgebra $\hat{h} = \sum_m \mathbb{C}y_1(m) \oplus \mathbb{C}c$. For $\hat{\mathfrak{a}}_2$ and $\hat{\mathfrak{a}}_3$ this restriction is still irreducible and therefore the representation space is isomorphic to $S(\hat{h}^-)$. For $\hat{\mathfrak{a}}_0$ and $\hat{\mathfrak{a}}_1$, to make it irreducible, we need to add the translation part of the Weyl group of $\hat{A}_1$. We define vertex operators by the formulas (2.11), (2.12), using a bilinear cocycle on $\mathbb{Z}(\frac{1}{4}\alpha)$ with values in $\langle \pm 1, \pm(-1)^{1/2} \rangle$ satisfying

$$(8.4)\quad \epsilon(\alpha, \tfrac{1}{4}\alpha) = (-1)^{1/2}.$$

This nontrivial choice of ϵ will be justified when we pass to higher rank lattices. The representation spaces have the following form:

$$(8.5)\quad \hat{\mathfrak{a}}_0:\ S(\hat{h}^-_{\mathbb{Z}}) \otimes \mathbb{C}[\mathbb{Z}\alpha \pm \tfrac{\alpha}{4} + \tfrac{\alpha}{4}]$$

$$(8.6)\quad \hat{\mathfrak{a}}_1:\ S(\hat{h}^-_{\mathbb{Z}}) \otimes \mathbb{C}[\mathbb{Z}\alpha \pm \tfrac{\alpha}{4}]$$

(8.7) $\hat{\alpha}_2$: $S(\hat{h}^-_{\mathbf{Z}+1/2}) \otimes \mathbb{C}e^{\pm\alpha/4+\alpha/4+\mathbf{Z}\alpha}$

(8.8) $\hat{\alpha}_3$: $S(\hat{h}_{\mathbf{Z}+1/2}) \otimes \mathbb{C}e^{\pm\alpha/4+\mathbf{Z}\alpha}$

where the one-dimensional factors in (8.7), (8.8) are denoted as they are for the sake of parallel definitions of e_α and e'_α on the second tensor factors. (The notations $\mathbb{C}[\mathbf{Z}\alpha\pm\frac{\alpha}{4}]$, etc., denote the obvious subspaces of the group algebra $\mathbb{C}[\frac{1}{4}\mathbf{Z}\alpha]$.)

Although S_3 does not act on each of the $\hat{\alpha}_i$ separately it permutes them and therefore their vertex representations. For example, there is a unique isomorphism[2]

(8.9) σ_2: $S(\hat{h}^-_{\mathbf{Z}}) \otimes \mathbb{C}[\mathbf{Z}\alpha\pm\frac{\alpha}{4}] \longrightarrow S(\hat{h}^-_{\mathbf{Z}+1/2}) \otimes \mathbb{C}e^{\pm\alpha/4+\mathbf{Z}\alpha}$

such that

$$\sigma_2: 1 \otimes e^{\pm\alpha/4} \mapsto 1 \otimes e^{\pm\alpha/4+\mathbf{Z}\alpha}$$

(8.10)

$$\sigma_2 \circ x(m) \circ \sigma_2^{-1} = (\sigma_2 x)(m),\ x = y_1,\ y_2 \text{ or } y_3,\ m\in\tfrac{1}{2}\mathbf{Z}.$$

It is interesting to note that comparison of the characters of the spaces in (8.9) gives a standard consequence of the Jacobi triple product identity:

$$(8.11)\quad \frac{\sum_{n\in\mathbf{Z}} q^{(n+1/4)^2}}{\eta(q)} = \frac{\eta(q)}{\eta(q^{1/2})}.$$

---

[2]This isomorphism was also discovered in a different context by R. Pfister [21].

9. TRIALITY FOR HIGHER RANK

Now we have the main tool for obtaining triality in our two cases of the E_8 and Leech lattices. The basic idea is to take the product of 8 or 24 copies of the affine Lie algebras of type $\hat{A}_1$ considered in the last section. Since the two cases are completely analogous, we shall explain triality for our main example of the Leech lattice. We first note that although the Leech lattice Λ has no elements of norm square 2, its double cover

(9.1) $\tilde{\Lambda} = \mathbb{Z}\text{-span } (\Lambda \cup \{\alpha_k\})$

has 48 such elements, which correspond to the root system of type $(A_1)^{24}$. Let

(9.2) $Q = \sum_{i=1}^{24} \mathbb{Z}\alpha_i$

and let N be the unique even unimodular sublattice of $\tilde{\Lambda}$ containing Q; N is a Niemeier lattice of type A_1^{24}. Then

(9.3) $\tilde{\Lambda} = N \cup N'$

where

(9.4) $N' = N + \alpha'$,

(9.5) $\alpha' = \frac{1}{4}\sum_{i=1}^{24} \alpha_i.$

We would like to define a bilinear cocycle ϵ on $\tilde{\Lambda}$ which has the properties (1.2), (1.3) for Λ and N. This forces us to use a cocycle with values in $<\pm 1, \pm(-1)^{1/2}>$. It will be convenient to have

(9.6) $\epsilon(\alpha, \beta+\gamma) = \epsilon(\alpha, \beta)$ for $\alpha, \beta \in \tilde{\Lambda}$, $\gamma \in Q$

(9.7) $\epsilon(\alpha_k, \alpha') = (-1)^{1/2}$ for $k=1,\ldots,24$.

Formula (9.6) will enable us to define explicit operators e'_α (cf. (2.7)) on the spaces $\mathbb{C}[N/Q]$ and $\mathbb{C}[N'/Q]$ below ((9.11), (9.12)). A cocycle with these properties can in fact be constructed (see [8]).

Now we consider the space

$$(9.8)\quad W = W_0 \oplus W_1 \oplus W_2 \oplus W_3$$

where

$$(9.9)\quad W_0 = S(\hat{h}^-_{\mathbf{Z}}) \otimes \mathbb{C}[N]$$

$$(9.10)\quad W_1 = S(\hat{h}^-_{\mathbf{Z}}) \otimes \mathbb{C}[N']$$

$$(9.11)\quad W_2 = S(\hat{h}^-_{\mathbf{Z}+1/2}) \otimes \mathbb{C}[N/Q]$$

$$(9.12)\quad W_3 = S(\hat{h}^-_{\mathbf{Z}+1/2}) \otimes \mathbb{C}[N'/Q].$$

One can see immediately that the Fourier components of the vertex operators $X(\pm\alpha_k, z)$ acting in W_i generate a copy of the algebra $\hat{\mathfrak{a}}_i$ (8.3), i=0,1,2,3. The 24 algebras of type $\hat{A}_1$, one for each k=1,...,24, commute pairwise, so that we have algebras of type $\hat{A}_1^{24}$ acting on the four pieces. Under these algebras each piece decomposes into 2^{12} irreducible components, in the first two according to the cosets of Q in N and N', respectively, and in the last two according to the natural basis of the second factor in the tensor product.

The highest weight vectors are the vectors

$$(9.13)\quad 1 \otimes e^{\alpha_C/2}, \qquad C\in\mathcal{C}_\Lambda$$

$$(9.14)\quad 1 \otimes e^{\beta}, \qquad \beta \text{ of the form } \frac{\pm\alpha_1\pm\dots\pm\alpha_{24}}{4}$$

$$(9.15)\quad 1 \otimes e^{\alpha_C/2+Q}, \qquad C\in\mathcal{C}_\Lambda$$

$$(9.16)\quad 1 \otimes e^{\beta+Q}, \qquad \beta \text{ of the form } \frac{\pm\alpha_1\pm\dots\pm\alpha_{24}}{4}$$

where $\alpha_C = \sum_i \epsilon_i \alpha_i$ if $C = (\epsilon_i)_i$, $\epsilon_i \in \{0,1\}$.

Taking a tensor product of 24 copies of the spaces described in Section 8 we obtain the action of the triality group S_3 on W, which permutes the W_i, i=1,2,3. For example, there is a unique isomorphism

$$(9.17) \quad \sigma_2 : S(\hat{h}^-_{\mathbf{Z}}) \otimes \mathbf{C}[N'] \longrightarrow S(\hat{h}^-_{\mathbf{Z}+1/2}) \otimes \mathbf{C}[N'/Q]$$

such that

$$(9.18) \quad \begin{aligned} &\sigma_2 : 1 \otimes e^{\beta} \longmapsto 1 \otimes e^{\beta+Q} \text{ for } \beta \text{ of the form } \frac{\pm\alpha_1 \pm \cdots \pm \alpha_{24}}{4} \\ &\sigma_2 \circ x(m) \circ \sigma_2^{-1} = (\sigma_2 x)(m) \end{aligned}$$

for $x = y_1$, y_2 or y_3 in one of the 24 $\hat{A}_1$s, and $m \in \frac{1}{2}\mathbf{Z}$.

Now let us consider products

$$(9.19) \quad y_i^{(r)}(z) y_i^{(s)}(z), \qquad i=1,2,3,$$

where r,s=1,...,24, r≠s; here each superscript designates a copy of $\hat{A}_1$. On each space W_j, each of the two factors in (9.19) has an expansion into powers z^n, $n \in Z$ ($Z=\mathbf{Z}$ or $\mathbf{Z}+\frac{1}{2}$). Thus (9.19) can always be expanded on W in integral powers of z. Each element of the triality group S_3 acting on W transforms each operator (9.19), by conjugation, into another such operator with the same r,s. When r=s, we must replace (9.19) by the corresponding normal ordered product (defined with respect to the Heisenberg algebra associated with the index i), which becomes the generating function of the Virasoro algebra associated with the corresponding copy of $\hat{A}_1$, and which is independent of i. Triality acts trivially on these operators.

For r,s = 1,...,24 and i=1,2,3, consider the expansion

$$(9.20) \quad :y_i^{(r)}(z) y_i^{(s)}(z): = \sum_{n \in \mathbf{Z}} y_i^{(r,s)}(n) z^{-n}.$$

Then all the operators $y_i^{(r,s)}(n)$, together with the identity operator, span a commutative algebra $\hat{\mathfrak{f}}_*$ under the cross-bracket operation, where

$$(9.21)\quad \mathfrak{f}_* = S^2(\mathfrak{h}) \oplus \Sigma \mathbb{C} k_\alpha,$$

α ranging through the elements (7.8) (see (3.2), (3.3), (3.4)). Note that

$$(9.22)\quad S^2(\mathfrak{h}) = \sum_{r,s=1}^{24} \mathbb{C}\alpha_r\alpha_s$$

breaks into the 24-dimensional space where r=s, corresponding to 24 Virasoro algebras, and the 276-dimensional space where r≠s, corresponding to (9.20) for i=1. Also, the second summand in (9.21) breaks up as

$$(9.23)\quad \bigoplus_{r<s} \mathbb{C}(k_{\alpha_r+\alpha_s} + k_{\alpha_r-\alpha_s}) \oplus \bigoplus_{r<s} \mathbb{C}(k_{\alpha_r+\alpha_s} - k_{\alpha_r-\alpha_s}),$$

two 276-dimensional spaces corresponding to (9.20) for i=2,3.

Thus we see that $\mathfrak{f}_*$ and $\hat{\mathfrak{f}}_*$ are preserved by the triality group S_3.

The structure $\mathfrak{f}_*$ does not exist in the case of the E_8-lattice Γ, and in fact will turn out to be the only difference between the algebras corresponding to Γ and Λ.

10. TRIALITY FOR VERTEX OPERATORS

Now we consider the vertex operators $X_Z(\alpha,z)$, $Z=\mathbb{Z}$ or $\mathbb{Z}+\frac{1}{2}$, for $\alpha \in N$. These operators preserve each space $W_0,\dots,W_3$ (see (9.9)-(9.12)). The triality group S_3 acting on W transforms each of these operators by conjugation. Since there are explicit formulas for the action of S_3 on W (cf. Section 9), we can in principle compute these transforms. We are interested in particular in the vertex operators $X_Z(\alpha,z)$, α as in (7.6), because the expansion coefficients of

$$(10.1)\quad X_Z(\alpha,z) + X_Z(-\alpha,z),$$

together with the operators corresponding to $\hat{\mathfrak{f}}_*$ (see (9.21)) span the subalgebra $\hat{\mathfrak{f}}_0$. It is an important fact that the family of operators $\hat{\mathfrak{f}}_0$ is preserved by triality.

Recall that

$$(10.2)\quad \sigma_2 : W_1 \longrightarrow W_3$$

in (9.17) induces the maps σ_2 for sl_2 (8.2) and σ_2 for $\hat{A}_1$ (8.9) (see (8.10), (9.18)). First we determine how the individual vertex operators $X_Z(\alpha,z)$, α as in (7.6), acting on W_1 (where $Z=\mathbb{Z}$) and W_3 (where $Z=\mathbb{Z}+\frac{1}{2}$), transform under σ_2. From the standard calculus of vertex operators, it follows easily that if we bracket an element of any one of the 24 $\hat{A}_1$'s with an operator $X_Z(\alpha,z)$, we obtain a linear combination of such operators. These commutation relations show that $\sigma_2 X(\alpha,z)\sigma_2^{-1}$ is a linear combination of vertex operators of the same type, and the coefficients are uniquely determined. More precisely,

$$(10.3)\quad \sigma_2(-i)^{|S|}X(\epsilon_S\tfrac{1}{2}\alpha_C,z)\sigma_2^{-1} =$$

$$=2^{-|C|/2}\sum_{T\subset C}(-1)^{|S\cap T|}(-i)^{|T|}X(\epsilon_T\tfrac{1}{2}\alpha_C,z),$$

for $S\subset C$, $C\in\mathcal{C}_\Lambda$, where $i=(-1)^{1/2}$, α_C is as in Section 9 and ϵ_S is the automorphism of $\mathfrak{h}$ which changes the sign of those α_k for which $k\in S$; here we view the code words in $\mathcal{C}_\Lambda$ as subsets of $\{1,\dots,24\}$.

So far, σ_2 has been defined only on W_1 and W_3. Postulating that (10.3) should hold on all of W, we are able (uniquely) to extend σ_2 to an involution σ_2 on W. This involution, however, does not preserve the subspace $V_{\mathbb{Z}}^{\natural}$ of W, and so we modify σ_2 slightly on one of the 24 coordinates to obtain an involution σ of W which does preserve $V_{\mathbb{Z}}^{\natural}$ (see [8]). Combining this with a suitable involution in C (see [8]), we obtain the triality group -- a copy of S_3 acting on $V_{\mathbb{Z}}^{\natural}$, and by conjugation, on the linear combinations of the vertex operators of the type $X(\alpha,z)$, α as in (7.6). The main content of this "triality for vertex operators" is the conjugation formula (10.3).

Now we recall that $\hat{\mathfrak{f}}_0$ is made up of $\hat{\mathfrak{f}}_*$ and the Fourier coefficients of the operators (10.1) for α as in (7.6). Combining the triality of $\hat{\mathfrak{k}}_*$ (see Section 9) with the triality for vertex operators, we obtain an action of S_3 on $\hat{\mathfrak{k}}_0$.

If we replace Λ by Γ and the Golay code $\mathcal{C}_\Lambda$ by the Hamming code $\mathcal{C}_\Gamma$, then we similarly obtain a triality for the corresponding vertex operators. The resulting triality of

$$\hat{\mathfrak{k}}_0 = (D_4 \times D_4)\hat{}$$

agrees with the usual Dynkin diagram triality discussed in Section 6.

This completes our discussion of triality. It is interesting to note that the principle of triality, which plays an important role in the construction of the exceptional Lie algebras, is applicable in a more complicated "sporadic" situation.

Using an analogue of (6.8), we obtain spaces of operators

(10.4) $\mathfrak{B} = \mathfrak{k}_0 \oplus p_1 \oplus p_2 \oplus p_3,$

(10.5) $\hat{\mathfrak{B}} = \bar{\mathfrak{k}}_0 \oplus \bar{p}_1 \oplus \bar{p}_2 \oplus \bar{p}_3 \oplus \mathbb{C}c,$

analogous to (6.9) and (6.10), respectively. However, in order to show that $\mathfrak{B}$ and $\hat{\mathfrak{B}}$ are algebras, we still must show

(10.6) $\bar{p}_1 \times \bar{p}_2 \subset \bar{p}_3$

and its cyclic permutations (see (3.4)). This cannot be proved by direct calculations since the operators involved act from $V_{\mathbf{Z}}^{+}$ to $V_{\mathbf{Z}+1/2}^{-}$ and vice versa; also, triality is not helpful here. In order to complete the proof we need to extend our space further. In particular, we need to consider a quadruple cover of the Leech lattice which contains the double cover already used. This quadruple cover arises naturally in the quaternionic construction of the Leech lattice, and we proceed to its definition.

11. QUATERNIONIC CONSTRUCTIONS OF LATTICES

The quaternionic constructions of the Leech and E_8 lattices [25] are in many respects analogous to their constructions from the binary codes. Now we consider codes over the field of four elements $\mathbf{F}_4$. Again there are two remarkable self-dual codes over $\mathbf{F}_4$ in dimensions two and six, which we shall denote by $\mathcal{C}_\Gamma(\mathbf{F}_4)$ and $\mathcal{C}_\Lambda(\mathbf{F}_4)$, respectively. The first is the unique self-dual code in dimension two, and the second is characterized by the property that the shortest word has 4 nonzero coordinates. In the division ring of quaternions, we set

(11.1) $P_4 = \mathbf{Z}[i,j,k,\frac{1}{2}(1+i+j+k)]$

(11.2) $Q_4 = \{x \in \mathbf{Z}[i,j,k] \mid x\bar{x} \in 2\mathbf{Z}\}$.

Then obviously we have

(11.3) $P_4/Q_4 \cong \mathbf{F}_4$,

where we choose the following representatives of the elements of $\mathbf{F}_4$ in P_4:

(11.4) $0, 1, \alpha = \frac{1}{2}(1+i+j+k),\ \alpha^2 = \frac{1}{2}(-1+i+j+k)$.

We also denote by θ_4 an element of Q_4 such that the ideal generated by θ_4 in P_4 is Q_4, e.g., $\theta_4 = i + j$, and set $\vec{a} = (a,a,\ldots,a)$ for $a \in P_4$. For $a,b \in \mathbf{F}_4$ and for $L=\Gamma$ or Λ, set

(11.5) $L^{ab} = \{\vec{a}+\theta_4 c+2x \mid c \in \mathcal{C}_L(\mathbf{F}_4), x \in P_4^{\ell}, \sum_{i=1}^{\ell} x_i \equiv b \pmod{Q_4}\}$.

In particular, when $L = \Lambda$, $\ell = 6$ and we have

(11.6) $N(D_4^6) = \Lambda^{00} \cup \Lambda^{01} \cup \Lambda^{0\alpha} \cup \Lambda^{0\alpha^2}$

(11.7) $\Lambda = \Lambda^{00} \cup \Lambda^{1\alpha} \cup \Lambda^{\alpha\alpha^2} \cup \Lambda^{\alpha^2 1}$

(11.8) $\bar{\Lambda} = \Lambda^{00} \cup \Lambda^{1\alpha^2} \cup \Lambda^{\alpha 1} \cup \Lambda^{\alpha^2\alpha}$, a Niemeier lattice of type D_4^6, a Leech lattice, and its conjugate, respectively. We consider the following quadruple cover of the Leech lattice:

$$(11.9) \quad \tilde{\tilde{\Lambda}} = \Lambda + N(D_4^6) = \bigcup_{a,b\in\mathbb{F}_4} \Lambda^{ab}.$$

12. $L_3(2)$ SYMMETRY

The construction of the Leech lattice and its double cover from the binary Golay code allowed us to introduce the action of 24 copies of $(sl_2)^\wedge$ in the extended vertex representation. Similarly the quaternionic construction of the Leech lattice and its quadruple cover give us an action of 6 copies of $\hat{D}_4$. The analogue of triality for sl_2 is now "$L_3(2)$-symmetry" of D_4. By $L_3(2)$ we mean the group of invertible 3×3 matrices over the 2-element field $\mathbb{F}_2$.

The simplest way to define this symmetry is to build it from two trialities of D_4: one of these trialities is induced from four sl_2 trialities (see Section 8) inside D_4, and the second follows from the symmetry of the Dynkin diagram (see Section 6). In more detail, fixing a Cartan subalgebra h of D_4, we have a natural decomposition of the root system,

$$(12.1) \quad \Delta(D_4) = \Delta^{(1)} \cup \Delta^{(2)} \cup \Delta^{(3)},$$

where $\Delta^{(i)}$, i=1,2,3, is a copy of the root system of type $(A_1)^4$. Thus the root system is given by $\pm\alpha_j^{(i)}$, i=1,2,3, j=1,2,3,4. Fixing i we get the decomposition

$$(12.2) \quad D_4 = (sl_2)^4 \oplus (\mathbb{C}^2)^{\otimes 4},$$

where $\mathbb{C}^2$ is the natural representation of sl_2. Then extending the triality of sl_2 described in Section 8 to its natural representation $\mathbb{C}^2$, and taking it four times we obtain the first triality of D_4. The second is obvious from (12.1). The two groups S_3 generate a group of automorphisms of D_4 which is an extension of $L_3(2)$ by $(\mathbb{Z}/2)^3$ and which we denote by $2^3L_3(2)$. It permutes the following seven subspaces according to the way that $L_3(2)$ permutes the nonzero elements of $(\mathbb{F}_2)^3$:

(12.3) $D_4 = \bigoplus_{i=1}^{7} \mathfrak{h}_i.$

Here $\mathfrak{h}_4$ is the fixed Cartan subalgebra $\mathfrak{h}$ of D_4,

(12.4) $\mathfrak{h}_i = \sum_{j=1}^{4} \mathbb{C}(x_{\alpha_j^{(i)}} - x_{-\alpha_j^{(i)}})$, i=1,2,3,

(12.5) $\mathfrak{h}_{4+i} = \sum_{j=1}^{4} \mathbb{C}(x_{\alpha_j^{(i)}} + x_{-\alpha_j^{(i)}})$, i=1,2,3.

Each subspace in the decomposition (12.3) is a Cartan subalgebra and any two Cartan subalgebras generate a copy of $(sl_2)^4$. It is sometimes convenient to identify the index set in (12.3) with the set of nonzero elements of $(\mathbb{F}_2)^3$. This group $L_3(2)$ is clearly transitive on the 2-dimensional subspaces of $(\mathbb{F}_2)^3$. Therefore any subalgebra $(sl_2)^4$ can be conjugated to the subalgebra $\mathfrak{h}_1 \oplus \mathfrak{h}_2 \oplus \mathfrak{h}_3$. This fact and its extension to Griess's algebra are crucial in our proof of the closure of $\hat{\mathfrak{B}}$.

We associate to the Lie algebra $\mathfrak{o}_8$ of type D_4 eight affine Lie algebras similar to the four algebras associated to sl_2 in Section 8.

(12.6) $\hat{\mathfrak{a}}_0 = \mathfrak{o}_8 \otimes \mathbb{C}[t,t^{-1}] \oplus \mathbb{C}c$

(12.7) $\hat{\mathfrak{a}}_i = \mathfrak{k}_i \otimes \mathbb{C}[t,t^{-1}] \oplus \mathfrak{p}_i \otimes t^{1/2}\mathbb{C}[t,t^{-1}] \oplus \mathbb{C}c$, i=1,...,8,

where $\mathfrak{k}_i = (sl_2)^4$, $\mathfrak{p}_i = (\mathbb{C}^2)^{\otimes 4}$. Then the group $2^3L_3(2)$ acts naturally on $\hat{\mathfrak{a}}_0$ and permutes the $\hat{\mathfrak{a}}_i$. We could then consider the basic representations of each of those algebras similar to (8.4)–(8.7) and define the action of $2^3L_3(2)$ on them.

Passing to our main picture, recall that in Section 11 we have defined the quadruple cover of the Leech lattice $\tilde{\Lambda}$, which contains the

Niemeier lattice N of type $\overline{D_4^6}$. Let

(12.8) $\tilde{\Lambda} = N^{(0)} \cup N^{(1)} \cup N^{(2)} \cup N^{(3)}, \qquad N^{(0)} = N$

be the coset decomposition with respect to N. We define

(12.9) $Y = \bigoplus_{i=0}^{7} Y_i$

where

(12.10) $Y_i = S(\hat{h}_{\mathbb{Z}}^{-}) \otimes \mathbb{C}[N^{(i)}], \; i=0,1,2,3$

(12.11) $Y_i = S(\hat{h}_{\mathbb{Z}+1/2}^{-}) \otimes \mathbb{C}[N^{(i-4)}/Q], \; i=4,5,6,7$

(with Q the root lattice of D_4^6). The group $2^3L_3(2)$ acts on Y and on its subspace V, which has a nontrivial intersection with every subspace in (12.9), $V_i = V \cap Y_i$. The spaces of operators $\mathbb{B}$ and $\hat{\mathbb{B}}$ have natural decompositions compatible with the decomposition (12.9):

(12.12) $\mathbb{B} = \bigoplus_{i=0}^{7} \mathbb{B}_i$

(12.13) $\hat{\mathbb{B}} = \bigoplus_{i=0}^{7} \bar{\mathbb{B}}_i \oplus \mathbb{C}c.$

In order to prove the closure of $\hat{B}$ we need to show that

(12.14) $\bar{\mathbb{B}}_i \times \bar{\mathbb{B}}_j \subset \bar{\mathbb{B}}_{i+j}$ (sum in $(\mathbb{F}_2)^3$).

Using the $L_3(2)$-symmetry we can conjugate the triple $\bar{\mathbb{B}}_i$, $\bar{\mathbb{B}}_j$, $\bar{\mathbb{B}}_{i+j}$ to $\bar{\mathbb{B}}_1$, $\bar{\mathbb{B}}_2$, $\bar{\mathbb{B}}_3$. But then we are inside $\hat{f}$, and the inclusion (12.14) is valid! This idea allows us to complete the proof that we have obtained a representation of the cross-bracket algebra $\hat{\mathbb{B}}$ in $V_{\mathbb{Z}}^{\natural}$.

13. THE MAIN THEOREM AND THE STRUCTURE OF GRIESS'S ALGEBRA

We now summarize what we have accomplished in the preceding sections:

We have constructed an infinite-dimensional graded commutative nonassociative algebra $\hat{\mathcal{B}}$ and an irreducible representation π of $\hat{\mathcal{B}}$ by cross bracket on a graded module $V = V_{\mathbb{Z}}^{\natural}$ with character $J(q)$. We have also constructed a group F_1 acting on V and as automorphisms of $\hat{\mathcal{B}}$, preserving their gradings, and such that

(13.1) $g\pi(x)g^{-1} = \pi(g \cdot x)$

for $g \in F_1$ and $x \in \hat{\mathcal{B}}$. The algebra $\hat{\mathcal{B}}$ is the affinization of a commutative nonassociative algebra $\mathcal{B}$ of dimension 196884, equipped with a nonsingular symmetric associative bilinear form $\langle \cdot, \cdot \rangle$.

Our construction of $\hat{\mathcal{B}}$ allows us to describe $\mathcal{B}$ very explicitly. Let us identify $\mathcal{B}$ with the subspace $\mathcal{B} \otimes 1$ in $\hat{\mathcal{B}}$. Then $\mathcal{B}$ acts on

$$V = \oplus_{n \leq 1} V_n = V_1 \oplus 0 \oplus V_{-1} \oplus V_{-2} \oplus \ldots$$

preserving the grading. The nontrivial homogeneous subspace V_{-1} "just below" the trivial one-dimensional space V_1 has the same dimension as $\mathcal{B}$, in analogy with the E_8 theory. We identify $\mathcal{B}$ with V_{-1} via the linear map

(13.2) $\iota(x) = \pi(x(-2))(1 \otimes 1), \; x \in \mathcal{B}.$

Again as in the E_8 case, V_{-1} turns out to be the adjoint representation of $\mathcal{B}$, in the sense that

(13.3) $\pi(x(0))\iota(y) = \iota(x \times y), \; x, y \in \mathcal{B}.$

This fact follows from the cross bracket identity

$$\pi((x \times y)(m+n)) =$$

(13.4)

$$= [\pi(x(m+\tfrac{1}{2})), \pi(y(n-\tfrac{1}{2}))] + [\pi(y(n+\tfrac{1}{2})), \pi(x(m-\tfrac{1}{2}))]$$

($m, n \in \mathbb{Z}+\frac{1}{2}$, $m+n \neq 0$) for the case $m = -\frac{1}{2}$, $n = -\frac{3}{2}$, applied to $1 \otimes 1$.

A similar argument shows that

(13.5) $\pi(x(2))\iota(y) = 2\langle x,y\rangle 1\otimes 1, \quad x,y\in\mathcal{B},$

which enables us to compute the form $\langle\cdot,\cdot\rangle$.

Now recalling that the module V is made up of two different vertex representations (see (5.5)), we see that

(13.6) $\mathcal{B} \simeq S^2(\mathfrak{h}(-1))\otimes 1 \oplus \sum_{\alpha\in\Lambda_4} \mathbb{C}\cdot 1\otimes(e^{\alpha}+e^{-\alpha}) \oplus \mathfrak{h}(-\frac{1}{2})\otimes T.$

One sees easily that

(13.7) $\iota(k_\alpha) = 1\otimes(e^{\alpha}+e^{-\alpha}).$

Then the action of $\mathfrak{f}$ on V_{-1} can be found explicitly using standard calculations with vertex operators. Finally, using the associativity of the form $\langle\cdot,\cdot\rangle$, we find the product of any two elements in $\mathfrak{h}\otimes T$. This gives us explicitly the structure of

(13.8) $\mathcal{B} = S^2(\mathfrak{h}) \oplus \sum_{\alpha\in\Lambda_4} \mathbb{C}k_\alpha \oplus \mathfrak{h} \otimes T.$

We first note that the action of F_1 decomposes V_{-1} and thus $\mathcal{B}$ into two orthogonal irreducible components, of dimensions 1 and 196883. The trivial one-dimensional module is spanned by the identity element, which we call I, of $\mathcal{B}$. The element I lies in $S^2(\mathfrak{h})$, and the operators I(n) ($n\in\mathbb{Z}$) in $\hat{\mathcal{B}}$, together with the identity operator on V, span a copy of the Virasoro algebra, under Lie brackets.

The product structure on $\mathcal{B}$ is given by:

$g^2\times h^2 = 4\langle g,h\rangle gh$

$g^2\times k_\alpha = \langle g,\alpha\rangle^2 k_\alpha$

$$k_\alpha \times k_\beta = \begin{cases} 0 & \text{if } \langle\alpha,\beta\rangle = 0, \pm 1 \\ \epsilon(\alpha,\beta) k_{\alpha+\beta} & \text{if } \langle\alpha,\beta\rangle = -2 \\ \alpha^2 & \text{if } \beta = \pm\alpha \end{cases}$$

$$g^2 \times h \otimes \tau = (\langle g,h\rangle g + \frac{1}{8}\langle g,g\rangle h) \otimes \tau$$

$$k_\alpha \times h \otimes \tau = \frac{1}{8}(h - 2\langle\alpha,h\rangle\alpha) \otimes e'_\alpha \tau$$

$$h_1 \otimes \tau_1 \times h_2 \otimes \tau_2 = \frac{1}{2}(2h_1 h_2 + \langle h_1,h_2\rangle I)\langle\tau_1,\tau_2\rangle +$$

$$+ \frac{1}{16} \sum_{\alpha \in \Lambda_4} (\langle h_1,h_2\rangle - 2\langle\alpha,h_1\rangle\langle\alpha,h_2\rangle)\langle\tau_1, e'_\alpha \tau_2\rangle k_\alpha,$$

where $g,h,h_i \in \mathfrak{h}$, $\alpha,\beta \in \Lambda_4$, $\tau,\tau_i \in T$, and $\langle\cdot,\cdot\rangle|_T$ is a nonzero $\bar{\Lambda}^{\wedge}$-invariant form on T (cf. Section 2).

The decomposition (13.8) is orthogonal with respect to the form $\langle\cdot,\cdot\rangle$ on $\mathfrak{B}$, which is given by:

$$\langle g^2,h^2\rangle = \langle g,h\rangle^2$$

$$\langle k_\alpha,k_\beta\rangle = \delta_{\bar{\alpha}\bar{\beta}}$$

$$\langle h_1 \otimes \tau_1, h_2 \otimes \tau_2\rangle = \langle h_1,h_2\rangle\langle\tau_1,\tau_2\rangle,$$

where $\bar{\alpha}, \bar{\beta}$ are the reductions of α, β mod 2Λ.

In a similar way we could explicitly describe the actions of C and S_3, and therefore the group F_1, on V_{-1} and $\mathfrak{B}$.

Combining our constructions with certain results in [15], or the recent simplification of these results in [28] (see also [27]), we can conclude that F_1 is a finite simple group with the group C as the centralizer of an involution. Thus F_1 is the Monster. By [28],

$$F_1 = \text{Aut } \mathfrak{B}.$$

Thus we have constructed a natural representation of the

Monster in the graded space V with character J(q). Recall that V is also an irreducible $\hat{\mathfrak{B}}$-module and that F_1 and $\hat{\mathfrak{B}}$ act compatibly on V (see (13.1)). We call the $(F_1,\hat{\mathfrak{B}})$-module V the *Moonshine module*.

Finally, one can explicitly check that up to an inessential one-dimensional subspace (the span of the identity element), our algebra $\mathfrak{B}$ and form $\langle\cdot,\cdot\rangle$ are isomorphic to Greiss's [15]. Another check shows that a certain involution in our triality group S_3 agrees with Griess's involution σ. Thus we obtain Griess's construction as a homogeneous "slice" of ours.

It is important to note that the infinite-dimensional vertex operator approach is more natural than the finite-dimensional approach, even though the primary object of study is a finite group. In particular, the six arbitrary parameters in Griess's construction are determined intrinsically in advance in our construction. Also, we have avoided the tedious verification that the involution σ preserves the algebra structure. Even simplification of Griess's work ([27], [28]) does not eliminate these complications. In our construction we use instead an entirely natural "calculus of vertex operators", which we consider standard, developed first for affine Lie algebras. In addition, we use a principle of triality which is based on the triality of a basis of the simplest simple Lie algebra $\mathfrak{sl}_2$.

14. THOMPSON SERIES

As another corollary of our construction, we obtain character formulas for elements of the Monster. By definition one has the Thompson series

$$(14.1)\quad T_g(q) = \sum_{n\geq -1} (\mathrm{tr}\ g|_{V_{-n}})\, q^n,\ g\in F_1.$$

We shall find a closed formula for the elements of C. For any $w\in W$ (see Section 4) we set

(14.2) $\eta_w(q) = \prod_{k>0} \eta(q^k)^{p_k}$

(see (4.6)) if w has characteristic polynomial

(14.3) $\det(1-xw) = \prod_k (1-x^k)^{p_k}$

on Λ. We also set

(14.4) $\theta_a^{\pm}(q) = \sum_{\substack{\alpha\in\Lambda \\ \bar{a}\alpha=\pm\alpha}} s(\alpha) q^{\langle\alpha,\alpha\rangle/2}$

where $a(e_\alpha) = s(\alpha)e_{\bar{a}\alpha}$, $a\in\tilde{W}$, and $\bar{a}\in W$ is the element induced by a. Then for $g\in C$ which has a preimage $(a,a_T)\in\hat{C}$ (see Section 5) we obtain

(14.5) $$T_g(q) = \frac{1}{2}\left[\frac{\theta_a^+(q)}{\eta_{\bar{a}}(q)} + \frac{\theta_a^-(q)}{\eta_{-\bar{a}}(q)} + \frac{tr(a_T)\eta_{\bar{a}}(q)}{\eta_{\bar{a}}(q^{1/2})} + \frac{tr(-a_T)\eta_{-\bar{a}}(q)}{\eta_{-\bar{a}}(q^{1/2})}\right].$$

Note that this expression is independent of the choice of preimage (a,a_T). It is interesting to note that Conway and Norton [4] obtained another formula for the trace of an element of C:

(14.6) $$T_g(q) = \frac{1}{2}\left[\frac{\theta_a^+(q)}{\eta_{\bar{a}}(q)} + \frac{\theta_a^-(q)}{\eta_{-\bar{a}}(q)} + \frac{tr(a_T)\theta_a(q^2)}{\eta_{\bar{a}}(q)} + \frac{tr(-a_T)\theta_{-a}(q^2)}{\eta_{-\bar{a}}(q)}\right] - c_g$$

where c_g is a constant and

$$\theta_a(q) = \sum_{\substack{\alpha\in\Lambda \\ \bar{a}\alpha=\alpha}} q^{\langle\alpha,\alpha\rangle/2}.$$

Thus, provided these formulas hold for the same Moonshine module, one obtains a whole collection of identities for modular functions enumerated by the conjugacy classes of elements of C. We have seen ((4.1), (4.4), (5.8)) that in the case when $g = e$ one has

$$(14.7)\quad T_e(q) = J(q) = \frac{\theta_\Lambda(q)}{\eta(q)^{24}} - 24.$$

Similarly, if one takes $g = g_2$ (resp., $g = g_3$) to be an involution (resp., an element of order 3) in F_1 whose centralizer is the double cover of the Baby Monster (resp., the triple cover of the Fischer group F'_{24}), then using the same argument as in the case $g = e$, namely, the isomorphism of suitable subspaces of the homogeneous and twisted vertex representations, one obtains

$$(14.8)\quad T_{g_2}(q) = J_2(q) = \frac{\theta_{\Lambda_2}(q)}{[\eta(q)\eta(q^2)]^8} - 8$$

$$(14.9)\quad T_{g_3}(q) = J_3(q) = \frac{\theta_{\Lambda_3}(q)}{[\eta(q)\eta(q^3)]^6} - 6$$

where Λ_2 (resp., Λ_3) are 16-dimensional (resp., 12-dimensional) even lattices without short vectors. Those lattices are certainly not unimodular; however they are "close" to unimodular lattices in the following sense: We call L an n-*modular lattice* if the dual lattice $L^* \simeq \frac{1}{n}L$. Then Λ_n, n=2,3, is an n-modular lattice. It follows immediately from the definitions (14.8), (14.9) that $J_n(q)$, n=2,3, is invariant with respect to the Atkin-Lehner involution ω: $z \longmapsto -\frac{1}{nz}$, where $q = e^{2\pi iz}$; it is also clearly invariant with respect to $\Gamma_0(n)$. This implies the invariance of $J_n(q)$ with respect to the group $\Gamma_0(n)^+ := \langle\Gamma_0(n),\omega\rangle$, and proves that $J_n(q)$, n=2,3, is a Hauptmodul for the group $\Gamma_0(n)^+$. The above argument certainly has a more general nature.

15. TWO DUAL MODELS AND TWO SPORADIC GROUPS

It was noticed from the first construction of the basic representation of $sl(2)^\wedge$ (see [19]) that it can be expressed in terms of operators similar to those appearing in the dual resonance models. Later this connection became more apparent [6]. However, the central result of the dual model theories, namely, the no-ghost theorem, remained untouched. This theorem implies that the dual models in rank 24 have the best behavior in terms of symmetry and unitary structure. This result seems to be deeply related to our construction, where rank 24 again plays a special role, providing a natural setting for the representation of the largest sporadic group F_1. Trying to reduce the critical dimension of the model, physicists introduced additional fermionic degrees of freedom. In this way the critical dimension reduces to 8, and the corresponding model is known as the Neveu-Schwarz model. In this case

(15.1) $$V^f_{\mathbf{Z}} = V_{\mathbf{Z}} \otimes \Lambda(\hat{h}^-_{\mathbf{Z}+1/2})$$

where $V_{\mathbf{Z}}$ is defined by (2.9). Then

(15.2) $$\mathrm{ch}\, V^f_{\mathbf{Z}} = \frac{\theta_{E_8}(q)\eta(q)^8}{\eta(q^{1/2})^8\eta(q^2)^8} = q^{-1/2} + 8 + 276\, q^{1/2} + \dots$$

One notices that this character has "moonshine" properties for the Conway group $\cdot 1$ similar to the properties of the character of $V_{\mathbf{Z}}$ with $L = \Lambda$ for the Monster. As was explained in Section 4, the natural automorphism group acting on $V^f_{\mathbf{Z}}$ is $\tilde{W}_{E_8} = 2^8 W_{E_8}$, and it is known that $\cdot 1$ has an involution centralizer C isomorphic to $2^{1+8}_+(W'_{E_8}/\langle\pm 1\rangle)$, where the prime denotes commutator subgroup. This situation is already familiar to us. In order to construct the right module we need to introduce an odd space

(15.3) $V^f_{\mathbf{Z}+1/2} = V_{\mathbf{Z}+1/2} \otimes \Lambda(\hat{h}^-_{\mathbf{Z}})$

and define $V^{f^\natural}_{\mathbf{Z}}$, $V^{f^\natural}_{\mathbf{Z}+1/2}$ as in Section 5. By analogy with the construction of F_1 one expects that

(15.4) $\mathrm{ch}\, V^{f^\natural}_{\mathbf{Z}} = \mathrm{ch}\, V^f_{\mathbf{Z}} - 8 = q^{-1/2} + 0 + 276\, q^{1/2} + \ldots$

which is in fact the case. Moreover the analogy between the two constructions exists at practically every step; this is also true in dual resonance theory. In particular one can repeat triality, $L_3(2)$ symmetry, etc. The analogue of the Conway-Norton formulas also exists, so that one can obtain a collection of identities for any conjugacy class of C. In particular for $e \in C$ we get the identity (cf. (14.6))

$$\mathrm{ch}\, V^{f^\natural}_{\mathbf{Z}} = \frac{1}{2}\left[\frac{\theta_{E_8}(q)\eta(q)^8}{\eta(q^{1/2})^8\eta(q^2)^8} + \frac{\eta(q^{1/2})^8}{\eta(q^2)^8}\right] + \tag{15.5}$$
$$+ \frac{1}{2}\, 2^4 \left[\frac{\theta_{E_8}(q^2)\eta(q)}{\eta(q^{1/2})^8\eta(q^2)} - \frac{\eta(q^{1/2})^8}{\eta(q^2)^8}\right] - 8\cdot 2^4.$$

It is interesting to note that besides several central ingredients of dual theories like the spaces $V_{\mathbf{Z}}$, $\Lambda(\hat{h}^-_{\mathbf{Z}})$, vertex operators and the Virasoro algebra, physicists have also used many other elements which occur in our construction, such as $V_{\mathbf{Z}+1/2}$ without the factor T, $\Lambda(\hat{h}^-_{\mathbf{Z}})$, triality, mixed spaces similar to V' (see e.g. [13]). Even the cross bracket appears ([13, page 526])!

The remarkable relationship of the two modules for sporadic groups and the two dual resonance models suggests a deep connection between two of the most distant theories in mathematics and physics. This also gives us some hints for the continuation of our work. In the physical models, besides the "noncovariant approach" which corresponds to what we have used in the paper, there exists a "covariant" one, which starts from the vertex representation of rank

26 with signature (25,1). In this approach the symmetries of the model are displayed in a more apparent way. This fact as well as the definition of the Monster Lie algebra of rank 26 in [1] suggest that our construction can be "embedded" into a more symmetric one. One can then hope to attach to each of the 23 constructions of the Leech lattice [5] a construction of the group F_1 (cf. [1]). Thus our present construction would correspond to the construction of the Leech lattice from the binary Golay code, the next one should correspond to the complex Leech lattice construction from the ternary code, and so on.

Among the finite-dimensional simple Lie algebras, E_8 is in a sense the richest. The proper understanding of its structure yields all the exceptional Lie algebras, via the "magic square", or all the simple algebras, via root system theory. We believe that the Monster -- really a uniquely beautiful creature, -- whose full understanding will involve Lie algebras, vertex operators, modular functions and physical models, should play an analogous role in a future unified theory of the finite simple groups.

REFERENCES

[1] R. E. Borcherds, J. H. Conway, L. Queen and N. J. A. Sloane, A Monster Lie algebra?, to appear.

[2] J. H. Conway, A group of order 8,315,553,613,086,720,000, Bull. London Math. Soc. 1 (1969), 79-88.

[3] J. H. Conway, A characterization of Leech's lattice, Inventiones Math. 7 (1969), 137-142.

[4] J. H. Conway and S. P. Norton, Monstrous moonshine, Bull. London Math. Soc. 11 (1979), 308-339.

[5] J. H. Conway and N. J. A. Sloane, Twenty-three constructions for the Leech lattice, Proc. Roy. Soc. London Ser. A 381 (1982), 275-283.

[6] I. B. Frenkel and V. G. Kac, Basic representations of affine Lie algebras and dual resonance models, Inventiones Math. 62 (1980), 23-66.

[7] I. B. Frenkel, J. Lepowsky and A. Meurman, An E_8-approach to F_1, Proceedings of the 1982 Montreal Conference on Finite Group Theory, ed. by J. McKay, Springer-Verlag Lecture Notes in Mathematics (1984).

[8] I. B. Frenkel, J. Lepowsky and A. Meurman, A natural representation of the Fischer-Griess Monster with the modular function J as character, Proc. Nat. Acad. Sci. U.S.A. 81 (1984), 3256-3260.

[9] H. Garland, The arithmetic theory of loop algebras, J. Algebra 53 (1978), 480-551.

[10] H. Garland, The arithmetic theory of loop groups, Publ. Math. IHES 52 (1980), 5-136.

[11] F. Gliozzi, D. Olive and J. Scherk, Supersymmetry, supergravity theories and the dual spinor model, Nuclear Physics B122 (1977), 253-290.

[12] D. Gorenstein, Finite simple groups, Plenum Press, New York, 1982.

[13] M. B. Green and J. H. Schwarz, Supersymmetrical dual string theory, Nuclear Physics B181 (1981), 502-530.

[14] R. L. Griess, Jr., A construction of F_1 as automorphisms of a 196,883 dimensional algebra, Proc. Nat. Acad. Sci. U.S.A. 78 (1981), 689-691.

[15] R. L. Griess, Jr., The Friendly Giant, Invent. Math. 69 (1982), 1-102.

[16] V. G. Kac, An elucidation of "Infinite-dimensional...and the very strange formula" $E_8^{(1)}$ and the cube root of the modular invariant j, Advances in Math. 35 (1980), 264-273.

[17] J. Leech, Notes on sphere packings, Canadian J. Math. 19 (1967), 252-267.

[18] J. Lepowsky, Euclidean Lie algebras and the modular function j, Amer. Math. Soc. Proc. Symp. Pure Math. 37 (1980), 567-570.

[19] J. Lepowsky and R. L. Wilson, Construction of the affine Lie algebra $A_1^{(1)}$, Comm. Math. Phys. 62 (1978), 43–53.

[20] D. Mitzman, Integral bases for affine Lie algebras and their universal enveloping algebras, Ph.D. thesis, Rutgers University, 1983 and Contemporary Math., Amer. Math. Soc., to appear.

[21] R. Pfister, Spin representations of $A_1^{(1)}$, Ph.D. thesis, Rutgers University, 1984.

[22] G. Segal, Unitary representations of some infinite-dimensional groups, Comm. Math. Phys. 80 (1981), 301–342.

[23] S. Smith, On the head characters of the Monster simple group, Proceedings of the 1982 Montreal Conference on Finite Group Theory, ed. by J. McKay, Springer-Verlag Lecture Notes in Mathematics (1984).

[24] J. Thompson, Some numerology between the Fischer-Griess Monster and the elliptic modular function, Bull. London Math. Soc. 11 (1979), 352–353.

[25] J. Tits, Four presentations of Leech's lattice, in "Finite Simple Groups II", Proceedings of a London Math. Soc. Research Symposium, Durham, 1978, pp. 303–307, Academic Press, London/New York, 1980.

[26] J. Tits, Résumé de cours, Annuaire du Collège de France, 1980–81, 75–87.

[27] J. Tits, Remarks on Griess' construction of the Griess-Fischer sporadic group, I–IV, preprints, 1983.

[28] J. Tits, Le monstre, Séminaire Bourbaki, 36e année, 1983/84, no. 620 (1983).

Rutgers University
New Brunswick, NJ 08903

Mathematical Sciences Research Institute
Berkeley, CA 94720

www.ingramcontent.com/pod-product-compliance
Lightning Source LLC
Chambersburg PA
CBHW081212220326
41598CB00037B/6762
* 9 7 8 9 9 7 1 5 0 4 2 0 5 *